LUMINESCENCE SPECTROSCOPY OF SEMICONDUCTORS

Further information for the reader's reference can be found at the end of the book: conversion of spectroscopic units, indispensable constants, semiconductor materials, overview of the most common excitation sources for luminescence spectroscopy, detectors for optical spectroscopy.

Luminescence Spectroscopy of Semiconductors

IVAN PELANT

Institute of Physics, v.v.i.
Academy of Sciences of the Czech Republic, Prague

JAN VALENTA

Department of Chemical Physics and Optics
Charles University, Prague

Great Clarendon Street, Oxford, OX2 6DP,
United Kingdom

Oxford University Press is a department of the University of Oxford.
It furthers the University's objective of excellence in research, scholarship,
and education by publishing worldwide. Oxford is a registered trade mark of
Oxford University Press in the UK and in certain other countries

First published 2012
First published in paperback 2016

Published in the United States of America by Oxford University Press
198 Madison Avenue, New York, NY 10016, United States of America

British Library Cataloguing in Publication Data
Data available

Library of Congress Cataloging in Publication Data
Data available

ISBN 978–0–19–958833–6 (Hbk.)
ISBN 978–0–19–875754–2 (Pbk.)

To our families:
Alena, Irena and Dita
Helena, Jakub and David

Preface

Luminescence of semiconductors has been widely used by many physicists, engineers and chemists for multiple purposes: For the study of the macroscopic optical properties of semiconducting materials as well as of their microscopic electronic processes, for the determination of purity and crystalline perfection of newly prepared compounds as well as of industrially grown single crystals, for the evaluation of the application prospects of novel active optoelectronic media, etc. The importance of luminescence techniques has considerably increased hand in hand with the recent rapid progress in low-dimensional semiconductor structures. The field of luminescence has been evolving rapidly during recent years also thanks to the advanced and ever developing spectroscopic instrumentation.

The purpose of this book is to introduce the reader to the study of the physical principles underlying semiconductor luminescence phenomena from the very beginning. Because, during the past 50 years, work on various types of luminescence radiation in semiconductors has been taking place in different streams, occurring mostly in different time periods, the relevant results are consequently scattered over many special monographs. The present book attempts to summarize the heart of the matter and to treat the whole subject in a unifying way. It is based on our one-semester course on 'Luminescence spectroscopy of semiconductors', read by us at the Faculty of Mathematics and Physics of Charles University in Prague for many years. For the purpose of the book we have extended our syllabus to cover a broader scope, by including especially a detailed treatment of experimental instrumentation (Chapter 2) and some theoretical parts, for example in Chapters 9, 10 and 12. Our own present interest has influenced the inclusion of Chapter 15; all other chapters, we believe, may serve as a general basis for many specialized and newly emerging branches of luminescence. It is understood that Chapters 12–17, devoted to luminescence phenomena of nanostructures, represent, on the one hand, an essential part of the book, and on the other hand, they should be regarded just as a 'snapshot' of the rapidly developing and versatile field of nanoscience that will very soon be outdated. The level of presentation throughout the book, being based predominantly on verbal and graphical descriptions of the phenomena rather than on a rigorous quantum-mechanical treatment, has been chosen to be comprehensible to graduate university students. Of course, a basic knowledge of solid state physics (the crystalline and amorphous states, **k**-space, energy band structure, elementary statistics of electrons and holes) is required.

We are indebted to many colleagues of ours for countless discussions over the years of cooperation, in particular to R. G. Elliman, B. Hönerlage, J. Kočka, J. Linnros, P. Malý, A. Mysyrowicz, J. Oswald and K. Vacek. We appreciate the creative atmosphere in the Group of Thin Films and Nanostructures of the Institute of Physics and the inspiring long-term tradition of luminescence research in the Department of Chemical Physics and Optics at the Faculty of Mathematics and Physics of Charles University in Prague, Czech Republic. We illustrate the text in part by our experimental results supported by the MŠMT Centre LC510 and the Grant IAA101120804 of GAAVCR. We should like to thank Academia, Prague, for granting us the electronic version of the Czech edition to work from. The book was translated from Czech by K. Herynková, K. Kůsová, J. Valenta, M. Žofka and I. Pelant under the editorial control of the last mentioned. Special thanks are due to V. Havlíček for his invaluable technical support. The book would never have been accomplished without the helpful cooperation of Oxford University Press, in particular S. Adlung, A. Warman and C. Charles.

Prague, May 2011
I. Pelant and J. Valenta

Contents

17 Spectroscopy of single semiconductor nanocrystals 455

Appendices 493

Introduction

1

Terminological notes

By the term 'luminescence of solids' we understand a surplus of the electromagnetic (light) radiation, emitted by a solid, over its equilibrium radiation that can be described by Planck's law.[1] At the same time this radiation has to have a decay time much longer than the period of light oscillations (10^{-14}–10^{-15} s). Above all, it follows from this definition that, from the thermodynamical point of view, the luminescence is a *non-equilibrium radiation*. This means that the solid needs to be supplied in some way with extra energy (with respect to that being exchanged by the solid with its surroundings through the equilibrium electromagnetic radiation). This extra energy is transformed inside the medium into the luminescence radiation. The supplied extra energy is called *excitation energy*, and the luminescence can be classified according to the way the excitation energy is applied, as follows:

Photoluminescence is excited with light (of wavelength λ_{ex}, which is usually shorter than the emission wavelength λ_{em}. The relation $\lambda_{ex} \leq \lambda_{em}$ is called Stokes' law).

Electroluminescence originates as a consequence of the application of an electric field and the relevant electric current flow through the material (do not confuse this with the thermal radiation due to the Joule heat!).

Chemiluminescence accompanies certain types of exothermic chemical reactions—the released heat or part of it is radiated in the form of light.

Bioluminescence coexists in a similar way with certain physiological biochemical reactions.

Cathodoluminescence arises when a high-energy electron beam (10^2–10^3 eV) impinges on a luminescent screen.

Mechanoluminescence is light (usually a short flash) emitted in some cases during mechanical deformation of a solid.

Thermoluminescence occurs when a solid is first cooled to a low temperature, then illuminated (excited) using short-wavelength electromagnetic

[1] The notion of light should be understood here in a wider sense, namely, not only visible radiation, but also the near-infrared and ultraviolet regions.

radiation and finally its temperature is allowed to increase slowly, which is accompanied by emission of the luminescence radiation.

Sometimes one can also encounter terms like *X-ray luminescence* (excitation by X-rays), *sonoluminescence* (excitation by acoustic or ultrasound vibrations), *triboluminescence* (luminescence due to friction when the material is scratched or crushed), etc. In recent years, light induced through injecting low-energy electrons or holes (energy of the order of 1 eV) into a semiconductor or metal in a scanning tunnelling microscope is being investigated. In the case of metals, however, the term luminescence seems inappropriate, as luminescence in solid-state physics is—for historical reasons—connected with non-metallic solids, insulators and semiconductors. Therefore, here the term *photon emission* is used instead.

The second part of the definition of luminescence, dealing with its finite decay time, differentiates luminescence from other types of so-called *secondary radiation*: reflected light, various types of scattered light (Rayleigh, Raman and Brillouin scattering) and Cerenkov radiation. The thing is that these kinds of radiation originate as an act of very fast photon–matter interaction and there is virtually no exchange of energy between the impinging photon and the electronic system of the solid. On the contrary, during the process of luminescence the electrons are excited to higher energy states, which entails genuine absorption of the excitation energy in the material and its subsequent gradual transformation. Consequently, the succession of all involved events lasts a relatively long time. For this reason, upon cessation of the excitation the luminescence continues to decay for some time. The lower bound of this time period gets shorter with the development of time-resolved spectroscopy techniques. Nowadays it can be considered to amount to hundreds of femtoseconds. (Of course, the scale of luminescence decay times is much broader and ranges from nanoseconds up to tens of hours.)

The study of luminescence has a long history, in the course of which the relevant terminology has also been developing. Radiating rotten stumps or certain kinds of luminescent insect and fish have been known in Nature from time immemorial. Probably more than a thousand years ago the Chinese and Japanese knew about luminescent dyes. In the seventeenth century, the 'Bolognian stone' was described, emitting red light upon prior exposure to sunlight (nowadays we know that it was barium sulphide, BaS). This and similar stones were given the name *phosphors* and the relevant effect, i.e. long-term light emission after cessation of the excitation radiation, was named *phosphorescence*.[2] Later on, in order to distinguish from phosphorescence, the term *fluorescence* was introduced to designate light emission with an immeasurably short decay after stopping the excitation; this was observed for the first time in fluorite, CaF_2. The general term *luminescence*, comprising both

[2] This has nothing to do with the chemical element P, phosphorus (in the sense that the luminescing substance would have to comprise P atoms), even if historical connection with phosphorus lighting itself can be traced; see Harvey E. Newton (2005). *A History of Luminescence. From the Earliest Times until 1900.* Dover Publications, Mineola.

phosphorescence and fluorescence, was finally introduced towards the end of the nineteenth century.

In the literature even the word *phosphor* occurs nowadays. Even if its meaning is not strictly specified, it usually denotes a luminescent solid in a wider sense. Also the meaning of the notions of phosphorescence and fluorescence has shifted in time—presently, fluorescence is sometimes understood as light emission occurring in the course of the excitation event, while in speaking of phosphorescence we have in mind non-equilibrium light emission observed during its decay, provided this decay is long enough to be observable by the naked eye. However, these terms are more exactly specified in the case of luminescence of organic substances.

Luminescence of organic and inorganic materials

Luminescence occurs in most organic substances (aromatic hydrocarbons like benzene, naphthalene, anthracene, organic dyes, etc.) as well as in many inorganic solids (ionic crystals, semiconductors). Its underlying origin, however, in both cases differs substantially. In organic matter, the role of a characteristic luminescence bearer (*luminescence centre*) is played by a molecule. This means that the essential features of the luminescence radiation (spectral content, decay time) in the solid state and in a solution are very much alike. This is because organic crystals are composed of molecules, the binding between them being mediated only by weak van der Waals forces. Therefore, the molecules keep their individuality to a large extent; the weak intermolecular forces are not strong enough to modify them substantially. The same, of course, is true in the case of a solution, thus the luminescence spectrum of organic matter has virtually the same shape in solution as well as in the solid state.

The basic features of organic luminescence can thus be derived from the electron energy level scheme of large organic molecules (the Jablonski diagram). As is well known, the ground state of such a molecule is represented by a singlet state S_0, higher lying electronic states being mainly excited singlet states (S_1, S_2, $S_3, \ldots$). The higher excited singlet states S_2, $S_3, \ldots$ relax very rapidly to the S_1 state and the transition $S_1 \rightarrow S_0$ is accompanied by emission of a luminescence photon. This radiative transition is spin-allowed and, consequently, is fast (the characteristic time is of the order of 10^{-9}s) and is called fluorescence. There are, however, also excited triplet states T_1, T_2, $T_3, \ldots$ The transition $T_1 \rightarrow S_0$ can also be accompanied with photon emission, but is spin-forbidden and thus slow (10^{-3}s). It is called phosphorescence.

As for inorganic solids, especially semiconductors, speaking of a 'molecule' loses its meaning to a large extent (a molecule of silicon—does it exist?). Physical properties here, above all the development of the forbidden energy gap, etc., are conditioned by the existence of a minimal ensemble of at least several tens of atoms (sometimes called a 'cluster'). However, real bulk properties can only be achieved in substantially larger objects with lateral dimensions at least 1 μm that contain approximately 10^{10} atoms. We shall treat these bulk semiconductor luminescence properties in Chapters 5–11. Low-dimensional semiconductor structures are examined in Chapters 12– 17.

Intrinsic and extrinsic luminescence

Solid-state luminescence can be divided into two basic types:

- *intrinsic* ('proper');
- *extrinsic* ('improper').

Intrinsic luminescence originates in an ideal, pure and defect-free crystalline lattice, while extrinsic luminescence has its origin in lattice defects or impurities. (A similar classification can be applied also to non-crystalline, disordered solids, provided we have in mind topological disorder of the amorphous network instead of the crystalline lattice.) A luminescence-active impurity (admixture) atom, ion or molecule is frequently called an *impurity luminescence centre*.

Now, a natural question arises. If—in the extrinsic case—the luminescence radiation originates in a microscopic impurity centre, what then is the role of the host solid itself, i.e. the crystalline or amorphous matrix? This matrix fulfils multiple functions: (i) First of all, it represents a host medium inside which the luminescence centres are fixed, statistically dispersed and mechanically isolated. However, this is by no means all; if we imagine these centres located in a similar way in a vacuum, they would either not emit any radiation at all or such radiation would be substantially weaker and of different spectral content. This means that (ii) the matrix also serves as an 'antenna' capturing the excitation energy and transferring it very efficiently to the luminescence centres. Next, (iii) owing to the interaction of the impurity electronic system with the matrix vibrations the electron energy levels undergo important modifications, which leads to substantial alterations of the optical spectra of the centre. Finally, (iv) in the special case of electroluminescence the matrix must ensure suitable electrical conductivity for exciting the centres.

At present we know that in the beginning of the era of the modern quantitative study of luminescence phenomena in solids (dating roughly from the first third of the nineteenth century to the 1950s), physicists and chemists were working exclusively with impurity, extrinsic luminescence. At that time, technology was not able to prepare sufficiently pure materials, while it is known today that most semiconductors owe their strong luminescence radiation to the presence of impurities. Even a negligible amount of residual impurity (1 ppm or less) can manifest itself in luminescence of enormous intensity, which masks or totally prevents intrinsic emission. This phenomenon, which will be discussed in Chapter 7, forms the basis of a sensitive qualitative and quantitative photoluminescence analysis of impurities in semiconductors. As a typical example of extrinsic emission we quote the luminescence of Cu^{+} or Mn^{++} ions in ZnS crystals, while the most characteristic example of intrinsic emission is luminescence due to the radiative decay of free excitons.

Examples of luminescence spectra

This book contains in its title the word *spectroscopy*. It follows that we shall largely deal with the interpretation of the emission spectral line shape of semiconductor luminescence. One of our goals is therefore aimed at analysing as much of the characteristic features of the experimentally acquired emission spectrum as possible, in order to assign the given spectral line to a specific microscopic luminescence centre or to some other channel of radiative recombination. Let us introduce here several examples of luminescence spectra.

Figure 1.1 shows near-infrared photoluminescence spectra of three crystalline Si samples, measured at liquid-helium temperature (4.2 K) [1]. The A sample represents extremely pure silicon without intentional doping (i.e. without introducing electrically active impurities), with a low concentration of residual unintentional impurities—in the order of $10^{12}\,\mathrm{cm}^{-3}$ only. The B sample contains impurities in the order of $\sim 10^{14}\,\mathrm{cm}^{-3}$ and the sample C has even higher impurity concentration, $> 10^{15}\,\mathrm{cm}^{-3}$. The notations of individual emission lines and bands will be clarified in Chapters 7 and 8. The spectra are rich in structure, comprising both very narrow (almost atomic) lines and relatively broad bands. It might be interesting to draw the reader's attention to the different characters of the spectral components: e.g. the lines denoted I_{LO}(FE) and I_{TO}(FE) are, as well as the band $I_{LO,TO}$(EHL), of intrinsic origin, while the lines B_{TO}(BE), P_{TO}(BE), Sb_{TO}(BE) are of extrinsic origin.

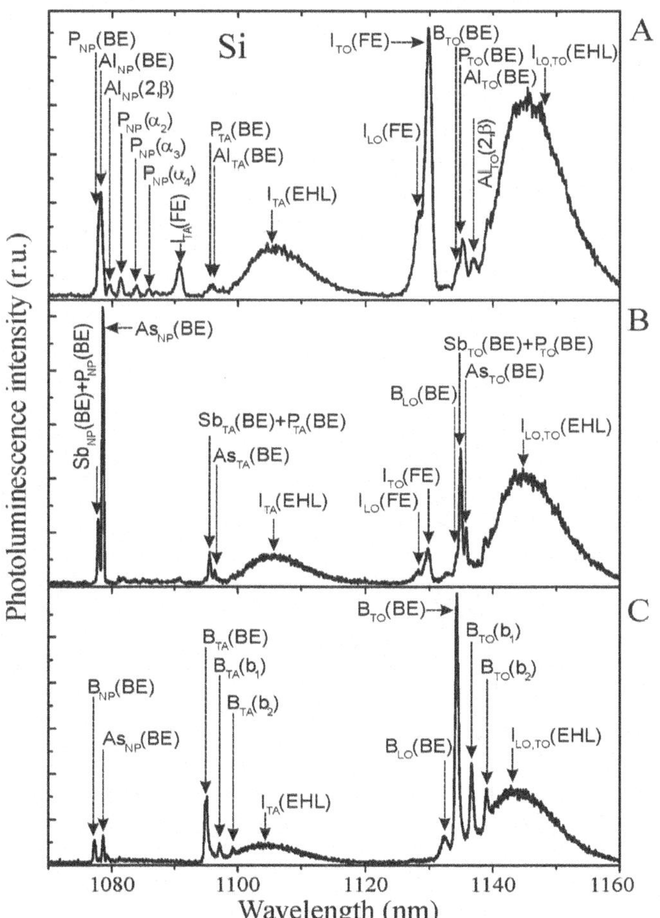

Fig. 1.1
Photoluminescence spectra of three samples of crystalline silicon close to the bandgap energy. The B and C samples were doped with (Sb + As) and (B + As), respectively; the A sample was not intentionally doped (nominally pure silicon). The samples were immersed in liquid helium ($T = 4.2\,\mathrm{K}$) and excited with a continuous-wave (cw) Ar^{+}-laser, $\lambda_{ex} = 488\,\mathrm{nm}$. After Valenta [1].

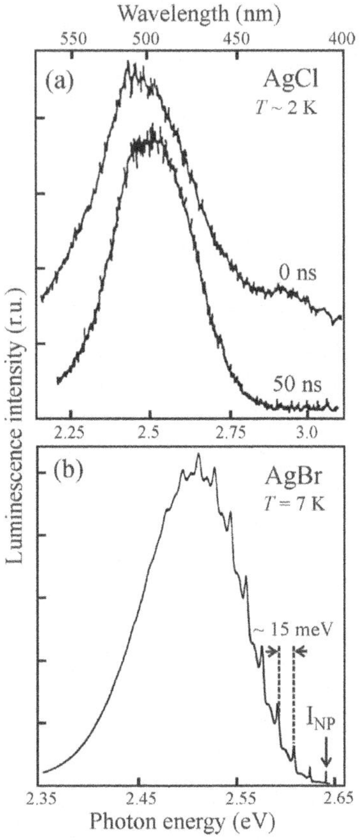

Figure 1.2(a) displays photoluminescence spectra of a pure silver chloride, AgCl, crystal, measured also at low temperature, 2 K, and under various modes of excitations [2]. Even if silver halides AgCl, AgBr are not typical semiconductors, their optical and, in particular, luminescence properties render them close to the family of wide-bandgap polar semiconductors. (Interestingly, at the same time, their mechanical properties are similar to those of metals.) In this context, they represent important model substances and throughout the book we shall often refer to their luminescence behaviour. The spectra contain a very broad emission band (full width at half maximum, FWHM, $\sim 0.4\,\text{eV}$) at $\sim 500\,\text{nm}$ (2.5 eV) and, besides this, an indication of a weaker band at $\sim$ 420 nm. Both bands are of intrinsic origin. The spectrum is situated in the visible region—under ultraviolet excitation and at low temperature AgCl exhibits very bright blue-green luminescence. Figure 1.2(b) shows the low-temperature (7 K) photoluminescence spectrum of nominally pure crystalline silver bromide, AgBr [3]. This material closely resembles AgCl both in chemical composition and in crystalline structure, and also the spectra displayed in Figs 1.2(a) and (b) are very much alike in their spectral position as well in their width. Despite these similarities, there is a fundamental difference in the origin of luminescence: Contrary to AgCl, the emission band at 2.5 eV in AgBr has extrinsic origin (!). (Another interesting observation is the 'fine structure' superimposed on the broad emission band in AgBr, which is completely absent in AgCl. The interpretation of these effects will be given in Chapter 7.) Finally, Fig. 1.3 demonstrates room-temperature emission spectra of zinc sulphide ZnS doped with various intentional impurities [4]. These broad extrinsic emission bands cover most of the visible region.

From these examples we can already draw several conclusions. First of all, it becomes quite obvious that—unfortunately—neither the shape nor the width of the emission spectrum can be used for judging the intrinsic or extrinsic origin of luminescence. Undoubtedly, this would be very profitable, but Nature does not reveal its mysteries so easily. Next, the reader has probably noticed

Fig. 1.2
(a) Photoluminescence spectra of pure AgCl at liquid-helium temperature under excitation with a pulsed N_2-laser ($\lambda_{ex} = 337\,\text{nm}$). Numbers near the curves denote temporal delays after the excitation pulse at which the spectra were recorded. After Pelant and Hála [2]. (b) Photoluminescence spectrum of nominally pure AgBr at $T = 7\,\text{K}$. The fine structure superimposed on the broad band is due to interaction with lattice vibrations—phonons. After Wassmuth *et al.* [3].

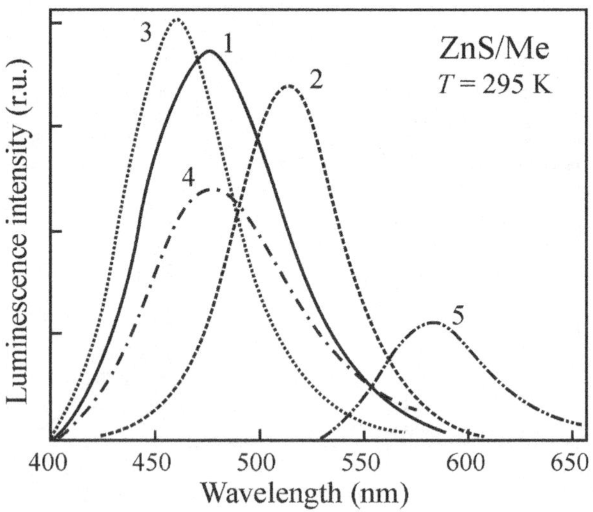

Fig. 1.3
Cathodoluminescence spectra of ZnS doped with various metals (Me), the so-called activators: 1. ZnS/Zn, 2. ZnS/Cu, 3. ZnS/Ag, 4. ZnS/Au and 5. ZnS/Mn. $T = 295\,\text{K}$. After Markovskii *et al.* [4].

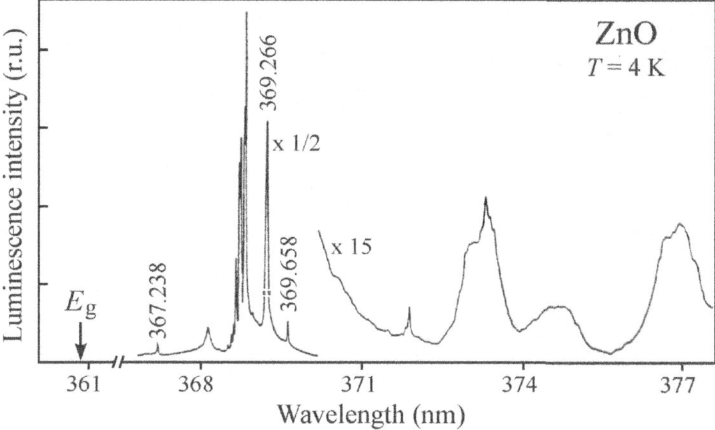

Fig. 1.4
Cathodoluminescence spectrum of edge emission in a nominally pure ZnO crystal. Excitation by bombardment with a 7 keV electron beam, $T = 4$ K. Adapted from Tomzig and Helbig [5]. The arrow labels the position of the bandgap $E_g = 3.436\,\text{eV}$ (~ 360.9 nm).

that most of the presented spectra are photoluminescence spectra, which anticipates the fact that the book is devoted predominantly to photoluminescence spectroscopy; the theory of electroluminescence and its applications are concentrated mainly in Chapter 11. Furthermore, the spectra displayed in Figs 1.1 and 1.2 were measured at very low temperatures. The reason for this is that the luminescence of most solids occurs only upon cooling them considerably below ordinary room temperature. And even if some materials luminesce at room temperature (Fig. 1.3), the intensity of this luminescence, as a rule, grows markedly in the course of cooling. This property contradicts flagrantly the ordinary behaviour of thermal radiation of solids (*incandescence*), described by the Planck and Stefan–Boltzmann laws, and that is why luminescence is sometimes known as *cold light*. The drop of luminescence intensity with increasing temperature is called *thermal quenching*.

Finally, in semiconductor luminescence one can encounter the term *edge emission*. This means luminescence radiation situated in the spectral range around the onset of interband optical absorption (absorption edge). More specifically, the energy of the luminescence photon $h\nu_{em}$ is only slightly less than the bandgap energy E_g. The edge emission spectrum usually consists of a large number of narrow (excitonic) lines, both intrinsic and extrinsic. An example of the low-temperature edge emission in zinc oxide, ZnO, is shown in Fig. 1.4 [5]. The bandgap energy $E_g = 3.436\,\text{eV}$ is marked by an arrow. The displayed wavelength range is very narrow, about 10 nm (90 meV); the individual lines are shifted with respect to E_g by 0.01–0.1 eV only. Similarly, the spectra shown in Fig 1.1 can also be specified as edge emission.

References

1. Valenta, J. (1994). *Photoluminescence characterization of selected semiconductors and insulators*. PhD Thesis, Charles University in Prague, Faculty of Mathematics and Physics, Prague.
2. Pelant, I. and Hála, J. (1991). *Solid State Com.*, **78**, 141.

3. Wassmuth, W., Stolz, H., and von der Osten, W. (1990). *J. Phys. C: Cond. Matter*, **2**, 919.
4. Markovskii, L. J., Pekerman, F. M., and Petoshina, L. N. (1966). *Phosphors* (in Russian: *Ljuminofory*). Chimija, Moskva, Leningrad.
5. Tomzig, E. and Helbig, R. (1976). *J. Luminescence*, **14**, 403.

Experimental techniques of luminescence spectroscopy

2

The aim of experimental luminescence spectroscopy is basically the same as that of other types of optical spectroscopy, namely, to disperse luminescence radiation into a spectrum, which is subsequently detected and recorded. Here, as a dispersion device, a monochromator or a polychromator is almost exclusively used; only exceptionally (in the case of Fourier luminescence spectroscopy) is an interferometer applied. Possible application of devices based on multiple-beam interference (Fabry–Perot interferometer, Lummer–Gehrke plate) brings no advantage as the luminescence spectra are usually broad and do not require extreme spectral resolution. Only exceptionally, e.g. in the case of narrow spectral lines of edge emission (Figs 1.1 and 1.4), a high-resolution monochromator or possibly a Fourier spectrometer with resolving power $R = (\lambda/\Delta\lambda)$ amounting at least to $\sim 5 \times 10^3$ ($\Delta\lambda$ denotes the minimum separation between two spectral lines around a central wavelength λ that are to be resolved) is indispensible.

Generally, what is characteristic for luminescence spectroscopy is typically an extremely low level of detected light flux rather than requirements for extreme spectral resolution. Therefore, the spectral device must have the highest possible throughput, and special care must be devoted to selecting a sensitive photodetector and, in particular, to setting up an efficient optical system to collect the luminescence signal. All these aspects will be discussed in this chapter.

2.1 Emission and excitation spectra

The purpose of monochromators in luminescence spectroscopy is, on the one hand, to observe the spectral content of luminescence emission and, on the other hand (in the case of photoluminescence) they can also be applied to select a suitable excitation wavelength from an optical excitation source. The luminescence signal can be measured either as a function of emission wavelength at a fixed excitation wavelength or vice versa. One speaks of *emission* and *excitation spectra*, respectively.

A general scheme of luminescence experiment is shown in Fig. 2.1(a). The excitation source supplies excitation energy to the sample, emitted

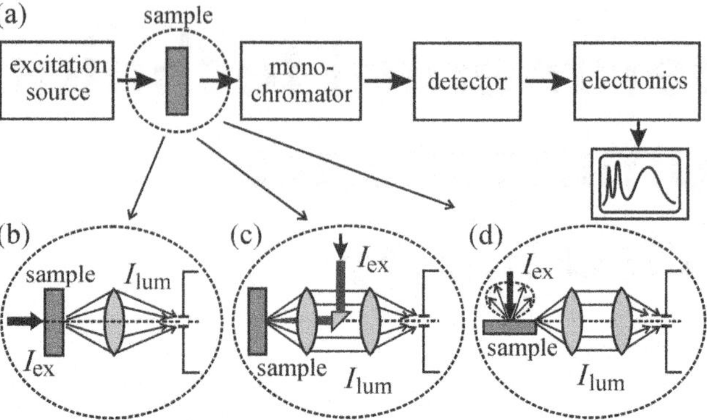

Fig. 2.1
(a) Block diagram of a luminescence experiment, (b) the 'transmission' geometry, (c) the 'back-reflection' geometry, (d) collection of luminescence from the sample edge using a waveguiding effect.

luminescence radiation is dispersed in a spectral device, a detector converts the optical signal into an electric signal that is then processed by electronic devices and finally by a control computer. The output in the form of a plot of luminescence intensity I_{lum} versus emission wavelength λ_{em} is called an *emission spectrum* (all the spectra shown in Figs 1.1–1.4 are emission spectra).

Spontaneous luminescence emission emerges from a sample practically in all directions, even if in many cases the sample might not be a perfect isotropic emitter. Various optical systems collecting the emitted light can be implemented. The simplest one is shown in Fig. 2.1(b). This is the so-called transmission geometry, where the optical axes of the excitation and luminescence rays coincide. Luminescence is thus detected from the opposite, non-excited side of the sample. This geometry is simple and easy to adjust but the luminescence signal could be significantly reduced by light scattering and especially by reabsorption in the sample. Since reabsorption may be spectrally dependent, the emission spectrum will be distorted. In addition, a significant fraction of non-absorbed excitation light could enter into the monochromator and cause problems by saturating or even damaging the detector. Therefore, the transmission geometry is seldom used.

More routinely applied is reflection geometry shown in Fig. 2.1(c). Luminescence is collected from the same spot on which the exciting radiation is focused by a lens (or an objective). The excitation beam is directed perpendicular to the optical axis of a collection system and the exciting light is sent towards the sample by means of a small totally reflecting prism (when the excitation is provided by a well-collimated narrow laser beam) or a dichroic filter (which reflects short wavelengths, i.e. excitation, and transmits longer wavelengths, i.e. luminescence). One and the same lens (objective) is employed to concentrate excitation light and to collect luminescence emission. The lens should have a very large numerical aperture in order to collect light from a wide solid angle. The collimated luminescence light is then focused by means of a second lens onto the entrance slit of a spectral device. Among merits of this geometry are the above mentioned large collection solid angle (remember that luminescence is weak secondary radiation, so the largest possible portion of the emitted power must be collected for successful spectral analysis, see Section 2.5), reduced reabsorption of luminescence radiation, as well as avoidance of stimulated emission (which will be explained later). As a possible drawback, complicated optical adjustment required for achieving both optimum excitation power density on the sample and simultaneous maximum luminescence collection can be mentioned. Another undesirable effect may be possible reflection of a significant part of the excitation light into a detection system, but that is easily remedied by slightly rotating the sample to reflect the exciting beam out of the detection cone.

In particular cases, when the sample takes the form of thin layer(s) with high-quality surfaces, the geometry sketched in Fig. 2.1(d) (an extreme case of the reflection geometry, excitation and emission axes being perpendicular to each other) may be advantageous. Such samples may act as a waveguide into which luminescence is coupled and thereby concentrated at the output edge. In comparison with the previous case, this scheme enables the excitation and emission optical paths to be adjusted independently, and it avoids reflection of the excitation beam into the detection system. The most serious problem that might appear is the possible occurrence of stimulated emission: if population inversion is (unintentionally) achieved, then in semiconductors with high probability of radiative recombination the luminescence (spontaneous emission) guided through the excited volume may be amplified by stimulated emission. This effect can be reduced or avoided by limiting the size of the excited area, but even a small portion of stimulated emission could significantly distort the shape of the emission spectrum [1, 2]. On the other hand, this geometry is convenient especially when stimulated emission and the presence of optical gain are the subjects of investigation. Details on such experiments and measurement techniques are provided in Chapter 10.

Two particular set-ups for measurements of photoluminescence emission spectra are presented in Fig. 2.2 [3, 4]. The experimental arrangement in Fig. 2.2(a) represents a particular case of the collection geometry sketched schematically in Fig. 2.1(c). A continuous-wave (cw) Ar^+-laser with the most intense lines at 488 and 514 nm is used as an excitation source. The sample is immersed in liquid helium inside a helium-bath cryostat (the temperature of the boiling point of helium at normal pressure is 4.2 K). Luminescence is spectrally dispersed in a double-grating monochromator and detected by a photomultiplier tube (PMT) placed behind the exit slit. The PMT is cooled down by liquid nitrogen (the temperature of the boiling point at normal pressure is 77 K). The signal from the PMT is processed by a photon counter and plotted, as a function of wavelength, by a computer.

Another frequently used set-up is shown in Fig. 2.2(b). The excitation source is represented by a high-pressure mercury lamp which has one of its dominant lines at 365 nm (in the near-UV spectral region) that is almost an ideal universal excitation wavelength for most semiconductors. (Nowadays, mercury lamps have been almost completely replaced by diode lasers or (UV adapted) Ar^+-lasers.) From the point of view of geometry, this set-up is halfway between Figs 2.1(c) and 2.1(d), so-called 'reflection at 45°'. The sample is, again, placed in a cryostat whose single window is replaced by a collection lens. The diameter of the lens defines the collection solid angle. Another lens focuses the collimated luminescence beam into a point where an optical chopper (a rotating disk with regularly arranged slots) periodically chops the beam.

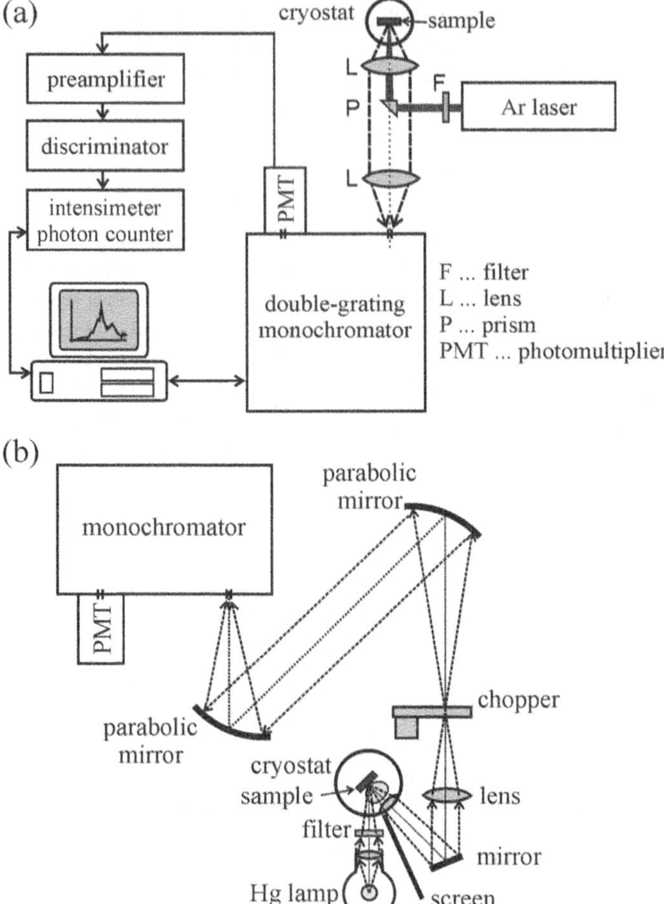

Fig. 2.2
(a) Experimental set-up for detection of photoluminescence emission spectra in a back-reflection geometry with double-lens collection system [3]. (b) Set-up for detection of emission spectra with the collection system combining lenses and two parabolic mirrors. This geometry is called 'reflection at 45°'. Adapted from Bebb and Williams [4]. The apparatus contains an optical chopper inserted into the luminescence beam, enabling application of a lock-in detection technique (Subsection 2.4.1).

The rectangular temporal trace of the luminescence signal is essential for treatment of the signal by a lock-in technique that will be discussed in Subsection 2.4.1 (in the case of chopping an unfocused luminescence beam the temporal profile of the signal would be trapezoidal and, consequently, the final signal-to-noise ratio will be degraded). Before entering the monochromator, the luminescence radiation must be refocused onto the entrance slit. This is done by a pair of parabolic mirrors; these cover a smaller area than a lens system that would play the same role, thus saving room on the optical table and, moreover, minimizing optical abberations. The detector behind the exit slit of the monochromator—again a photomultiplier tube—provides the output signal that is processed by a lock-in amplifier.

The photoluminescence *excitation spectrum* is a plot of the photoluminescence intensity I_{PL} (at a fixed emission frequency $\nu_{em} = c/\lambda_{em}$) as a function of the excitation photon energy $h\nu_{ex}$:

$$I_{PL}(\nu_{em}) = f(h\nu_{ex});\ \nu_{em} = \text{const.}$$

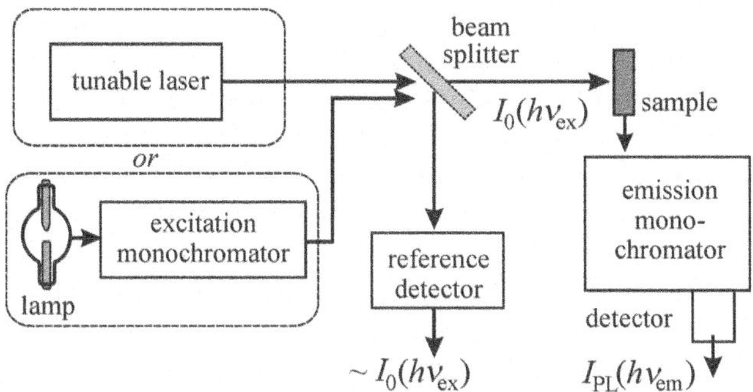

Fig. 2.3
Block diagram of an apparatus for excitation spectra measurements.

A schematic sketch of an experimental set-up for recording excitation spectra is shown in Fig. 2.3. In order to understand the usefulness of the excitation spectra, let us investigate firstly the dependence of the photoluminescence intensity $I_{PL}(\nu_{em})$ on the excitation light intensity. Suppose we have a sample in the form of a plane-parallel plate with thickness d and an excitation intensity $I_0(h\nu_{ex})$ impinging on it. The transmitted excitation intensity is given by the Lambert–Beer law

$$I(h\nu_{ex}) = I_0(h\nu_{ex}) \exp[-\alpha(h\nu_{ex})d], \tag{2.1}$$

where $\alpha(h\nu_{ex})$ is the absorption coefficient. If we consider the correction for reflectivity R on the front face of the sample (for the sake of simplicity R is assumed to be independent of wavelength), the absorbed part of the excitation intensity is equal to $I_0(h\nu_{ex})(1 - R)(1 - \exp[-\alpha(h\nu_{ex})d])$.

The photoluminescence intensity, being proportional to the absorbed energy (possible nonlinear effects are neglected here) then reads

$$I_{PL}(\nu_{em}) \cong \eta I_0(h\,\nu_{ex})(1 - R)(1 - \exp[-\alpha(h\nu_{ex})d]). \tag{2.2}$$

The proportionality coefficient $\eta \leq 1$ introduced here is called the *efficiency* or *quantum yield of luminescence*. In general it depends on many variables: $\eta = \eta(h\nu_{ex}, h\nu_{em}, T, \ldots)$. Now we should distinguish two extreme cases:

1. Weakly absorbing sample $\alpha(h\nu_{ex})d \ll 1$. By applying the common approximation of the exponential function $\exp(-\alpha d) \approx (1\text{–}\alpha d)$, we obtain from (2.2)

$$I_{PL}(\nu_{em}) \cong \eta\, I_0(h\nu_{ex})(1 - R)\, d\, \alpha(h\nu_{ex})$$

and, after dividing by a reference spectrum of the excitation source $I_0\,(h\nu_{ex})$, we find

$$\frac{I_{PL}(\nu_{em})}{I_0(h\nu_{ex})} \sim \alpha(h\nu_{ex}). \tag{2.3}$$

This means that the excitation spectrum, normalized in this way, reproduces the shape of the absorption spectrum. This can be applied to obtain (relative) absorption spectra in samples in which standard transmission spectroscopy fails, e.g. in thin semiconductor layers on an opaque (i.e. strongly absorbing or scattering) substrate, in measuring low concentrations of luminescent dopants in a strongly absorbing non-luminescent host material, in quantum wells or superlattices, etc. This is one of the motivations for investigating photoluminescence excitation spectra. (It is worth pointing out that such measurements of optical absorption spectra can also be performed in samples of a very irregular shape, in small fragments or even in powder materials. This is a general merit of most luminescence measurements—they can be performed without paying special attention and effort to preparing a sample of a perfectly defined shape, size, surface quality, etc.)

The second motivation for measuring excitation spectra is purely practical. As mentioned above, luminescence radiation is generally very weak; hence all means of optimizing the luminescence signal should be used. In this sense the excitation spectrum, once acquired, gives a clear hint about the choice of an optimum excitation wavelength—this is a wavelength (or excitation photon energy) at which the excitation spectrum is peaked.

As an example of a photoluminescence excitation spectrum we present that of crystalline quartz, containing a pronounced structure (see Fig. 2.4(a) [5]). A narrow excitonic peak around 8.7 eV as well as onsets of indirect and direct absorption edges around 9 eV and 10.7 eV can be identified. Note that the emission spectrum is significantly red-shifted from the deep-UV excitation peaks to the near-UV and blue spectral regions (between 2 and 5 eV, see Fig. 2.4(b)). Such a red-shift between excitation and emission peaks is often called the *Stokes shift*; a more exact definition will be given in Section 4.5. A sizeable Stokes shift, of several electron-volts like in Fig. 2.4, indicates low power efficiency of luminescence, and most likely also the occurrence of a complicated mechanism of excitation energy transfer from a light-absorbing system to luminescence centres. This is additional interesting information revealed by the excitation spectra.

Let us recall that all of the above discussion is valid in the case of weak absorption of excitation light, when relation (2.3) holds true. Now we are going to discuss the opposite case.

2. Strongly absorbing sample $\alpha(h\nu_{ex})d \gg 1$, i.e. $\alpha \approx 10^3$–$10^5\,\mathrm{cm}^{-1}$ and $d \geq 10^{-2}$ cm. In this case $\exp[-\alpha(h\nu_{ex})\,d] \ll 1$ and (2.2) is reduced to

$$I_{PL}(\nu_{em}) \approx \eta\, I_0(h\nu_{ex}). \tag{2.4}$$

This means that the excitation spectrum basically mimics the shape of the emission spectrum of the excitation source while information about the sample under study is lost. The condition $\alpha(h\nu_{ex})d \gg 1$ is valid for most semiconductors under deep band-to-band excitation, i.e. when $h\nu_{ex}$ is much larger than the

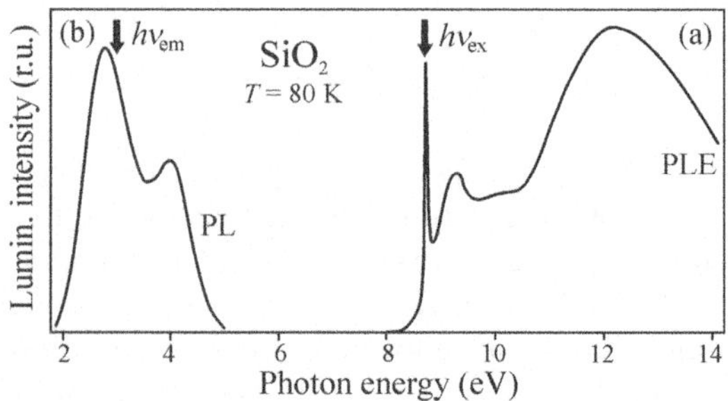

Fig. 2.4
(a) Photoluminescence excitation spectrum (PLE) of crystalline SiO_2 detected at $h\nu_{em} = 3.0$ eV and $T = 80$ K. (b) Photoluminescence emission spectrum (PL), excited by photons with an energy of $h\nu_{ex} = 8.7$ eV. Adapted from Itoh *et al.* [5].

bandgap energy E_g. Here, excitation spectra no longer bring useful information. In a number of experiments between the extremes 1 and 2, the usefulness of excitation spectra measurements must be assessed from case to case.

Note: From the experimental point of view eqn (2.4) can be exploited to monitor excitation intensity variations. When the quantum efficiency η of a luminescent material (or its solution—e.g. Rhodamin B in ethylenglycol) strictly satisfies the requirement of being independent of excitation wavelength, $\eta \neq \eta(h\nu_{ex})$, then this material (the so-called quantum counter) may be applied in combination with a photomultiplier to measure the emission spectrum of an excitation source. The advantage of this mode of measurement against using detectors with a spectrally flat response (thermal detectors) is its much higher sensitivity.

2.2 Types of photodetectors

In Fig. 2.2 a photomultiplier tube (PMT) is used to detect the spectrally dispersed luminescence signal. This is the most common highly sensitive photodetector for luminescence spectroscopy, even if nowadays it is often replaced by an avalanche photodiode or by an array of small semiconductor devices—usually by a charge-coupled device (CCD) camera. We shall briefly describe these three types of detectors.

Photomultiplier tube

Detection of optical radiation in a photomultiplier tube is based on the external photoelectric effect and secondary emission of electrons. A photomultiplier is an evacuated glass (or quartz) tube containing a photocathode (from which incident photons eject electrons—with quantum efficiency $\eta_{ph} < 1$—due to the external photoelectric effect), an array of dynodes (where primary electrons are multiplied through the emission of secondary electrons) and an anode (the collector of electrons, see Fig. 2.5(a)). The outlets of all electrodes pass through a socket of the photomultiplier. A photon can eject an electron from the photocathode only when the photon energy $h\nu$ exceeds the work function ϕ of the given photocathode material, $h\nu \geq \phi$. To detect (optical) photons from the visible range ($h\nu \approx$ 1–4 eV), the photocathode must be made of either an alloy containing alkali metals with low work function ϕ (e.g. $\phi = 1.8$ eV in Cs,) or a semiconductor with relatively low values of both the work function ϕ and bandgap E_g (GaAs). The applied photocathode material determines the spectral sensitivity of the detector, especially its long-wavelength cut-off, while the short-wavelength edge is determined mostly by the material of the input window.

Photoelectrons are accelerated and focused onto the first dynode, where each of these primary electrons can release δ (between 3 and 10) secondary electrons. These secondaries, upon being accelerated, hit the next dynode, and the secondary emission of electrons is repeated there so that an avalanche increase in the number of electrons is achieved: if n is the number of dynodes, the number of electrons that hit the anode will be equal to $G = \delta^n$. The quantity G is called the current amplification or photomultiplier gain and attains values

(a)

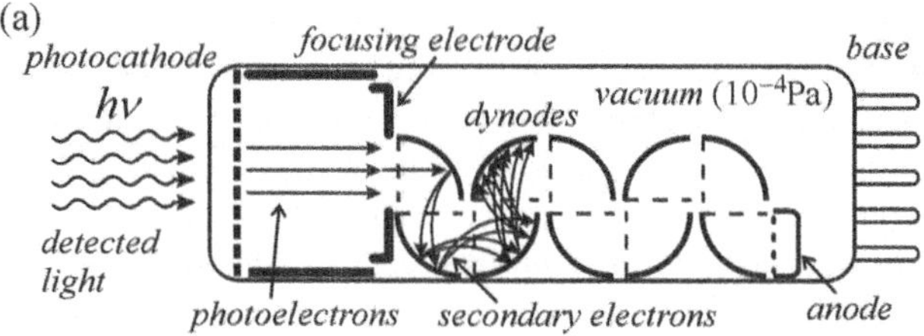

(b)

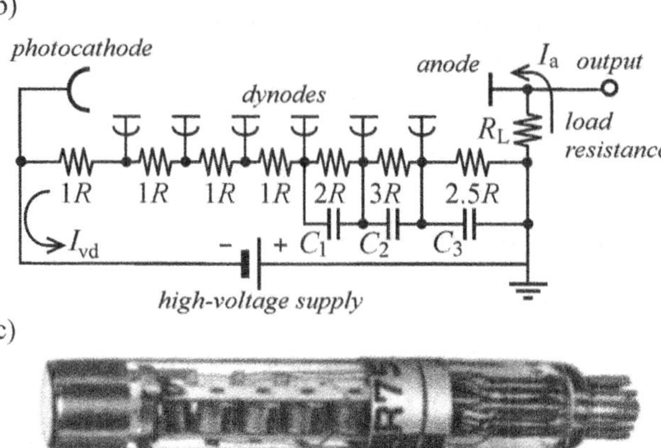

(c)

Fig. 2.5
(a) General scheme of a photomultiplier tube, (b) circuitry of the resistor voltage divider between dynodes, where I_a denotes the anode current and I_{vd} is the current flowing across the voltage divider, (c) a Hamamatsu R750 photomultiplier tube (the tube length and diameter are 88 and 20 mm, respectively). Adapted from a Hamamatsu brochure.

of 10^5–10^7. Dynodes are mostly fabricated from nickel or BeCu and are covered by a material with high coefficient δ of secondary emission (BeO, GaP, or KCsSb). The electron flux, collected by the anode, flows across a load resistance and gives an output electric signal, which is then treated by electronic circuits.

In order to accelerate the primaries as well as the secondaries, an electric field must be supplied between the relevant electrodes. This is done by applying a high voltage (usually between 1000 and 2000 V) between the photocathode and anode through a resistor voltage divider (Fig. 2.5(b)). The resistors are connected in series; the ratios of their values, defining how the applied voltage is divided between electrodes, are specified by the manufacturer and must be strictly maintained. Nowadays most manufacturers offer sockets with a built-in resistor divider for any type of photomultiplier tube. A typical value of a single resistor is of the order of 100 kΩ, determining the current through the divider to be in the order of mA (in the case of $n = 10$ dynodes and an applied voltage of 1000 V).

There are many various construction types of photomultiplier tubes. The photomultiplier sketched schematically in Fig. 2.5(a) and an overall view of which is shown in Fig. 2.5(c), has a so-called linear focus multiplier; other common types of multiplier structures are called 'Venetian blinds', boxes and grids, and circular cages. Besides, the photocathode may be deposited onto the inner surface of either the front side of the cylindrical tube (head-on type, Fig. 2.5(c)) or its side wall (side-on type). A more detailed description and comparison of different types of photomultipliers goes beyond the scope of this book and can be found in the dedicated literature, e.g. [6, 7].

We shall describe concisely the technical parameters of photomultiplier tubes. Photomultipliers owe their widespread use to, firstly, the already mentioned high gain G and also to their *spectral range of sensitivity*. This range extends between about 150 and 1100 nm; in special cases it can go up to 1400 nm. It should be stressed, however, that the entire spectral domain cannot be covered by a single type of photomultiplier. The selection of an appropriate photomultiplier type thus has to be based primarily on considerations about the spectral range of the detected light. The *anode sensitivity* $k(\lambda)$ is given as the anode current produced by 1 watt (in energy units) or 1 lumen (in photometric units) of incident radiant or luminous flux Φ, respectively:

$$k(\lambda) = \frac{\text{output signal}}{\text{incident radiant power}},$$

$$k(\lambda) = \frac{S}{\Phi} = \frac{S}{I\left(\mathrm{W/cm^2}\right)\mathscr{A}\left(\mathrm{cm^2}\right)} \quad (\mathrm{A/W}) \text{ or } (\mathrm{A/lm}), \qquad (2.5)$$

where $\mathscr{A}$ denotes the sensitive area of the photomultiplier. Typical values of $k(\lambda)$ are 10^3–10^5 A/W.

The above mentioned *quantum efficiency* η_{ph} of the photomultiplier tube is defined as the probability of ejecting from the photocathode one electron per incident photon (η_{ph} is typically about 0.1–20%). A very important parameter for properly operating a photomultiplier is its *maximum anode current* I_{am}, whose typical value is usually 10–100 μA and must not be exceeded (i.e. the incident radiant flux must be kept below a certain level), otherwise the excessive heating of the electrodes will induce permanent damage to the photomultiplier (due, e.g., to partial evaporation of the sensitive layer of the photocathode) which manifests itself as a decrease in sensitivity or an increase in background noise. In addition, for proper functioning of a photomultiplier tube (linearity of its response) the resistors in the voltage divider must be chosen so that the current flowing across the divider is at least an order of magnitude higher than I_{am} (Problem 2/1).

This definition of sensitivity does not take into account *noise* and the so-called *dark current* of the photomultiplier tube. The dark current I_{d} is a current due to the thermionic emission of electrons from the photocathode, which takes place even in darkness when no light strikes the photocathode. The thermionic current density is described by the well-known Richardson–Dushman equation [8], and thus increases exponentially with increasing temperature and decreasing work function ϕ of the photocathode. Photomultiplier tubes sensitive in the red and near-infrared spectral regions suffer from the highest dark currents, as the work function of their photocathodes must be low. Typical values of the anode dark current lie between a few nA up to hundreds of nA. Dark current at the output of a photomultiplier cannot be distinguished from the real (light-induced) signal, which is troublesome especially for very low signal levels—dark current obviously increases the limit of the lowest detectable signal (the detection limit of a photomultiplier tube). It is therefore desirable to suppress the dark current as much as possible. The strong (exponential) dependence of the dark current on temperature enables us to reduce it significantly by cooling the photomultiplier. The lowest allowed operational temperature is given by the manufacturer and must not be exceeded because mechanical deformations at lower temperatures can induce permanent mechanical damage. Cooling can be provided by means of a thermoelectric (Peltier) element (down to about −25°C with a single-stage device) or by liquid nitrogen (boiling point −195°C = 77 K). The cooling housings for different types of photomultiplier tubes are commonly offered by manufacturers or they can be home-built (which is not an easy task, though). Sometimes, 'for less demanding applications', photomultipliers can be used even without cooling, but that depends on the photomultiplier type and even the particular photomultiplier piece of choice.

The dark current I_d can thus be suppressed to a large extent by cooling. Alternatively, because the dark current contribution to the resulting output signal has the form of a direct current (DC) component, it can be filtered out when measuring alternating current (AC) signals. A similar approach cannot, unfortunately, be applied to reduce noise in the output signal—see, e.g., Figs 1.1 and 1.2, where we reproduce luminescence spectra as recorded by a plotter after being treated by the electronic part of the detection channel. The spectra in Fig. 1.1 were detected by using a cooled (77 K) photomultiplier tube with a near-infrared sensitive photocathode while the spectra in Fig. 1.2(a) were obtained by an uncooled photomultiplier tube with a semiconductor-type photocathode. In order to characterize the noise properties of a photodetector, the quantity termed *noise equivalent power* (*NEP*) (or equivalent noise input—*ENI*) is introduced. This means the incident light power which would induce an output signal equal to the root mean square of the intrinsic (dark) noise N (Fig. 2.6). The detected noise increases, of course, with increasing transmission bandwidth Δf of the entire measurement system; it turns out that the noise amplitude scales with the bandwidth like $(\Delta f)^{1/2}$. So the detected noise N can be written down as

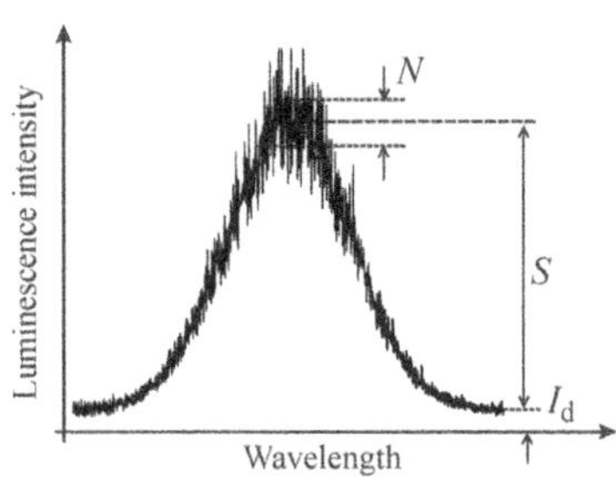

Fig. 2.6
Schematical representation of noise conditions in a DC luminescence emission spectrum. N stands for the root mean square noise value, i.e. $N = \left[\left(\int_0^{t_0} n^2(t)\,dt\right)/t_0\right]^{1/2}$, where $n(t)$ is the instantaneous time-variable value of the noise related to the mean signal value S.

$$N = k(\lambda)\,NEP\sqrt{\Delta f}, \tag{2.6}$$

where $NEP = (2e\,I_d G)^{1/2}/k(\lambda)$ [7] is a parameter already independent of the measurement system, i.e. solely characterizing the photodetector. According to the definition, the signal-to-noise ratio (S/N) is equal to unity when the incident light power is equal to *NEP*. In other words, *NEP* represents the lowest signal level (detection limit) which can be detected by a given detector; weaker signals will remain buried in noise. Referring to eqn (2.6), *NEP* is expressed in $\mathrm{W/Hz^{1/2}}$. Thus this is not directly power (in watts), but one usually supposes that $\Delta f = 1$ Hz and then $NEP(\Delta f)^{1/2}$ (W) and $NEP(\mathrm{W/Hz^{1/2}})$ in (2.6) are numerically equal to each other. A typical value of *NEP* for photomultipliers amounts to 10^{-16}–10^{-15} W.

One can also introduce related quantities: the *detectivity D* and *normalized detectivity D^**. The detectivity D is simply the reciprocal value of $NEP (D = 1/NEP)$, and the normalized detectivity is defined as

$$D^* = \sqrt{\mathscr{A}}/NEP = D\sqrt{\mathscr{A}} \quad (\mathrm{cm\,Hz^{1/2}/W}). \tag{2.7}$$

A definition such as this is introduced because for most detectors the *NEP* scales with the sensitive area $\mathscr{A}$ of the detector like $\sqrt{\mathscr{A}}$. Therefore the normalized detectivity D^* enables us to compare detectors of different sizes. A photomultiplier with a photocathode area of $\mathscr{A} \cong 5\,\text{cm}^2$, $NEP = 10^{-16}\,\text{W/Hz}^{1/2}$, and $\Delta f = 1\,\text{Hz}$ would feature $D^* \cong 2.3 \times 10^{16}\,\text{cmHz}^{1/2}/\text{W}$. Let us note, however, that the detectivities D and D^* are most often used to characterize semiconductor detectors rather than photomultipliers.

A more detailed analysis of noise is beyond the scope of this book. We should only note that the total photodetector noise has several components: the shot noise of the photocathode, the noise due to the statistical character of the secondary emission from dynodes, Johnson's noise in the load resistor, the so-called $1/f$ noise, and in the case of semiconductor detectors, generation-recombination noise is added, etc. The reader is referred to [6, 9].

On the other hand, the crucial task in luminescence spectroscopy is to optimize the signal-to-noise ratio, S/N. This will be discussed in detail in Section 2.5.

Here we shall briefly deal with one type of noise only—shot noise. This noise is a direct result of the quantum (corpuscular) nature of both the photons in the input optical signal and the electric charge generated in a detector. The number of incoming photons per unit time fluctuates around a certain average value. Statistical fluctuations in the stream of photons emitted by a classical source obey Poissonian photon statistics (some special cases violating Poissonian statistics are mentioned in Chapter 17 and Appendix L). Obviously, this photon-related noise component is independent of the particular detector, and represents an inevitable property of the incoming signal and thus also the lowest possible noise level of any optical detection system. An analogous statistical description applies to the internal detector process of photoelectron emission upon absorption of a photon. The current generated by a detector thus fluctuates around an average value (i.e. the signal S) and—in analogy with similar statistical ensembles—the relevant fluctuations are equal to the square root of the signal average. This means that the mean square value of the shot noise N reads

$$N = \sqrt{S}. \tag{2.8}$$

This kind of noise, increasing at higher signal levels, is shown in Fig. 2.6 and is noticeable also in Figs 1.1 and 1.2 (the noise of the background level is of different origin).

To conclude our account of photomultipliers we give several practical hints. A photomultiplier is a sensitive, fragile, and expensive electronic tube that must be handled with care. When not in use, the photomultiplier tube should be stored in a light-tight box. Its installation into a housing must be done under dimmed light and exposure of the photocathode to daylight should be as short as possible. The operational high voltage must be applied at least half an hour prior to measurement in order to stabilize the sensitivity $k(\lambda)$, the dark current value I_{d} and the noise properties. Varying or even switching on and off the applied high voltage during an experiment (e.g. for fear of exceeding the maximum allowed anode current) is a quite common mistake, which could lead to erroneous results—any change in voltage may disturb the stabilized operation conditions and cause a temporary variation in photomultiplier sensitivity, lasting up to tens of minutes (the so-called photomultiplier hysteresis). The correct approach to avoid overflow of I_{am} is decreasing the incoming light flux (using, e.g., calibrated neutral density filters). In the case of a cooled photomultiplier, the best practice is to keep cooling and permanently applying the high voltage even between measurements, the photocathode being kept in darkness. In such a way the dark current and the noise amplitude are minimized in the long term.

There might be two additional causes of irreversible noise increase (besides the above mentioned overflow of I_{am}): exceeding the maximum allowed voltage between the photocathode and anode U_{am} (this value is given by the manufacturer, as well as I_{am}) and penetration of moisture into the cooled photomultiplier housing. We recommend the maximum allowed voltage U_{am} should never be applied to a photomultiplier. First of all, the optimum operational voltage always lies below U_{am} (sometimes even by hundreds of volts) and increasing the voltage closer to U_{am} does not lead to better performance. Secondly, a possible inaccuracy of the voltmeter monitoring the applied high voltage can result in an unintentional overflow of U_{am}, entailing consequences much like the case of overflowing of I_{am}. Moisture penetrating towards a cooled photomultiplier can condense in the socket and leakage currents could cause irreversible changes of the properties of the photomultiplier. A good commercially supplied housing should be leakproof, anyhow.

Finally, let us remark that a magnetic field can be detrimental to the functioning of a photomultiplier as it alters the optimized trajectories of electrons between electrodes. The location of a photomultiplier (and of all auxilliary experimental apparatus) in a laboratory must be chosen to be free of disturbing electromagnetic fields. A special magnetic shield can possibly be used (these are offered by photomultiplier manufacturers).

Avalanche photodiode

The basic principle of an avalanche photodiode is the same as in other photovoltaic semiconductor photodetectors, namely, the *internal photoelectric effect*—the generation of an electron–hole pair through absorption of a photon. The photogenerated charge is then separated in an electric field established across the depletion layer of a p-n junction, and a photovoltage is created at the detector output, which is subsequently processed by electronics. Semiconductor photodiodes have a much wider spectral range compared to photomultipliers, especially toward the near-infrared region; their long-wavelength cut-off is determined by the bandgap width of the photosensitive semiconductor material. For example, the range of sensitivity of a Ge and InAs diode is 600–1800 nm and 1000–3200 nm, respectively. Despite their relatively low detectivities ($D^* \leq 10^{12}$ cm Hz$^{1/2}$/W), the p-n or p-i-n photodiodes[1] are

[1] An ultrasensitive germanium detector introduced by Edinburgh Instruments, Ltd., represents an exception suitable to detect low luminescence light fluxes. It is based on a p^+-i-n^+ diode made of ultrapure germanium and cooled by liquid nitrogen. The detection range is 800–1700 nm and its normalized detectivity of 5×10^{14} cmHz$^{1/2}$/W is comparable to photomultipliers. However, high sensitivity is counterbalanced by a very slow response, big size, considerable weight, and a relatively complicated operation.

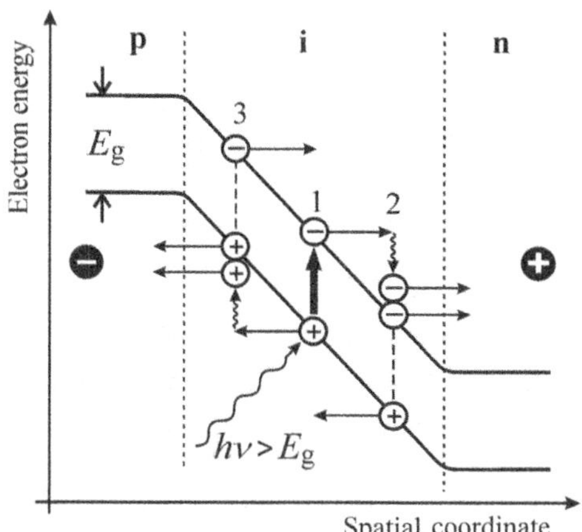

Fig. 2.7
The principle of an avalanche photodiode. Primary electron–hole pairs are created by absorption of photons $h\nu > E_g$ (1). A pair is dissociated and both the electron and hole are accelerated in opposite directions by the strong electric field in the intrinsic layer of the reverse biased p-i-n structure. After gaining enough energy the accelerated electron (hole) can create by impact ionization a new electron–hole pair (2) or (3) and the process can continue in an avalanche-like manner.

applied in luminescence spectroscopy for the detection in the infrared region beyond ~ 1000 nm, i.e. beyond the red cut-off in sensitivity of the common photomultipliers. However, current photodiodes lack the effect of photocarrier multiplication. Such multiplication is only possible in avalanche photodiodes that have internal *gain* in the range of $G_{AP} \sim 10$–500. The avalanche photodiode detectivity D^* is then increased but usually does not reach values comparable to photomultipliers.

The principal functional layout of an avalanche photodiode is shown in Fig. 2.7. Basically one deals with a p-i-n semiconductor structure under high reverse bias. Upon absorption of a photon $h\nu > E_g$ an electron–hole pair is created. This pair is dissociated in the strong electric field of the intrinsic layer (i) and both the electron and hole are accelerated in opposite directions. If the energy acquired by the accelerated carriers exceeds E_g (which is not always the case, because carriers are losing their energy in collisions with the crystal lattice), another electron–hole pair can be generated by *impact ionization*. Then the whole process may be repeated and an avalanche-like increase in number of carriers takes place.

Even if the principle of an avalanche photodiode looks very simple, there are several fundamental and technical obstacles making fabrication of an efficient and reliable device difficult. First of all, contrary to what might seem profitable at first sight, simultaneous amplification of both the electrons and holes (Fig. 2.7) is not favourable for obtaining high gain. In fact, such a situation is undesirable because this process is unstable and can result in avalanche breakdown and damage to the diode. In addition, the symmetrical amplification as a random process gives rise to a higher noise level of the detector and, moreover, prolongs the risetime of the avalanche diode, thereby reducing the frequency bandwidth of the diode. Therefore, avalanche diodes should be made of a material in which only one type of carrier is amplified. Let us characterize the probability of ionization by electrons (or holes), i.e. the number of ionization events per unit distance, by the *ionization coefficient* α_e (or α_h). Then an important avalanche photodiode parameter can be defined, namely, the ionization ratio $\kappa = \alpha_h/\alpha_e$ and it is required that κ tends either to zero $\kappa \to 0$ (electron multiplication) or to infinity $\kappa \to \infty$ (hole multiplication). The best commercial avalanche photodiodes based on silicon have $\kappa \approx 0.006$. In the extreme case of $\kappa = 0$ the photodiode gain is simply given by $G_{\mathrm{AP}} = \alpha_{\mathrm{e}} w$, where w is the width of the multiplication region.

Another difficulty one is faced with when designing an avalanche photodiode arises from this simple reasoning: to achieve high diode sensitivity (i.e. to maximize the number of primary electron–hole pairs), the intrinsic (i) layer should be made as thick as possible. On the other hand, the inevitable presence of a high electric field across the layer requires this layer to be thin. The last requirement is of a technical rather than essential nature: the volume of the amplifying material should be limited in order to minimize the possible occurrence of material inhomogenities that could cause a local overrun of the breakdown electric field, thus creating a microplasma and eventually leading to photodiode damage. This conflict of opposite demands is solved by the special design of the avalanche photodiode—the wide photon-absorption layer is separated from the thin electron-multiplication layer ('separate-absorption-multiplication'). Evidently, fabrication of good avalanche photodiodes involves the use of both a perfect homogeneous material and advanced technology. The first avalanche photodiodes produced in the 1970s suffered from excess noise and easy breakdowns but subsequent radical development of semiconductor technologies removed those constraints.

Finally, the requirement to minimize the multiplication volume (as well as to achieve fast response) leads to a small size of the sensitive area $\mathscr{A}_{\mathrm{AP}}$ of the avalanche photodiode (of the order of mm^2 or smaller).

The structure of an avalanche photodiode together with the profile of the internal electric field are presented in Fig. 2.8 (in this case hole multiplication occurs—photon absorption takes place in an i-InGaAs layer from which the holes drift in the weak electric field to an intrinsic InP multiplication layer) [10]. Obviously, the highest allowed bias applied to an avalanche photodiode must not be exceeded, similarly to photomultipliers.

Let us briefly compare photomultipliers with avalanche photodiodes from the point of view of luminescence spectroscopy. An advantage of avalanche photodiodes as solid-state devices over fragile vacuum tubes is their small compact size and robustness. On the other hand, noise properties of avalanche photodiodes are generally worse as expressed by the mean value of the dark noise [11]

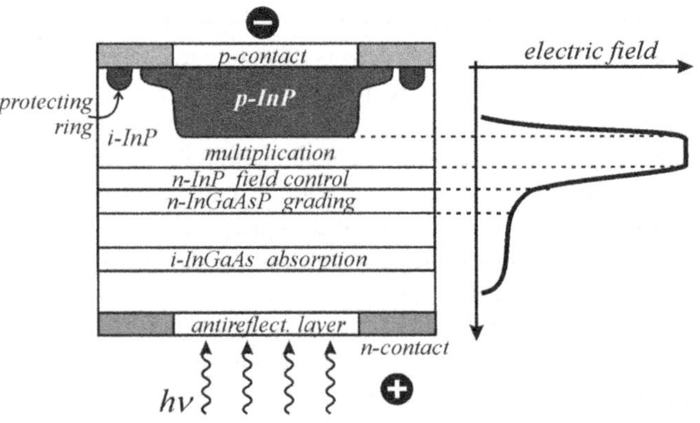

Fig. 2.8
Internal structure of an InGaAs/InP avalanche photodiode with separated absorption and multiplication layers and the corresponding profile of the electric field (right). Adapted from Ferrari *et al.* [10].

$$N_{\mathrm{AP}} = \left(2eI_{\mathrm{dAP}}\bar{G}_{\mathrm{AP}}^2 F(G_{\mathrm{AP}})\Delta f\right)^{1/2}. \tag{2.9}$$

When compared to eqn (2.6) which is valid for photomultipliers we notice firstly in (2.9) the *squared* mean value of the gain $\bar{G}_{\mathrm{AP}}^2$ and, secondly, an additional noise factor $F(G_{\mathrm{AP}})$. The squared factor arises as a consequence of the contribution of both types of carriers (electrons and holes) to the photodiode current (even if only one type of carrier is multiplied). The factor $F(G_{\mathrm{AP}})$ arises from the stochastic nature of the ionization process. For multiplication of electrons this factor is given by

$$F(G_{\mathrm{AP}}) = \kappa\, G_{\mathrm{AP}} + (2 - 1/G_{\mathrm{AP}})(1 - \kappa). \tag{2.10}$$

In the case of an Si avalanche photodiode with parameters $\kappa = 6 \times 10^{-3}$ and $\bar{G}_{\mathrm{AP}} = 100$ we obtain $F(G_{\mathrm{AP}}) \cong 2.6$. The *NEP* of an avalanche photodiode can be expressed, using (2.9), as

$$NEP_{\mathrm{AP}} = \frac{\sqrt{2eI_{\mathrm{dAP}}\,\bar{G}_{\mathrm{AP}}^2\,F(G_{\mathrm{AP}})}}{k_{\mathrm{AP}}(\lambda)}$$

and for the normalized detectivity it follows from (2.7) that

$$D_{\mathrm{AP}}^* = \frac{\sqrt{\mathscr{A}_{\mathrm{AP}}}\,k_{\mathrm{AP}}(\lambda)}{\sqrt{2eI_{\mathrm{dAP}}\,\bar{G}_{\mathrm{AP}}^2\,F(G_{\mathrm{AP}})}}.$$

Finally, the ratio of the normalized detectivity of a typical photomultiplier D_{PM}^* (with gain $G_{\mathrm{PM}} \approx 10^5$) to that of an avalanche photodiode D_{AP}^* is found to be

$$\frac{D_{\mathrm{PM}}^*}{D_{\mathrm{AP}}^*} = \sqrt{\frac{\mathscr{A}_{\mathrm{PM}}}{\mathscr{A}_{\mathrm{AP}}}}\sqrt{\frac{I_{\mathrm{dAP}}}{I_{\mathrm{dPM}}}}\,\frac{k_{\mathrm{PM}}(\lambda)}{k_{\mathrm{AP}}(\lambda)}\left(\frac{\bar{G}_{\mathrm{AP}}\sqrt{F(G_{\mathrm{AP}})}}{\sqrt{G_{\mathrm{PM}}}}\right), \tag{2.11}$$

where I_{dAP} and I_{dPM} stand for the dark currents of the avalanche photodiode and photomultiplier, respectively, and $k_{\mathrm{AP}}, k_{\mathrm{PM}}$ are the corresponding sensitivities.

If we try to compare, using eqn (2.11), two particular detectors—a Hamamatsu R316 photomultiplier with a Perkin Elmer InGaAs avalanche photodiode (the spectral sensitivities of both detectors overlap in the range 700–1100 nm), we find $D_{\mathrm{PM}}^* \cong 50D_{\mathrm{AP}}^*$. However, a comparison like this must be considered with care, because all device parameters are strongly dependent on the radiation wavelength as well as on operational conditions. Sometimes even the datasheets for detectors, supplied by the manufacturers, do not contain all the necessary parameters.

Nevertheless, in general it holds true that the normalized detectivity of avalanche photodiodes does not attain the photomultiplier values. Moreover, the small size of the sensitive area of the avalanche photodiode is hardly appropriate for luminescence spectroscopy—to ensure tight focusing of luminescence radiation at the exit slit of a monochromator onto a small area of $1\,\mathrm{mm}^2$ is basically impossible; a large part of the signal would thus be lost. Here, the optimum choice is indeed a photomultiplier because its photocathode of typically rectangular shape with size about $3 \times 12\,\mathrm{mm}^2$ fits the cross-section of the light beam emerging from the exit slit quite well. Therefore, avalanche photodiodes are applied in unconventional spectroscopic techniques only, like time-correlated single-photon counting and the determination of photon autocorrelation (see Subsection 2.4.2 and Chapter 17). They can also be used for the detection of extremely low-level luminescence signals from small sources—nanoobjects, like single molecules or nanocrystals, in a microspectroscopy set-up (Chapter 17). Single-photon counting avalanche photodiode modules are produced, e.g., by Perkin Elmer, ID Quantique and Hamamatsu.

In these special luminescence measurements, the avalanche photodiode operates in the so-called Geiger mode, the applied bias being higher than the breakdown voltage and the diode gain being very high ($G_{\mathrm{AP}} \cong 10^5 - 10^6$). A particular diode design simultaneously ensures that no current flows across the diode except in the case when an avalanche event is initiated by a released photoelectron; after generating a current pulse the diode is closed again (in analogy with the Geiger–Müller counter of radioactive radiation). In this way individual photons may be resolved. Normalized detectivity of photon counting modules based on silicon avalanche photodiodes can be as high as $5 \times 10^{13}\ \mathrm{cmHz}^{1/2}/\mathrm{W}$. Obviously, in this case cooling of the avalanche photodiode is of particular importance in order to suppress the dark counts (the dark counts of the best diodes can be below 20 counts/s).

Multichannel detectors

Photomultipliers and avalanche photodiodes are single-channel photodetectors, which means that they are able to detect, at a given time, the photon flux at a single point (area) in space only. Alternatively, the so-called multichannel detectors, in the form of linear or two-dimensional arrays of tiny photosensitive solid-state devices, are able to detect the photon flux simultaneously at many spatial points. In luminescence spectroscopy, when the single-channel detector is placed behind the exit slit of a monochromator, only a single point from the spectrum is recorded. The entire spectrum is then acquired sequentially, step by step, by rotating the dispersion element (grating or prism) and thereby scanning the spectral image across the exit slit. Such a measurement is, of course, quite lengthy and the obtained spectrum can be distorted owing to, for instance, the long-term instability of both the excitation source and/or the sample (photochemical changes, generation of defects, etc.).

On the other hand, multichannel detectors, fixed in the output plane of a spectrograph, simultaneously detect the intensity (photon flux) at many spectral points within some spectral range, or even the whole spectrum without moving the dispersion element (Fig. 2.9). This allows for significant shortening of the signal acquisition time, avoiding instability effects, monitoring the entire spectrum in real time, eliminating inaccuracy of the dispersion element scanning mechanism, etc. In addition, this approach considerably simplifies those experiments that apply a pulsed mode of luminescence excitation. It may be worth evoking memories of the first type of multichannel detector used in spectroscopy, which was a photographic plate. Its replacement by photomultipliers and scanning monochromators during the 1950s and 1960s

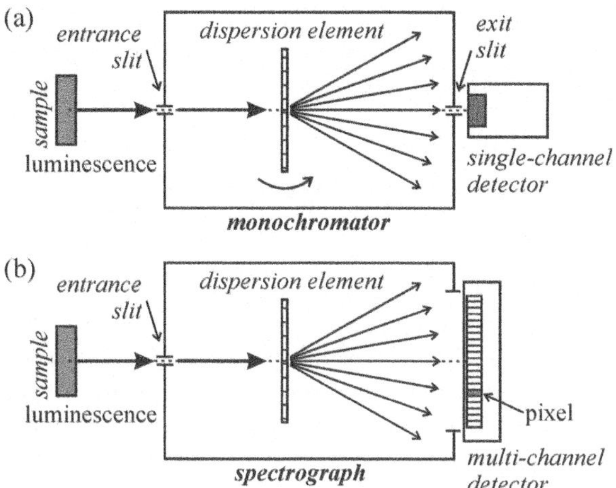

Fig. 2.9
Scheme of (a) a monochromator equipped with a rotating dispersion element (prism or grating) and a single-channel detector, (b) a spectrograph with a fixed dispersion element and a multichannel detector.

introduced a qualitative improvement to optical, and especially luminescence, spectroscopy. Even if the photomultiplier is a single-channel detector only, its many times higher sensitivity and reproducibility compared to photographic emulsions enabled researchers to investigate many new semiconductor phosphors. This finally resulted in the discoveries and inventions of, e.g., injection electroluminescence, light-emitting diodes, semiconductor lasers, and the accelerated development of optoelectronics and its applications.

Present-day multichannel detectors combine detection sensitivity—comparable to that of photomultipliers—with the advantages of a 'photographic plate' and the mechanical robustness of a solid-state detector. The better noise and detection properties of multichannel detectors arise from the so-called *Fellgett* (multiplex) *advantage*, which makes it possible to integrate a weak luminescence spectrum for an arbitrarily long time.

Let us give a more detailed explanation. Suppose we have a linear array detector containing n elements (called pixels, an abbreviation of 'picture element'), the sensitivity of each of them being the same as that of a photomultiplier. In the case of a scanning monochromator and a single-channel detector, the detector (photomultiplier) 'sees' any of the n spectral elements separately, say for one second of time, therefore the entire spectrum is recorded within n seconds, featuring a certain signal-to-noise ratio $(S/N)_0$. A multichannel detector records all of the n pixels simultaneously, therefore, supposing the integration time to be also one second, the resulting signal-to-noise ratio will be the same, i.e. $(S/N)_0$. The spectrum, nevertheless, is recorded within $1/n^{\text{th}}$ of the time required to be measured (to the same 'quality') by the photomultiplier. If we now record (integrate) the spectrum also for the time span of n seconds, we obviously obtain an n-times higher signal nS but—according to eqn (2.8)—only $\sqrt{nS}$-times higher noise level. Consequently, the spectrum recorded in this way will feature the signal-to-noise ratio

$$\left(\frac{S}{N}\right)_{\text{Fellgett}} = \frac{nS}{\sqrt{nS}} = \sqrt{n}\left(\frac{S}{N}\right)_0. \tag{2.12}$$

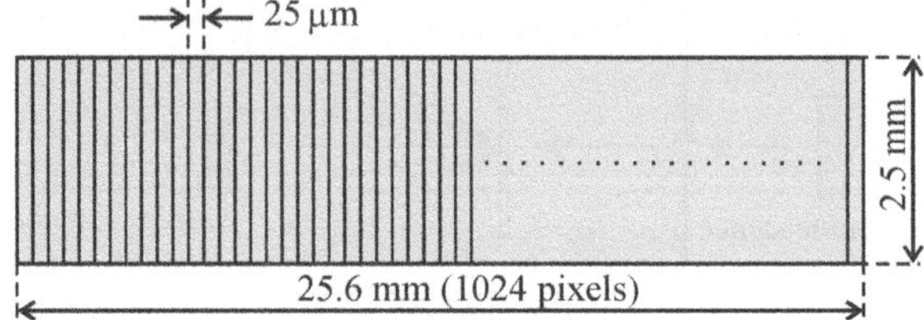

Fig. 2.10
Standard array of silicon photodiodes.

This increase in signal-to-noise ratio is known as Fellgett advantage. Let us consider $n \cong 1000$ (indeed, the number of pixels is often $1024 = 2^{10}$); then according to eqn (2.12) the increase in the signal-to-noise ratio is $\sqrt{1000} \approx 32\times$ (!). The foregoing example is valid strictly in the case of shot noise only but, as we shall see shortly, most other noise components can be successfully suppressed in multichannel detectors; in particular the thermal noise is effectively suppressed by cooling.

Of course, a further increase in S/N can in principle be achieved by further prolonging the acquisition time, as is immediately evident from eqn (2.12). With an acquisition time of m seconds ($m > n$) the signal-to-noise ratio is improved $\sqrt{m}$-times against $(S/N)_0$. It must be remembered, however, that in this way we deliberately renounce other benefits of multichannel detectors, namely, the elimination of possible sample 'fatigue' or of other unwanted variations in time of both the sample and the whole experimental set-up; on the other hand, such a measure is often beneficial because it facilitates successful recording of ('recovered from the noise') the spectra of extremely low light fluxes, undetectable by making use of photomultipliers.

Two of the most common types of multichannel detectors are *photodiode arrays* and *charge-coupled-device* (CCD) *cameras*. A linear array of photodiodes is made up of a row of tiny (silicon) photodiodes (Fig. 2.10). Every photodiode (pixel) is reverse biased and charged like a capacitor. The incident photons generate electron–hole pairs, increasing electric conductivity in the capacitor and consequently its discharging. Then, a read-out period follows, during which the diodes are charged up by a current that is proportional to the number of absorbed photons. The usual duration of the read-out period in the case of the 1024 diode array shown in Fig. 2.10 is 13 ms. The spectral range of the sensitivity is, similarly to common single Si photodiodes, about 300–1100 nm, and the spectrum acquisition time can be arbitrarily varied. The discrete structure of the photodiode array seemingly imposes certain limits to the achievable *spectral resolution*. However, luminescence spectra are usually relatively broad and so this fact does not usually pose serious problems.

The dark signal in a diode array, given by thermally generated electron–hole pairs, determines the sensitivity threshold of the array but can be substantially reduced by cooling—mostly thermoelectric cooling down to −20°C is applied.

The sensitivity of the diode array is not necessarily sufficient for the purposes of luminescence spectroscopy. Nowadays, the most sensitive photodetectors, aimed at detecting extremely weak signals and broadly exploited especially in luminescence spectroscopy, are CCD cameras. Their basic principle again consists in the generation of charge carriers by incident photons, but

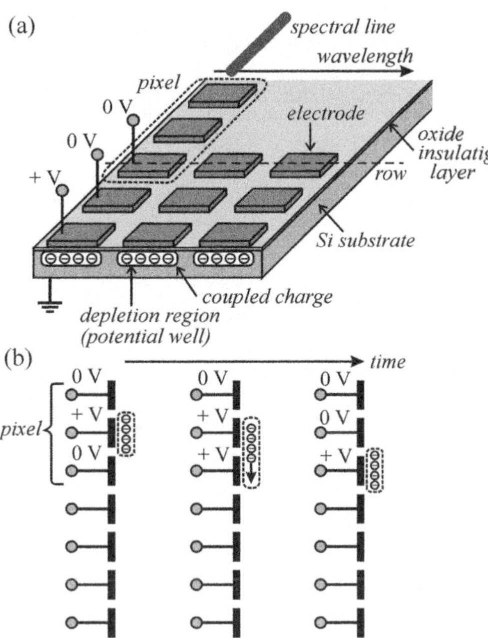

Fig. 2.11
(a) Schematics of a CCD detector. (b) Charge transfer in a CCD detector. Adopted from ISA Jobin-Yvon-Spex flyers [12].

their construction and read-out scheme differ substantially from the photodiode arrays.

One deals with a system of discrete small components, in which the individual pixel is usually defined as a set of three electrodes with a variable applied bias [12]. The electrodes are made of conductive (highly doped) polycrystalline silicon and deposited on a photosensitive silicon wafer of lower conductivity; the said materials are separated by a thin layer of insulating SiO_2 (Fig. 2.11(a)). The photoelectrons generated by incident light may, depending on the applied bias, be trapped and stored in potential wells that are created in silicon beneath the electrodes. The light-detection process involves three steps:

1. Absorption of incident photons in photosensitive silicon and generation of electron–hole pairs. The number of pairs in each pixel is proportional to the incident light intensity. When positive bias is applied to the pixel central electrode, only the electrons are trapped in the potential well while holes are drained to earth.
2. Transfer of the collected charge from a given pixel to the next one. By applying the bias to the outer electrode of the pixel, the electron cloud is shifted to the pixel edge and eventually to the next pixel in the column, i.e. along the particular spectral line or a specific wavelength. This process is repeated until the charge from each pixel of a row reaches a read-out register located along the edge of the CCD chip, i.e. in parallel with the wavelength axis. This is accomplished by gradually varying the bias on individual pixels (see Fig. 2.11(b)). It is now clear why a pixel is composed of three electrodes. A great advantage of this mode of information read-out is the extreme reduction of the read-out noise due to the use of a single

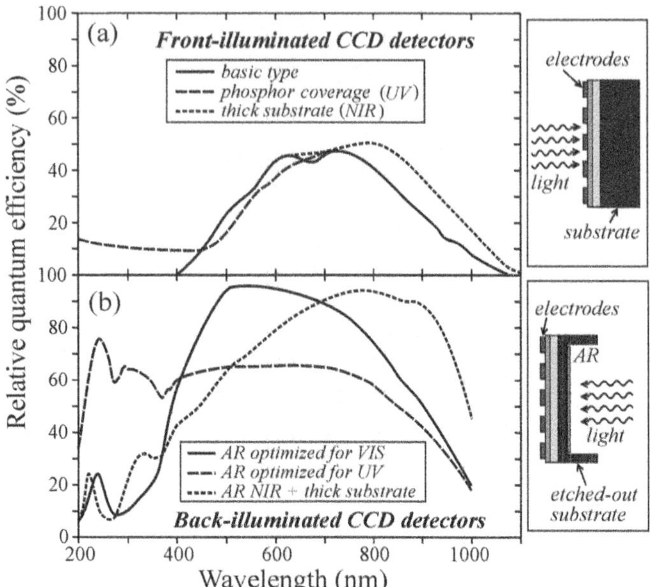

Fig. 2.12
Spectral response of a silicon CCD camera. (a) Sensitivity of a standard front-illuminated CCD detector (extension of the detection range into the IR region—short dashed line—can be achieved by using a thicker substrate that absorbs effectively longer wavelengths). (b) Spectral sensitivity of a back-illuminated chip where absorption of incident light in electrodes is avoided. Extension into the UV region can be obtained for example by depositing a phosphor layer (emitting visible light under illumination by UV light) directly onto the detector. AR means antireflection coating. Adapted from the ISA Jobin-Yvon-Spex flyer [12].

common read-out register for all of the pixels instead of employing a read-out system in every pixel.

3. Conversion of electric charge into voltage, amplification and digitization. The stored charge is transferred row by row into the read-out register and then into an output node, converted to an appropriate voltage, amplified and turned into a digital signal. Finally, the number of detected photoelectrons in every pixel of a CCD is stored, and the obtained picture (spectrum) can be visualized on a display, numerically treated, etc.

The typical area of a pixel in a spectroscopic CCD detector is about $20 \times 20\,\mu\text{m}^2$, and the full size of the sensitive area is, e.g., $26.6 \times 6.6\,\text{mm}^2$ (1024×256 pixels with a pixel size of $26 \times 26\,\mu\text{m}^2$).

The spectral sensitivity of a CCD detector is determined by its material—usually silicon. The typical range of about 400–1000 nm can be modified to some extent by adjusting the detector surface and thickness and the mode of illumination (front or back-illuminated), see Fig. 2.12. The experimental *spectral coverage* depends on the physical length of the detector and on the monochromator dispersion, and it can obviously be expressed as

$$\Delta\lambda_{\text{cov}} = (\text{reciprocal linear dispersion[nm/mm]}) \times (\text{CCD detector length [mm]}).$$

We have to stress that multichannel detectors should not be coupled to a common scanning monochromator, by merely removing its output slit and replacing it by a CCD camera or a diode array. An ordinary monochromator is able to image an aberration-free spectrum (i.e. to create a spectrally resolved projection of the entrance slit into the exit slit plane) of a very narrow wavelength range only, covering roughly the slit width (~ 1 mm). For application with multichannel detectors, special *imaging* spectrographs are available. Their optical aberrations must be corrected over the whole size (height and width) of the detector. For example, combination of a Horiba Jobin Yvon iHR 320 imaging spectrograph having a dispersion of 2.64 nm/mm with a CCD chip of 1024 × 256 pixels (pixel size 26 × 26 μm^2) shows a spectral coverage of $\Delta\lambda_{\text{cov}} = 70.3$ nm. In order to measure a wider spectral range the ultimate spectrum must be glued together from several segments with a shifted central wavelength.

The sensitivity of the CCD detector is extremely high. Thermal noise can be almost eliminated by cooling down to −90°C or even −140°C when the dark signal can be suppressed down to an unbelievable 1 electron/pixel/h only (!) (In practise, the recommended temperature is usually above −110°C because lower temperatures would decrease the quantum efficiency of the detector; efficient cooling is provided either by liquid nitrogen or a multistage thermoelectric Peltier element.) The lowest detectable signal is then limited by the so-called read-out noise, or possibly only by the shot noise originating mostly in the signal itself.

Let us put forward a few examples:

1. Suppose the total root mean square noise value N_{CCD} of the detector is given as a combination of three components

$$N_{\text{CCD}} = \sqrt{|n_{\text{shot}}|^2 + |n_{\text{dark}}|^2 + |n_{\text{read}}|^2}, \qquad (2.13)$$

with n_{shot}, n_{dark} and n_{read} standing for the shot noise, dark signal and the read-out noise, respectively. Assuming a relatively low flux of incident photons of 10 photon/pixel/s entails (in the visible range) a power flux density of $\sim 4 \times 10^{-13}$ W/cm^2. Considering the quantum efficiency of the detector to be 50% we obtain 5 photoelectron/pixel/s. Setting the acquisition time to be 600 s (10 min), the total integrated signal is 5 photoelectron/pixel/s ×600 s = 3000 photoelectron/pixel. Shot noise of such a signal is $n_{\text{shot}} = (3000)^{1/2} = 54.77$. Suppose further this cooled detector has a dark signal of 1 electron/pixel/min, then the total dark signal accumulated during 600 s is 10 electron/pixel and the corresponding noise component is $n_{\text{dark}} = 10^{1/2} = 3.16$. Finally, the typical read-out noise of a 1024 × 256 pixel CCD detector with slow reading is about $n_{\text{read}} = 4$ electrons (root mean square value).

According to (2.13) the overall noise of the detection system is

$$N_{\text{CCD}} = \sqrt{(54.77)^2 + (3.16)^2 + 4^2} = 55.$$

The signal-to-noise ratio is $S/N = 3000/55 = 54.54$. Note that, were the detector completely free from both the dark current and read-out noise, then the best achievable signal-to-noise ratio would be $S/N = 3000/54.77 = 54.77$, i.e. only negligibly higher. We conclude that under an incident photon flux of 10 photon/pixel/s the intrinsic detector noise does not play any detrimental role—such measurement is referred to as 'shot noise limited'. The same conclusion would hold true even for a ten times lower incident flux of 1 photon/pixel/s!

2. Let us now consider a complelety different situation when an extremely low signal of only 1000 photon/s hits the *whole area* of a CCD detector

(1024 × 256 pixels^2 = 676 μm^2). This means, at a photon energy of 2 eV, an incident power density of 1.7×10^{-17} W/cm^2. Such extremely low photon fluxes take place, for example, when detecting luminescence excited by the tip of a scanning tunnelling microscope (STM) or photoluminescence of single semiconductor quantum dots (see Chapter 17). Suppose we have at our disposal the same CCD detector as in the previous example. This means with quantum efficiency of 50%, dark signal of 1 electron/pixel/min, read-out noise of $n_{\mathrm{read}} = 4$ electrons, and we also leave unchanged the acquisition time 600 s.

The average photon flux incident on a single pixel is 1000 (photon/s)/ 2.62×10^5 pixels = 3.8×10^{-3} photon/pixel/s; the corresponding number of photoelectrons is $0.5 \times 3.8 \times 10^{-3} = 1.9 \times 10^{-3}$ photoelectron/pixel/s, and after an integration time of 600 s we obtain the total signal as 1.9×10^{-3} photoelectron/pixel/s × 600 s = 1.14 photoelectron/pixel. The corresponding shot noise is $n_{\mathrm{shot}} = \sqrt{1.14} \approx 1.07$. From the preceding example we remember that $n_{\mathrm{dark}} = 3.16$, $n_{\mathrm{read}} = 4$ and thus the total CCD noise reads

$$N_{\mathrm{CCD}} = \sqrt{(1.07)^2 + (3.16)^2 + (4)^2} = 5.21.$$

The signal-to-noise ratio is thus $S/N = 1.14/5.21 \approx 0.22$. Taking $S/N = 1$ as the detection limit criterion, the light flux under consideration is not detectable and will be 'buried in noise'. Here the experiment is already clearly limited by the detector's noise itself!

Does this example imply that even sensitive CCD detectors are not able to detect photon fluxes of the order of 10^3 photon /detector/s? No, it does not, as we shall see. The solution could be to set a longer detection time and/or apply a less noisy detector.

3. Let us keep all the parameters from example 2 fixed, except the integration (acquisition) time, which we increase now from 600 s to 1 hour (3600 s). The signal will be increased to 6.84 photoelectron/pixel and the overall noise to $\sqrt{(2.61)^2 + (7.75)^2 + 4^2} = 9.1$. We obtain $S/N = 6.84/9.1 \cong 0.75 < 1$. We can see that even quite substantial prolongation of the acquisition time does not lead to $S/N = 1$, and further extension of the experimental time is usually out of the question (due to possible instabilities of sample properties or experimental conditions).
4. Let us select the best available CCD detector with a high quantum efficiency (90%) and almost zero dark signal (1 electron/pixel/h)—these are typical parameters of a back-illuminated liquid-nitrogen-cooled silicon CCD chip. Let the incident photon flux be the same as in example 2, i.e. 3.8×10^{-3} photon/pixel/s, and integration time is 600 s. Then the signal is $0.9 \times 3.8 \times 10^{-3} \times 600 = 2.05$ photoelectron/pixel, the shot noise is $\sqrt{2.05} = 1.43$, the dark signal is 1/6 electron/pixel and the corresponding noise is $\sqrt{1/6} = 0.41$. The read-out noise remains equal to 4 electrons. Now we have $N_{\mathrm{CCD}} = \sqrt{(1.43)^2 + (0.41)^2 + 4^2} = 4.27$, $S/N = 2.05/4.27 \cong 0.48 < 1$. Even now the spectrum remains undetectable, due to the relatively high read-out noise. The last chance is to extend the acquisition time; 40 min (2400 s) is on the limit but still acceptable. The reader then easily

finds that the resulting S/N ratio finally breaks the detection limit, being $S/N = 8.21/4.98 \cong 1.65 > 1$. The spectrum can thus be recorded within a 40-minute exposure time, even if quite noisy.

In conclusion, CCD detectors manage to display the whole spectrum of comparatively strong signals in real time (which, by the way, appears to be an excellent tool to adjust the experimental set-up in this case). The whole spectral range of such signals is detected almost instantaneously (no monochromator scanning is needed), all the wavelengths being recorded simultaneously. The CCD detectors have very low dark signal and their sensitivity is comparable to photomultipliers or even better. For detection of extremely low signals, the S/N ratio can be improved by increasing acquisition times or by binning signals from several neighbouring pixels. One of the drawbacks of a CCD is the relatively narrow spectral coverage, as another possible weakness may be their low read-out speed given by operating a single read-out register ('serial read-out'). CCD detectors must be handled with care, avoiding strong overexposure by light (like photomultipliers) and, in particular, discharge of static electricity into both the disconnected CCD camera and the control electronics connectors.

2.3 Monochromators and spectrographs

A *monochromator* serves to disperse the analysed light signal into a spectrum using a dispersion element (grating or prism), and then to cut out a narrow (quasi)-monochromatic part of the spectrum by the exit slit, located at the rear (output) focal plane. A monochromator can be tuned manually (e.g. for selecting the appropriate excitation wavelength) or scanned automatically by rotating the dispersion element—in this case the spectrum moves continuously across the exit slit and can be recorded using a single-channel detector.

A *spectrograph* projects the spectrum onto a fixed position in its output focal plane. The exit slit is replaced by a photographic plate or a multichannel detector. Most modern spectrographs are equipped with two output channels—a lateral output (selectable by a moving mirror), being fitted with a slit, enables application of the spectrograph in a scanning monochromator mode (Fig. 2.13).

An *imaging spectrograph* is a special type of spectrograph, equipped with purpose-built optics that reduces optical aberrations. The spectral image of the entrance slit is perfectly sharp over a large area at the output image plane (typically 1 cm high and 2 cm wide), not only in its central part (close to the optical axis) as was the case of a monochromator.

The terms *polychromator* and *spectrometer* can also be encountered in the literature. A *polychromator* has several parallel monochromatic outputs provided by multiple output slits. A *spectrometer* is a complete device designed for measurements of optical emission spectra, consisting of a monochromator (or spectrograph) coupled with a detector.[2]

[2] This nomenclature of spectral devices is not employed systematically in the scientific literature. The terms monochromator, spectrograph or spectroscope are often used arbitrarily like

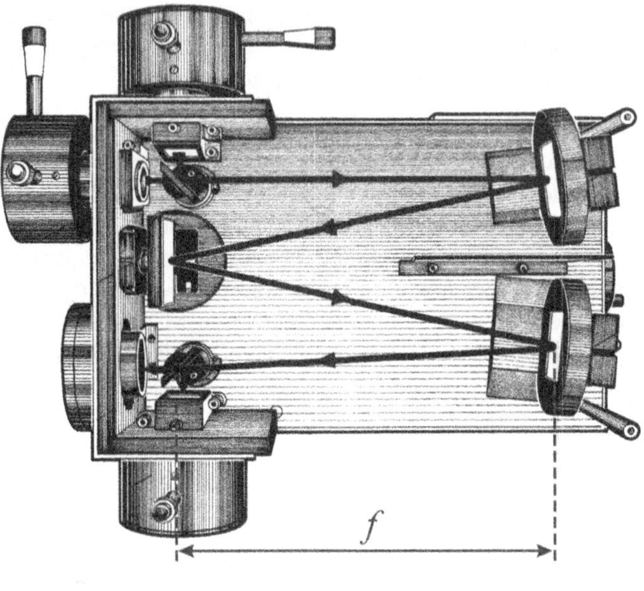

Fig. 2.13
A HR 320 monochromator–spectrograph. Adapted from Horiba Jobin Yvon flyer.

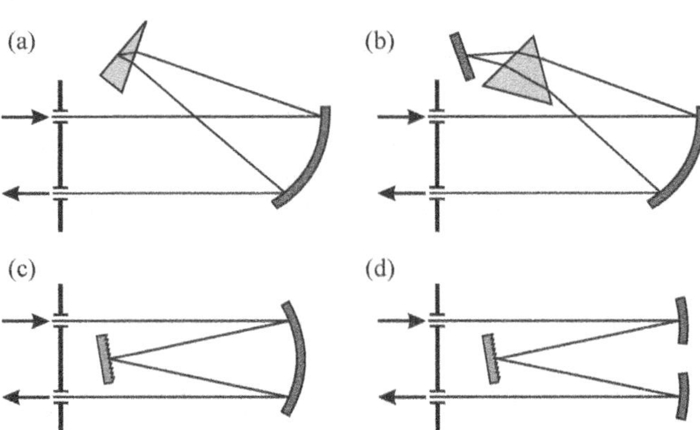

Fig. 2.14
The most common optical designs of spectral devices. (a) Littrow type with an autocollimating prism, (b) Walsch type with a prism, (c) Ebert construction with a grating, (d) Czerny–Turner type with a grating. The quality of spectral imaging increases from (a) to (d).

There are various designs of optical spectral devices with different levels of compensation for spherical aberration, astigmatism and coma, observed when displaying the spectral elements. Some of these designs are shown in Fig. 2.14. Most of the currently fabricated monochromators suitable for luminescence spectroscopy are of the Czerny–Turner type (Fig. 2.14(d))—see for example the HR 320 monochromator–spectrograph in Fig. 2.13. Other, somewhat less familiar optical schemes are those by Ebert–Fastie, Gillies, Wadsworth, the Rowland's construction with a concave grating and many others; for more details the reader is referred to the special literature [13, 14, 15].

The basic quantities describing the properties of a monochromator or spectrograph are dispersion, resolving power, and throughput (aperture). We shall

synonyms (originally a *spectroscope* denoted a device producing an optical spectrum for visual observation).

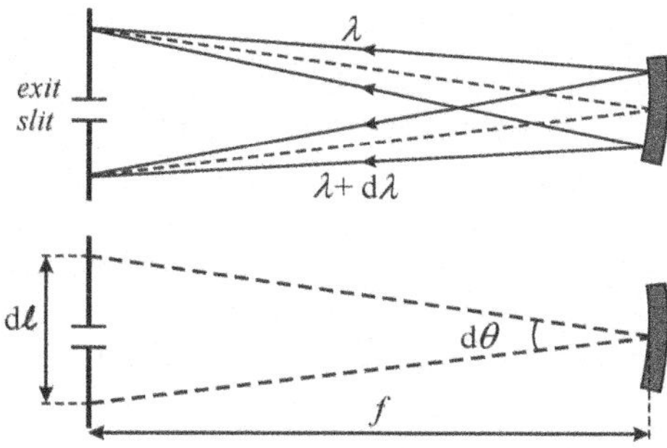

Fig. 2.15
Illustration of the definition of angular and linear dispersion.

describe them only to the limited extent necessary for the experimental foundations of luminescence spectroscopy.

2.3.1 Dispersion and resolving power

Let us consider two light beams whose wavelengths differ by $\mathrm{d}\lambda$. They leave the dispersion element at angles differing by $\mathrm{d}\theta$. We can define the *angular dispersion* as

$$\mathscr{D} = \frac{\mathrm{d}\theta}{\mathrm{d}\lambda}\ (\mathrm{rad/nm}). \tag{2.14}$$

The spatial separation $\mathrm{d}\ell$ between two spectral lines λ, $\lambda + \mathrm{d}\lambda$ in the exit slit plane depends on the focal length f of the spectral device, determined by a focusing mirror or lens, see Figs 2.13 and 2.15:

$$\mathrm{d}\ell = f\,\mathrm{d}\theta. \tag{2.15}$$

The *linear dispersion L* is then defined as

$$L = \frac{\mathrm{d}\ell}{\mathrm{d}\lambda} = f\,\frac{\mathrm{d}\theta}{\mathrm{d}\lambda} = f\mathscr{D}. \tag{2.16}$$

More frequently its inverted value, the so-called *reciprocal linear dispersion*, is used:

$$L^{-1} = \frac{\mathrm{d}\lambda}{\mathrm{d}\ell} = \frac{1}{f}\frac{\mathrm{d}\lambda}{\mathrm{d}\theta} = \frac{1}{f\,\mathscr{D}}\ (\mathrm{nm/mm}). \tag{2.17}$$

To make an order-of-magnitude estimate, let us consider a simple diffraction grating and light incident at an angle $\alpha \neq \alpha(\lambda)$. Light diffraction is described by the the grating equation $d(\sin\alpha + \sin\theta) = k\lambda$ (d is the grating constant, k is the diffraction order). By differentiating with respect to λ we obtain

$$\mathscr{D} = \mathrm{d}\theta/\mathrm{d}\lambda = \frac{k}{d\,\cos\theta}. \tag{2.18}$$

In order to estimate $\mathscr{D}$ we approximate $\theta \approx 0°$ (corresponding to the situation when the analysed light impinges on the grating at an angle $\alpha \neq 0$ and the spectrum is observed in a direction almost perpendicular to the grating plane). Then for a grating with 1200 grooves/mm operating in the first diffraction order we obtain an angular dispersion $\mathscr{D} = 1/d = 1.2 \times 10^3$ rad/mm $= 1.2 \times 10^{-3}$ rad/nm. A monochromator with $f = 500$ mm features a linear dispersion

$L = 500 \times 1.2 \times 10^{-3} = 0.6\,\text{mm/nm}$ and reciprocal linear dispersion $L^{-1} \approx 1.66\,\text{nm/mm}$. Therefore, in the case of luminescence spectroscopy, when—as we shall specify below—neither very high dispersion nor excessive spectral resolution is required, the spectral devices possess reciprocal linear dispersion in the range of 1–10 nm/mm. The importance of the reciprocal linear dispersion L^{-1} results from the fact that the spectral bandwidth $\Delta\lambda$ of a given device is simply determined by L^{-1} (declared by the manufacturer) and the mechanical slit width $\Delta\ell$ (set by the experimenter) via the relation

$$\Delta\lambda = L^{-1}\,\Delta\ell \tag{2.19}$$

that follows immediately from (2.17). In our example (a monochromator with $L^{-1} = 1.66\,\text{nm/mm}$) for a chosen slit width of $\Delta\ell = 0.5\,\text{mm}$ we thus find $\Delta\lambda = 0.83\,\text{nm}$. Note that eqn (2.19) is strictly valid only when both the entrance and exit slits are of the same width $\Delta\ell$. This setting is commonly used in spectroscopy for optimizing spectral brightness and facilitating spectral correction procedures (see Section 2.8).

The angular dispersion of a prism can be found by differentiating the condition of minimum deviation (see e.g. [14, 16]) which yields

$$\mathscr{D} = \frac{2\sin\,(\varphi/2)}{\sqrt{1 - N^2\sin^2\,(\varphi/2)}}\,\frac{\mathrm{d}N}{\mathrm{d}\lambda}, \tag{2.20}$$

where ϕ is the prism apex angle, and N and $\mathrm{d}N/\mathrm{d}\lambda$ are the refractive index and material dispersion of the glass (or other material) from which the prism is fabricated, respectively.

In the case of a monochromator, the measurement of the spectra is performed by rotating the dispersion element, thereby scanning the spectral image across the exit slit. Most grating monochromators are equipped with a so-called *sine-drive* scanning mechanism that ensures linear scanning of the wavelength in time, which means $\Delta\lambda = \text{const}\,\Delta t$. The principle of the sine-drive mechanism is shown in Fig. 2.16. There is a triangle ABC; one of its sides (a) is allowed to vary linearly in time t by means of a precise motor-driven screw, $a = a_0 + vt$. The length of the hypotenuse c, which is anchored by a ball-joint at point B, is constant. The second side b is carried in a sliding assembly allowing its length to be varied. Then the angle γ is a function of time, $\gamma = \gamma(t)$. By differentiating the goniometric relation $\sin\gamma(t) = (a_0 + vt)/c$ we find

$$\cos\gamma\,\mathrm{d}\gamma = (v/c)\mathrm{d}t. \tag{2.21}$$

A grating can be coupled with this mechanical system in such a way that $\gamma = \theta$. (Spectroscopic gratings have an asymmetric profile of grooves—the so-called blaze, which enables efficient diffraction of a selected spectral range into the desired angle and diffraction order.)

Then eqn (2.21) is replaced by

$$\cos\theta\,\mathrm{d}\theta = (v/c)\mathrm{d}t$$

and using (2.18) we obtain

$$(k/d)\mathrm{d}\lambda = \cos\theta\,\mathrm{d}\theta \quad \text{or} \quad (k/d)\mathrm{d}\lambda = (v/c)\mathrm{d}t,$$

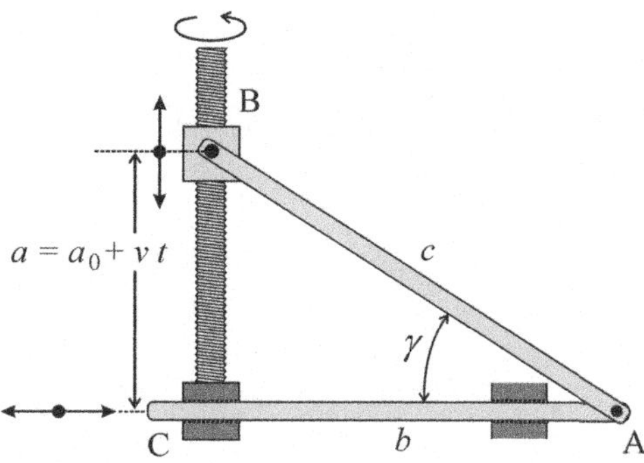

Fig. 2.16
Schematic of a sine-drive mechanism.

therefore, in the given diffraction order k we have $\Delta\lambda = \text{const}\ \Delta t$; the wavelength scan is linear in time.

A simple modification of the sine-drive (Fig. 2.16) is possible: the length of the hypotenuse c can be varied in time using the screw mechanism while a is kept constant. Then the reciprocal wavelength $\nu^* = 1/\lambda$, called the wavenumber, is a linear function of time $\Delta\nu^* = \text{const}\Delta t$. This design is referred to as a *cosecant-drive* (or just a *cosec-drive*) since here one deals with a linear temporal variation of the function $1/\sin\gamma = \text{cosec}\ \gamma$. This drive makes the spectrum to be scanned linear in the photon energy $h\nu = hc\nu^*$ instead of in wavelength, which could be advantageous for many experiments as well as their physical interpretation. Indeed, it is used in some Raman spectrometers. In the past few decades, most monochromators were provided with sine-drives but modern spectrographs are equipped with direct drives, where a motor rotates the grating directly (through some gear wheels). The number of motor steps per unit wavelength in the course of scanning is not constant and is driven by control software. The benefits of this design are a very high speed of grating rotation from one position to another, possible integration of more selectable gratings in one holder, and use of cheaper mechanical components.

Unlike in gratings, there is no simple mechanical way to ensure linear scanning of λ or ν^* in a prism monochromator. The angular dispersion of a prism as given by eqn (2.20) contains the dispersion relation $N = N(\lambda)$ which is strongly nonlinear in glasses and other optical materials. Therefore, spectra measured using a prism monochromator usually have a nonlinear wavelength scale; compare, e.g., Fig. 2.28 (a prism monochromator) with Fig. 1.1 (a grating monochromator).

Another important fact contained in relations (2.18) and (2.20) is worth pointing out: the dispersion of a grating or prism is independent of its size—e.g. two prisms made of the same material and having equal apex angles φ have equal dispersion regardless of size (Fig. 2.17). In other words, the dispersion of a spectral device is defined in the framework of geometrical optics, where diffraction due to finite beamwidth is not taken into account.

On the other hand, the *resolving power* R, defined with the aid of the smallest resolvable difference between two closely spaced spectral lines $\Delta\lambda$ as $R = \lambda/\Delta\lambda$, is limited just by diffraction of the light beam, whose diameter is driven by the lateral size of the dispersion element. A small prism or grating with a small effective area gives rise to sizeable broadening of spectral lines in

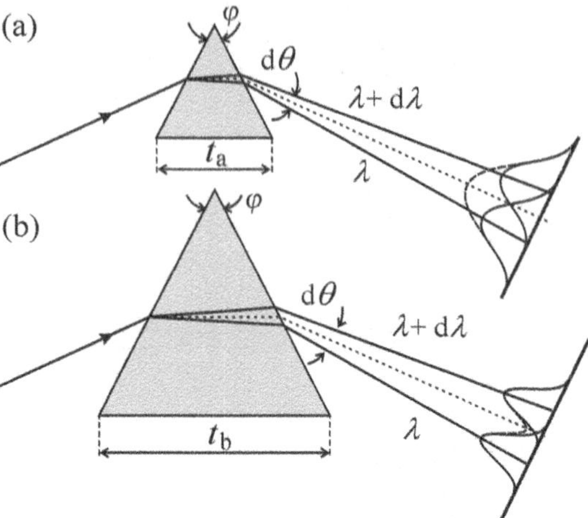

Fig. 2.17
Illustrating the distinction between dispersion and resolving power. Two prisms (a) and (b) have equal (angular and linear) dispersion but the bigger one (b) has higher resolving power, as shown by the profiles of two closely spaced spectral lines in the exit focal plane.

the exit slit plane due to light diffraction. Consequently, taking into account Rayleigh criterion for the resolution of two peaks, the resolving power is decreased (Fig. 2.17). But even having a sufficiently large dispersion element, the effective cross-section of the beam to be analysed may happen to be limited by the insufficient diameters of the lenses or mirrors used in the spectral device, and possibly also by improperly collimating the input radiation.

Along with diffraction due to the limited diameter of the light beam, another diffraction effect enters the play—diffraction at the entrance slit, which could negatively influence the output spectral image when the slit width is set too narrow, below some 'reasonable' limit.[3] Let us discuss both of these phenomena in greater detail.

To begin with, suppose we can neglect diffraction at the entrance slit. It is obvious that reducing the width of the monochromator slits leads, according to eqn (2.19), to narrowing of the transmitted spectral bandwidth $\Delta\lambda$ and, consequently, to increased spectral resolution. Let us employ the term *theoretical* or *ultimate resolving power* for $R_t = \lambda/\Delta\lambda_{opt}$, where $\Delta\lambda_{opt}$ is the (minimum) spectral bandwidth at which the input slit diffraction does not yet become evident, and thus an ultimate resolution R_t is achieved. Let us consider this situation, as occurring in a prism monochromator, displayed in Fig. 2.18. We remember that the diffraction angle on a circular aperture of diameter W (here the 'effective aperture stop of the monochromator') is expressed by the well-known relation

$$\delta\theta \approx \lambda/W, \tag{2.22}$$

if we omit the numerical factor 1.22 on the right-hand side for the sake of simplicity. Now, should two closely spaced lines be resolved ($\Delta\lambda = \Delta\lambda_{opt}$), the condition $\Delta\theta = \delta\theta$ must be satisfied according to Fig. 2.18. The angle $\Delta\theta$

[3] Note that in spectral devices equipped with a diffraction grating, the interplay between three diffraction effects is taking place—diffraction on the input slit, on the grating and diffraction due to the finite beam size.

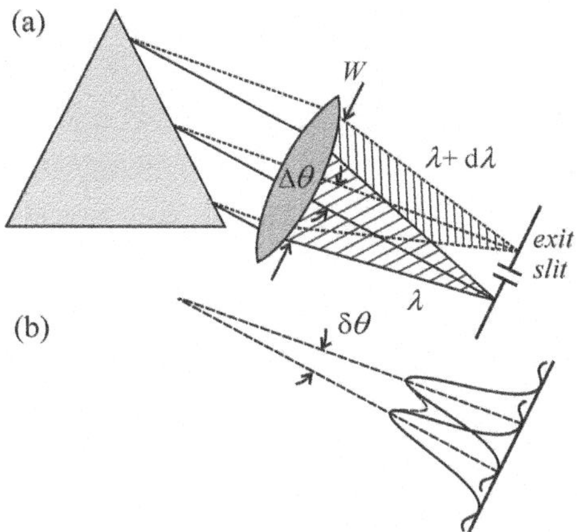

Fig. 2.18
Illustrating the resolving power of a simple prism monochromator. (a) Rays leaving the prism and propagating through an exit focusing lens. The lateral size of the beam W is basically determined by the prism dimensions. The images of two spectral lines in the plane of the exit slit have finite width due to light diffraction. (b) Profiles of two closely spaced spectral lines in the plane of the exit slit that meet Rayleigh resolution criterion (i.e. the centre of one linear diffraction pattern falls on the first minimum of the diffraction profile of the other line).

is determined by the angular dispersion of the prism; by combining eqns (2.14) and (2.22) we obtain

$$R_t = \lambda/\Delta\lambda_{opt} = \frac{\delta\theta\, W}{\Delta\lambda_{opt}} = \frac{\Delta\theta\, W}{\Delta\lambda_{opt}} \approx \frac{d\theta}{d\lambda} W = W\,\mathscr{D}. \tag{2.23}$$

Obviously, this relation is valid for any dispersion element. Importantly, it shows that there is a close relation between the angular dispersion $\mathscr{D}$ and the ultimate resolution R_t of a dispersion element, and also supplies a qualitative explanation for Fig. 2.17.

By inserting the angular dispersion into eqn (2.23) for the grating (2.18) or prism (2.20), respectively, we obtain, after some simple algebra, for the ultimate resolving power of the grating

$$R_t = mk \tag{2.24}$$

and for that of a prism (valid under the assumption that the whole prism entrance face is illuminated)

$$R_t = t\,\frac{dN}{d\lambda}. \tag{2.25}$$

In eqn (2.24) m stands for the number of illuminated grating grooves (it is easy to show that $m = W/d\cos\theta$), and t in eqn (2.25) denotes the width of the prism base, see Fig. 2.17. For example, a grating with 1200 grooves/mm and width of 5 cm has in the first diffraction order $R_t = 6 \times 10^4$; a prism made of a heavy flint glass ($dN/d\lambda \approx 2.7 \times 10^{-4}$/nm for blue light) with a base of $t =$ 5 cm features $R_t \approx 1.35 \times 10^4$. In reality, the ultimate resolution power may be decreased by one-third or even more due to imperfections in the mechanical and optical elements, imperfect homogeneity of the optical materials, etc.

Coming back to the problem of diffraction at the entrance slit and the determination of the optimum slit width $\Delta\ell_{opt}$, we start with the scheme in Fig. 2.19. The detected light is focused onto the entrance slit under a top angle of 2φ and adjusted so that the whole collimating element (lens) of diameter W is covered. Along with the angle φ, determined basically by the rules of geometrical optics, the deviation angle due to diffraction at the slit emerges

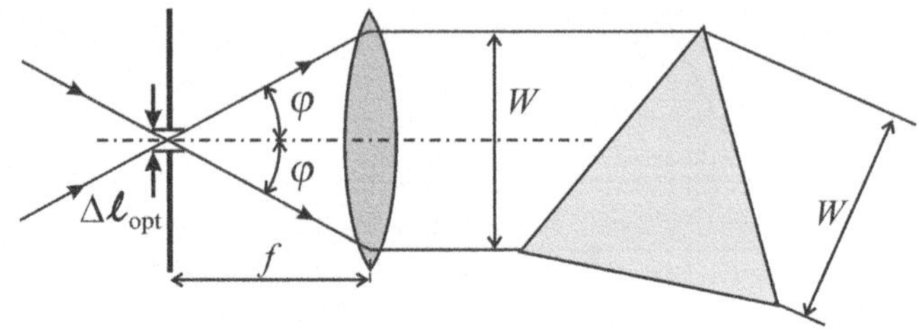

Fig. 2.19
The entrance slit of a spectral device should not be set narrower than $\Delta\ell_{\mathrm{opt}}$, otherwise, due to diffraction, the incoming light fills up an angle larger than 2φ and part of the signal is lost.

as an independent factor. This diffraction angle $\varphi_{\mathrm{d}} \sim \lambda/\Delta\ell$ increases with decreasing $\Delta\ell$. When setting $\Delta\ell$ too small (with the best of intentions to improve resolution), below some optimum value $\Delta\ell_{\mathrm{opt}}$, the angle φ_{d} becomes larger than φ and part of the input light fails to meet the lens and is lost for spectral detection. Such a situation has several negative consequences: (i) first of all, the brightness of the spectrum drops which might be fatal in studying low-level luminescence signals; in any case the signal-to-noise ratio deteriorates (see Section 2.5); (ii) light passing through the lens hits the internal walls or other parts of the device, thereby producing parasitic reflections and thus increasing the background signal (the so-called stray light) which, eventually, (iii) results in futher suppression of the S/N ratio. Moreover, mechanical imperfections, e.g. microscopic damage to the slit blades, can substantially degrade the sharpness of the spectra when setting the slit very narrow.

The optimum value $\Delta\ell_{\mathrm{opt}}$ of the slit width is thus achieved if the diffraction angle

$$\varphi_{\mathrm{d}} \sim \lambda/\Delta\ell_{\mathrm{opt}} \tag{2.26}$$

is just equal to the angle φ given by the geometrical condition

$$\varphi \sim \frac{W}{2f}, \tag{2.27}$$

imposed by Fig. 2.19. By comparing (2.26) and (2.27) we obtain

$$\Delta\ell_{\mathrm{opt}} \approx \frac{2f\lambda}{W}. \tag{2.28}$$

The reader is, undoubtedly, aware of some oversimplification of the discussion given above; in particular relation (2.26) valid for Fraunhofer diffraction of plane waves must be considered as an approximation only. Thus, to make an order-of-magnitude estimate, we neglect also the numerical factor 2 in (2.28), we shall consider a monochromator with focal length of $f = 60\,\mathrm{cm}$ and effective width of the dispersion element $W = 6\,\mathrm{cm}$, we put $\lambda = 1\,\mu\mathrm{m}$ and then we get for the optimum slit width

$$\Delta\ell_{\mathrm{opt}} \approx f\lambda/W \approx 10\,\mu\mathrm{m}.$$

This is a very important message for luminescence measurements: Decreasing the input slit width below approximately 10 μm makes no sense—resolution will not be improved; on the contrary the quality of the spectrum will deteriorate and the S/N ratio will decrease.

Of course, in most practical luminescence experiments much wider slits (100 μm or more) are used, partly due to low light fluxes and also because the luminescence spectra are often composed of broad bands that do not require an extremely high resolution, as already stressed above. Nevertheless, one can meet situations where the luminescence emission spectrum of a semiconductor contains a series of narrow, sometimes closely spaced lines (see, e.g., the spectrum of bulk silicon in Fig. 1.1). To resolve such lines, the width of the slits must be chosen very carefully. The optimum width is found as a trade-off between an acceptable noise level and the requirement to distinguish the closely spaced lines. Here, it appears useful to realize that by opening slits to a real value $\Delta\ell_{\rm real} > \Delta\ell_{\rm opt}$, the theoretical limit of the resolving power $R_{\rm t}$ is relaxed to a real value $R_{\rm real}$ called the 'effective-slit-width-limited resolving power': $R_{\rm real} = \lambda/\Delta\lambda_{\rm real} = \lambda/L^{-1}\Delta\ell_{\rm real}$. An approximate relation between these two resolving powers may be written as

$$R_{\rm real} \leq \left(\frac{\Delta\ell_{\rm opt}}{\Delta\ell_{\rm real}}\right) R_{\rm t}. \tag{2.29}$$

All this is illustrated schematically in Fig. 2.20.

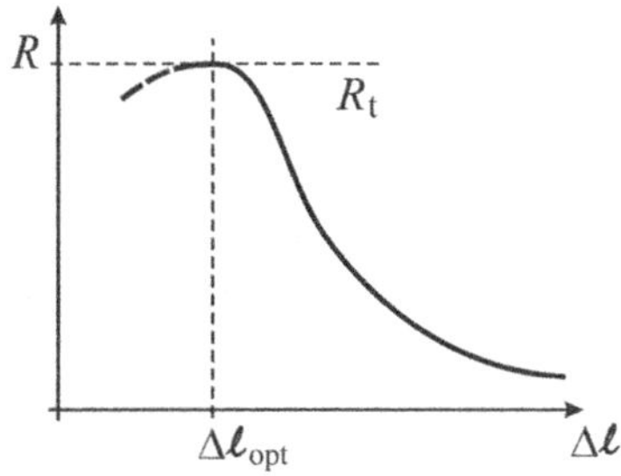

Fig. 2.20
Real resolving power of a spectral device as a function of the slit width $\Delta\ell$. The quantity $R_{\rm t}$ stands for the ultimate resolving power. When $\Delta\ell$ is large enough, diffraction effects on the slits may be neglected and the simple relation $(R_2/R_1) = (\Delta\ell_1/\Delta\ell_2)$ holds true.

To conclude our discussion of the dispersion properties we shall attempt to compare briefly prism- and grating-based spectral devices. A diffraction grating gives a linear sweep in λ or $\nu^* = 1/\lambda$, which is one of its main merits. Another advantage, as we have seen, is—as a rule—the higher resolving power compared to a prism of comparable size. On the other hand, a grating monochromator or spectrograph has at least two unpleasant properties: Firstly, different wavelengths can be sent to the same diffraction angle, in other words, spectra of different diffraction orders may partially overlap one another. An inexperienced or careless experimenter can consequently come to a completely erroneous interpretation of his observation. The problem is usually solved by inserting appropriate band-pass or edge filters in front of the input slit in order to restrict the spectral width of the detected signal. A second disadvantage of gratings is less familiar—the diffraction efficiency of a grating can often exhibit a sharp local minimum at certain wavelengths. This is known as Wood's anomaly (the physical origin of which is based presumably on the generation of surface plasmons in the metal coating, but is not known in all its details). Such a minimum is projected onto the measured spectrum and might be misinterpreted as a spectral feature. Therefore, a correction for the spectral response of the apparatus is often necessary; this will be treated in Section 2.7.

A prism spectrometer also has its pros and cons. The main advantage over a grating device is that the apparatus is absolutely free from any problems connected with the overlap of different order spectra. In general, prisms may also have a wider dispersion range than gratings; depending on the material the prism is made of (various glasses, ionic crystals, etc.), the application range can extend from the ultraviolet to the far-infrared spectral regions. On the other hand, as the main drawback of the prism, its nonlinear dispersion along with the relevant nonlinear sweep of the spectra inducing a significant decrease in resolving power at longer wavelengths—as follows from eqn (2.25)—is usually noted.

Even if none of these two types of dispersion elements is ideal, most commercial spectrometers produced during the last 20 or 30 years are based on gratings. The advanced technology of holography has led to striking improvements in the quality of holographic diffraction gratings, which are now available in a wide range of groove densities and sizes for reasonable prices. In order to overcome the disadvantage of the relatively narrow spectral range, several interchangeable gratings are mounted on a revolving table (turret) inside the spectrometer.

We shall conclude this subsection with a short note on wavelength calibration of monochromators and spectrographs. The current commercially available spectral devices have an internal control system for rotating the grating. The exit (detected) wavelength is calculated and indicated by control software. Nevertheless, the correctness of the wavelength calibration must be checked from time to time, especially after moving the device or before some high-resolution measurements. Wavelength calibration is done using a light source of well-defined wavelengths. For a fast check we can use some lines of gas lasers, e.g. the red line of a He-Ne laser at 632.8 nm. Low-pressure discharge lamps (filled with Hg, Ar, Kr, Ne, etc.), whose emission wavelengths are tabulated, are considered to be the best sources of calibration spectral lines. Calibration of a monochromator is checked by scanning the spectrum of a calibration lamp using narrow slits and comparing the result with the tabulated wavelength values. Calibration of a spectrograph is more sophisticated: in the first step the 'monochromator-like' calibration at the centre of a multichannel detector (more precisely, at the point of intersection of the optical axis of a spectrograph with the exit image plane) must be checked, and the second step calibrates the dispersion of wavelengths on the multichannel detector (off-axis points). The appropriate calibration procedures should be integrated in the control software and described in a user manual.

2.3.2 Throughput of monochromators and spectrographs

In order to define the term *throughput* we begin by recalling the definition of the *brightness* (*radiance*) of an extended emitter, Fig. 2.21(a). The radiant flux $\Delta\phi_{\mathrm{i}}$, emitted by a surface element ΔS of the emitter, is proportional to the size of this element, to the cosine of the angle ϑ between the direction of radiation and the normal to ΔS, and to the magnitude of the pertinent solid angle $\Delta\omega$, i.e. $\Delta\phi_{\mathrm{i}} \sim \Delta S \cos\vartheta\,\Delta\omega$. The proportionality factor in this equation is just the brightness of the emitter B:

$$\Delta\phi_{\mathrm{i}} = B\,\Delta S \cos\vartheta\,\Delta\omega. \tag{2.30}$$

Suppose the element ΔS is represented by the uniformly illuminated (effective) area of a monochromator entrance slit. Due to the narrow field of view of the collimator optics we can take $\cos\vartheta \approx 1$ and approximate the angle $\Delta\omega$ via the relation $\Delta\omega \approx W^2/f^2$, see Fig. 2.21(b). Equation (2.30) then reads

$$\phi_{\mathrm{i}} \approx B\,S\,(W^2/f^2). \tag{2.31}$$

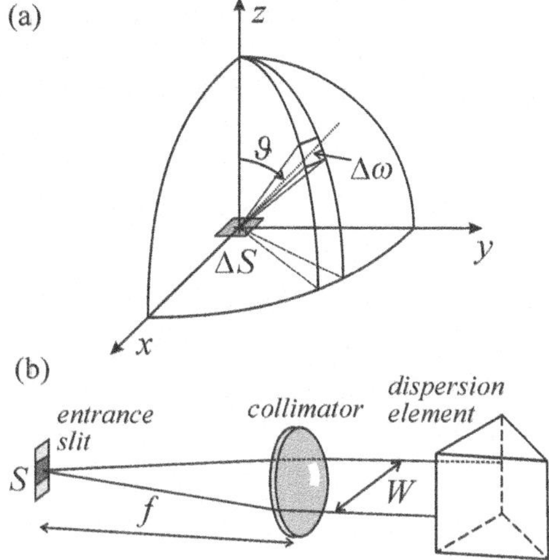

Fig. 2.21
Illustrating the definitions of (a) the brightness of an extended flat emitter and (b) the aperture ratio of a spectral device.

Let us first consider the case of a *monochromator*. Here an essential quantity is the (quasi-)monochromatic luminous or radiant flux ϕ_λ at the exit slit that induces the photoelectric response of a single-channel detector. The efficiency and/or transmittance of all optical elements inside the monochromator (mirror reflectivity, lens transmittance, efficiency of the diffraction grating, or prism transmittance) at a wavelength λ can be included in the coefficient $\kappa(\lambda) < 1$. Then, based on eqn (2.31), we can write for the output flux

$$\phi_\lambda = \kappa(\lambda)\,\phi_\mathrm{i} = \kappa(\lambda)\,B\,S\left(W^2/f^2\right). \tag{2.32}$$

The *throughput* of the monochromator is then defined as

$$\sum\nolimits_\mathrm{mono} = \frac{\kappa(\lambda)\,\phi_\mathrm{i}}{B} = \kappa(\lambda)\,S\,(W^2/f^2) \approx \kappa(\lambda)\,S(A/f^2). \tag{2.33}$$

Here A stands for the effective area of the dispersion element. Note that the throughput is proportional to the maximum solid angle that just illuminates the whole area of the collimator. It should be noted that the definition of the throughput is not standardized. Sometimes it is taken simply as (A/f^2) or even $\sqrt{A/f^2} \approx W/f$. Here, in analogy with a photographic camera, the aperture ratio is given by 1/#, where (aperture number) # = f/(diameter of the entrance pupil). Producers of spectral devices usually give the throughput[4] in the form of $(f/\#)$, with $\# = f/W$ ranging from about 3.5 to 20 (the bigger this number the lower is the throughput of the device!). This simplified (commercial) definition of the throughput does not take into account the factor S in eqn (2.33), i.e. the variability of the slit width.

In luminescence spectroscopy, dealing mostly with weak signals, spectral devices with a large throughput are usually preferred. However, similar to the mutual interconnection between dispersion and resolving power, there exists a simple relation between the throughput and resolving power of a

[4] Unfortunately, the terminology is not universal. Different terms like luminosity, throughput, aperture, aperture ratio, acceptance angle, etc. are used.

device. In what follows we shall derive this relation and discuss its practical consequences.

Let w and h be the width and height of the illuminated part of the entrance slit, respectively. Then $S = wh$ and the quantity $\alpha = w/f$ stands for the angular width of the entrance slit as 'seen' from within the device. Analogously, $\beta = h/f$ means the angular height of the slit. Consequently, $\alpha\beta = S/f^2$ and instead of eqn (2.32) we write

$$\phi_\lambda = \kappa\, B\, \alpha\, \beta\, W^2. \tag{2.34}$$

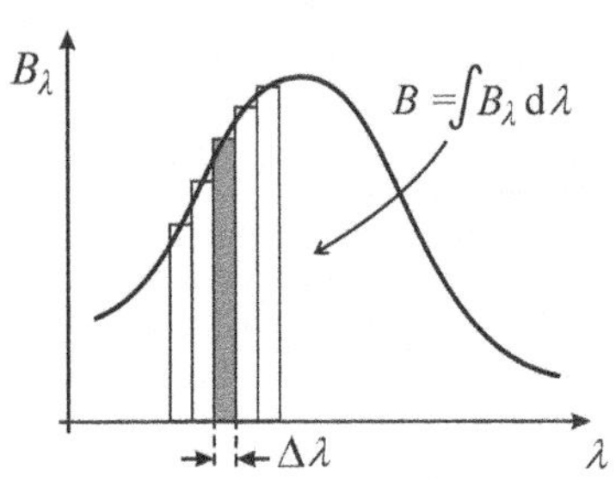

Fig. 2.22
A continuous spectrum characterized by its spectral brightness B_λ. The entrance slit is illuminated by the spectrally integrated brightness $B = \int B_\lambda d\lambda$.

Using the relations for the linear and angular dispersions (2.19) and (2.17), respectively, the transmitted bandwidth $\Delta\lambda$ determined by the slit width[5] w reads

$$\Delta\lambda = L^{-1}\, w = \frac{1}{f\,\mathscr{D}}\, w = \alpha\, \mathscr{D}^{-1}$$

and eqn (2.34) takes the form

$$\phi_\lambda = \kappa\, B\, \beta\, \Delta\lambda\, \mathscr{D}\, W^2. \tag{2.35}$$

In case of a sufficiently broad continuous spectrum (common in luminescence) we can introduce the 'spectral brightness' of the entrance slit B_λ by replacing $B \to B_\lambda \Delta\lambda$ (see Fig. 2.22). Then, the output flux is given by

$$\phi_\lambda = \kappa\, B_\lambda\, \beta\, \mathscr{D}\, W^2 (\Delta\lambda)^2, \tag{2.36}$$

which can be further modified by introducing the real resolving power R_{real}

$$\phi_\lambda = \frac{\kappa\, B_\lambda\, \beta\, \mathscr{D}\, W^2 \lambda^2}{R_{\text{real}}^2}. \tag{2.37}$$

The final relation between resolving power and throughput Σ_{mono} we are looking for is derived from (2.37) using eqns (2.32) and (2.33) and reads

$$R_{\text{real}} = \frac{\kappa\, \beta \mathscr{D} W^2\, \lambda}{\sum_{\text{mono}}}. \tag{2.38}$$

This relation shows that the real resolving power and the (real) throughput of a monochromator are, in agreement with intuition, inversely proportional to each other—widening the slits boosts the output signal but reduces the spectral resolution. Again, the experimenter must find a viable compromise using his/her experience. However, more interesting conclusions can probably be drawn from eqn (2.37):

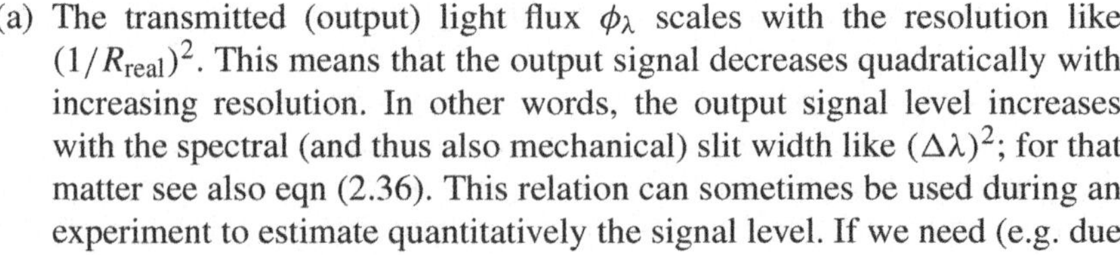

(a) The transmitted (output) light flux ϕ_λ scales with the resolution like $(1/R_{\text{real}})^2$. This means that the output signal decreases quadratically with increasing resolution. In other words, the output signal level increases with the spectral (and thus also mechanical) slit width like $(\Delta\lambda)^2$; for that matter see also eqn (2.36). This relation can sometimes be used during an experiment to estimate quantitatively the signal level. If we need (e.g. due

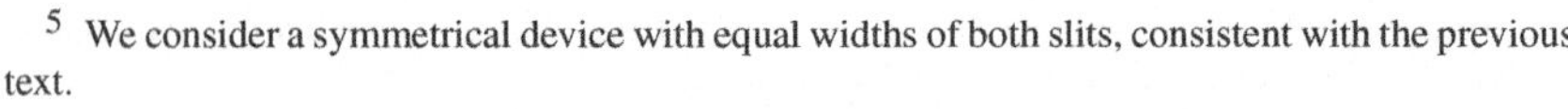

[5] We consider a symmetrical device with equal widths of both slits, consistent with the previous text.

to low signal levels at the emission band edges) to increase the 'detection sensitivity', one option is to set wider slits. Now, we know that doubling the slit width entails a fourfold increase in signal level. We have to keep in mind, however, that the above quadratic relation holds well for broad structureless spectra only (as follows from the introduction of the spectral brightness B_λ, see Fig. 2.22). Besides, there are plenty of more precise means to increase sensitivity.

(b) The output flux ϕ_λ scales linearly with β, the angular height of the illuminated slit. Therefore, it appears highly desirable to illuminate the whole height of the entrance slit, which, however, is difficult to arrange because the emitting spot is commonly of circular shape. Nevertheless, when the signal is brought by an optical cable, its termination can be shaped to fit the entire surface of the entrance slit. Also in cases of weakly absorbed or homogeneous two-photon excitation of a sufficiently voluminous sample by a laser beam (Chapter 5, Fig. 5.15(b)), the elongated luminescence spot can be oriented so that its image covers the entrance slit.

(c) The relation $\phi_\lambda \sim W^2$ implies that a high-throughput monochromator has to be equipped with a large grating or prism.

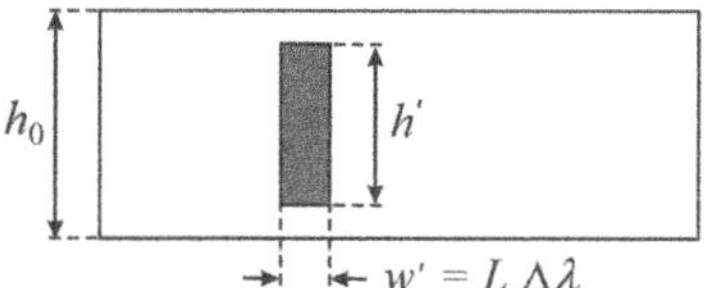

Fig. 2.23
Image of the entrance slit on a multichannel detector located in the spectrograph exit plane. h_0 is the height of the sensitive area of the detector, and h' and w' stand for the height and width of the entrance slit image, respectively.

In the case of a *spectrograph* it is the so-called *irradiance* of a multichannel detector (the radiant flux at the spectrometer output divided by the illuminated area) that has a crucial role to play, rather than the radiant or luminous flux itself. This quantity determines the exposure time and the signal-to-noise ratio. Luminescence radiation, spatially distributed by a dispersion element, projects the entrance slit area $S = wh$ characterized by the spectral brightness B_λ to its image $S' = w'h'$ (in the plane of the multichannel detector, see Fig. 2.23), the dimensions of the slit and its image being identical ($w = w', h = h'$) in the case of a symmetrical device. In analogy with definition (2.33) we define the throughput of the spectrograph Σ_{spectro} using the irradiance $(\kappa(\lambda)\phi_i/S')$ on the area $S' = w'h'$ of the slit image created by quasi-monochromatic light $\lambda \pm \Delta\lambda/2$ as

$$\sum\nolimits_{\text{spectro}} = \frac{\phi_\lambda}{B_\lambda\,\Delta\lambda\,S'} = \frac{\phi_i\,\kappa(\lambda)}{B_\lambda\,\Delta\lambda\,S'} = \frac{S}{S'}\kappa(\lambda)\left(\frac{W}{f}\right)^2 = \kappa(\lambda)\left(\frac{W}{f}\right)^2. \tag{2.39}$$

Now we can express the 'spectral irradiance' of the detector with respect to its full height $h_0 \geq h'$ as $E_\lambda = \phi_\lambda / h_0 w$, as follows from Fig. 2.23. This expression for E_λ can then be easily modified using eqn (2.32) and the definition of the aperture number # as

$$E_\lambda = \frac{\phi_\lambda}{h_0\,w} = \frac{\kappa(\lambda)\,B_\lambda\,\Delta\lambda\,S\,(W/f)^2}{h_0\,w} = \left(\frac{h}{h_0}\right)\kappa(\lambda)\,B_\lambda\left(\frac{1}{\#}\right)^2\frac{\lambda}{R_{\text{real}}}. \tag{2.40}$$

Relation (2.40), valid for a spectrograph, is analogous to expression (2.37) relevant to a monochromator. Here, again, are some conclusions important for experiments:

(a) The irradiance E_λ decreases proportionally to $(1/R_{\text{real}})$, unlike the transmitted flux at the exit slit which scales as $(1/R_{\text{real}})^2$.

(b) The irradiance E_λ is proportional to (h/h_0). Therefore, it is desirable to choose a combination of spectrograph, multichannel detector and a method to illuminate the entrance slit so that $h = h_0$, in order to maximize E_λ. Naturally, the case $h > h_0$ should be avoided because part of the signal flux fails to meet the detector.
(c) The irradiance E_λ scales like $(1/\#)^2$, i.e. the irradiance of the detector increases with the square of the aperture ratio. This factor is actually the most important one among those aimed at maximizing the detector irradiance.

2.4 Signal detection methods in luminescence spectroscopy

Following the discussion of spectral devices and detectors we are now going to describe briefly methods of signal treatment in spectroscopy. Essentially this means dealing with the principles of electronic devices that treat the electric signal from the photodetector. Here we concentrate on single-channel detectors because multichannel ones are quite straightforward and user friendly—the accumulated photoinduced charges, after being digitized, are simply transferred to a computer to be processed by software.

For the sake of simplicity we consider in this section only continuous-wave (cw) luminescence signals, i.e. signals excited by a continuously emitting lamp or laser. In particular, we are going to describe two methods: phase-synchronous detection and the photon-counting technique. Experimental methods of pulsed luminescence signals are left to Section 2.9.

2.4.1 Phase-synchronous detection

The simplest arrangement of optical cw signal detection is obviously direct current detection—time invariant radiant flux (at the monochromator exit slit) induces in the attached photomultiplier a constant photocurrent, which is subsequently directly measured by a voltmeter connected to a load resistance (Fig. 2.5(b)). In the case of extremely low luminescence photon fluxes a DC preamplifier could be inserted between the photomultiplier and the voltmeter. However, in reality such DC measurements are rarely applied, because low photocurrents are commonly comparable in magnitude with the photomultiplier dark current or only slightly higher. The dark current amplitude must be subtracted in this case from the detected signal, which leads to important errors like in any other indirect measurements where two comparable quantities are to be subtracted. The inevitable presence of broad-frequency noise makes the quality of the resulting output even worse because there is no electronic component available for noise suppression.

It therefore becomes obvious that weak continuous optical signals can be successfully detected only when applying some electronic device that is able to extract reliably a signal buried completely in the dark current and/or in various types of noise whose amplitude may be higher by several orders of magnitude. This is possible, first and foremost, by employing the method of phase-synchronous detection through a device called a *lock-in amplifier*. This device is able to measure alternating voltage or currents and to provide a corresponding DC output voltage that scales linearly with the input signal level (the output signal level is usually much higher than the input level, hence the term 'amplifier'). Here, only certain frequencies of the input signal are detected while virtually all the others are suppressed (mainly all noise with a broad frequency spectrum), hence the term 'lock-in'. The ability of a synchronous detector to extract and amplify a low synchronous signal buried in noise is characterized by a quantity called the *dynamic reserve*. This is the ratio of the maximum acceptable amplitude of the asynchronous (noise) signal (i.e. the amplitude value which just starts to induce nonlinearity and saturation of input circuits) to the peak synchronous signal giving a full-scale DC output. The dynamic reserve of up-to-date lock-in amplifiers is extremely high, routinely reaching values of 100 dB (!) or even higher.

Any lock-in amplifier contains two key parts: a mixer (phase detector) and an integrator. The principle of operation of a lock-in amplifier is illustrated in Fig. 2.24. The input to the mixer (left side of Fig. 2.24) receives two electric signals—the measured sine wave signal e_1 from a photodetector and a square wave so-called reference signal e_2 (the origin of these particular forms of signals will be described later). The output of the mixer e_3 is the product of the input signals: $e_3 = e_1e_2$. The technical solution consists in alternating usage of a direct and an inverting amplifier in the mixer input. The amplitude of the reference signal E_2 can be put equal to unity. Then Fig. 2.24(a) shows immediately that for equal frequencies of both the signals ($f_1 = f_2$) and provided their relative phase shift is equal to 0, 2π, 4π, etc. the output mixer signal e_3 contains a pulsating voltage of positive polarity only. The measured signal is said to be synchronous with the reference. Enabling e_3 to pass through the integrator having a time constant $\tau = RC$ (right-hand side of Fig. 2.24), a DC output voltage with amplitude E_{DC} proportional to the amplitude of the measured input signal E_1 is obtained.

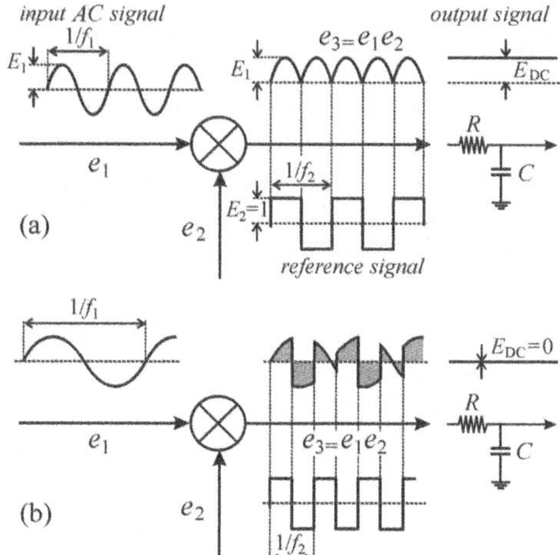

Fig. 2.24
Principle of a lock-in amplifier. The left-hand side shows a mixer and the right-hand side an integrator (low-pass filter). (a) For a synchronous input signal $f_1 = f_2$ we obtain a non-zero DC output signal $E_{DC} \neq 0$. (b) Asynchronous input signal $f_1 \neq f_2$ along with a sufficiently long time constant $\tau = RC$ result in a zero DC output level.

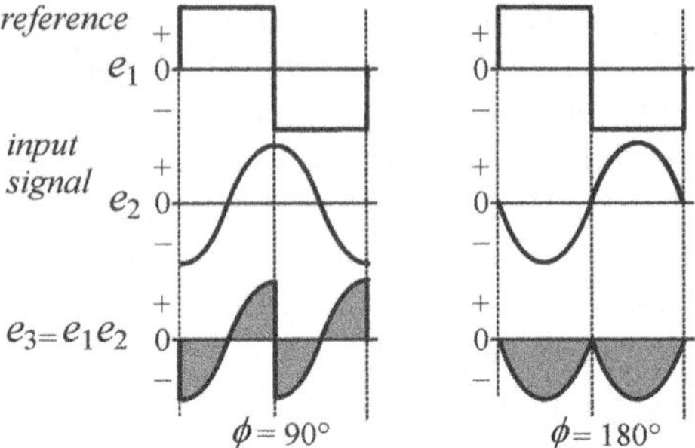

Fig. 2.25
Demonstration of the phase sensitivity of a lock-in amplifier. The input signal from a photodetector e_1 and the reference signal e_2 have equal frequencies but the DC output is strongly dependent on the phase shift between these two signals.

In the case of an asynchronous signal $f_1 \neq f_2$ (Fig. 2.24(b)) the signal e_3 contains both positive and negative components compensating each other if a long enough time constant τ is applied, and thus zero DC output results. It can be seen that the lock-in amplifier acts as an extremely selective filter processing and amplifying only the 'meaningful' signal component, i.e. that corresponding to the reference frequency; this is the succinct essence of the principle of the lock-in amplifier.

There are, however, additional useful features of synchronous detection that remain to be explained. First of all, these is the *phase sensitivity*, which strongly influences the magnitude of the output signal—for certain values of the phase shift between the signal and the reference, the output $E_{\rm DC}$ may be zero even if the measured and reference signals have the same frequency $f_1 = f_2$(!). Two examples ($f_1 = f_2$ while the phase shift is different from $2\pi n$, n being an integer) are demonstrated in Fig. 2.25: the phase shift $\phi = 90°$ (or $(2n+1)\pi/2$) induces zero DC output, while under a shift of $\phi = 180°$ (or $(2n+1)\pi$) the output signal E_{DC} is of the same amplitude as for $\phi = 0°$ but of negative polarity. This is why a lock-in amplifier is sometimes also called a phase-sensitive detector.

The phase sensitivity has several interesting consequences. But prior to entering the relevant discussion we should describe how cw luminescence radiation is converted into a sine-modulated input signal and how the reference signal is generated. Everything is based on simple mechanical chopping of the light beam (as a rule, the excitation beam is chopped, less commonly chopping of the photoluminescence signal is applied—see Fig. 2.2(b)) by a mechanical chopper, which is a rotating disk with periodically arranged slots. (Of course, mechanical chopping itself cannot create the sinusoidal signal shown in Figs 2.24 and 2.25, nevertheless, the input circuits of a lock-in amplifier are able to adapt a chopped cw signal into a sine wave.) The reference signal is generated by an optical pair (a LED and a photodiode) attached to the chopper. In this way the reference and measured signal are modulated by the same mechanical system and undoubtedly $f_1 = f_2$ holds true. However, there is certainly a non-zero phase shift ϕ between both signals due to the different response times of the photomultiplier and the reference photodiode along with different delays introduced by wiring and circuits on the path of both signals towards the mixer input.

Here, the phase sensitivity of the lock-in amplifier appears to be very useful. The phase shift ϕ between the two input signals can be continuously tuned

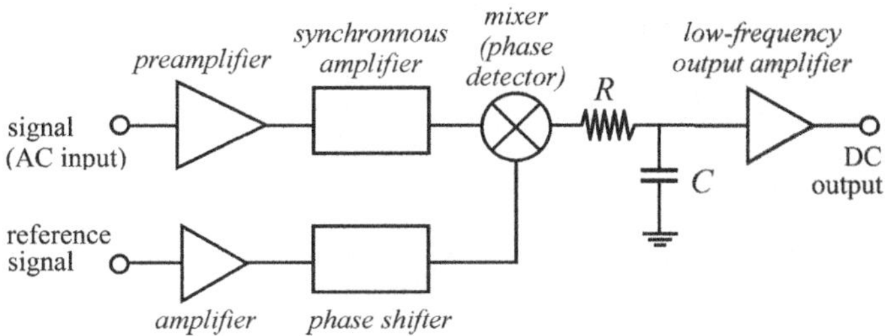

Fig. 2.26
Block diagram of a phase-sensitive detector. Both the modulated input and the reference signal are first amplified and possibly shape-adapted. The reference signal branch contains an adjustable phase shifter. After passing through the RC circuit (low-pass filter) the signal is amplified again.

to maximize the output signal amplitude E_{DC}. Up-to-date lock-ins can do this phase tuning automatically. Another useful feature is the possibility of alternating the polarity of the output simply by setting the phase shift to $\pm 180°$. The converse is also true: if the device, on its own, automatically changes the phase by $\pm 180°$ during measurement, this indicates that the polarity of the input signal has changed, which may carry important experimental information. (While one can hardly meet such a situation in luminescence spectroscopy, it is, however, quite common in photoelectric measurements.) Finally, the phase sensitivity proves very useful in filtering out all kinds of noise, even if their frequencies may happen to be equal $f_1 = f_2$, because they have no fixed phase relation to the reference signal, i.e. their phases fluctuate randomly.

The time constant $\tau = RC$ also deserves a short discussion. The illustrations in Figs 2.24 and 2.25 are oversimplified. It can be shown that the output of the mixer can contain not only pulsating components with frequency $f_1 = f_2$, but also with $f_1 = (2n + 1) f_2$ where n is an integer. Moreover, asynchronous input frequencies reveal themselves at the mixer output as sum and difference frequency components $(f_1 + f_2)$, $(f_1 - f_2)$ and higher harmonics. These frequencies could possibly appear as noise superimposed on the output voltage E_{DC} but they are mostly filtered out by the RC circuit, which then acts not only as an integrator but also as a low-pass filter. A block diagram of a lock-in amplifier is shown in Fig. 2.26.

Concerning the optimum choice of the time constant τ, smoothing of the DC output is evidently improved with increasing time constant (available values of τ range from microseconds to tens of seconds) but, on the other hand, very high values of τ make the overall response of the device slower. This means that the speed of scanning a spectrum across the exit slit must be adapted to the selected time constant, otherwise the acquired spectral shape may be distorted. Very long time constants can lead to intolerably long acquisition times. This problem will be discussed in more detail in Section 2.5.

We shall conclude our discussion of the synchronous detection technique by mentioning the limits of the method. One limit arises from luminescence kinetics and its relation to the chopping rate. If the mechanical chopper modulates the excitation beam, the expected square wave photoluminescence signal may be distorted owing to the finite speed of the rise and decay of luminescence (Fig. 2.27). If the relevant characteristic time constants are comparable to or even longer than the chopping period $T_2 = 1/f_2$, the lock-in amplifier input signal loses its modulated shape and can no longer be processed by the device (Fig. 2.27(d)). Possible solutions are obvious—either decrease the chopping

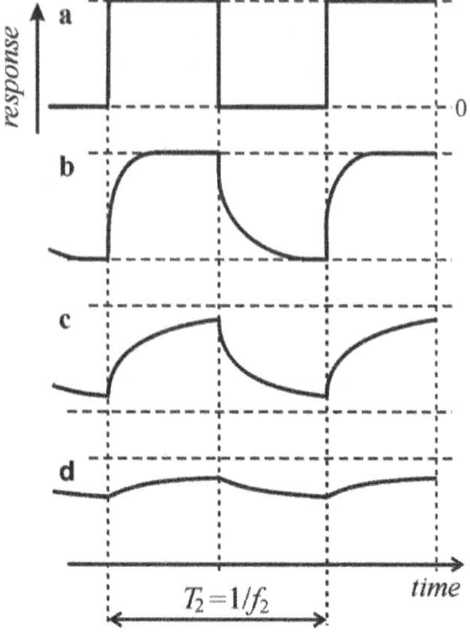

Fig. 2.27
Temporal trace of the photoluminescence response excited by a mechanically chopped beam with frequency f_2. Curve (a) pertains to luminescence with very fast kinetics, i.e. the rise and decay times are short compared to $T_2 = 1/f_2$; curves (b)–(d) illustrate luminescence with slower and slower response. A lock-in amplifier will perceive case (d) as almost an unmodulated DC signal and will not respond to it.

frequency f_2 (which is not always possible) or modulate the luminescence instead of the excitation beam (which is experimentally not so easy to do). On the other hand, the above drawback can sometimes be turned into an advantage and applied as a simple tool for time-resolved luminescence spectroscopy.

Suppose that the material under study shows two luminescence emission bands, partially overlapping spectrally but having substantially different kinetics. Then these spectral peaks may be cleanly separated by performing two measurements with a lock-in amplifier under two different chopping frequencies $f_2' \ll f_2''$. While the measurement at frequency f_2' reveals a spectrum comprising both the bands, the measurement at frequency f_2'' shows only the 'fast' band provided this frequency is high enough to eliminate the slow component. An example is shown in Fig. 2.28 [17, 18].

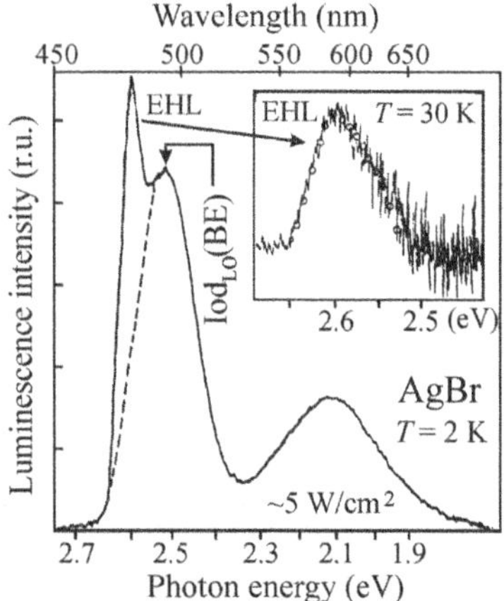

Fig. 2.28
Low-temperature photoluminescence spectrum of AgBr under excitation with a cw UV laser and using low-frequency synchronous detection ($f_2' = 500$ Hz). The spectrum contains three bands: luminescence of the electron–hole liquid (EHL) at ~2.6 eV, a band Iod_{LO}(BE) due to radiative recombination of excitons bound to ions of an isoelectronic unintentional dopant (iodine) at ~2.5 eV, and a band at ~2.1 eV due to a residual impurity. The inset shows the spectrum obtained under high chopping frequency $f_2'' = 100$ kHz, which eliminates the iodine-related band Iod_{LO}(BE) with a decay time of ~ 25 μs and retains only the fast band of EHL with a decay time of about 20 ns. Adapted from Pelant *et al.* [17] and Hulin *et al.* [18].

The second limit beyond which application of the lock-in technique serves no useful purpose is detection of very low light fluxes, where the discrete character of both light (photons) and detection events (photoemission from the photomultiplier photocathode) becomes apparent. This situation is nicely illustrated by Fig. 2.29. The lock-in amplifier cannot detect very weak signals since they are formed by an array of random anode pulses (Fig. 2.29(c)) rather than by a DC component processable through chopping into a form suitable for phase-sensitive treatment. Detection of such low photon fluxes requires another method called photon counting.

2.4.2 Photon counting

Every current pulse at the output of a photomultiplier, shown in Fig. 2.29(c), corresponds to the impact of one photon on the photocathode, accompanied by emission of one photoelectron. Therefore, the number of pulses N_r registered per unit time is proportional to the number of photons N_h hitting the photocathode during the same time interval, i.e. to the incident light flux or

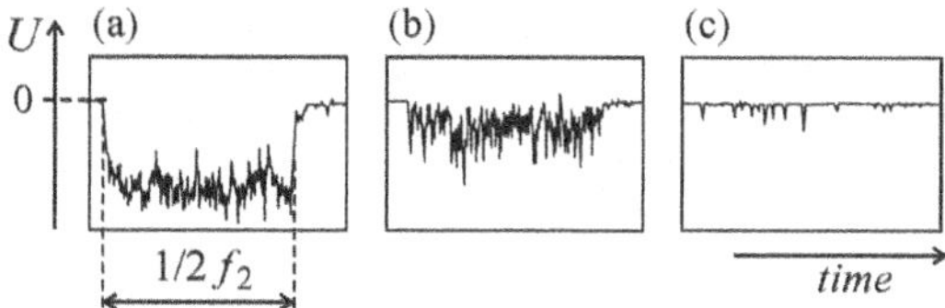

Fig. 2.29
Visualization of the output signal of a photomultiplier (chopped at frequency f_2) on an oscilloscope screen for different signal levels. (a) The light flux hitting the detector is high enough, single output voltage pulses are merged together. (b) The number of incoming photons is lower; pulses corresponding to the detection of single photons can already be resolved. (c) Very low signal photon flux. The signal from the detector is composed of discrete voltage pulses; there is almost no continuous component. Therefore the lock-in sees only incoherent noise and the DC output level E_{DC} is zero. The signal has negative polarity, which corresponds to the most common photomultiplier wiring with a grounded anode (Fig. 2.5b). Modified from Hamamatsu [7].

intensity. (In quantitative measurements, e.g. those of the quantum efficiency of luminescence, we have to keep in mind that these quantities are only proportional to each other, not equal. The number of registered pulses N_r is always lower than the number of incoming photons N_h, the proportionality coefficient being the product of the photocathode quantum efficiency η_{ph} and the collection efficiency of dynodes α_d, which means $N_r = \eta_{ph}\alpha_d N_h$.)

A block diagram of a photon counting apparatus is presented in Fig. 2.30. The output signal from the photomultiplier is composed of single voltage spikes (as already illustrated in Fig. 2.29(c)), generally superimposed on various low- and high-frequency noise and instabilities—circle A. Big pulses are due to photoelectrons emitted from the photocathode while smaller ones come from the thermionic emission in dynodes. This signal is treated by an input amplifier, which multiplies the signal approximately one thousand times and filters out the low-frequency ripple—circle B. The amplified pulses enter a discriminator (circle C), which eliminates pulses that are below a certain adjustable discrimination level (dashed line in the circle B). In this way the parasitic pulses originating in the thermionic emission of dynodes are ruled out. Each pulse passing through the discriminator generates at the output a standard TTL voltage pulse (circle C) and these pulses are counted by a common counter. The counter can send out either digital information about the number of pulses per unit time or it can produce an analogue signal (for a plotter) employing a digital-to-analogue converter.

The principle of the photon-counting method looks simple but the apparatus requires excellent temporal and noise characteristics of all of its components. The most critical is selection of a suitable photomultiplier tube meeting stringent criteria imposed on its parameters, as well as the selection of its high anode–cathode voltage and of the appropriate discrimination level of the discriminator. These issues deserve to be discussed in more detail.

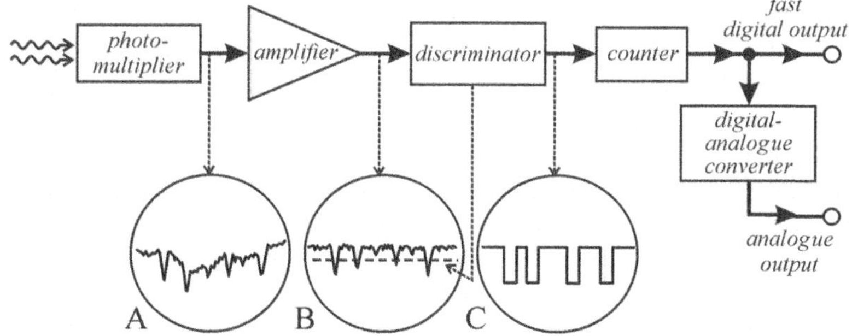

Fig. 2.30
Block diagram of a simple version of a photon counter.

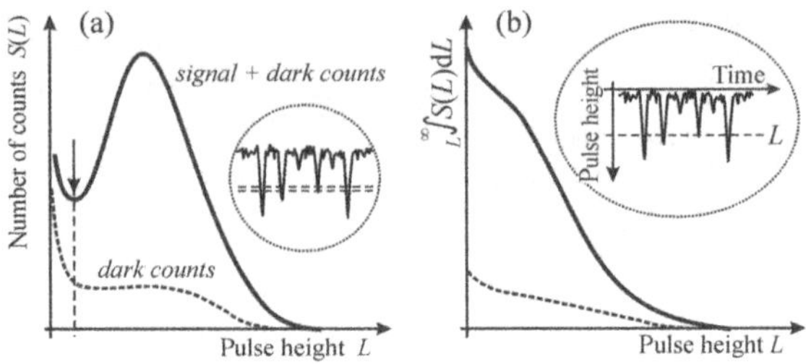

Fig. 2.31
(a) Qualitative illustration of a (differential) distribution of photomultiplier pulse heights. The dashed and solid curves are the height distributions for 'dark' pulses (photocathode not illuminated) and for 'light' pulses (signal photons hit the photocathode), respectively. Such distributions can be measured by making use of a multichannel analyser or a discriminator with adjustable upper and lower levels (see inset). (b) An integral distribution of pulse heights, which can be experimentally acquired with a single-level discriminator. Adapted after a Hamamatsu brochure [7].

By no means is every photomultiplier applicable to the photon-counting method. In selecting a suitable photomultiplier the following factors have a crucial role to play:

1. Temporal width of the output voltage pulses. This width must be sufficiently short in order to distinguish the height of pulses due to photoelectrons from that of pulses originating in thermionic emission in dynodes. Excess temporal blurring of the electron packets propagating through the photomultiplier could reduce the contrast in amplitude between the two kinds of pulses so that their separation in the discriminator would be impossible. Moreover, shorter pulses enable the photon-counting technique to be applied to higher pulse rates—the upper limit of pulse counting corresponds to the situation in which adjacent pulses become significantly overlapped and thus indistinguishable. A common value of the temporal pulse halfwidth in photomultipliers designed for photon counting is about 10–20 ns.
2. A high quantum efficiency η_{ph} of photocathode emission and a low anode dark current. It is evident from the very principle of the method that a high signal-to-noise ratio requires high yield of generation of primary photoelectrons from photons hitting the photocathode, along with minimization of the number of thermally released electrons. Quantum efficiencies of suitable photomultipliers are usually around $\eta_{\mathrm{ph}} = 20$–30%.
3. Appropriate distribution of pulse heights. Firstly, we have to explain what we understand by the distribution of pulse heights. One can hardly expect that all the pulses or counts originating from a single photoelectron event on the photocathode (or 'dark' counts initiated by thermally released electrons) would have the same amplitude on the anode; the statistical character of secondary electron emission at the dynodes, individual variations in electron trajectories, etc. make the anode pulse heights of both the signal and 'dark' counts fluctuate. The average amplitude of signal counts is expected, naturally, to be higher compared to that of dark counts. This phenomenon is illustrated in Fig. 2.31(a) which shows a histogram $S(L)$ of the occurrence of pulses with height L; this is called the distribution of photomultiplier pulse heights. The centre of gravity of the dark-count distribution (dashed curve) is clearly left-shifted compared to the pulse distribution relevant to an illuminated photomultiplier (signal+dark counts, solid curve). An optimum pulse distribution for the photon-counting technique should have a pronounced—and as narrow as possible—peak with strongly suppressed occurrence of small pulses. The discrimination level is then adjusted close

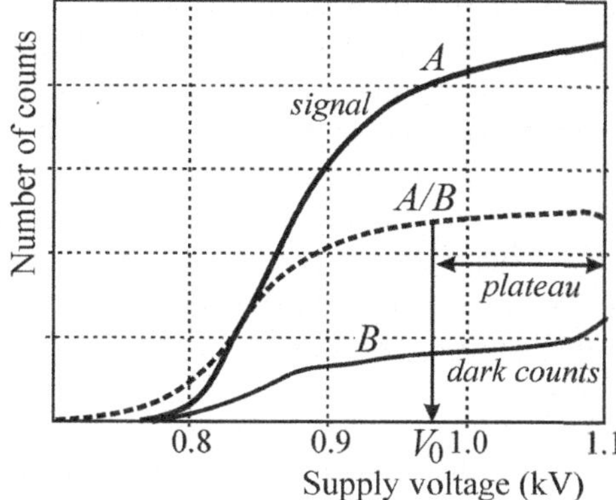

Fig. 2.32
Number of counts as a function of voltage applied to a photomultiplier in the photon-counting technique (important is the occurrence of a plateau on the curve A/B). Adapted after a Hamamatsu brochure [7].

to the local minimum of the distribution curve (see the arrow in Fig. 2.31(a)) in order to reject efficiently the dark pulses.

Here, however, the experimenter is faced with a problem. Photomultiplier manufacturers usually do not provide curves of pulse height distributions in their catalogues. Therefore, it appears advisable for the user to perform his/her own measurement of these characteristics prior to application of a particular photomultiplier[6] in photon-counting detection. A direct measurement of distributions as shown in Fig. 2.31(a) is possible with a multichannel analyser of pulse heights or using a discriminator that has two adjustable discrimination levels (such a narrow discrimination 'window' is scanned from low- to high-amplitude pulses and the occurrence of pulses belonging to this span of heights is recorded). Commonly, however, the discriminator has only one discrimination level. In this case all pulses with amplitude higher (in absolute value) than the adjusted discrimination level are recorded, resulting in the curve shown in Fig. 2.31(b). One can easily infer that this curve can be converted to the peak height distribution via numerical differentiation. Therefore, the curve (b) displayed in Fig. 2.31 is sometimes called an *integral representation*, while the curve (a) is called a *differential representation* of the pulse height distribution.

4. The choice of the photomultiplier anode–cathode high voltage and of the discrimination level. Suppose we have already acquired the differential curve of the height distribution and adjusted the discrimination level approximately to the local minimum (the arrow in Fig. 2.31(a)). We are now in a position to find a proper operational high voltage. The procedure is as follows: the applied voltage is varied in small steps (allowing the operation conditions to be stabilized adequately), and the number of pulses is recorded at each step. This is done for both non-illuminated and illuminated photomultipliers. Curves similar to those plotted in Fig. 2.32 are obtained. The number of signal pulses (upper curve) increases with increasing voltage faster than the number of dark pulses (lower curve) because the latter are

[6] There are photomultiplier types especially fabricated for use in photon-counting applications. But the manufacturer can also select an exeptionally suitable exemplar from other ranges of photomultipliers.

already substantially restricted by the discriminator. The ratio of these two curves (dashed curve in the middle) increases rapidly as the high voltage raises but then saturates, achieving a plateau. (The occurrence of a broad plateau is another prerequisite imposed upon the photomultiplier.) The correct applied high voltage is then situated in the plateau region that starts at a voltage V_0 at which the ratio of signal to dark pulses reaches saturation (Fig. 2.32) and that extends basically up to the maximum allowed voltage. The output signal is thus to a large extent independent of the anode–cathode voltage and its fluctuations, featuring an obvious benefit of the photon-counting technique.

It is important to realize that proper choice of the operational high voltage may comprise some ambiguities. The distribution of pulse heights shown in Fig. 2.31 was obtained with a certain pre-selected photomultiplier high voltage. If, later, we found that this bias was, unfortunately, outside of the plateau region, another operational voltage must be chosen, which may influence the shape of the pulse height distribution. Then, in principle, we shall have to remeasure the curves shown in Fig. 2.31, find a new discrimination level and determine a new plateau characteristic. Possibly, the operational voltage will then have to be shifted again, etc. and we risk ending up in a closed loop. The solution is a compromise based both on the experimenter's experience and on the requirements of the particular experiment. Obviously, shifting the discrimination level down (to the left side from the local minimum of the height distribution) increases the overall efficiency of pulse counting, while shifting it to higher pulse amplitudes (roughly towards the middle between the local minimum and the distribution peak) rejects more noise pulses and improves the signal-to-noise ratio.

We shall close our discussion of photon counting with a few final notes. First of all, let us remind ourselves that not only a photomultiplier but also an avalanche photodiode can be applied in the photon-counting method (Section 2.2). In fact, we have also already mentioned the natural limits of application of the technique. The high count rate limit is set by the resolution of two closely following pulses, where both the temporal width of the pulse and the dead time of the detector and discriminator are significant. A typical value of the maximum pulse rate is $1\text{–}3 \times 10^6\ \mathrm{s}^{-1}$, which corresponds to a photon flux of about 10^7 photon/s (i.e. optical power ~ 3 pW) hitting a photomultiplier with a photocathode quantum efficiency of $\eta_{\mathrm{ph}} = 20\%$. The low count rate limit is determined by various noise characteristics and by the rate of dark counts. Using an up-to-date cooled photomultiplier one can easily detect signals of tens of photons per second, but if the entire spectrum is to be recorded in this case, then the overall measurement time already has a decisive role to play, as will be analysed in Section 2.5.

A troublesome peculiarity of the photon-counting technique is its excessive receptiveness to external perturbations. A photon-counting apparatus can register, for example, spurious pulses generated by a stepping motor that drives the dispersion element located inside the monochromator or high-frequency perturbations from other laboratory equipment, originating possibly even in neighbouring rooms. The only remedy is perfect shielding of all instruments including usage of well screened cables, good earthing, and sometimes finding an optimum location of the apparatus within the laboratory. The spheres of applicability of the lock-in and photon-counting techniques overlap over a certain range of photon fluxes. If both techniques are usable, then synchronous detection is usually preferred as it is simpler and less sensitive to external perturbations.

For more information on the photon-counting technique, noise analysis, dead times, accuracy, etc. the reader is referred to, e.g., [6, 19].

2.5 Signal-to-noise ratio in a scanning monochromator

In the preceding sections we examined the basic characteristics of spectral devices, introduced parameters describing the properties of optical detectors, and analysed the two basic techniques of signal detection in luminescence spectroscopy—synchronous detection and photon counting. Now, we are ready to deduce a relatively simple expression for the signal-to-noise (S/N) ratio [20] of luminescence spectra. This is of basic importance for laboratory practice.

Consider a scanning monochromator, provided at the exit slit with a detector of sensitivity $k(\lambda)$, normalized detectivity D^* and sensitive area $\mathscr{A}$. The detector is illuminated by a monochromatic radiant flux ϕ_λ. The definition (2.7) of D^* can be modified using (2.6) to

$$D^* = \sqrt{\mathscr{A}}/NEP = \frac{\sqrt{\mathscr{A}}\sqrt{\Delta f}\,S}{\phi_\lambda N} = k(\lambda)\,\frac{\sqrt{\mathscr{A}}\sqrt{\Delta f}}{N}. \tag{2.41}$$

To be specific, let the output from the detector be treated by a lock-in amplifier. Then the frequency bandwidth Δf of the detection path is driven by the integrator time constant τ as $\Delta f \approx 1/2\pi\tau$. This can be understood as a straightforward result of a Fourier transform, or—in a more illustrative way—we can imagine the signal as being recorded by a classical plotter (or traced in real time on an oscilloscope screen) when the time constant τ determines the speed of movement of the recorder pen. Suppose the spectrum contains very sharp features, for example an intense narrow line; then the fast upward movement of the pen is damped by the time constant τ. In the case of excessively long τ, the peak is not well reproduced, is smaller, broader and shifted along the direction of wavelength scanning. It may happen that less intense spectral peaks even disappear entirely from the record. This implies that, in order to detect a spectrum without distortion, the wavelength scanning speed must be adapted to the applied time constant τ.

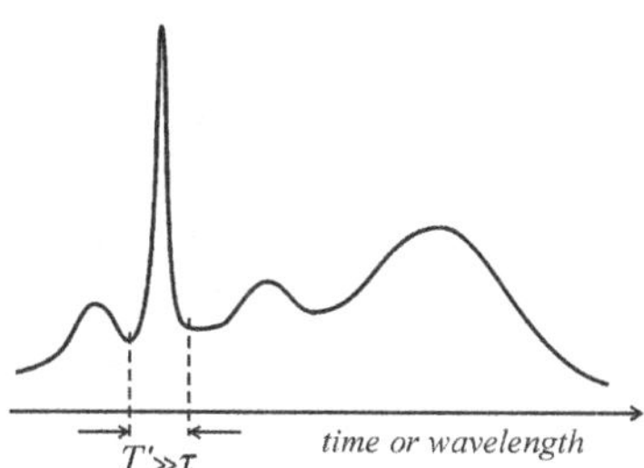

Fig. 2.33
Scanning of emission spectrum in real time. The recording time T' of the narrowest peak must be much longer than the time constant τ of the detection system.

The mathematical formulation of this requirement is simple. Crucial for the scanning speed are the sharpest features in the spectrum. The narrowest peak, if it is to be recorded perfectly, must be scanned for a time T' significantly longer than τ (Fig. 2.33), which means

$$\tau = \varepsilon T'\ (\varepsilon \ll 1),\ \ \Delta f \approx 1/2\pi\tau. \tag{2.42}$$

Now, the noise level N in eqn (2.41) can be expressed using eqn (2.42) as

$$N = k(\lambda)\,\frac{\sqrt{\mathscr{A}}\,\sqrt{\Delta f}}{D^*} = \frac{k(\lambda)}{D^*}\sqrt{\frac{\mathscr{A}}{2\pi\varepsilon T'}}. \tag{2.43}$$

The magnitude of the signal S as registered by the detector for the output radiative flux ϕ_λ (2.37) is given by

$$S = k(\lambda)\,\phi_\lambda = k(\lambda)\,\frac{\kappa\, B_\lambda\, \beta\, \mathscr{D}\, W^2\lambda^2}{R^2_{\text{real}}}. \tag{2.44}$$

Finally, the signal-to-noise ratio we are looking for follows from the last two equations as

$$\frac{S}{N} = \frac{\kappa\, B_\lambda\, \beta\, \mathscr{D}\, W^2\lambda^2}{R^2_{\text{real}}}\sqrt{\frac{2\pi\varepsilon T'}{\mathscr{A}}}\,D^*. \tag{2.45}$$

This relation hold true for the narrowest spectral line. Obviously, recording any other spectral feature, like a wider peak or band, will run for even longer time. For the detection time of the whole spectrum thus the relation $T > T' \gg \tau$ holds.

Expression (2.45) for S/N contains several interesting pieces of information and is worth analysing in detail. First of all, the detector sensitivity $k(\lambda)$ does not enter eqn (2.45). This surprising fact confirms that the S/N ratio is determined predominantly by the detectivity D or D^*. Therefore, a responsible choice of a suitable detector should not be just a matter of a cursory glance at the catalogue's data on sensitivity.

Suppose now the entire experimental set-up is ready, which means the detector and monochromator parameters (detectivity D^*, transmittance κ, effective area of the dispersion element proportional to W^2, and angular dispersion $\mathscr{D}$) have been fixed. Then, for the experimenter just four remaining parameters are left to be adjusted in order to maximize S/N: the spectral brightness of the entrance slit B_λ, the angular height of the illuminated slit β, the time constant τ (or acquisition time T), and the resolving power R_{real}. In Subsection 2.3.2 we already saw that homogeneous illumination of the whole slit height is, except in special cases, hardly possible in luminescence spectroscopy. Thus, only three adjustable parameters remain, i.e. B_λ, T, and R_{real}.

First we pay attention to the resolving power and detection time T; they are closely related. An increase in resolving power R_{real} (e.g. by setting narrower slits) is accompanied, according to eqn (2.45), by a decrease in the S/N ratio. If in this situation we want to keep the S/N constant, we can attempt to increase the time constant τ and simultaneously to set a longer acquisition time T, which means a slower spectrum scan. This method is, however, very limited. Imagine we have recorded a spectrum within $T_1 = 10$ min and found that, in order to resolve better two neighbouring spectral lines, we have to increase the resolving power (i.e. to narrow down the slits), say, 3.3 times. According to eqn (2.45), the S/N ratio remains unchanged provided we prolong the acquisition time to $T_2 \approx 100T_1 = 1000$ min ≈ 16 hours (!). This is not feasible. In other words, the influence of the detection time on the S/N ratio is strongly limited. Obviously, this is because of the weak relationship $(S/N) \sim \sqrt{T}$ and, on the

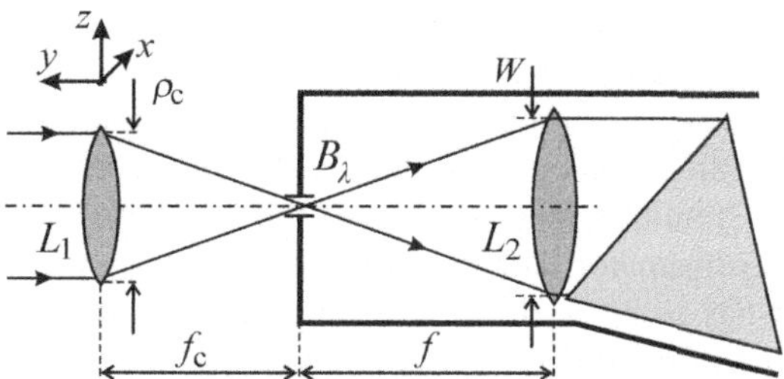

Fig. 2.34
Optical system for illuminating the spectrometer entrance slit. A collimated beam of luminescence emission comes from the left-hand side. The parameters of the focusing lens L_1 (condenser) are chosen so that the light cone inside the spectrometer just covers the whole area of the collimating element (here the lens L_2, but most of today's spectral devices utilize mirrors). At the same time this design ensures plane parallelism of the beam entering the dispersion element.

other hand, to the strong dependence $(S/N) \sim 1/R^2_{\text{real}}$. Let us note that a similar weak square root relation $(S/N) \sim \sqrt{T}$ holds true also for the photon-counting technique.

We come to the conclusion that the crucial parameter available to the experimenter is the entrance slit brightness B_λ. This can be optimized by setting up an effective optical system to collect luminescence from as wide as possible solid angle (see Figs 2.1 or 2.2, and the relevant discussion). Futhermore, the optical system focusing the luminescence signal on the entrance slit must be well designed. The basic principle is illustrated in Fig. 2.34. Luminescence emission is to be focused by a lens L_1 with minimized spherical and chromatic aberration (an achromatic doublet is preferred) exactly onto the centre of the entrance slit. Moreover, this imaging system should be designed so that the light cone entering the spectroscope just covers the whole area of the collimating lens (or mirror) L_2. Let the diameter and focal length of the lens L_1 be ρ_c and f_c, respectively; then simple geometrical similarity (Fig. 2.34) implies

$$\frac{\rho_c}{f_c} = \frac{W}{f}\left(= \frac{1}{\#}\right), \tag{2.46}$$

in other words, the aperture ratios of lenses L_1 and L_2 must be equal to one another. This equation represents a recipe for choosing the ratio ρ_c/f_c of the lens L_1 (condenser). In case relation (2.46) is not fulfilled, either part of the luminescence emission is lost for spectral analysis (if $\rho_c/f_c > 1/\#$), or the resolving power of the spectrometer is not fully exploited (if $\rho_c/f_c < 1/\#$) as the incoming light cone does not cover the whole area of the dispersion element. It is also desirable to ensure precise positioning of the properly chosen condenser L_1 with respect to the optical axis of the whole system by fixing it to a suitable $\{x, y, z\}$ translation stage equipped with micrometer adjustment screws.

Nowadays, there are many advanced numerical methods (often being part of commercial software packages) that can be applied to 'improve' noisy spectra *a posteriori*, having finished the experiment. But it is always advisable to pay attention to optimizing the luminescence signal during the experiment, especially to maximize the entrance slit brightness and simulatenously to exploit properly the throughput of the spectral device in the described way.

One way to adjust the whole optical path from the source of luminescence emission to the entrance slit is based on the principle of reversibility of light rays, a law known from geometrical optics. Here, the detector is removed

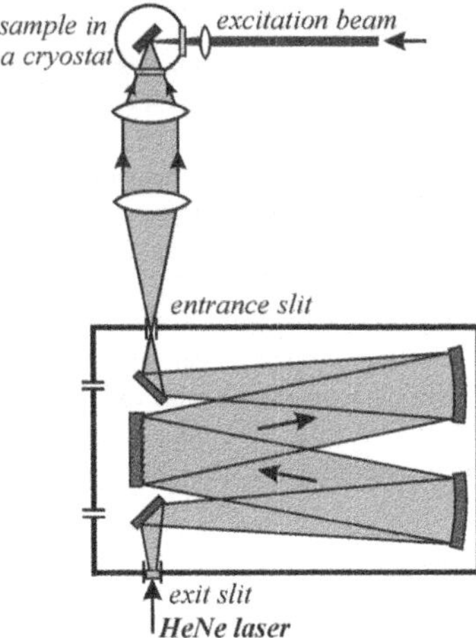

Fig. 2.35
Adjustment of the optical system for luminescence collection by reversing the direction of light propagating through a monochromator. A piece of ground glass is inserted into the exit slit.

and the exit slit is illuminated from outside, preferentially with a HeNe laser (633 nm) or other gas or diode laser emitting visible light, as shown in Fig. 2.35. A small piece of ground glass or calking paper is inserted into the plane of the exit slit in order to scatter the low-divergence laser beam into many directions, simulating the passage of the luminescence emission beam through the monochromator. The laser light, after leaving the entrance slit (the monochromator has to be set to the laser wavelength, obviously), maintains its direction going through the collection optical system and is finally focused onto the sample, where a visible red spot appears. This is the place onto which the excitation light beam is to be directed. Such an arrangement ensures correct focusing of the luminescence emission onto the entrance slit along with the optimized path through the monochromator and the right illumination of the detector. Do not forget to remove the ground glass from the exit slit before remounting the detector (!). A final signal maximization prior to recording the spectrum is done via fine adjustment of the positions of the lenses using $\{x, y, z\}$ translation stages.

Achieving maximum brightness at the entrance slit represents the experimental priority also when using a spectrograph in conjunction with a multichannel detector, because B_λ naturally figures in eqn (2.40), standing for the detector irradiance. Other experimental factors, the spectrum acquisition time and the related S/N ratio, were already described in that part of Section 2.2 dealing with multichannel detectors.

Finally, it is worth noting that, nowadays, optical waveguides (fibres) are increasingly applied to guide light signals towards spectrometers. Among numerous benefits, for example, is their capacity to bring the signal from distant and/or badly accessible places. In order to minimize signal losses, one has to choose an appropriate waveguide material with maximum transmittance over the spectral region of interest (common waveguides are made of various glasses, silica, polymers, and liquids) and then correctly couple light in and out of the waveguide.

A glass waveguide can be formed by a single fibre or a bundle of fibres. The single-fibre waveguide is mostly of cylindrical shape, the central part of which is called the core (refractive index n_1) and is surrounded by a cladding layer with lower refractive index $n_2 < n_1$ (the change in refractive index may be either step-like or gradual, hence the so-called step-index or graded-index fibres). Coupled light is guided within the core thanks to total reflection on the interface between the core and cladding, provided the entrance angle is equal to α_a or smaller (see Fig. 2.36(a)). The maximum entrance angle α_a is given by

$$\sin \alpha_a = \frac{1}{n_0}(n_1^2 - n_2^2)^{1/2}, \tag{2.47}$$

where n_0 is the relative refractive index of the ambient medium (for air we can put $n_0 = 1$). The same value of maximum angle holds for light leaving the waveguide. Typical values for glass and silica fibres are $\alpha_a = 34°$ and $12°$, respectively. The angle α_a is called *the acceptance* or *aperture angle* and it determines the so-called *acceptance* (and output) *cone of the fibre*. The sine of the aperture angle is the *numerical aperture* (NA), and sometimes also the term

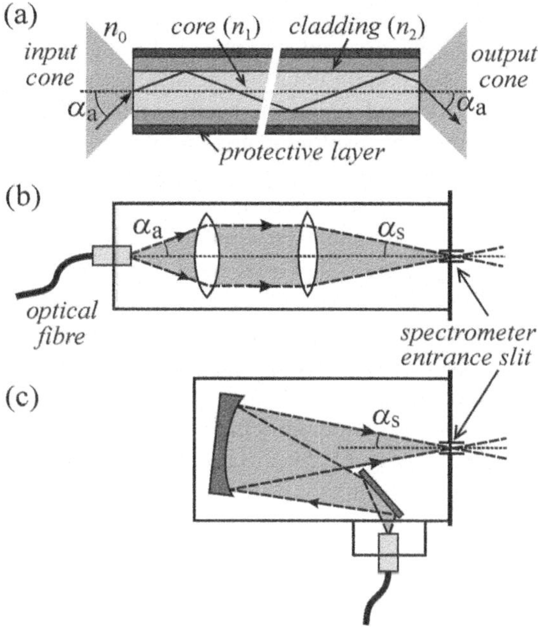

Fig. 2.36
(a) Schematic of light rays guided in an optical waveguide, (b) and (c) examples of two optical systems designed to couple correctly light from a fibre to a spectrometer using lenses (a) or mirrors (b). Based on documents by Horiba Jobin Yvon, Ltd.

aperture ratio $1/\# \approx 2\text{NA}$ is used (in analogy with photographic cameras), see Subsection 2.3.2.

The numerical apertures of a waveguide and that of a spectrometer only exceptionally happen to match one another. Therefore, a special optical system is required to adapt the output cone of a fibre to the acceptance cone of a spectrometer described by an angle α_S (Fig. 2.36(b),(c)). Such coupling adapters for various optical waveguides and spectrometers are usually designed and supplied by spectrometer manufacturers.

2.6 Fourier luminescence spectroscopy

We have shown that increasing the resolving power of a scanning monochromator leads either to a decrease in the S/N ratio or to impractically long spectral acquisition times. One possible way to solve this problem—the multichannel detectors—was also mentioned; these devices are indeed increasingly used in modern luminescence spectroscopy. However, if a need for relatively very high resolution ($\Delta\lambda \leq 0.01$ nm) arises, we shall be faced with a fundamental limitation: the ultimate resolution of the multichannel detectors is strictly limited by the finite size of their individual pixels. For example, a typical detector with pixels of $26\,\mu\text{m} \times 26\,\mu\text{m}$ attached to a spectrometer with a dispersion of $L^{-1} \approx 3$ nm/mm has the resolution limit $\Delta\lambda \approx 3\,\text{nm/mm} \times 26 \times 10^{-3}\,\text{mm} \approx 0.08$ nm. Even if the need for such a very high spectral resolution is not typical for luminescence spectroscopy, it may sometimes appear, as we have already indicated several times. In this case one has no choice but to resort to Fourier luminescence spectroscopy.

A Fourier spectrometer, widely applied in infrared absorption spectroscopy, is basically a scanning Michelson interferometer. The radiation to be analysed (luminescence in our case), upon entering the spectrometer, is split into two arms and, after being reunited again, the beams interfere due to the varying length of one of the arms. The resulting temporal interference signal undergoes Fourier analysis that transforms it into a frequency (or wavelength) spectrum. There are two fundamental differences compared to standard monochromators: (i) all spectral components are analysed simultaneously (the so-called Fellgett (or multiplex) advantage which, however, takes place also in spectrographs equipped with multichannel detectors—see eqn (2.12)), and (ii) the interferometer has no slits and, consequently, its throughput/aperture is much higher than that of a traditional monochromator that shows the same resolving power (the so-called Jacquinot advantage).

It is possible to derive the following relation comparing the signal-to-noise ratio of a Fourier spectrometer $(S/N)_{\mathrm{F}}$ to that of a scanning monochromator $(S/N)_{\mathrm{scan}}$ utilizing the same detector [20]

$$\frac{(S/N)_{\mathrm{F}}}{(S/N)_{\mathrm{scan}}} \approx \frac{2\pi f}{h}\sqrt{M}, \tag{2.48}$$

where M is the number of narrow spectral components (lines) that are analysed by a Fourier spectrometer during the same time interval that a scanning monochromator needs to analyse a single component. Other variables have their usual meaning: f is the focal length of the monochromator collimator and h is the illuminated slit height (the first factor on the right side of (2.48) thus reflects the Jacquinot advantage while the second one results from the Fellgett advantage).

The right-hand side of eqn (2.48) has a typical value of 10^3. If we consider the highest real resolution of a typical scanning monochromator to be $\Delta\lambda \approx 0.1$ nm, then, keeping the same S/N ratio, a Fourier spectrometer can in principle attain a resolution of about three orders of magnitude better, i.e. $\Delta\lambda \approx 10^{-4}$ nm. However, since, as a rule, semiconductor photodiodes rather than photomultipliers are used in a Fourier luminescence spectrometer, it appears more realistic to speculate that the right-hand side of eqn (2.48) is one order of magnitude lower. Then, the practically achievable resolution of a Fourier spectrometer is about $\Delta\lambda \approx 10^{-3}$ nm; expressed in wavenumbers ν^*, which are frequently used here, this corresponds to $\Delta\nu^* \approx 10^{-2}\,\mathrm{cm}^{-1}$. We shall mention a benefit of this unique property in Subsection 7.2.2. Nevertheless, in general Fourier spectrometers are still rather expensive and their exploitation is as yet worthwhile only in very special applications (e.g. high-resolution near-infrared spectroscopy of weak luminescence signals).

2.7 Spectral corrections

Suppose we have optimized our optical collection system and our spectral apparatus consists of a high-throughput monochromator and a high-quality cooled detector. Even then, however, the experimental emission spectra of the sample under study, in spite of being minimally noisy, do not correspond—

as far as their shape is concerned—to the actual luminescence spectrum. The reason for this is due to the fact that the photodetector sensitivity $k(\lambda)$ as well as the characteristics of other optical elements—the reflectivity or transmittance of the dispersion element, mirrors and lenses—are spectrally dependent. Thus, if we denote the luminescence intensity as function of wavelength (which means the actual emission spectrum) as $I(\lambda)$, then the measured spectrum $i(\lambda)$ is equal to

$$i(\lambda) = Q(\lambda)\, I(\lambda), \tag{2.49}$$

where $Q(\lambda)$ is usually called the spectral response of the whole detection system.[7] Thus $Q(\lambda)$ includes both the transmittance of the spectral device $\kappa(\lambda)$ and the detector sensitivity $k(\lambda)$. Sometimes $Q(\lambda)$ is simply referred to as a correction function (of the detection channel). The function $Q(\lambda)$ must be determined experimentally, separately for each particular configuration of the detection system (various combinations of a dispersion element and a detector unit). A faithful spectral shape $I(\lambda)$ is then obtained, according to eqn (2.49), by dividing the measured spectrum $i(\lambda)$ by the correction function $Q(\lambda)$.

The correction function $Q(\lambda)$ is determined with the aid of a light source with well-known spectral emittance. A commonly applied calibrated source for the visible and near-infrared regions is a tungsten strip (filament) lamp, whose emission spectrum bears a close resemblance to blackbody radiation. Such calibrated lamps are supplied, together with a detailed description of their spectral energy density, either in graphical or tabular form (or simply colour temperature), valid under a specific heating current. Light is emitted by a narrow tungsten strip in a vacuum bulb; this design ensures the distribution of temperature over the entire surface of the light-emitting body to be as homogeneous as possible. (A traditional tungsten spiral in a bulb or other incandescent lamp types do not fully satisfy this condition but even special sources of this type—filament lamps—can be calibrated and applied for correction purposes.) If an experimentally acquired spectrum of the calibrated source is $w(\lambda)$ and its actual spectral emittance is $W(\lambda, T)$, the required correction function is calculated according to eqn (2.49) simply as

$$Q(\lambda) = \frac{w(\lambda)}{W(\lambda, T)}. \tag{2.50}$$

Then, to obtain undistorted results, any measured luminescence spectrum has to be divided by this correction curve.

The procedure for determining $Q(\lambda)$ as described above seems to be quite simple but the opposite is true. First of all, it is not easy to keep the prescribed lamp current stabilized over the whole measurement time, while even small

[7] A somewhat more exact formulation of the above debate is that the luminescence intensity means (usually) the photon flux (power) per unit interval of wavelengths, $[I(\lambda)] =$ photon/s/unit interval of λ. From the point of view of radiometry this corresponds to radiant flux $\phi_L(\lambda)$ per unit interval of λ, $[\phi_L(\lambda)] =$ power/unit interval of λ. This flux impinges on the entrance slit creating its irradiance $E_\lambda = \phi_L(\lambda)/S$. Then the slit becomes a planar light source for the spectrometer, featuring brightness $B_\lambda = E_\lambda$. According to eqn (2.37), the radiant flux at the exit slit is $\phi_\lambda \sim \kappa(\lambda) B_\lambda$ and the detector signal reads $i(\lambda) \sim k(\lambda)\phi_\lambda$, which means $i(\lambda) \sim k(\lambda)\kappa(\lambda)$ $B_\lambda \sim k(\lambda)\kappa(\lambda)\, I(\lambda)$.

deviations from the specific temperature (usually in the range of 2800–3000 K) cause significant changes of the $W(\lambda, T)$ spectrum. Too high lamp intensity may appear as another problem, because of possible detector oversaturation. Next, ageing of such lamps may be quite fast and can cause harmful alteration of the original spectrum $W(\lambda, T)$. Often, to determine $Q(\lambda)$ correctly, a diffusion reflector (made of, e.g., MgO or $BaSO_4$) is recommended, but variations in its spectral reflectivity can affect the result. In addition, problems could also arise from polarization effects—light becomes partially polarized or depolarized by reflective surfaces, by diffraction gratings, etc. In particular the last mentioned effects are often ignored even if their impact could be significant. The correction curve thus has to be determined with due care. Some difficulties in determining $Q(\lambda)$ can be eliminated by using an integration sphere fixed at the entrance of the spectrometer. This ensures correct filling of the collimator by light and also depolarizes possible partial polarization of the incoming lamp radiation.

Nowadays, performing the correction of luminescence spectra for spectral response of the detection path should be commonplace in every laboratory. Any spectrum submitted for publication in international refereed journals may be automatically supposed to have been corrected. Indeed, the difference between a corrected and an uncorrected spectrum can be considerable, as illustrated in Fig. 2.37 [21]. Naturally, the correction becomes imperative when the spectrum is going to be fitted with a theoretical curve. Actually one of the aims of this book is to teach the reader how to identify the microscopic origin of the luminescence centre, based on the corresponding spectral shape and its changes due to controlled variations of experimental parameters: temperature, excitation intensity, etc.

Nevertheless, one can admit that sometimes the correction in question is not indispensable—for example, if only a qualitative demonstration of spectral changes (e.g. one spectral band grows at the expense of another during an increase in temperature) is sufficient for a given purpose. Also in the case of very narrow emission lines the spectra can often be left uncorrected and even fitting its shape with a theoretical model is possible. The reason is that the correction curve is usually only slowly varying with wavelength and its change

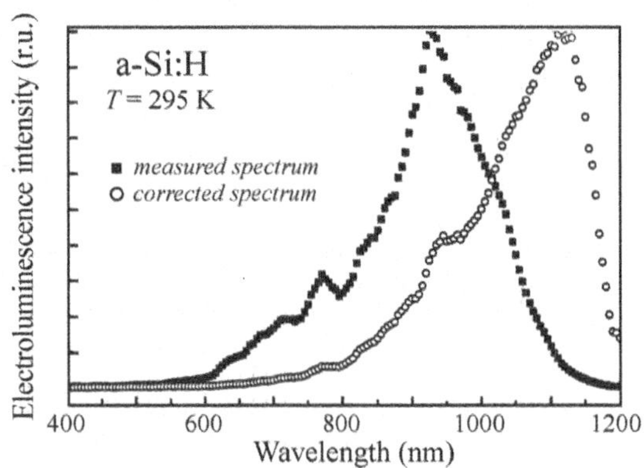

Fig. 2.37
As measured (full squares) and corrected (empty circles) electroluminescence spectra of amorphous hydrogenated silicon. The pronounced deformation of the measured spectrum at long wavelengths is mainly due to the rapid decrease in photomultiplier sensitivity in this spectral region (S1 photocathode), while the dip at $\sim$ 790 nm is due to the so-called Wood's grating anomaly. The spectra are normalized. Adapted from Fojtik [21].

in a narrow wavelength range (a few nanometres) can be neglected. But always be vigilant!

We should stress that this correction does not serve for establishing the absolute number of emitted photons or determining the absolute luminescence quantum yield η—this is a domain of very special and quite difficult techniques, which are described for example in [22] and in Appendix K. Any spectrum corrected using (2.50) 'only' faithfully represents (in relative units) the number of photons emitted by the luminescent sample per unit interval of wavelengths or photon energies. It remains to be noted that the described correction procedure is basically the same irrespective of whether a monochromator or spectrograph is used.

In the context of corrections of optical emission spectra we must mention another experimental problem, which used to be unjustly neglected. It concerns two alternative modes of plotting the emission spectra—against either wavelengths λ or photon energy $h\nu$. Both representations are possible and widely used (sometimes even both modes are applied to the same graph, one at the bottom and the other at the upper axis or vice versa, see e.g. Fig. 1.2), conversion between wavelengths and photon energies being performed simply via the trivial relation

$$h\nu = h\frac{c}{\lambda}. \tag{2.51}$$

However, such an approach is not always fully justified. One has to take into account what type of dispersion element and wavelength scan was used during the experiment.

Let us consider firstly a grating monochromator with a linear scan in wavelengths λ (the sine-drive, see Section 2.3). The spectral measurement is carried out using constant slit widths $\Delta\lambda$, the result being either a raw spectrum $i(\lambda)$ or, after having performed the correction, the actual spectrum $I(\lambda)$. However, the *elemental step (interval) in photon energy* is not kept constant during the measurement, as is clearly seen by differentiating (2.51),

$$|\mathrm{d}(h\nu)| = \frac{hc}{\lambda^2}|\mathrm{d}\lambda|. \tag{2.52}$$

Obviously, when increasing the wavelength (keeping $\mathrm{d}\lambda = \text{const}$), the energy interval $|\mathrm{d}(h\nu)|$ drops, because it scales like λ^{-2}. Therefore, if one wishes to plot the spectrum against the photon energy scale like $I(h\nu)$, the decrease in the magnitude of the energy interval $|\mathrm{d}(h\nu)|$ should be corrected[8] by multiplying each value of $I(\lambda)$ by λ^2:

$$I(h\nu) = \lambda^2 I(\lambda). \tag{2.53}$$

This multiplication obviously boosts the long-wavelength wing of the spectrum, causing a red-shift of spectral peaks; see Fig. 2.38(a) [23]. Thus, in principle it is incorrect to convert simply λ into $h\nu$ while keeping the spectral curve unchanged. However, at the same time one can see that the correction

[8] Let us recall that $I(\lambda)$ means the photon flux per unit interval of wavelength, and by analogy $I(h\nu)$ is the photon flux per unit (therefore constant) interval of photon energies.

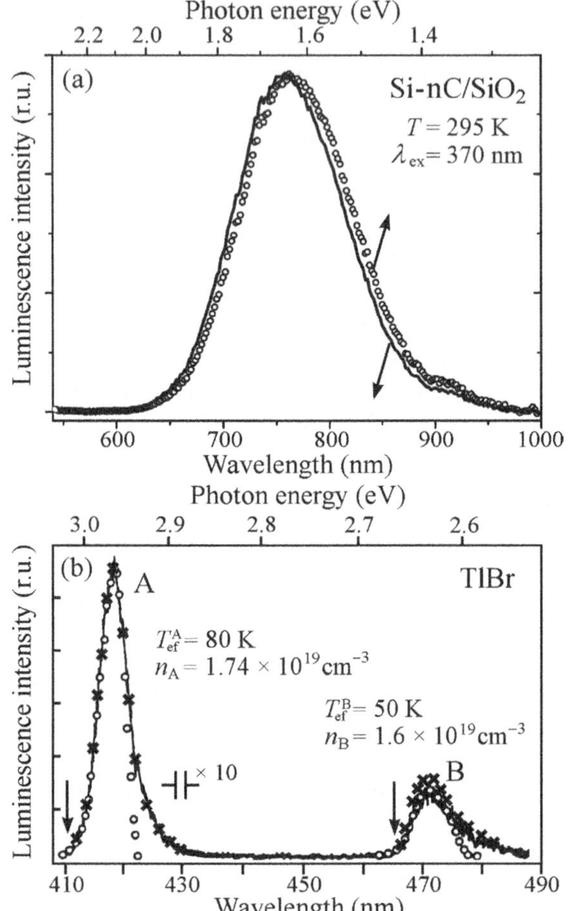

Fig. 2.38
Correction of emission spectra when transforming the wavelength-to-photon energy scale. (a) Photoluminescence spectrum of silicon nanocrystals embedded in an SiO_2 matrix. The spectrum was recorded using a grating monochromator with a linear scan of λ (solid line, the spectrum is related to the lower axis). Correction pertinent to transformation into photon energies $h\nu$ was performed by multiplying by λ^2 (open circles, the spectrum is related to the upper axis). Both curves are normalized. Adapted from Dian *et al.* [23]. (b) Emission spectra of TlBr crystals due to radiative decay of the electron–hole plasma. The solid line is the experimental spectrum measured by making use of a linear scan in λ, and the crosses show the recalculated spectrum connected with axis transformation to photon energies (circles are related to a theoretical fit, which is irrelevant to the present discussion). Both the A and B emission lines being relatively narrow, the correction as regards their spectral shape and position is of minor importance—just the relative intensities of these bands are slightly changed (the intensity of the B line increases relative to the A line). Adapted from Pelant *et al.* [24].

effect is not large and becomes significant for sufficiently broad spectra only (full width at half maximum about 100 nm or more). This is illustrated by Fig. 2.38(b) [24]: for emission lines only a few tens of nanometres in width or narrower, the distinction in shape between the recalculated curve and the spectrum as recorded becomes negligible, usually buried in noise.

In the case of a grating monochromator with a linear sweep in wavenumbers $1/\lambda$ or energy $h\nu$ (a cosec-drive, which is a much less frequent case) we obtain an experimental spectrum against photon energy like $I(h\nu)$. For transforming it into wavelengths we again use relation (2.53), but in the opposite direction: the spectrum $I(h\nu)$ will be divided by λ^2.

Monochromators equipped with a prism should in principle require two corrections. Nevertheless, the following discussion is intended to show that the situation there is, in fact, not so serious. First of all, the reciprocal linear dispersion L^{-1} is a nonlinear function of λ which, according to (2.19), causes an increase in the spectral slit width $\Delta\lambda$ (when keeping the mechanical slit width $\Delta\ell$ constant) with increasing λ.

Consequently, the shape of the measured spectrum $I(\lambda)$ does not exactly correspond to reality. In principle, for each prism device, knowing its particular $L^{-1}(\lambda)$ dependence, we should have to determine a specific correction function[9] $\xi(\lambda) = \Delta\lambda_0/\Delta\lambda < 1$ to multiply the experimental spectrum $I(\lambda)$; since $\xi(\lambda)$ is a decreasing function, the long-wavelength wing of $I(\lambda)$ would have been slightly suppressed. However, a second correction takes place during the transition from $I(\lambda)$ to $I(h\nu)$, namely, multiplication by λ^2 according to (2.53), which relatively enhances the long-wavelength side. All in all, both corrections go in opposite directions and may roughly compensate each other. This means that we can practically omit both of them within reasonable accuracy.

To conclude, this somewhat arbitrary way of interchanging between the λ and $h\nu$ axes, found frequently in the literature, is not usually at the expense of the correctness of presentation, except, perhaps, for very broad spectra. For completeness, let us also mention spectrographs, where wavelength scanning is replaced by multichannel detection. Most of these devices employ a diffraction grating whose dispersion is linear in wavelength, i.e. $\Delta\lambda = \text{const}$. This situation is analogous to that of a sine-drive monochromator—transformation from $I(\lambda)$ to $I(h\nu)$ is done using eqn (2.53), if necessary.

2.8 Influence of slit opening on the shape of emission spectra

The last important correction of emission spectra we are going to treat here concerns the effect of the finite width of the instrument slits on the spectral shape. Intuitively, we feel that setting too wide slits can bring about loss of spectral details and blurring or broadening of the spectra. In this section, the influence of the slit will be described quantitatively. Firstly, the case of a monochromator (a dispersion device with two slits) and then that of a spectrograph (having one entrance slit and a multichannel detector at the output) will be analysed.

Monochromator

In Appendix A we give an illustrative explanation of the convolution of two functions in terms of a shift of the first function over the second one along the independent variable axis, integrating simultaneously the product of the two functions (Fig. A.1). The functioning of a scanning monochromator equipped with a rotating dispersion element fits the above scheme well: the emission spectrum, created with the cooperation of the entrance slit with the dispersion element (and being affected by the spectral variations in reflectivity (transmittance) of all elements contained within the device) is projected onto the exit slit plane. Obviously, this is not the true spectrum $I(h\nu) = I(E)$ but it can be described as $\kappa(E)I(E)$, where $\kappa(E)$ comprises all the above mentioned spectral factors (see also (2.32)). If, moreover, we include now also

[9] Here $\Delta\lambda_0$ means the spectral slit width at the short-wavelength boundary λ_0 of the spectral range covered by the device, and $\Delta\lambda = L^{-1}(\lambda)\Delta\ell$ is taken for $\lambda > \lambda_0$.

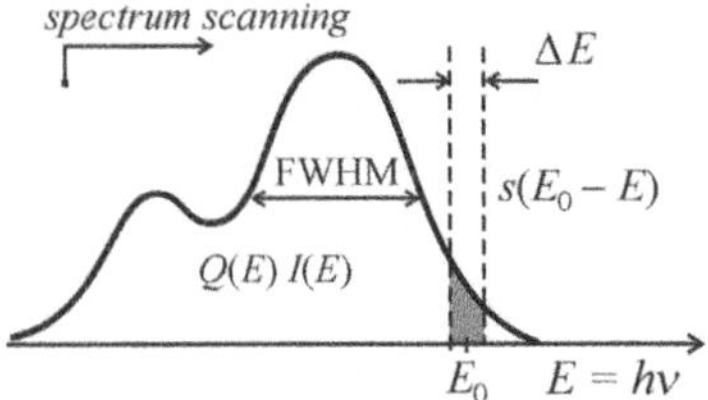

Fig. 2.39
The emission spectrum in a scanning monochromator is scanned across the exit slit (its spectral width being ΔE). The joint activity of the entrance and exit slits is described by a function $s(E_0-E)$. This means that the function $Q(E)I(E)$ is convoluted with $s(E_0-E)$; FWHM means full width at half maximum of the spectral band.

the detector spectral sensitivity $k(E)$, then the spectrum projected onto the exit slit corresponds to $\kappa(E)k(E)I(E) \equiv Q(E)I(E)$.

This spectrum is now 'swept' or scanned across the exit slit, consistently with rotation of the grating or prism (Fig. 2.39). The exit slit of spectral width ΔE transmits (and so 'integrates') slices of the spectrum towards the detector. If we denote, according to eqn (A.1) and Fig. A.1, the product $Q(E)I(E)$ as f_2, and the function describing the effect of slits as $f_1 = s(E)$, we see that the measured spectrum is given by the convolution of the functions QI and s:

$$i(E_0) = \int_{-\infty}^{+\infty} s(E)Q(E_0-E)I(E_0-E)\mathrm{d}E = \int_{-\infty}^{+\infty} Q(E)I(E)s(E_0-E)\mathrm{d}E,$$

where we have applied commutativity of the operation of convolution according to eqn (A.3). Thus the measured spectrum $i(E_0)$ is, taking into account the effect of the finite slit opening, given by the convolution

$$i(E_0) = \int_{-\infty}^{+\infty} Q(E)I(E)s(E_0-E)\mathrm{d}E. \tag{2.54}$$

The function $s(E_0-E)$ is usually called *the apparatus (instrumental) function* of the spectral device. It must comply with the normalization condition

$$\int_{-\infty}^{+\infty} s(E_0-E)\mathrm{d}E = 1. \tag{2.55}$$

We are now in a position to decide under what conditions the convolution (2.54) is to be taken unavoidably into consideration when treating the experimental data; we shall discuss three typical situations.

(a) The spectral slit width ΔE is much narrower than the full width at half maximum (FWHM) of the measured spectrum, $\Delta E \ll$ FWHM. Then the function $s(E_0-E)$ can be approximated by the delta function $\delta(E_0-E)$ and (2.54) is reduced straightaway to

$$i(E_0) = Q(E_0)I(E_0). \tag{2.56}$$

Or, for a broad spectrum and narrow slits the effect of apparatus function need not be taken into account. The true emission spectrum is obtained simply by dividing the experimental result by the correction function $Q(E)$.

(b) The spectral slit width ΔE remains smaller than the FWHM of the spectrum but the apparatus function cannot be considered infinitely narrow. Then the measured spectrum and the actual spectrum are not identical, and in order to extract the actual spectrum $I(E)$ from the convolution integral (2.54), one needs to know the function $s(E_0-E)$.

Let us restrict our discussion to the standard case of equal entrance and exit slit widths ΔE, large enough to neglect diffraction effects. Then the geometrical image of the entrance slit in the output focal plane is represented by a homogeneously illuminated area (rectangle). Then an *ideal monochromator* (i.e. a device free of imperfections and aberrations of optical imaging) features the apparatus function of a triangle shape as described by

$$s(E_0 - E) = \begin{cases} \frac{1}{\Delta E}\left[1 - \frac{|E_0-E|}{\Delta E}\right] & \text{for} |E_0 - E| \le \Delta E \\ 0 & \text{for} |E_0 - E| > \Delta E, \end{cases} \quad (2.57)$$

which is illustrated in Fig. 2.40(b) and whose FWHM is equal to ΔE. Figure 2.40(a) helps us to understand why this is so. When the dispersion element is turning, the monochromatic image of the entrance slit is moving across the exit slit, and the detector records a signal proportional to the overlap area (just the convolution of the entrance slit image with the exit slit). This signal increases linearly from zero to its maximum value (both areas are fully overlapping) and then drops back. As a result, the triangular shape of the function $s(E_0-E)$ appears.

In a *real monochromator*, however, various imperfections, inhomogeneities of all optical elements, parasitic diffraction effects, imperfect absorption of the internal blackened surfaces of the monochromator, etc. produce a weak light background that can enter the exit slit at wavelengths different from the nominally set value. This effect leads to making the triangle vertices round and the real apparatus function can be approximated by a Gaussian curve, as shown in panel (c) of Fig. 2.40. Its analytical expression takes the form

$$s(E_0 - E) = N(\Delta E') \ \exp\left[-\frac{4 \ln 2 \ (E_0 - E)^2}{(\Delta E')^2}\right], \quad (2.58)$$

where $\Delta E'$ is the FWHM of the curve and $N(\Delta E') = 2\sqrt{\ln 2}/(\sqrt{\pi}\Delta E')$ is a normalizing factor.

The relatively simple form of the apparatus function (2.58) is advantageous for performing a deconvolution of eqn (2.54) in order to extract the real spectrum $I(E)$ from the experimental one $i(E)$. (We should not forget, however, that also the correction function $Q(E)$ must be known.) There are a large number of deconvolution methods [25] and current progress in computational techniques makes them easily available. Nevertheless, it seems that an inverse approach, the so-called iterative convolution, remains among the most widely applied methods. The principle is based on setting down a suitable model function for the actual spectrum $I(E)$ with some initial parameters and calculating the convolution (2.54). The parameters of $I(E)$ are then iterated until the best correspondence between the calculated spectrum $i(E)$ and the experimental one is achieved. Evidently, the initial selection of an appropriate microscopic model of luminescence centre (and related spectral shape) is crucial and depends strongly on both the experience and intuition of the experimenter.

Application of the apparatus function in the form of (2.58) includes one difficulty which is not evident at first glance, namely, the unknown value of $\Delta E'$. Unlike the triangular function (2.57), the halfwidth of the Gaussian peak $\Delta E'$ (2.58) is not equal to the spectral width of the entrance or exit slit ΔE. This is because of the normalization (2.55) and is illustrated in Fig. 2.40(c): the postulate of a constant area below the curve yields $\Delta E' > \Delta E$. Therefore, knowledge of the mechanical (and, consequently, also spectral) widths of slits does not enable us to determine the particular shape of the Gaussian instrumental function (2.58). How, then, can the apparatus function shape be obtained?

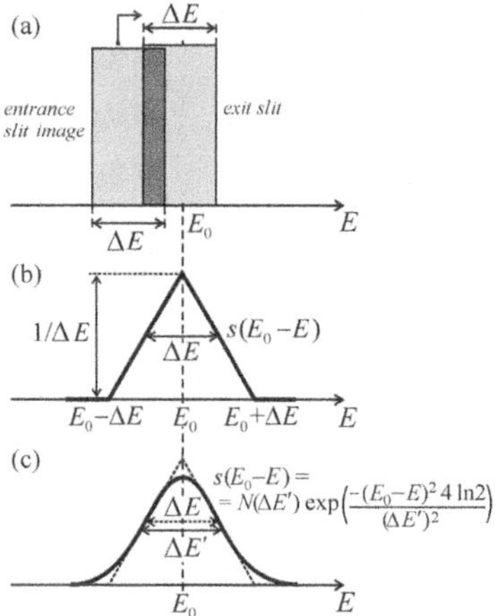

Fig. 2.40
Illustration of the apparatus function of a monochromator. (a) The monochromatic image of the entrance slit is moving to the right across the exit slit in the focal plane of an ideal device. The detector 'sees' an intensity proportional to the overlap area (dark grey rectangle) whose variation as a function of E is the apparatus function $s(E_0-E)$ plotted in panel (b). This function has a triangular shape with unit area, in order to fulfil the condition (2.55). Panel (c) shows the shape of this function in a real monochromator, as influenced by imperfections in the instrument and optical inhomogeneities. $N(\Delta E')$ stands for a normalizing factor.

The answer to this question is related to our last situation to be discussed:

(c) The spectral slit width ΔE is much wider than the width of the measured spectrum, $\Delta E \gg$ FWHM. Then we put $I(E) \approx \delta(E - E_0')$, $Q(E) \approx$ const, and eqn (2.49) is reduced to

$$i(E_0) \approx \int_{-\infty}^{+\infty} \delta(E - E_0')\, s(E_0 - E)\, \mathrm{d}E \approx s(E_0 - E_0'). \tag{2.59}$$

Thus, paradoxically, the result of this experiment is the apparatus function rather than the spectrum (spectral peak) itself. Such a situation can be easily realized for example by measuring narrow emission lines from a low-pressure spectral lamp. If such a measurement is performed with the same slit widths as applied during the luminescence experiment, we can thereby use the obtained shape of the Gaussian apparatus function (2.58) for deconvolution (or iterative convolution) to find the actual spectral shape $I(E)$.

Let us add a final technical note: All the functions in the integral (2.54) are non-zero within relatively narrow energy intervals only; then the limits of numerical integration may be strongly restricted. In fact they are given by the narrowest function. Therefore, in the case of (b), applying the Gaussian apparatus function (2.58), reasonable integration limits are $\langle E_0 - 5\Delta E',\ E_0 + 5\Delta E' \rangle$.

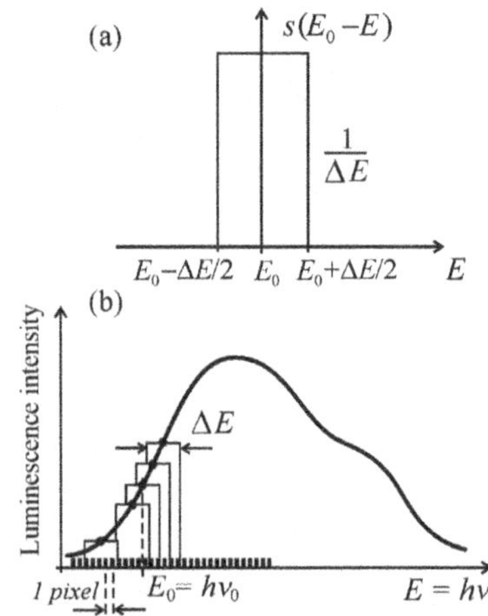

Fig. 2.41
(a) Apparatus function of a spectrograph. (b) Detection of a broad luminescence spectrum by a multichannel detector and a spectrograph with a wide entrance slit (ΔE means the spectral width of the entrance slit image).

Spectrograph

Here, assessment of the influence of the slit opening on the spectral shape is somewhat easier for two reasons. Firstly, there is only a single (entrance) slit. Secondly, the option of acquiring any spectrum for a very long time enables us to set the slit width narrow enough so as not to distort the spectral shape. Nevertheless, we are going to analyse several potential situations.

First of all, we should ask about the shape of the apparatus function $s(E)$. The answer is obvious: in any spectrograph, a uniformly illuminated monochromatic image of its entrance slit is created in the exit focal plane. Therefore, the relevant apparatus function is of a rectangular shape described by

$$s(E_0 - E) = \begin{cases} \frac{1}{\Delta E} & \text{for } |E_0 - E| \le \frac{\Delta E}{2} \\ 0 & \text{for } |E_0 - E| > \frac{\Delta E}{2}, \end{cases} \tag{2.60}$$

where ΔE is again the spectral slit width (Fig. 2.41(a)). Then the convolution (2.54) takes on the form

$$i(E_0) \approx \int_{E_0 - \Delta E/2}^{E_0 + \Delta E/2} Q(E) I(E) \mathrm{d}E \approx Q(E_0) \int_{E_0 - \Delta E/2}^{E_0 + \Delta E/2} I(E) \mathrm{d}E, \tag{2.61}$$

if we exploit the fact that the correction function $Q(E)$ is only slowly varying across the slit width. Application of relation (2.61) to deconvoluting a measured spectrum is then relatively simple.

However, a closer inspection of the preceding procedure can raise an objection: we have introduced the convolution (Appendix A) as a relative shift of two functions, which is not completely true here—the grating is not rotating, and the spectrum is not moving. Is then the proposed application of eqn (2.61) correct? The response is yes. Although the 'monochromatic' images of the entrance slit hit the detector simultaneously, not in sequence like in a monochromator, the final effect is the same (see Fig. 2.41(b)). The mechanism of the generation of the spectral image is, however, different, as we shall indicate.

Figure 2.41(b) shows the situation when the experimenter has chosen the entrance slit width of a spectrograph[10] several times wider than the size of a single pixel. This is a standard situation; in Section 2.3 we have shown that setting the slit width narrower than about 10 μm serves no useful purpose. Therefore, a slit width of $\Delta\ell = 100$–200 μm is a reasonable choice (also in relation to an acceptable acquisition time of the spectrum), $\Delta\ell$ being really much larger than the typical CCD pixel size 25 μm. Let us select a fixed photon energy E_0. If the slit image width exceeds that of a single pixel, the signal level on the pixel at $E_0 = h\nu_0$ may 'get saturated', but this does not prevent the signal at E_0 from attaining the correct level, because 'monochromatic' images of neighbouring spectral elements join the signal. Naturally, this is accompanied by spectral broadening; everything is comprised in eqn (2.61).

We shall close Sections 2.7 and 2.8 with a brief evaluation of the effects and corrections just discussed in view of their practical importance. The reader, possibly confused and upset, might have come to the conclusion that any experimental luminescence spectrum is only a relatively useless raw product, which must undergo many mathematical operations prior to being presented and interpreted. This is not the case, though. Among the above mentioned corrections only one is really essential and almost always indispensable, namely, to divide the experimental spectrum by the spectral correction function $Q(\lambda)$ given in eqn (2.50), in order to correct the effect of spectral sensitivity of the whole detection system. The 'presentation problem' of the spectral data, i.e. care about whether to present spectra against wavelength λ or photon energy $E = h\nu$, takes place only exceptionally, in the case of very broad spectra (and if one wants to be very careful). Similarly, the correction through making use of the apparatus function $s(E_0{-}E)$ in order to remove the effect of the finite slit width is worth doing only in the case of the measurement of very narrow emission lines, like for instance those typical of radiative recombination of free excitons and biexcitons (Chapters 7 and 8). To a certain extent the corrections $Q(\lambda)$ and $s(E_0{-}E)$ may even be considered disjunctive, because the spectral correction function $Q(\lambda)$ is important for wide spectra, when the slit-width effect is negligible, while the function $s(E_0{-}E)$ is important for narrow lines when, to a good approximation, $Q(\lambda_0) \approx$ const.

2.9 Time-resolved luminescence measurements

Knowledge of the temporal behaviour of the luminescence signal, especially the shape of the luminescence decay curve after the excitation is switched off, carries valuable information contributing to the identification of the recombination process and the microscopic nature of the relevant luminescence centre (along with an analysis of the excitation and emission spectral shapes

[10] We again and again assume equal widths of the entrance slit and of its image.

and of the dependence of the pump intensity). However, the techniques of time-resolved luminescence measurements, in particular under pumping with ultrashort laser pulses, represent a largely specialized research field that cannot be fully treated within the limited space of this section. Therefore, we shall describe just the basic principles of the most common experimental methods and instruments. A comprehensive treatment of this subject can be found for example in [26] and [27].

2.9.1 Direct imaging of the luminescence response

The simplest set-up to study luminescence decay consists of a single-channel detector, usually a photomultiplier, of which the output is connected to the vertical deflection plates of an oscilloscope and swept in time by its horizontal time base. The result is either observed on a screen or transferred to a computer. Obviously, in order to record, in this way, the decay of the luminescence radiation in time, either a cw excitation beam must be chopped (then curves similar to those plotted in Fig. 2.27 are obtained) or a pulsed excitation source supplying sufficiently short pulses (a lamp or laser) must be used. Since the character of luminescence decay may be strongly dependent on the emission wavelength, the decay curves $i_\lambda(t)$ of a quasi-monochromatic luminescence signal are usually measured after inserting a monochromator or at least narrow-band interference filters in front of the detector.

The next step consists in analysing the measured curves $i_\lambda(t)$. In the simplest case, one deals with a single-exponential function $i_\lambda(t) = i_\lambda(0)\exp(-t/\tau)$ where τ is referred to as the luminescence decay time or lifetime. However, many other shapes of decay, e.g. double-exponential, hyperbolic, etc. may occur in luminescence spectroscopy. Their detailed study will be presented in Chapter 3.

Generally, the mean decay time $\bar{t}$ can be defined from the point of view of statistics as the mean value of a continuous stochastic varible by

$$\bar{t} = \frac{\int_0^\infty t\, i_\lambda(t)\,\mathrm{d}t}{\int_0^\infty i_\lambda(t)\mathrm{d}t},$$

where $i_\lambda(t)$ plays the role of the occurrence probability density of a time interval t in which an elementary emitter is found in an excited state. In the case of the single-exponential decay $i_\lambda(t) = i_\lambda(0)\exp(-t/\tau)$, simple evaluation of the integrals leads to $\bar{t} = \tau$.

For practical purposes and in analogy with the single-exponential decay, as the 'decay time' in the case of more complex decay curves $i_\lambda(t)$, the time at which the initial intensity $i_\lambda(0)$ decreases to the value $i_\lambda(0)/\mathrm{e} \approx i_\lambda(0)/2.72$, is often (somewhat arbitrarily) taken.

Although the method of direct decay imaging looks very simple, its practical implementation challenges the quality of the instrumentation and the experimenter's knowledge of the limits of applicability of the method. First of all, the whole experimental set-up (a pulsed excitation source, a photomultiplier

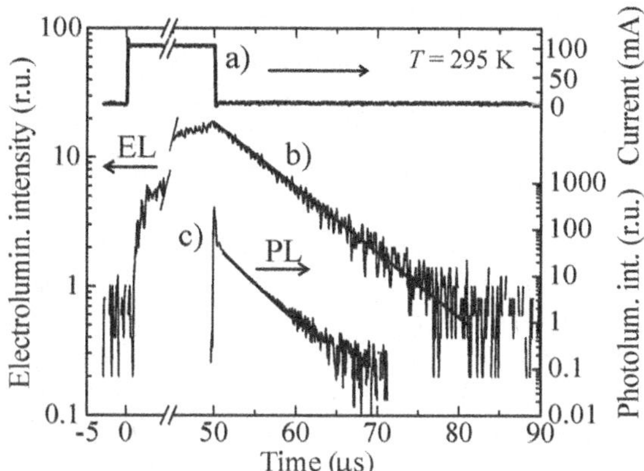

Fig. 2.42
An example of direct imaging of the luminescence response by a digital oscilloscope (the signal comes from red electroluminescence (EL) and photoluminescence (PL) of silicon nanocrystals). The upper trace (a) shows a $50\,\mu$s-long electric injection pulse, which excites electroluminescence, and curve (b) is part of the onset of EL and the subsequent decay. Curve (c) is the photoluminescence response of the same sample to excitation with 3 ns laser pulses (the third harmonics from a Nd:YAG laser). For both EL and PL, the excitation switch-off is fast enough which enables direct determination of decay times without any mathematical treatment (deconvolution). The decay curves (b), (c) are fitted by single-exponential functions with decay times of $\tau_{EL} \approx 8\,\mu$s and $\tau_{PL} \approx 2.5\,\mu$s. After Luterová *et al.* [28].

or photodiode, and an oscilloscope) should be fast enough, highly sensitive and perfectly linearly responding to the optical signal, which requirements are difficult to be satisfied simultaneously. (For example, a highly sensitive photomultiplier must contain many dynodes, which in turn makes its time resolution worse due to statistical fluctuations of secondary emitted electrons and the long trajectory of the electron packet.) One can easily imagine a number of potential limitations of the method: (i) the insufficient response speed of the photomultiplier; if (ii) a fast p-i-n photodiode is applied, it is inherently less sensitive and a preamplifier must be used in connection with the oscilloscope, whose limited bandwidth may pose a problem; (iii) similarly, the frequency bandwidth of the oscilloscope itself may be a limiting factor, etc. The direct measurement of the luminescence response may be easily applied (without any complementary mathematical treatment) to study relatively slow decays only (under a pulsed excitation, for roughly $\tau \geq 0.1$–$1\,\mu$s). An example is shown in Fig. 2.42 [28]. If photoluminescence is excited by a chopped beam, this limit is pushed to even longer times ($\tau \geq 50\,\mu$s), considering that the typical maximum frequency of mechanical chopping is of the order of 1 kHz and the excitation pulses are of trapezoidal rather than of perfectly rectangular shape (the slope of the pulse edges is given by the speed of chopper rotation as well as by the diameter of the chopped beam).

Under the complementary (*a posteriori*) mathematical treatment, *deconvolution* of the measured decay curve is to be understood. In analogy to the influence of the monochromator slit opening on the shape of the measured emission spectra (Section 2.8), we can deduce that the experimentally determined luminescence time response $R_\lambda(t)$ is given by a convolution of the actual time response $i_\lambda(t)$ (i.e. the luminescence response to an infinitely short excitation pulse) with the temporal profile of the excitation pulse $L(t)$ measured by the same detection system as that used to obtain $R_\lambda(t)$:

$$R_\lambda(t) = \int_0^t L(t - t')i_\lambda(t')\mathrm{d}t'. \tag{2.62}$$

Here the integration goes from zero to t. This follows from analogy with the convolution example given in Appendix A, where the actual decay shape $i_\lambda(t)$, which has non-zero values only within an interval $\langle 0,\infty)$, substitutes now for the function defined by eqn (A.4). Convolution with the excitation pulse is defined by eqn (A.5), from which eqn (2.62) follows via simple application of the commutativity of convolution (A.3).

Often the duration of the excitation pulse $L(t)$ happens to be comparable to the decay time of the luminescence response $i_\lambda(t)$. In this case the actual luminescence response must be determined via deconvolution of relation (2.62).[11] This can also be simply formulated as follows: Due to the finite temporal width of the excitation pulse $L(t)$, the exact (sharp) initial point of decay $t = 0$ cannot be found and, consequently, the time at which the intensity drops to $i_\lambda(0)/e$ cannot be determined. Only in case the pulse width $L(t)$ is much shorter than the decay time, can we approximate $L(t\text{–}t') \to \delta(t\text{–}t')$ and we obtain $R_\lambda(t) = i_\lambda(t)$; this means that the experiment gives the actual shape of the decay curve, in complete analogy with the situation when the spectral slit width is negligible compared to the width of spectral lines and measured emission spectra do not need to be deconvoluted to recover their actual shape.

It is worth stressing that the function $L(t\text{–}t')$ appearing in (2.62) does not represent the actual shape of the excitation pulse but its 'projection' as seen by the whole detection system. Thus, it is not necessary to apply an extremely fast detection set-up that would be able to detect a distortion-free pulse shape, which might pose problems, especially in experiments with nanosecond and picosecond pulses. On the other hand, the detection system must be fast enough to distinguish reliably the shape of the excitation pulse from the shape of the relevant luminescence response, anyway. The pulse shape $L(t)$ can be most easily measured by detecting reflected or scattered excitation light.

Note that the above discussion concerns photoluminescence experiments. In the case of electroluminescence the response is determined not only by optical transitions and recombination phenomena but also by electric effects related to charge transport or capacitance of the device under study. Then, the observed electroluminescence kinetics is a combination of electric and optical effects and its analysis may be considerably more difficult.

2.9.2 Phase-shift method

The principle of this method consists in excitation of the sample with light, the intensity of which is sinusoidally modulated in time with an angular frequency ω. The luminescence response is obviously expected to be also modulated with the same frequency, but the finite value of the excited state lifetime τ induces two effects: Firstly, the luminescence response is phase shifted by an angle φ with respect to the excitation function and, secondly, its modulation depth is decreased.

[11] The relevant methods were mentioned in Section 2.8 when discussing relation (2.58). For the case of luminescence decay curves, a critical review is given by O'Connor, D. V., Ware, W. R., and Andre, J. C. (1979). *J. Phys. Chem.*, **83**, 1333.

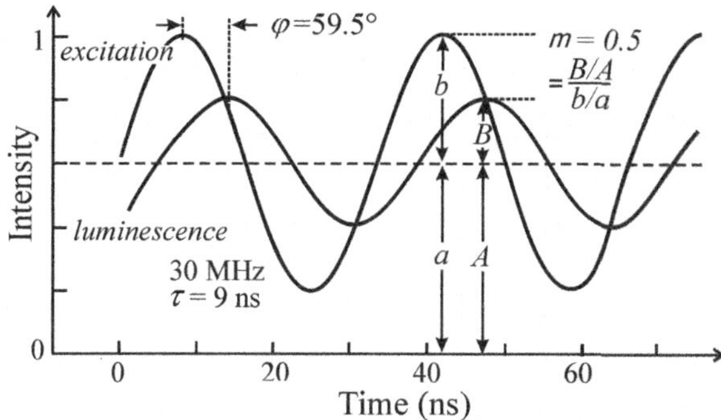

Fig. 2.43
Modulation of excitation light and luminescence response in the phase-shift method. Important parameters: modulation frequency $\omega = 2\pi \times 30\,\text{MHz}$, phase angle $\varphi = 59.5°$, demodulation coefficient $m = 0.5$ and extracted (single-exponential) decay time $\tau = 9\,\text{ns}$. Adapted from Lakowicz [27].

The overall picture is displayed in Fig. 2.43. It can be shown that [27]

$$\tan\varphi = \omega\tau, \tag{2.63a}$$

$$m = \left(1 + \omega^2\tau^2\right)^{-1/2} = \cos\varphi, \tag{2.63b}$$

where $m = (B : A)/(b : a)$ is the ratio of modulation depths of the luminescence to that of the excitation light (Fig. 2.43). As measured quantities thus φ and, possibly, m are obtained; on the basis of these the decay time τ is calculated using (2.63a). It should be stressed that the simple relations (2.63) are strictly valid for pure *single-exponential* decays only. If this is not the case, eqns (2.63a) and (2.63b) provide values of τ differing slightly from one another. They offer an order-of-magnitude estimate of the decay kinetics but their physical meaning is not evident.

To apply this method successfully, the frequency ω must be comparable to $1/\tau$. In the case of $\omega \ll 1/\tau$, the fast luminescence response does follow perfectly the modulations of the excitation light, i.e. $\varphi \to 0$, $m \to 1$, and thus analysis of the experimental curves yields no information about the decay time; if, on the other hand, $\omega \gg 1/\tau$, luminescence is not able to follow fast modulation variations, $m \to 0$, φ cannot be reliably determined and the method fails again. The applicability of the method usually lies within the modulation frequency range $\omega/2\pi \approx 1$–$30\,\text{MHz}$, which allows decay times from 10^{-1} s to $\sim 10^{-9}$ s to be measured.

The phase-modulation method has been given considerable attention for a long time. Its main benefit consists in the capacity to measure relatively fast luminescence kinetics without the need for any source of short or ultrashort pulses. Excitation can be provided by a cw lamp whose output is modulated at high frequency using, for example, light diffraction on ultrasound waves in crystals (the technique of ultrasound waves in the frequency range of 10 MHz is well mastered and a large variety of modulators are available).

Due to the development of reliable picosecond and femtosecond lasers with high repetition rate and good pulse-height stability, the phase-shift method has been put largely in the shade. Also its restriction to a simple single-exponential decay as outlined above had a role to play. A wider spectrum of information on luminescence decay can be obtained using other methods, based on high repetition rate pulsed excitation, like the time-correlated single–photon counting or sampling methods, which are the subject of the following subsections.

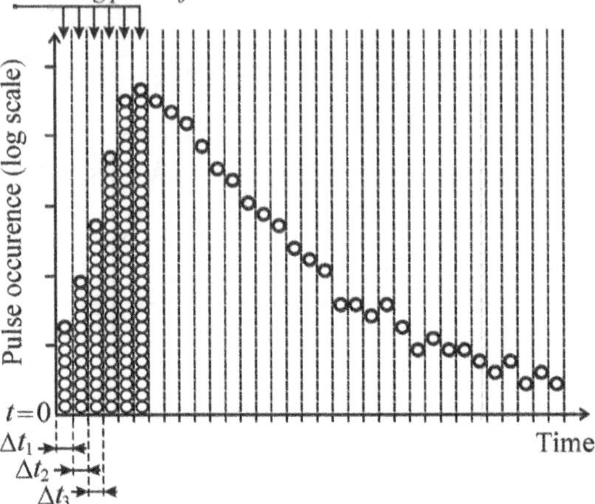

Fig. 2.44
Schematic representation of a luminescence response measured by the time-correlated photon-counting method. A narrow excitation pulse (not shown) is emitted at time $t = 0$.

2.9.3 Time-correlated photon counting

This method is based on the statistical character of luminescence photon emission. Upon excitation of a luminescent material by a very short light pulse, the luminescence emission builds up and then decays. During the decay, the number of emitted photons decreases in time. Photons are emitted randomly in all directions. Suppose we select a narrow solid angle (e.g. using a circular aperture) and record (by a photomultiplier) the arrival time of single photons within this angle, i.e. the time when a pulse appears at the photomultiplier output corresponding to a single photoelectron released from the photocathode (see Subsection 2.4.2); the moment when the excitation pulse was emitted is taken as the time origin $t = 0$. The pulse excitation of the sample is repeated many times.

The luminescence rise and decay curve we are interested in is obtained by plotting a histogram of the pulse occurrence within certain time intervals Δt_i (called time slots), as illustrated in Fig. 2.44. This is based on the fact that the probability of emitting a photon into the selected solid angle at any time is proportional to the number of luminescence centres actually being in an excited state.

One should be aware that the curve shown in Fig. 2.44 is accumulated randomly during the experiment, which means not by shifting regularly from the first time slot towards the next one in sequence Δt_1, Δt_2, $\Delta t_3, \ldots$, etc. The detection events arrive randomly into any of the time slots.

In order to perform such an experiment, two basic requirements must be satisfied:

1. The luminescence intensity must be very low to enable the arrival of single photons to be recorded.[12]

[12] In fact there are two kinds of time-correlated photon counting (TC-PC) experiments: The first one is called time-correlated single-photon counting. Here only one (or no) count is detected following each excitation event (actually, the correct proportion is one detected photon per 10 or 100 excitation pulses—if the rate is higher, the shape of the temporal response is distorted). This

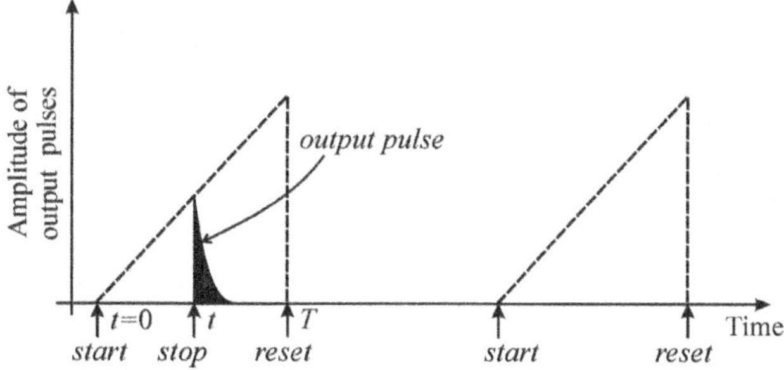

Fig. 2.45
Functioning of a time→voltage amplitude converter.

2. A high number of photons must be detected to obtain conclusive statistics. The repetition rate of special pulsed lamps emitting nanosecond pulses is of the order of 10–100 kHz; repetition rates of mode-locked pulsed lasers can be of the order of 10–100 MHz. Even for such high-repetition excitation, the accumulation of the entire rise/decay curve may take from minutes up to hours.

On the instrumental side, the essential part of this apparatus is a time → voltage converter. Its functioning is illustrated in Fig. 2.45. The converter is initiated by the excitation pulse at time $t = 0$ (start pulse) and begins to generate a DC voltage linearly increasing in time. The voltage rise is stopped by a stop pulse coming from the photomultiplier (which detects single luminescence photons). The resulting voltage amplitude is proportional to the time delay between $t = 0$ and the stop pulse. This voltage is sorted according to its amplitude and then ranked into the corresponding channel of a multichannel analyser. The immediate result of this measurement thus has the form shown in Fig. 2.44, where the horizontal axis represents channel numbers, not directly the time intervals Δt_i.

The time-correlated photon-counting method is very sensitive and accurate, but requires special electronic devices; besides, measurements are sometimes time-consuming. It is mainly applied to the detection of luminescence kinetics at the time-scale of $\sim 10^{-6}$–10^{-10} s. Obviously, the finite width of the excitation pulses $L(t)$ must also be taken into account here and the experimental result, if necessary, deconvoluted in order to extract the actual luminescence response, according to (2.62). A special deconvolution method dedicated to the case of very high repetition rates of the excitation pulses (larger than the reciprocal decay time) is described in [29].

2.9.4 Boxcar integrator

All of the above mentioned methods enable us to detect the luminescence time response at a fixed wavelength (or a range of wavelengths). Now we are going

is used to measure fast nanosecond decays under high-repetition (> MHz) pulsed lasers (in order to accumulate enough statistics in a reasonable time). The second type of TC-PC is applied to measure slow (microsecond) decays. Here more than one count should be detected at each (low-repetition) excitation pulse, otherwise the statistics cannot be accumulated within a relatively short time period.

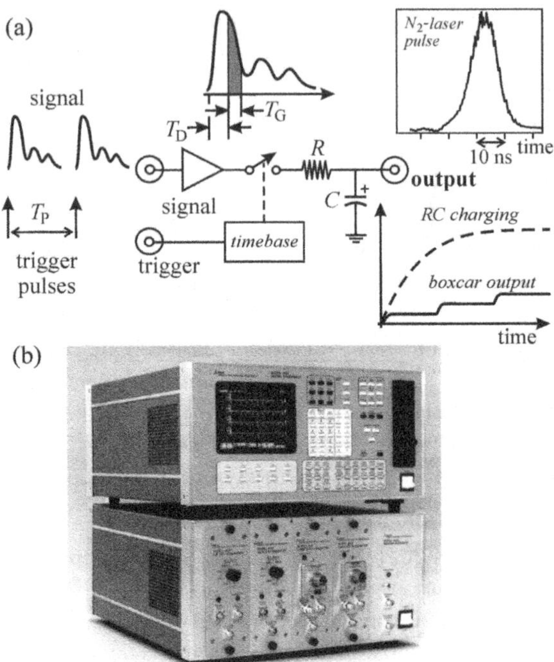

Fig. 2.46
(a) Schematic of boxcar integrator operation. The signal input is loaded by periodic short voltage pulses from a photomultiplier. The trigger input receives synchronous trigger pulses that turn on (via a time base and after a defined delay T_D) a fast sampling switch (gate) for a period T_G. The repetition rate of the whole process is $f = 1/T_P$. After Letzter [30]. The attached graph demonstrates the measured profile of N_2-laser pulses (337 nm) with a temporal FWHM of about 10 ns (detection by a photomultiplier, pulse repetition rate $f = 50$ Hz). (b) Digital EG&G PAR Model 4400 boxcar integrator. Reproduced with permission by Ametek Signal Recovery (formerly EG&G Signal Recovery).

to describe an electronic device called a boxcar integrator (or boxcar averager) which has a much greater diversity of experimental options; of practical importance for us is its capacity to record complete time-resolved luminescence spectra (i.e. emission spectra when the luminescence intensity is registered at a predefined delay after the excitation event). To begin with, we describe the principle of the device, then its application to luminescence kinetics and finally we shall address the measurement of time-resolved emission spectra.

The boxcar integrator was developed by Princeton Applied Research in the middle of the 1960s, mainly in response to demands for the detection of a wide variety of fast optical signals with high repetition rate, emerging due to the rapid development of laser techniques at that time. It can also be classified as a device aimed at detecting weak repetitive signals unprocessable by the lock-in detection technique.

A simplified scheme of boxcar functioning is shown in Fig. 2.46(a) [30]. There are two inputs: (i) the investigated signal (pulsed output from a photomultiplier or other detector) that is repeated with a period T_P, and (ii) a trigger pulse, synchronous with the signal. The signal channel contains a broadband preamplifier, a fast sampling unit, and a low-pass *RC* filter. With the incoming trigger pulse, the sampling unit (gate) is switched on (with a defined delay T_d) for a time interval T_G. Therefore, during this time interval a capacitor *C* is being charged and its voltage rises. The whole process is repeated with a frequency $f = 1/T_P$.

As far as the delay T_D is concerned, two modes of operation are feasible. Either T_D is kept constant, or is left to increase linearly in time. The latter mode is applied to measure the temporal shape of the incoming pulse, i.e. the

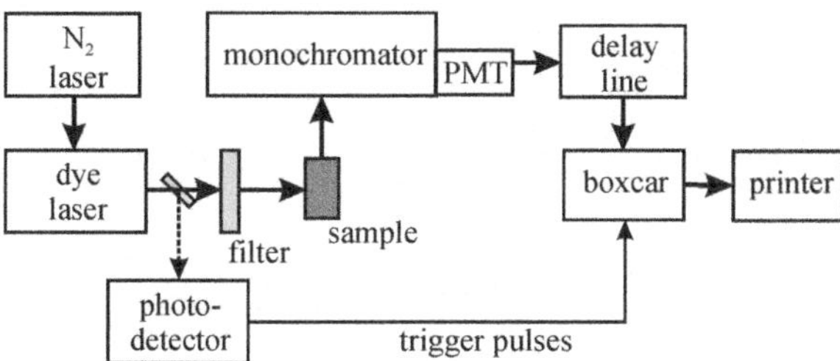

Fig. 2.47
Single-channel luminescence spectrometer with a pulsed laser excitation. Pulses from a dye laser (pumped by a nitrogen laser) periodically excite luminescence of a sample, a small fraction of the excitation pulse energy (reflected on a beam splitter) being applied to generate trigger pulses for a boxcar integrator. The delay line, inserted between the photomultiplier and the boxcar signal input, serves to compensate for internal delay in the triggering circuits of the integrator. Adapted from Letzter [30].

luminescence kinetics. Here the gate samples the temporal profile of signal pulses by shifting the delay by an increment Δt after each incoming pulse.[13]

The RC time constant is usually much larger than T_G, thus the voltage on the capacitor C increases in small steps (Fig. 2.46(a)). Due to the simultaneous slow discharging of the capacitor the output signal asymptotically approaches the average value of the input pulse amplitude within the given time interval. In other words, the latest signal acquisitions have predominant weight over the earlier ones that fade away gradually and the input signal shape is replicated at the output. This mode of signal treatment is referred to as an exponential averaging.

At the same time, an important condition is to find an appropriate relation among several experimental parameters—too fast gate scanning distorts the pulse shape while too slow scanning leads to unnecessary prolongation of the experiment and possible errors due to drift of experimental parameters. This means that an optimum time of signal scanning T_M must exist; it can be shown that this optimum scanning time is given approximately as $T_M \approx 5RCX/T_G^2 f$, where X denotes the overall sampled time interval. The inset in Fig. 2.46(a) shows an example: a N_2-laser pulse shape measured in this way ($X = 60$ ns, $T_G = 1$ ns, $T_M = 60$ s, $f = 50$ Hz).

Time-resolved emission spectra are measured in the $T_D =$ const mode (to be more specific, using several discrete values of T_D). The experiment can be performed in the set-up shown in Fig. 2.47: for any selected delay value T_D the monochromator is let to scan the spectrum. This means that this time the same portion of the temporal profile of luminescence pulse is sampled for a time T_G, while the wavelength is being scanned. On condition of a proper setting of all experimental parameters, i.e. the scanning speed in relation to RC, T_G and f, exponential averaging takes place and the luminescence spectrum is continuously recorded under a pulsed excitation. (This ability of the boxcar integrator contributed considerably to the fast development of laser spectroscopy in the 1960s and 1970s; till that time, there was practically only one way to detect emission spectra under pulsed excitation, namely, lengthy and uncertain registration on a photographic plate.) Moreover, the signal-to-noise ratio is improved when using a boxcar, because the input noise is not coherent with the trigger pulse rate and its contribution to the integrated output signal is therefore small, likewise in the lock-in technique.

[13] In principle, one deals with a stroboscopic method of measurement of periodic events, exploited also in, for example, a sampling storage osciloscope.

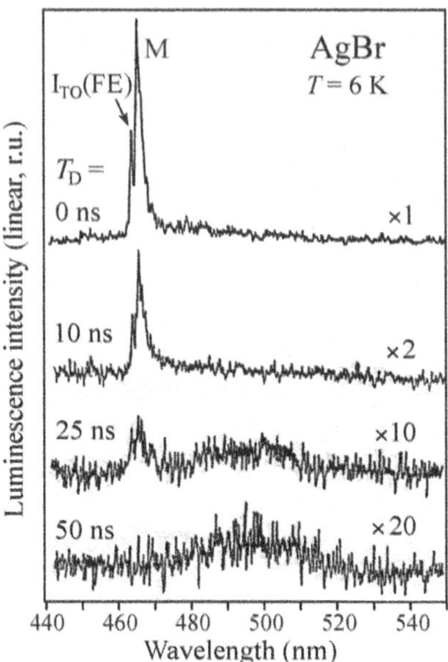

Fig. 2.48
Time-resolved luminescence spectra of AgBr at a temperature of $T = 6$ K, measured in an experimental configuration similar to that shown in Fig. 2.47 (only the dye laser was omitted, the sample was excited directly with N_2-laser pulses 337 nm, 2 MW/cm^2, 10 ns). The gate width is $T_G \sim 3$ ns and the values of the delay T_D behind the excitation pulse maximum are indicated on the left-hand side of each spectrum. With increasing delay T_D the emission lines I_{TO}(FE) and M are rapidly fading out and a slow band around 490 nm appears (this emission band is shown also in Fig. 1.2(b), but excited with cw UV light). Adapted from Baba and Masumi [31].

Upon changing the time position of the gate, a new spectrum is registered at another delay T_D after the excitation pulse—this is the principle of time-resolved luminescence spectroscopy. An example is presented in Fig. 2.48, showing low-temperature (6 K) time-resolved emission spectra of pure AgBr under pulsed excitation with a N_2-laser (337 nm) [31]. Clearly, the spectra in different time delays (windows) behind the excitation event can differ substantially. Immediately after excitation only two narrow emission lines ('fast component'), labelled I_{TO}(FE) and M, are observed. They rapidly fade out with increasing delay T_D and a new broad emission band ('slow component') appears around $\sim$ 490 nm. The origin of these spectral features will be discussed in Sections 7.1, 7.2, and 8.2.

2.9.5 Streak camera

A boxcar integrator is an excellent instrument to study luminescence kinetics and decay times from about 500 ps to hundreds of ns but does not serve this purpose for faster processes taking place on shorter time-scales of 10–100 ps. This constraint is due to the limited speed of the boxcar electronic circuits as well as the relatively slow response of the photomultipliers (even the special fast photomultipliers do not have the signal onset faster than $\sim$ 500 ps). Similar limitations take place also in oscilloscopes, more specifically in their input amplifiers. For luminescence transients occurring on the time-scale of picoseconds and femtoseconds, either special methods of nonlinear optics (up-conversion) or the so-called streak camera should be used. This device enables the measurement of both time evolutions of ultrafast light signals and time-resolved emission spectra, likewise the boxcar integrator. The development of the streak camera at the beginning of the 1970s (Hamamatsu Co. Ltd.) was

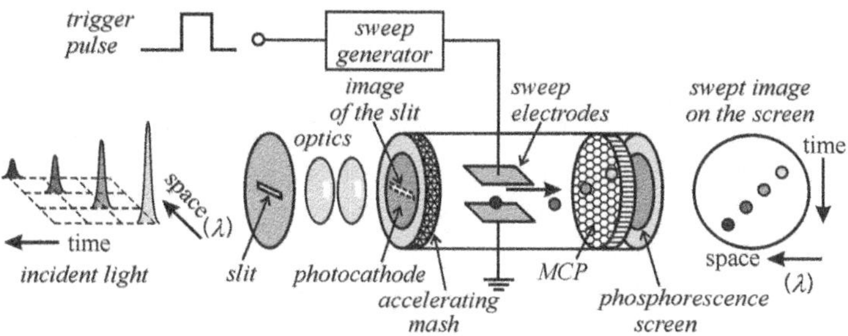

Fig. 2.49
Operation schematic of a streak camera. The light pulse passing through the entrance slit and hitting a photocathode induces emission of photoelectrons inside an evacuated tube. Photoelectrons are accelerated along the horizontal direction (i.e. the direction of the incoming light) by a voltage applied between the photocathode and a metal mesh and subsequently deflected by a high voltage in the vertical direction. Upon impinging on a phosphorescence screen, the photoelectrons produce a phosphorescent trace with spatial distribution of intensity corresponding to the temporal profile of the incoming pulse. After materials by Hamamatsu Co. Ltd.

again motivated by the fast dissemination of lasers emitting picosecond and femtosecond pulses.

The streak camera is a device which converts temporal information, originating in a light-emission event, to spatial information. Roughly speaking, the streak camera combines, in a single device, the principles of a photomultiplier with the basic components of an oscilloscope: the external photoelectric effect, deflection of an electron beam and a phosphorescence screen. The principle of a streak camera is shown in Fig. 2.49. The analysed light pulse is focused on the entrance (mechanical) slit, whose image is projected onto a photocathode deposited on the internal front face of an evacuated tube. Emitted photoelectrons (whose number is proportional to the intensity of the incident light) are accelerated in an electric field and then deflected by a high voltage applied in the direction perpendicular to the acceleration voltage. This deflection voltage is applied synchronously with the incoming light pulses, the photoelectrons thus being swept with a defined speed and known trajectory (like in an oscilloscope, but without input amplifier here) by which means the time information is converted to spatial information. Then these electrons undergo multiplication via secondary electron emission in the so-called microchannel plate, and finally they bombard a phosphorescence screen on the opposite end of the vacuum tube. This creates the phosphorescence image, rendering the temporal distribution of intensity of the analysed luminescence pulses in the form of a vertical spatial distribution of phosphorescence intensity on the screen. The phospherescence image is then recorded (e.g. by a photodiode array or a CCD detector), treated and displayed on a computer screen.

An example of a measured luminescence transient is presented in Fig. 2.50: luminescence of Hg_2Cl_2 (calomel) crystals excited by $\sim$ 35 ps pulses of second harmonic frequency from a passively mode-locked Nd:YAG laser (532 nm) [32]. As Hg_2Cl_2 is an anisotropic uniaxial crystal, its luminescence response exhibits a slightly different temporal shape under polarization of the excitation light perpendicular to the optical axis **c** (electric field vector $\mathscr{E}\perp\mathbf{c}$) and parallel with this axis ($\mathscr{E}||\mathbf{c}$). On the left-hand side of the lower curve in Fig. 2.50 one can also see the shape of the excitation pulse; this was enabled by the appropriate setting of an optical delay line for both luminescence and the reflected fraction of the excitation light and their simultaneous focusing onto the camera entrance slit.

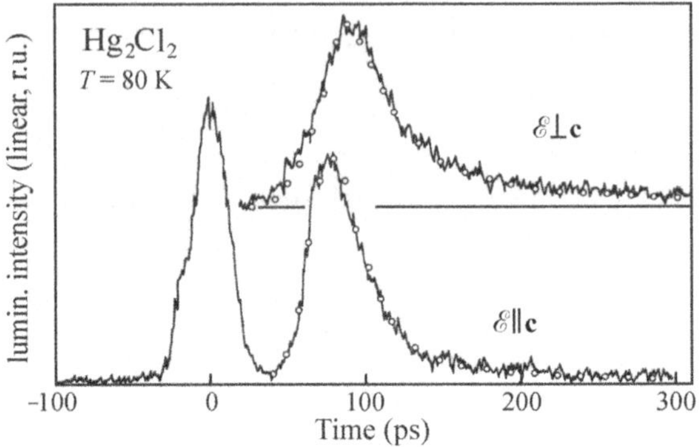

Fig. 2.50
Luminescence transients measured and displayed by a streak camera (luminescence of Hg_2Cl_2 crystals spectrally situated around $\sim$ 396 nm, excited at $T \sim 80$ K via two-photon absorption of second harmonic pulses from a Nd:YAG laser with temporal FWHM $\sim$ 35 ps). The lower curve shows also (on the left) the excitation pulse profile. Full lines represent experimental data while points show the calculated convolution of the excitation pulse profile with a double-exponential decay. After Svoboda *et al.* [32].

The open circles in Fig. 2.50 represent the best numerical fit using the model of double-exponential decay $i_\lambda(t) = a_1 \exp(-t/\tau_1) + a_2 \exp(-t/\tau_2)$ with parameters:

$$a_1 = 0.93,\ \tau_1 = (17 \pm 5)\,\text{ps},\ a_2 = 0.07,\ \tau_2 = (87 \pm 30)\,\text{ps}\ (\mathscr{E}\|\mathbf{c}),$$
$$a_1 = 0.82,\ \tau_1 = (25 \pm 7)\,\text{ps},\ a_2 = 0.18,\ \tau_2 = (77 \pm 26)\,\text{ps}\ (\mathscr{E}\perp\mathbf{c}). \quad (2.64)$$

Figure 2.50 indicates that the excitation pulse FWHM and the characteristic times τ_i are of the same order of magnitude, therefore to fit the experiment properly, numerical convolution of the measured system response to the excitation pulses with a presumed model shape of luminescence decay $i_\lambda(t)$ according to eqn (2.62) was applied. The numerical values given in (2.64) give an idea of the attainable accuracy in extracting the decay times τ_i by this method.

The principle of the measurement of time-resolved luminescence spectra follows from Fig. 2.49. The streak camera can be attached to the spectrograph output in such a way that the spectrum of the luminescence signal is imaged onto the entrance slit. Then the two-dimensional image displayed on the phosphorescence screen represents the temporal evolution of the pulse spectral content while the photoelectrons are vertically swept.

Various models of streak cameras are commercially available for various time regimes, with time resolution as short as ≤ 1 ps. Streak cameras can operate in a single-shot mode but also at high repetition rates of the order of 100 MHz (the so-called synchroscan camera).

2.10 Problems

2/1: *Photomultiplier nonlinearity.* When illuminating a high-gain photomultiplier by too high photon flux, the photocurrent between the last dynodes may attain such high values (without yet inducing irreversible damage to the photomultiplier) that the condition I_a (anode current) $\ll I_{vd}$ (current through the voltage divider) is no longer fulfilled—see Fig. 2.5(b). Consequently, the last stages of the voltage divider are partially 'short-circuited' by the current I_a flowing inside the photomultiplier. Then the voltage from an external high-voltage supply is concentrated predominantly in several first stages of the divider (close behind the photocathode). Thus a positive feedback appears, the photomultiplier gain increases and the linearity of the photomultiplier response is violated. Discuss the consequences of this effect for measurements of optical gain according to the paper by Webb [33].

2/2: Show that the relation between the photomultiplier anode sensitivity $k(\lambda)$ (expressed in units of A/W) and the photocathode quantum efficiency η_{ph} has the form

$$\eta_{ph}(\%) = \frac{h\,c}{\lambda e}\frac{k}{G} \times 100 = \frac{1240}{\lambda}\frac{k}{G} \times 100,$$

where h is Planck's constant (Js), c is the speed of light (m/s), e is the electron charge (C), λ denotes the wavelength of the incident light, and G is the photomultiplier gain.

2/3: You are in a position to investigate the emission spectrum of a semiconductor with low luminescence yield. This spectrum is expected to be situated between 600 and 800 nm and composed of spectral bands and lines, the FWHM of the narrowest ones being about 1 nm and that are separated from one another by 2.5 nm. Decide which of the following monochromators is more convenient for this task and justify your selection:

1. Focal length $f = 1000$ mm, aperture $f/7.5$, holographic grating $120 \times 140\,\mathrm{mm}^2$ (blaze at 630 nm), reciprocal linear dispersion $L^{-1} = 0.8\,\mathrm{nm/mm}$, resolving power $R = 7\times10^4$ (at the slit width of 10 μm).
2. Focal length $f = 320$ mm, aperture $f/4.2$, holographic grating $120 \times 140\,\mathrm{mm}^2$ (blaze at 750 nm), reciprocal linear dispersion $L^{-1} = 2.4\,\mathrm{nm/mm}$, resolving power $R = 6\times10^3$ (at the slit width of 10 μm).

What is the maximum slit width of the selected monochromator you may apply in the experiment?

2/4: A scanning monochromator is used to measure one of the emission lines of the material mentioned in the previous problem. The signal from a photomultiplier is treated by a lock-in amplifier. What is the maximum time constant you can apply (in order to obtain the spectrum as smooth as possible but without distortions), if the wavelength scan speed is 10 nm/min? What is the total acquisition time of the spectrum? Note: The values of the time constant can usually be set in steps 1–3–10, e.g. 1 ms–3 ms–10 ms.

2/5: (a) Show that the FWHM of a bell-shape curve $f(x) = x\exp(-x/a)$ is FWHM $\sim$ 2.45 a.

(b) Suppose you excite photoluminescence with optical pulses of temporal FWHM = 5 ns and measure the corresponding luminescence decay. Let the decay curve $i(t)$ be of single-exponential type with the characteristic time constant $\tau = 15$ ns. Estimate the uncertainty in determining τ if no deconvolution of the experimental curve is applied. For the sake of simplicity suppose that the excitation pulse is described by a function $L(t) = t\exp(-t/a)$ and use the result of part (a).

References

1. Voos, M., Leheny, R. F., and Shah, J. (1980). *Radiative recombination*. In: *Handbook on Semiconductors* (ed. T. S. Moss), Vol. 2 (ed. M. Balkanski), p. 329. North Holland, Amsterdam.
2. Kulakovskii, V. D. and Timofeev, V. B. (1983). *Thermodynamics of electron–hole liquid in semiconductors*. In: *Modern Problems in Condensed Matter Sciences* (general ed. V. M. Agranovich and A. A. Maradudin), Vol. 6, *Electron–Hole Droplets in Semiconductors* (ed. C. D. Jeffries and L. V. Keldysh), p. 95. North Holland, Amsterdam.
3. Valenta, J. (1994). *Photoluminescence characterization of selected semiconductors and insulators*. PhD. thesis, Charles University in Prague, Faculty of Mathematics & Physics, Prague.
4. Bebb, H. B. and Williams, E. W. (1972). *Photoluminescence I. Theory*. In: *Semiconductors and Semimetals* (ed. R. K. Willardson and A. C. Beer), Vol. 8, p. 181. Academic Press, New York.
5. Itoh, C., Tanimura, K., Itoh, N., and Itoh, M. (1989). *Phys. Rev. B*, **39**, 11183.
6. Engstrom, R. W. (1980). *Photomultiplier Handbook*. RCA/Burle, USA.
7. *Photomultiplier Tube. Principle to Application*. Hamamatsu Photonics K. K. 1994.
8. Rieke, G. H. (2003). *Detection of Light. From Ultraviolet to Submillimeter*, 2nd edn, p. 194. Cambridge University Press, Cambridge.
9. Saleh, B. E. A. and Teich, M. C. (2007). *Fundamentals of Photonics*, 2nd edn. Wiley-Interscience, New York.
10. Ferrari, M., Pavesi, L., and Righini, G. C. (eds) (2002). *Micro-Optoelectronics: Materials, Devices and Integration*, p. 26. Italian Society of Optics and Photonics, Centro Editoriale Toscano.
11. Mac Intyre, R. J. (1966). *IEEE Trans. Elect. Dev.*, **13**, 164.
12. *Spectrum One, Spectroscopic CCD Detectors*. ISA Jobin Yvon-Spex, Horiba.
13. Michel, P. (1953). *La spectroscopie d'emission et ses applications*. Collection Armand Colin, Paris.
14. Sawyer, R. A. (1951). *Experimental Spectroscopy*. Prentice Hall. New York.
15. James, J. and James, J. F. (2007). *Spectrograph Design Fundamentals*. Cambridge University Press, Cambridge.
16. Demtröder, W. (2003). *Laser Spectroscopy: Basic Concepts and Instrumentation*, p. 109. Springer, Heidelberg.
17. Pelant, I. , Hála, J., Parma, L., and Vacek, K. (1980). *Solid State Comm.*, **36**, 729.
18. Hulin, D., Mysyrowicz, A., Combescot, M., Pelant, I., and Benoit à la Guillaume, C. (1977). *Phys. Rev. Lett.*, **39**, 1169.
19. *Photon Counting using Photomultiplier Tubes* (1998). Technical information, Hamamatsu Photonics K. K.
20. Bagdanskis, N. I., Bukreev, V. S., Žižin, G. N., and Popova, M. N. (1982). High resolution infrared spectrometers (in Russian: *Infrakrasnyje spektrometry vysokogo razrešenija*). In *Current Trends in Spectroscopy Techniques* (in Russian: *Sovremennyje tendencii v technike spektroskopii*, ed. S. G. Rautman), p. 153. Nauka, Novosibirsk.
21. Fojtík, P. (1999). *Selected photoelectrical properties of amorphous and microcrystalline silicon with respect to lateral resolution* (in Czech: *Vybrané fotoelektrické vlastnosti amorfního a mikrokrystalického křemíku s přihlédnutím k laterálnímu rozlišení*). Diploma thesis, Charles University in Prague, Faculty of Mathematics & Physics, Prague.
22. Shionoya, S. and Yen, W. M. (eds) (1999). *Phosphor Handbook*, Chap. 14. CRC Press, Boca Raton.
23. Dian, J., Valenta, J., Luterová, K., Pelant, I., Nikl, M., Muller, D., Grob, J. J., Rehspringer, J.-L., and Hönerlage, B. (2000). *Mater. Sci. Engin. B*, **69–70**, 564.
24. Pelant, I., Kohlová, V., Hála, J., Ambrož, M., and Vacek, K. (1987). *Solid State Com.*, **64**, 939.

25. Bertero, M. and Boccacci, P. (1998). *Introduction to Inverse Problems in Imaging*. Institute of Physics Publishing, Bristol.
26. Shah, J. (1999). *Ultrafast Processes of Semiconductors and Semiconductor Nanostructures*. Springer Series in Solid-State Sciences, Vol. 115. Springer, Berlin.
27. Lakowitz, J. R. (1983): *Principles of Fluorescence Spectroscopy*, Chap. 3. Plenum Press, New York and London.
28. Luterová, K., Pelant, I., Valenta, J., Rehspringer, J.-L., Muller, D., Grob, J. J., Dian, J., and Hönerlage, B. (2000). *Appl. Phys. Lett.*, **77**, 2952.
29. Sakai, Y. and Hirayama, S. (1988). *J. Luminescence*, **39**, 145.
30. Letzter, S. (1983). *Laser Focus*, December 1983.
31. Baba, T. and Masumi, T. (1977). *Nuovo Cimento B*, **39**, 609.
32. Svoboda, A., Večeř, J., Barta, Č., and Pelant, I. (1990). *phys. stat. sol. (b)*, **157**, K69.
33. Webb, C. (2003). *Optics & Photonics News*, May 2003, 14.

3 Kinetic description of luminescence processes

A multitude of characteristic features of luminescence can described and understood using simple kinetic equations, which, from the point of view of mathematics, represent first-order differential equations in time. In this chapter, we shall first introduce important parameters called the probability of radiative and non-radiative transition and the luminescence quantum yield. We shall discuss their relation to the experimentally established luminescence decay time. Next, we shall formulate kinetic equations for several simple luminescence phenomena. Solving the equations, we shall proceed firstly to explicate transient luminescence phenomena (i.e. the form of luminescence decay or possibly rise in time) and, secondly, we shall obtain the functional dependence of emission intensity on the optical excitation intensity in the steady state.

3.1 Radiative and non-radiative recombination. Luminescence quantum yield

An excited luminescence centre is in a non-equilibrium thermodynamic state and it loses its electron excitation energy after a certain relaxation time and goes into the ground state. The transition can occur in two ways. The excitation energy is either radiated away in the form of a luminescence photon after the mean lifetime τ_r or the excess energy is transmitted in the form of heat (vibrations) to the crystal lattice after time τ_{nr} or it can possibly cause photochemical changes in the crystal matrix or generate a lattice defect. The first kind of transition is called a *radiative transition*, the time τ_r is the radiative recombination time (radiative lifetime), transitions of the second kind are called *non-radiative* and τ_{nr} is the non-radiative recombination time (non-radiative lifetime). The inverse values $(\tau_r)^{-1}$ and $(\tau_{nr})^{-1}$ then represent the probability of the corresponding transition per unit time (the recombination rate). The total probability of transition to the ground state is given by

$$\frac{1}{\tau} = \frac{1}{\tau_r} + \frac{1}{\tau_{nr}}; \tag{3.1}$$

we can imagine this in analogy to electromagnetism as two parallel conductors, the resulting conductivity of which is given by the sum of the inverse values of their individual resistances (Fig. 3.1).

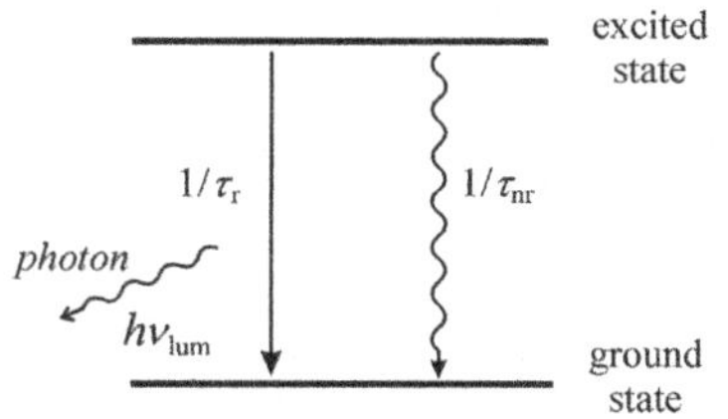

Fig. 3.1
Illustration of the definition of luminescence quantum efficiency.

We define the luminescence *quantum yield* or *quantum efficiency*, already mentioned in Section 2.1, as

$$\eta = \frac{1/\tau_r}{1/\tau_r + 1/\tau_{nr}} \leq 1 \tag{3.2}$$

i.e. as the ratio of the radiative recombination rate to the total recombination rate. For brightly luminescent materials $\tau_r \ll \tau_{nr}$ holds true, which means the given excited centre is much more likely to go into the ground state by radiating a photon $h\nu_{lum}$ than by a non-radiative transition. Neglecting $1/\tau_{.nr}$ in (3.2), we get $\eta \rightarrow 1$ (we also say the quantum yield is 100%). However, such materials almost never occur in nature. In fact, we speak of strongly luminescent materials already if η is of the order of 0.1.

The word 'quantum' in the definitions given here implicitly points to the fact that both luminescence excitation and light emission take place through one photon elementary steps; e.g. $\eta = 0.5$ means there is one emitted photon to every two absorbed photons. Thus, multiple-photon phenomena, although feasible in general, are not considered here. Likewise, definition (3.2) does not allow for the energy difference between the exciting $h\nu_{ex}$ and emitted $h\nu_{lum}$ photons—in this sense, we sometimes introduce the *power efficiency* $\eta_P = (h\nu_{lum}/h\nu_{ex})\eta$. Due to Stokes' law, we have $\eta_P \leq \eta$. For injection electroluminescence, η relates not to the number of photons absorbed but to the number of electron–hole pairs injected.

What is the physical meaning of the time τ given by expression (3.1)? It characterizes the emptying of the excited-state level in a large ensemble of luminescence centres by radiative and non-radiative transitions simultaneously. Yet, the population decay of an excited level is reflected just in a decrease of the luminescence intensity as it decays after the excitation is turned off. Therefore, the time τ is (in accordance with the notation introduced in Section 2.9) an experimentally accessible quantity—the luminescence decay time. By means of this decay time, we can rewrite the expression for quantum efficiency (3.2) in the form

$$\eta = \frac{\tau}{\tau_r}(\leq 1). \tag{3.3}$$

Let us point out straightaway the occasional erroneous use of the terms we have just introduced. The radiative lifetime (τ_r) is not the same as the luminescence decay time (τ) although the admitted similarity of the terms may tempt us to confuse them! (On the contrary, we shall show shortly that, in a special case, the luminescence decay time may be identical to the non-radiative recombination time!)

From the point of view of mathematics, relations (3.1), (3.2), and (3.3) are just trivial fractions but, surprisingly enough, they involve a wide spectrum of physical information. So firstly, it is interesting that expression (3.3) relates two measurable quantities (η, τ) to an important quantum-mechanical parameter characterizing optical transitions from a theoretical perspective (the reciprocal value of the radiative lifetime τ_r^{-1} is proportional to the square of the modulus of the corresponding transition matrix element [1–3]). The magnitude of the matrix element basically cannot be obtained by direct measurements and it is neither simple nor reliable to calculate it. Therefore, the simple formula (3.3) appears to provides a welcome opportunity to obtain τ_r^{-1} from two experimentally available quantities and to compare the result to the theory (in a semiconductor with a direct bandgap, τ_r is of the order of 10^{-9} s, and with an indirect bandgap, $\tau_r \approx 10^{-4}$ to 10^{-3} s). However, while it is very easy to measure τ—and we have discussed several of the most common experimental methods in Section 2.9—it is, in contrast, extremely difficult to measure the quantum efficiency η with sufficient accuracy, as we have, for that matter, also mentioned before. Hence, we should not overestimate the significance of expression (3.3) in this sense; nonetheless, it is very important that by measuring the temperature dependence of the luminescence intensity (which we can consider, to a good approximation, proportional to the quantum yield η) and of the luminescence decay time τ we can obtain from (3.3) information on the temperature dependence of the radiative recombination time τ_r, albeit in relative units.

In addition, relations (3.1)–(3.3) enable us to asses quickly a number of situations we may encounter in practice. Let us first consider the case $\tau_{nr} \ll \tau_r < \infty$. Then the non-radiative recombination rate, τ_{nr}^{-1}, is much higher than the radiative transition rate; nonetheless, some weak luminescence radiation may still be present. Naturally, the quantum yield η will be very low ($\eta \approx \tau_{nr}/\tau_r \ll 1$ as follows directly from (3.2)) and what is particularly interesting is that—as an immediate consequence of (3.1)—we get $\tau \cong \tau_{nr}$. Thus, measuring the luminescence decay time, we are now actually measuring the *non-radiative lifetime*. Or, put in still another way, if we reveal in an experiment a weak luminescence decaying very quickly then there will very likely be a strong influence of non-radiative transitions.

The opposite case reads $\tau_r \ll \tau_{nr}$. The quantum yield η is close to unity according to (3.2) and $\tau \cong \tau_r$ holds. Although we encounter this situation much less often than the previous case we must still mention the following possibility. Let us consider a material with a relatively large τ_r, which, in itself, is thus not a very efficient phosphor. Yet, if we are able to 'forbid' somehow all non-radiative transitions then $\tau_{nr} \to \infty$, $\tau_r \ll \tau_{nr}$ and the electronically excited system has no other choice but to return—albeit slowly—to the ground state exclusively via radiative transitions. Hence, in this way, we can obtain a material featuring efficient luminescence ($\eta \cong 1$). The total number of photons emitted per unit of time will be relatively low; nevertheless, this mechanism is thought to be one of the factors contributing to the intense luminescence of so-called porous silicon and of silicon nanocrystals. Silicon is a poor phosphor as a result of its indirect bandgap. The radiative lifetime is long ($\tau_r \approx 10^{-3}$ s) and almost all transitions to the ground state occur via much more efficient non-radiative processes. In nanocrystals, however, the non-radiative transitions take place mostly on the surface of the nanocrystal and they can be prevented through efficient passivation of unoccupied surface bonds by attaching, e.g., hydrogen or oxygen atoms.

3.2 Monomolecular process

Let us consider a *localized luminescence centre*[1] in a crystal or possibly in an amorphous matrix, typically represented by an impurity atom. Let us assume the centre can be brought into an electronic excited state by either absorbing light directly or via energy transfer from the optically excited surrounding matrix. Let both processes be described by a quantity G, which is the number of photons absorbed in unit volume per unit time. Subsequently, centres in an excited state return to the ground state, partly radiatively and partly non-radiatively. If the concentration of the excited centres is $n(t)$, the processes under consideration are described by the equation

$$\frac{\mathrm{d}n}{\mathrm{d}t} = G - \frac{n}{\tau_r} - \frac{n}{\tau_{nr}} = G - \frac{n}{\tau}, \qquad (3.4)$$

where $\mathrm{d}/\mathrm{d}t$ denotes the derivative with respect to time and G is called the generation term. A schematic is shown in Fig. 3.2(a). We shall now analyse eqn (3.4) in two cases: (a) luminescence decay and (b) steady state.

(a) Luminescence decay is characterized by the excitation being switched off ($G = 0$) at time $t = 0$. We are interested in the fall-off of the luminescence intensity $i(t)$ at times $t \geq 0$. Equation (3.4) has the form

$$\frac{\mathrm{d}n}{\mathrm{d}t} = -\frac{n}{\tau}$$

and its solution thus reads $n(t) = n(0)\exp(-t/\tau)$. The luminescence intensity $i(t)$ (here, the number of photons emitted from unit volume per unit time) is given by the ratio $n(t)/\tau_r$, and hence the desired intensity fall-off is

$$i(t) = \frac{n(t)}{\tau_r} = \frac{n(0)}{\tau_r}\mathrm{e}^{-t/\tau} = i(0)\mathrm{e}^{-t/\tau}, \quad t \geq 0. \qquad (3.5)$$

Here, $i(0)$ denotes the intensity at time $t = 0$. We can see that in this case (the so-called *monomolecular process* or a localized centre) the luminescence exhibits strictly single-exponential decay governed by the total lifetime τ, as we have already described in Section 3.1 and as plotted in Fig. 3.2(b).

(b) The steady state—when the excitation G is turned on and when, after the initial increase in $i(t)$, the intensity of luminescence settles at a constant value—is determined by the condition $\mathrm{d}n/\mathrm{d}t = 0$. From eqn (3.4) and with regard to Fig. 3.2(b), we have

$$G = \frac{n(0)}{\tau_r} + \frac{n(0)}{\tau_{nr}} = i(0) + \frac{n(0)}{\tau_{nr}}. \qquad (3.6)$$

The generation term G is closely related to the excitation radiation intensity I_{ex} (here, the number of photons incident upon unit surface area per unit time). Let us consider the common case when all excitation radiation

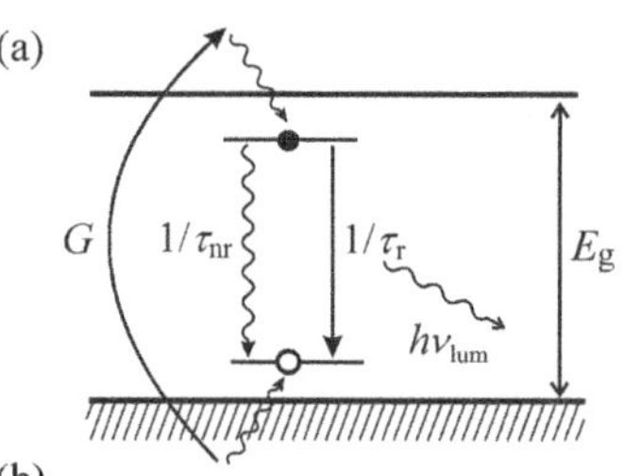

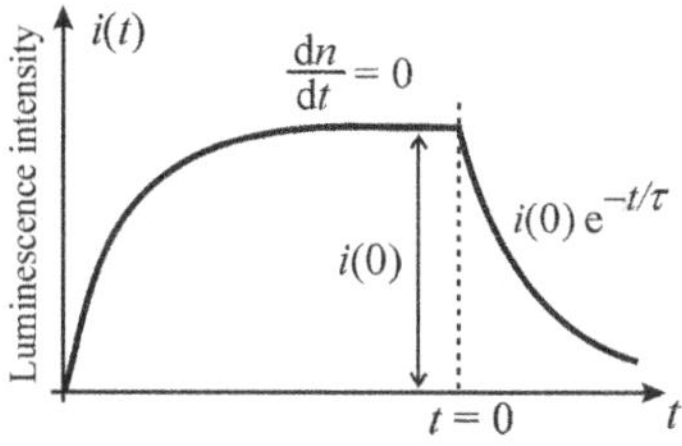

Fig. 3.2
(a) The monomolecular luminescence process. Wavy lines represent non-radiative thermalization of the carriers, their capture, and possibly non-radiative recombination. (b) The kinetics of monomolecular luminescence upon turning the excitation source on and off.

[1] By this we shall mean an electron excitation localized at a certain impurity or defect. Free quasi-particles do not participate in recombination phenomena.

is absorbed in the sample (we neglect reflection for the sake of simplicity). Absorption occurs in a thin subsurface layer, the depth of which is roughly equal to the reciprocal value of the absorption coefficient of the excitation light α^{-1}. If the area of the excited luminescence surface is ΔS then there are $I_{ex}\Delta S$ photons incident per unit time and the photons are thus absorbed in a volume $\sim \Delta S\alpha^{-1}$. The density of the absorbed photons, or the number of centres excited per unit volume per unit time, is thus $G \approx (I_{ex}\Delta S)/\Delta S\alpha^{-1} = \alpha I_{ex}$.

From eqn (3.6) we then get

$$i(0) = G - \frac{n(0)}{\tau_{nr}} = \alpha\, I_{ex} - \frac{\tau_r}{\tau_{nr}} i(0)$$

and after some simple algebra and considering definition (3.2), we immediately have

$$i(0) = \eta\alpha I_{ex}. \tag{3.7}$$

In the case of monomolecular recombination, the stationary value of the luminescence intensity, $i(0)$, is thus directly proportional to the excitation intensity I_{ex}.

Hence, we have obtained two simple and important guidelines facilitating the interpretation of experimental data. Observing a single-exponential decay together with a linear increase in luminescence intensity are rather strong indications that the microscopic origin of the examined luminescence might lie within a localized centre. This is not unambiguous evidence—similar features can be seen in, for example, the luminescence of a kind of free quasi-particle, the free exciton—and yet, such an observation helps us to exclude a number of other types of recombination processes.

Let us note further that the presence of two kinds of localized luminescence centres will show itself via a double-exponential decay

$$i(t) = a_1 e^{-t/\tau_1} + a_2 e^{-t/\tau_2}, \tag{3.8}$$

where $a_i (i = 1, 2)$ denotes the amplitudes of the individual components. Naturally, the prerequisite of observing a dependence of the form (3.8) is that we detect spectrally unresolved luminescence of both centres, e.g. when the emission wavelengths of both centres are close to each other and their spectra overlap. Today, computer-aided fitting of the experimental decay curve enables us to determine the parameters a_1, a_2, and τ_1, τ_2 quickly and reliably even in case we need to use deconvolution methods (see (2-64) and Fig. 2.50).

One cannot help thinking of generalizing relation (3.8) to a sum of multiple exponential functions. However, when attempting to fit the luminescence decay curve using the expression $i(t) = \sum_{i=1}^{j} a_i e^{-t/\tau_i}$ $(j \geq 3)$, we must be very careful. Of course, from the point of view of mathematics, such a fit is feasible and we usually achieve very good agreement between our calculation and experiment, yet, from the point of view of physics, such a result in itself can hardly be considered as evidence for the presence of three or more luminescence centres. This is because, in principle, any of the characteristic forms of decay curve, which we shall discuss below, can be modelled very nicely as a sum of three exponentials with six fitting parameters (a_i, τ_i, $i = 1, 2, 3$); the informative value of such a fit, however, is highly questionable.

3.3 Bimolecular process

Let us consider the process shown in Fig. 3.3. If a semiconductor absorbs a photon of energy above the bandgap, E_g, then free electron–hole pairs are produced. Let us denote the concentration of electrons by n and the concentration of holes by p. If there is no impurity atom or possibly a lattice defect nearby where localization of the created photocarriers could occur then electrons and holes will diffuse along the edges of the respective bands and, eventually, electrons will recombine with the nearest holes (as their wavefunctions overlap most), whether radiatively or non-radiatively. The above processes are described by the equation

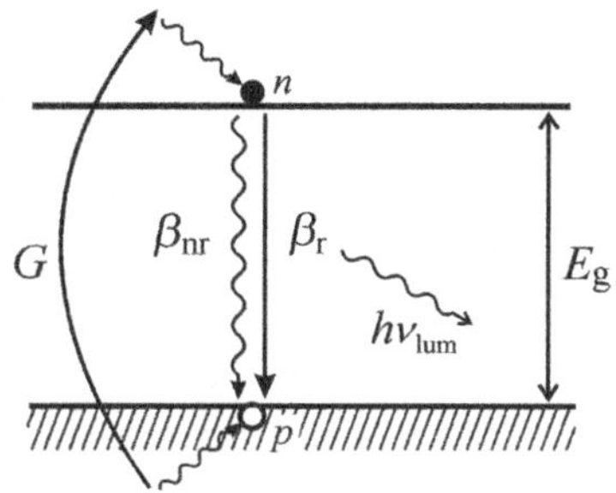

Fig. 3.3
An example of a bimolecular luminescence process.

$$\frac{dn}{dt} = G - \beta\, n\, p, \tag{3.9}$$

where G again denotes the generation term. The recombination term proportional to the product np represents the fact that recombination requires a free electron to encounter a free hole. The coefficient β is called the *bimolecular recombination coefficient* and can be expressed as the sum of bimolecular radiative (β_r) and bimolecular non-radiative (β_{nr}) recombination coefficients: $\beta = \beta_r + \beta_{nr}$.[2] If we consider, for the sake of simplicity, an intrinsic semiconductor at low temperatures, then $n = p$ and eqn (3.9) can be rewritten as

$$\frac{dn(t)}{dt} = G - \beta_r n^2 - \beta_{nr} n^2 = G - \beta n^2. \tag{3.10}$$

Here, the term $\beta_r n^2$ clearly represents the intensity of luminescence generated by bimolecular recombination (order-of-magnitude values of the coefficients β_r are $\beta_r \sim 10^{-10}\,\mathrm{cm^3\,s^{-1}}$ for direct bandgap semiconductors and $\beta_r \sim 10^{-15}$–$10^{-14}\,\mathrm{cm^3\,s^{-1}}$ in indirect bandgap semiconductors). We shall again analyse both the luminescence decay and steady state.

(a) Similarly to the monomolecular process, let us turn off the excitation ($G = 0$) at time $t = 0$ and follow the decay. Then, relation (3.10) reduces to

$$\frac{dn(t)}{dt} = -\beta n^2,$$

[2] It should be clear by now that the terms 'monomolecular' and 'bimolecular' historically originated from the analogous chemical reactions.

which is an equation we can solve easily if, for example, we separate the variables, use the substitution $1/n = x$, and apply the boundary condition $n(t = 0) = n(0)$. We get

$$n(t) = \frac{n(0)}{(\beta n(0)t + 1)}$$

and the luminescence intensity $i(t) = \beta_r n^2(t)$ comes out immediately as

$$i(t) = \frac{\beta_r n^2(0)}{(\beta n(0)t + 1)^2} = \frac{i(0)}{(\beta n(0)t + 1)^2} = \frac{i(0)}{(\gamma\sqrt{i(0)}t + 1)^2}, \tag{3.11}$$

where $i(0)$ is the intensity at time $t = 0$ and $\gamma = \beta/\sqrt{\beta_r}$. We can see that the form of the bimolecular luminescence decay differs substantially from the exponential function. It is a power-law function of time and, for late times, we can approximate $i(t) \approx t^{-2}$.

(b) The steady state is characterized by the condition $dn/dt = 0$. From (3.10) we get

$$i(0) = \beta_r n^2(0) = G - \beta_{nr} n^2(0) = G - \beta_{nr}\frac{i(0)}{\beta_r},$$

which can be transformed into the form

$$i(0) = \left(\frac{\beta_r}{\beta_r + \beta_{nr}}\right) G = \left(\frac{\beta_r}{\beta_r + \beta_{nr}}\right) \alpha I_{ex}. \tag{3.12}$$

Therefore, in the case of bimolecular recombination, the steady-state luminescence intensity remains a linear function of the excitation intensity I_{ex}.

However, let us stress again that this result is valid only if the steady state has been reached, that is, in particular when using continuous excitation radiation of an incandescent lamp, or possibly of a gas or a semiconductor laser. Nevertheless, bimolecular recombination requires a rather high density of electron–hole pairs since its probability scales like the square of this density. This is more likely to occur when using excitation by powerful laser pulses. And under a pulsed excitation, the intensity dependence of bimolecular luminescence can be governed by a law different from the linear relation of the type (3.12) as we now proceed to show.

Let us first solve a general differential equation of the form (3.10). It is easy to check that the solution reads

$$n(t) = \sqrt{\frac{G}{\beta}\frac{1 - e^{-2\sqrt{G\beta}t}}{1 + e^{-2\sqrt{G\beta}t}}} + \text{const.} \tag{3.13}$$

Let us now choose a boundary condition different from the one we have used up to now. Let $n(0) = 0$ at time $t = 0$ and now we promptly turn on the excitation G in the form of a rectangular pulse of duration t_p (Fig. 3.4(a)). If $2\sqrt{G\beta}t \ll 1$ holds for all $0 \leq t \leq t_p$ (i.e. the excitation pulse is sufficiently short), then we can approximate the exponential terms in (3.13) using a series expansion $\exp(-x) \approx 1 - x$ which yields

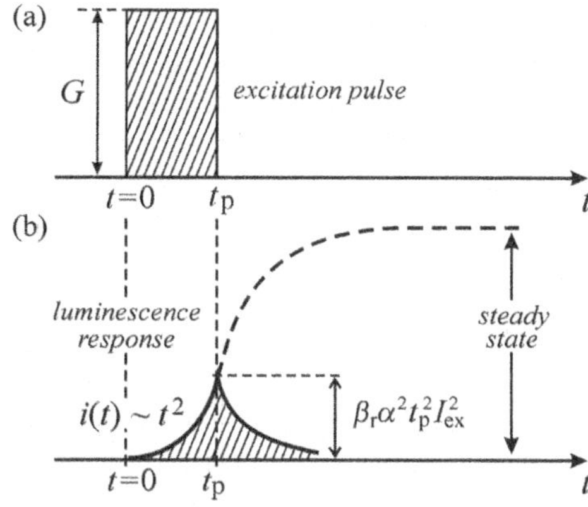

Fig. 3.4
Time-dependent bimolecular luminescence under a pulsed excitation. (a) A short excitation pulse $t_p \ll 1/2\sqrt{G\beta}$; (b) luminescence response pulse. The dashed curve represents the evolution of luminescence intensity towards the steady state provided the excitation lasts long enough ($t_p \to \infty$).

$$n(t) \approx \sqrt{\frac{G}{\beta}} \frac{2\sqrt{G\beta}t}{2(1-\sqrt{G\beta}t)} \approx Gt. \tag{3.14}$$

Then, the luminescence intensity $i(t) = \beta_r n^2(t)$ is given by

$$i(t) = \beta_r n^2(t) \approx \beta_r G^2 t^2 \approx \beta_r \alpha^2 t^2 I_{ex}^2. \tag{3.15}$$

Hence, the amplitude of the luminescence response pulse is proportional to the square of the excitation intensity I_{ex} (Fig. 3.4(b)). We leave it to the reader to confirm that we can never get a similar quadratic dependence for a monomolecular process even if we excite using an arbitrarily short pulse $t_p \ll \tau$ (Problem 3/2).

When is the excitation pulse 'short enough' so that the quadratic dependence (3.15) can reveal itself? To get a numerical estimate, let us consider, for example, excitation by using a rather weak pulsed lamp with an excitation power density of $100\,\text{W/cm}^2$, bimolecular recombination coefficient $\beta = 10^{-15}\,\text{cm}^3\,\text{s}^{-1}$, and the excitation photon energy $h\nu_{ex} = 3.2\,\text{eV}$. Then the excitation intensity in terms of the relevant photon flux is $I_{ex} = 100\,(\text{J/cm}^2\,\text{s})/h\nu_{ex} \approx 2 \times 10^{20}\,\text{photon/cm}^2\,\text{s}$. With absorption coefficient $\alpha = 10^4\,\text{cm}^{-1}$ we have $G = I_{ex}\alpha \approx 2 \times 10^{24}\,\text{cm}^{-3}\,\text{s}^{-1}$ and $\sqrt{G\beta} = \sqrt{2 \times 10^{24} \times 10^{-15}} \approx 4.5 \times 10^4\,\text{s}^{-1}$. Thus, the relation $t_p \ll (1/2\sqrt{G\beta}) \approx 1 \times 10^{-5}\,\text{s}$ will hold for pulses of duration of roughly 1 μs and less. Of course, under a higher-intensity excitation G and higher values of β, the duration of pulses leading to the quadratic dependence (3.15) will get substantially shorter; nevertheless, it is still easily achievable with excitation by common laser pulses with durations of the order of nanoseconds or tens of picoseconds.

3.4 Stretched exponential

In practice, we observe certain cases when the luminescence decay curve is neither governed by the exponential law (3.5) nor by a dependence of the type (3.11) but, instead, it can be described by the so-called stretched exponential

$$i(t) = i(0)\exp[-(t/\tau)^\delta], \tag{3.16}$$

where τ is called the decay time and δ is the dispersion factor with $0 < \delta < 1$. It is obvious that for the limiting value $\delta = 0$, eqn (3.16) represents a constant function, and for $\delta = 1$ relationship (3.16) reduces to the standard exponential decay. Between the limiting values of the parameter δ, the function (3.16) interpolates these two extremes as shown in Fig. 3.5. This also hints at the origin of the name—as if we 'stretched' an exponentially decreasing function towards later times.

This decay law is often observed in disordered systems and it is believed to be due to the so-called dispersion diffusion of photoexcited carriers. In particular, eqn (3.16) describes very precisely luminescence decay in amorphous semiconductors or silicon nanostructures. Nevertheless, it is important to realize that, in itself, the presence of the stretched exponential function in luminescence decay does not allow us to identify the luminescence centre in microscopic detail. There exist several plausible explanations—e.g. in a silicon nanocrystal system, this can be the case for photocarrier migration between various localized states either within a single nanocrystal or along multiple closely packed nanocrystals. Moreover, the diffusion itself can occur due to carriers being excited from localized to free states (multiple trapping and release) or hopping directly between the localized states. The degree of the dispersion of motion is measured by δ, which depends on the trap density, the distribution of their depths, etc. Moreover, both δ and τ depend also on the temperature and the wavelength at which the decay is measured.

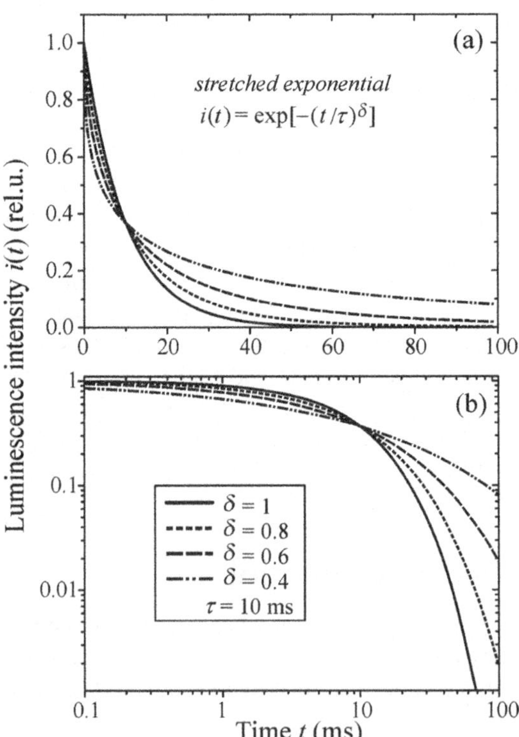

Fig. 3.5
A plot of the stretched exponential $i(t) = i(0)\exp[-(t/\tau)^{\delta}]$ for $i(0) = 1$, $\tau = 10$ ms and several values of the parameter δ. (a) A lin–lin scale; (b) a log–log scale often used to plot the stretched exponential.

3.5 Multiple processes present simultaneously

Let us now consider a somewhat more complex situation that, however, comes closer to that from a large variety of possible experimental events. We shall deal with two parallel recombination channels: bimolecular (intrinsic) recombination of free electron–hole pairs produced by optical excitation $G = \alpha I_{ex}$ and monomolecular (extrinsic) recombination of impurity atoms present in a total concentration N_0. These impurity centres are excited primarily via the same optical excitation that produces free electrons and holes. All processes involved are shown in Fig. 3.6. We shall be mainly interested in the character of the dependence of both luminescence lines on the excitation intensity in the steady state. We shall show that in case there are both monomolecular and bimolecular luminescence processes present in the material simultaneously, they can be distinguished by their typical intensity dependences.

We divide the free photoelectrons and holes into the so-called 'optical', n_0, and 'thermal', n_t, electrons and 'optical', p_0, and 'thermal', p_t, holes. Optical electrons and holes represent a population of carriers occupying almost all available states in the conduction and valence bands.[3] They can mutually recombine radiatively with a bimolecular recombination coefficient β, emitting a photon $h\nu_B \geq E_g$. By thermal photocarriers n_t or possibly p_t, we shall mean electrons and holes in the lowest energy states just above the bottom of the conduction or below the top of the valence band, respectively. These quasi-particles originate from the population of optical carriers via thermalization, which process is described using a rate constant k. They have the highest chance to be captured by an impurity centre, promoting this into an excited state. Let us denote the concentration of the excited centres by N. Let the process of both electron and hole capture be described by the same rate constant a; its occurrence probability is proportional to the product of the

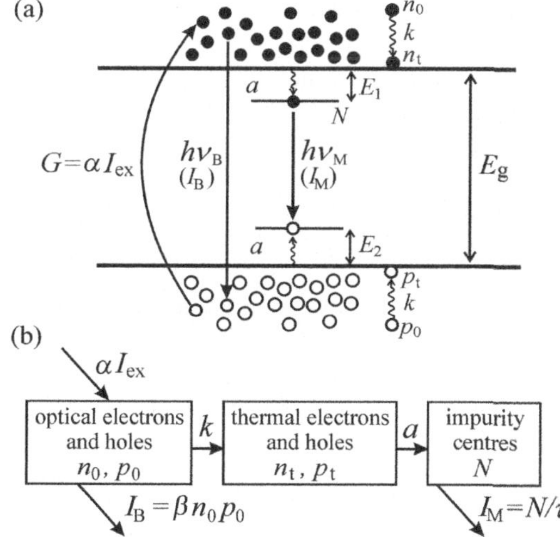

Fig. 3.6
(a) The band scheme of a semiconductor showing the processes of bimolecular (intrinsic) and monomolecular (extrinsic) photoluminescence—see the main text and eqns (3.17)–(3.19). The respective photon energies are $h\nu_B \approx E_g$, $h\nu_M \approx E_g-(E_1 + E_2)$. (b) The corresponding block diagram of the transitions.

[3] We can presume their energy distribution to be described by the classical Boltzmann tail of the Fermi–Dirac distribution function with the Fermi energy level being roughly amid the bandgap.

concentration of impurity centres in the ground state (N_0–N) with the concentration of thermal electrons n_t, or holes p_t, respectively. The excited impurity atoms then luminesce (emitting a photon of energy $h\nu_M \approx E_g-(E_1+E_2)$), the number of recombination acts per unit volume being given by N/τ. Let us note that, for the sake of simplicity and without loss of universality of our arguments, we shall assume 100% quantum efficiency in both luminescence processes, i.e. $\beta = \beta_r$ and $\tau = \tau_r$.

All the above processes can be described by the following kinetic equations:

$$\frac{dn_0}{dt} = \alpha I_{ex} - \beta n_0 p_0 - k\, n_0, \tag{3.17}$$

$$\frac{dn_t}{dt} = k\, n_0 - a(N_0 - N)n_t, \tag{3.18}$$

$$\frac{dN}{dt} = a(N_0 - N)n_t - \frac{N}{\tau}; \quad N \neq N_0, N \neq 0, \tag{3.19}$$

$$n_0 = p_0,\ n_t = p_t.$$

In the steady state, $d/dt = 0$, we obtain

$$\beta n_0^2 + k\, n_0 - \alpha\, I_{ex} = 0, \tag{3.20}$$

$$a(N_0 - N)n_t - k\, n_0 = 0, \tag{3.21}$$

$$a(N_0 - N)n_t = N/\tau. \tag{3.22}$$

We rewrite equation (3.22) in the form

$$\frac{N}{\tau} = \frac{a\, n_t N_0}{(1 + a\, n_t \tau)}, \tag{3.23}$$

having thus obtained a parametric expression for the monomolecular luminescence intensity, $I_M = N/\tau$; for the bimolecular luminescence intensity, we have in accordance with our previous notation $I_B = \beta\, n_0^2$.

Solving the quadratic equation (3.20), we get the equilibrium concentration of optical electrons, n_0, and hence immediately also I_B; using eqns (3.21) and (3.23), we obtain n_t and thus also I_M.

The solution to (3.20) reads

$$n_0 = \frac{k}{2\beta}\left(\sqrt{1 + \frac{I_{ex}}{I_0}} - 1\right), \qquad I_0 = \frac{k^2}{4\alpha\beta}. \tag{3.24}$$

Substituting this expression into (3.21) and taking advantage of (3.23), we get after some simple algebra

$$a\, n_t \tau = \frac{\frac{k^2}{2\beta}\left(\sqrt{1 + \frac{I_{ex}}{I_0}} - 1\right)}{\frac{N_0}{\tau} - \frac{k^2}{2\beta}\left(\sqrt{1 + \frac{I_{ex}}{I_0}} - 1\right)}. \tag{3.25}$$

In relations (3.24) and (3.25), $I_0 = k^2/4\alpha\beta$ represents a sort of characteristic excitation intensity governed exclusively by the material parameters. We now distinguish two excitation modes, weak excitation and strong excitation, characterized by the relations $I_{ex} \ll I_0$ and $I_{ex} \gg I_0$, respectively.

(a) Weak excitation ($I_{ex} \ll I_0$). In this case we apply the familiar approximation $\sqrt{1+x} \approx 1 + x/2$ valid for $x \ll 1$. From eqn (3.24) we then get

$$n_0 \approx \frac{\alpha\, I_{ex}}{k}, \qquad I_B = \beta\, n_0^2 = \frac{\alpha^2 \beta}{k^2} I_{ex}^2. \tag{3.26}$$

Therefore, bimolecular luminescence varies quadratically with excitation intensity. From eqn (3.25), we have

$$a\, n_t \tau \approx \frac{\alpha\, I_{ex}}{\frac{N_0}{\tau} - \alpha\, I_{ex}}, \qquad a\, n_t N_0 \approx \frac{N_0 \alpha I_{ex}}{N_0 - \alpha \tau I_{ex}} \tag{3.27}$$

and, finally, substituting the above relations into (3.23) yields, for the monomolecular luminescence,

$$I_M = \alpha\, I_{ex}; \tag{3.28}$$

thus, I_M increases linearly with the excitation.

Unlike the situation when there is just a single active luminescence process in a semiconductor (Sections 3.2 and 3.3), the steady-state bimolecular and monomolecular luminescences now depend on the excitation intensity in a markedly different way: I_B scales quadratically while I_M scales linearly with excitation intensity. This corresponds to the intuitive and rather widespread scheme; we stress, however, that the necessary prerequisite is the parallel occurrence of both phenomena, which is sometimes overlooked just as the weak excitation condition. (Short-pulse excitation, discussed in Section 3.3, is another example of a quadratic demonstration of bimolecular recombination.)

It is not difficult to understand what goes on in our model material during weak excitation. Comparing (3.28) with the starting eqn (3.17), we see that basically all the absorbed energy αI_{ex} is radiated in the form of monomolecular impurity luminescence. Thus, under weak excitation, bimolecular emission is substantially weaker than monomolecular emission ($I_B \ll I_M$), which makes sense since the density of the 'optical' electron–hole pairs is low and it is difficult for the electrons and holes to find one another to recombine. Almost all the photoelectrons are trapped at the impurity centres, which subsequently return to the ground state, emitting an extrinsic luminescence photon $h\nu_M \approx E_g{-}(E_1 + E_2)$.

This enables us to formulate now the weak excitation condition alternative to the relationship $I_{ex} \ll I_0$. Under the circumstances, the low excitation intensity is physically equivalent to the fact that by far not all impurity atoms are brought into the excited state. In other words, the number of all available recombination impurity events, N_0/τ, is much larger than the density of photoexcitation events per second: $N_0/\tau \gg \alpha I_{ex}$. This, of course, then reduces relation (3.27) to the form

$$a\, n_t \tau \approx \frac{\alpha \tau\, I_{ex}}{N_0} \ll 1. \tag{3.29}$$

Indeed, neglecting $a n_t \tau$ as compared to unity in the denominator of eqn (3.23), we immediately obtain $I_M = N/\tau \approx a n_t N_0$. This, considering (3.29) again yields $I_M \approx (\alpha I_{ex}/N_0) N_0 = \alpha I_{ex}$, which is in accordance with (3.28).

Therefore, inequality (3.29) is an alternative expression for the weak excitation condition.

(b) Strong excitation. It appears beneficial to formulate the strong excitation condition as a relation opposite to (3.29):

$$a n_t \tau \gg 1.$$

Equation (3.23) then immediately informs us of the monomolecular luminescence intensity

$$I_M = \frac{N}{\tau} = \frac{a\, n_t\, N_0}{(1 + a\, n_t \tau)} \approx \frac{a\, n_t\, N_0}{a\, n_t \tau} = \frac{N_0}{\tau}. \tag{3.30}$$

This is clearly the maximum achievable intensity in the case when all impurity centres are excited. Its value no longer depends on the excitation intensity. Therefore, the extrinsic radiative recombination channel becomes *saturated*.

What dependence of the bimolecular intrinsic luminescence on the excitation intensity do we get now? Using the strong excitation condition in the form $I_{ex}/I_0 \gg 1$, we obtain from eqn (3.24)

$$n_0 = \frac{k}{2\,\beta}\sqrt{\frac{I_{ex}}{I_0}}$$

or

$$I_B = \beta\, n_0^2 \approx \frac{\beta\, k^2}{4\,\beta^2} \frac{I_{ex}}{I_0} = \alpha\, I_{ex}. \tag{3.31}$$

Hence, bimolecular luminescence increases linearly in the strong excitation limit and, in a sense, it basically swaps roles with monomolecular luminescence as can be seen by comparing with eqn (3.28). The overall behaviour of the system on a wide scale of excitation intensities is summarized by the plot in Fig. 3.7(a).

We can formulate a system of kinetic equations analogous to eqns (3.17)–(3.19) for a large number of various luminescence centres and photocarrier trap levels. An example of a very successful model is the kinetic model of exciton and biexciton optical creation, which introduces, among other things, exactly the concept of 'optical' and 'thermal' excitons to explain the experimental data [4]. An older monograph by Antonov-Romanovskij [5] even attempts to cover the basis of all luminescence phenomena in semiconductors by making use of various combinations of unspecified radiative and non-radiative transitions and a large number of trapping levels. From today's perspective, however, this approach appears obsolete.

To conclude this section, a few more general notes may be in order. The kinetic approach to luminescence phenomena may be useful when analysing experimental data, in the sense that it is able to suggest a basic idea or a hint about what type the examined luminescence radiation might be. Here, both the character of the decay curve and the measured intensity dependence can help us. A specific model can give us a very good picture of the experimental

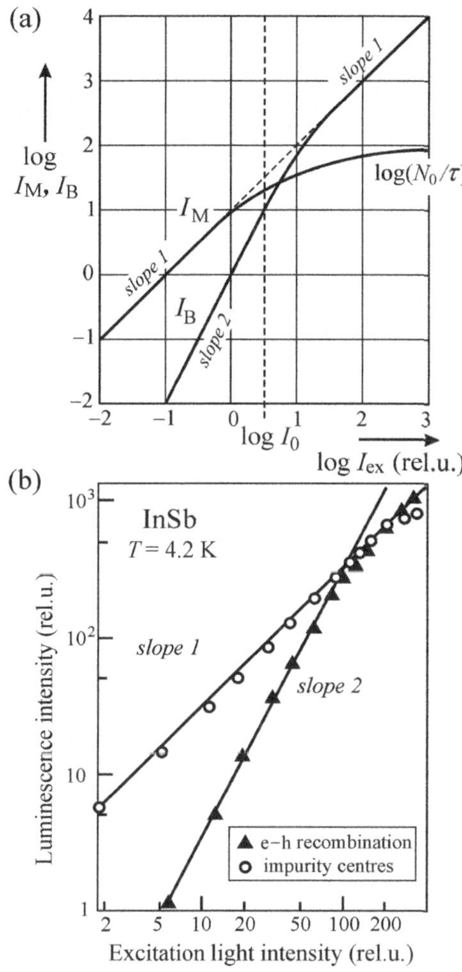

Fig. 3.7
(a) A schematic intensity dependence of monomolecular (I_M) and bimolecular (I_B) luminescence for the model shown in Fig. 3.6. (Using the log-log scale, the plot of a linear function is a straight line with slope equal to 1, the plot of a quadratic function is a straight line with slope equal to 2.) (b) The intensity dependence of bimolecular luminescence (electron–hole intrinsic recombination, triangles) and monomolecular luminescence (localized impurity centre, circles) in InSb at temperature $T = 4.2$ K. Adapted from Mooradian and Fan [6]. Compare to panel (a).

results in a number of cases. For example, the basic features of the curves in Fig. 3.7(b), representing the intensity dependence of intrinsic and impurity low-temperature luminescence in InSb [6], are qualitatively absolutely identical to the attributes of our model curves from Fig. 3.7(a) (the quadratic dependence of intrinsic luminescence and the linear dependence of impurity luminescence in the case of weak excitation; a marked decrease in the slopes and possibly a tendency to saturation at high excitation intensities).

On the other hand, we must not overestimate the significance of kinetic equations. First of all, they contain no information about the spectral form of the individual luminescence lines. In addition, based on these equations alone we cannot usually draw unambiguous conclusions in the sense of a bilateral mathematical implication. For example, an experimentally established quadratic intensity dependence is evidence (if we disregard possible nonlinear optical phenomena and very high excitation effects) for the presence of a bimolecular recombination mechanism, yet the absence of such a dependence does not necessarily imply the absence of a bimolecular process—see Fig. 3.7(a) on the side of high excitation intensities. Finally, we must keep in mind that the effects of very high excitation accompanied by the bands filling (being reflected e.g. in spontaneous emission in an electron–hole system of high density, n_0, $p_0 \geq 10^{17}\,\text{cm}^{-3}$, occurring for instance in LED operation or near the threshold condition of a semiconductor laser) may modify the formulation of the kinetic equations very substantially; e.g. the simple relation (3.10) is no longer valid, etc. [7]. After all, in the case of very high excitation, brand-new luminescence systems appear (they are discussed in Chapter 8), for which it often does not make much sense to distinguish mono- and bimolecular luminescence phenomena.

3.6 Problems

3/1: Show that even if we introduce quantum yields of monomolecular and bimolecular luminescence ($\eta = \tau/\tau_r < 1$, $\beta_r/(\beta_r + \beta_{nr}) = (\beta_r/\beta) < 1$) into eqns (3.17)–(3.19), all the salient features of the model shown in Fig. 3.6 remain qualitatively preserved.

3/2: Check that for a monomolecular process we cannot (as against a bimolecular process) obtain a quadratic dependence of the luminescence response to the excitation intensity even if we excite the system by an arbitrarily short pulse $t_p \ll \tau$, where τ is the luminescence decay time.

3/3: Prove that the solution to the equation $dn/dt = -\beta n^2 - n/\tau$ is a decay curve $n(t)$ of the form

$$n(t) = \frac{n(0)e^{-t/\tau}}{1 + n(0)\,\beta\,\tau\,(1 - e^{-t/\tau})}.$$

Show that at later times after the excitation was turned off (when the density of the recombining photocarriers, $n(t)$, has already dropped substantially) the total luminescence decay is governed by a sum of exponential curves. What condition must be satisfied for the decay to be described by a single exponential $\sim \exp(-t/\tau)$?

3/4: Show that (a) the solution to eqn (3.10) is (3.13) and that (b) for long excitation pulses and after a transient initial increase in luminescence intensity, $i(t)$, a steady-state value of intensity is established, which is in accord with (3.12). Plot the curve $i(t)$, from the time we turn on the excitation until the emission intensity settles.

3/5: Discuss the kinetic model of exciton and biexciton creation according to [4].

References

1. Bebb, H. B. and Williams, E. W. (1972). *Photoluminescence I. Theory*. In *Semiconductors and Semimetals* (ed. R. K. Willardson and A. C. Beer), Vol. 8, p. 181. Academic Press, New York.

2. Rice, T. M. (1977). *The electron-hole liquid in semiconductors: Theoretical aspects*. In *Solid State Physics* (ed. H. Ehrenreich, F. Seitz, and D. Turnbull), Vol. 32, Chap. 1. Academic Press, New York.
3. Yu, P. Y. and Cardona, M. (1996). *Fundamentals of Semiconductors*. Springer, Berlin
4. Knox, R. S., Nikitine, S., and Mysyrowicz, A. (1969). *Optics Comm.*, **1**, 19.
5. Antonov-Romanovskij, V. V. (1966). *Kinetics of photoluminescence of crystalline phosphors* (in Russian: *Kinetika fotoljuminescencii kristalofosforov*). Nauka, Moskva.
6. Mooradian, A. and Fan, H. Y. (1966). *Phys. Rev.*, **148**, 873.
7. Bourdon, G., Robert, I., Sagnes, I., and Abram, I. (2002). *J. Appl. Phys.*, **92**, 6595.

4 Phonons and their participation in optical phenomena

Optical and thus also luminescence transitions in a semiconductor occur primarily within its electron system originating from the outer electron shells of the atomic constituents of the crystal lattice. However, interaction of the system with atomic nuclei binding the inner shell electrons is by far no negligible in optical phenomena and reveals itself in a number of effects very easy to observe. This affects the shape of the absorption edge, the width of both the absorption and emission spectral lines, gives rise to their possible fine structure, and it often results in thermal luminescence quenching, etc. In this chapter, we first recap concisely the standard facts about crystal lattice vibrations, recalling the notion of phonons. Then we mention the microscopic mechanisms responsible for interactions of excited electronic states with semiconductor lattice vibrations, the so-called exciton–phonon interaction. Next, we point out the possible local vibrations of impurity atoms. Finally, we discuss a model of the interaction between a localized optical centre and lattice vibrations (the concept of configurational coordinate) and related effects and terms: the Franck–Condon principle, thermal luminescence quenching, the Huang–Rhys factor, and weak and strong interactions with lattice.

4.1 Lattice vibrations—phonons

Atomic nuclei in a crystal keep oscillating around their equilibrium positions, determined by lattice point coordinates. The amplitudes of these oscillations rise with increasing temperature; however, they remain small compared to the lattice constant a at temperatures related to the luminescence processes (these range basically from the pumped liquid He temperature, about 1.5 K, up to room temperature of 295 K). In the simplest case of a one-dimensional crystal made of an infinite chain of identical atoms of mass m, we can write the equation of motion in the form

$$m\ddot{x}_n = -f(x_n - x_{n-1}) - f(x_n - x_{n+1}), \tag{4.1}$$

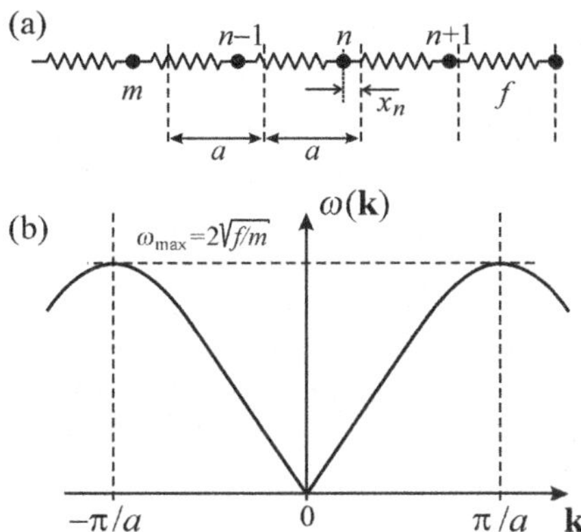

Fig. 4.1
(a) An infinite one-dimensional chain of identical oscillating atoms, and (b) the corresponding dispersion relation $\omega = \omega(\mathbf{k})$.

where x_n is the instantaneous deflection of the nth atom from its equilibrium position (Fig. 4.1(a)). In (4.1), we only consider interactions of the nearest neighbours. As a further consequence of the small amplitude of the nuclei, we can use a harmonic approximation assuming, similarly to a harmonic oscillator, that the force restoring the nuclei to their equilibrium positions depends linearly on the instantaneous deflection with the force constant (or 'spring constant') f.

It is natural to look for a solution to eqn (4.1) in the form of a wave of frequency ω, running along the chain [1, 2]

$$x_n = A\,\mathrm{e}^{\mathrm{i}\omega(t-na/c_s)} = A\,\mathrm{e}^{\mathrm{i}(\omega t-kna)}, \tag{4.2}$$

where $c_s = c_s(\omega)$ is the velocity of the wave motion, $k = |\mathbf{k}| = \omega/c_s$ is the magnitude of the corresponding wavevector $\mathbf{k}$, and A is the amplitude of the vibrations.

Similarly to other cases in wave mechanics, we are interested in the so-called *dispersion relation* for the given system, i.e. the dependence $\omega = \omega(\mathbf{k})$. We obtain it simply by substituting (4.2) into (4.1); the result is plotted in Fig. 4.1(b).

Here, several facts are worth mentioning.

(a) $\omega = \omega(\mathbf{k}) \equiv \omega_{\mathbf{k}}$ is a periodic function of k and hence it is enough to consider the range $k \in (-\pi/a, \pi/a)$ only, which is nothing but the first Brillouin zone that we encounter, in complete analogy, with the electron band structure of semiconductors.

(b) There is a maximum oscillation frequency, ω_{max}, at which the vibrations can still propagate along the chain. It follows from the solution to eqn (4.1) that $\omega_{max} = 2(f/m)^{1/2}$. This is markedly different from a continuum (represented by an oscillating string) where no such restriction on the transmitted frequency occurs.

(c) Near the centre of the Brillouin zone ($k = 0$), i.e. for $k \ll \pi/a$, the dispersion relation becomes almost linear: $\omega_{\mathbf{k}} = c_s k$. This can be easily understood. If, for $k \ll \pi/a$, we write the wavevector as $k = 2\pi/\lambda$ and combine both expressions, we get the condition $\lambda \gg a$, which means that the discrete chain structure vanishes in comparison to the wavelength.

Therefore, we are in a continuum-type situation for which, as is well known, a linear dependence governs the relationship between ω and k with c_s denoting the speed of sound.

Regarding the mechanical wave motion discussed above, we can introduce quantization, which is to a large extent analogous to the quantization of an oscillating electromagnetic field. If we restrict, as a certain approximation to reality, the number of atoms in the chain to a large but finite number, N, we can view the chain, regarding the energy, as an ensemble of N harmonic oscillators. The mean energy of a quantized harmonic oscillator $\bar{\varepsilon}$ at a given temperature T is given by the expression that constitutes part of Planck's blackbody radiation law

$$\bar{\varepsilon} = \frac{\hbar\omega_{\mathbf{k}}}{e^{\hbar\omega_{\mathbf{k}}/k_B T} - 1}. \tag{4.3}$$

Here, k_B denotes the Boltzmann constant. Thus, the energy of the entire one-dimensional crystal will be

$$\bar{E} = N\bar{\varepsilon} = \frac{N\,\hbar\omega_{\mathbf{k}}}{e^{\hbar\omega_{\mathbf{k}}/k_B T} - 1}. \tag{4.4}$$

We can now grasp expression (4.4) as stating that the total vibrational energy of the one-dimensional crystal is given by the sum of the energies of $N_{\mathbf{k}}$ quasi-particles, when

$$N_{\mathbf{k}} = \frac{\bar{E}}{\hbar\omega_{\mathbf{k}}} = \frac{N}{e^{\hbar\omega_{\mathbf{k}}/k_B T} - 1}; \tag{4.5}$$

we assign energy $\hbar\omega_{\mathbf{k}}$ to every quasi-particle. We call these quasi-particles—quanta of the vibrational energy of a solid—*phonons.* According to statistical physics, the statistical distribution (the distribution function) of the phonons is defined as the mean number of phonons in one quantum state. For a linear chain of N oscillating atoms, the number of allowed quantum states is equal to N [1], hence, from (4.5) we obtain for the phonon distribution function

$$n_{\mathbf{k}} = \frac{N_{\mathbf{k}}}{N} = \frac{1}{e^{\hbar\omega_{\mathbf{k}}/k_B T} - 1}. \tag{4.6}$$

Various vibrationally excited states of the crystal are characterized by different values of $n_{\mathbf{k}}$ or, in other words, each vibrational state of the lattice can be described as a state of a perfect gas of non-interacting particles—phonons. Equation (4.6) tells us that the phonons are subject to Bose–Einstein statistics and that the chemical potential of an ensemble of phonons is equal to zero. It also follows from the same equation that the phonon population depends very strongly on temperature T—as the temperature drops, the phonon population decays quickly.

The analogy to the quanta of electromagnetic radiation—photons—is rather obvious; it is, however, important to realize there is a certain difference. The expression $h\nu$ for the energy of photons holds exactly, while $\hbar\omega_{\mathbf{k}}$ for the phonons only holds in the harmonic approximation. With large amplitudes of atomic vibrations, we can no longer apply the harmonic approximation and phonons can undergo mutual interactions. There are well-known macroscopically observable effects that cannot be explained within the context of harmonic oscillations of the lattice (e.g. the thermal expansion of solids).

Although the one-dimensional chain of identical atoms provides us with the basic concept of the phonon, it cannot encompass the full reality of a three-dimensional crystal. It is astonishing, however, that it suffices to modify the one-dimensional model just slightly to be able to represent all the basic features of vibrations in a three-dimensional crystal. This modification consists in introducing two kinds of atoms into the chain as shown in Fig. 4.2(a). We denote their masses by m and M ($m < M$). The lattice constant is then equal to $2a$ and there are two atoms per unit cell. Using an approach analogous to the previous case of identical atoms, we can formulate equations of motion for both kinds of atoms and, by solving them, we can obtain the dispersion relations shown in Fig. 4.2(b). The most important new feature as compared to Fig. 4.1 is the fact that there are now two branches of phonons separated by a frequency gap.[1] The lower branch, called the *acoustic branch*, is characterized by the relation $\lim_{\mathbf{k}\to 0} \omega(\mathbf{k}) = 0$ while on the boundary of the Brillouin zone, it has its maximum value of $\omega_{\max}^{(a)} = (2f/M)^{1/2}$. On the contrary, the upper branch, called the *optical branch*, has the highest allowed frequency in the centre of the Brillouin zone (equal to $\omega_{\max}^{(o)} = [2f(1/m + 1/M)]^{1/2}$) and the magnitude of the optical frequency gently decreases towards the boundary of the zone $k = \pm\pi/2a$ where it reaches the value of $\omega_{\min}^{(o)} = (2f/m)^{1/2}$.

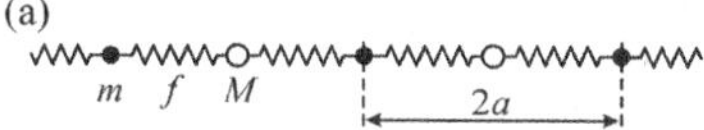

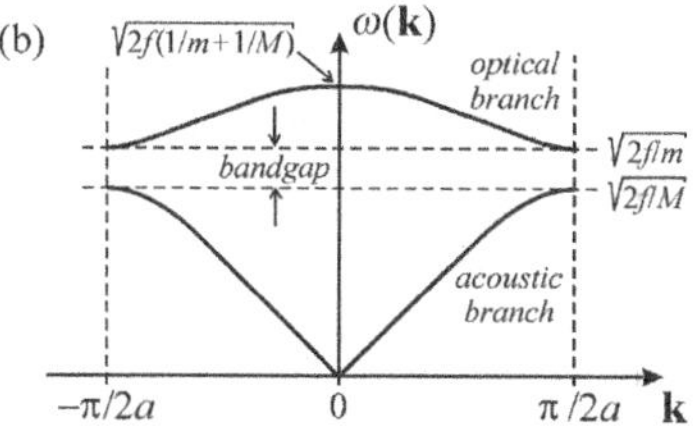

Fig. 4.2
(a) A one-dimensional chain with two kinds of oscillating atoms of masses M and m ($M > m$); (b) the corresponding dispersion relations $\omega = \omega(\mathbf{k})$.

Therefore, we speak of optical and acoustic vibrations or phonons. From the microscopic perspective, the difference between them consists in the fact that—as can be easily shown based on the equations of motion [1, 2]—light and heavy atoms move in counter-phase during optical vibrations, see also Fig. 4.5 below, while for acoustic vibrations both kinds of atoms vibrate in phase. In this case, however, even the centre of mass in each cell moves and there is a deformation wave of 'compression and expansion' travelling through the crystal, as for sound propagation. Conversely, optical vibrations—just due to the atoms moving in counter-phase—leave the centre of mass of the unit cell at rest, which allows for a non-zero frequency $\omega_{\max}^{(o)}$ even at $k = 0$. The name 'optical vibrations' stems from the fact that if there are two kinds of atoms of opposite charge per unit cell (a partially ionic bond), then the counter-phase motion means electric dipole oscillations, which can also be excited by resonance absorption of photons from the mid-infrared region.

Besides explaining the fundamental difference between the optical and acoustic vibrations, Fig. 4.2 shows several other facts that are worth emphasizing because of their importance in optical and, particularly, in luminescence phenomena:

(a) Near the first Brillouin zone boundary ($k = \pm\pi/2a$), both phonon branches are basically dispersionless, i.e. $\mathrm{d}\omega/\mathrm{d}k \approx 0$.

[1] Again, we can perceive a certain similarity to the electron energy band structure in semiconductors: the existence of allowed energy (frequency) bands and gaps. This is the common characteristic feature of waves propagating in periodic structures.

(b) Near the Brillouin zone centre ($k = 0$), the optical branch reveals itself dispersionless while the acoustic branch has its highest dispersion here.
(c) Figure 4.2 considers only atomic oscillations along the chain axis. These are the so-called *longitudinal vibrations*. In addition to these, however, there are also *transverse vibrations* when the atoms move perpendicularly to the chain. (Yet, in both cases, the vector **k** points in the same direction, that is, along the chain.) Therefore, we speak of transverse acoustic (TA), longitudinal acoustic (LA), transverse optical (TO) and longitudinal optical (LO) phonons.

Up to this point, we have explored only the single dimension of the linear model. It turns out that in a real three-dimensional crystal, the concept of TA-, LA-, TO-, and LO-phonons remains fully valid, but the phonon dispersion curves acquire somewhat more variable forms as against Fig. 4.2, due to numerous vibrational waves travelling through the crystal lattice in various directions featuring different electron cloud densities, different effective spring constants, etc. An example of actual phonon dispersion curves in a particular semiconductor—crystalline silicon—is shown in Fig. 4.3 [3]. Such curves are most often obtained using inelastic scattering of neutrons as they interact with phonons in various directions within a crystalline sample. These then correspond to different directions of **k** in the Brillouin zone.

Generally, in the three-dimensional case, if a crystal has n atoms per unit cell then there are exactly 3 types of acoustic vibrations characterized by the condition $\lim_{\mathbf{k}\to 0} \omega(\mathbf{k}) = 0$, and $3n$-3 types of optical vibrations with $\omega(\mathbf{k}) \neq 0$ for $k = 0$. The transverse vibrations may be doubly degenerate; the degeneration is generally removed in an anisotropic medium.

As regards Fig. 4.3, let us firstly notice that the typical phonon energy in a semiconductor is of the order of 10 meV. It is further interesting to realize that in silicon the gap between the optical and acoustic branches disappears in a certain direction in the Brillouin zone. This is due to the fact that there are two Si atoms per unit cell in the silicon (diamond) structure, and the atoms—naturally—have the same mass. According to Fig. 4.2(b), the acoustic and

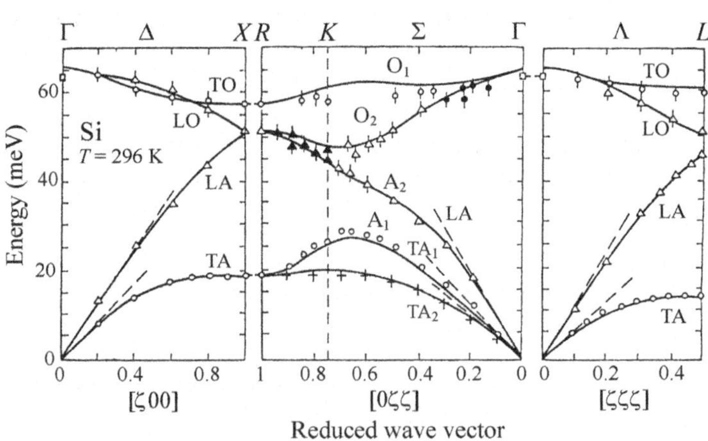

Fig. 4.3
Phonon dispersion curves of crystalline silicon in the principal directions of the Brillouin zone at room temperature. The reduced wavevector is plotted in units of $\zeta = ka/2\pi$, where a is the lattice constant. Adapted from Dargys and Kundrotas [3].

optical curves indeed coincide for $m = M$ at the boundary of the Brillouin zone.[2]

Finally, a remark on the terminology: Despite their name, acoustic phonons can participate in optical processes in a crystalline solid just like the optical phonons do! For instance, at the onset of the optical absorption edge in a semiconductor with an indirect bandgap (Si, Ge, AgBr), the type of phonons ensuring the quasi-momentum conservation law is determined by selection rules following from the symmetry of lattice vibrations, the symmetry of electron wavefunctions, and the symmetry of the corresponding matrix elements. Luminescence processes in these semiconductors are in a fully analogous situation (Subsection 7.1.4). Primarily, it is thus not of crucial importance whether the phonons are optical or acoustic. However, each of these phonon types may have a different mechanism of interaction with excited electronic states; these mechanisms are briefly treated in Section 4.2.

4.2 Electron–phonon and exciton–phonon interactions

In the preceding section, we dealt with phonons as with an isolated system. However, phonon participation in optical processes naturally requires interaction with the system of electrons responsible for the optical properties, which are basically the electrons occupying energy levels in the valence or possibly conduction band. In that case, we speak of the so-called electron–phonon interaction. Clearly, we can expect a similar type of interaction also in case the semiconductor electronic system is in an excited state and returns to the ground state, the phonon system thus participating in emission of luminescence radiation. We shall be primarily interested in this process. If, in this respect, we call the excited electronic system of the crystal an 'exciton' then we may call the corresponding interaction an exciton–phonon interaction. We shall discuss a somewhat more detailed view of the exciton–phonon and electron–phonon interactions in an excited semiconductor in Subsection 7.2.4.[3]

The following example [4] is a clear illustration of how we can imagine the charge carriers may influence the lattice vibrations. Let us consider a charge,

[2] One might then ask whether it actually makes any sense to speak of acoustic and optical vibrations in silicon if the lattice is made up of a single kind of atom. Yes, it does make sense, exactly because there are two atoms per unit cell. Then $n = 2$ and we have exactly three acoustic and $3n$–$3 = 3$ optical vibration branches. Although the optical branches cannot be excited by infrared radiation here as the bond is fully covalent, this requirement is not a condition necessary to call the lattice vibrations optical.

[3] The original meaning of the word 'exciton' in a semiconductor was clearly specified and strictly defined (Wannier, G. H. (1937). *Phys. Rev.*, **52**, 191; Elliot, R. J. (1957). *Phys. Rev.*, **108**, 1384; Nikitine, S. (1959). *Phil. Magazine*, **4**, 1) as the lowest excited electronic state of a perfect, pure crystal that can propagate freely through the lattice and transfer excitation energy (not electric charge). We shall discuss this quasi-particle, also called a Wannier 'free exciton', in connection with its important role in luminescence properties of semiconductors in Chapter 7. At a later time, the term exciton relaxed substantially from its rigour and, at present, it is used rather loosely also to denote an arbitrary bound electron–hole pair generated by light (e.g. an excited state of an impurity atom, a localized electronic excitation in an amorphous material, an electron–hole pair in a semiconductor nanocrystal, and suchlike).

whether bound or free, in a harmonically oscillating lattice made up of two kinds of ions (an ionic crystal or a semiconductor with an ionic contribution to the lattice binding energy). We shall focus on a single mode of these vibrations and we shall write its total energy (in the absence of charge that would introduce interactions) in the well-known form

$$W_0 = \frac{1}{2} m \dot{Q}^2 + \frac{1}{2} f Q^2, \tag{4.7}$$

where Q is the normal coordinate and $(f/m)^{1/2}$ denotes the angular frequency of the mode. Now we insert a charge into the lattice, which, due to Coulomb forces, introduces an additional term, $-FQ$, into (4.7), where F is the force the charge exerts on the mode. Here, we neglect higher-order terms in Q and, likewise, a possible change of the spring constant, f, due to the inserted charge. Hence, the total energy is then equal to

$$W = W_0 - F\,Q,$$

which, after introducing a new coordinate $\tilde{Q} = Q - F/f$, can be rewritten in the form

$$W = \frac{1}{2} m \dot{\tilde{Q}}^2 + \frac{1}{2} f \tilde{Q}^2 - \frac{1}{2}\frac{F^2}{f}.$$

Physically, this means a shift of the oscillator vibrations to a new equilibrium position $Q_0 = F/f$ accompanied by the release of relaxation energy $W_R = F^2/2f$, as shown in Fig. 4.4.

On the whole, there are three main microscopic mechanisms mediating the electron–phonon and exciton–phonon interactions: the deformation potential, the piezoelectric mechanism, and the Fröhlich mechanism.

The deformation potential

The energy of an electron and a hole, and thus also that of an exciton, is a sensitive function of the interatomic distance in the crystal lattice. It is well known that, for example, a deformation caused by an external force changes the bandgap value of a semiconductor. We can thus expect a similar effect to occur also due to crystal lattice vibrations. The simplest picture corresponds to long-wavelength, longitudinal acoustic vibrations when atomic displacements parallel to the direction of wave propagation occur, inducing periodic expansions and compressions of the medium on a macroscopic distance. These

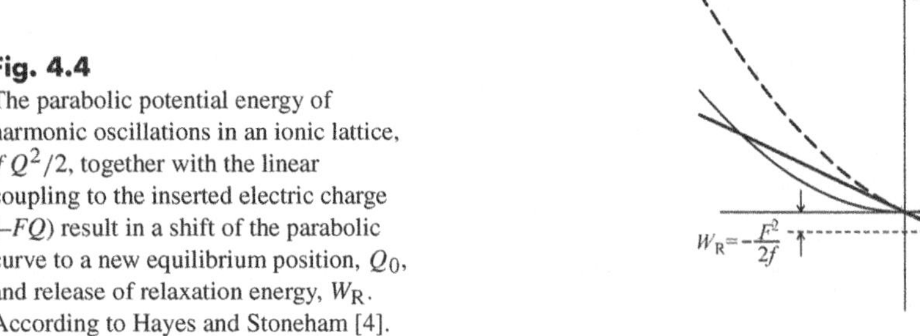

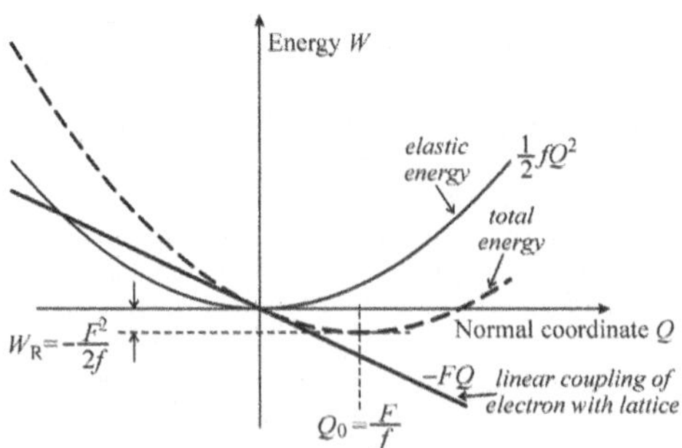

Fig. 4.4
The parabolic potential energy of harmonic oscillations in an ionic lattice, $fQ^2/2$, together with the linear coupling to the inserted electric charge ($-FQ$) result in a shift of the parabolic curve to a new equilibrium position, Q_0, and release of relaxation energy, W_R. According to Hayes and Stoneham [4].

can be expressed using the relative change in the volume of a sample, $\delta V/V$. The relative change then yields a shift in the extreme of the electron energy band $\delta E_{\mathbf{k}n}$ ($\mathbf{k}$ denotes the electron wavevector, n indexes the band). To a first approximation, the dependence reads [5]

$$\delta E_{\mathbf{k}n} = a(\mathbf{k}, n)\ (\delta V/V), \tag{4.8}$$

where $a(\mathbf{k}, n)$ is the *volume deformation potential* of the energy level $E_{\mathbf{k}n}$. It follows from this equation that the deformation potential has the dimension of energy and it is measured in eV.

However, it is neither simple to calculate such a (absolute) deformation potential theoretically nor to determine it experimentally. It is much easier to establish the difference between deformation potentials of two energy levels (the relative deformation potential). One of the commonly used experimental techniques is an optical measurement performed under external hydrostatic pressure applied to the sample. This determines the energy difference in optical transitions near the onset of the absorption edge, that is, near the extremes of the valence and conduction bands. To illustrate the situation, Table 4.1 gives the values of the relative deformation potential between extremes of the valence and conduction bands at the Γ point ($\mathbf{k} = 0$) of the first Brillouin zone. In semiconductors with a direct bandgap, this energy difference corresponds to the onset of an interband absorption edge; in semiconductors with an indirect bandgap (Si, GaP, AgBr), we in addition need to consider the deformation potential at the relevant points at the boundary of the first Brillouin zone. Table 4.1 thus also lists the available values of two absolute deformation potentials Σ_u and Σ_d, necessary to describe the energy shift according to the generalized relation (4.8) at the point $\Delta = [k/k_{max}, 0, 0]$ in the direction [100], where the conduction band minima are located in silicon and GaP.

Based on the data in Table 4.1, we can infer the following facts as regards the electron–phonon interaction. Values of the relative deformation potential given in the table decrease slightly with increasing bond ionicity, i.e. from Si towards AgBr; nevertheless, the dependence is weak and all the deformation potential values are of the same order. Therefore, this electron–phonon

Table 4.1 Deformation potentials for extremes of conduction and valence bands in selected semiconductors. The symbols $a(\Gamma_{1c})$–$a(\Gamma_{15v})$ or possibly $a(\Gamma_{15c})$–$a(\Gamma'_{25v})$ denote the relative volume deformation potential for the lowest minimum of the conduction band and the highest maximum of the valence band at the Γ point ($\mathbf{k} = 0$). Σ_d and Σ_u denote deformation potentials at the minimum of the conduction band in Si and GaP. The volume deformation potential along the [100] direction in the Brillouin zone of silicon is equal to ($\Sigma_d + \Sigma_u$). Most data adopted from Yu and Cardona [5]. Values are given in eV.

	$a(\Gamma_{1c}) - a(\Gamma_{15v})$	$a(\Gamma_{15c}) - a(\Gamma'_{25v})$	Σ_d	Σ_u
Si		−10	5	8.77
GaP	−9.3			13
GaAs	−9			
ZnS	−4			
ZnSe	−5.4			
CdTe	−3.4			
AgBr	−2.31			

interaction mechanism has roughly the same efficiency in all semiconductors. Important from the luminescence point of view is that, in addition to mediating the coupling between electrons or excitons and the acoustic phonon field that we have mentioned several times already, both the magnitude and sign of the deformation potential have an immediate impact also on the behaviour of luminescence in semiconductors subject to a static mechanical deformation. In this sense, we must understand the data in Table 4.1 as follows.

A negative value of the relative deformation potential $a(\Gamma_{1c}) - a(\Gamma_{15v})$ means that in hydrostatic pressure experiments ($\delta V/V < 0$), the energy separation $(\Gamma_{1c}) - (\Gamma_{15v})$, i.e. the width of the bandgap, E_g, becomes larger according to (4.8). This means that, as the pressure increases in these experiments, semiconductors with a direct bandgap will show a shift in the wavelength of edge emission luminescence lines towards lower values (a 'blue-shift'). Experiments confirm this. For indirect semiconductors Si and GaP, Table 4.1 gives positive signs of the deformation potentials at the minimum of the conduction band. As the values of the deformation potential at the conduction band edge commonly turn out to be about one order of magnitude larger than at the valence band maximum, the values shown here define the character of the indirect bandgap shift and thus also that of the edge emission in Si and GaP; the applied hydrostatic pressure causes a red-shift in the edge emission here.

It should be emphasized that the account given above was considerably simplified and only applies to the long-wavelength longitudinal acoustic phonons. A more detailed discussion of the deformation potentials—additionally including also anisotropic uniaxial deformation of the crystal (or shear deformation of the TA-phonons), which lifts the degeneracy of the conduction band minima in reciprocal space of Si and Ge crystals, and the participation of *short-wavelength phonons* near the first Brillouin zone boundary in optical transitions in Si, Ge, and AgBr—can be found in [5].

An important note: The deformation potential mechanism may be partially related to optical phonons as well, particularly the long-wavelength ones ($k \ll \pi/a$). These phonons from around the centre of the Brillouin zone can also induce band edge shifts due to a relation analogous to (4.8) because a relative displacement of two ions in the unit cell induces on macroscopic distances a quickly changing local deformation (Fig. 4.5(a)), which the electron system can feel as well.

The exciton–phonon interaction due to the deformation potential occurs in all crystalline solids.

The piezoelectric mechanism

In materials of lower lattice symmetry, an electric voltage associated with a macroscopic electric field inside the material may develop, in particular between opposite sample surfaces under mechanical deformation; this is known as the piezoelectric effect. The same mechanism of generating an electric field also works in the case of deformations varying in time and resulting from the relative shift between adjacent atomic layers; that is, most

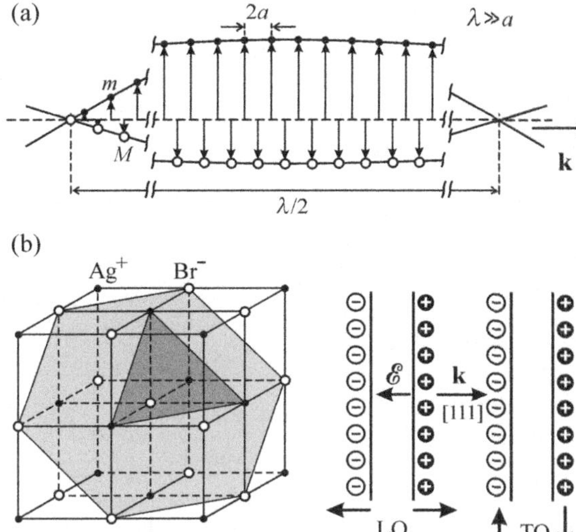

Fig. 4.5
(a) A long-wavelength TO vibration with a rather strong local deformation between adjacent atoms oscillating in counter-phase with large amplitude.
(b) In a cubic crystal with the rock salt structure (e.g. AgBr), planes occupied only by positive ions alternate with planes of negative ions perpendicular to the [111] direction. LO vibrations then represent a vibrating capacitor with a constant charge but varying plate separation. This produces a varying macroscopic electric field $\mathcal{E}$ along the wave propagation direction **k** || [111]. TO vibrations do not possess this property. We can also imagine an analogous situation to occur in type II-VI and III-V semiconductors with partially ionic bonds.

importantly again in the case of longitudinal acoustic modes of lattice vibrations. The piezoelectric potential Φ_{PE}, describing the generated electric field, then changes the energy of a carrier of charge q, an electron or a hole, by the quantity $q\Phi_{\mathrm{PE}}$; this is the essence of this electron–phonon coupling. Although the exciton, as a bound electron–hole pair, is electrically neutral, it is affected by the electric field, too, as the field modulates its energy levels and binding energy. Since the piezoelectric electron–phonon interaction is underlain by (long-range) Coulomb forces, the interaction is strongest for phonons with a small wavevector **k** (long wavelength), that is, for long-wavelength LA-phonons again.

This exciton–phonon interaction only occurs in crystals whose crystallographic point group does not contain the centre of inversion (in semiconducting materials, these are particularly the point groups T_{d} and $C_{6\mathrm{v}}$ that govern the crystal structure of the following materials showing significant luminescence: GaAs, GaP, GaN, CdS, CdSe, ZnO, ZnS, and ZnTe). It can be rather easily screened by free carriers and it is thus relatively weak; nonetheless, it is stronger than the deformation potential mechanism.

The Fröhlich mechanism

This is the most important exciton–phonon interaction. However, it only occurs in ionic crystals and in semiconductors with an ionic contribution to the lattice energy (also called polar semiconductors); it is thus completely absent in elemental semiconductors like Ge and Si. Its principle consists in the fact that, due to the opposite orientation of oscillations by oppositely charged ions, long-wavelength LO-phonons induce in the vibrating crystal lattice a long-range macroscopic electric field along the direction of the wavevector **k** (Fig. 4.5(b)). This field—or equivalently the macroscopic crystal polarization—again enters the Coulomb interaction with charge carriers or excitons. It is worth mentioning that TO-phonons do not produce similar crystal polarization and thus the energy of long-wavelength LO-phonons (i.e. phonons in the neighbourhood of

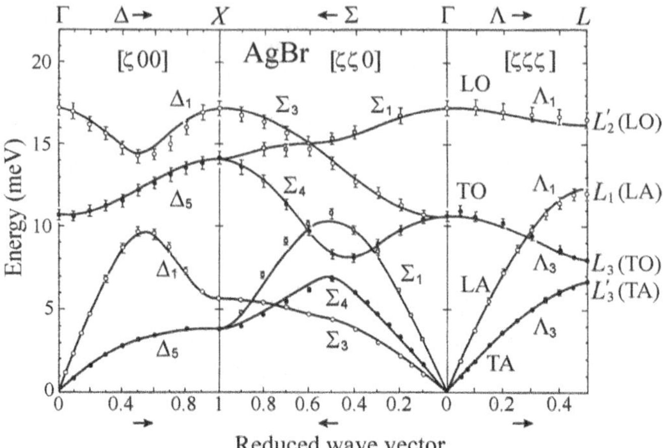

Fig. 4.6
Phonon dispersion curves in a polar semiconductor AgBr at $T = 4.4$ K. The reduced wavevector is plotted in units of $\zeta = ka/2\pi$, where a is the lattice constant. Symbols denote experimental points from neutron diffraction measurements, and lines represent the theoretical model. Symmetries of the individual phonon branches are marked by the irreducible representations of the corresponding point groups. It is particularly worth noting that at the Γ point ($\mathbf{k} = 0$), the LO-phonon energy is greater than the TO-phonon energy, which is common behaviour in polar semiconductors. Besides, however, the plot also shows some unexpected features. At the Lpoint, the energy of the TO-phonon L_3(TO) is lower than the energy of the LA-phonon L_1(LA) which contradicts the simple Fig. 4.2. This basically means that the heavier ions $M(\mathrm{Ag}^+)$ oscillate with a higher frequency here than the lighter ones m (Br^-). This can only be explained by the effective force constants, f, being considerably different for both modes. This and related effects are discussed in more detail in Problem 4/4. The irreducible representation notation at the L point holds for a negative ion (Br^-) at the origin of coordinates. Adapted from Song and Williams [6].

$\mathbf{k} = 0$) in a polar lattice is always greater than that of TO-phonons, as illustrated, for example, by Fig. 4.6 for AgBr [6].

The Fröhlich mechanism can be well quantified using the so-called polaron coupling constant α. An electric charge in an ionic lattice polarizes its neighbourhood. The polarization involves two components: electronic and ionic. The potential at a distance r from a point charge e is screened by the dielectric constant. The electronic polarization component (the shift of the electron cloud with respect to the nuclei) and the ionic component (the shift of ions from their equilibrium positions) are jointly associated with a static dielectric constant, ε_s, that adequately describes screening at low field frequencies. The potential energy at a distance r from a point charge then reads

$$\frac{1}{4\pi\varepsilon_0\varepsilon_s}\frac{e^2}{r}$$

where ε_0 is the dielectric constant or permittivity of free space. Electric field screening by the electronic component of the lattice constituents alone is described by a high-frequency or optical dielectric constant $\varepsilon_\infty (< \varepsilon_s)$ and leads to the potential energy

$$\frac{1}{4\pi\varepsilon_0\varepsilon_\infty}\frac{e^2}{r}.$$

Ionic polarization itself, which is due to the relative motion of anions and cations (representing the mass of the vibrating lattice), is then characterized by the difference

$$\frac{e^2}{4\pi\varepsilon_0 r}\left(\frac{1}{\varepsilon_\infty} - \frac{1}{\varepsilon_s}\right).$$

Obviously, $(\varepsilon_\infty^{-1} - \varepsilon_s^{-1})$ is a crucial factor for the description of the Fröhlich mechanism of electron–phonon and exciton–phonon interactions. In practice, one introduces the above-mentioned dimensionless coupling constant α in the form [7]

$$\alpha = \frac{e^2}{8\,\pi\varepsilon_0\, r_p}\,(\varepsilon_\infty^{-1} - \varepsilon_s^{-1})\,\frac{1}{\hbar\omega_{LO}}. \tag{4.9}$$

In (4.9), $r_p = \hbar\,/\sqrt{2m_p\,\hbar\omega_{LO}}$ stands for the so-called *polaron radius.* By the polaron term we denote a quasi-particle—an electron or a hole of mass m_p—that polarizes and thus locally deforms the ionic lattice as it propagates through it. This manifests itself in an increased mass of the electron or hole—as if they 'pulled' along the polarized surrounding lattice that slows them down in their motion. (We speak of an electron–polaron with a coupling constant α_e and of a hole–polaron with a constant α_h.) Thus, the effective polaron mass, m_p, is always larger than the effective mass of a 'bare' quasi-particle.

The radius r_p denotes the very size of the polarization-deformed lattice region around the charge carrier—the polaron; this is basically of the same magnitude in all polar crystals ($r_p \approx$ 1–1.5 nm). The crucial factor when characterizing the Fröhlich interaction using (4.9) is thus the value of the coupling constant, quoted for selected semiconductors in Table 4.2. As expected, the values of α_e, α_h increase with the fraction of bond ionicity (from homopolar semiconductors Ge, Si towards type I-VII semiconductors). Therefore, we can expect the strongest Fröhlich electron–phonon coupling, i.e. polaron effects, in silver halides AgBr, AgCl and in thallous halides. Experimental results confirm this reflection. However, in the case of the exciton–phonon coupling, the situation becomes somewhat different. It is expected that the exciton–phonon interaction will become stronger here, like in the case of the deformation mechanism, with increasing radius of the Wannier exciton a_x. (We can infer this on the basis of an analogy between the Wannier exciton and a free atom—the polarizability of an atom increases with its radius.) The exciton Bohr radius decreases, although not completely monotonically, with increasing bandgap E_g, as we shall discuss in Chapter 7. Thus, what comes into play here is not only the coupling constant α abut also the last column in Table 4.2. It turns out that the most prominent effects of the exciton–phonon coupling can be

Table 4.2 Polaron coupling constants α_e (electron–polaron), α_h (hole–polaron), and the bandgap width, E_g, in selected crystalline semiconductors; (i) denotes the indirect bandgap.

Semiconductor		α_e	α_h	E_g (eV)
Ge		0	0	(i) 0.745 (T = 0 K)
Si		0	0	(i) 1.17 (T = 0 K)
GaAs		0.03		1.519 (T = 0 K)
GaP	III-V	0.13		(i) 2.35 (T = 0 K)
GaN		0.4–0.5		3.49 (T = 0 K)
CdTe		0.39		1.606 (T = 4.2 K)
CdSe	II-VI	0.46		1.829 (T = 80 K)
CdS		0.65		2.583 (T = 4.2 K)
ZnS		0.71		3.78 (T = 19 K)
AgBr		1.6	2.8	(i) 2.7 (T = 4.2 K)
AgCl	I-VII	1.86	–	(i) 3.3 (T = 4.2 K)
TlBr		2.05	3.2	(i) 2.663 (T = 4.2 K)

observed in the luminescence of some II-VI compounds, in particular CdS. This is also partially due to the fact that the piezoelectric interaction here joins the Fröhlich interaction.

4.3 Lattice vibrations associated with point defects

In our considerations of phonons and their interactions with charge carriers we had, until now, always had in mind a perfect, pure crystal lattice without impurities and defects. However, we have already emphasized several times that a number of solid-state luminescence processes, very important as regards their application, are conditioned just by the presence of atomic-type impurities. It is thus in order to attempt to analyse now the influence of point defects in the crystal lattice on its vibrational properties [4].

Let us first clarify the differences in approaches used to describe perfect crystal vibrations and those of a crystal containing point defects. Firstly, as we have seen in Section 4.1, the vibrational modes of a perfect, defect-free crystal are characterized by the wavevector **k**. Secondly, these modes are fully delocalized due to the translational symmetry of the lattice. This means that, knowing the oscillation amplitude in a certain part of the crystal, we find the same oscillation amplitude—without being enhanced or damped—in any other macroscopically separated region.

We usually assume that the imperfect lattice of crystals containing point defects retains its harmonic character of vibrations and hence we may speak of its vibrational modes and phonons. Yet, we may no longer use the term wavevector and employ fully the concept of the translational symmetry of the lattice. The amplitude of the vibrational mode will be generally different in different sections of the crystal.

Let us now assume that the lattice contains a single point defect, e.g. an impurity atom of mass M' different from the masses of atoms or ions constituting the host crystal. One can imagine three different types of vibrational modes.

1. Certain crystal vibrational modes may happen not to be affected by the presence of the defect at all. If the defect atom indeed only has a different mass M' as we assumed and the harmonic force ('spring') constant, f, remains unchanged in its neighbourhood, then the modes with a node located at the impurity atom will not feel the presence of the defect. This is because the different mass will stay at rest (Fig. 4.7(a)).
2. The second type of mode can be called a *resonance mode*. If the mass M' is large (a heavy impurity atom) or the force constant f' in the neighbourhood of the impurity decreases due to a different type of interatomic bond ($f' < f$), or possibly if both effects are present simultaneously, then there appears a mode of a rather low frequency $\omega_R = (f'/M')^{1/2}$ (Fig. 4.7(b)). The frequency ω_R is usually situated within the band of allowed phonon frequencies of the unperturbed lattice and a resonance transfer of vibrational energy from the unperturbed area leads to an amplification of the local

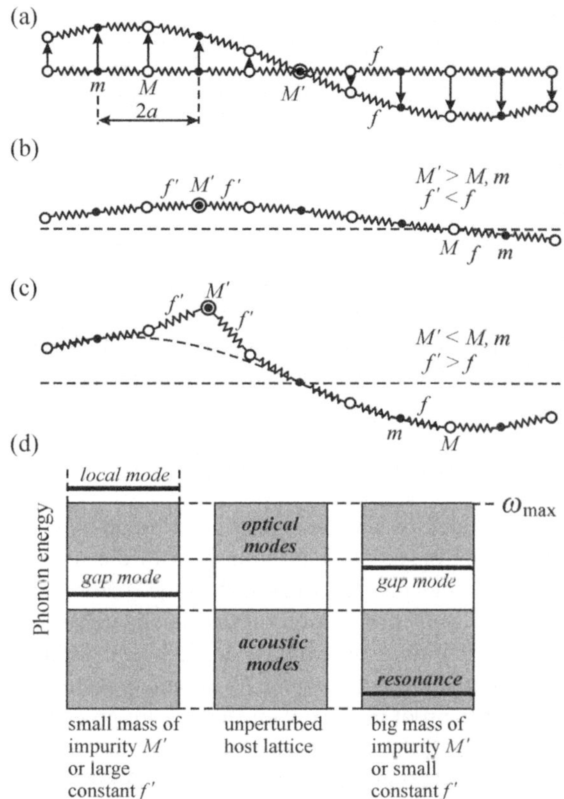

Fig. 4.7
Various types of localized vibrations of impurity atoms for the one-dimensional case: (a) an unaffected mode of the basic lattice; (b) a low-frequency resonance mode; (c) a local mode in superposition with the basic lattice mode; (d) a schematic picture of the defect modes in the energy band scheme of phonon branches.

amplitude as compared to the amplitude of vibrations in areas far away from the defect atom; hence the name resonance mode.

3. The third type of mode occurs in the neighbourhood of an impurity atom, the mass M' of which is lower than the masses of atoms in the unperturbed crystal (specifically, e.g. $M' < M$, $M' < m$ for a two-atom chain), and the neighbouring force constant f' is high, $f' > f$. The frequency of such a mode $\omega_L = (f'/M')^{1/2}$ is then located above the highest allowed frequency of perfect lattice vibrations ω_{max}. It is clear that the resonance energy transfer cannot occur here and the amplitude of such a mode decreases rather quickly away from the defect; the mode is called a *local mode* (Fig. 4.7(c)). An analogous type of defect vibration can also occur (for various ratios of the values f, f', M', M, and m) within the frequency gap between the acoustic and optical phonon branches. We can call this a *gap mode*. The frequency position of the modes discussed above in the band scheme of phonon energies is shown in Fig. 4.7(d).

It remains to assess the degree of localization of the modes under discussion in a quantitative way, if possible. The mean quadratic value of the oscillation amplitude of a given atom, in a given mode, and how this quantity varies with the distance from an impurity atom, suggest themselves to be suitable parameters. Let us assume there are N atoms oscillating in this mode, with the same amplitudes u. The potential energy $\langle V \rangle$ of N oscillating atoms in the

Table 4.3 Oscillation amplitude $\langle u^2\rangle^{1/2}$ of various types of modes.

Mode	Close to a defect	Far from a defect	Figure
Perfect crystal mode	$\sim N_\infty^{-1/2}$	$\sim N_\infty^{-1/2}$	4.7(a)
Resonance mode	$\sim N_{\mathrm{L}}^{-1/2}$	$\sim N_\infty^{-1/2}$	4.7(b)
Local mode	$\sim N_{\mathrm{L}}^{-1/2}$	0	4.7(c)

harmonic approximation is, expressed classically, equal to $\langle V\rangle = Nfu^2/2$. At the same time, the total energy of a *quantum harmonic oscillator* (representing a given mode) is equal to $2\langle V\rangle$ and is given by relation (4.3), independent of N. Even if we add to it the zero-point energy $h\omega/2$ (which is not included in (4.3)), we obtain

$$2\langle V\rangle = \frac{1}{2}\hbar\omega \coth\frac{\hbar\omega}{2k_{\mathrm{B}}T}, \tag{4.10}$$

which is fully determined by the frequency ω and temperature T and is also independent of N. The only option to comply with the requirement that the expression ($Nfu^2/2$) be independent of N is thus the scaling $\langle u^2\rangle^{1/2} \sim 1/\sqrt{N}$. This then supplies us with information on the localization of various modes. If we denote by N_∞ the number of atoms in the host crystal and if the point defect affects the oscillation amplitude of N_{L} atoms in its neighbourhood then the relative values of the root mean square $\langle u^2\rangle^{1/2}$ are given in Table 4.3.

Finally, let us note that the defect-induced vibrations modify the selection rules for infrared and Raman spectroscopy. Translational symmetry of the unperturbed crystal leads to the quasi-momentum conservation law and only optical phonons around $\mathbf{k} \sim 0$ are active in infrared spectra and Raman scattering. The presence of a defect breaks the symmetry and, therefore, modes from almost the entire Brillouin zone (with frequencies from the allowed bands) can participate in infrared and Raman experiments. At the same time, the presence of the defect itself does not appear much in the resulting spectra; they reflect primarily the phonon density of states of the perfect host crystal.

The typical impurity concentrations to which we can apply the approximation of a single isolated defect discussed above (without the need to consider defect pairs or even clusters) are of the order of 1000 ppm and less. This complies fully with the requirements for extrinsic luminescence in semiconductors, which is characterized by impurity concentrations mostly within the range 10 ppb–100 ppm.

4.4 A localized optical centre in a solid matrix—the configurational coordinate model

In the present section, we shall continue investigating the properties of a point defect (an impurity atom) in a solid. However, we shall now be primarily interested in the electronic properties of such an impurity centre related to its optical excitation and subsequent radiative transition to the ground state—

the extrinsic luminescence.[4] In Section 3.2 we gave such an optically active area in a solid the name of a *localized luminescence centre*, in order to distinguish the phenomenon from a process where emission of a luminescence photon is connected with quasi-particles moving freely through the matrix or crystal lattice (e.g. radiative recombination of a free electron–hole pair). We now introduce the concept of a configurational coordinate, which will enable us to understand how an excited luminescence centre enters into interaction with the defect modes of lattice vibrations in its neighbourhood, how the exciton–phonon interaction affects the shape of the optical spectra, and how this interaction can be quantified using optical measurements.

Let us consider an impurity atom of mass M_P embedded in a host solid; the simplest case of a one-dimensional chain with a single type of atom is shown in Fig. 4.8(a). Such a defect, as we know, leads to a distortion in the equilibrium positions of the nearest neighbours—the host atoms of mass $m \neq M_P$. Let the defect be in its ground electronic state. Let us denote by Q_{g0} the equilibrium distance between the impurity and its nearest neighbours of

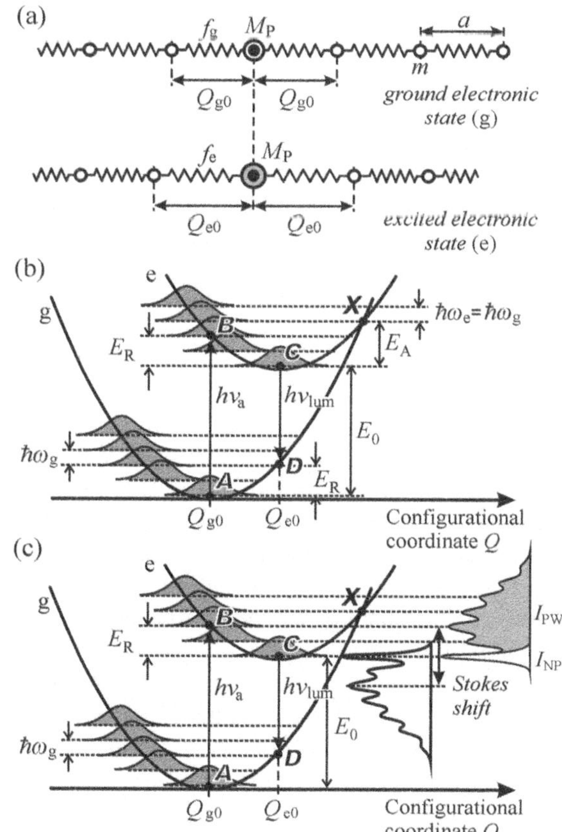

Fig. 4.8
(a) A one-dimensional chain of atoms ('matrix') containing an optically active impurity atom of mass M_P. (b) Energy of the ground (g) and excited (e) electronic states of a localized centre as a function of the configurational coordinate Q for the Huang–Rhys factor $S = 2$. (c) Schematic of the origin of optical absorption and luminescence spectra within the framework of the Franck–Condon principle at $T = 0K$. I_{NP} and I_{PW} denote the integral intensities of a no-phonon line and a phonon wing, respectively.

[4] In certain cases, the formalism discussed here can also be applied to intrinsic processes (see Problem 7/4 or the luminescence of pure AgCl in Subsection 7.2.4).

mass m; this distance generally differs from the unperturbed 'lattice constant' a, occurring sufficiently far away from the defect. The impurity atoms, just like the host atoms, oscillate around their equilibrium positions at any temperature and (assuming harmonic oscillations) we can write for the total energy of the oscillating impurity

$$E_g = E_{g0} + \frac{1}{2} f_g (Q - Q_{g0})^2. \tag{4.11}$$

Here f_g is the effective force constant determining the oscillation frequency $\omega_g = (f_g/M_p)^{1/2}$, and $(Q-Q_{g0})$ represents the instantaneous displacement from the equilibrium position Q_{g0}. We shall call Q a *configurational coordinate* and the overall situation is presented in Fig. 4.8(b). The diagram also shows the quantum-mechanical energy levels of a harmonic oscillator, represented by the oscillating atom M_p, as being equidistant with spacing $h\omega_g$. Let us remind ourselves that the quantum-mechanical probability of finding the harmonic oscillator in one of these vibronic levels reaches its highest value at the classical turning points where the corresponding vibronic level intersects the parabolic curve. Figure 4.8(b) shows this schematically, plotting the square of the modulus of the wavefunction (shaded curves). The ground vibrational state of energy $h\omega_g/2$ is an exception to this rule, with its occurrence probability highest for $Q = Q_{g0}$ (point A).

Let us now imagine the impurity atom in an excited electronic state. The classical picture is as follows: one of the valence electrons jumps to a higher energy level with a larger orbital radius; this means that the effective atom radius increases and the excited atomic 'pushes' away its nearest neighbours somewhat further off. The equilibrium distance thus grows from the value Q_{g0} to a value $Q_{e0}(> Q_{g0})$, the impurity will oscillate around a new equilibrium position, and we can write for its total energy

$$E_e = E_{e0} + \frac{1}{2} f_e (Q - Q_{e0})^2, \tag{4.12}$$

where $E_{e0} = E_{g0} + E_0$ and we usually assume that the force constant f_e remains unchanged as the atom goes to the excited electronic state, thus $f_e = f_g$. Equation (4.12) is represented by the upper parabola in Fig. 4.8(b), which, under the assumptions, arises by translating the parabola (4.11) into a new origin $[Q_{e0}, E_{g0} + E_0]$. The energy E_0 is equal to the electron excitation energy of the impurity atom and is called the *zero-phonon* or *no-phonon energy* for reasons that will become clear hereafter.

Before entering the discussion on optical transitions with Fig. 4.8 as promised, let us attempt to specify more closely the notion of the configurational coordinate Q. We introduced it by a visual demonstration—as a coordinate describing in a simple way the relative motion of a point defect and its nearest neighbours. Admittedly, such a picture very often corresponds to reality; on a more general level, however, Q may represent a configuration of multiple atoms as well as of a three-dimensional host matrix in the neighbourhood of a luminescent impurity atom or molecule. In another words, the coordinate Q is generally a combination of normal vibrational modes of a defective periodic atomic structure. The reduction of a three-dimensional configuration to a single coordinate represents a considerable technical simplification of the problem. When introducing the term of configurational coordinate, we nowhere explicitly invoked the translational symmetry of the crystal lattice; we were only speaking of a defect atom and its immediate neighbourhood. Hence, only the short-range order is important here and the configurational coordinate model can thus be applied to both crystalline and amorphous solids.

Let us now consider how the impurity atom can be raised from the ground to an excited electronic state by absorbing a photon of visible radiation. In its ground electronic state, the system is located on the lower parabola around the minimum A. The absorption of a suitable photon proceeds extremely fast, within a time interval of about 10^{-15} s. During this time, all atoms in the solid keep their momentary positions unchanged since the typical frequencies of lattice vibrations are comparatively low, $\omega \sim 10^{12}$–10^{13}s^{-1}. Therefore, the absorption event takes place much faster than nuclei around the impurity atom can rearrange into a new equilibrium configuration and, thus, in the diagram of Fig. 4.8(b), it is represented by a vertical transition between points A and B. (Not to be confused with direct optical transitions in the Brillouin zone!) Point B, as the final transition state, is determined by that vibronic level of the excited electronic state that has the highest occurrence probability at $Q = Q_{\text{g0}}$ (the 'turning point'). Therefore, a photon of energy $h\nu_{\text{a}}$ equal to the energy difference between B and A is absorbed. However, point B represents an excited vibrational state of the upper energy curve, which is a non-equilibrium position. Hence, by rearranging the configuration of its adjacent nuclei, the system gradually reaches an equilibrium given by the minimum C of the upper parabola at $Q = Q_{\text{e0}}$. In addition it releases its excess energy by emitting phonons $h\omega_{\text{e}}(= h\omega_{\text{g}})$, belonging to one of the vibrational modes associated with point defects as discussed in Section 4.3. Therefore, the ultimate outcome of the relaxation is to hand over a certain fraction of the electron excitation energy $E_{\text{R}} = (E_{\text{B}} - E_{\text{C}})$ to the matrix (lattice) in the form of heat. The energy E_{R} is called the *relaxation energy*. At point C, the system is still in an excited electronic state of finite lifetime (10^{-9}–10^{-8} s for the allowed dipole transitions); after this time, it drops to point D of the ground electronic state, radiating a luminescence photon of energy $h\nu_{\text{lum}} = E_{\text{C}}–E_{\text{D}}$. The verticality of the transition $C \to D$ and the choice of point D are driven by the same rules as applied in the case of the absorption transition ($A \to B$). The entire process is again completed by releasing the relaxation energy equal to the energy difference between D and A (which can be easily shown to be equal to $E_{\text{R}} = E_{\text{B}} - E_{\text{C}}$) and returning to point A.

The approach assuming that atomic nuclei remain at rest in a solid matrix during optical excitation of the electron, and thus leading to verticality of the resulting optical transitions in the configurational coordinate diagram, is called the *Franck–Condon principle*. We can immediately see that the principle inherently explains the Stokes' luminescence law: $h\nu_{\text{a}} \geq h\nu_{\text{lum}}$. Thus, the localized luminescence centres are generally 'transparent' to their own emitted radiation.

Considering Fig. 4.8(b), we define several additional important terms. First of all, the two parabolic curves intersect at point X. Putting aside zero-point oscillations, the energy separation E_A of point X from the minimum C of the upper parabola is equal to

$$E_A = \frac{(E_0 - E_R)^2}{4E_R}. \tag{4.13}$$

This is the *activation energy*, the meaning of which will be discussed in Section 4.6. Next, we usually express the relaxation energy E_R, supposing $h\omega = h\omega_g = h\omega_e$, as

$$E_R = S\hbar\omega, \tag{4.14}$$

where the dimensionless parameter S is called the *Huang–Rhys* factor. By eqn (4.14), this factor hence represents the mean number of phonons $h\omega$ emitted as the centre relaxes along the path $B \to C$ or $D \to A$. (If $\omega_e \neq \omega_g$, which is experimentally observed in certain cases, then it is basically possible to define two different values of the Huang–Rhys factor: S_e a S_g.) This factor, S, is an important indicator of the strength of coupling between the centre and the matrix. The larger the parameter S, the larger is the shift $(Q_{e0} - Q_{g0})$ and thus the stronger is the interaction between the excited centre and its surroundings. Finally, the meaning of the term zero-phonon energy for E_0 should be clearer now: it is the lowest excitation energy to be supplied to the centre to bring it to the excited electronic state. Optically, this will certainly be possible if the axes of both parabolas merge, i.e. $Q_{g0} \approx Q_{e0}$ and $E_R \to 0$, or $S = 0$. We shall show shortly, however, that this is not the only configuration in which such a transition can occur. No phonons are released during such an optical excitation and hence the term zero-phonon energy for E_0.

4.5 The shape of absorption and emission spectra of a localized centre

By making use of Fig. 4.8(b) and employing qualitative arguments, we can now proceed to get an insight into the shape of optical absorption and emission (luminescence) spectra of a localized centre. At first glance, it may seem that both the absorption and emission are formed by narrow lines corresponding to the transitions $A \to B$ and $C \to D$. The truth is, however, somewhat more complicated. The 'harmonic oscillators' of the ground as well as the excited electronic states oscillate incessantly, and so both the initial and the final point of an optical transition must respect the spatial distribution of the oscillator probability density. Thus, for instance, although an absorption transition starting from point A will most likely end at point B, it may also—with just a slightly lower probability—terminate on a vibrational level (of the excited electronic state) with an energy higher or lower by $h\omega$. Transition to the vibrational levels $\pm 2h\omega$ will even have a somewhat lower probability, etc. Obviously, even the transition with zero-phonon energy $h\nu_a = E_0$ will occur with a certain probability (which, in Fig. 4.8(b), is just the transition $-2h\omega$), and it will happen all the more so the smaller the difference $Q_{e0} - Q_{g0}$ is.

The above considerations thus lead to the following picture of the absorption spectrum: it will consist of a series of narrow lines spaced equidistantly by the phonon energy $h\omega$, and their intensities will be modulated by a certain probability distribution. Since analogous reasoning may be applied to the emission spectrum as well, only reversing the transition directions, the shape of the emission spectrum is expected to be mirror-like with respect to the absorption spectrum, both spectra sharing a common line at the zero-phonon energy E_0. All this is shown in the right part of Fig. 4.8(c).

What has been said up to now can be, naturally, represented quantitatively by calculating the transition probability [8]; it is proportional to the square of the dipole moment matrix element. Assuming independence of the dipole moment operator of the configurational coordinate Q, the spectral shape will be determined—in accordance with the above reflections—by the overlap integral of the wavefunctions of the vibrating nuclei.

At absolute zero temperature, this yields the shape of the absorption spectrum to be given by the Poisson distribution

$$I(n) \approx e^{-S} S^n / n!, \quad n = 0, 1, 2, \ldots, \tag{4.15}$$

where n numbers the vibrational level of the excited electronic state (the up-shifted parabola); as $T \to 0\,\text{K}$, the transition begins from the ground vibrational level (point A) of the ground electronic state.

Figure 4.9 represents schematically several spectra of the type (4.15) for various values of the parameter S. One usually distinguishes three cases: weak ($S \leq 1$), medium ($S \approx 5$–10), and strong ($S \geq 20$) exciton–phonon coupling. It immediately follows from relation (4.15) that in the case of an extremely weak bond ($S = 0$), the spectrum will consist of a single line corresponding to the zero-phonon transition to the $n = 0$ state (the so-called *zero-phonon* or *no-phonon line*). With increasing values of S, gradually more and more lines with a higher n (the so-called *phonon replicas* or satellites) appear and the intensity of the no-phonon line $I(0) = I_{\text{NP}}$ decreases; the maximum of the spectral 'envelope' occurs roughly at $n = S$. The energy separation between the maxima of absorption and emission is called the *Stokes shift* and it is obviously given by $2Sh\omega$, as demonstrated by Fig. 4.8(c).

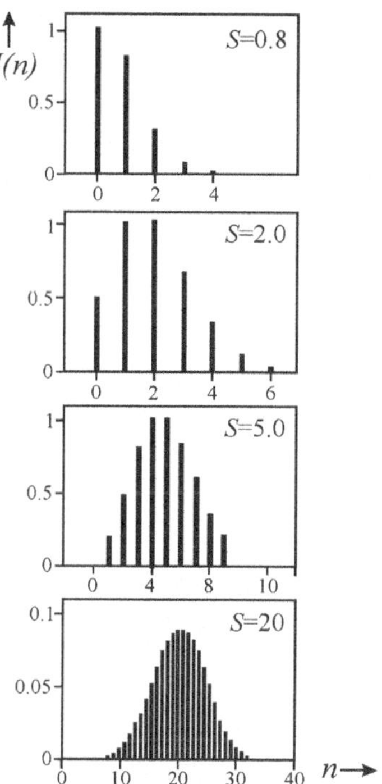

Fig. 4.9
Schematic shape of the (absorption) spectrum given by the Poisson distribution (4.15) for several values of the Huang–Rhys parameter S.

Figure 4.9 considers the individual lines to be infinitely narrow. In fact, the lines are already broadened at $T = 0$ K, as actually shown in Fig. 4.8(c). This happens as a result of the zero-point oscillations and other mechanisms not included in the Franck–Condon principle: homogeneous broadening due to the finite excited state lifetime and inhomogeneous broadening due to the more or less variable matrix parameters in the vicinity of any localized centre, i.e. because of possible variability in the values of energy $h\omega$. Therefore, as the parameter S increases the satellite phonon lines $n \geq 1$ gradually merge into a continuum, which is sometimes also called the *phonon wing*. Let I_{PW} denote the wing integral intensity, shown as the shaded area on the right in Fig. 4.8(c). The ratio of the no-phonon line intensity I_{NP} to the total intensity of the absorption (or emission) band ($I_{\text{NP}} + I_{\text{PW}}$) is called the *Debye–Waller factor*

$$u_{\text{DW}} = \frac{I_{\text{NP}}}{I_{\text{NP}} + I_{\text{PW}}} \leq 1. \tag{4.16}$$

The value of u_{DW} drops quickly with increasing S as one can infer immediately from Fig. 4.9. Furthermore, it also decreases considerably with increasing temperature (here we can refer to the complete analogy with the Debye–Waller factor known from X-ray diffraction on the vibrating lattice).

Before discussing typical experimental examples, we shall briefly summarize the typical spectral features for various strengths of the exciton–phonon coupling:

- Weak coupling: a distinct no-phonon line and a negligible phonon wing, zero Stokes shift and a large Debye–Waller factor $u_{DW} \approx 1$.
- Medium coupling: an inconspicuous no-phonon line, a number of phonon satellites (up to $n \approx 10$) in the phonon wing, a small factor $u_{DW} \leq 10^{-2}$.
- Strong coupling: the no-phonon line missing completely, phonon satellites merging into a broad phonon wing, the spectrum takes the form of a broad, structureless Gaussian band with a large Stokes shift (for $n \geq 10$, the Poisson distribution turns into the Gaussian form, see Fig. 4.9), factor $u_{DW} \approx 0$.

Examples of experimental spectra are given in Fig. 4.10. We shall mainly concentrate on emission spectra since often it is not quite straightforward to measure the absorption spectrum of a localized centre in semiconductors. (Either the absorption measurements are not sufficiently sensitive as compared to the luminescence measurement, as we have mentioned already, or the absorption spectrum overlaps with the onset of the matrix absorption edge itself, and so on. In this respect, the situation is much more favourable in wide-bandgap optical materials such as alkali halides, Al_2O_3, CaF_2, CaO.)

As an example of weak exciton–phonon coupling in Fig. 4.10(a), the low-temperature (4.2 K) luminescence spectrum of phosphorus impurity atoms embedded with a concentration of $10^{16}\,cm^{-3}$ into crystalline silicon is displayed [9]. This is a pure no-phonon, narrow line ($S = 0$, $u_{DW} \approx 1$). If we give a little more thought to this observation, it may seem strange that in a semiconductor with an indirect bandgap, such as silicon, an optical recombination act without phonon participation can be found; the quasi-momentum conservation law should not enable similar processes to occur. This is a justified objection, we have not been considering a situation like this so far; in both the present section and Section 4.4, we have tacitly assumed that the semiconductor—matrix—had a direct bandgap and from this point of view, phonons were not indispensable to ensure optical transitions. Now, we really should need them; however, it is the relatively heavy phosphorus doping that breaks the perfect translational symmetry of the lattice and partially removes the ban on no-phonon transitions. We can also immediately ask the question why is it just silicon where the experiment points to a weak exciton–phonon interaction. The answer is simple. The only mechanism of this interaction in non-polar, centrosymmetric silicon is the deformation potential and the two most efficient mechanisms of this interaction, the Fröhlich and piezoelectric mechanisms, are thus missing.

An example of medium exciton–phonon coupling is displayed in Fig. 4.10(b): the low-temperature (7 K) luminescence spectrum of iodine ions I^- in a crystalline AgBr matrix [10]. We can clearly see a number of narrow phonon lines for $n = 0$ to $n = 10$, superimposed on a rather broad phonon wing.

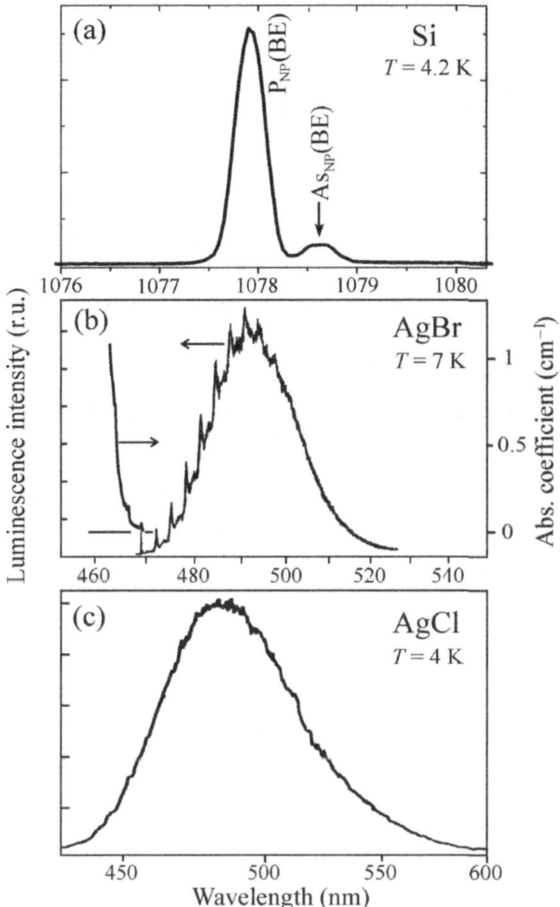

Fig. 4.10
Emission spectra of localized optical centres in semiconductors. (a) Weak exciton–phonon coupling: The (no-phonon) luminescence line of phosphorus atoms in crystalline silicon. We can also see a weak no-phonon line associated with arsenic atoms present at a lower concentration. After Pelant *et al.* [9]. (b) Medium exciton–phonon coupling: $AgBr/I^-$ luminescence. To suppress coupling between the iodine-related I^- centre and the lattice vibrations (in order to emphasize the phonon structure), an external hydrostatic pressure was used here. After Wassmuth *et al.* [10]. The left part shows the corresponding absorption spectrum after Kanzaki and Sakuragi [11]. (c) Strong exciton–phonon coupling: luminescence of a self-trapped exciton in AgCl. After Pelant and Hála [13]. Note the different wavelength scales.

Therefore, we can estimate the values of the interaction parameters: $S \approx 8$ and $u_{DW} < 10^{-2}$ (because, although the no-phonon line is present, its intensity is minute). Also here, we notice the presence of the no-phonon line in the spectrum even though this is again the case of a semiconductor with an indirect bandgap. The reason for this is analogous to that for doped silicon. The left part of Fig. 4.10(b) also shows the onset of the absorption spectrum as measured in AgBr samples of high optical quality [11]. Comparison of the spectra confirms the correct identification of the individual lines—we can clearly see the mirror-symmetry of the spectra together with the no-phonon lines overlapping in absorption and emission. Let us again note that the presence of no-phonon lines at a localized centre in an indirect semiconductor is not commonplace. For example, $ZnSiP_2$ (with an unknown impurity as the centre) is usually cited as an indirect semiconductor where the no-phonon spectral line is missing and there is an energy separation of $2\hbar\omega$ between the onsets of absorption and emission [12].

The low-temperature (4 K) luminescence spectrum of AgCl serves as an example of strong exciton–phonon coupling in Fig. 4.10(c): a smooth, broad curve without any phonon structure [13]. Here, the localized centre is represented by the so-called self-trapped hole in the vicinity of an Ag^+ ion, which attracts a photoelectron via Coulomb forces, thus producing the so-called self-trapped or auto-localized exciton. Its radiative annihilation then gives rise to a broadband luminescence spectrum. We shall consider these excitons in rather more detail in Subsection 7.2.4. This is an untypical *intrinsic localized centre*.

There is not much difference between the emission spectra of AgBr and AgCl apart from the total absence of both the no-phonon line and fine phonon structure in AgCl. In fact, the interaction of excitons with the lattice vibrations is very strong in both cases due to the efficient Fröhlich mechanism.

4.6 Thermal quenching of luminescence

Let us return to Fig. 4.8(b) and explore point X in more detail. At a temperature close to absolute zero, the excited centre is located near point C (and thermal vibrations are negligible). With increasing temperature T, however, the amplitude of vibrations of the centre around its position Q_{e0} increases, and, consequently, so does the chance of the system to reach point X. In solid-state physics terminology, this means overcoming the potential barrier of height E_A. The probability of such a thermally activated process is governed by the well-known relation

$$p = p_0 \, e^{-E_A/k_B T}, \tag{4.17}$$

where $p_0(s^{-1})$ is the so-called frequency factor, acquiring values of the order of the lattice vibration frequency. After reaching point X, the excited centre now has the option of a non-radiative return to the ground electronic state: at point X, it can pass over the curve of the ground state potential energy and, following the path $X \to A$ while emitting phonons, it can reach equilibrium at point A. Thus, all the energy gained via absorption of the excitation photon $h\nu_a$ is converted into the matrix vibrational energy, i.e. into heat.

Realizing that the probability (4.17) is thus nothing but the probability of a non-radiative transition $p = \tau_{nr}^{-1}$ and assuming that the probability τ_r^{-1} of a radiative transition $C \to D$ is independent of temperature, the luminescence quantum efficiency (3-2) can be expressed as

$$\eta = \frac{1}{1 + \xi \exp(-E_A/k_B T)}, \tag{4.18}$$

where $\xi = \tau_r p_0$. Therefore, η drops with increasing temperature and this effect is called *thermal quenching* of the luminescence.

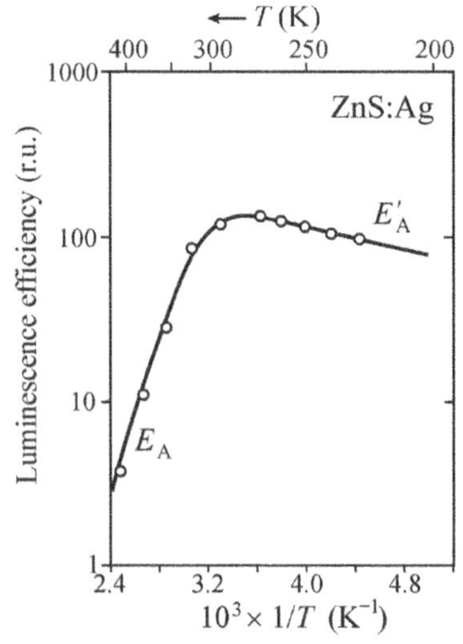

Fig. 4.11
Luminescence efficiency of ZnS:0.015% Ag as a function of temperature. Points—experiment; solid line—calculation according to (4.19) with $E'_A = 3.9$meV, $E_A = 132$meV. After Williams [14].

An example of the temperature dependence of the luminescence efficiency, established experimentally in a classical semiconductor phosphor ZnS doped with Ag (which is well-known from the Second World War era when it was used for radar screens), is shown in Fig. 4.11. This is an often-used type of graph where we plot η as a function of reciprocal temperature, $1/T$; this is because for sufficiently large T we can usually approximate (4.18) by $\eta \sim \xi^{-1} \exp(E_A/k_B T)$, and from the slope of such a logarithmic-scale plot we can immediately infer the activation energy E_A. At the same time, it can be seen in Fig. 4.11 that the thermal quenching only occurs here for temperatures $T \geq 310$ K (i.e. for $1000/T \leq 3.2$ K^{-1})—hence, this phosphor radiates most at room temperature which is highly unusual and which explains why the material is so attractive as regards its applications. It is thus obvious that relation (4.18) does not fully grasp the overall form of the experimental dependence since, when temperature increases from about 200 K to approximately 280 K, the luminescence efficiency does not drop but grows instead. A similar increase is generally observed quite often and the simple configurational coordinate model, as discussed above, can no longer encompass this fact. In order to describe the experimentally found temperature dependence η, eqn (4.18) is sometimes modified by adding another factor that describes a weakly thermally activated process resulting in an increase of the radiative transition probability. For example, the experimental points in Fig. 4.11 can be very well fitted by the equation

$$\eta \sim \frac{T\,\mathrm{e}^{-E'_{\mathrm{A}}/k_{\mathrm{B}}T}}{1+\xi\,\mathrm{e}^{-E_{\mathrm{A}}/k_{\mathrm{B}}T}} \tag{4.19}$$

(see the curve in Fig. 4.11 with $E'_{\mathrm{A}} \approx 3.9$ meV and $E_{\mathrm{A}} \approx 132$ meV) [14]. However, in this case, it is just a phenomenological model while currently there are already more exact models based on microscopic concepts. Let us emphasize, however, that any such increase in luminescence intensity with increasing temperature is only temporary and, ultimately, thermal quenching always prevails and luminescence thus fades away completely.

And finally a note regarding the relative position of points B and X in Fig. 4.8(b). Our analysis of thermal quenching was based on the assumption that the intersection point X is at a higher energy than point B. If both points are located on the same level or if X is even lower than B then there will be a steep drop in luminescence intensity since a non-radiative transfer of the electronic excitation energy to the matrix will prevail already at very low temperatures.

4.7 Problems

4/1: Show that the vibrational energy of a one-dimensional 'crystal' with N atoms is at high temperatures ($h\omega_{\mathbf{k}} \ll k_{\mathrm{B}}T$) equal to $\bar{E} = N\,k_{\mathrm{B}}\,T$. (Hint: Transform relation (4.4) into the form

$$\frac{\bar{E}}{k_{\mathrm{B}}T} = \frac{N\hbar\omega_{\mathbf{k}}}{k_{\mathrm{B}}T}\left[\exp\left(\frac{\hbar\omega_{\mathbf{k}}}{k_{\mathrm{B}}T}\right)-1\right]^{-1}.)$$

4/2: Show that the total energy of a harmonic oscillator in quantum physics is equal to $(h\omega/2)\coth(h\omega/2k_{\mathrm{B}}T)$.

4/3: Prove that in the harmonic approximation and assuming $f_{\mathrm{e}} = f_{\mathrm{g}}$, the relaxation energies E_{R} in Fig. 4.8(b) in the ground and excited electronic states are the same. Show that the activation energy E_{A} is given by eqn (4.13).

4/4: The lattice vibrations in crystalline silver bromide show a number of peculiarities. This is caused by the deformability of the Ag^{+} ion and the considerable fraction of both ionic and covalent bonds in the crystal.

These effects lead to very dissimilar force constants for different vibration types, to a low energy of TO-phonons at the L point (Fig. 4.6), and to the inverted character of TO- and TA-phonons near the L point. Discuss these phenomena according to Dorner *et al.* [15].

4/5: Confirm that at $T = 0\,\mathrm{K}$ the Debye–Waller factor u_{DW} and the Huang–Rhys factor S are related by $u_{\mathrm{DW}} = \exp(-S)$. (Hint: Use eqns (4.15) and (4.16).)

References

1. Dekker, A. J. (1963). *Solid State Physics*. Chap. 2. Prentice-Hall, Englewood Cliffs, N.J.
2. Peyghambarian, N., Koch, S. W., and Mysyrowicz, A. (1993) *Introduction to Semiconductor Optics*. Chap. 4. Prentice Hall, Englewood Cliffs, N.J.
3. Dargys, A. and Kundrotas, J. (1994). *Handbook on Physical Properties of Ge, Si, GaAs and InP*. Science and Encyclopedia Publishers, Vilnius.
4. Hayes, W. and Stoneham, A. M. (1985). *Defects and Defect Processes in Non-metallic Solids*. Chap. 1. John Wiley, New York.
5. Yu, P. Y. and Cardona, M. (1996). *Fundamentals of Semiconductors*. Springer, Berlin.
6. Song, K. S. and Williams, R. T. (1996). *Self-Trapped Excitons* (Springer series in solid state sciences Vol. 105). Springer, Berlin.
7. Böer, K. W. (1990). *Survey of Semiconductor Physics*. Van Nostrand Reinhold, New York.
8. Vij, D. R. (ed.) (1998). *Luminescence of solids*. Plenum Press, New York; Yamamoto, H. (1999). *Fundamentals of Luminescence*. In *Phosphor Handbook* (ed. S. Shionoya and W. M. Yen), p. 35. CRC Press, Boca Raton.
9. Pelant, I., Hála, J., Ambrož, M., Vácha, M., Valenta, J., Adamec, F., Kohlová, V., and Matoušková, J. (1990). *Impurity assessment in Si wafers by photoluminescence method V*. Research report for Tesla Rožnov. Charles University in Prague, Faculty of Mathematics & Physics, Prague.
10. Wassmuth, W., Stolz, H., and von der Osten, W. (1990). *J. Phys. C: Cond. Matter*, **2**, 919.
11. Kanzaki, H. and Sakuragi, S. (1969). *J. Phys. Soc. Japan*, **27**, 109.
12. Shah, J. (1972). *Phys. Rev. B*, **6**, 4592.
13. Pelant, I. and Hála, J. (1991). *Solid State Comm.*, **78**, 141.
14. Williams, F. E. (1948). *The mechanism of rate processes in the luminescence of solids*. In *Preparation and Characteristics of Solid Luminescent Materials*. Cornell Symposium 1946, p. 337. John Wiley, New York; Chapman & Hall, London.
15. Dorner, B., von der Osten, W., and Bührer, W. (1976). *J. Phys. C: Solid State Phys.*, **9**, 723.

Channels of radiative recombination in semiconductors

Various radiative recombination processes have already been mentioned several times in the previous chapters. These processes were, however, not discussed from the point of view of the underlying physics; more emphasis was put on the introduction (or reminder) of the basic concepts and terminology, which will be indispensable later in the book. The configurational coordinate model, which was described in the previous chapter, for example, is broadly applicable. In this chapter, we are going to show in which particular aspects of luminescence in semiconductors the application of this model is completely straightforward, when, on the other hand, it does not make any sense at all, or when it can be used only formally and with prudence.

Now that everything is ready we can move on to giving systematic explanations of known channels of radiative recombination in semiconductors. First, we will list an overview of these channels. Secondly, we will focus in more detail on the spectral shape of the emission line of recombining free electron–hole pairs, and of related recombinations of a free electron with a neutral acceptor or a free hole with a neutral donor. Then we will talk about the shape of the emission spectra of recombining donor–acceptor pairs and finally we will mention the particularity of exciting luminescence via a two-photon absorption process.

5.1 Overview of luminescence processes in crystalline semiconductors

Let us start with the widely accepted classification of luminescence processes according to the intensity of excitation. The *low-fluence* or *weak excitation* processes (i.e. taking place when a gas-discharge lamp, an incandescent lamp or a continuous-wave gas laser with the output power of the order of 0.01–10 W/cm^2 are used for excitation) comprise:

- radiative recombination of free excitons (often denoted as FE for a free exciton, or X);
- radiative recombination of free excitons with simultaneous emission of an LO phonon (FE–LO, X–LO);

- radiative recombination of a bound exciton (BE), which can be further subdivided into
  - radiative decay of an exciton bound to a neutral donor (D^0, X) or (D^0–X);
  - radiative decay of an exciton bound to a neutral acceptor (A^0, X) or (A^0–X);
  - radiative decay of an exciton bound to an ionized donor (D^+, X) or (D^+–X);
  - radiative decay of an exciton bound to an ionized acceptor (A^-, X) or (A^-–X);
  - radiative decay of an exciton bound to an isoelectronic impurity;
- radiative recombination of donor–acceptor pairs (D^0–A^0);
- radiative recombination of a free hole with a neutral donor (h–D^0) or a free electron with a neutral acceptor (e–A^0);
- radiative recombination of free electron–hole pairs (e–h), which can also be referred to as band-to-band recombination.

At *high-fluence (strong) excitation*, usually by means of a pulsed laser with excitation intensity of the order of 1 kW–10 MW/cm^2, additional processes occur:

- radiative decay of an excitonic molecule (EM or XX);
- luminescence coming from inelastic collisions of excitons (X–X collisions);
- luminescence of electron–hole liquid (EHL);[1]
- luminescence of electron–hole plasma (EHP);
- Bose–Einstein condensation of excitons or excitonic molecules and luminescence coming from the condensate.

Obviously, two or more types of these processes can occur simultaneously in a material. In addition, most of these processes can also be observed in low-dimensional semiconductor structures, although the emission of light from low-dimensional structures is in some aspects specific. The emission of light from amorphous semiconductors, on the other hand, is governed by significantly different processes.

Later in this chapter, the most important spectral features connected with luminescence processes occurring at low-fluence excitation levels will be treated, deliberatly neglecting any luminescence manifestation of excitons, to which Chapter 7 is devoted. The high-excitation luminescence spectral features will be discussed in Chapter 8 and the luminescence of amorphous and low-dimensional semiconductors will be described in Chapters 9 and 12.

5.2 Recombination of free electron–hole pairs

Even though, in reality, this process is not encountered very often, the explanation of the shape of a luminescence emission spectrum will begin here; the

[1] This process is sometimes not considered typical for high-level excitations because it can often be observed even when a continuous source of radiation is used.

reasons for choosing this particular process will become evident later. (We would like to note here that this process was called 'bimolecular' in Chapter 3, where it was treated from the point of view of its kinetics.)

5.2.1 Direct bandgap

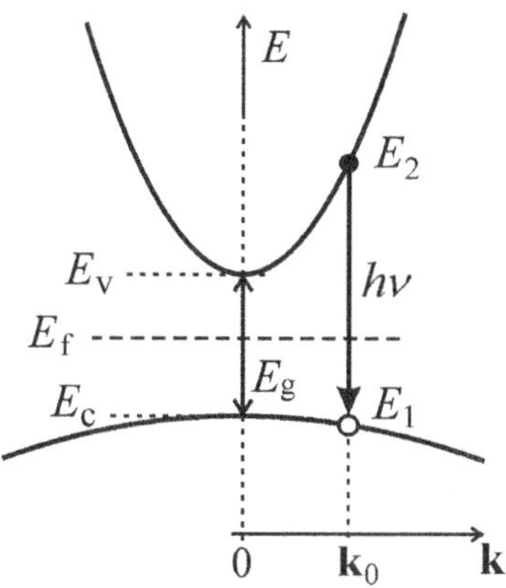

Fig. 5.1
Free electron–hole recombination in a direct bandgap semiconductor. E_f stands for the Fermi level.

The process is schematically shown in Figure 5.1. An electron with energy E_2 situated close to the minimum of the conduction band spontaneously recombines with a hole situated at the top of the valence band with energy E_1 and wavector $\mathbf{k}_0$ via a direct vertical transition. The physical quantity of interest here is the probability density $I_{sp}(h\nu)$ of the spontaneous emission of a photon with energy $h\nu = E_2–E_1$. This is obviously influenced by three factors:

1. the probability of a suitable configuration of occupancy of conduction and valence band states $f_e(\nu)$;
2. the quantum-mechanical probability of this transition in the form of the square of the absolute value of the matrix element of the transition (in the dipole approximation) $|M|^2$;
3. the joint density of electron and hole states $\rho(\nu)$ in the corresponding bands.

Thus

$$I_{sp}(h\nu) \approx |M|^2\, f_e(\nu)\rho(\nu) \approx \frac{1}{\tau_r} f_e(\nu)\,\rho(\nu). \tag{5.1}$$

Ad 1. For the radiative recombination to take place, the corresponding energy state with energy E_2 in the conduction band has to be occupied by an electron while the state with energy E_1 in the valence band has to be empty (i.e. occupied by a hole), as is indicated in Fig. 5.1. Consequently, $f_e(\nu)$ is given as the product of the occupancy factors f_c and f_v:

$$f_e(\nu) = f_c\, f_v = f(E_2)[1 - f(E_1)], \tag{5.2}$$

where $f(E) = \{\exp[(E - E_f)/k_B T] + 1\}^{-1}$ is the Fermi–Dirac distribution function. Let us assume that the semiconductor is intrinsic, has a sufficiently wide bandgap E_g and the process takes place at low temperature. At low-fluence excitation, the semiconductor can be treated as if it is still in thermal equilibrium, i.e. the Fermi level stays localized in the middle of the bandgap at low temperature (Fig. 5.1). As long as the condition of low temperature and wide bandgap $k_B T \ll (E_g/2) \approx E_2–E_f \approx E_f–E_1$ is fulfilled, one can write

$$f(E_2) = \frac{1}{e^{(E_2-E_f)/k_B T} + 1} \approx \exp\left[-(E_2 - E_f)/k_B T\right],$$

$$[1 - f(E_1)] = 1 - \frac{1}{e^{(E_1-E_f)/k_B T} + 1} \approx \frac{e^{(E_1-E_f)/k_B T}}{e^{(E_1-E_f)/k_B T} + 1}$$

$$\approx \exp[-(E_f - E_1)/k_B T],$$

which transforms (5.2) into the simple form

$$f_e(\nu) \approx \exp\left(-h\nu/k_B T\right). \tag{5.3}$$

Ad 2. The matrix element of the optical transition M can be considered to be constant to a very good approximation over a wide range of wavevectors close to the extremes of the bands. In relation to luminescence, the radiative lifetime τ_r will be treated as inversely proportional to the square of the matrix element $\tau_r \sim |M|^{-2}$. (This simple relation is strictly valid only for atomic transitions. When it comes to semiconductors, the application of this relation is a bit less straightforward; the parameters of the particular band structure as well as the bandgap width come into play. A detailed theoretical discussion can be found in [1]. This approximation, however, is fully sufficient within the framework of this textbook.)

Ad 3. The density of states $\rho(\nu)$ is pertinent to the difference energy band $E_c(\mathbf{k})$–$E_v(\mathbf{k})$, which is obtained—mathematically speaking—by subtracting $E_v(\mathbf{k})$ from $E_c(\mathbf{k})$. Since these transitions are vertical the subtraction makes sense in this context. The densities of states in the individual bands, i.e. in the conduction band $\rho_c(E)$ and the valence band $\rho_v(E)$, are well known (see, e.g., [2, 3]):

$$\rho_c(E) = \frac{(2m_e)^{3/2}}{2\pi^2\hbar^3}(E - E_c)^{1/2}, \; E \geq E_c, \tag{5.4a}$$

$$\rho_v(E) = \frac{(2m_h)^{3/2}}{2\pi^2\hbar^3}(E_v - E)^{1/2}, \; E \leq E_v, \tag{5.4b}$$

where m_e and m_h stand for the effective masses of an electron and a hole, respectively, connected with the curvatures of the corresponding bands close to the extremes. Thus, one can easily get (using the incremental relation $\rho_c(E_2)\mathrm{d}E_2 = \rho(\nu)\mathrm{d}\nu$)

$$\rho(\nu) = \frac{(2m_r)^{3/2}}{\pi\hbar^2}(h\nu - E_g)^{1/2}; \quad h\nu \geq E_g. \tag{5.5}$$

Relation (5.5) includes the reduced mass of the electron and hole given by the expression

$$\frac{1}{m_r} = \frac{1}{m_e} + \frac{1}{m_h}. \tag{5.6}$$

To derive $\rho(\nu)$ by making use of the aforementioned incremental relation, the following expressions, which stem directly from Fig. 5.1 (see Problem 5/1), are employed:

$$E_2 = E_c + \frac{m_r}{m_e}(h\nu - E_g), \tag{5.7}$$

$$E_1 = E_v - \frac{m_r}{m_h}(h\nu - E_g). \tag{5.8}$$

One more thing worth mentioning is that the density $\rho(\nu)$ is given per unit frequency, not per unit energy, which causes different fractional prefactors to appear in (5.4) and (5.5).

Now, it is possible to get the desired formula for the spectral shape of luminescence coming from the radiative recombination of free electron–hole pairs in a semiconductor with direct bandgap by combining (5.3), (5.5) and (5.1):

$$I_{sp}(h\nu) \approx D_0(h\nu - E_g)^{1/2} \exp\left[-(h\nu - E_g)/k_B T\right], \qquad (5.9a)$$

where $D_0 = [(2m_r)^{2/3}/\pi\hbar^2\tau_r]\exp(-E_g/k_B T)$ is a parameter independent of the photon energy $h\nu$. Disregarding this parameter, $I_{sp}(h\nu)$ is given by the product of a factor characterizing the joint density of states $(h\nu–E_g)^{1/2}$ with a term describing basically the occupancy of these states, i.e. $\exp[-(h\nu - E_g)/k_B T]$. The shape of the luminescence spectrum (5.9a) is sometimes also called the *Maxwell–Boltzmann distribution*. It is interesting to note that the formula describing the radiative band-to-band recombination contains solely the reduced electron–hole mass m_r, not the individual effective masses of an electron and a hole m_e, m_h. In other words, it is impossible to determine the values of the effective masses of the carriers using only luminescence (or absorption) measurements.

The spectral shape $I_{sp}(h\nu)$ is shown in Fig. 5.2(a). The curve is asymmetric with a tail towards the high-energy photon side. The asymmetry stems from the kinetic energy of recombining free electrons and holes, which can be described by an exponential distribution function.[2] The full-width

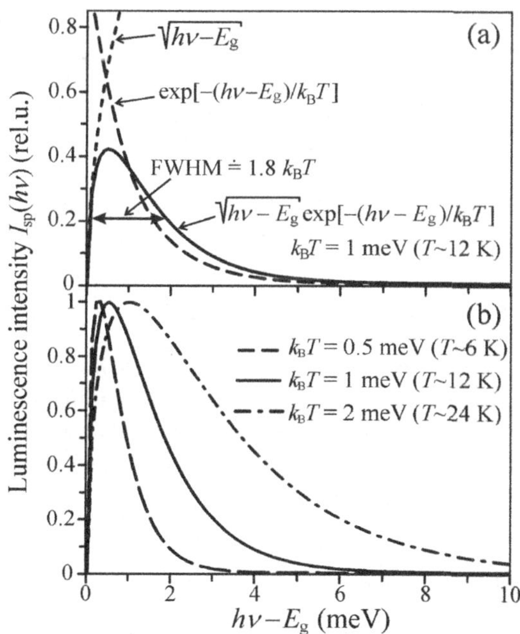

Fig. 5.2
(a) Maxwell–Boltzmann lineshape $I_{sp}(h\nu) \approx \sqrt{h\nu - E_g}\exp[-(h\nu - E_g)/k_B T]$ at $T \approx 12$ K. (b) Asymmetric linewidth broadening with increasing temperature. Curves are normalized.

[2] It may be of interest to recall that the formula for the number of molecules of an ideal gas Δn (out of the total number n), whose kinetic energy falls in the interval $(E_k, E_k + \Delta E_k)$, reads

$$\frac{\Delta n}{n\,\Delta E_k} = \frac{2}{\sqrt{\pi}\,(k_B T)^{3/2}}\sqrt{E_k}\exp(-E_k/k_B T).$$

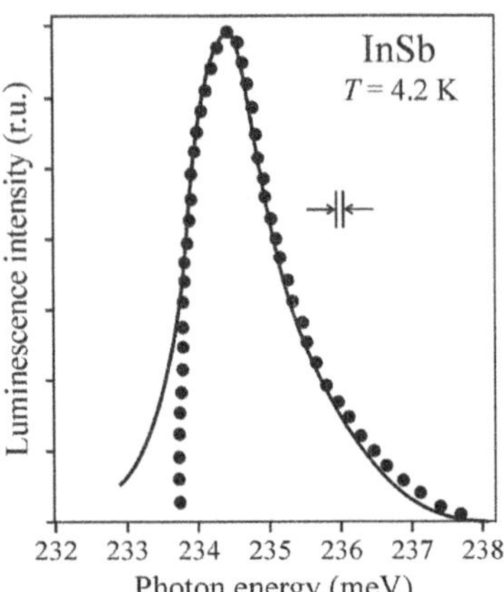

Fig. 5.3
Emission spectrum of the recombination of free electron–hole pairs in InSb at $T = 4.2$ K. Solid curve—experiment; circles—theoretical lineshape given by (5.9). After Mooradian and Fan [4].

at half maximum (FWHM) of this curve can then be shown to be directly proportional to temperature

$$\text{FWHM} \sim 1.8\, k_B T.$$

Thus, the linewidth broadens with rising temperature, as can be seen in Fig. 5.2(b). On the whole, however, the line is very narrow: temperatures T between 4 and 100 K correspond to FWHM $\approx$ 1–20 meV, which amount to only tenths of nanometres, or nanometres at most, in the visible spectral region. The physical reason underlying the linewidth broadening is obvious: the Boltzmann tail of the Fermi–Dirac distribution broadens with increasing temperature, or in other words, at higher temperatures photoexcited electrons and holes occupy states further from the extremes of the bands. Thus, photons with higher energy are generated by these higher-energy electron–hole pairs, while the low-energy onset of the spectral line is determined by E_g and its position remains fixed. (The possible shift of E_g with changing temperature leads to a shift of the emission line as a whole.)

Figure 5.3 demonstrates the experimentally observed luminescence of electron–hole pairs in InSb at 4.2 K [4]. The dotted curve indicates the theoretical lineshape given by (5.9a) and agrees very well with experiment except for the low-energy side. This slight difference can be due to several factors, including the influence of the slit of the monochromator as described in Section 2.8, which smears the sharp onset of luminescence. As has already been mentioned, the radiative recombination of free electrons and holes is a scarcely encountered process. Indium antimonide InSb is the exception that proves the rule. Its very narrow bandgap ($E_g \sim 0.23$ eV at $T = 4.2$ K) and consequently also the low exciton binding energy ensure that even at low temperatures excitons are thermally dissociated. In nearly all other semiconductive compounds, low-temperature luminescence is dominated by excitonic effects, which means that free electrons and holes and their luminescence are practically absent.

In the literature, formula (5.9a) can sometimes be written in the slightly modified form

$$I_{sp}(h\nu) \approx \nu^2 (h\nu - E_g)^{1/2} \exp\left[-(h\nu - E_g)/k_B T\right]. \tag{5.9b}$$

The factor ν^2 comes from the *photon density of states* in three dimensions, which equals $4\nu^2/c^3$ in free space. Taking into account the already mentioned narrow spectral width and high optical frequencies (FWHM$/h\nu$) ≤ 0.01, however, makes it possible to consider ν^2 a constant, and thus to neglect the influence of this term on the spectral lineshape.

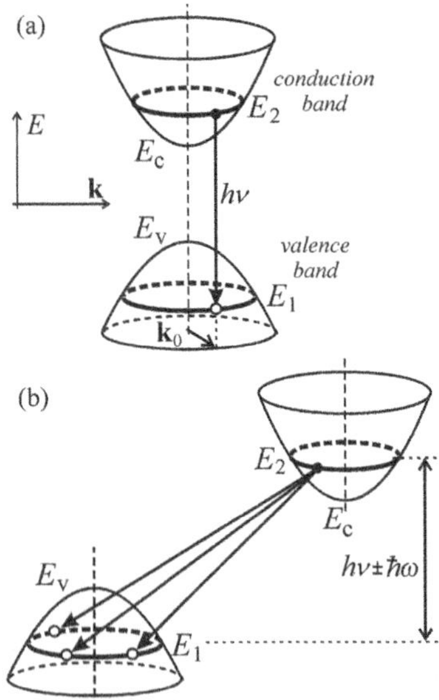

Fig. 5.4
Schematic drawing of radiative recombination of electron–hole pairs in the energy band struture diagram $E(\mathbf{k})$ for a semiconductor with (a) a direct and (b) an indirect bandgap. $h\nu$ stands for the energy of a photon, $\hbar\omega$ for the energy of a phonon.

5.2.2 Indirect bandgap

A fundamental difference between the recombination of electron-hole pairs in a semiconductor with a direct and indirect bandgap is shown in Fig. 5.4.

This formula implies that the Maxwell–Boltzmann distribution is a fully valid name for (5.9a) and, moreover, the comparison of these two formulas suggests that at low concentrations of free quasiparticles ($= 10^{17}$ cm^{-3}, i.e. for weak optical excitation) one can speak of a 'gas' of free electrons, holes or excitons in semiconductors.

In a direct-bandgap semiconductor, the recombination takes place between electrons and holes with the same wavevector $\mathbf{k}_0$, whereas in an indirect-bandgap semiconductor, an excited electron with energy E_2 can recombine with any hole with energy E_1 and varying values of $\mathbf{k}$, as long as the energy-conservation rule $E_2–E_1 = h\nu \pm \hbar\omega$ is fulfilled. The energy $\hbar\omega$ then denotes the energy of a phonon, which takes part in the process in order to ensure the conservation of quasi-momentum: $k_{\hbar\omega} = |\mathbf{k}(E_2)–\mathbf{k}(E_1)|$. Considering the fact that phonon dispersion relations are defined for any $\mathbf{k}$ from the first Brillouin zone (see, e.g., Fig. 4.3), it is obvious that an appropriate phonon can always be found, unless the transitions are forbidden due to the symmetry of electron and/or phonon states. For the sake of simplicity, only low-temperature cases will be treated hereafter. At low temperatures, only a small number of phonons are present in the semiconductor and the recombination process is accompanied by the emission of a photon and the *emission of a phonon*, i.e. the electron excitation energy is released both in the form of luminescence and as an additional vibration energy of the lattice: $E_2–E_1 = h\nu + \hbar\omega$.

The calculation of the probability density of spontaneous emission $I_{\mathrm{sp}}^{\mathrm{in}}$ of a luminescence photon with energy $h\nu = (E_2–E_1)–\hbar\omega$ is then carried out by summing all of the contributions over the curves marked in bold in Fig. 5.4(b), subsequently 'infinitesimally shifting' the difference $(E_2–E_1)$ on the energy axis, repeating the summation, etc. Factors which play important roles in the calculation are thus, naturally, not only the densities of states in the bands (5.4), but also the occupancy factors f_{c}, f_{v}.[3] In analogy with (5.1), the described procedure leads to the integration over all pairs of states in the conduction and valence bands, whose energies relative to the corresponding band edges will be denoted as $E_{\mathrm{e}} = E_2–E_{\mathrm{c}}$ for the electron and $E_{\mathrm{h}} = E_{\mathrm{v}}–E_1$ for the hole, respectively (see Fig. 5.5(a)). The mathematical formulation of the described procedure then reads

$$\begin{aligned} I_{\mathrm{sp}}^{\mathrm{in}}(h\nu) &\approx |M|^2 \int_0^{h\nu+\hbar\omega-E_{\mathrm{g}}} \rho_{\mathrm{c}}(E_{\mathrm{e}})\, f_{\mathrm{c}}(E_{\mathrm{e}})\, \rho_{\mathrm{v}}(E_{\mathrm{h}})\, f_{\mathrm{v}}(E_{\mathrm{h}})\mathrm{d}E_{\mathrm{e}} \\ &= |M|^2 \int_0^{\overline{h\nu}} \rho_{\mathrm{c}}(E_{\mathrm{e}})\, f_{\mathrm{c}}(E_{\mathrm{e}})\, \rho_{\mathrm{v}}(h\nu + \hbar\omega - E_{\mathrm{g}} - E_{\mathrm{e}}) \\ &\qquad f_{\mathrm{v}}(h\nu + \hbar\omega - E_{\mathrm{g}} - E_{\mathrm{e}})\mathrm{d}E_{\mathrm{e}} \\ &= |M|^2 \int_0^{\overline{h\nu}} \sqrt{E_{\mathrm{e}}}\, f_{\mathrm{c}}(E_{\mathrm{e}})\sqrt{\overline{h\nu} - E_{\mathrm{e}}}\, f_{\mathrm{v}}(\overline{h\nu} - E_{\mathrm{e}})\mathrm{d}E_{\mathrm{e}}, \end{aligned} \tag{5.10}$$

where, as can be seen from Fig. 5.5(a), $h\nu = E_{\mathrm{g}} + E_{\mathrm{e}} + E_{\mathrm{h}}–\hbar\omega$ or $E_{\mathrm{h}} = h\nu + \hbar\omega–E_{\mathrm{g}}–E_{\mathrm{e}}$. The upper integration limit $\overline{h\nu} = h\nu + \hbar\omega–E_{\mathrm{g}}$ reflects the above-mentioned summation of all contributions to the flux of emitted photons $h\nu$.

[3] The joint density of states (5.5) cannot be applied, since it has a reasonable meaning only for vertical transitions.

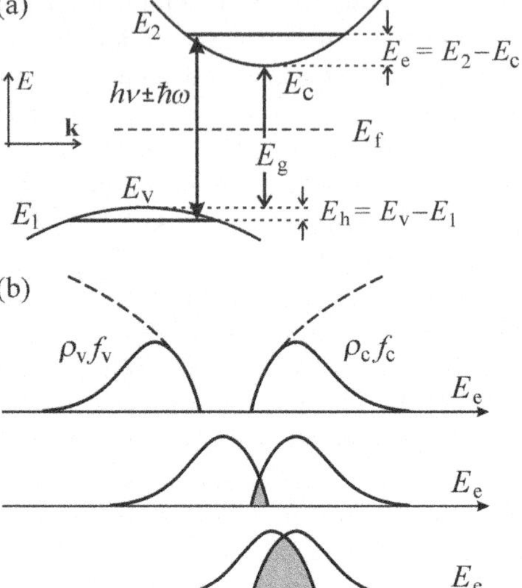

Fig. 5.5
Illustrating the calculation of the spectral lineshape of luminescence coming from the recombination of free electron–hole pairs in an indirect-bandgap semiconductor; (a) detail of band structure; (b) convolution of occupied electron and hole states according to (5.10).

On closer inspection, one can easily recognize that what (5.10) represents is the convolution of states occupied by electrons in the conduction band with the states occupied by holes in the valence band. (Let the reader compare formula (5.10) with eqn (A-5) in Appendix A.) The convolution-based calculation (5.10) is schematically shown in Fig. 5.5(b). The matrix element $|M|^2$ this time contains both vertical transitions (via virtual states with higher energies of electrons and holes at the same quasi-momentum) and also electron–phonon scattering ensuring the indispensable alteration of quasi-momentum in the recombination event. If both the photon and phonon parts of the matrix element are symmetry-allowed, this matrix element can be in a very good approximation treated as independent of both E_e and E_h and its influence on the emission lineshape can be neglected.

The calculation (5.10) can be carried out analytically without being too complicated, although one might not think so at first sight. First, we can realize that the product $f_c(E_e)\, f_v(\overline{h\nu} - E_e)$ under the integration sign is completely analogous to (5.2), but with $h\nu$ being replaced with $(h\nu + \hbar\omega)$. Similarly to (5.3) the term thus can be approximated

$$f_c(E_e) f_v(\overline{h\nu} - E_e) \approx \exp\left[-(h\nu + \hbar\omega)/k_B T\right]. \qquad (5.11)$$

Since the product $f_c f_v$ is thus independent of E_e, it can be placed outside the integration sign in (5.10). The formula then reads

$$I_{sp}^{in}(h\nu) \approx e^{-(h\nu+\hbar\omega)/k_B T} \int_0^{\overline{h\nu}} \sqrt{E_e}\sqrt{\overline{h\nu} - E_e}\, dE_e, \qquad (5.12)$$

where the definite integral is of the type $\int_0^a \sqrt{x}\sqrt{(a-x)}\, dx \approx a^2$ (Problem 5/4). The resulting formula for the spectral lineshape of luminescence due to recombination of free electron–hole pairs in a semiconductor with an indirect bandgap thus acquires the form

$$I_{\mathrm{sp}}^{\mathrm{in}}(h\nu) \approx \left[h\nu - (E_{\mathrm{g}} - \hbar\omega)\right]^2 \exp\left[-(h\nu - (E_{\mathrm{g}} - \hbar\omega))/k_{\mathrm{B}}T\right]. \quad (5.13)$$

Finally, this formula is not very different from the analogous relation for a direct-bandgap semiconductor (5.9a). The first difference is represented by the red-shift of the low-energy side of the spectrum by $\hbar\omega$, the second then consists of the replacement of the square-root dependence of the pre-exponential factor in (5.9a) by a quadratic dependence in (5.13). This change in the pre-exponential factor, however, does not influence the lineshape radically; it just slightly modifies the low-energy onset of the spectrum (a precise experimental determination of which is, moreover, problematic due to spectral broadening by the slit of the dispersion device as well as by other mechanisms), while the shape of the high-energy tail is most significantly influenced by the exponential factor, which is of identical form in both (5.9) and (5.13). This similarity is documented in Fig. 5.6.

Now, it is very important to note that this luminescence process scarcely occurs in reality. The fact that the excited electrons and holes in an indirect-bandgap semiconductor live much longer than in direct-bandgap materials causes nearly all electrons and holes at low temperatures to be bound in excitons. The emission spectrum then features only the luminescence line due to the radiative recombination of excitons. Emission coming from the recombination of free electron–hole pairs can only be observed in pure crystals at slightly elevated temperatures (at about 100 K), when some of the excitons are already ionized and when, simultaneously, thermal luminescence quenching is not yet sufficiently efficient. The intensity of such luminescence is, however, very low, since the matrix element in (5.10) is of second-order correction in perturbation theory.

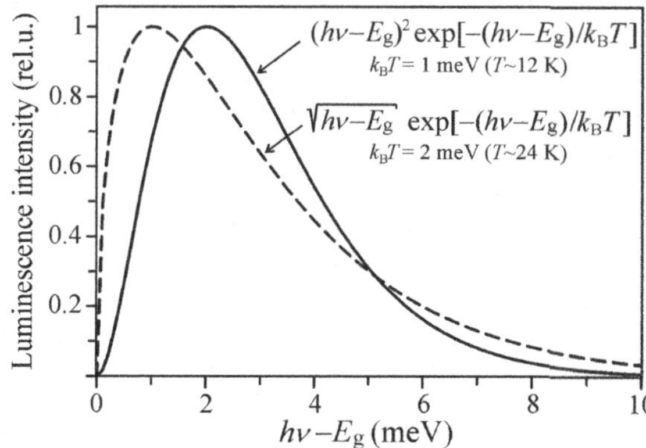

Fig. 5.6
Comparison of emission lineshapes of the recombination of free electron–hole pairs (5.9a)—dashed curve—with the lineshape described by (5.13)—solid curve. The curves are normalized both vertically and relative to the origin of the energy axis (i.e. $\hbar\omega = 0$). Typical asymmetry in the high-energy tail is present in both cases.

5.3 Recombination of a free electron with a neutral acceptor (e–A^0) and of a free hole with a neutral donor (h–D^0)

An impurity atom in a semiconductor, whose valence is smaller (larger) by one electron than that of the main constituent of the crystal lattice, is referred to as a shallow acceptor (donor). Classic examples of a shallow acceptor (donor) are boron (phosphorus) atoms in silicon. The additional hole or electron arising from the replacement of the main constituent's atom in its position in the crystal lattice by the impurity atom is weakly bound and can be thermally excited into the valence (conduction) band, thus boosting the electrical conductivity of the material. The corresponding amount of energy needed for the release of the hole or electron is referred to as the *ionization (binding) energy* of the acceptor E_A or donor E_D. The acceptor and donor states are usually represented in the band structure diagram by short segments drawn inside the bandgap, just above the maximum of the valence band or below the minimum of the conduction band. As the definition implies, the difference in energy between the acceptor (donor) level and the extreme of the valence band equals E_A (or E_D in the case of the conduction band). A neutral acceptor (donor) is a term used to describe an acceptor (donor) with a non-ionized hole (electron); in the opposite case, the impurity will be referred to as an ionized acceptor or donor. The short segment in the band structure demonstrates the fact that the hole or the electron is localized at the neutral acceptor or donor and their quasi-momentum $\hbar\mathbf{k}$ is practically zero. This situation is depicted in Fig. 5.7 for an acceptor and a donor in a direct-bandgap semiconductor.

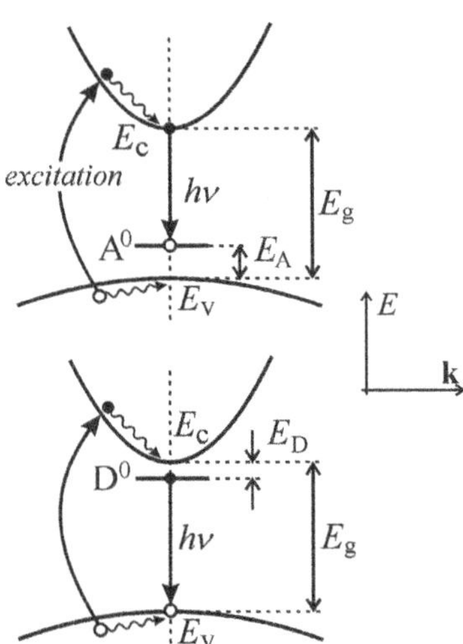

Fig. 5.7
(e – A^0) recombination (top panel) and (h – D^0) recombination (bottom panel) in a direct-bandgap semiconductor.

Figure 5.7 implies that the presence of a neutral acceptor or donor in the semiconductor can give rise to specific luminescence. Hereafter, we will focus only on acceptors (Fig. 5.7(a)). If a free electron is present at the bottom of the conduction band, it can recombine with a hole localized on the acceptor level, which leads to the emission of a luminescence photon with approximate energy $h\nu \approx E_g - E_A$. The luminescence process would not proceed further, however, if it were not for excitation, either optical or by employing other means, resulting in the liberation of other free electron–hole pairs, as is also demonstrated in Fig. 5.7. Excited electrons and holes quickly thermalize and 'diffuse' close to the extremes of the bands. The electron remaining after the radiative recombination at the acceptor level recombines (usually non-radiatively) with a hole at the top of the valence band, which restores the neutral state of the acceptor and the whole process of the emission of a luminescence photon can happen again.

Next, the spectral shape of luminescence (e–A^0) in a direct-bandgap semiconductor will be described. Let us start with (5.9a); this relation can be adapted so that it explicitly contains the effective masses m_e and m_h:

$$I_{sp}(h\nu) \approx (h\nu - E_g)^{1/2} \exp\left(-\frac{m_e}{m_e + m_h}\frac{h\nu - E_g}{k_B T}\right) \exp\left(-\frac{m_h}{m_e + m_h}\frac{h\nu - E_g}{k_B T}\right). \tag{5.14}$$

Assuming a high hole effective mass $m_h \to \infty$ and thus $m_h \gg m_e$ immediately yields, by making use of (5.14),

$$I_{sp}(h\nu) \approx (h\nu - E_g)^{1/2} \exp\left(-\frac{h\nu - E_g}{k_B T}\right). \tag{5.15}$$

We get back to a formula identical to (5.9a) and the whole procedure seems pointless. In reality, however, the transition from (5.14) to (5.15) only once again stresses the fact that optical transitions across the bandgap do not carry information regarding the values of effective masses m_e, m_h themselves. Importantly now, in the case of recombination of the type (e–A^0) the hole bound to an acceptor can be viewed as an immobile quasi-particle with an effective mass $m_h \to \infty$, which immediately suggests that the lineshape will again correspond to (5.9a), or (5.15), except for the fact that the low-energy onset is determined by (E_g–E_A) instead of E_g, which can also be easily deduced from Fig. 5.7(a). Consequently, the formula will read

$$I_{sp}^{(e-A^0)} \approx \left[h\nu - (E_g - E_A)\right]^{1/2} \exp\left(-\frac{h\nu - (E_g - E_A)}{k_B T}\right). \tag{5.16a}$$

Since similar reasoning (but for the substitution of electrons with holes) applies to the recombination of a free hole with a neutral donor (h–D^0), one can instantaneously write the formula for the corresponding lineshape

$$I_{sp}^{(h-D^0)} \approx \left[h\nu - (E_g - E_D)\right]^{1/2} \exp\left(-\frac{h\nu - (E_g - E_D)}{k_B T}\right). \tag{5.16b}$$

All in all, the Maxwell–Boltzmann lineshape (5.9a), (5.16) (or the analogous formula (5.13)) is shared by all luminescence processes for which at least one of the recombining quasi-particles is free. This similarity in behaviour reflects the *kinetic energy* of a free quasi-particle, which manifests itself in the asymmetry of the spectral line and the broadening of the high-energy tail with increasing temperature.[4]

An example of an experimentally observed luminescence spectrum (e–A^0) coming from a high-purity epitaxial GaAs layer at $T = 1.9$ K is shown in Fig. 5.8 [5]. Both aspects of the experiment, i.e. the high purity and low temperature, are essential for this observation. If the semiconductor is not highly pure (high purity for GaAs stands for the concentration of unintentional (residual) acceptors and donors of the order less than 10^{15} cm^{-3}), several different acceptors and donors with slightly varying ionization energies are present in the sample, their emission lines thus being, according to (5.16), slightly shifted with regard to one another. This may result in the occurrence of two inhomogeneously broadened lines, one representing the unresolved acceptor transitions and the second one unresolved donor transitions, with the typical asymmetry given by (5.16) smeared and consequently lost. The sample whose spectrum is

[4] From the point of view of the configurational coordinate model, the processes (e–A^0) and (h–D^0) can be viewed as weakly localized excitation (i.e. only the zero-phonon line is observed), and broadening of the line due to kinetic energy of the quasi-particles taking part in the process goes beyond the scope of this model.

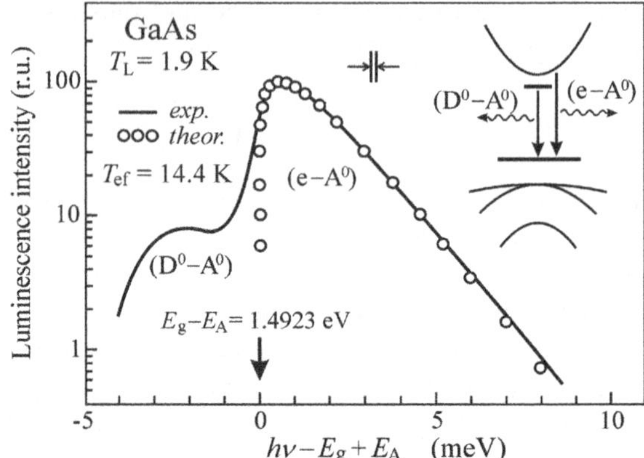

Fig. 5.8
Photoluminescence spectrum (e–A^0) of high-purity GaAs containing a single residual shallow acceptor with ionization energy $E_A \approx 27$ meV. The solid curve denotes the experimental data, and circles stand for the theoretical lineshape (5.16a). The low-energy edge reveals the occurrence of donor–acceptor pair emission (D^0–A^0), which is, however, more than an order of magnitude lower in intensity (note the logarithmic intensity scale). The bath temperature of liquid helium was $T_L = 1.9$ K, and the effective temperature of the electron gas for the fit of eqn (5.16a) was $T_{ef} = 14.4$ K. After Ulbrich [5].

shown in Fig. 5.8 contains a single type of acceptor only, which is why the typical asymmetry of the experimentally obtained curve towards the high-energy wing is well evidenced. Yet, at the low-energy edge near $h\nu \cong E_g - E_A$ another emission band (D^0–A^0), which will be treated in Section 5.4, clearly stands out. The second experimental aspect, i.e. low temperatures, plays an important role not only due to the effect of thermal quenching, which has already been mentioned several times, but even more importantly to prevent the ionization of donors by the thermal energy $k_B T$.

Figure 5.8 contains our very first notion of two different temperatures in semiconductors: the lattice temperature (T_L) and the so-called *effective temperature* of a gas of free quasi-particles T_{ef}, introduced in photoluminescence measurements. The concept of the effective temperature indicates that, although the energy distribution of photoexcited free electrons and holes (or excitons) can be described by the Fermi–Dirac distribution function or by its Boltzmann tail, the temperature T_{ef} in the relevant formula is higher than the bath temperature T_L at which the sample is kept (liquid helium, nitrogen, etc.). In other words, even if the free quasi-particles are in thermal equilibrium with one another, they are not in thermal equilibrium with the subsystem of atoms (ions) vibrating in the lattice. This effect was found to be quite general and occurs not only at high-fluence excitation with high-power laser pulses, but also at conventional low-fluence excitation levels. The different temperatures of the carriers and the bath are a natural outcome of the relatively short photocarrier or exciton lifetime, the term ‘relatively short’ meaning the fact that the lifetime of carriers and excitons is shorter than the relaxation time necessary for the establishment of full thermodynamic equilibrium between the two described subsystems, i.e. the gas of free quasi-particles and the crystal lattice.[5] Naturally, the difference between the temperatures T_{ef} and T_L increases if a high-power excitation source is used or when the energy difference between the exciting photons and the bandgap E_g rises. Ultrafast semiconductor spectroscopy [6] is the branch of optics dealing with, among other things, the study of the rate of dissipation of this ‘hot carrier’ excess

[5] The exception to the rule $T_{ef} > T_L$ can sometimes be encountered in indirect-bandgap semiconductors, in which the lifetime of excitons may be sufficiently long.

energy (i.e. the decrease of T_{ef} as a function of time after the excitation pulse has ended) after picosecond or femtosecond excitation.

The last remark concerns the practical importance of (e–A^0) luminescence: it enables the ionization energy E_A to be spectroscopically determined as the energy difference between the low-energy onset of the line (E_g–E_A) and the energy corresponding to the width of the forbidden gap E_g. The bandgap energy E_g, however, needs to be obtained with sufficient accuracy from another type of measurement, since luminescence measurements rarely allow for its determination (but for the case of band-to-band recombination). It is of the essence, moreover, that a single type of acceptor is present; the alternative would inevitably lead to the smearing of the lineshape. A characteristic feature confirming the proper identification of a line being due to (e–A^0) or (h–D^0) can, besides the Maxwell–Boltzmann lineshape, be the saturation tendency of the intensity dependence, as discussed in Section 3.5 for extrinsic luminescence.

Radiative recombination (e–A^0), (h–D^0) is difficult to observe in an indirect-bandgap semiconductor due to the negligible concentration of free electrons and holes with respect to the concentration of excitons.

5.4 Recombination of donor–acceptor pairs (D^0–A^0)

If both neutral donor and acceptors are present in a semiconductor simultaneously, another type of radiative recombination might occur, as depicted in Fig. 5.9. In the approximation of effective mass, the neutral donor D^0 can be considered as an 'extra' electron orbiting all the remaining parts of the atom, i.e. the atomic nucleus, the inner atomic shells and electrons taking part in the covalent bonding with the nearest constituents of the crystal lattice. A neutral acceptor A^0 can then be treated analogously, if the extra electron is substituted with a hole. This electron–hole pair can recombine radiatively while emitting a photon $h\nu$ (Fig. 5.9(a)). The very same process is sketched in Fig. 5.9(b) in the band structure diagram, although the (E, x) system of coordinates, x being the real space coordinate, is used instead of the conventional $(E, \mathbf{k})$ layout.

Immediately after the recombination, both the donor and the acceptor become ionized (D^+, A^-), but free electron–hole pairs generated by the band-

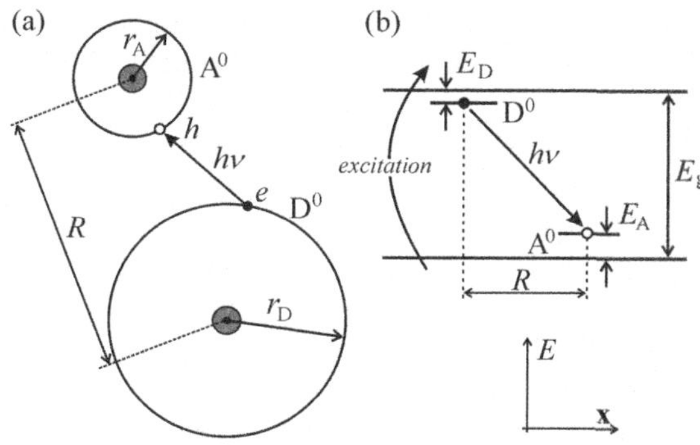

Fig. 5.9
The concept of luminescence arising from the recombination of donor–acceptor pairs (D^0–A^0): (a) in real space, (b) in the electron energy band structure as a function of the real-space coordinate x. Donor and acceptor atoms are located at different positions of the crystal lattice, separated by the distance R. The quantities r_D and r_A stand for the radius of the donor and acceptor, and E_D and E_A are their ionization energies, respectively.

to-band excitation quickly neutralize these ionized donors and acceptors and the whole process resumes.

This process is sometimes schematically described as

$$D^0 + A^0 \rightarrow D^+ + A^- + h\nu.$$

Glancing at Fig. 5.9(b) one could deduce that the energy of the emitted photon is determined as

$$h\nu(D^0 - A^0) = E_g - (E_A + E_D). \quad (5.17)$$

Nevertheless, this formula is not fully valid. Since a charged pair (D^+, A^-) is generated as a result of the recombination process, the additional energy of its mutual electrostatic attraction $e^2/4\pi\varepsilon_0\varepsilon_r R$ needs to be added to (5.17):

$$h\nu(D^0 - A^0) = E_g - (E_A + E_D) + \frac{e^2}{4\pi\varepsilon_r\varepsilon_0 R}(-J(R)). \quad (5.18)$$

In this formula, R denotes the distance between the donor and the acceptor and r_D stands for the radius of the electron shell in D^0 (and, analogously, r_A for the hole in A^0), see also Fig. 5.9(a). Formula (5.18) holds only provided the donor and the acceptor represent point charges, i.e. $R \gg r_D, r_A$. The last term $J(R)$ in (5.18) constitutes a non-Coulomb term, correcting for the overlap of the wavefunctions in the case of small distance R and decreasing the additional Coulomb energy. This term disappears when R is sufficiently large.

The (D^0–A^0) recombination manifests itself in the emission spectrum in various ways, depending on the strength of the electron–phonon coupling. First we shall discuss semiconductors with weak electron–phonon coupling (which means, according to Table 4.2, the fourth group and III-V group semiconductors, taking into account the polaron coupling constant α_e as a measure of this interaction). Since the donors and acceptors replace the main constituents of the crystal lattice in their sites, the values of R can only be discrete and, thus, so will the energies of the emitted photons, according to (5.18). When the electron–phonon interaction is weak, one can consequently expect the emission spectrum to comprise a number of narrow (zero-phonon) lines corresponding to the allowed values of R, or, in other words, to virtual spherical 'shells' of partners for radiative recombination around the particular donor or acceptor.

A prime example of such an emission spectrum is shown in Fig. 5.10 [7], in which one immediately recognizes an overwhelming number of narrow lines. The depicted luminescence spectrum is that of silicon- and sulphur-doped gallium phosphide, GaP, at liquid helium temperature (1.6 K). In this case, silicon and sulphur are the so-called substitution donors and acceptors of type I, which substitute atoms only in one of the sublattices of the crystal, in this case phosphorus: a phosphorus atom is replaced with a sulphur (donor S_P) or silicon (acceptor Si_P) atom. Zinc and oxygen, on the other hand, is a pair that can serve as an example of substitution donors and acceptors of type II, which are built into both sublattices, i.e. Zn_{Ga}–O_P.

Both the density of lines and their intensity obey general rules. The lines located at the high-energy side of the spectrum are relatively sparse and of low intensity, and in the direction towards lower energies they become more

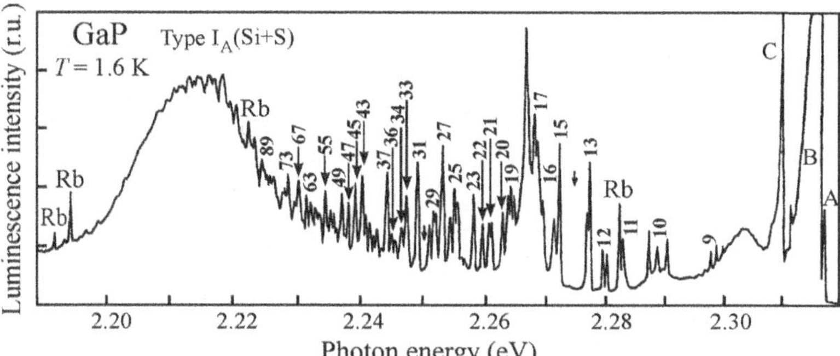

Fig. 5.10
Photoluminescence spectrum of silicon- and sulphur-doped GaP (the doping atoms are the so-called substitution impurities of type I, S_P donor and Si_P acceptor). The doping level is $|N_D–N_A| \sim 10^{18}\,\text{cm}^{-3}$, and measurement temperature 1.6 K. The numbers next to the lines denote the number of the shell, lines marked with Rb come from a rubidium calibration lamp. Lines A, B, C are due to bound excitons. The whole spectrum is located between $E_g = 2.35$ eV and $E_g–(E_D + E_A) \approx 2.044$ eV. After Thomas *et al.* [7].

densely packed and their intensity increases, whereas at the very low-energy edge the intensity decreases again and the lines merge into a 'fine-structure-modulated' wider band. Such behaviour is easily understandable considering relation (5.18): high photon energies correspond to small R, the number of suitable neighbours in the nearest shell being low, though; the corresponding discrete values of R are relatively scarce, resulting in sparse and weak lines on the high-energy side of the spectrum. When the distance R increases, however, more partners can be found in the given shell, and the shells themselves are located closer to one another, which means that the lines become more intense and more densely packed. The situation changes once again when R becomes even larger, as the suitable shells nearly merge and a decrease in the probability of radiative recombination $W(R)$ comes into play owing to the fact that the electron and a hole are too far apart, for one can prove that

$$W(R) = W(0)\exp(-2R/a), \tag{5.19}$$

where $a = \max(r_D, r_A)$. This is why the low-energy end of the spectrum gets smeared into a broad band and drops to zero somewhere above the energy limit of $E_g–(E_A + E_D)$, see Fig. 5.10.

This type of emission spectrum contains so many characteristic features that the experimental identification of the (D^0–A^0) radiative channel should not pose any problems (even though some spectral similarity with the luminescence of an electron–hole liquid in a doped semiconductor can appear, see Chapter 8). Nevertheless, two additional characteristic features of this type of luminescence spectrum can be listed, the first one being an overall *blue-shift* of the broad band with increasing excitation intensity, as displayed in Fig. 5.11. This blue-shift is easily noticeable and can be qualitatively understood again on the basis of eqn (5.18): when increasing the excitation intensity, more D^0, A^0 atoms closer to one another, i.e. with lower R, get excited, which increases the influence of the Coulomb term in (5.18). The second attribute of the (D^0–A^0) luminescence spectrum is then typical behaviour in time-resolved luminescence experiments: the increasing delay after the excitation impulse gives rise to a pronounced *red-shift* of the emission spectrum. The reason for the behaviour lies in the fact that, according to (5.19), the closest pairs with the lower R are the first ones to recombine, while pairs further and further apart start to send out luminescence in longer time delays. This, in agreement with

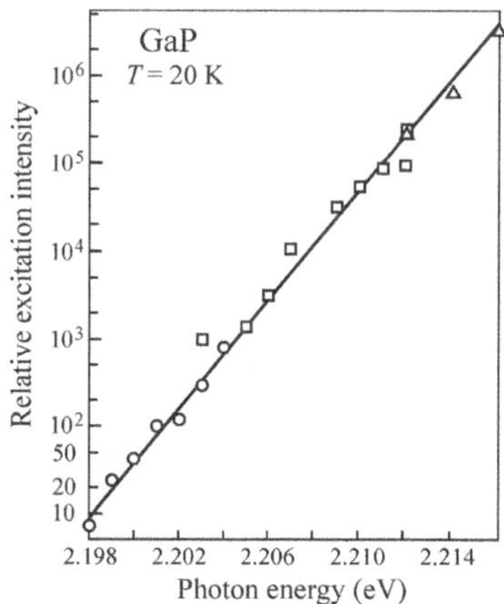

Fig. 5.11
Position of the maximum of the broad ($D^0 - A^0$) band from Fig. 5.10 as a function of excitation intensity. Bath temperature of 20 K. Different types of symbols denote different samples and different excitation wavelengths. After Thomas *et al.* [7].

(5.18), signifies the shift of $h\nu(D^0\text{–}A^0)$ towards lower energies and evidently also the decrease of the overall emission intensity.

In the foregoing discussion we have put the $(D^0\text{–}A^0)$ radiative recombination in the framework of the configurational coordinate model as a special example of 'localized excitation' without the contribution of the kinetic energy of free carriers. The narrow spectral lines in Fig. 5.10 were referred to as 'zero-phonon' lines. The arguments that justify the application of such terminology are summarized in Fig. 5.12, which represents the emission spectrum of the very sample of Figs 5.10 and 5.11 measured, however, in a much wider spectral range and at slightly higher temperature (20 K). The spectrum contains an intense zero-phonon component near 2.20 eV and several satellites with decreasing intensities, shifted by multiples of the LO-phonon energy. Such behaviour is a typical manifestation of weak electron–phonon coupling. This time, the densely packed lines from Fig. 5.10 are hidden in the unresolved shoulder at 2.25 eV and the wide low-energy band originally located at 2.215 eV (Fig. 5.10) is shifted to 2.20 eV in Fig. 5.12, as a result of the thermally induced decrease in E_g.

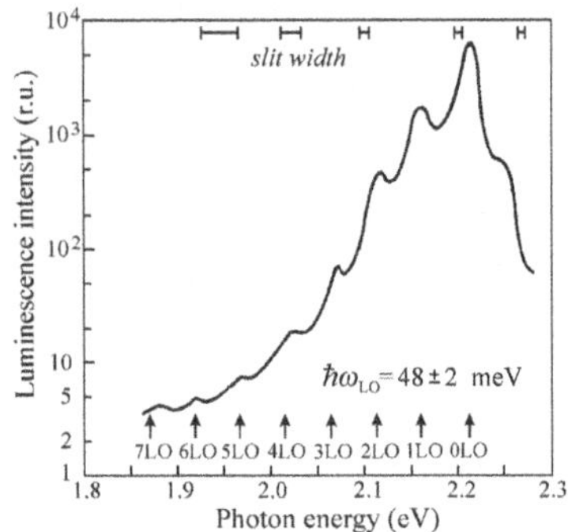

Fig. 5.12
$(D^0\text{–}A^0)$ luminescence spectrum of GaP at $T = 20$ K (the very same sample as in Figs 5.10 and 5.11). The high-energy tail at ~ 2.25 eV includes the unresolved lines of $D^0\text{–}A^0$ pairs (wide slit, higher temperature) and the peak at 2.20 eV corresponds to the broad band from Fig. 5.10. The structure is periodically repeated with period $\hbar\omega_{LO}$. Note the gradual widening of the monochromator slit for higher-energy phonon replicas and the logarithmic intensity scale. After Thomas *et al.* [7].

Under what experimental conditions can the characteristic $(D^0\text{–}A^0)$ recombination luminescence with enormous numbers of narrow lines be observed in a semiconductor with weak electron–phonon coupling? The most important parameter will be a suitable level of concentration of both dopants. If the number of donors and acceptors is too low, D^0 and A^0 in the pairs will be too far apart from one another and no recombination will take place; on the other hand, when the concentration is too high, the simple Coulomb interaction of D^0 and A^0 ceases to be valid and, moreover, concentration quenching can set in. Therefore, the typical emission with densely packed lines can be observed in moderately or highly doped materials (e.g. in GaP at $|N_D\text{–}N_A| \sim 10^{17}\text{–}10^{18}\ \text{cm}^{-3}$) and under slightly elevated excitation (otherwise, only donors and acceptors far apart from one another will be excited and the spectrum will feature only a band). Furthermore, low temperatures are absolutely essential and, if possible, the number of competing radiative recombination channels should be minimized. The importance of fulfilling all of these conditions becomes more obvious if we realize that GaP is, in fact, the only material in which a large number of narrow $(D^0\text{–}A^0)$ lines can be observed. For instance in silicon, whose electron–phonon coupling is even weaker than in GaP (and which, similarly to GaP, is an indirect-bandgap semiconductor), luminescence lines of this origin have never been experimentally observed. In GaAs, just to give another example, whose electron–phonon interaction is comparable to that in GaP, the $(D^0\text{–}A^0)$ emission manifests itself as a single unresolved line. There are, obviously, other factors coming into play; the problem will be further discussed in Section 7.2.

Now, let us briefly focus on semiconductors with strong electron–phonon coupling. A typical example of such a semiconductor is, taking into account the Fröhlich constant α_e in Table 4.2, direct-bandgap zinc sulphide ZnS with $E_g \sim 3.8$ eV. A particular case of doping by Cu_{Zn} acceptors and Cl_S donors leads to intense green emission, which has been well known for many years [8]. The most recent interpretation of this luminescence consists in the $D^0\text{–}A^0$ recombination. As the electron–phonon coupling in this material is particularly

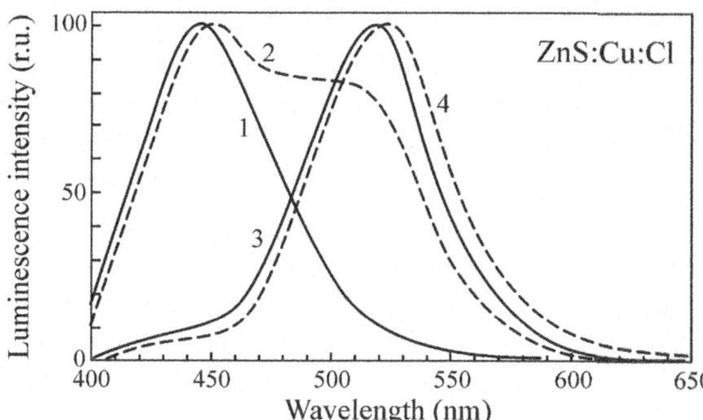

Fig. 5.13
Spectral distribution of cathodoluminescence of ZnS doped (activated) by copper with the addition of a 'fluxing agent' in the form of NaCl. This results in the luminescence of donor (Cl_S)–acceptor (Cu_{Zn}) pairs, giving rise to the emission band at ~ 520 nm. Curves enumerated 1–4 correspond to increasing concentration of copper: 1 – nominally pure ZnS, 2–10^{-3} % Cu, 3–5 × 10^{-3} % Cu, 4–10^{-2} % Cu. Measured at room temperature, curves are normalized. After Espe [8].

strong, the luminescence band is broad, lacks any structure and is located far from edge emission (i.e. with large Stokes shift), near 520 nm (Fig. 5.13). The existence of other dopants, the introduction of which makes it possible to cover an important part of the visible spectral region, has been of great importance for applications in the development of phosphors for cathodoluminescence.

Here, it might be of interest to make a small historical detour: the fact that efficient luminescence in zinc sulphide can be achieved only if the material is co-doped with two different elements was empirically discovered as long ago as in the 1940s. Since the D^0–A^0 recombination mechanism was not yet known to be responsible for the luminescence, the metal dopant was not called an 'acceptor', but an 'activator', and the Cl^- or Br^- ions were not known as 'donors', but as 'co-activators'.

A similar type of broadband D^0–A^0 emission can also be encountered in ZnSe and CdS and some attempts were also made to interpret the intense low-temperature blue-green luminescence of AgCl by means of the same mechanism. From the application perspective, D^0–A^0 recombination is exploited as a mechanism of efficient emission from GaP light-emitting diodes. One of the proposed models describes the red electroluminescence at 1.8 eV as being due to the recombination of a compact O_P–Zn_{Ga} pair in highly doped material.

5.5 Luminescence excited by two-photon absorption

In the text so far, we have taken for granted that the excitation of the photoluminescence processes takes place via the conventional absorption of a single (suitable) photon with energy $h\nu_{ex} > E_g$ which leads to the generation of electron–hole pairs or excitons and is followed by radiative recombination via one of the above-mentioned mechanisms. This kind of absorption is governed by the well-known Lambert–Beer law (2.1)

$$I_{h\nu_{ex}}(d) = I_0 \exp\left(-\alpha\left(h\nu_{ex}\right) d\right), \tag{5.20}$$

where $\alpha(h\nu_{ex})$ is the absorption coefficient and d stands for the depth under the surface of the sample to which the light penetrates. Unless the experiment has some specific requirements, the usual energy of the excitation photons substantially exceeds E_g, which guarantees high values of the absorption coefficient and in turn boosts the luminescence intensity.

The typical values of absorption coefficient for these cases are in the order of $\alpha(h\nu_{ex}) \approx 10^4$–$10^5\,\text{cm}^{-1}$, which allows the penetration of light only into depths of the order of $d \approx \alpha^{-1} = 10^{-5}$–$10^{-4}\,\text{cm} = 100\,\text{nm}$–$1\,\mu\text{m}$. Such low values can pose a problem, because radiative recombination can be strongly influenced by surface and subsurface states, the existence of which is well known in solid-state physics. These states can either quench the luminescence, or, if the semiconductor is prone to oxidation, the surface oxide layer (or a surface layer of adsorbed molecules or radicals) can be the source of luminescence. Under such conditions, the obtained spectra *might not* in fact reflect the properties of the studied material—or, at least, an argument of this kind can be raised by any reviewer of a thesis or manuscript dealing with luminescence of semiconductors.

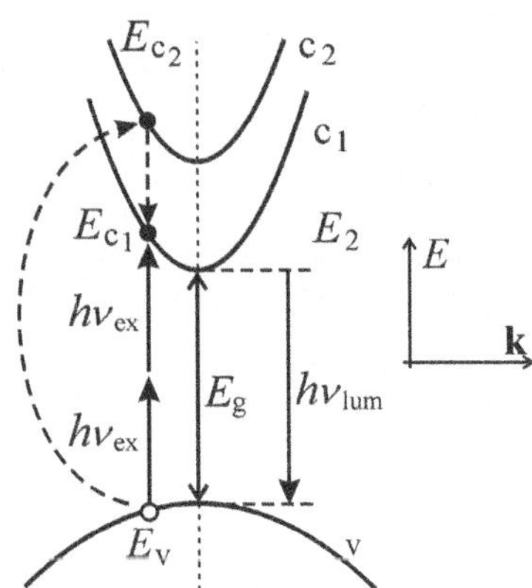

Fig. 5.14
Schematic illustration of two-photon absorption in semiconductors. The process is mediated by the presence of a higher lying conduction band c_2, but the energy-conservation law reads $E_{c1} - E_v = 2h\nu_{ex}$.

A simple but reliable way to confirm or refute such an argument may consist in comparing emission spectra excited via a one-photon and two-photon process. In order to understand why, it is important to give an explanation of two-photon absorption first.

This is a nonlinear optical phenomenon, in which two photons are absorbed at the same time *in a single elemental act*. The special case of absorption of two identical photons $h\nu_{ex}$ in a semiconductor is schematically represented in Fig. 5.14. A condition to be fulfilled if this process is to occur reads

$$E_g > h\nu_{ex} \geq E_g/2. \tag{5.21}$$

The right part of the inequality can be easily understood, as it reflects the energy-conservation law. The left part of eqn (5.21) results from the nonlinear nature of this process, since two-photon absorption is described in second-order perturbation theory [9] of quantum mechanics, implying that this process is several orders of magnitude weaker than the conventional one-photon absorption (first-order perturbation theory). Thus, were the condition $h\nu_{ex} > E_g$ fulfilled, intense single-photon absorption over the bandgap would set in and two-photon absorption would cease to be detectable.

The formalism of second-order perturbation theory implicitly comprises the so-called 'virtual intermediate level', which mediates the transition between the ground (E_v) and final (E_{c1}) states. The role of the virtual intermediate state in a semiconductor is taken on for example by one of the higher lying conduction bands (c_2 in Fig. 5.14). The energy-conservation law, though, is a relation between the ground and final states: $2h\nu_{ex} = E_{c1} - E_v$.

Now, we will derive a relation describing the decrease in light intensity as it passes through a semiconductor in the case when eqn (5.21) holds; in other words, we will derive a two-photon analogue of the Lambert–Beer law (5.20). Let us denote by $\alpha^{(2)}(I)$ the two-photon absorption coefficient, which evidently depends, unlike its one-photon counterpart, on the intensity of light I. It can be shown that this dependence is linear: $\alpha^{(2)}(I) = aI$ [9], where the coefficient a is a material constant. The decrease in the intensity of the light beam $\mathrm{d}I$ as a function of travel distance $\mathrm{d}x$ in a sample can then be described by the differential equation

$$dI = -\alpha^{(2)} I\, dx = -a\, I^2\, dx. \tag{5.22}$$

This equation can be easily solved using separation of variables with the initial condition $I(x = 0) = I_0$. The result can be written as

$$I(d) = \frac{I_0}{(a\, I_0\, d + 1)}, \tag{5.23}$$

which is the formulation of the law we were looking for; d again stands for the depth of penetration under the surface of the sample, or its thickness. This relation was derived for the first time by Basov [10].

In contrast to the fast exponential decrease as in the case of linear one-photon absorption (5.20), the decrease of light intensity during two-photon absorption is only hyperbolic. This fundamental difference in the rate of decrease is documented in Fig. 5.15(a); the hyperbolic decrease is incomparably slower. In order to induce nonlinear optical phenomena, obviously, the application of a high-power source of light is necessary. Two-photon absorption can only be observed when relatively high-power laser beams are used as excitation sources. Taking into account the typical value of coefficient $a(\sim 10^{-3}$ cm/MW) and the power density of the incident beam $I_0 =$ 10 MW/cm^2, one can immediately see that—according to (5.23)—a sample length of 1 cm(!) is needed so that the intensity could decrease by a single per cent (i.e. $I(d)/I_0) \approx 0.99$)! Two-photon excited luminescence coming from thick samples is then, without doubt, a phenomenon connected with the bulk of the material. (If semiconductor nanostructures with typical dimension of units and tens of nanometres are used as samples, however, the difference in the depth of penetration between one- and two-photon excitation is no longer so significant.)

An experimental verification of the described behaviour can be found in Fig. 5.15(b). The figure demonstrates a view through a cryostat window (dark circle) of a bulk AgCl sample at $T \cong 80$ K (grey rectangle in the upper part of the window). The ruby laser beam ($h\nu_{ex} = 1.78$ eV) is coming from the left side and excites the luminescence in AgCl; the forbidden gap width $E_g \cong 3.2$ eV of the silver chloride signifies that condition (5.21) is fulfilled. This luminescence can be observed as a bright trace coming through the sample in its middle part along its whole length, i.e. about 1.5 cm. Its brightness only minutely decreases towards the right side of the sample, reflecting a very modest decrease of laser intensity during its passage through the sample.

Which brings us back to the initial question: 'What does a two-photon excited, i.e. inherently coming from the bulk, luminescence spectrum look like in comparison with one-photon excitation?' Although a general and unambiguous answer to this question does not exist, a large number of experiments carried out at different laboratories (including that of the authors of this book) suggest that in the vast majority of cases the emission spectra are almost identical, at least when it comes to the salient features. This similarity is documented in Fig. 5.16, in which emission spectra of (a) a bulk AgCl sample at low temperatures [11] and (b) a thin CdS film (thickness of 1.6 μm) at room temperature [12] are compared. Similar comparisons were obtained using a large number of other semiconductors under diverse experimental conditions,

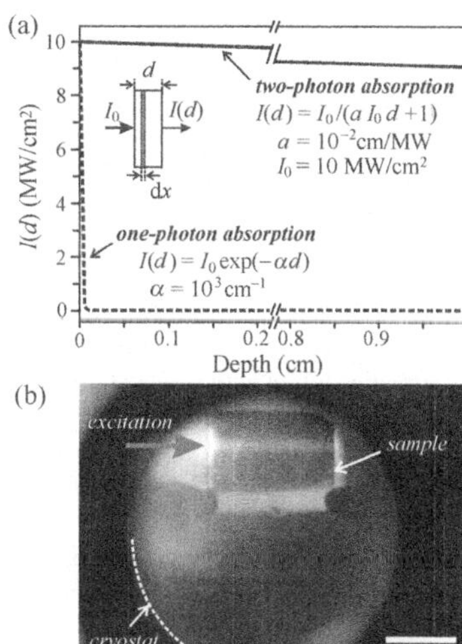

Fig. 5.15
(a) Numerical comparison of the decrease in intensity of a light beam as a function of the penetration depth during one- and two-photon absorption. An unrealistically high value of $a = 10^{-2}$ cm/MW was chosen in order to make the decrease of intensity during the two-photon absorption process observable to the naked eye. (b) Experimental demonstration: a view through the window of a cryostat (dark circle) of a bulk AgCl sample with a horizontal trace of two-photon excited blue-green luminescence across the whole length of the sample (grey rectangle with a light strip in the middle). The sample temperature was $T \cong 80$ K. The focused beam of the excitation ruby laser 694 nm is coming from the left side, and three green optical filters with high optical densities were applied when the photograph was taken in order to block the scattered laser light.

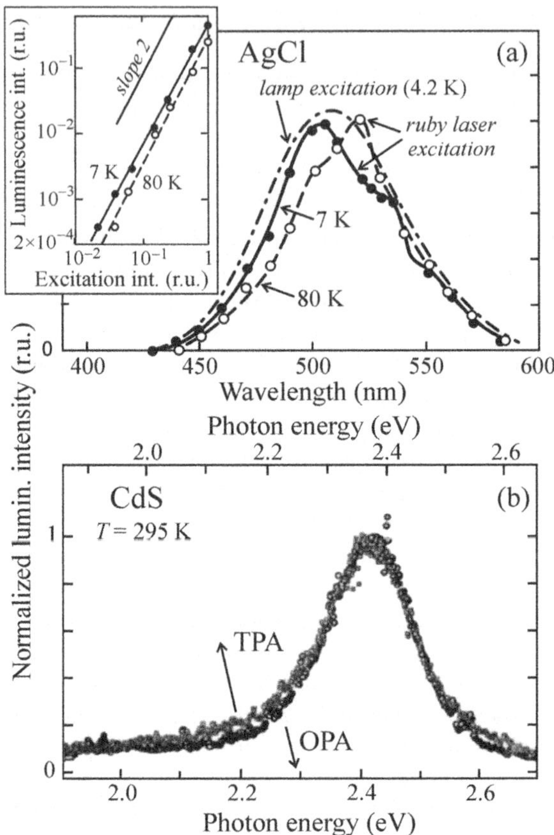

Fig. 5.16
Comparison of emission spectra acquired with one- and two-photon excitations. (a) AgCl single crystal, excited by one-photon absorption with a cw high-pressure lamp ($h\nu_{ex} = 3.4\,eV$) and by two-photon absorption with a pulsed ruby laser ($2h\nu_{ex} \cong 3.56\,eV$). After Laibowitz and Sack [11]. (b) Thin-film CdS excited at room temperature by one-photon absorption (OPA) with nanosecond pulses of the third harmonic of a Nd:YAG laser ($h\nu_{ex} \cong 3.49\,eV$) and by two-photon absorption (TPA) with 200-fs pulses of a Ti:sapphire laser ($2h\nu_{ex} \cong 3.23\,eV$). The spectra are shifted with respect to one another because the penetration depth of the femtosecond laser was larger than the thickness of the sample (1.58 μm); that is why the luminescence in this case is strongly influenced by reabsorption and consequently shifted to lower energies. After Yano and Ulbrich [12].

both in air and under vacuum. This indicates that the influence of surface states on luminescence is not so pronounced as was commonly supposed in the past. In some cases, some new features, e.g. phonon replicas, appear in the two-photon excited spectrum; these new features, however, are again more closely related to the properties of the material itself than to its surface. When in doubt, though, comparative experiments should be carried out on the material.

The last question needing clarification here is how to recognize undoubtedly that the observed luminescence is, indeed, two- and not one-photon excited. Two characteristic features can be applied, namely the quadratic intensity dependence and an 'anti-stokes' shift of the two-photon excited luminescence. The term 'anti-stokes shift' actually indicates that the spectral position of the emission spectrum violates Stokes' law, which states that the luminescence wavelength λ_{lum} is always longer than the excitation wavelength λ_{ex}. This law was formulated in the nineteenth century when one-photon excitation was the only conceivable way to excite luminescence (the very first laser was designed as late as in 1960); this mode of excitation, obviously, always leads to a Stokes shift. This shift clearly renders nothing less than the energy-conservation law $h\nu_{ex} \geq h\nu_{lum}$. During two-photon excitation, however, the law reads $2h\nu_{ex} \geq h\nu_{lum}$ or $\lambda_{lum} \geq \lambda_{ex}/2$ and, in particular, the condition $\lambda_{ex} \geq \lambda_{lum} \geq \lambda_{ex}/2$ is very likely to occur; this happens very often indeed in semiconductors. Such behaviour is demonstrated in Fig. 5.14 depicting the excitation that takes place not too high above the bandgap, and luminescence photons have an energy very close to E_g, meaning that $\lambda_{lum} < \lambda_{ex}$. Thus, when looking at the photograph in Fig. 5.15(b), one should realize that the trace of luminescence across the sample is blue-green ($\lambda_{lum} \approx 500\,\text{nm}$), while the exciting laser beam is red ($\lambda_{ex} = 694\,\text{nm}$).[6]

The fact that two-photon excited luminescence depends quadratically on the excitation intensity can be derived from simple energy-based reasoning. The density of the excitation power absorbed in the sample of thickness d equals (if reflection losses are neglected), according to eqn (5.23),

$$\left(I_0 - \frac{I_0}{(ad\,I_0 + 1)}\right) = \frac{ad\,I_0^2}{(ad\,I_0 + 1)} \approx ad\,I_0^2; \qquad (5.24)$$

the last modification in (5.24) is justified by the fact that even when applying the highest power densities of the order of $I_0 = 10\,\text{MW/cm}^2$, still low enough not to destroy the material with the application of nanosecond laser pulses, and considering typical values of $a \leq 10^{-3}\,\text{cm/MW}$, the condition $adI_0 \ll 1$ easily holds. Making just the natural assumption, namely, that luminescence is proportional to the absorbed power, we see that luminescence is proportional to the squared incident laser intensity I_0^2 according to (5.24). This dependence is illustrated in the inset in Fig. 5.16(a). The reader surely recalls the case of bimolecular (one-photon excited) recombination which can, under certain circumstances, exhibit quadratic dependence on the excitation intensity, as was discussed in Section 3.5. Consequently, even though the experimental determination of such a dependence on its own does not prove that two-photon excitation takes place, the combination of this quadratic excitation-intensity dependence with an anti-stokes luminescence shift cannot be interpreted in any other way.

It is important to keep in mind that the realization of a two-photon experiment for a given value of bandgap is to some extent limited by the availability of high-power pulsed lasers complying with (5.21). Due to the growing numbers of tunable parametric generators this requirement is no longer as limiting as it used to be. The very last remark in this section relates to the fact that two-photon excitation can sometimes be applied to ensure excellent homogeneity of photoexcited carriers or excitons in order to compare experimental results with theoretical models, as will be mentioned later on in Chapter 8.

[6] To detect two-photon excited luminescence, the 'transmission geometry' cannot be applied. The direct laser light is many orders of magnitude more powerful than the luminescence and a direct detection of the beam could easily destroy the detector. The two-photon excited luminescence signal is rarely detected in a different way than at right angles to the excitation beam.

5.6 Luminescence from transition metal and rare earth ion impurities

Donors and acceptors are introduced in semiconductors as impurities in order to control their electrical transport properties. A number of impurities, however, are introduced into solids primarily in order to boost the luminescence efficiency and achieve a particular emission wavelength. Some of these impurities will be discussed in Chapter 7 (the so-called isoelectronic impurities); this section will deal with two types of specific impurities active in luminescence, which are exploited in countless commercially produced phosphors (e.g. for various types of reflecting surfaces, displays, luminescent signs, etc.): transition metal ions from the fourth period of the periodic table and rare earth ions.

These impurities, which cannot be easily classified from the point of view of their electrical activity as donors or acceptors, are active also in the luminescence of semiconductors. The corresponding channels of radiative recombination, however, will be mentioned only briefly, despite being both efficient and very commonly applied; they are not typical for semiconductors. They can be much more frequently encountered in various glasses, wide-bandgap ionic compounds, polymer-based materials, etc. Anyone more deeply interested in them can consult the chapters in the textbooks [13, 14].

Transition metal ions

Transition metal ions exhibit luminescence as a result of transitions of electrons between the levels of the partially filled 3d shell. Ions having the greatest importance in applications are Cr^{3+}, Mn^{4+} and Mn^{2+}. For example, $ZnS : Mn^{2+}$ has been a material commonly exploited for the preparation of electroluminescent thin film for quite some time; Cr^{3+} ions present in sapphire, Al_2O_3, were, on the other hand, the active medium for the very first ruby laser operating in 1960.

It is important to realize that $3d^{n*} \rightarrow 3d^n$ (* stands for an excited state) transitions are, from the atomic point of view, dipole-forbidden, since they do not comply with the condition $\Delta\ell = \pm 1$. When the ions are embedded into the crystal lattice, the interaction with the electric field of the crystal makes the transitions partially allowed. This interaction can be described in the framework of so-called ligand-field theory, which, among other things, shows that in some crystal lattices the emission spectra of the transition metal impurity ions can consist of both discrete lines (atomic-like emission) and broad bands. The crucial factor determining the spectral shape is the arrangement of the neighbouring ligands, the radius of the d orbitals and the distances between the built-in metal ion and the ligand. The shape of the spectrum can be predicted on the basis of the so-called Tanabe–Sugano diagram [15].

Recently, much effort has been devoted to the study of magnetic semiconductors, i.e. semiconductors with impurities of paramagnetic ions (Mn^{2+}), with the view towards applications in spintronics. Spintronics deals with the potential exploitation of spin polarization of charge carriers in informatics. Luminescence spectroscopy is presumed to be an excellent instrument for the study of these materials.

Rare earth ions

The most commonly applied rare earth ions are trivalent ions of lanthanides Ce^{3+}–Yb^{3+}, in which luminescence comes from $4f^{n*} \rightarrow 4f^{n}$ transitions. Similarly to the previous case, these transitions are also dipole-forbidden, while the crystal field makes them partially allowed. An important difference with respect to the previous case of $3d^{n*} \rightarrow 3d^{n}$ transitions, however, lies in the fact that the 4f shell is strongly screened from the influence of the crystal field by the external filled orbitals $5s^2 5p^6$. The influence of the crystal field is, on the one hand, strong enough to partially alter the character of the transitions, and on the other hand, the screening manifests itself by keeping the atomic-like character of the emission spectrum. In the framework of the configurational coordinate model in Fig. 4.8 this weak interaction with the lattice can be described as $\Delta Q = (Q_{e0} - Q_{g0}) \rightarrow 0$. This description implies two important consequences:

Firstly, the weak effect of the matrix suggests that the energy levels in ions, and thus also the emission (absorption) spectra, will be independent of the chemical nature and periodic arrangement of the matrix. Secondly, the activation energy E_A of the thermal quenching of luminescence (Fig. 4.8(b)) will be high, giving rise to more or less temperature-independent emission. Consequently, the emission can be expected to occur even at room temperature unless a different quenching mechanism, such as an Auger process (see Chapter 6), is present.

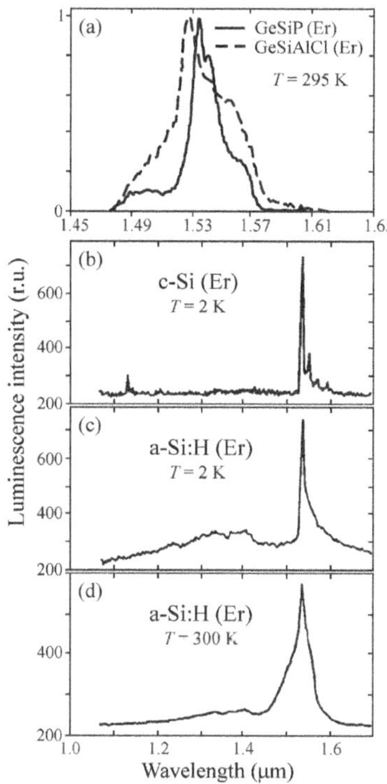

Fig. 5.17
Emission spectra of Er^{3+} ions built into various matrices and measured at different temperatures. (a) A glass GeO_2–SiO_2 matrix doped with Al_2O_3 (dashed curve) and P_2O_5 (solid curve) at $T = 295$ K. After Martucci *et al.* [16]. (b) Crystalline silicon matrix at $T = 2$ K. (c) Hydrogenated amorphous silicon (a-Si:H) matrix at $T = 2$ K. (d) a-Si:H matrix at $T = 300$ K. Except for the temperature, the experimental conditions and intensity scales are the same for (c) and (d). Graphs (b)–(d) after Bresler *et al.* [17].

These aspects of rare-earth-doped semiconductor luminescence are nicely demonstrated in Fig. 5.17, depicting the luminescence of Er^{3+} ions built into various matrices [16, 17]. The emission is due to $4f^{11*} \rightarrow 4f^{11}$ transitions from the excited state $4I_{13/2}$ to the ground state $4I_{15/2}$. Comparison of the emission spectra in Fig. 5.17(a), (b) and (c) immediately confirms that the emission of these ions is represented by a relatively narrow line peaking approximately at 1530 nm, regardless of whether they are incorporated in a glass GeO_2–SiO_2 matrix or in crystalline or amorphous silicon. Comparison of the emission spectra in Fig. 5.17(c) and (d) further suggests that the intensity of luminescence of Er^{3+} in hydrogenated amorphous silicon decreases by only about 20% when the temperature increases from 2 K to 300 K. Recently, there has been considerable interest in the luminescence of Er^{3+} in silicon as the exploitation of such luminescence would allow for the operation of an optoelectronic light source on a silicon basis, a necessity for silicon optoelectronics or photonics. The emission wavelength near 1.53 μm is highly attractive because it lies in the so-called second transmission window of the glass fibres used in optical communications.

Apart from the trivalent form, rare earths can be built into the lattice also in the form of divalent ions. In the Eu^{2+} ion, for example, which is an important activator in a number of phosphors used in applications, luminescence occurs via dipole-allowed $5d \rightarrow 4f$ transitions. The reason underlying this difference against the case of Er^{3+} ions is as follows: the Eu^{2+} ground state is $4f^7$, and it can have for partners in optical transitions both the $4f^{7*}$ excited state and the $4f^6 5d$ excited state. The intensity of the crystal field then determines which one of these excited states lies lower in energy; the $4f^6 5d$ level turns out to be the lower one in most Eu^{2+}-ion-activated phosphors, resulting in

$4f^6 5d \rightarrow 4f^7$ being the allowed transition underlying the luminescence in this case. One more difference can be found: the 4f shell is, similarly to trivalent ions, screened from the influence of the crystal field by the filled $5s^2 5p^6$ shell, but the excited 5d orbital is spatially located outside this shell and as such feels the interaction of the matrix much more strongly. This implies that, on the one hand, the emission spectrum can contain broad bands and, on the other hand, that its spectral position may vary with the chemical composition and spatial symmetry of the matrix.

For these reasons, Eu^{2+} ions provide a wide range of luminescence and absorption bands. The $CaF_2 : Eu^{2+}$ crystal can serve as an interesting example, since it was chosen as the very first material for the experimental demonstration of the occurrence of two-photon absorption in solids in 1961. The only laser then available was the ruby laser $\lambda = 694$ nm ($h\nu = 1.78$ eV); the cubic $CaF_2 : Eu^{2+}$ crystal is transparent at this wavelength and at the same time has a strong absorption band $4f \rightarrow 5d$ at $2h\nu = 3.56$ eV. In addition, $CaF_2 : Eu^{2+}$ was known to exhibit intense blue luminescence with a band around 420 nm (~ 2.95 eV), being due to radiative transitions after the relaxation of the excited electrons. The clever employment of these properties by Kaiser and Garrett [18] led to the actual detection of blue luminescence excited with red ruby laser pulses, with the intensity of this luminescence quadratically rising with the excitation intensity according to (5.24), see Fig. 5.18 (compare with the inset in Fig. 5.16(a)).

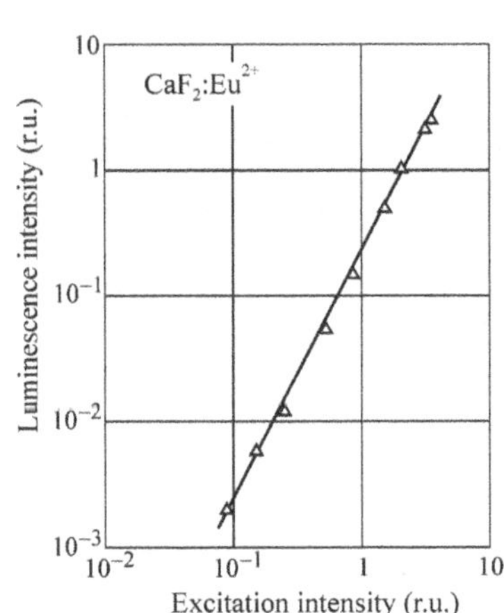

Fig. 5.18
Quadratic dependence of the intensity of blue luminescence on red laser excitation in $CaF_2 : Eu^{2+}$ in an original experiment demonstrating two-photon absorption in solids; $T = 295$ K. After Kaiser and Garret [18].

5.7 Problems

5/1: Derive relations (5.7) and (5.8) by making use of Fig. 5.1. Employ the parabolic approximation $E_2 = E_c + \hbar^2 k_0^2/2m_e$, $E_1 = E_v - \hbar^2 k_0^2/2m_h$, along with $E_2 - E_1 = h\nu$.

5/2: Relation (5.9) for the luminescence lineshape of electron–hole pair recombination includes a parameter $D_0 \sim \exp(-E_g/k_B T)$, which increases with rising temperature. Does this mean that the intensity of this luminescence will also rise with temperature? Which important effect was neglected when deriving eqn (5.9)?

5/3: Prove (e.g. graphically) that the halfwidth of the Maxwell–Boltzmann lineshape (5.9) approximately equals $1.8\,k_B T$.

5/4: Calculate the integral $I = \int_0^a \sqrt{x}\sqrt{(a-x)}\mathrm{d}x$, $a \geq 0$, in eqn (5.12). Hint: Apply integration by parts and then the substitution $x = a\sin^2 t$. (Result: $I = (\frac{2}{3}\int_0^{\pi/2} \sin^4 t\,\mathrm{d}t)a^2$.)

5/5: Estimate what concentration of donor–acceptor pairs should occur in crystalline silicon so that the discrete D^0–A^0 lines appear in its low-temperature luminescence spectrum. The spectral position of such lines would be determined by the Coulomb term in (5.18). Using Fig. 1.1, find the corresponding spectral positions of these lines. Typical binding energies of donors and acceptors in silicon are E_D(phosphorus) $\cong E_A$(boron) $= 45$ meV, and consider typical radii $r_A \cong r_D \cong 3$ nm.

References

1. Bebb, H. B. and Williams, E. W. (1972). *Photoluminescence I. Theory*. In *Semiconductors and Semimetals* (ed. R. K.Willardson and A. C. Beer), Vol. 8, p. 181. Academic Press, New York.
2. Kittel, C. (1976). *Introduction to Solid State Physics*, 5th edn. John Wiley, New York.
3. Saleh, B. E. A. and Teich, M. C. (1991). *Fundamentals of Photonics*. John Wiley, New York.
4. Mooradian, A. and Fan, H. Y. (1966). *Phys. Rev.*, **148**, 873.
5. Ulbrich, R. (1973). *Phys. Rev. B*, **8**, 5719.
6. Shah, J. (1999). *Ultrafast Spectroscopy of Semiconductors and Semiconductor Nanostructures*. Springer, Berlin.
7. Thomas, D. G., Gershenzon, M., and Trumbore, F. A. (1964). *Phys. Rev. A*, **133**, 269.
8. Espe, W. (1954). *Luminescent Substances in Electrical Engineering* (in Czech: *Luminiscenční látky v elektrotechnice. Výroba, vlastnosti a použití*). SNTL, Prague.
9. Yariv, A. (1967). *Quantum Electronics*. John Wiley, New York.
10. Basov, N. G. (1966). *International Conference on Physics of Semiconductors*, Kyoto 1966. *J. Phys. Soc. Japan*, **21**, Supplement, p. 277.
11. Laibowitz, R. L. and Sack, H. S. (1966). *phys. stat. sol.*, **17**, 353.
12. Yano, S. and Ulbrich, B. (2003). *Thin Solid Films*, **444**, 295.
13. Shionoya, S. (1998). *Photoluminescence*. In *Luminescence of Solids* (ed. D. R.Vij), p. 95. Plenum Press, New York.
14. Shionoya, S. and Yen, W. M. (eds.) (1999). *Phosphor Handbook*, p. 141. CRC Press, Boca Raton.
15. Tanabe, Y. and Sugano, S. (1954). *J. Phys. Soc. Japan*, **9**, 753.
16. Martucci, A., Guglielmi, M., Strohhöfer, C., Fick, J., Pelli, S., and Righini, G. C. (1999). *Germania based sol-gel waveguides doped with* Er^{3+}. In *Innovative Light Emitting Materials*. Advances in Science and Technology (ed. P. Vincenzini and G. C. Righini), Vol. 27, p. 197. Techna, Faenza.
17. Bresler, M. S., Gusev, O. B., Kudoyarova, V. Kh., Kuznetsov, A. N., Pak, P. E., Terukov, E. I., Yassievich, I. N., Zakharchenya, B. P., Fuhs, W., and Sturm, A. (1995). *Appl. Phys. Lett.*, **67**, 3599.
18. Kaiser, W. and Garrett, C. G. B. (1961). *Phys. Rev. Lett.*, **7**, 229.

Non-radiative recombination

Luminescence efficiency or quantum yield almost never reaches $\eta = 1$; even if this were the case, Stokes' law implies that a portion of the supplied excitation energy is not transformed into luminescence radiation but, during relaxation to the system ground state, changes into other types of energy instead. The relevant transitions are called *non-radiative transitions* and the relevant electron–hole recombination is called *non-radiative recombination*. It is mostly an undesirable competing effect, which can even result, under strong pumping, in local overheating and mechanical damage of the material.

Depending on the final form of the dissipated excitation energy, one can recognize three basic types of non-radiative recombination in semiconductors:

1. recombination when the excitation energy transforms into heat (phonons);
2. recombination leading to a creation of new point defects in the lattice;
3. recombination transforming the excitation energy into photochemical changes of the material under study.

The first type of non-radiative recombination, which consequently slightly increases the sample temperature, is found most commonly. Within its framework, multiphonon transitions and Auger recombination can be distinguished. Photochemical changes induced by excitation radiation are restricted to only a limited number of compounds such as halides of silver and thallium. Similarly, the formation of lattice defects occurs quite rarely, mainly in some wide-bandgap materials at the border between semiconductors and ionic crystals.

Even though the theoretical foundations of non-radiative recombination processes in semiconductors are covered in the monograph by Abakumov *et al.* [1], in general non-radiative recombination is not as widely studied and understood as the diverse types of radiative recombination.

6.1 Transformation of the excitation energy into heat

6.1.1 Multiphonon recombination

The simplest idea that comes to mind under the concept of 'multiphonon non-radiative recombination in semiconductors' is probably as follows: an excited electron near the bottom of the conduction band, a hole near the valence band maximum and the entire excitation energy is—instead of being emitted in the form of photons $h\nu \cong E_g$ owing to radiative e–h recombination—handed over to the lattice through n phonons with energy $\hbar\omega$: $E_g = n\hbar\omega$, as depicted in Fig. 6.1(a). Actually, it is a non-radiative analogy of bimolecular recombination, discussed in Section 3.3. However, this process is highly improbable. From the perturbation theory point of view, it is an nth order process and for its probability $1/\tau_{nr}$ a relation like $1/\tau_{nr} \approx \exp(-E_g/\hbar\omega) \approx \exp(-n)$ can be foreseen. For a typical semiconductor with $E_g \approx 2$ eV and $\hbar\omega \approx 25$ meV we obtain $n = 80$, therefore, the corresponding probability is negligibly small and decreases with increasing forbidden gap E_g. Moreover, expression (5-9b) implies that the probability of the radiative transition, considering its proportion to ν^2, increases strongly in this case with the value of $h\nu(\approx E_g)$. Therefore, there are two reasons why wide-bandgap semiconductors outperform, as far as the luminescence efficiency is concerned, those with a narrow bandgap.

The probability of such free electron–hole pair non-radiative bimolecular recombination would increase substantially if a deep energy level occurs

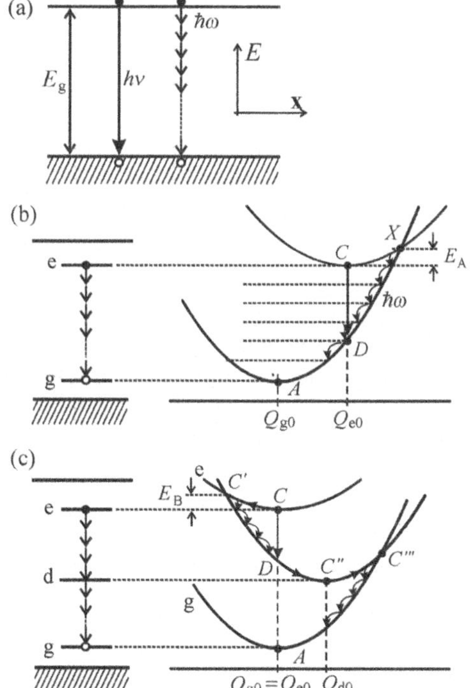

Fig. 6.1
(a) Non-radiative bimolecular recombination of free electron–hole pairs across the bandgap via multiple phonon $\hbar\omega$ emission. This process in fact does not occur in reality. (b) Non-radiative transition in a localized luminescence centre $C \to X \to A$ through the emission of many local phonons $\hbar\omega$('cooling transitions'). The centre is assumed to be in strong interaction with the lattice (high local change in the coordinate Q_{e0}–Q_{g0}). (c) Non-radiative transition into the ground electronic state $C \to C' \to C'' \to C''' \to A$ inside a localized centre that has only a weak interaction with the lattice ($Q_{e0} \approx Q_{g0}$). The non-radiative recombination is mediated by a deep level d. The thermal activation energy of the transition is denoted by E_B.

approximately in the middle of the bandgap; this process will be discussed in the next subsection. Now we shall pay attention to how non-radiative multiphonon transitions can be applied in a *localized luminescence centre*. Actually, we have already addressed the problem for the case of strong exciton–phonon coupling in Section 4.6, with the help of Fig. 4.8(b); its salient features are reproduced here as Fig. 6.1(b). The ordinate of the intersection point X lowers with increasing difference $Q_{g0} - Q_{e0}$ (i.e. with increasing exciton–phonon binding) and so does the activation energy E_A. The excited centre then reaches point X easily and then, down the trajectory $X \rightarrow A$, it may pass non-radiatively to the ground electronic state, emitting local phonons. Notice that the phonon emission here has a character substantially different from the (hypothetical) non-radiative bimolecular recombination depicted in Fig. 6.1(a). While in Fig. 6.1(a) multiphonon emission would be mediated by 'virtual' electron levels (much like, for instance, two-photon absorption) and therefore it would be a pure process of nth order perturbation theory, in the case of a localized centre, on the other hand, it proceeds by sequential application of one-phonon processes with participation of a real intermediate level.[1] The probability of such processes is then incommensurately higher, and non-radiative recombination of the localized centre via local phonon emission becomes very efficient, provided the centre is in strong enough interaction with the lattice.

If the localized centre interacts with the surrounding lattice only very weakly (e.g. rare earth ions), then $Q_{g0} \approx Q_{e0}$ (Fig. 6.1(c)) and the point of intersection X in fact does not exist. In this case, non-radiative recombination with many-phonon emission also becomes improbable, similarly to the case of a free electron–hole pair depicted in Fig. 6.1(a). Apart from other things this is one of the reasons why rare earth ions represent efficient luminescent centres if embedded into any lattice, as we have already discussed in Section 5.6.

Here, a very important factor mediating non-radiative recombination proves to be a *deep level* lying approximately in the middle of the bandgap, which corresponds to a defect located in the vicinity of the luminescent centre. It is significant that a defect associated microscopically with the deep level is always in strong interaction with the surrounding lattice, therefore the configurational coordinate changes from Q_{e0} to Q_{d0} during transfer of the excitation energy to this defect (Fig. 6.1(c)). This allows for a gradual dissipation of the electron excitation energy via 'cooling transitions' $C \rightarrow C' \rightarrow C''$. From here, the remaining energy 'spills' into the matrix by a pathway of similar character $C'' \rightarrow C''' \rightarrow A$, thus completing the non-radiative relaxation of the excited luminescence centre via multiphonon emission. The non-radiative recombination mediated by a defect is sometimes called Shockley–Read recombination.

It is worth realizing that, at the same time, the radiative channel of this centre may not be completely blocked ($\tau_r \rightarrow \infty$). The transition $C \rightarrow D$ in Figs 6.1(b) and 6.1(c) comprises the emission of a luminescence photon; transitions to the deep level can therefore be accompanied by weak broadband

[1] Such transitions are sometimes called 'cooling transitions' because they lead to a decrease of the non-equilibrium amplitude of the vibrations in the excited electronic state, therefore to a decrease of the high effective temperature.

luminescence. However, the intensity of such luminescence strongly decreases with decreasing thermal activation energy E_B (Fig. 6.1(c))—in other words, the stronger the interaction of the deep defect with the surrounding matrix, the weaker is the luminescence connected with this deep level.

We assumed tacitly that the luminescence centre was photoexcited via a path $A \rightarrow C$, see Fig. 6.1(c). The excitation energy, however, can be supplied to the centre also via capturing injected free carriers (the case of injection electroluminescence). This presumes, however, a sufficient electric conductivity of the surrounding lattice. Luminescence quenching of this type does indeed occur in light-emitting diodes [2].

What is the microscopic basis of the non-radiative centre constituting the deep level? In many cases this is not known precisely. It may be dislocations or vacancies in the bulk of the semiconductor, or recombination in the proximity of the interface between two materials may be involved (dislocations caused by different lattice constants, aggregated impurity atoms). Nevertheless, what is especially worth mentioning is the so-called *surface recombination*.

Surface recombination is a specific non-radiative recombination in solid-state phosphors, which takes place only in a thin subsurface layer and is conditioned by the occurrence of deep localized levels located within the forbidden gap (sometimes called Tamm levels or Tamm surface states). It is known that these levels, related to surface electron states, owe their origin to perturbation of the ideal translation symmetry of the infinite lattice by the crystal surface. More exactly, from the microscopic point of view it is unoccupied (dangling) bonds of surface atoms in a semiconductor that give rise to these levels.

For the sake of simplicity let us assume an intrinsic semiconductor, excited with a generation rate G. The situation is depicted in Fig. 6.2(a). As a consequence of the existence of the deep levels under discussion, the non-radiative recombination (see Fig. 6.1(c) or possibly Fig. 6.5) will be strongly inhomogeneous along the x coordinate normal to the surface, assuming a maximum at $x = 0$. This means, however, that the photocarrier concentration $n(x)$ just under the surface will decrease significantly, thus creating a concentration gradient driving photocarriers to the surface. Kinetic equations describing the generation and recombination processes then will, in comparison with previous considerations, incorporate an additional term arising from the *diffusion current* density $|\mathbf{j}| = -D|\,\mathrm{grad}\, n(x)| = -D\partial n(x)/\partial x$, where D stands for the diffusion coefficient of the carriers.[2]

We now analyse a steady-state regime, i.e. we apply the kinetic equations for $\partial/\partial t = 0$. The diffusion term, containing a derivative with respect to the space coordinate x, thus enters the kinetic equations through Fick's second law in the form of $-\,\partial/\partial x(-D\partial n(x)/\partial x)$. Thus instead of equation of the type (3.4) we have

$$D\,\frac{\mathrm{d}^2 n(x)}{\mathrm{d}x^2} - \frac{n(x)}{\tau} = -G, \qquad (6.1)$$

[2] Carrier diffusion is in this case governed by slower (heavier) carriers which, via Coulomb interaction, slow down the motion of their faster partners and ensure the electric neutrality of the photoexcited electron–hole system.

where, considering the steady state conditions, we substituted the total derivative d/dx for the derivative partial $\partial/\partial x$. Let us remind ourselves that τ is the (total) non-equilibrium carrier lifetime. When solving eqn (6.1) we apply the following boundary condition which tells us that the surface recombination described by the recombination rate S determines the value of the derivative of $n(x)$ at $x = 0$ according to relation

$$D \left. \frac{dn(x)}{dx} \right|_{x=0} = S\ n(x)\,|_{x=0}\,. \tag{6.2}$$

Note that S is expressed, as follows from (6.2), in cm/s.

An example of the solution of eqn (6.1) using the boundary condition (6.2) is graphically demonstrated in Figs 6.2(b) and (c). Panel (b) shows a strongly inhomogeneous photoexcitation profile $G = \alpha I_0 \exp(-\alpha x)$ under the common case of one-photon excitation. The second boundary condition here reads $n(x \to \infty) = 0$. Panel (c) refers to the case of a constant excitation profile $G = \text{const}$ connected with the second boundary condition $n(x \to \infty) = G\tau$. This may occur, e.g. under two-photon laser excitation or in electric injection in luminescent diodes. It is observed that in both cases a marked decrease of the concentration $n(x)$ close beneath the surface, indeed, occurs which may

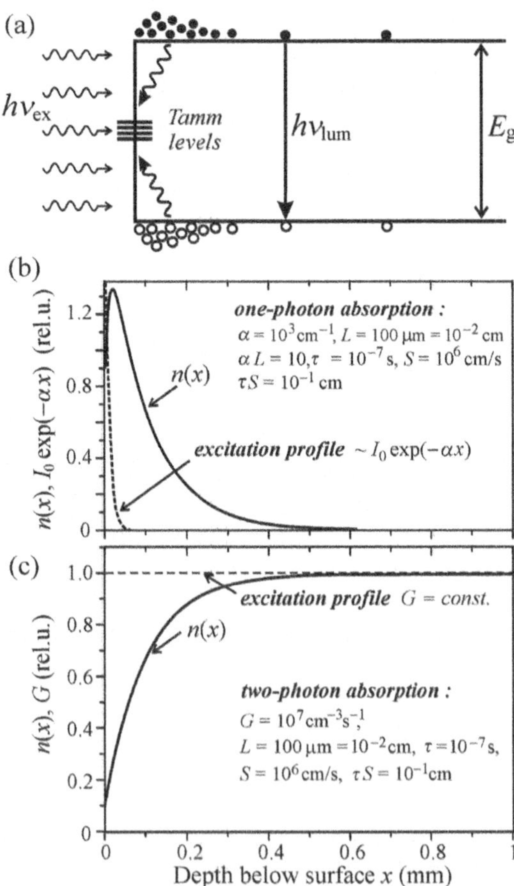

Fig. 6.2
Illustration of surface non-radiative recombination. (a) Schematic of photoexcitation and deep levels within the forbidden gap in the proximity of the surface. (b) Profiles of the generation rate $G(x) \sim I_0\exp(-\alpha x)$ and of the excited carrier concentration $n(x)$ under one-photon excitation. (c) Carrier concentration profile $n(x)$ under a constant generation rate $G = \text{const}$ (e.g. two-photon excitation). Parameters $L = (D\tau)^{1/2}$ (diffusion length), τ and S were chosen the same for both (b) and (c).

give rise to a luminescent 'dead layer' with strongly reduced luminescence intensity. The width of this layer can range—depending on the values of the parameters α, τ, S, D—from units to hundreds of micrometres. It should be mentioned that in order to demonstrate clearly the effects depicted in Figs6.2(b) and (c) a rather high surface recombination velocity ($S = 10^6$ cm/s) had to be chosen.

Under the condition G = const the subsurface decrease of $n(x)$ is more prominent and penetrates deeper. This fact can, on the one hand, be favourable in two-photon luminescence spectroscopy because it contributes to suppressing the spurious surface extrinsic radiative channels (Section 5.5). On the other hand, in applications—light-emitting semiconductor devices—the subsurface dead layer should be regarded as an undesirable effect which decreases the luminescence efficiency. In order to suppress the surface recombination special arrangements of the diode active layer are used. In general, the surface non-radiative recombination rate can be reduced by application of various passivation techniques which can decrease substantially the concentration of the subsurface recombination centres in the bandgap.

6.1.2 Auger and bimolecular recombination

Non-radiative Auger recombination in semiconductors can be considered parallel to the Auger effect known from electron spectroscopy of atoms and solids or from X-ray generation. In this process, an incident high-energy electron ejects another electron from an atomic core level; if some electron from an outer shell falls into the created vacancy, one X-ray photon is emitted. However, the filling process of the vacant state can also result in another outcome: the excitation energy is not released in the form of an emitted X-ray photon but, instead, is handed over to another electron from the same outer shell, ejecting it out of the atom. It is obviously a competing non-radiative process accompanying the X-ray generation.

As regards electron–hole pair recombination in semiconductors, an analogous Auger process is depicted in Fig. 6.3. It requires the presence of three quasi-particles, either two electrons and one hole (process *eeh*, see Fig. 6.3(a)) or two holes and one electron (process *ehh* from Fig. 6.3(b)). The energy released by one electron–hole pair recombination is transferred to the third quasi-particle which is then 'catapulted' either higher to the conduction band (e) or lower to the valence band (h). The energy of the recombining pair is passed to the third quasi-particle as its kinetic energy, and for this reason the Auger recombination is sometimes classified as an independent category of non-radiative transitions, characterized by the fact that the electronic excitation energy is—instead of being emitted as a luminescent photon—delivered to another electronic excitation. In the end, however, the kinetic energy of the third quasi-particle spills very rapidly (in picoseconds or their fractions) into the crystalline or amorphous matrix by multiphonon emission (Fig. 6.3, right column); the ultimate form of energy is therefore again heat. The ejected 'hot quasi-particle' (Auger quasi-particle) releases energy by relaxing to lower allowed energy levels in the respective band, taking advantage of the quasi-continuous character of these states.

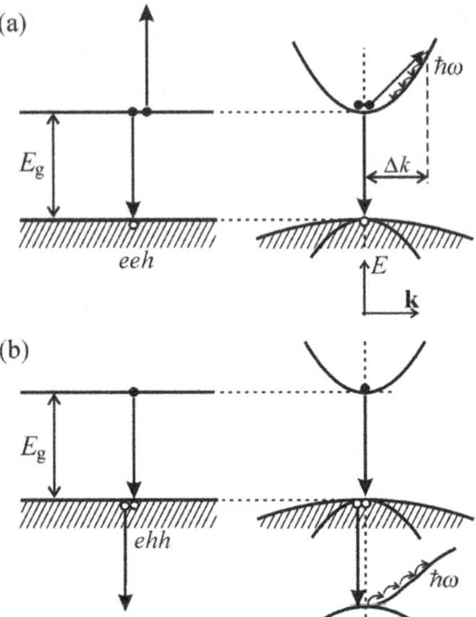

Fig. 6.3
Intrinsic non-radiative Auger recombination: (a) electron–electron–hole (*eeh*); (b) electron–hole–hole (*ehh*).

During the Auger process, obviously both energy and quasi-momentum conservation laws have to be satisfied. The schematics on the right-hand side of Fig. 6.3 show that the Auger process can occur both with (a) and without (b) involvement of a quasi-momentum conserving phonon $\hbar\Delta\mathbf{k}$; the no-phonon Auger process becomes feasible provided the ejected quasi-particle has to its disposal a suitable energy band, attainable without the alteration of wavevector of the (i.e. $\Delta\mathbf{k} = 0$). The latter alternative does not happen frequently; the first type of Auger process ($\Delta\mathbf{k} \neq 0$) is more common. The Auger mechanisms depicted in Fig. 6.3 can be called *intrinsic* because their occurrence is not conditioned by the presence of a defect or impurity.

Let us choose a rate of the non-radiative process $(\tau_{\mathrm{nr}}^{\mathrm{A}})^{-1}$ to be characteristic of the Auger recombination. Then we remember the bimolecular recombination described by eqn (3.9)

$$\frac{\mathrm{d}n(t)}{\mathrm{d}t} = G - \beta_{\mathrm{r}}\, n\, p - \beta_{\mathrm{nr}}\, n\, p, \tag{6.3}$$

where G is the generation term and $\beta_{\mathrm{r}}(\beta_{\mathrm{nr}})$ stands for the radiative (non-radiative) bimolecular recombination coefficient, see also Fig. 3.3. The Auger effect is a three-particle process, unlike two-particle recombination depicted in Fig. 3.3. We can therefore expect that Auger recombination will be described by a term $\sim np^2$ and instead of (6.3) it is possible to write

$$\frac{\mathrm{d}p(t)}{\mathrm{d}t} = G - A_{\mathrm{n}}n^2 p \qquad (eeh), \tag{6.4a}$$

$$\frac{\mathrm{d}n(t)}{\mathrm{d}t} = G - A_{\mathrm{p}}n\, p^2 \qquad (ehh), \tag{6.4b}$$

depending on whether the process (a) or (b) in Fig. 6.3 is under consideration. The so-called Auger coefficients A_{n}, A_{p} in eqn (6.4) will be related to the rate of non-radiative Auger recombination if we rewrite this equation in the form

$$\frac{\mathrm{d}p}{\mathrm{d}t} = G - \frac{p}{\tau_{\mathrm{nr}}^{A_{\mathrm{n}}}}, \tag{6.5a}$$

$$\frac{\mathrm{d}n}{\mathrm{d}t} = G - \frac{n}{\tau_{\mathrm{nr}}^{A_{\mathrm{p}}}}. \tag{6.5b}$$

By comparing eqns (6.4a), (6.4b) with (6.5a), (6.5b) we obtain

$$\frac{1}{\tau_{\mathrm{nr}}^{A_{\mathrm{n}}}} = A_{\mathrm{n}}\, n^2, \qquad \frac{1}{\tau_{\mathrm{nr}}^{A_{\mathrm{p}}}} = A_{\mathrm{p}}\, p^2. \tag{6.6}$$

Typical values of the Auger coefficient in semiconductors are of the order of $A \sim 10^{-31}$–$10^{-29}\mathrm{cm}^6/\mathrm{s}$, exceptionally (for the no-phonon process $\Delta\mathbf{k} = 0$) up to $10^{-26}\mathrm{cm}^6/\mathrm{s}$. The quadratic dependence of $(\tau_{\mathrm{nr}}^{A})^{-1}$ on the carrier concentration in eqn (6.6) signifies an important fact: the intrinsic Auger effect occurs mainly for high concentrations of free carriers. The *(eeh)* is type is expected to be found mainly in n-type semiconductors, while *(ehh)* is predominantly in p-type materials. Non-radiative Auger recombination obviously plays an important role in luminescence processes under high excitation and also whenever multiple free electron–hole pairs are localized in a small volume, e.g. in nanocrystals. Auger lifetimes τ_{nr}^{A} for $n, p = 10^{18}\mathrm{cm}^{-3}$ are of the order of 10^{-6} s, but for $n, p = 10^{19}\mathrm{cm}^{-3}$ they drop to only 10^{-8} s (!). It is worth noting that intrinsic Auger recombination contributes to the fact that luminescence of narrow-bandgap semiconductors is weak. The reason is that easy thermal generation of free carriers across the bandgap leads to a relatively high free carrier concentration already in unexcited material.

In addition to intrinsic Auger recombination, there also exists *extrinsic Auger recombination* when carriers localized at impurities also play a role. An example of such a process is displayed in Fig. 6.4: a non-radiative process accompanying luminescence originating in the transitions (e–A^0) and (h–D^0) and therefore decreasing its efficiency. Again a high concentration of free carriers or impurity atoms is essential here because the interaction of, for example, a free electron with an electron localized on a donor must occur, which is conditioned by the close proximity to each other. A slightly different type of extrinsic Auger process (accompanying bound exciton recombination) will be discussed in Chapter 7.

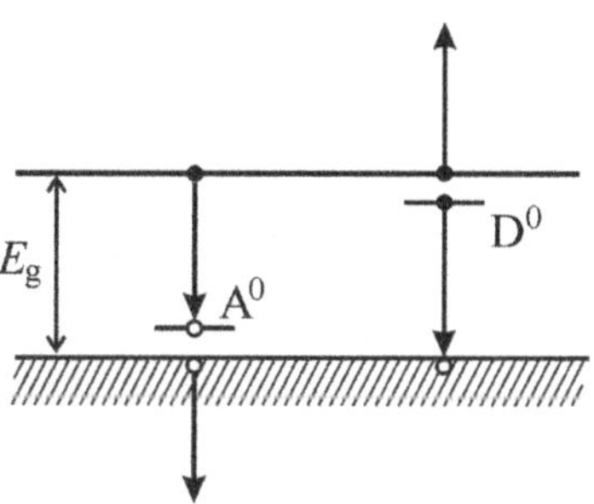

Fig. 6.4
Extrinsic nonradiative Auger recombination competing with radiative transitions $(e - A^0)$ and $(h - D^0)$. Compare with Fig. 5.7.

Let us now return to the question of non-radiative bimolecular recombination of free carriers. We know that bimolecular recombination, both radiative and non-radiative, is characterized by the quadratic nature of the recombination terms in kinetic equations of the type (6.3) or (3.9). We have already noted that the probability of such non-radiative processes in an ideal pure semiconductor (Fig. 6.1(a)) is in fact zero. In spite of this, it has been discovered experimentally that bimolecular non-radiative recombination occurs in a wide range of materials. We will show that the reason for this may lie in the existence of strongly localized states approximately in the middle of the bandgap (a suitably located 'deep level') and their co-actions with Auger recombination. (One therefore deals with the extrinsic effect because deep levels are due either to a defect or an impurity.) This mechanism has been proposed by Juška *et al.* [3].

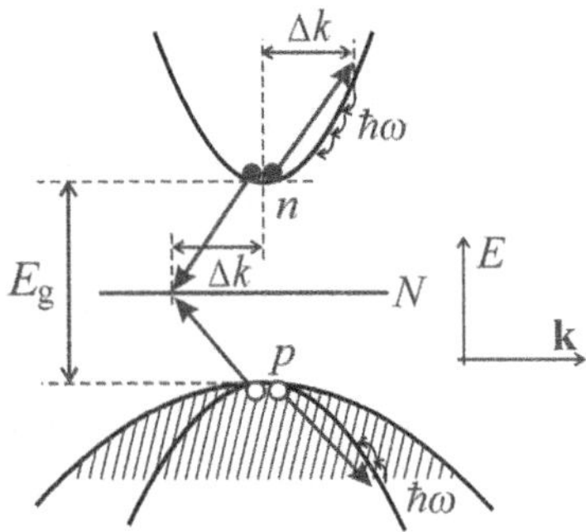

Fig. 6.5
Bimolecular nonradiative recombination mediated by a deep level (with concentration N) in the forbidden band in cooperation with Auger recombination of electrons (n) and holes (p) in respective extremes of the valence and conduction bands. $\hbar\Delta\mathbf{k}$ stands for quasi-momentum change, $\hbar\omega$ is phonon energy.

The process is clarified by the schematic sketch in Fig. 6.5. Its basis consists in Auger recombination of free electrons and holes in relevant bands; however, the final state of the transition of the particle releasing energy becomes the deep level within the forbidden gap. At the same time, if this energy state is strongly localized in **r**-space, it is strongly delocalized in **k**-space, which facilitates automatic satisfaction of the quasi-momentum conservation law. The probability of such processes is therefore high. Furthermore, its rate in comparison with the Auger recombination across the full bandgap (Fig. 6.3) is enhanced because of the smaller required change $\Delta\mathbf{k}$ in the Auger particle wavevector (for the Auger recombination matrix element M_A we have $M_A \approx 1/(\Delta\mathbf{k})^2$).

Let N be the concentration of the localized states, $n = p$ the concentration of the excited free carriers and n_N the momentary concentration of electrons on the localized level; the corresponding kinetic equations for electrons read

$$\frac{dn}{dt} = -A_n n^2 (N - n_N) \tag{6.7}$$

and for holes

$$\frac{dp}{dt} = -A_p p^2 n_N = -A_p n^2 n_N.$$

We eliminate n_N:

$$\frac{dn}{dt} = -A_n n^2 N - \frac{A_n}{A_p}\frac{dn}{dt},$$

$$\frac{dn}{dt}\left(1 + \frac{A_n}{A_p}\right) = -(A_n N)\, n^2,$$

$$\frac{dn}{dt} = -\left[\frac{A_n N}{(1 + A_n/A_p)}\right] n^2 = -\beta_{nr} n^2. \tag{6.8}$$

Therefore, the process depicted in Fig. 6.5 indeed has a bimolecular character, as reflected in the quadratic dependence $\sim n^2$. If $A_n = A_p = 10^{-27}\text{cm}^6/\text{s}$ ($\Sigma\Delta\mathbf{k} = 0$, the Auger coefficients therefore have high values) and $N = 10^{18}\text{cm}^{-3}$, we get from (6.8) for the bimolecular coefficient of non-radiative recombination $\beta_{nr} = 5 \times 10^{-10}\,\text{cm}^3/\text{s}$, a value in order-of-magnitude agreement with both experiment and the typical magnitude of the radiative bimolecular coefficient β_r, see Section 3.3.

Let us emphasize that the effective functioning of this model assumes a relatively high concentration of the deep traps, because there has to be a positive value $(N - n_N)$ on the right side of eqn (6.7). This model was originally proposed for amorphous silicon where the role of the deep levels is played by valence band tail states, which fulfil the above condition. Also a relatively high excitation is indispensable because it is in principle necessary to fulfil another condition n, $p > N$ in order to ensure a sufficient concentration of electrons and holes in the extremes of the bands during rapid carrier capture in the deep trap.

6.2 Creation of lattice defects

The principle of this excitation energy dissipation lies in the following mechanism: An electronic excitation is localized at some atom occupying a regular lattice position. Relaxation of such an excited atom into the ground state can proceed in the following way: the released excitation energy is transferred to the relevant atom as a whole, which is thus 'kicked off' into an interstitial lattice position, creating a so-called Frenkel defect (a vacancy plus an interstitial atom). This defect can be either transient or permanent depending on whether the interstitial atom does or does not return into its original lattice site once the excitation is terminated.

In order to make this non-radiative recombination process possible, the following criteria must be fulfilled, [4]:

(a) *Localization of the electronic excitation* must occur. Expressed quantitatively, the lifetime of such an excitation at the atom in question must be higher than the effective vibration period of the lattice, because the creation of the Frenkel defect is conditioned by relatively slow displacements of surrounding heavy atoms or ions. Strong localization of the excitation happens most easily if photoexcited electrons and holes or excitons move slowly throughout the crystal lattice. This occurs most probably when they have high effective masses, thus *polarons* with non-zero coupling constants α_e, α_h (Table 4.2) appear to be good candidates. It can be shown that the polaron mass m_p^* is connected with the effective mass of a 'bare' quasi-particle m^* by the expression

$$m_p^* = \frac{1 - 0.08\alpha^2}{1 - \alpha/6 + 0.0034\,\alpha^2}\, m^* \approx \left(1 + \frac{\alpha}{6}\right) m^*, \tag{6.9}$$

therefore $m_p^* > m^*$(the approximation on the right side holds for $\alpha \leq 0.5$). Such a localized excitation is usually called a self-trapped exciton; we will address self-trapped excitons in the next chapter.

One can immediately infer from Table 4.2 that the creation of self-trapped excitons is to be expected in II–VI and I–VII semiconductors, which have considerable ionic binding, a strong exciton–phonon interaction and a relatively wide forbidden gap.

(b) *Energy criterion.* The energy of the self-trapped exciton E_e (which does not differ too much from the bandgap width E_g) must of course be higher than the energy E_d of formation of the defect

$$E_g \geq E_e > E_d. \tag{6.10}$$

The energy E_d required by an atom or ion to jump to an interstitial position is very weakly dependent on the type of semiconductor and is mostly in the range of 1.5–3.5 eV. Relation (6.10) tells us, in accordance with criterion (a), that the creation of a lattice defect is mainly a feature of wide-bandgap semiconductors with a high fraction of ionic binding.

(c) *Protection against degradation to heat—phonons.* Local vibrations in the vicinity of an atom around which the self-trapped exciton gets localized must have a frequency ω_ℓ higher than the highest LO-phonon frequency $\omega_{\max}^{\mathrm{LO}}$, i.e. $\omega_\ell > \omega_{\max}^{\mathrm{LO}}$, which guarantees that phonons do not 'spread' the localized energy over the whole crystal. The phonon mode concerned must thus belong, referring to Fig. 4.7(d), among the so-called local modes, resulting from an increase of the effective spring constant $f = M\,\omega_\ell^2$ around the localized excitation; redistribution of the charge density accompanied by changes of the equilibrium atom positions may indeed often lead to the required increase of f.

(d) *Orientation criterion.* The self-trapped exciton (its charge density) should have a suitable symmetry in the crystal lattice so that the creation of the interstitial atom is energetically favourable.

An example of a semiconductor material in which such defect formation owing to the localized electronic excitation very probably occurs is AgCl. Here, the exciton–phonon interaction is so strong that a self-trapped exciton (in the form of an autolocalized hole on a silver ion $(AgCl_6)^{4-}$ in the lattice, which subsequently attracts a photoelectron via Coulomb interaction) is created at low temperatures very rapidly (in the order of picoseconds). In addition to radiative recombination, presenting itself as a broadband blue-green luminescence, this localized exciton can recombine also non-radiatively causing the relevant Ag^+ ion to be pushed out (probably only transiently) into an interstitial position. Therefore a Frenkel defect is created.

Interstitial silver ions Ag^+ in silver halides can also be created by heating; these materials are characterized by a rather high concentration of these interstitial ions, about 10^{13}–$10^{14}\,\mathrm{cm}^{-3}$ at room temperature. Their presence, hand in hand with their mobility, plays a fundamental role in photolysis and the formation of latent photographic images.

6.3 Photochemical changes

In silver halides AgCl, AgBr and AgI at sufficiently high temperatures ($T \geq 200\,\mathrm{K}$) there exists, besides defect formation, another channel dissipating energy of the optically created electron–hole pairs. A photoelectron becomes captured at a certain 'photosensitivity centre' while a hole gets localized on a Cl^- ion. The captured electron then attracts an interstitial Ag^+ ion which is able to diffuse rather freely through the lattice (the diffusion coefficient of the Ag^+ ions at room temperature is of the order of $10^{-12}\,\mathrm{cm}^2/\mathrm{s}$, while that of Cl^- ions is by several orders lower, basically unmeasurable). A neutral Ag^0 atom results, and successive trapping of other electrons and Ag^+ ions may lead to the growth of a metallic silver cluster—a seed of the latent photographic image:

$$\mathrm{Ag}^+ \underset{e}{\rightarrow} \mathrm{Ag}^0 \underset{e}{\rightarrow} \mathrm{Ag}^- \underset{\mathrm{Ag}^+}{\rightarrow} 2\mathrm{Ag}^0 \underset{\mathrm{Ag}^+}{\rightarrow} 3\mathrm{Ag}^+ \underset{e}{\rightarrow} 3\mathrm{Ag}^0 \underset{e}{\rightarrow} 3\mathrm{Ag}^- \ldots \qquad (6.11)$$

Sufficient mobility both of the electrons and of the interstitial metal ions, Ag^+ (the specificity of silver halides resides just in this) is therefore essential, and also a sufficiently high temperature is required because the ion mobility grows exponentially with T. The captured holes neutralize the Cl^- ions and chlorine gas leaks out, at least from the surface layers of the crystal. Therefore, a photochemical dissociation of the solid occurs.

At low temperatures ($T \leq 100\,\text{K}$) the motion of the interstitial Ag^+ obviously becomes strongly limited ('freezes'), the formation of the latent image fades away and silver halides become, on the contrary, relatively efficient phosphors. Their properties under UV irradiation are then very stable. (Hardly anybody is aware of the impossibility of taking photographs at cryogenic temperatures.) There exists a rather wide range of temperatures (100–200 K) where all three actions, i.e. radiative recombination, defect formation and photochemical reaction, may coexist. It is interesting that, although silver halides have been widely used in photography for as long as 150 years, a detailed understanding of all the steps in the reaction described by scheme (6.11) is still lacking. In particular, what is not fully clear is the microscopic nature of the primary photosensitivity centres.

Similar photographic processes occur, nevertheless with much lower efficiency, in halides of lead ($PbCl_2$), thallium (TlCl, TlBr) and mercury (Hg_2Cl_2, Hg_2Br_2); all these materials show luminescence in the visible region at low temperatures.

6.4 Problems

6/1: If a bimolecular radiative recombination $\sim \beta n^2$ and Auger non-radiative recombination $\sim An^3$ are simultaneously active in a semiconductor, the non-equilibrium carrier concentration $n(= p)$ is driven by the kinetic equation

$$\frac{dn}{dt} = -\beta\, n^2 - A\, n^3.$$

In the case when non-radiative recombination prevails (i.e. $\beta \ll An$), it will be possible to detect a very weak luminescence, whose kinetic behaviour will be described by the equation $dn/dt \cong -An^3$. Show that the luminescence decay curve can then be expressed as $I(t) = I(0)/[1 + 2AI(0)t/\beta]$, where $I(0) = \beta n^2(0)$ is the intensity at time $t = 0$.

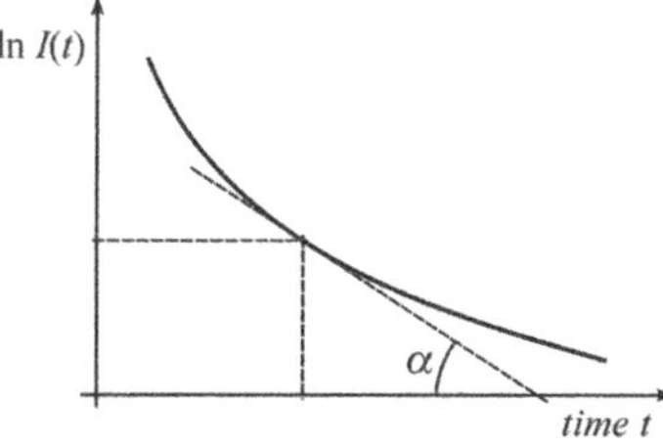

Fig. 6.6

6/2: *Quick graphical test of the type of recombination mechanism.* From the experimentally measured curve of a luminescence decay $I(t)$ (Fig. 6.6) it is possible to deduce the dominating type of carrier recombination by making use of the following method: We plot the curve on a semi-logarithmic scale and at several points we draw a tangent, thereby obtaining several values of its slope $\tan\alpha$. Then we plot $\ln|\tan\alpha|$ as a function of $\ln I(t)$. If this plot is constant, independent of $I(t)$, the dominating recombination is monomolecular, if it constitutes a straight line with slope 1/2 then a bimolecular recombination dominates, and a straight line with slope 1 indicates that non-radiative Auger recombination prevails. Prove this. (Hint: Start from the explicit expressions for $I(t)$. Employ the results of Problem 6/1.)

6/3: Discuss the experimental methods applied to study the Auger effect in semiconductors according to [2]; see also [5].

References

1. Abakumov, V. N., Perel, V. I., and Yassievich, I. N. (1991). *Nonradiative recombination in semiconductors*. In *Modern Problems in Condensed Matter Sciences*, Vol. 33 (ed. V. M. Agranovich and A. A. Maradudin). North Holland, Amsterdam.
2. Pilkuhn, M. H. (1981). *Light emitting diodes*. In *Handbook on Semiconductors* (ed. T. S. Moss), Vol. 4 (ed. C. Hilsum), p. 539. North Holland, Amsterdam.
3. Juška, G., Viliunas, M., Arlauskas, K., and Kočka, J. (1995-I). *Phys. Rev. B*, **51**, 16668.
4. Luščik, Č. B. and Luščik, A. Č. (1989). *Decay of electronic excitations with defect formation in solids* (in Russian: *Raspad elektronnych vozbužděnij s obrazovaniem defektov v tverdych telach*). Nauka, Moskva.
5. Dziewior, J. and Schmid, W. (1977). *Appl. Phys. Lett.*, **31**, 346; Benz, G. and Conradt, R. (1977). *Phys. Rev. B*, **16**, 843.

Luminescence of excitons

In an ideal pure semiconductor, the primary electronic excitation is a free electron–hole pair, the energy required for its creation (supplied, for example, by an incident photon) being equal—at the very minimum—to the bandgap value E_g. In a simplified way, an exciton may be visualized as a couple consisting of an electron and the associated hole, attracted to each other via Coulomb forces. Therefore, such a bound electron–hole pair no longer represents two independent quasi-particles and its internal energy is lower than E_g. The exciton is thus a quasi-particle representing the lowest electronic excitation in a semiconductor. There exist three basic types of excitons:

1. *Frenkel exciton* or a small-radius exciton. The spatial extension of the excitation is approximately restricted to a single unit cell. These excitons are to a large extent localized at a specific atom or molecule, and their movement through the crystal is limited to a hopping mechanism. They occur in molecular crystals.
2. *A charge transfer exciton* occurs primarily in ionic crystals. One can imagine its creation as follows: An electron is transferred from a lattice anion to a nearest neighbour cation, thereby creating there a maximum of the electron charge density. The radius of the charge transfer exciton can therefore be somewhat larger than that of the Frenkel exciton.
3. *Wannier exciton* or a large-radius exciton. The electron and hole are separated over many lattice constants, the exciton wavefunction is strongly delocalized and the exciton can move freely inside the crystal. Such a quasi-particle is also called a *free exciton*. The free exciton transfers the excitation energy, however, not the electric charge, because—as a whole—it is electrically neutral. Wannier excitons occur mainly in semiconductors.

The annihilation of an exciton is accompanied by a characteristic luminescence due to radiative recombination of the electron with the hole. In this chapter, we shall explain in detail the concept of the Wannier exciton and expound how it manifests itself in luminescence. Afterwards, we shall discuss the characteristic features of luminescence of the so-called bound excitons (i.e. excitons localized at impurity atoms or self-trapped owing to the strong exciton–phonon interaction).

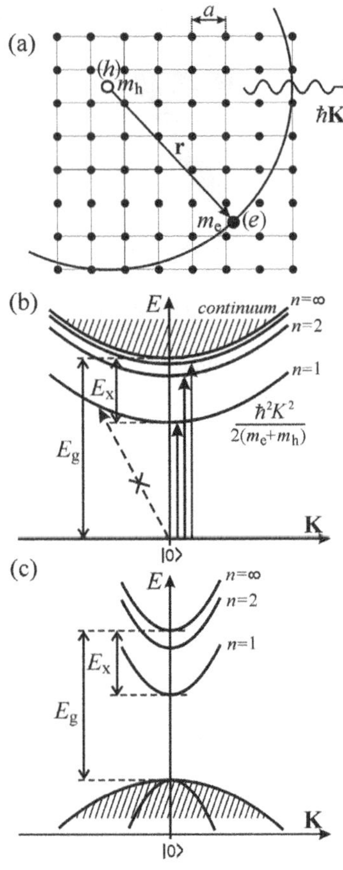

Fig. 7.1
(a) Schematic of the Wannier exciton in a two-dimensional crystal lattice with lattice constant a. Here $\hbar\mathbf{K}$ stands for the quasi-momentum belonging to the translational movement of the exciton centre of mass, and the effective masses of the electron (e) and hole (h) are denoted m_e and m_h, respectively. (b) The exciton dispersion relations $E = E(\mathbf{K})$. Optical absorption transitions are also marked; 'non-vertical' transitions are not allowed. (c) An incorrect plot of the exciton levels within the energy band scheme of a semiconductor (see text).

7.1 Concept of the Wannier exciton

The Wannier exciton can be conceived, in a first approximation, as a weakly bound electron–hole pair at which an electron and a hole circulate around each other; the attractive force results from the Coulomb potential

$$U(r) = -\frac{e^2}{4\pi\varepsilon_0\varepsilon\, r}, \tag{7.1}$$

where r is the electron–hole distance and ε stands for the dielectric constant of the substance. A striking resemblance with the hydrogen atom is already evident at first sight; the role of the proton is played here by the hole. If we regard the electron and hole as point charges characterized by their charge and effective masses (the so-called effective mass approximation), we can apply a modified Bohr model of the hydrogen atom. We shall see that this illustrative approximation can explain the majority of the principal features observed in the optical spectra of Wannier excitons in semiconductors.

Let us rotate the electron around the hole (which is always heavier, see Fig. 7.1(a)), thus circulating an orbit with radius r given by eqn (7.1). With regard to the Bohr radius of the hydrogen atom $a_B \approx 5 \times 10^{-2}$ nm, however, we encounter here two factors which impose the necessity of rescaling this characteristic length. Firstly, now we have a substantially different ratio of the (effective) masses. Unlike the pair of a light electron and a very heavy proton, the exciton is composed of two light quasi-particles with comparable masses m_e, m_h which entails a lower stability of the exciton in comparison with the hydrogen atom, and thus to a larger radius of the electron orbit. Secondly, the attractive electrostatic force between the electron and hole in a semiconductor is shielded by the dielectric constant ε,[1] which reduces further the attractive force and results in an increase in the orbital radius. It turns out that we may write

$$r_n = \frac{\varepsilon}{(m_r/m_0)} a_{Bn} = \frac{\varepsilon}{(m_r/m_0)} n^2 a_B, \tag{7.2}$$

where $n = 1, 2, 3, \ldots$ is the principal quantum orbit number, $m_r = (m_e m_h)/(m_e + m_h)$ is the reduced mass of the exciton and m_0 stands for the free electron mass. The commonly understood Bohr radius of the hydrogen atom a_B corresponds to the ground state, i.e. to the quantum number $n = 1 (a_B \equiv a_{B1})$. Considering typical values $\varepsilon \approx 10$ and $(m_r/m_0) \approx 0.1$ for a semiconductor, an order-of-magnitude estimate of the Wannier exciton ground state radius thus yields $a_X \equiv r_1 \approx 100 a_B \approx 5$ nm.

Let us rescale in a similar way the binding (or ionization) energy of the hydrogen atom. It is well known that this energy, commonly known as one Rydberg, amounts to 13.6 eV and refers to the ground state of the hydrogen atom. Let us denote $Ry(\mathrm{H}) = 13.6$ eV. As noted above, the low value of the

[1] A completely independent issue is whether the static ε_s or high-frequency ε_∞ dielectric constant is to be applied. This question has not yet been clarified up to now. It depends on whether the volume occupied by the exciton is large enough in order to be regarded as a homogeneous dielectric. Most frequently, the so-called Haken correction which interpolates between ε_s and ε_∞ is used.

Table 7.1 The binding energy and Bohr radius of free excitons in selected semiconductors. The width (E_g) and type (d–direct, i–indirect) of the bandgap are indicated.

Semiconductor	Binding energy E_X (meV)		Radius a_X (nm)	E_g (eV) i/d
	experimental	calculated		
CuCl	190		0.7	3.395 d
CuBr	108		1.25	3.077 d
Cu_2O	150		0.5	2.17 d
ZnS	36			3.78 d
AgBr	~ 22		~ 3	2.71 i
GaN	28	23	3.1	3.49 d
ZnO	61	~ 58		3.44 d
TlCl	23^i, 11^d			3.22i/3.42d
AgCl	23			3.27 i
TlBr	19^i, 9^d		2.6^i 4.1^d	2.66i/3.02d
GaP	18	21		2.35 i
ZnSe	19	20	4.5	2.8 d
CdS	30	28	2.7	2.58 d
ZnTe	13	13	5.5	2.39 d
CdSe	15	15	5.4	1.83 d
CdTe	10.5	12	6.7	1.60 d
GaAs	4.2	4.9	13	1.52 d
InP	5.1	4.8	12	1.42 d
Si	14.7	14.7	4.9	1.17 i
Ge	4.15	4.17	11.4	0.74 i
GaSb	2.8	2	23	0.81 d
InSb		0.4	100	0.23 d
SiC	28			2.42i

reduced mass (m_r/m_0) along with the high dielectric constant reduces the stability of the exciton. The exciton energy levels are therefore expected to lie markedly closer to the ionization continuum in comparison with the hydrogen atom:

$$E_{X(n)} = \frac{(m_r/m_0)}{\varepsilon^2}\frac{1}{n^2}Ry(H) = \frac{E_X}{n^2}. \tag{7.3}$$

The binding energy, denoted E_X in eqn (7.3), corresponds to the exciton ground state $n = 1$, therefore

$$E_X \equiv E_{X(1)} = \frac{(m_r/m_0)}{\varepsilon^2}Ry(H). \tag{7.4}$$

A typical value of E_X thus amounts to $E_X \approx (0.1/10^2) \times 13.6\,\text{eV} \approx 13\,\text{meV}$. The Wannier exciton *binding energy* is therefore of the order of tens of meV. The exciton is stable only if the attractive potential (7.1) is strong enough to prevent the exciton from breaking up owing to collisions with phonons. The binding energy E_X must thus be higher than $\sim k_BT$. Because the value $k_BT = 10\,\text{meV}$ is associated with $T \approx 110\,\text{K}$, one comes to the important conclusion that excitons in semiconductors occur only at low temperatures. Table 7.1 summarizes the basic exciton parameters in selected semiconductors.

To understand the next properties of the Wannier exciton, we have to go a step beyond the effective-mass approximation. From the point of view of wave–particle duality, it is possible to describe an exciton by its wavefunction and to tackle the problem of finding its stationary energy states using the standard methods of solid-state quantum theory [1, 2]; they lead to expressions (7.2) and (7.3). The exciton wavefunction, which is constructed as a linear combination of atomic functions of the crystal, has the same translational symmetry as the crystal lattice. The larger or smaller extent of the exciton's spatial localization (corresponding to its Bohr radius a_X) is described in quantum mechanics by the so-called envelope wavefunction.

Free excitons may thus move throughout the lattice. The following quantities are associated with such a translational movement: the exciton wavevector **K**, quasi-momentum $\hbar\mathbf{K}$ and kinetic energy $E_{kin} = \hbar^2 K^2/2(m_e + m_h)$. Considering eqn (7.3), the total exciton energy can then be written in the form

$$E_{(n)}(\mathbf{K}) = E_g - E_{X(n)} + E_{kin} = E_g - \frac{(m_r/m_0)}{\varepsilon^2}\frac{Ry(H)}{n^2} + \frac{\hbar^2 K^2}{2(m_e + m_h)}. \tag{7.5}$$

For the ground state, $n = 1$, energy we thereby get

$$E_{(1)}(\mathbf{K}) = E_g - E_X + \frac{\hbar^2 K^2}{2(m_e + m_h)} = E_g - E_X + \frac{\hbar^2 K^2}{2m_{exc}}, \tag{7.6}$$

where $m_{exc} = m_e + m_h$ is the total (or effective) exciton mass. Here, it is necessary to stress that, in addition to low temperatures, also sufficient *purity of the crystal* is essential for free excitons to exist. That is, the impurity atoms represent very efficient traps for the free excitons propagating through the crystal, which then get localized at the impurities (losing their kinetic energy) and *bound excitons* are created.

Equations (7.5) or (7.6) are nothing but the dispersion relations of a free exciton. They are depicted in Fig. 7.1(b). This figure looks very similar to a commonly used representation of the semiconductor band structure, just some extra discrete exciton levels within the bandgap appear. However, there is a substantial difference—the common energy band scheme of a semiconductor follows from the so-called one-electron approximation in which one chosen test electron moves in the effective potential of electric forces exerted by all of the other electrons and periodically arranged atomic cores. The exciton concept goes beyond this one-electron approximation because it takes into consideration the extra influence of a positive hole. Therefore, the scheme in Fig. 7.1(b) no longer belongs to the one-electron approximation and the $\mathbf{K} = |0\rangle$ point cannot be identified with the top of the valence band; it represents the ground state of the whole ideal pure crystal where all the spin and orbital electron momenta are compensated.

What may be somewhat misleading is that these exciton levels can in fact be formally drawn also into the one-electron scheme, which is depicted in Fig. 7.1(c). Even if one can meet similar pictures in renowned textbooks, this approach is not fully accurate and might lead to an incorrect interpretation of such schemes, as we shall see in Chapter 8.

7.1.1 Absorption spectrum of the Wannier exciton

Before we set about the exposition of exciton luminescence (i.e. light emission) properties accompanying the annihilation of the exciton, we describe how the creation of an exciton due to the absorption of an appropriate incident photon manifests itself in the optical absorption spectra of semiconductors.

Let us first consider a direct-bandgap semiconductor. It is well known that the onset of the absorption edge here is described (in the so-called dipole approximation and without considering the exciton effects) by the square-root law [2–4]

$$\alpha(h\nu) \approx (h\nu - E_g)^{1/2}, \quad h\nu \geq E_g. \tag{7.7}$$

The photons possessing energy $h\nu$ smaller than the bandgap width E_g therefore should not be absorbed at all. In fact, however, very distinctive absorption lines at certain discrete energies $h\nu < E_g$ are observed experimentally at low temperatures. Their origin is depicted by the upward arrows in Fig. 7.1(b), namely, they arise from transitions from the $|0\rangle$ state to the $n = 1$ state (creation of the exciton in its ground state) and to the higher exciton levels $n = 2, 3, \ldots$ This is thus the characteristic and famous hydrogen-like series in the absorption spectra, comprising lines at energies

$$h\nu = E_g - \frac{E_X}{n^2}, \quad n = 1, 2, 3, \ldots \infty; \tag{7.8}$$

the band-to-band absorption to the states labelled in Fig. 7.1(b) as 'continuum' begins by transitions to $n = \infty$.

It becomes clear by comparing (7.5) and (7.8) that the excitons created by this kind of light absorption have zero kinetic energy. In other words, the excitons at $\mathbf{K} = |0\rangle$ are created by vertical transitions. The question arises as to whether also 'non-vertical' transitions as indicated in Fig. 7.1.(b) can occur. The answer is no; they cannot, because the relation $\lambda \gg a$ holds between the optical wavelength $\lambda = c/\nu$ and the lattice constant a. The magnitude of the photon wavevector $2\pi/\lambda$ is therefore negligible compared with the characteristic dimension of the Brillouin zone $2\pi/a$ and, consequently, the quasi-momentum of the absorbed photon, is not sufficient to ensure the necessary change of $\mathbf{K}$ required for the non-vertical transition. Such a transition would thus have to incorporate a third quasi-particle—phonon—and the absorption act would then be classified as a higher-order process, the probability of which is many orders of magnitude lower.

A textbook example of the direct-bandgap absorption spectrum in the exciton region is given in Fig. 7.2 (pure GaAs at $T = 1.2$ K) [5]. Two additional important facts follow from this figure. Firstly, the intensity of the absorption lines drops with increasing n (proportionally to $\sim 1/n^3$), and, secondly, the optical absorption for $h\nu \geq E_g$ becomes enhanced under the influence of excitons (the absorption edge shape without considering the exciton effects, as given by eqn (7.7), is depicted by the dashed line).

In a direct-bandgap semiconductor in which the dipole transitions in the vicinity of the absorption edge are not allowed owing to the symmetry of the wavefunction in the band extrema (Cu_2O), the absorption spectrum has

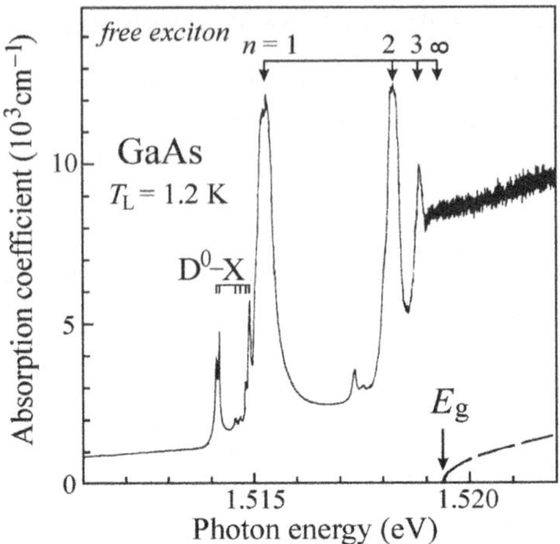

Fig. 7.2
Optical absorption spectrum of a very pure GaAs sample at $T = 1.2$ K in the vicinity of the absorption edge. Marked absorption lines, leading to the creation of the free exciton in the state $n = 1, 2$ and 3, dominate the spectrum. Absorption into higher exciton states merges gently into the absorption continuum; the value of the forbidden gap $E_g = 1.5194$ eV is determined by extrapolating the series to $n = \infty$. The dashed line at the bottom right represents the absorption edge shape $(h\nu - E_g)^{1/2}$ expected in the absence of the electron–hole interaction. D^0–X stands for the absorption on residual impurities. After Weisbuch and Benisty [5].

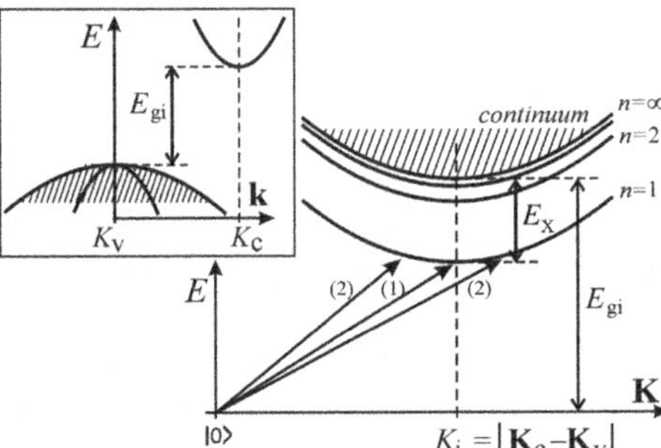

Fig. 7.3
Exciton dispersion relations in a semiconductor with indirect bandgap of width E_{gi}. The arrows indicate optical absorption transitions from the $|0\rangle$ point. The inset shows the relevant electronic band structure in the one-electron approximation.

a somewhat different character (the line intensity decreases proportionally to $(n^2-1)/n^5$ and the $n = 1$ line is missing). However, the marked absorption lines still dominate the spectrum.

In indirect-bandgap semiconductors, however, a substantially different spectral dependence around the onset of the intrinsic absorption edge occurs. The situation is schematically depicted in Fig. 7.3. The $\mathbf{K} = |0\rangle$ point again represents the crystal ground state, however, the exciton dispersion curves have a minimum at $K_i = |\mathbf{K}_c - \mathbf{K}_v|$, where $\mathbf{K}_c(\mathbf{K}_v)$ stands for the wavevector of the conduction (valence) band extremum, respectively; see the inset in Fig 7.3. A stable exciton can be created only when the group velocities of the electron and hole are the same; only in this case can the electron and hole move together as a bound pair. Since the electron or hole group velocity in the band is equal to

$$\mathbf{v}_g = \frac{1}{\hbar}\frac{\partial E}{\partial \mathbf{k}}, \tag{7.9}$$

the claim for equality of the group velocities (zero gradient) will undoubtedly be fulfilled in the valence $\mathbf{K}_v$ and conduction $\mathbf{K}_c$ band extrema, therefore the exciton created with a minimum energy will be characterized by the wavevector $K_i = |\mathbf{K}_c - \mathbf{K}_v|$.[2]

The onset of optical absorption should thus turn up at the photon energy $h\nu = E_{gi} - E_X$ (transition (1) in Fig. 7.3). However, because now a phonon $\hbar\omega$ with a wavevector $|\mathbf{K}_c - \mathbf{K}_v|$ must obviously participate in the absorption process in order to meet the quasi-momentum conservation law, the threshold value of the photon energy will shift to $h\nu = E_{gi} - E_X + \hbar\omega$ (we consider only low temperatures and hence the *phonon emission* into the lattice reservoir). Besides, the values of the absorption coefficient will be much lower compared with direct-gap semiconductors and—what is most important—there is no reason to observe any absorption lines, because the transitions (2) in Fig. 7.3 will have the same probability as the transition (1), both types of process being of the same order. Then the density of states in the exciton bands, proportional to $[E - (E_{gi} - E_{X(n)})]^{1/2}$, enters the play and the absorption coefficient will increase with increasing $h\nu$ starting from the threshold values $E_{gi} - E_{X(n)} + \hbar\omega$. Such a spectral course of the absorption coefficient is schematically depicted in Fig. 7.4(a). Instead of discrete lines, it contains characteristic 'shoulders' at the energies $E_{gi} - E_{X(n)} + \hbar\omega$. An example of a simple experimental exciton absorption spectrum of an indirect-bandgap semiconductor (AgBr) is shown Fig. 7.4(b) [6]. In reality, the experimental spectrum often shows a somewhat richer structure than that displayed in Fig. 7.4(b)—for instance in Si and Ge—as a consequence of the higher number of various types of participating phonons. Then, however, the analysis of such a spectrum is not an easy task.

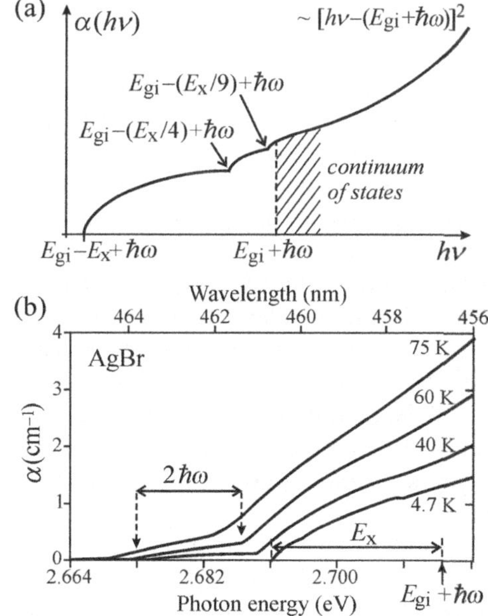

Fig. 7.4
(a) Schematic drawing of the low-temperature exciton absorption spectrum in an indirect-bandgap semiconductor. (b) Measured temperature dependence of the exciton absorption spectrum in AgBr. The drawing (a) corresponds to the curve recorded at $T = 4.7$ K. The value of $(E_{gi} + \hbar\omega) \approx 2.714$ eV (determined with the help of the binding energy value $E_X \approx 22$ meV established from other measurements) is indicated. As the temperature rises, a shift of E_{gi} towards lower energies occurs on the one hand and, on the other, absorption processes with phonon annihilation begin to operate, resulting in the occurrence of another shoulder red-shifted by $2\hbar\omega$. Thence it is possible to determine the energy of the quasi-momentum conserving phonon: $\hbar\omega \cong 8$ meV (TO(L)). After Joesten and Brown [6].

[2] In direct-bandgap semiconductor $\mathbf{K}_c = \mathbf{K}_v = 0$ holds and thus the exciton is created at $\mathbf{K} = |0\rangle$, in accordance with the previous discussion.

It is worth noting that the transitions of type (2) result in the appearance of excitons with non-zero kinetic energy, unlike direct-bandgap semiconductor where the resonant condition (7.8) leads to the creation of excitons with zero kinetic energy. However, it is important to realize that in both the types of semiconductors, excitons with non-zero kinetic energy can be created easily by optical absorption in the 'non-resonant' way. By means of photons $h\nu > E_g$ one creates free electron–hole pairs which are very quickly, in the order of picoseconds, bound to excitons. These thermalize rapidly, getting rid of the excess energy through the emission of phonons into the lattice. Such an exciton gas can be described statistically by its kinetic energy distribution with a certain effective temperature, as was explained in Section 5.3.

7.1.2 Direct bandgap: resonant luminescence of free exciton–polaritons

Looking at Fig. 7.1(b), an idea about exciton luminescence may cross our mind, namely, that luminescence transitions accompanying exciton annihilation can be depicted simply by turning upside down the arrows pointing up from $|0\rangle$ to $n = 1, 2, \ldots$. In the *emission spectrum* thereby a series of lines fully analogous to the lines shown in the *absorption spectrum* in Fig. 7.2 will appear (at least, it works like this in the case of a hydrogen atom). Unfortunately, this is not quite true. Here, the difference between a gas of non-interacting hydrogen atoms and the collective properties of the solid state will reveal itself, as well as the influence of residual impurities which are present even in nominally pure undoped semiconductor materials. The situation around the luminescence of free excitons thus becomes rather complicated.

The luminescence photon, created during exciton annihilation due to the radiative recombination of the electron with a hole, will be resonantly reabsorbed while travelling through the lattice (the absorption coefficient for exciton absorption lines is very high – see Fig. 7.2) and transformed back into an exciton; this exciton, when its lifetime is out, annihilates emitting a photon which is again reabsorbed, and so on. Such an oversimplified 'mechanistic' view is evidently untenable; however, it points to the fact that the concepts of the exciton and photon inside a solid can hardly be separated from each other. We speak of a mixed electronic-polarization and optical wave propagating through the crystal, for which the term *polariton* (not to be confused with polaron!) or more strictly *exciton–polariton* is used.[3]

A more solid idea about polaritons and their luminescence can be obtained from the graphical representation of their dispersion curves $E(\mathbf{K})$. The dispersion curves of the exciton $E(\mathbf{K}) = E_g - E_X + \hbar^2 K^2/2m_{exc}$ and of a bare photon $E(K) = h\nu = \hbar\, c\, K/\sqrt{\varepsilon_\infty}$ (i.e. the straight line with slope $\hbar\, c\,/\sqrt{\varepsilon_\infty}$) without considering the above mentioned interaction are schematically depicted in Fig. 7.5(a). Both curves must necessarily cross each other. However, it is known that in a degenerate quantum system, the degeneracy is

[3] Analogously, one can encounter a *phonon–polariton* in the infrared part of the spectrum in polar crystals.

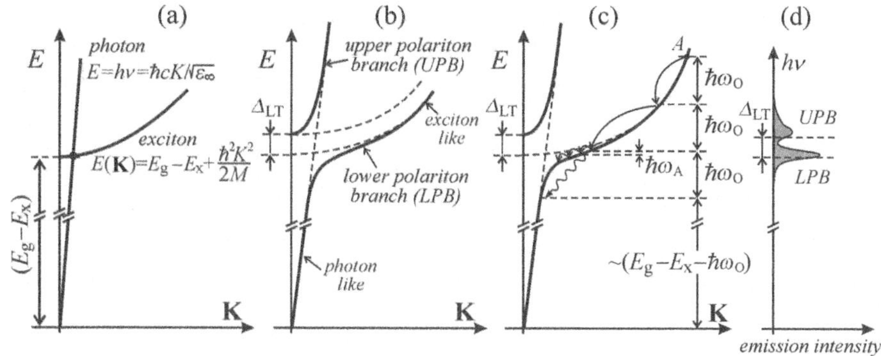

Fig. 7.5
(a) Dispersion curve of a 'bare' exciton and 'bare' photon (straight line with slope of $\hbar c(\sqrt{\varepsilon_\infty})^{-1}$), i.e. without considering their mutual interaction. (b) Schematic of the exciton–polariton dispersion curves. Δ_{LT} stands for the longitudinal-transverse exciton splitting.
(c) Thermalization of LPB polaritons owing to the cascade emission of optical ($\hbar\omega_{\mathrm{O}}$) and acoustic ($\hbar\omega_{\mathrm{A}}$) phonons. (d) Schematic of the luminescence spectrum of polaritons.

lifted owing to the interaction. Here, it means simply that a splitting of the curves in the vicinity of this point of intersection will appear, as suggested in Fig. 7.5(b). Moreover, the so-called longitudinal-transverse splitting Δ_{LT} of the exciton gains importance in polar cubic semiconductors, in analogy with the energy difference between LO- and TO-phonons (Section 4.2). The resulting polariton dispersion curves are drawn by solid lines in Fig. 7.5(b). One can indeed recognize around the intersection point a region having the character of mixed exciton-photon states, while far from this point the curves have a character very similar to a bare photon (the so-called 'photon-like' straight lines with a slope close to the value of $\hbar$ multiplied by the light velocity) or to a bare exciton with the characteric parabolic dependence $E \sim K^2$. The entire dispersion curve is then composed of two branches, namely, the upper polariton branch (UPB) and the lower polariton branch (LPB).

The existence of these two branches and, in particular, of the strong LPB curvature around the point where the 'photon-like' straight line turns into the 'exciton-like' parabola $\sim \hbar^2K^2/2m_{\mathrm{exc}}$, is essential for understanding the shape of the polariton emission spectra. This region in the $(E, \mathbf{K})$ plane is often termed a *bottleneck*.

Let us consider a common situation during which the polaritons are created (non-resonantly) via absorption of photons with energy high above the bandgap as, for example, at A point in Fig. 7.5(c). Thermalization of these polaritons runs initially—far away from the 'bottleneck'—very efficiently via emission of the optical phonons possessing a high energy $\hbar\omega_0$ at $\mathbf{K} \approx 0$; close to the bottleneck, however, the scattering to lower energy states can be mediated only via emission of the acoustic phonons with very low energy $\hbar\omega_{\mathrm{A}}$. Naturally, the thermalization process is slowing down significantly and a considerable accumulation of the polariton population on the LPB in the vicinity of the bottleneck will occur. A similar accumulation is likely to happen also near the bottom of the UPB. After a lapse of their lifetime of the order of

10^{-9} s, these accumulated polaritons 'recombine' to give rise to a no-phonon[4] exciton–polariton luminescence that, localized spectrally in the close vicinity of $h\nu \approx (E_g - E_X)$, emerges from the crystal. The emission spectrum thus consists, according to what has just been said, of two lines separated by $\sim \Delta_{LT}$. This is schematically depicted in Fig. 7.5(d). The line originating from the LPB branch is usually more intense, as may be inferred from thermodynamic considerations. It is necessary to emphasize, however, that no simple analytic formula describing the corresponding spectral lineshape like, e.g., eqns (5.9) or (5.13) exists, in spite of numerous attempts to deduce such a lineshape formula theoretically.

It may be interesting to mention that this spectral shape actually complies with the original simple concept of the excitonic luminescence reabsorption—the central part of a narrow emission line is (seemingly?) reabsorbed and only both of the wings survive (see also Appendix H).

Figure 7.6 shows an example of the experimental emission spectrum of the exciton–polariton in GaAs at $T = 4.2$ K [7]. The thick arrow marks the position of the $n = 1$ absorption line (i.e. $h\nu = E_g - E_X$), taken from Fig. 7.2. A qualitative agreement with Fig. 7.5(d) is evident.

The above discussed model of the exciton–polariton luminescence was refined by Koteles *et al.* [8] who noticed that the occurrence of the 'UPB/LPB doublet' was affected to a large extent by the concentration of residual impurities, in particular donors. The authors assume that all the excitons, after having thermalized, accumulate exclusively on the LPB branch and, accordingly, their luminescence spectrum should consist of a single line only. However, if donors are present at a relatively high concentration ($\geq 10^{15}$ cm^{-3}), the exciton–polaritons are scattered on their way from the bulk by elastic collisions just with these donors (not by acceptors because their Bohr radius—which means, in this case, their effective scattering cross-section—is

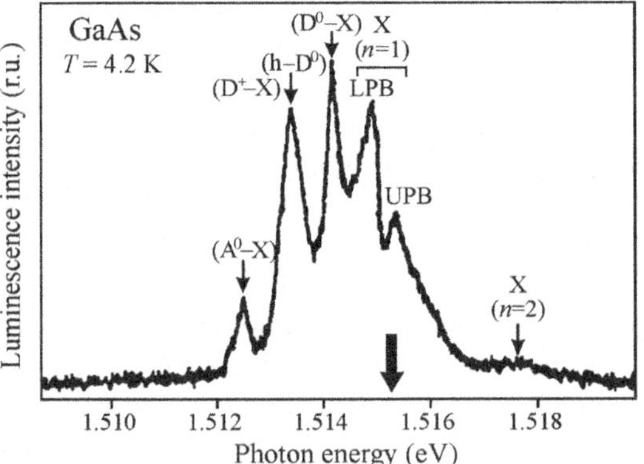

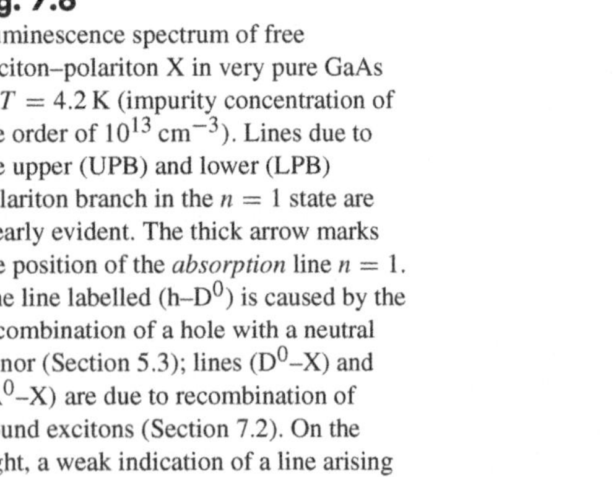

Fig. 7.6
Luminescence spectrum of free exciton–polariton X in very pure GaAs at $T = 4.2$ K (impurity concentration of the order of 10^{13} cm^{-3}). Lines due to the upper (UPB) and lower (LPB) polariton branch in the $n = 1$ state are clearly evident. The thick arrow marks the position of the *absorption* line $n = 1$. The line labelled (h–D^0) is caused by the recombination of a hole with a neutral donor (Section 5.3); lines (D^0–X) and (A^0–X) are due to recombination of bound excitons (Section 7.2). On the right, a weak indication of a line arising from recombination of the $n = 2$ state of the exciton–polariton can also be seen. After Razeghi *et al.* [7].

[4] That is, phonons do not participate in the actual emission process. However, it is possible to conceive of a process (depicted by the lower wavy line in Fig. 7.5(c)), during which emission of another phonon $\hbar\omega_0$ occurs and, at the same time, the polariton moves to the photon part of the dispersion curve. This means that the radiative decay of the exciton–polariton is accompanied by emission of an optical phonon. We shall treat this mechanism shortly.

substantially lower!). This scattering leads, according to detailed calculations, to a 'hole burning' in the luminescence line profile and to the emergence of the doublet. In sufficiently pure samples polariton scattering does not occur. Very convincing support for this theory is given in Fig. 7.7, which depicts three photoluminescence spectra of nominally undoped epitaxial GaAs layers with different residual donor concentrations. Incontestably, the free exciton (X) line splitting fades away with increasing sample purity. The effect seems to be confirmed by more and more frequent observation of a simple and intense luminescence line at the $n = 1$ position of the free exciton also in other materials, e.g. InP (Fig. 7.8) [9]. This tendency is likely to be closely related to the incessant improvement of a wide range of preparation techniques of ultrahigh-purity semiconductors, such as molecular beam epitaxy (MBE) or metal organic vapour phase epitaxy (MOVPE), etc. Figures 7.6–7.8 can serve as examples of a photoluminescence assessment of the semiconductor purity that shows a considerable application impact, and is *de facto* a unique and irreplaceable method used even in the microelectronics industry. Application of this photoluminescence impurity analysis in the case of silicon, where it even gives quantitative information, will be discussed in detail in Subsection 7.2.2.

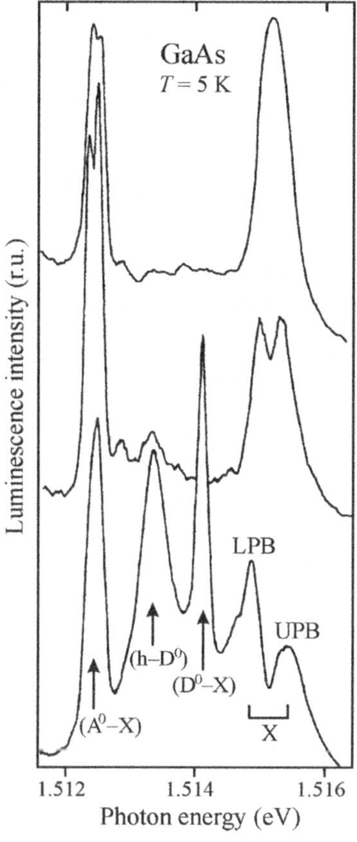

Fig. 7.7
Emission spectra of three undoped epitaxial layers of GaAs with different residual donor concentrations. The bottom spectrum belongs to the sample with the highest concentration of donors, and this concentration decreases towards the middle and upper spectra. At the same time it can be seen that the free exciton-polariton line (X) is losing its splitting into the UPB and LPB. Bath temperature is 5 K. Labelling of the remaining lines is the same as in Fig. 7.6. After Koteles *et al.* [8].

7.1.3 Direct bandgap: luminescence of free excitons with emission of optical phonons

The previous subsection might have evoked an impression that the entire spectral manifestation of free exciton luminescence in a direct-bandgap semiconductor is characterized by a kind of not-very-exactly defined lineshape which often almost disappears in the background of plenty of other emission lines. Nevertheless, there exists another intrinsic free exciton recombination channel in direct-bandgap semiconductors, significant especially in II-VI type semiconductors, namely, luminescence with simultaneous emission of LO-phonons (possibly TO-phonons, as well). Specific for this channel is a pronounced emission line (or even a series of lines) with a well-defined spectral shape.

This process, as we have already mentioned, is depicted by the lower wavy line in Fig. 7.5(c). From here we can at once estimate the spectral position of the relevant line as

$$h\nu_{\mathrm{X}-m\,\mathrm{LO}} \approx (E_{\mathrm{g}} - E_{\mathrm{X}}) - m\,\hbar\omega_0, \quad m = 1, 2, 3, \dots \tag{7.10}$$

This relation formulates nothing but the law of energy conservation, expressing that the lines are shifted, with respect to the exciton–polariton resonant line (or 'doublet'), towards lower energies by a multiple of the phonon energy $\hbar\omega_0$. (Figure 7.5(c) shows this recombination for $m = 1$ only.) In the literature these lines are sometimes marked as A–LO, A–2LO, ..., or X–LO, X–2LO, etc. Their intensity is relatively high but, at the same time, the reason for this is not obvious at first sight—after all, one deals with a process incorporating the participation of a third quasi-particle and thus its probability to occur should

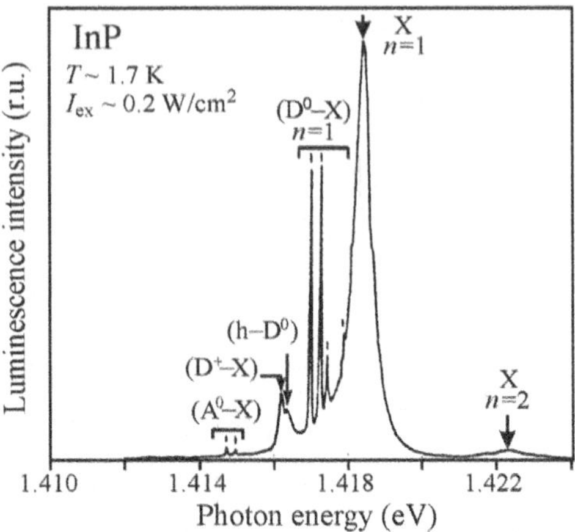

Fig. 7.8
Emission spectrum of a pure epitaxial layer of InP (residual impurity concentration below 10^{14} cm^{-3}). The spectrum is dominated by an unsplit line of the exciton–polariton ground state ($X_{n=1}$); a weak trace of the first exciton–polariton excited state ($X_{n=2}$) is also present. Other lines on the low-energy side of the spectrum are attributed to recombination of an exciton bound to a neutral donor in the ground state $(D^0–X)_{n=1}$, an exciton bound to an ionized donor (D^+–X), exciton bound to a neutral acceptor (A^0–X) and to recombination of a hole with a neutral donor (h–D^0). Notice that the bound exciton linewidth is much narrower in comparison with the linewidth of the free exciton–polariton; the latter reflects the kinetic energy distribution in the free exciton gas. After Bose *et al.* [9].

be substantially lower than that of the resonant emission itself. The reasons for this are in fact two:

1. The emitted phonon delivers part of the energy of the decaying exciton–polariton into the lattice, thereby lowering the energy of the luminescence photon by multiples of $\hbar\omega_0$. The photon may thus escape to a large degree (but not completely) from the bottleneck region and circumvent number the resonant polariton complications (reabsorption and scattering). Therefore, a large number of photons finally reaches the sample surface and leaves the semiconductor.
2. The participation of phonons makes it possible for *all* the free excitons residing at the LPB (which have a certain distribution of their kinetic energy $\hbar^2 K_{\text{exc}}^2/2m_{\text{exc}}$ characterizing their propagation through the crystal) to take part in the radiative recombination process. This is because the participating phonons ensure the quasi-momentum conservation law $\mathbf{K}_{\text{exc}} \approx \mathbf{q}_{\text{phonon}}$ to be satisfied for the whole exciton population, while it is only exciton–polaritons with virtually zero kinetic energy, i.e. a zero wavevector $\mathbf{K}_{\text{exc}} = \mathbf{K}_{\text{photon}} \approx 0$, that can undergo resonant no-phonon radiative recombination. Such excitons, however, represent only a small fraction of the whole free exciton population, as shown in Fig. 7.9(a) (see also Problem 7/1). This is therefore the second reason why phonon participation in the emission process substantially increases the $(X - m\text{LO})$ line intensity. (Sometimes one speaks of 'recoil' phonons released during the radiative annihilation of excitons.)

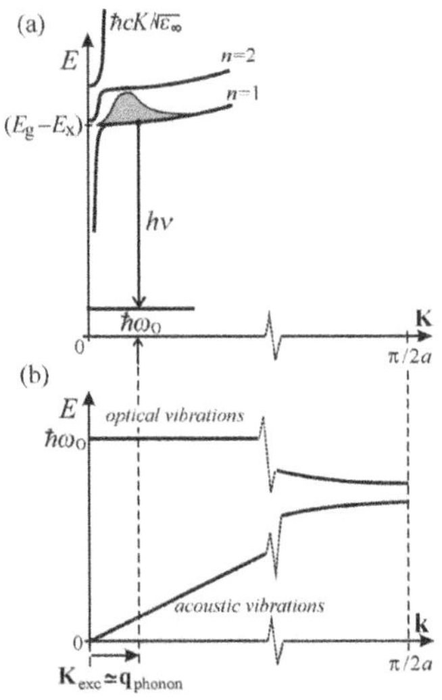

Fig. 7.9
(a) Radiative recombination of a free exciton accompanied by emission of one optical phonon. The photon energy is $h\nu \approx (E_g - E_X) - \hbar\omega_0$, the quasi-momentum conservation law $\mathbf{K}_{\text{exc}} \cong \mathbf{q}_{\text{photon}}$ is satisfied. The shaded area labels the Maxwell–Boltzmann distribution of the exciton kinetic energies. In the vicinity of $\mathbf{K} = 0$, a splitting of polariton states into UPB and LPB for both $n = 1$ and $n = 2$ is indicated. (b) Corresponding phonon dispersion curves.

Now, one can anticipate intuitively that shapes of the emission lines with phonon participation will reflect the Maxwell–Boltzmann distribution of the free exciton kinetic energy and will thus be analogous to eqn (5.9). Basically this is the case; nevertheless, some small differences appear.

The lineshape should reflect in some way the probability of phonon(s) creation $W^{(m)}(m\mathbf{q}_{\text{phonon}})$; this is a new factor that we did not have to take into account while deriving the relation (5.9); phonons did not participate there. We thus write formally

$$I_{\mathrm{sp}}^{(m)}(h\nu) \approx (h\nu - [(E_{\mathrm{g}} - E_{\mathrm{X}}) - m\hbar\omega_0])^{1/2}$$
$$\times \exp\left[-\frac{h\nu - [(E_{\mathrm{g}} - E_{\mathrm{X}}) - m\hbar\omega_0]}{k_{\mathrm{B}}T}\right]$$
$$W^{(m)}(m\mathbf{q}_{\mathrm{phonon}} \approx K_{\mathrm{exc}}), \tag{7.11}$$

considering simultaneously that the low-energy threshold of the line is given by eqn (7.10). Let us pay attention only to the (X–LO) and (X–2LO) lines, i.e. $m = 1, 2$. The quasi-momentum conservation law for one- and two-phonon assisted radiative annihilation of a free exciton is depicted in Fig. 7.10. During the one-phonon process (Fig. 7.10(a)), each exciton with a given $\mathbf{K}_{\mathrm{exc}}$ can create only phonons with essentially a discrete spectrum of magnitudes of their wavevectors $|\mathbf{q}_{\mathrm{phonon}}|$. The important question now reads: will the probability $W^{(1)}(\mathbf{q}_{\mathrm{phonon}})$ be the same for all excitons with different kinetic energy, i.e. with a different vector $\mathbf{K}_{\mathrm{exc}} \approx \mathbf{q}_{\mathrm{phonon}}$? The answer is no, because the moduli of all these vectors are small in comparison with the dimensions of the first Brillouin zone, and any of their relative variations $|\Delta\mathbf{q}|/|\mathbf{q}|$, however small they may be, is important. It can be expected that phonons whose *wavelengths* are comparable with the linear *dimension of the exciton* will be emitted most probably; therefore, the modulus of their wavevector will correspond to the modulus of the exciton wavevector. More accurate theoretical considerations lead to a scaling of the type $W^{(1)}(\mathbf{q}_{\mathrm{phonon}} \approx \mathbf{K}_{\mathrm{exc}}) \sim K_{\mathrm{exc}}^2$. In other words, $W^{(1)}$ is proportional to the exciton kinetic energy which

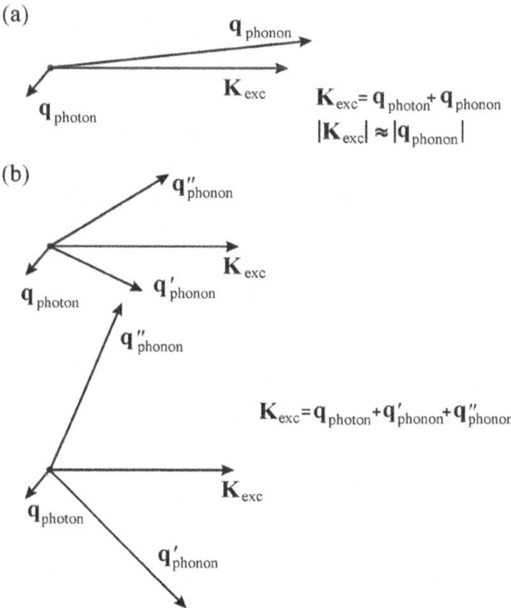

Fig. 7.10
Vector diagrams depicting the quasi-momentum conservation law during radiative annihilation of a free exciton: (a) with emission of one phonon, (b) with emission of two phonons.

is $(h\nu - [(E_g - E_X) - \hbar\omega_0])$. Therefore, for the (X–LO) process we obtain from (7.11)

$$I_{sp}^{(1)}(h\nu) \approx (h\nu - [(E_g - E_X) - \hbar\omega_0])^{3/2} \exp\left[-\frac{h\nu - [(E_g - E_X) - \hbar\omega_0]}{k_B T}\right]. \quad (7.12)$$

We emphasize once more that the scaling $W^{(1)} \sim |\mathbf{K}_{exc}|^2$ follows from the fact that the moduli of the exciton and phonon wavevectors are in this case comparable.

Two-phonon radiative recombination (X–2LO) satisfies the quasi-momentum conservation law in a slightly different way. The number of corresponding combinations of wavevectors $(\mathbf{q}'_{phonon}, \mathbf{q}''_{phonon})$ for each recombining exciton is high, and their moduli may be much larger than $|\mathbf{K}_{exc}|$, as can be seen in Fig. 7.10(b). Such phonons no longer 'feel' the spatial extent of the exciton, and one can expect $W^{(2)}(\mathbf{q}'_{phonon}, \mathbf{q}''_{phonon})$ to be no longer dependent upon the energy or upon the wavevector of the recombining exciton; thus $W^{(2)} = \text{const}$. It is this fact which causes the important difference between the one- and two-phonon process, and from (7.11) it follows immediately that the two-phonon lineshape ($m = 2$) will be Maxwell-like:

$$I_{sp}^{(2)}(h\nu) \approx (h\nu - [(E_g - E_X) - 2\hbar\omega_0])^{1/2} \exp\left[-\frac{h\nu - [(E_g - E_X) - 2\hbar\omega_0]}{k_B T}\right]. \quad (7.13)$$

These qualitative considerations were put forward for the first time by Gross *et al.* [10] and confirmed by quantum-mechanical calculations by Segall and Mahan [11]. Figure 7.11 reproduces (a) one of the first published emission spectra by Gross *et al.* in the spectral range of the (X − m LO) lines in CdS [10], and (b) a comparison of the experiment with the theoretical lineshape (7.12) in CdSe [12].

The second important feature of the emission lines (X–LO), (X–2LO),..., contributing to their identification, is their intensity dependence. The free exciton recombination, as a typical intrinsic 'monomolecular' process, increases with the excitation linearly in a wide interval of excitation intensities (no saturation is observed).

We have not yet given a full account of why just the *optical phonons* have to participate in phonon-assisted free exciton luminescence. In fact acoustic phonons could not make the relevant effect possible because at low values of $|\mathbf{q}_{phonon}|$, their energy is close to zero (Fig. 7.9(b)) and would not be sufficient to release photons from the bottleneck region, or—if you like—to enable them to escape from the resonant reabsorption. The optical phonons have a constant non-zero energy $\hbar\omega_0$ in the vicinity of $|\mathbf{q}_{phonon}| = 0$ and do not introduce any additional factor which could complicate the emission lineshape (7.12) or (7.13). But then the 'legal' question arises: Is the participation of the 'welcome' optical phonons allowed also by selection rules imposed by the crystal symmetry?

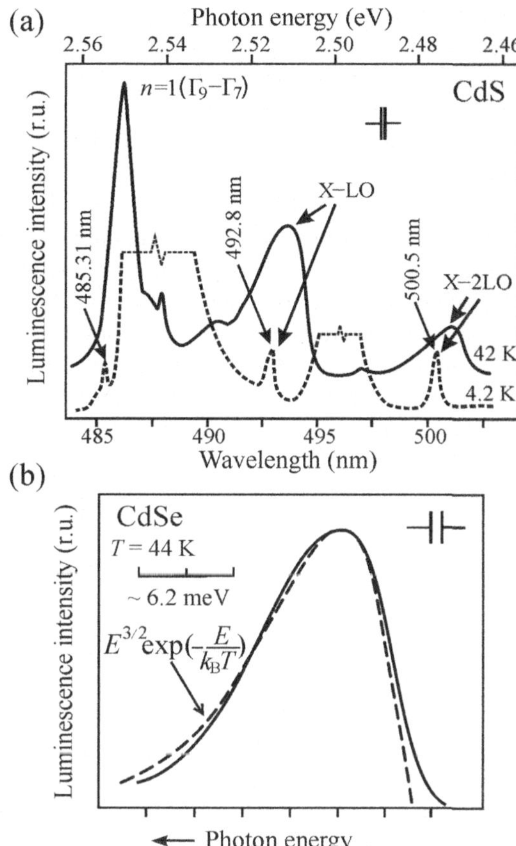

Fig. 7.11
(a) Emission spectrum of $n = 1$ free excitons (Γ_9–Γ_7) in CdS at $T = 4.2$ K (lower curve) and $T = 42$ K (upper curve). The lower curve from the left: a weak resonant line of the exciton–polariton (485.3 nm), a very intense and broad line (486–492 nm, intensity out of scale) due to an exciton bound to a residual impurity, a X–LO line at 492.8, once more an intense line of bound excitons (494–498) nm and finally a X–2LO line at 500.5 nm. At a temperature of $T = 42$ K, the lines of bound excitons almost disappear, the free exciton emission dominates the spectrum and the lines X–LO, X–2LO show the characteristic asymmetry described by relations (7.12) and (7.13). After Gross *et al.* [10]. (b) Comparison of the experimental lineshape X–LO, solid line, with the theoretical one (7.12), dashed line. Crystal CdSe at a temperature of $T = 44$ K. After Gross *et al.* [12].

In order to answer this question, a short excursion into the irreducible representations of point groups will be of use [3, 13]. Let Γ^{α}, Γ^{β}, Γ^{μ} be the irreducible representations (in general, of three different point groups of symmetry), corresponding to the wavefunction of the initial and final electron states and to the matrix element operator of the corresponding optical transition, respectively. Matrix elements of the type $\langle\Gamma^{\beta}|\Gamma^{\mu}|\Gamma^{\alpha}\rangle$ are then non-zero (i.e. the transition is allowed) if and only if

$$\text{the direct product}(\Gamma^{\alpha}\otimes\Gamma^{\mu})\text{includes}\Gamma^{\beta} \tag{7.14a}$$

or, equivalently,

$$\text{the direct product}(\Gamma^{\alpha}\otimes\Gamma^{\beta}\otimes\Gamma^{\mu})\text{includes a fully symmetric representation}\Gamma_1. \tag{7.14b}$$

Next, by using (7.14), we shall find the selection rule concerning LO-phonon participation in the excitonic luminescence of a commonly occurring hexagonal wurtzite modification of CdS crystals. Let us denote as **c** the six-fold rotation axis of the highest symmetry. The band structure is schematically shown in Fig. 7.12. The corresponding point group of crystal symmetry is C_{6v}.

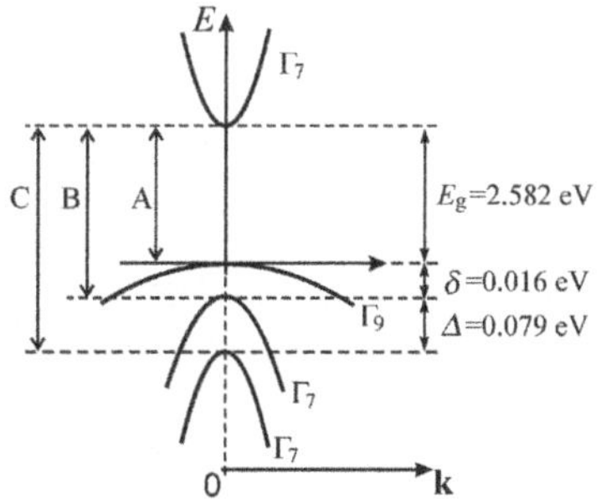

Fig. 7.12
Schematic of the energy band structure of hexagonal wurtzite CdS crystals. Δ stands for the spin–orbit splitting and δ is the splitting of the highest valence band by the crystal field. The transitions between three valence bands and the conduction band are usually marked as A, B, C. The emission lines in Fig. 7.11 belong to the A transitions.

Because the spin–orbit splitting of the valence band at the Γ point plays an important role here, irreducible representations of the double group are used for labelling the symmetry of the electron states. Let us first study the onset of the absorption edge $\Gamma_9 \rightarrow \Gamma_7$, neglecting for a while the exciton states.

The CdS crystals have a direct bandgap. In the relevant notation, the highest valence band maximum at $\mathbf{k} = 0$ has a symmetry $\Gamma^{\alpha} = \Gamma_9$, and the conduction band minimum at the same value of $\mathbf{k}$ has a symmetry $\Gamma^{\beta} = \Gamma_7$. In crystals with the point group C_{6v}, the photon operator for light polarization $\mathscr{E} \perp \mathbf{c}$ (in the dipole approximation) transforms according to the irreducible representation $\Gamma^{\mu} = \Gamma_5$. If we calculate the direct product $\Gamma^{\alpha} \otimes \Gamma^{\mu} = \Gamma_9 \otimes \Gamma_5$ employing the character tables of the irreducible representations (or we may consult previously published multiplication tables [3]), we obtain $\Gamma_9 \otimes \Gamma_5 = \Gamma_7 \oplus \Gamma_8$. This product contains the representation $\Gamma_7 (= \Gamma^{\beta})$ and, according to (7.14a), this means the interband transitions in the vicinity of the absorption edge are dipole allowed.[5]

Finally, let us pursue the luminescence of excitons with phonon participation. Because the interband transitions are dipole allowed, the exciton envelope function must be s-like ($\lambda = 0$) with a full Γ_1 symmetry. Therefore, the initial electronic state of the transition is

$$\Gamma^{\alpha} = \Gamma_1 \otimes \Gamma_7 \otimes \Gamma_9 \tag{7.15}$$

(i.e. a free exciton), the final state being the crystal ground state $|0\rangle$, which is certainly spherically symmetric: $\Gamma^{\beta} = \Gamma_1$. The overall operator of the transition matrix element transforms like $\Gamma^{\mu} = \Gamma_{\text{photon}} \otimes \Gamma_{\text{phonon}} = \Gamma_5 \otimes \Gamma_{\text{phonon}}$. In CdS, the LO-phonon has an energy of $\hbar\omega_0 = 37\,\text{meV}$ and symmetry Γ_5. Taking into consideration (7.15) and using the multiplication table we therefore obtain

$$\begin{aligned}
\Gamma^{\alpha} \otimes \Gamma^{\beta} \otimes \Gamma^{\mu} &= (\Gamma_1 \otimes \Gamma_7 \otimes \Gamma_9) \otimes \Gamma_1 \otimes (\Gamma_5 \otimes \Gamma_5) \\
&= (\Gamma_7 \otimes \Gamma_9) \otimes (\Gamma_5 \otimes \Gamma_5) \\
&= (\Gamma_5 \oplus \Gamma_6) \otimes (\Gamma_1 \oplus \Gamma_2 \oplus \Gamma_6) \\
&= \Gamma_1 \oplus \Gamma_2 \oplus \Gamma_3 \oplus \Gamma_4 + 3\Gamma_5 + 3\Gamma_6.
\end{aligned}$$

From the last expression it is evident that the product $\Gamma^{\alpha} \otimes \Gamma^{\beta} \otimes \Gamma^{\mu}$ includes Γ_1, therefore, according to condition (7.14b), the corresponding transition is allowed. The LO-phonon assisted radiative recombination of free excitons in crystals of CdS type is thus allowed by the selection rules.

We shall close this subsection with a brief summary: The participation of optical phonons in free exciton radiative annihilation in direct semiconductors removes polariton effects and leads to the appearance of luminescence lines with a characteristic shape described by relations (7.12) and (7.13). Their intensity dependence is linear in a wide range of excitation intensities. Prominent participation of LO-phonons in exciton luminescence, which cannot be overlooked particularly in II-VI materials, follows from the fact that these phonons have in polar semiconductors the strongest coupling—Fröhlich

[5] Sometimes the term 'first-class dipole-allowed' direct transitions is used; see the book by Peyghambarian, N., Koch, S. W., and Mysyrowicz, A. (1993), *Introduction to Semiconductor Optics*. Prentice Hall, Englewood Cliffs N.J.

coupling—with excitons (Section 4.2). In I-VII semiconductors, which are even more polar, the LO-phonon assisted luminescence of free excitons also occurs (for example in CuCl), but some of them have an indirect bandgap (AgBr, TlBr) and the mechanism of exciton luminescence is rather different there; we will explain this now.

7.1.4 Luminescence of free excitons in indirect-bandgap semiconductors

The first important fact affecting quite essentially (and basically in a positive sense) luminescence processes in indirect-bandgap semiconductors is the absence of polariton states. This is documented in Fig. 7.13. The energy minimum of the exciton states is located at the $\mathbf{K}_i$, a point sufficiently distant from $|0\rangle$ and the exciton and photon dispersion curves therefore do not cross each other. This phenomenon markedly simplifies the discussion of exciton luminescence in these materials.

A second important factor, following from the very nature of the indirect bandgap and also from Fig. 7.13, is the indispensability of participation of non-zero $\mathbf{q}_{\text{phonon}} \cong \mathbf{K}_i$ phonons in the radiative recombination process. This is, as we already know, a positive aspect—it enables the whole population of free excitons to recombine radiatively because a suitable 'recoil phonon' provides the momentum mismatch $\sim \hbar\mathbf{K}_i$. This time, however, the approximate equality $\mathbf{q}_{\text{phonon}} \cong \mathbf{K}_{\text{exc}}$ is no longer valid, unlike the case of a direct semiconductor. The reason for this consists in the fact that the $\mathbf{K}_i$ point is *de facto* in all known indirect semiconductors situated at the first Brillouin zone boundary or in its close vicinity, thus $|\mathbf{q}_{\text{phonon}}| \cong |\mathbf{K}_i| \gg |\mathbf{K}_{\text{exc}}|$, as indicated in Fig. 7.13. In this case, as we discussed in the preceding section, the probability $W^{(1)}$ of phonon creation does not depend on $\mathbf{q}_{\text{phonon}}$, and it follows immediately from eqn (7.11) that the spectral lineshape of the one-phonon assisted free exciton luminescence will reflect exactly the Maxwell–Boltzmann distribution of exciton kinetic energies. Now we can rewrite expression (7.11) as

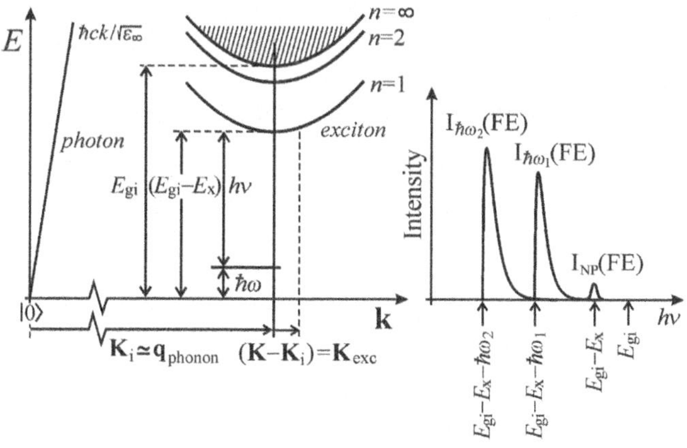

Fig. 7.13
Dispersion curves of a free exciton and a photon in an indirect-bandgap semiconductor. The polariton states do not exist. On the right: schematic of the exciton luminescence spectrum (I_{NP} stands for a no-phonon line).

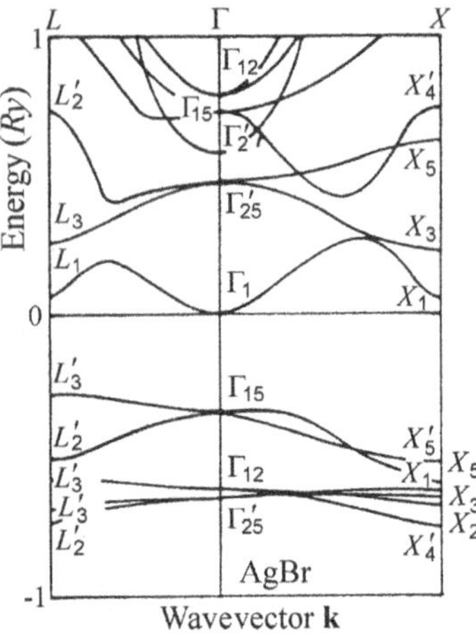

Fig. 7.14
Band structure of AgBr. The valence band maximum at the L point has a symmetry L_3', and the conduction band minimum with a symmetry Γ_1 is located at the Brillouin zone centre (Γ point, $\mathbf{k} = 0$). An ion of Br^- is placed at the origin of the coordinate system. Energy is plotted in units of $Ry(\mathrm{H}) \approx 13.6\,\mathrm{eV}$. After Kunz [14].

$$I^{(1)}_{\mathrm{sp(i)}}(h\nu) \cong (h\nu - [(E_{\mathrm{gi}} - E_{\mathrm{X}}) - \hbar\omega])^{1/2} \exp - \frac{h\nu - [(E_{\mathrm{gi}} - E_{\mathrm{X}}) - \hbar\omega]}{k_{\mathrm{B}}T}, \tag{7.16}$$

where the index (i) indicates the indirect bandgap. Appendix B features in detail the way in which the shape (7.16) or (7.12) can be modified in experiment by the influence of a finite slit width of the spectral apparatus and by so-called phonon broadening.

What remains to be considered is what kind of a phonon will take part in the recombination process. In strongly polar crystals of AgBr one would *a priori* expect participation of an LO-phonon. Surprisingly, this is not the case (that is why we have also omitted the index (0) in the phonon energy $\hbar\omega$ in (7.16)). Why?

Selection rules arising from the symmetry of phonon and electron states will apply, analogously to Subsection 7.1.3. Crystals of AgBr have a cubic structure belonging to the point group O_{h}. The valence band maximum is situated at the boundary of the first Brillouin zone and has a L_3' symmetry, while the conduction band minimum is at the Brillouin zone centre and has a Γ_1 symmetry (see Fig. 7.14) [14]. A photon in the crystal structure O_{h} transforms according to the irreducible representation Γ_{15}. (Relativistic effects and thus the spin–orbit splitting of the highest valence state in AgBr can be disregarded; thus the electronic states are classified according to a single point group.) Considering the symmetry of the exciton envelope to be Γ_1, the initial state of the transition related to the exciton recombination has a symmetry $\Gamma^{\alpha} = \Gamma_1 \otimes \Gamma_1 \otimes L_3'$. The crystal ground state Γ_1 represents the transition final state Γ^{β} (thus $\Gamma^{\beta} = \Gamma_1$), and the transition matrix element belongs to the representation $\Gamma^{\mu} = \Gamma_{\mathrm{photon}} \otimes L_{\mathrm{phonon}} = \Gamma_{15} \otimes L_{\mathrm{phonon}}$. Then, in accordance with (7.14a), the direct product

$$(\Gamma^{\alpha} \otimes \Gamma^{\mu}) = (\Gamma_1 \otimes \Gamma_1 \otimes L_3') \otimes (\Gamma_{15} \otimes L_{\mathrm{phonon}}) = L_3' \otimes \Gamma_{15} \otimes L_{\mathrm{phonon}} \tag{7.17}$$

must contain $\Gamma^{\beta} = \Gamma_1$ if the phonon-assisted transition is to be allowed.

Because L_3' and Γ_{15} are odd representations, L_{phonon} must be an even representation if the direct product (7.17) is to contain an even (Γ_1) representation. At the L point, AgBr has two types of phonons transforming according to an even representation: L_1(LA) and L_3(TO), see Fig. 4.6 (and also Problem 4/4).

We perform the multiplication indicated in (7.17), using the tables of irreducible representations of points with symmetries L and Γ [13], for both types of phonons in question:

phonon L_1:

$$\begin{aligned}(L_3' \otimes \Gamma_{15}) \otimes L_1 &= L_3' \otimes (\Gamma_{15} \otimes L_1) = L_3' \otimes (L_1' \oplus L_3') \\ &= \Gamma_1 \oplus \Gamma_2 \oplus 2\Gamma_{12} \oplus 3\Gamma'_{25} \oplus 3\Gamma'_{15},\end{aligned} \tag{7.18a}$$

phonon L_3:

$$\begin{aligned}(L_3' \otimes \Gamma_{15}) \otimes L_3 &= L_3' \otimes (\Gamma_{15} \otimes L_3) = L_3' \otimes (L_1' \oplus L_2' \oplus 2L_3') \\ &= 2\,(\Gamma_1 \oplus \Gamma_2 \oplus 2\Gamma_{12} \oplus 3\Gamma'_{25} \oplus 3\Gamma'_{15}).\end{aligned} \tag{7.18b}$$

Indeed, both eqns (7.18a) and (7.18b) contain Γ_1 on their right side, therefore the participation of both the L_1(LA) and L_3(TO) phonons in the radiative recombination of an indirect free exciton in AgBr is allowed. On the contrary, the LO-phonon in AgBr has symmetry L_2' at the L point and therefore its participation in the exciton luminescence is forbidden.

Selection rules derived in this way are in perfect agreement with experiment [15], as documented in Fig. 7.15. The dominant line in the low-temperature edge emission of pure AgBr crystals is a TO(L)-phonon replica of free exciton luminescence ($\mathrm{I_{TO}}$(FE)),[6] and, in addition, also a markedly weaker line $\mathrm{I_{LA}}$(FE) is present. This is due to the higher density of states of the TO-phonons in the L point along with a stronger coupling of these phonons with excitons over that of TA-phonons. In Fig. 7.15, also the theoretical lineshape (7.16) calculated using $\hbar\omega_{\mathrm{TO}} \cong 8\,\mathrm{meV}$, $\hbar\omega_{\mathrm{LA}} \cong 12\,\mathrm{meV}$ and an effective temperature of $T = 10\,\mathrm{K}$ is depicted (symbols; the height of the theoretical curves is always normalized to the maximum of the measured line). Agreement between theory and experiment, as far as the lineshape is concerned, appears also to be very good. We point out the characteristic asymmetry towards higher photon energies. In this context, we refer the reader also to Fig. 8.20.

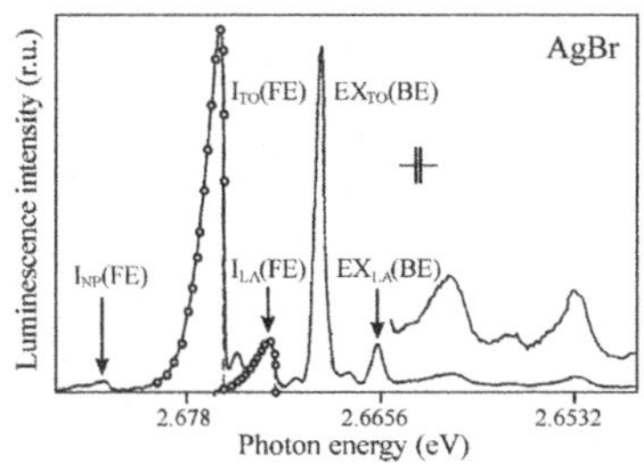

Fig. 7.15
Photoluminescence edge emission spectrum of pure AgBr (residual impurities ≤ 1 ppm) at a bath temperature of $T = 4.2\,\mathrm{K}$. Starting from the highest photon energy: a line of no-phonon recombination of a free exciton $\mathrm{I_{NP}}$(FE), a TO-phonon $\mathrm{I_{TO}}$(FE) and LA-phonon $\mathrm{I_{LA}}$(FE) assisted luminescence of a free exciton, respectively. Normalized emission lineshapes, calculated using (7.16) with an effective temperature of 10 K, are denoted by circles. (Compare with Fig. 7.13.) The $\mathrm{EX_{TO}}$(BE) and $\mathrm{EX_{LA}}$(BE) lines originate in the radiative annihilation of an exciton bound to an unknown residual impurity. The last three weak lines are two-phonon replicas. After von der Osten and Weber [15].

Figure 7.15 contains another two interesting pieces of information. Firstly, also a weak line $\mathrm{I_{NP}}$(FE) corresponding to the radiative recombination of an indirect free exciton without phonon participation is present. This process is theoretically forbidden in an ideal periodic lattice, however, due to the presence of residual impurities and lattice defects breaking the strictly periodic arrangement of atoms, it becomes partially allowed. Secondly, these residual impurity atoms manifest themselves in the luminescence spectrum very distinctly: free excitons get localized at the impurities, creating bound excitons (BE) which also decay radiatively. The corresponding lines are, contrary to free exciton luminescence, of extrinsic origin and this is reflected in their labelling: $\mathrm{EX_{TO}}$(BE), $\mathrm{EX_{LA}}$(BE). It is expected that in this case the intensity ratio $\mathrm{I_{TO}}$(FE)/$\mathrm{EX_{TO}}$(BE) varies significantly from sample to sample.

In silicon and germanium, similar symmetry-driven selection rules apply like in the case of AgBr. It turns out that in phonon-assisted free exciton radiative recombination in silicon all types of phonons (LO, TO, LA, TA) are allowed. Experimentally, though, only three lines $\mathrm{I_{TA}}$(FE), $\mathrm{I_{LO}}$(FE) and $\mathrm{I_{TO}}$(FE) are observed, as can be seen in Fig. 1.1. The reason for this consists in the degeneration of the LA, LO phonon dispersion curves at the X point of the Brillouin zone (Fig. 4.3), in the close vicinity of which the conduction band minimum in silicon is situated (Fig. 7.16). The weak $\mathrm{I_{LA}}$(FE) line thus *de facto* merges with the $\mathrm{I_{LO}}$(FE) line. In germanium, two allowed free exciton luminescence lines, namely $\mathrm{I_{LA}}$(FE) and $\mathrm{I_{TO}}$(FE), are observed, again in accord with the selection rules. Here, the spectral lineshape is more complicated due to the exciton ground state splitting (interaction of the anisotropic conduction band minima with degenerate hole states) and non-parabolicity of the density of states. If the theoretical description of the lineshape (7.13) is to be used to fit exactly the experiment, it needs substantial corrections. Further information on this issue can be found in [16, 17].

[6] This labelling is sometimes used in order to stress the intrinsic (I) character of the free exciton (FE) emission.

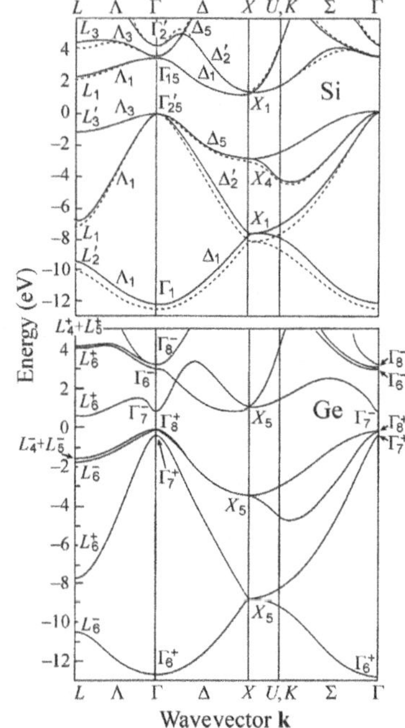

Fig. 7.16
Band structure of Si and Ge. In silicon, an indirect forbidden gap $E_{gi} \cong 1.17\,\mathrm{eV}$ is formed between the top of the valence band (Γ'_{25}) and the conduction band minimum situated along the Δ line, close to the X_1 point. In germanium, there is an indirect bandgap $\Gamma_8^+ - L_6^+$ of the magnitude of $E_{gi} \cong 0.745\,\mathrm{eV}$. Because Ge is heavier than Si, the spin–orbit splitting of the valence band maximum is more distinct and a double group notation is used for labelling the symmetry of the electronic states.

Finally, one important note: Similarly to the case of (X − m LO) lines (Section 7.1.3), a significant additional guideline to identify various free exciton luminescence lines in indirect semiconductors is their linear dependence on pump intensity.

We shall close this subsection by stressing once more what has already been mentioned repeatedly: a single microscopic luminescence centre in an indirect-bandgap semiconductor may manifest itself through several emission lines of comparable intensities (phonon replicas).

7.2 Bound excitons

We have already noted that lattice defects, and especially impurity atoms, present in the crystal either intentionally (doping) or as residual impurities which the preparation technology fails to get rid off, make very efficient potential wells—traps—for free excitons. The exciton loses its kinetic energy owing to localization at these traps and such excitons are then called *bound excitons* (BE), to be distinguished from free excitons (FE).[7] Afterwards, a radiative or non-radiative recombination of the bound exciton, accompanied in the former case by emission of a characteristic luminescence photon, occurs again. It is a typical *extrinsic luminescence*.

As a rule, the efficiency of the bound exciton luminescence is substantially higher than that of free excitons. Emission spectra of medium-level doped or sometimes also nominally pure semiconductors are usually dominated by the bound exciton luminescence. Why? Two reasons are of importance.

1. First of all, it is the large radius of the Wannier exciton a_X itself, representing in this context a large capture cross-section σ_X for capturing of an exciton at the potential well ($\sigma_X \approx \pi a_X^2 \approx 10^{-12}\,\mathrm{cm}^2$ if we consider a typical Bohr radius $a_X = 5\,\mathrm{nm}$). Let us estimate the mean free exciton lifetime τ_{tr} before trapping. Obviously, it is expected that the higher the cross-section σ_X, the higher the impurity concentration N and the faster the free exciton movement, the smaller is τ_{tr}:

$$\tau_{tr} \approx \frac{1}{\sigma_X N v}, \tag{7.19}$$

where v is the thermal velocity of exciton diffusion through the lattice. Let us assume $N = 10^{15}\,\mathrm{cm}^{-3}$ (corresponding to a nominally pure semiconductor, except silicon perhaps, where purity of $N \le 10^{11}\,\mathrm{cm}^{-3}$ can be reached), $\sigma_X = 10^{-12}\,\mathrm{cm}^2$ and $v = 10^6\,\mathrm{cm/s}$ (corresponding to a temperature of approximately 2 K); from (7.19) we obtain $\tau_{tr} \approx 10^{-9}\,\mathrm{s}$.

This value of τ_{tr} is comparable with the radiative recombination lifetime τ_r in direct-bandgap semiconductors. This means the chance of a free exciton in such a semiconductor to emit its characteristic luminescence is roughly the same as that of being trapped at an impurity and then to recombine with emission of BE luminescence.

[7] In the case of exciton localization at an impurity atom, the term 'bound exciton' is in fact synonymous with the term 'excited state of the impurity atom'.

In an indirect-bandgap semiconductor $\tau_r \geq 10^{-4}$ s holds, thus we get $\tau_{tr} \ll \tau_r$. Therefore, the chance of a free exciton to be trapped, forming in this way a bound exciton, is far higher than to recombine radiatively with emitting the characteristic free exciton luminescence ($h\nu \approx E_{gi} - E_X - \hbar\omega$). This simple estimate implies immediately that the low-temperature free exciton luminescence—for instance in silicon at a doping level of the order of 10^{15} cm^{-3}—should not appear at all. Indeed, Fig. 1.1 corroborates this conjecture: The bottom spectrum (C) originates in a sample with a total concentration of boron and arsenic atoms of approximately 8×10^{15} cm^{-3} and it is evident that the free exciton lines I_{TO}(FE)/I_{LO}(FE) at $\sim$ 1130 nm and I_{TA}(FE) at $\sim$ 1090 nm (i.e. the intrinsic emission) are totally missing, while the extrinsic emission of excitons localized at boron atoms (e.g. B_{TO}(BE), B_{LO}(BE)) is very intense. We will return to this point in more detail in Subsections 7.2.1 and 7.2.2.

2. The bound exciton luminescence is characterized by the so-called 'giant oscillator strength', as shown by Rashba and Gurgenishvili [18]. In principle this means that all the unit cells inside a 'volume' of approximately $\sim a_X^3$ around the impurity atom or defect contribute to the radiative recombination of the bound exciton. The probability of this radiative recombination (and, consequently, the relevant oscillator strength f) is thus increased by several orders of magnitude, and the radiative lifetime $\tau_r \sim 1/f$ decreases proportionally.

These two effects can also be expressed in the following way: The crystal lattice, serving as a 'matrix' for impurity atoms or defects, plays, during photoluminescence processes, the role of an 'antenna' that 'harvests' the excitation radiation and transfers the excitation energy efficiently—with the help of the motion and trapping of free excitons—to the extrinsic luminescence centres. Expressed concisely, photoluminescence is a very sensitive tool to study impurities and defects in semiconductors.

How do we identify the bound exciton luminescence in the emission spectrum? One deals with a localized excitation, characterized very often (but not always!) by a small Huang–Rhys factor. This is owing to the large radius of a typical bound exciton, or owing to the fact that the excited area is spread in the host lattice over a considerable volume. Therefore, the spectra are composed of very narrow emission lines (FWHM of the order of 0.1 meV) which, moreover, do not broaden with increasing temperature (because the bound exciton lacks kinetic energy). Figure 7.17 represents the first emission spectrum of a bound exciton published in 1960 [19], which clearly demonstrates these characteristic features. A very narrow As_{TO}(BE) line originates in the TO-phonon assisted radiative recombination of an exciton bound to an arsenic atom (donor impurity in silicon). Its narrow width contrasts sharply with the neighbouring free exciton line $I_{TO/LO}$(FE), broadened considerably at the experimental temperature of $T = 25$ K. A narrow no-phonon line of the bound exciton As_{NP}(BE) is also present.

There are two additional important features contributing to the reliable identification of the extrinsic bound exciton luminescence. The first one is the

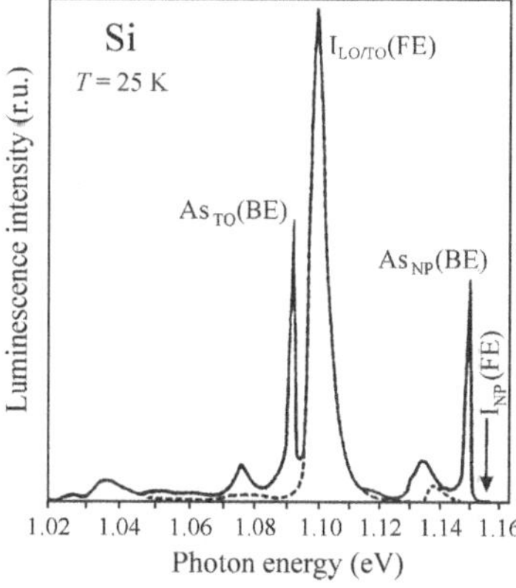

Fig. 7.17
The first published emission spectrum of a bound exciton: crystalline silicon containing an arsenic impurity with a controlled concentration of 8×10^{16} cm^{-3} (full line) in comparison with a silicon sample containing negligible concentration of impurities (dashed line). Silicon doped with arsenic shows a new line As_{TO}(BE) due to radiative decay of excitons bound to As atoms. An intense $I_{LO/TO}$(FE) line is a fingerprint of the free exciton luminescence; As_{NP}(BE) denotes no-phonon emission of the bound exciton. After Haynes [19].

linear intensity dependence with the expected tendency to *saturation* at higher excitation power densities, as outlined in Section 3.5. Next, the specific spectral position of the bound exciton line may be considered to be the second feature. Because the free exciton during its localization hands over part of its energy to the surroundings (and the rest is later emitted), the bound exciton emission line must be shifted—with respect to the corresponding free exciton replica—towards *lower photon energies*. This is clearly demonstrated in Fig. 7.17. In the next subsections we shall explain in more detail the physical origin of this energy shift.

7.2.1 Excitons bound to shallow impurities

Shallow donors and acceptors, controlling in a principal way the electric conductivity of semiconductors, are of fundamental importance in semiconductor technology. These impurities can reveal themselves in a characteristic way also in luminescence radiation; this is naturally profitable because it enables the donors and acceptors to be investigated also by optical means. We shall point to the slightly different luminescence behaviour of excitons bound to ionized donors or acceptors, and that of excitons bound to these impurities in a neutral state.

Excitons bound to ionized donors (D^+–X) or acceptors (A^-–X)

This is the simplest type of bound exciton. To describe the processes of free exciton capture and subsequent radiative annihilation of the bound exciton, it is advantageous to use quantum-chemistry notation. Let us thus denote

e.................. electron,
h.................. hole,
$\oplus e$............... neutral donor; also D^0,
$\ominus h$............... neutral acceptor; also A^0,
FE............... free exciton,
$\oplus$................. ionized donor; also D^+,
$\ominus$................. ionized acceptor; also A^-,
$\oplus\, eh$............. exciton bound to ionized donor; also (D^+–X) or (D^+, X),
$\ominus\, eh$............. exciton bound to ionized acceptor; also (A^-–X) or (A^-, X),
$\oplus\, eeh$............ exciton bound to neutral donor; also (D^0–X) or (D^0, X),
$\oplus\, heh$............ exciton bound to neutral acceptor; also (A^0–X) or (A^0, X).

The capturing event of a free exciton by an ionized donor can be described schematically as

$$\oplus + \mathrm{FE} \rightarrow \oplus\, eh + D_1, \tag{7.20a}$$

where D_1 is the dissociation or binding energy of the exciton in the relevant complex. It is thus the energy released by the exciton during localization or, alternatively, the energy required to tear the exciton off the ionized donor and make it free again, i.e. the energy necessary to start the reaction (7.20a) going in the opposite direction. (D_1 is not to be confused with the binding energy E_X of an electron–hole pair in a free exciton!) The radiative decay of such a bound exciton $\oplus eh$, resulting in emission of a photon $h\nu_{\mathrm{BE}}$, then reads

$$\oplus eh \rightarrow \oplus + h\nu_{\mathrm{BE}}. \tag{7.20b}$$

By combining both eqns (7.20) we obtain immediately

$$h\nu_{\mathrm{BE}} = \mathrm{FE} - D_1 \cong (E_g - E_X) - D_1, \tag{7.21}$$

where the meaning of the spectral shift of the bound exciton line is clear, i.e. the energy separation between the low-energy threshold of the free exciton line and the maximum of the bound exciton line is equal to the dissociation (binding) energy D_1 of the exciton at the impurity. Luminescence measurements therefore provide a simple way to find out this crucial parameter via optical experiments. Among others, this is important also because theoretical calculations are not able to cope with D_1 completely.

The problem of fundamental importance related to excitons bound to ionized impurities concerns their existence depending on the ratio of the effective masses of the carriers $\sigma = m_e/m_h$. It appears that neither (D^+–X) nor (A^-–X) can exist (i.e. be a stable complex) at arbitrary values of σ. The following qualitative consideration may help us to unravel this claim. In order for the exciton bound to an ionized donor $\oplus eh$ to be stable, it is required that the hole is sufficiently heavy (has small kinetic energy), otherwise it will break away from the neutral remainder $\oplus e$. In other words, the kinetic energy of the hole will surmount the gain in potential energy accomplished during localization of the exciton and the complex (D^+–X) will disappear. The theoretical dependence of the dissociation energy D_1 on σ is shown qualitatively in Fig. 7.18. It is seen that there exists a critical ratio $\sigma_c \cong 0.43$ and the complex (D^+–X) is stable only for $\sigma \leq \sigma_c$. Similar consideration can be repeated also for an exciton bound to an ionized acceptor (A^-–X); merely the electron and hole exchanging their roles. Such an exciton will thus be stable for $\sigma > 1/\sigma_c \approx 2.33$, as is also shown in Fig. 7.18. (In reality, theoretical computations of the complex stability calculate, in analogy with a H_2^+ molecular ion, a slightly different dissociation energy, namely, the energy needed to break away only a hole, thus a neutral donor remains there, or only an electron, when a neutral acceptor remains.) A more detailed discussion can be found, e.g. in [20].

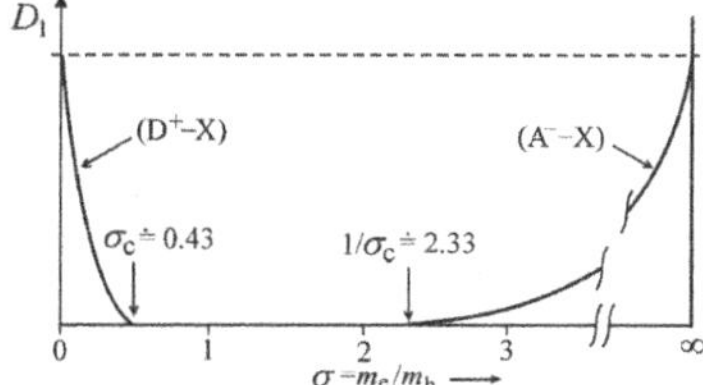

Fig. 7.18
Qualitative illustration of the dissociation energy of an exciton bound to an ionized donor (acceptor)—i.e. the energy needed to tear the free exciton off the complex—as a function of the effective masses ratio $\sigma = m_e/m_h$. The exciton at an ionized donor (D^+–X) may exist only for $\sigma \leq \sigma_c \cong 0.43$, the exciton at an ionized acceptor (A^-–X) only for $\sigma > 1/\sigma_c \cong 2.33$.

Several interesting consequences can be extracted from the above exposition:

(a) In a given material, there cannot exist simultaneously both (D^+–X) and (A^-–X).
(b) The existence of (A^-–X) in general is very unlikely because, as a rule, the hole is usually heavier than the electron ($\sigma < 1$).
(c) In silicon ($\sigma \cong 0.61$), no excitons bound to ionized impurities can exist at all, neither (D^+–X) nor (A^-–X). This is illustrated for instance in Fig. 1.1; lines of this origin occur in none of the three panels.
(d) In GaAs ($\sigma \cong 0.11$) and InP ($\sigma \cong 0.0944$) we find $\sigma < \sigma_c$, therefore the (D^+–X) exciton should exist in these materials, unlike the (A^-–X) one. Also this conclusion is confirmed by experiment. The (D^+–X) line can be identified in the emission spectrum of InP shown in Fig. 7.8; this line

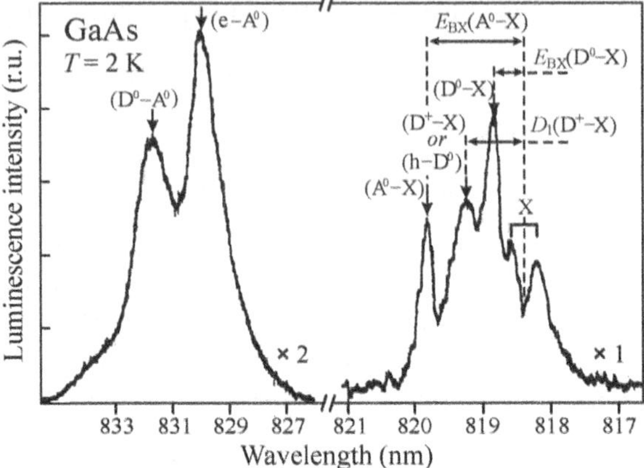

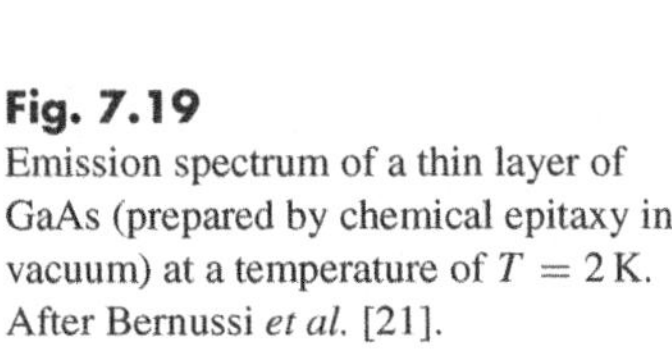
Fig. 7.19
Emission spectrum of a thin layer of GaAs (prepared by chemical epitaxy in vacuum) at a temperature of $T = 2$ K. After Bernussi *et al.* [21].

occurs also in GaAs but coincides spectrally with the (h–D^0) line (see, e.g., Figs 7.6 and 7.19 [21]). No line which could be attributed to (A^-–X) occurs in the mentioned figures.

Excitons bound to neutral donors (D^0–X) or neutral acceptors (A^0–X)

Numerous theoretical computations and experiments have shown that excitons bound to a neutral impurity are stable at an arbitrary σ ratio. This is a substantial difference in regard to the previous case. One can speculate qualitatively about the reasons in the following way.

An exciton bound to a neutral donor $\oplus eeh$ is—in the limit of a heavy hole $\sigma \to 0$ (i.e. $m_h \to \infty$)—an analogy to the hydrogen molecule H_2, which is a very well known and stable particle with a dissociation energy (the energy necessary for decomposing H_2 into two hydrogen atoms) equal to 0.33 Ry(H) = 4.5 eV. By analogy, we can conclude that the dissociation energy E_{BX} of an exciton localized at a neutral donor will be $E_{BX} \approx 0.33\, E_D$, where E_D is the ionization energy of the relevant donor. In the opposite limit of a light hole $\sigma \to \infty$, only the electron part of the exciton is bound to the impurity, which is actually an analogy to the H^- ion. This ion is also a stable object with dissociation energy of approximately 0.055 Ry (H); the corresponding exciton binding energy will thus be $E_{BX} \approx 0.055\, E_D$. A variety of interpolations between these two limit cases exist, nevertheless, it appears that E_{BX}, despite passing through a minimum, stays permanently above zero.

To describe the creation and subsequent radiative decay of a bound exciton, equations analogous to eqn (7.20) can be written:

$$\oplus e + \mathrm{FE} \to \oplus eeh + E_{BX}, \tag{7.22a}$$

$$\oplus eeh \to \oplus e + h\nu_{BE}. \tag{7.22b}$$

By eliminating $\oplus eeh$ we obtain

$h\nu_{BE} = \mathrm{FE} - E_{BX} = (E_g - E_X) - E_{BX}$ (in a direct-bandgap semiconductor),

$h\nu_{BE} = \mathrm{FE} - E_{BX} = (E_{gi} - E_X - \hbar\omega) - E_{BX}$ (in an indirect-bandgap semiconductor).

The emission line of a bound exciton is thus again red-shifted by an amount equal to the binding (dissociation) energy E_{BX}.

As an example we give in Fig. 7.19 the emission spectrum of GaAs at $T = 2$ K which, being measured in a sufficiently wide range of wavelengths, incorporates a number of lines. We have already discussed previously the origin of all of them, starting from the (D^0–A^0) band at the longest wavelengths and terminating with the polariton emission (X) on the opposite side. At this moment, let us pay attention to the (D^0–X), (A^0–X) and (D^+–X) lines. The binding energy E_{BX} of the (D^0–X) exciton is, as can be deduced from Fig. 7.19, substantially smaller than that of (A^0–X). This is understandable because m_h ($\approx 0.61\ m_0$) is significantly larger than m_e ($\approx 0.066\, m_0$). What is less understandable at first sight is why the (D^+–X) line is more distant from the free exciton line than the (D^0–X) one, or—in our notation—why $D_1 > E_{BX}$. Then, a related question arises as whether a similar relation holds universally in all materials.

The answer is: No, it does not. The relation between D_1 and E_{BX} depends strongly on the ratio σ, as demonstrated qualitatively in Fig. 7.20. The exciton bound to an ionized donor is, as noted above, a stable complex only for $\sigma < \sigma_c$; however, D_1 increases rather abruptly with further decrease of σ. On the other hand, it is true that $E_{BX} > 0$ holds for arbitrary σ but the increase in E_{BX} is rather slow with decreasing σ and both curves cross each other somewhere. Therefore, the relation between D_1 and E_{BX} is material-dependent; σ is sufficiently small in GaAs as well as in most direct semiconductors and therefore $D_1 > E_{BX}$ applies there.

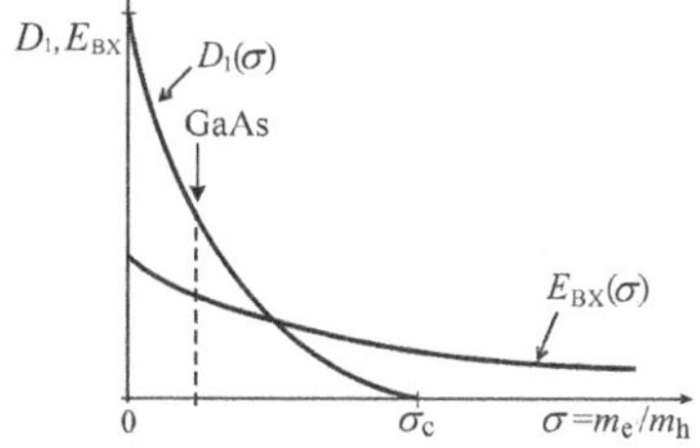

Fig. 7.20
The ratio of the binding energies of an exciton localized at an ionized (D_1) and neutral (E_{BX}) donor.

There is a very important law related to excitons bound to neutral donors and acceptors, usually referred to as *Haynes' rule*. It can be seen in Figs 7.6, 7.8 or 7.19 that the binding energy of excitons localized at acceptors differs in general from that of excitons localized at donors, as we have already discussed. Nevertheless, Haynes in 1960 found out empirically—just with the help of luminescence spectroscopy—an even more delicate relation, namely, that E_{BX} depends on the chemical nature of the donor or acceptor atom. Or, the energy E_{BX} depends (slightly) on whether the exciton is bound to an atom of, for example, phosphorus or arsenic in silicon (both the impurities being donors in silicon). Till now, we have not taken this fact into account in our exposition, in other words, we used the effective mass approximation and we were interested only in the ratio $\sigma = m_e/m_h$ of the host material, but the possible influence of the chemical nature of the impurity was disregarded. Haynes' rule thus points to the fact that the localized exciton 'feels' the chemical identity of the impurity atom and can bear witness to it via the specific energy of the emitted luminescence photon. This phenomenon is sometimes referred to as the 'central cell correction'.

The standard quantitative formulation of Haynes' rule in case of donors reads

$$E_{BX} = a + b\, E_D, \tag{7.23}$$

which, expressed verbally, tells that us the binding energy E_{BX} of an exciton localized at a neutral donor increases linearly with increasing binding energy

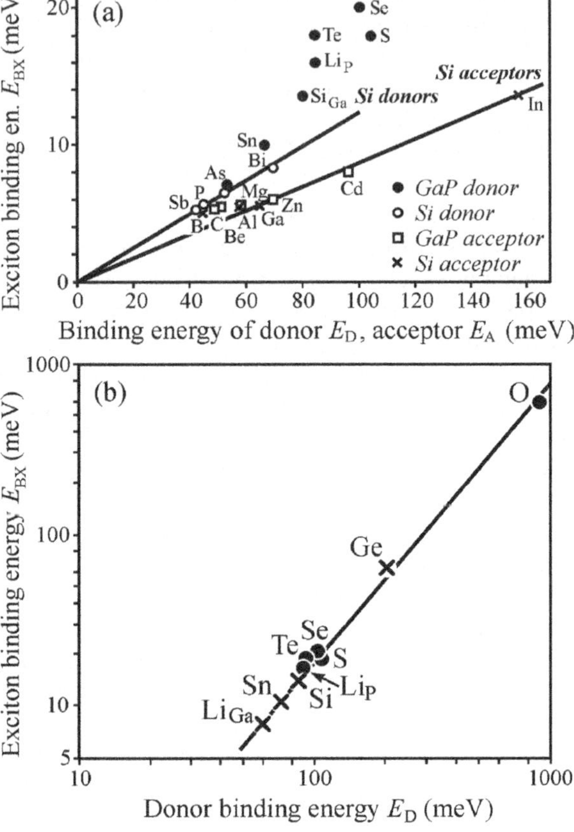

Fig. 7.21
(a) Haynes' rule: the binding energy E_{BX} of an exciton localized at a neutral impurity as a function of the binding energy of the corresponding donor (E_D) or acceptor (E_A) in Si and GaP. Full lines apply for the linear relation (7.23) in silicon. After Hayes and Stoneham [22]. (b) Modified Haynes' rule for donors in GaP: $E_{BX} \sim E_D^{1.6}$, after Herbert [24].

of the donor E_D itself. As for donors in silicon, $a = 0$ and $b \approx 0.1$, thus the binding energy of a localized exciton is approximately equal to one-tenth of the binding energy of the donor itself, which corresponds reasonably to the above mentioned estimates based on the analogy with a molecule H_2 or ion H^-. The dependence (7.23) is plotted in Fig. 7.21(a) [22]. An equation fully analogous to (7.23) holds also for the acceptors in silicon, the only difference being that there appears E_A instead of E_D and values of the constants a, b will be slightly different. A corresponding plot is also shown in Fig. 7.21(a), together with data relevant to GaP where eqn (7.23) also holds true but $a \neq 0$ and, moreover, this parameter has the opposite sign for donors and acceptors.

The simple Haynes' rule in the form of (7.23) was later explained theoretically through a chemical shift of the short-range force potential [23]. Unfortunately, this rule is not of universal validity. It turned out to be inapplicable for example to some III-V compounds such as GaAs or InP. The binding energy E_{BX} differs there for different donors and acceptors non-systematically and only very little. Therefore, if the line due to an exciton bound to a neutral donor is marked simply (D^0–X) in Figs 7.6, 7.7 or 7.19, this means that it may incorporate spectrally unresolved contributions from several donor species. A modified Haynes' rule $E_{BX} \sim E_D^{1.6}$ was proposed for GaP, as shown in Fig. 7.21(b) [24]. This modification resulted in particular from extending the

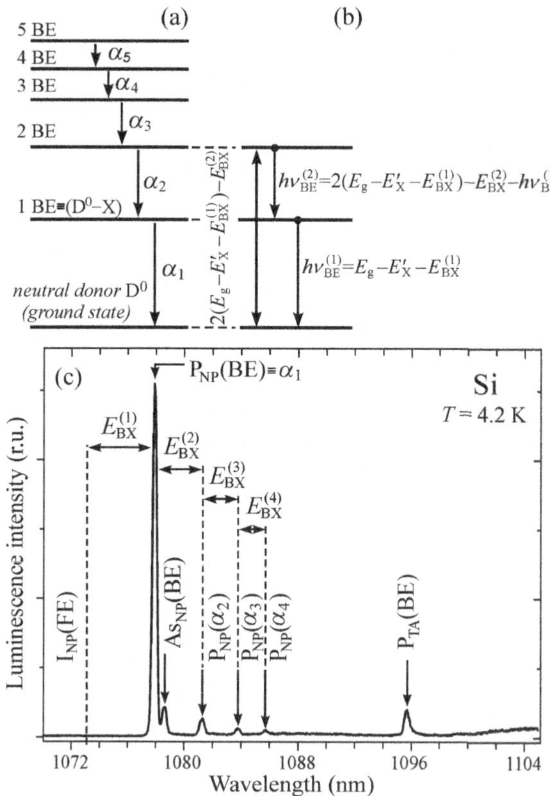

Fig. 7.22
(a) Schematic shell model of a bound multiexciton complex localized at a neutral donor. Transitions denoted α_m result from radiative decay of the mth exciton, while the exciton number in the complex decreases to $(m-1)$. (b) Energy balance for luminescence of a BMEC consisting of two bound excitons. E'_X stands for either the binding energy of the bound exciton in the case of a direct-bandgap semiconductor ($E'_X \equiv E_X$) or at no-phonon emission, or $E'_X = E_X + \hbar\omega$ for phonon replicas in a semiconductor with an indirect bandgap. (c) Low-temperature ($T = 4.2$ K) emission spectrum demonstrating the occurrence of α_1–α_4 lines in a BMEC localized at a phosphorus atom in silicon. The lines are no-phonon (NP) replicas. Also the hypothetic spectral position of a no-phonon free exciton luminescence I_{NP}(FE) is depicted. Correlation with the notation used in panel (a) reads $P_{NP}(BE) \equiv \alpha_1$, $P_{NP}(\alpha_2) \equiv \alpha_2$, $P_{NP}(\alpha_3) \equiv \alpha_3$, $P_{NP}(\alpha_4) \equiv \alpha_4$. In addition to the said lines, the spectrum comprises also a weak no-phonon line $As_{NP}(BE) \equiv (D^0\text{–}X)$ originating in exciton annihilation at another neutral donor (arsenic), and also a TA-replica of luminescence of a single exciton bound at phosphorus P_{TA}(BE). The silicon sample was doped by P and As to a total donor concentration exceeding $10^{16}\,\mathrm{cm}^{-3}$.

data in Fig. 7.21(a) to deep donors Ge and O. The validity of Haynes' rule (7.23) is also commonly not accepted in II-VI semiconductors.

Bound multiexciton complexes (BMEC)

At the beginning of the 1970s, Kaminskii and Pokrovskii discovered experimentally with the aid of luminescence spectroscopy that more than one exciton can be localized at a neutral donor or acceptor [25]. Later the term *bound multiexciton complex* (BMEC) became common to denote such objects. Here, the situation differs a bit against single excitons bound to neutral impurities (D^0–X) and (A^0–X), because the existence of (D^0–X), (A^0–X) was predicted theoretically (together with predicting the existence of excitonic molecules—Section 8.2 [26]), while the discovery of bound multiexciton complexes was, from the theoretical point of view, unexpected.

A BMEC is created in a photoexcited semiconductor in such a way that one, two, three or more excitons with step by step diminishing binding (localization) energy are successively 'wrapped around' a neutral donor D^0 or acceptor A^0. They later annihilate radiatively emitting narrow luminescence lines as depicted in Fig. 7.22(a), drawn specifically for a BMEC at a neutral donor.

If m stands for the number of bound excitons, then the energy level of the mth bound exciton is denoted as m BE ($m = 1, 2, 3, \ldots$) in Fig. 7.22.

Conformingly to previous notation 1 BE ≡ (D^0–X). During the radiative annihilation of m BE, an emission line α_m arises and the number of excitons in the BMEC lowers by one (m BE → (m–1)BE). The energy balance of the multiexciton trapping and subsequent radiative recombination can be obtained by generalizing relations (7.22):

$$\oplus e + \mathrm{FE} \to \oplus e(eh) + E_{\mathrm{BX}}^{(1)}$$

$$\oplus e(eh) - E_{\mathrm{BX}}^{(1)} + \mathrm{FE} \to \oplus e(2eh) + E_{\mathrm{BX}}^{(2)}$$

$$\oplus e(2eh) - \left(E_{\mathrm{BX}}^{(1)} + E_{\mathrm{BX}}^{(2)}\right) + \mathrm{FE} \to \oplus e(3eh) + E_{\mathrm{BX}}^{(3)}$$

$$\vdots$$

$$\oplus e\,((m-1)\,eh) - \sum_{j=1}^{(m-1)} E_{\mathrm{BX}}^{(j)} + \mathrm{FE} \to \oplus e(m\,eh) + E_{\mathrm{BX}}^{(m)}. \quad (7.24a)$$

In eqns (7.24a), describing the gradual trapping of free excitons FE, $E_{\mathrm{BX}}^{(m)}$ stands for the binding energy of the mth localized exciton ($E_{\mathrm{BX}}^{(1)} \equiv E_{\mathrm{BX}}$ in our previous notation).

The radiative decay of the mth exciton leading to emission of a photon $h\nu_{\mathrm{BE}}^{(m)}$ (α_m line) proceeds in accordance with the scheme

$$\oplus e(meh) \to \oplus e\ [(m-1)\,eh] + h\nu_{\mathrm{BE}}^{(m)}, \quad (7.24b)$$

which, by means of the last equation of (7.24a), gives immediately

$$h\nu_{\mathrm{BE}}^{(m)} = \mathrm{FE} - \sum_{j=1}^{m} E_{\mathrm{BX}}^{(j)}.$$

This relation tells us that the α_m lines are shifted with respect to the position of the free exciton emission line FE (= $E_{gi} - E_X - \hbar\omega$ in an indirect-bandgap semiconductor) by a sum of binding energies of all of the localized excitons. Figure 7.22(b) illustrates the energy balance (7.24) for the $m = 1$ and $m = 2$ lines; the notation $E'_X = E_X + \hbar\omega$ is introduced there for brevity's sake. Considering that $E_{\mathrm{BX}}^{(m)} < E_{\mathrm{BX}}^{(m-1)}$, the α_m lines make a series densifying with increasing m. At the same time, the intensity of these lines decreases, because the number of localized excitons with higher m due to the condition $E_{\mathrm{BX}}^{(m)} < E_{\mathrm{BX}}^{(m-1)}$ decreases monotonically. This is nicely illustrated in Fig 7.22(c) which displays the low-temperature emission lines α_1–α_4 originating from a BMEC at a phosphorus atom in silicon.

The schematic displayed in Fig. 7.22(a) represents a considerably simplified version of the so-called *shell model* which was put forward by Kirczenow for explaining the luminescence in BMECs [27]. In reality, the levels in Fig. 7.22(a) are split as a result of the higher number of electron and hole states differing slightly in energy. Using a very sensitive detection system with

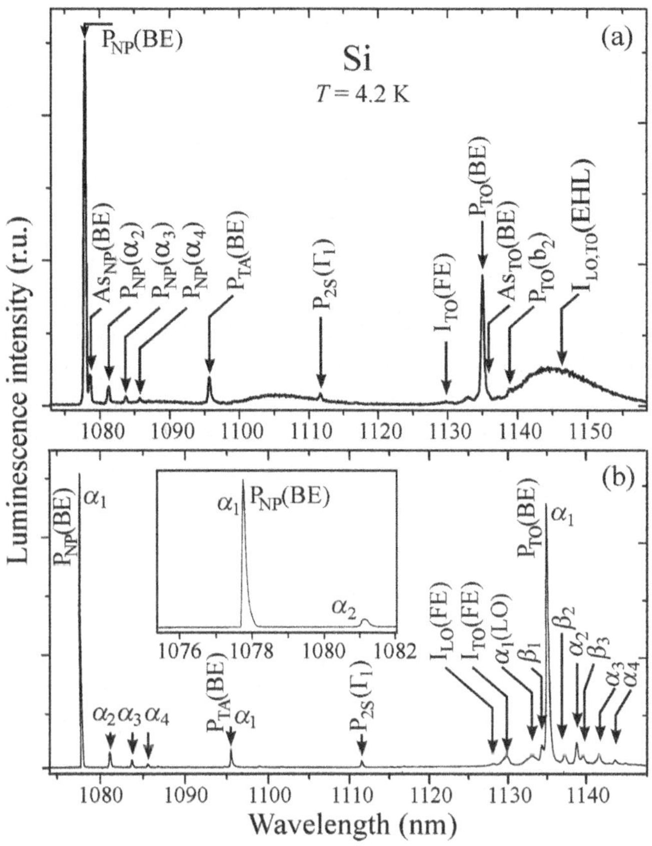

Fig. 7.23
Photoluminescence spectra of two samples of crystalline Si doped with phosphorus to approximately the same donor concentration $N_P \approx 3 \times 10^{14}\,\text{cm}^{-3}$. Bath temperature $T = 4.2$ K. (a) Continuous-wave excitation 488 nm from an Ar^+-laser (excitation intensity $\sim 20\,\text{W/cm}^2$), a scanning detection monochromator HRD-1 and a cooled photomultiplier with a photocathode of S1 type. After Pelant *et al.* [29]. (b) Continuous-wave excitation 514 nm from an Ar^+-laser (estimated excitation intensity $\sim 2\,\text{W/cm}^2$), detection with the help of a Nicolet Fourier spectrometer with a cooled North Coast Ge detector. The inset shows the enlarged area of NP replicas. After Colley and Lightowlers [30].

a high S/N ratio, one can, in addition to the series of α_m lines, observe also weaker emission lines β_m—they can be seen for instance in Fig. 7.23(b).[8]

It is worth noting that the α_m lines in Fig. 7.22(c) represent no-phonon luminescence and, surprisingly, are more intense than the corresponding series of lines in the region of TO/LO- or TA-replicas, as one can make sure of by looking at Fig. 1.1(A) where a similar series of phosphorus-related α_1–α_4 lines occurs, too. At this point it may appear suitable to make a small deviation in a more general sense: when discussing recombination processes in indirect-bandgap semiconductors we encounter certain competition between the configurational coordinate model and the principle of quasi-momentum conservation (the configurational coordinate model was introduced in Sections 4.4 and 4.5 by assuming tacitly a direct-bandgap material). As far as silicon donors are concerned, there is a very weak exciton–phonon interaction, thus the Huang–Rhys factor is small ($S \cong 0.1$) implying that the no-phonon lumines-

[8] True values of the binding energies then differ from those of the quantities $E_{BX}^{(m)}$ introduced by us, and need not satisfy the condition of monotonic decrease with increasing m. Details can be found in Thewalt, M. L. W. (1983). *Bound multiexciton–impurity complexes*. In *Excitons* (ed. E. I. Rashba and M. D. Sturge). *Modern Problems in Condensed Matter Sciences*, Vol. 2, Chap. 10. North Holland, Amsterdam.

cence lines of a localized impurity centre (phosphorus atom) are expected to be relatively intense, as indeed is observed experimentally; the first phonon replicas are substantially weaker. However, we cannot forget that in an indirect semiconductor this phenomenon is conditioned by relaxation of the quasi-momentum conservation law owing to the phoshorus-atom-induced breaking of the lattice translational symmetry. Hereby, the influence of the indirect bandgap is *de facto* annulled in this way.[9]

A similar scheme of energy levels as in Fig. 7.22 (shell model) can be introduced also in the case of a BMEC at a neutral acceptor. A corresponding series of emission lines, characteristic for a BMEC localized at boron in silicon is clearly noticeable in Fig. 1.1(C). Somewhat stronger exciton–phonon coupling in an acceptor-related BMEC causes the first phonon replicas to be more intense than no-phonon lines here. The B_{TO}(BE) line, is analogous to a TO-replica of the donor α_1 line, $B_{TO}(b_1)$ corresponds to a TO-replica of the α_2 line, $B_{TO}(b_2)$ corresponds to a TO-replica of the α_3 line, etc. Fully analogous notation is used for TA-replicas in Fig. 1.1(C).

BMECs were studied in detail mainly in silicon. Lines that can be attributed to radiative decay of a BMEC were observed also in Ge, β-SiC and GaP. These are all indirect-bandgap semiconductors where the occurrence of a BMEC is quite understandable, for two reasons. First, the excitons have a relatively long lifetime in indirect semiconductors (which enhances the probability for trapping of several excitons at one centre), and second is the N-fold degeneracy of the conduction band minimum ($N = 4$ in Ge and $N = 6$ in Si) allowing up to $2N$ electrons to be placed in one molecular orbital without violating Pauli exclusion principle. Some emission lines, observed in the direct-bandgap semiconductor GaAs, were also attributed to a BMEC; however, definitive confirmation of the existence of BMECs in these materials is missing.

The radiative recombination of various types of excitons bound to donors and acceptors represents a strong rival channel to the (D^0–A^0) luminescence.

7.2.2 Quantitative luminescence analysis of shallow impurities in silicon

The utilization of low-temperature photoluminescence spectroscopy over the exciton spectral range to determine the concentration of shallow donors and acceptors in crystalline silicon was proposed in 1982 by Tajima [28]. In Table 7.2 we give the wavelengths of the luminescence lines originating in various complexes (D^0–X), (A^0–X) in silicon. These are TO-phonon assisted lines of excitons bound to commonly applied donors and acceptors. As will become clear hereafter, the TO-replicas are those used most frequently for the quantitative luminescence analysis of silicon.

[9] Strictly speaking, however, the presence of impurity atoms should also allow the intrinsic no-phonon luminescence of free exciton I_{NP}(FE) to appear, like in AgBr, see Fig. 7.15. In spite of this, no line like I_{NP}(FE) has ever been reported in silicon; this difference between AgBr and Si is probably given by the much higher concentration of both point and line defects (dislocations) in AgBr.

It follows from Table 7.2 that the spectral positions of the lines characteristic of donors and acceptors vary—according to Haynes' rule—with the chemical nature of the dopants; the variations range from fractions of nanometres to several nanometres. Such differences are in principle large enough for the individual lines to be distinguished, provided a scanning monochromator with resolving power of $R \approx 5 \times 10^3$ is used. However, if there are several types of impurities in a given sample, their lines will very probably merge together due to their natural linewidth and due to possible line broadening by the finite monochromator slit, and will form an almost irresolvable band. Any effort to identify the individual lines then would not probably attain the target. Fortunately, silicon technology is highly advanced so that a very low level of residual (introduced unintentionally) impurities—down to $10^{10}\,\text{cm}^{-3}$—can be reached by zone-refining. In order to achieve the desired type and value of electric conductivity, mostly phosphorus (donor) and boron (acceptor), possibly also As, Sb, Al and Ga, are used as dopants. In practise, their concentrations usually range from $10^{12}\,\text{cm}^{-3}$ to $10^{18}\,\text{cm}^{-3}$ and knowledge of the exact values of the concentration is basic information for both the producers and purchasers of Si ingots and wafers.

Photoluminescence analysis enabling us to determine such a concentration in the range of approximately 10^{11}–$10^{15}\,\text{cm}^{-3}$ was developed primarily for boron and phosphorus. In this concentration range—i.e. for nominally pure or weakly doped Si wafers—this technique has an important role to play in supplementing the wide range of diagnostic methods (temperature-dependent Hall effect, resistivity measurements, infrared absorption spectroscopy, etc.) being commonly applied at higher concentrations of donors and acceptors. It can be stated that photoluminescence diagnostics is nowadays the only analytic method of weakly doped Si crystals that is used routinely in practice.

The essence of the method is very simple and is based on the following use of the emission lines $B_{TO}(BE)$, $P_{TO}(BE)$ and $I_{TO}(FE)$, shown for example in Fig. 1.1 or Fig. 7.23 [29, 30]. It is, undoubtedly, reasonable to assume the

Table 7.2 Spectral positions of luminescence lines (TO-replicas) due to excitons bound at various impurities in Si; $T = 4.2\,\text{K}^{*)}$

Impurity	Type	Wavelength (nm)
P	SD	1135.13
As	SD	1135.97
Sb	SD	1135.04
Bi	SD	1138.32
Li	ID	1133.74
B	SA	1134.39
Al	SA	1135.60
Ga	SA	1136.15
In	SA	1144.83
Tl	SA	1177.59
C	I	1164.36
free exciton	–	1129.76

*)SD stands for a substitutional donor, ID interstitial donor, SA substitutional acceptor, I isoelectronic impurity.

intensities of the lines B_{TO}(BE) and P_{TO}(BE) to be proportional to the relevant concentrations of boron N_B or phosphorus N_P, respectively. On the other hand, the intrinsic emission I_{TO}(FE) line, as we have already stressed several times, is characteristic of those parts of the crystal where these dopants are not present. The ratio of the line intensities B_{TO}(BE)/I_{TO}(FE) or P_{TO}(BE)/I_{TO}(FE) is surely an increasing function of the concentrations N_B, or N_P, respectively. If we determine experimentally this dependence (the so-called calibration curve) via luminescence measurements performed on samples with a known content of boron and phosphorus, the dopant concentrations N_B and N_P in unknown samples can then be determined easily from the relevant luminescence spectra using these calibration curves.

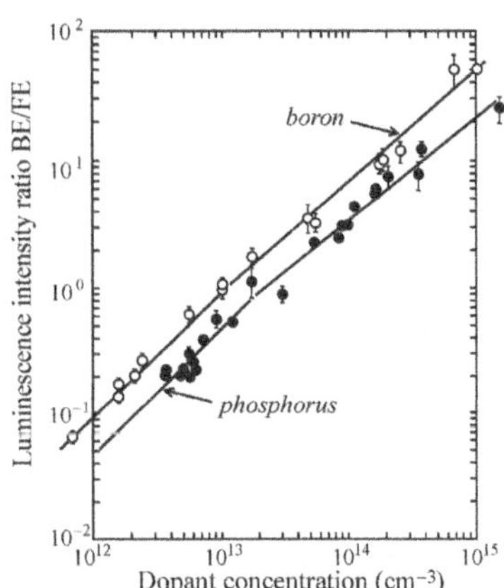

Fig. 7.24
Calibration curves for the quantitative determination of B and P concentrations in silicon from photoluminescence spectra at $T = 4.2$ K (dependence of the ratio of intensities P_{TO}(BE)/I_{TO}(FE) and B_{TO}(BE)/I_{TO}(FE) on phosphorus and boron concentrations, respectively). After Pelant *et al.* [31].

An example of such calibration curves is displayed in Fig. 7.24 [31]. It should be stressed that these curves—as shown—are not applicable universally because the intensity ratio of the bound to free exciton lines depends slightly on the excitation power density. Measurements of calibration curves and experimental investigation of samples in which the dopant concentration is to be determined thus have to be always performed using the same experimental setup and keeping all the experimental parameters constant. In this case, it is not even necessary to take into account spectral corrections. The calibration curves in Fig. 7.24 were obtained using the experimental set-up drawn in Fig. 2.2(a).

It is worth stressing that the measurement of intensities of the B_{TO}(BE) or P_{TO}(BE) lines alone is not sufficient to determine the unknown concentrations because the overall luminescence intensity is sample-dependent (being affected by surface treatment, reproducibility of the sample position with respect to the optical collecting system, etc.); it is thus always necessary to determine the ratio BE/FE. This fact then specifies the above mentioned limits of applicability of the method: in strongly doped samples (N_B, $N_P \geq 10^{15}$ cm^{-3}) the I_{TO}(FE) line becomes undetectable, as clearly demonstrated in Fig. 1.1(C); in very pure samples (N_B, $N_P \leq 10^{11}$ cm^{-3}), on the other hand, the B_{TO}(BE) and P_{TO}(BE) lines disappear.

While the essence of the method is simple, its experimental realization is not quite trivial. There are several reasons for this. First of all, the luminescence of silicon is very weak, several orders of magnitude below that of direct-bandgap materials (GaAs). Moreover, judged from the point of view of detectors of weak photon fluxes, this luminescence is situated in a very unsuitable spectral range. As for photomultipliers, only types with an S1 photocathode are sensitive in the range 1100–1200 nm, but also for them this wavelengths represent the red limit of their sensitivity. Similar arguments can be applied as for silicon CCD detectors. It is possible to use a germanium detector, however, only special high-sensitive types with efficient noise suppression are suitable (Edinburgh Instruments, formerly North Coast). The problem is all the more difficult because it is necessary to measure the spectra with a high resolution R_{real}, therefore with a relatively narrow slit of the spectral apparatus ($\Delta\lambda = L^{-1}\Delta\ell \approx 0.25$ nm). The possibility to circumvent the mentioned difficulties by using a very intense excitation source (e.g. a pulsed laser) is also out of the question because, under strong pumping, a qualitative change of the emission spectrum occurs—a novel broad band at ~1143 nm labelled $I_{LO,TO}$(EHL) in Fig. 1.1 begins to prevail, whose origin, consisting in radiative recombination of an electron–hole liquid, is discussed in detail in Section 8.4. This band to a large degree masks the lines of bound excitons, particularly in very pure samples.

It is therefore necessary to choose a certain optimal level of (continuous) pumping and pay attention to the proper choice of monochromator, detector and also of efficient collection of an optical system. What follows from the above discussed difficulties is the particular choice of TO replicas for the method under discussion, because in their spectral range, both the lines originating from the impurities and the free exciton line I_{TO}(FE) are relatively intense.

This method can serve as an example to demonstrate clearly some advantages of Fourier luminescence spectroscopy. Figure 7.23 shows a comparison of (a) an emission spectrum obtained in a conventional experimental set-up [29] with (b) an emission spectrum measured by a Fourier spectrometer [30]. It becomes evident that the Fourier-measured spectrum exhibits a markedly higher *S/N* ratio, even under lower excitation intensity (as reflected among others in the total absence of the $I_{LO,TO}$(EHL) band). The accuracy achieved in the Fourier approach is substantially better (± 10%) compared with that obtained with a scanning monochromator (± 30%). The sensitivity in the Fourier arrangement can attain almost unbelievable levels—it enables us to distinguish safely a tiny spectral shift and fine structure in the emission lines of bound excitons in Si crystals with different content of silicon isotopes [32]! This opens further diagnostic possibilities.

Another experimental aspect deserves special attention, namely, the method of sample cooling. It is mandatory that the samples must be immersed in a cooling medium (liquid He), an optical bath He cryostat is thus a necessity. We emphasize this because there is sometimes a tendency to replace the bath cryostat by a cheaper option—a continuous flow cryostat with a closed He cycle where the cooling is mediated only by the heat transfer in He vapour. In this case, however, the cooling power is incomparably lower and the samples are strongly heated by the absorbed excitation radiation (the reader is reminded that silicon has low luminescence efficiency—the prevailing part of the excitation energy is converted, owing to non-radiative transitions, into heat!). This results in 'evaporating' bound excitons from the impurity atoms and in a corresponding increase of the free exciton concentration, which entails an appreciable drop in intensity of the B_{TO}(BE) and P_{TO}(BE) lines and, vice versa, an increase in the I_{TO}(FE) line intensity. The significance of this effect is demonstrated in Fig. 7.25 [33, 34]: upon an increase in temperature by a mere three kelvin the intensity ratio B_{TO}(BE)/I_{TO}(FE) is decreased by more than one order of magnitude! Insufficient cooling therefore leads to a rapid deterioration of the measurement sensitivity, loss of the required spectral resolution and the method finally declines in importance.

Upon introducing the photoluminescence analysis of silicon by Tajima, the method was independently developed and confirmed at several other laboratories [30, 31, 35, 36]. The results were identical as far as the principal features were concerned, even if Colley and Lightowlers [30] proposed a modification to consider the NP replicas instead of TO ones for the bound excitons, and the TO replica to employ for free excitons only. Besides calibration curves for boron and phosphorus, they presented calibration curve also for aluminium; Schumacher and Whitney [36] in addition for gallium and Broussel *et al.* [37] for arsenic.

An effort is being made to extend similar photoluminescence diagnostics of dopants to other semiconductor materials. However, serious problems are often encountered owing either to high residual impurity concentrations or insufficient resolved lines from different donor species. Nevertheless, the determination of luminescence of shallow impurity concentrations in CdTe has been published [38] and attempts to develop characterization methods of GaAs with the help of Fourier photoluminescence and magneto-photoluminescence spectroscopy have been reported, e.g. [39].

7.2.3 Excitons bound to isoelectronic impurities

The exciton can be trapped—besides the above discussed cases of localization at ionized or neutral donors and acceptors—also at *isovalent* or *isoelectronic impurities*. By an impurity like this we understand a substitutionally built-in atom from the same group (column) of the periodic table of elements which the original constituent of the host lattice comes from. Unlike donors or acceptors, we do not encounter here a weakly bound 'extra' electron or hole that could influence the electric conductivity. Typical examples to be discussed in more detail are a nitrogen atom in place of phosphorus in GaP (GaP/N) and an iodine ion I^- substituting bromine in AgBr (AgBr/I^-). In most cases, as we shall see, the exciton bound to an isoelectronic trap presents a highly efficient channel of radiative recombination. (Also thallium activated alkali halides (KCl/Tl, CsI/Tl), often quoted in older literature as examples of efficient phosphors and scintillators, belong to a certain extent to this category [40].)

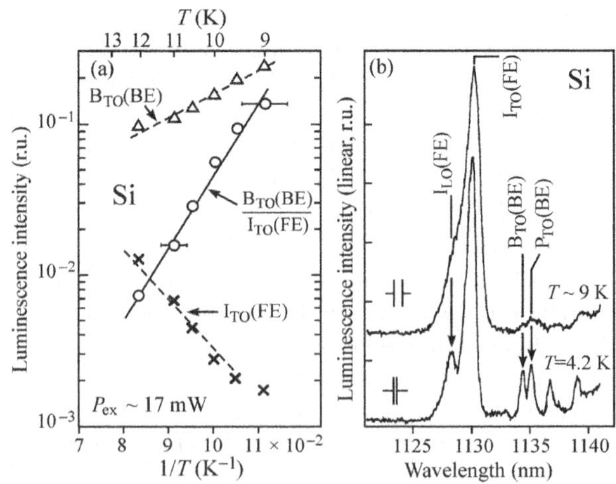

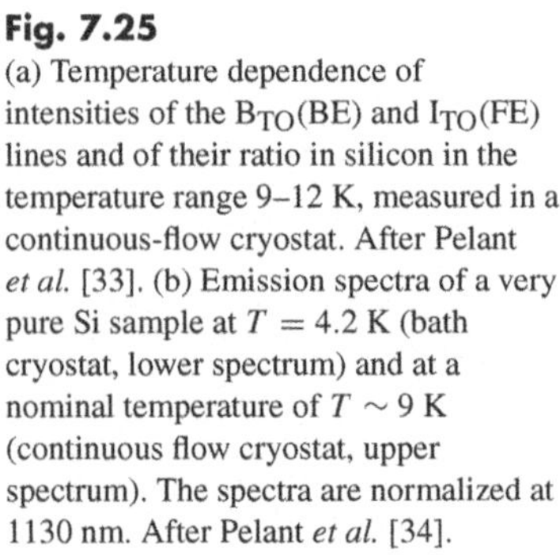

Fig. 7.25
(a) Temperature dependence of intensities of the B_{TO}(BE) and I_{TO}(FE) lines and of their ratio in silicon in the temperature range 9–12 K, measured in a continuous-flow cryostat. After Pelant *et al.* [33]. (b) Emission spectra of a very pure Si sample at $T = 4.2$ K (bath cryostat, lower spectrum) and at a nominal temperature of $T \sim 9$ K (continuous flow cryostat, upper spectrum). The spectra are normalized at 1130 nm. After Pelant *et al.* [34].

The physical mechanism of free exciton trapping is based on the electronegativity difference between the original and substituting atom. For example, a nitrogen atom N has higher electronegativity X (i.e. the ability to attract an electron) than a phosphorus atom P ($X_N = 3.0 > X_P = 2.2$). Therefore, if free photoelectrons and photoholes are created in GaP by an external excitation, the N atom may capture a nearby photoelectron with the help of short-range forces. The negatively charged N atom then attracts—now via long-range Coulomb forces—a free photohole, thereby a bound exciton localized at the nitrogen atom is created. Its radiative recombination gives a characteristic luminescence emission. Similarly, one can imagine the process to take place in AgBr crystals doped with an iodine impurity I^-; the iodine electronegativity is lower than that of bromine ($X_I = 2.7 < X_{Br} = 3.0$) and thus iodine has a higher capability to capture a photohole. A bound exciton is created again, owing to the subsequent Coulomb interaction of the positively charged centre with a photoelectron.

What is the spectral shape of the luminescence emission accompanying the recombination of an exciton bound to an isoelectronic trap? The Maxwell–Boltzmann lineshape can be certainly excluded since possible broadening due to the kinetic energy of either a free carrier or a free exciton is missing. One deals with a localized excitation; a qualitative answer can thus be obtained by applying the configurational coordinate model. It is essential in this context that the initial photocarrier capture is by short-range forces. Localization of the excitation is thus basically mediated by alteration in the occupation of the electronic shell in a single (impurity) atom. The corresponding change in the configurational coordinate $r_Q = (Q_{e0}–Q_{g0})/Q_{g0}$ (see Fig. 4.8) can then vary according to the strength of the exciton–phonon interaction. If this strength is relatively weak, which is the case of III-V semiconductors and thus also of GaP, r_Q is small (small Huang–Rhys factor $S \cong 0$) and the emission line will be narrow.

This is demonstrated in Fig. 7.26(a) where sharp lines A, B can be seen in the edge emission of a nominally pure sample of GaP [41]; both the lines are attributed to an exciton localized at a residual impurity N, being present in a concentration of $\sim 10^{15}\,cm^{-3}$. The A line originates from an exciton with a total quantum number $J = 1$, the B line belongs to the exciton in the $J = 2$ state (in principle, a forbidden transition $\Delta J = 2$). The splitting is caused by the electron–hole exchange interaction; also the terms singlet and triplet exciton are often used. Both no-phonon lines and phonon replicas involving emission of optical phonons are present. (Looking at Fig. 5.10 we may find out that the lines in question are very distinct also in a GaP sample doped with Si and S.)

In strongly polar AgBr, the exciton–phonon coupling is much stronger, the surrounding host lattice reacts to the hole trapping by a sizable change of r_Q and the emission due to the iodine presence reveals itself through a broad band rather than a narrow line. The emission spectrum of the exciton localized at I^- (which is again a residual impurity in a nominally pure material, similarly to N in GaP) was already mentioned briefly in the context of our discussion of the exciton–phonon interaction (see Figs 1.2(b) and 4.10(b)). The Huang-Rhys factor S can be estimated, based on the serial number of phonon replicas located at the band maximum, to be $S = 8–9$. A similar spectrum is shown

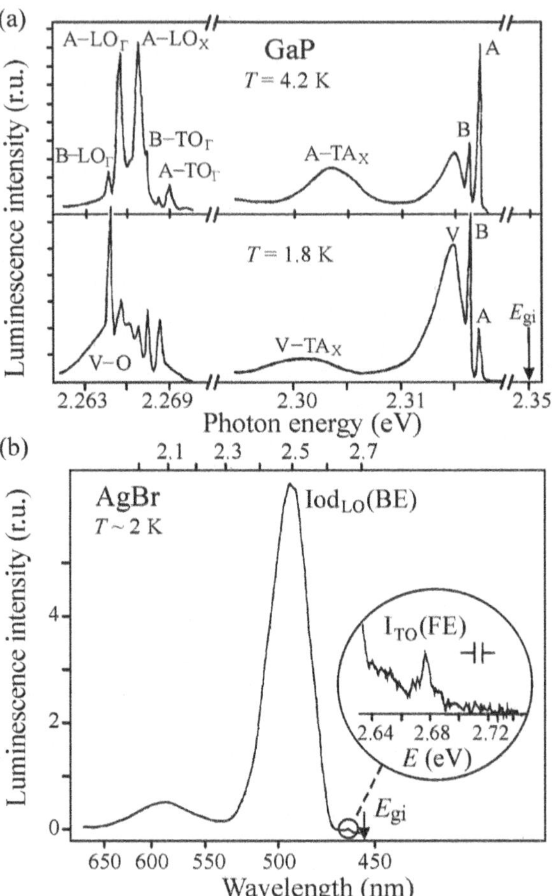

Fig. 7.26
(a) Emission spectra of nominally pure GaP under cw pumping with a dye laser 2.409 eV. The A and B lines (no-phonon ones) and their TA-, TO- and LO-phonon replicas, occurring in the edge emission range, are due to radiative decay of the excitons localized at the residual isoelectronic impurity of nitrogen ($\sim 10^{15}$ cm^{-3}). Band V originates from a similar exciton which is, moreover, influenced by the presence of distant donors and acceptors. After Gershoni *et al.* [41]. (b) Emission spectrum of nominally pure AgBr under cw pumping with a Kr^+-laser 3.53 eV. A weak indication of edge emission in the vicinity of 2.7 eV (enlarged in the circle) contains the free exciton line I_{TO}(FE). An intense broad band at $\sim$ 2.5 eV is due to an exciton bound to a residual isoelectronic impurity of iodine ($\sim 10^{16}$cm^{-3}). After Pelant [42]. The position of the indirect bandgap E_{gi} is marked. Note the different energy scales in (a) and (b).

again in Fig. 7.26(b) [42]. This figure is, however, rather exceptional in two respects. Firstly, unlike the spectra in Figs 1.2(b) and 4.10(b), the bound exciton band at ~2.5 eV lacks any fine structure. This is, however, an experimental artefact due to a not-entirely-suitable choice of the time constant of the detecting lock-in amplifier (over-damped state). On the other hand, however, this facilitated the discovery of the second—and more interesting—aspect: a weak emission of the free exciton I_{TO}(FE) at ~2.678 eV. What made recording this extremely weak line possible was just the strong noise suppression, together with a certain exceptionality of the investigated sample. The result enables us now to estimate in an illustrative way the extraordinary efficiency of the radiative recombination of excitons localized at an isoelectronic trap.

From Fig. 7.26(b), we can estimate the ratio of the spectrally integrated intensities of the bound exciton line Iod$_{LO}$(BE) to the free exciton line I_{TO}(BE) as [Iod$_{LO}$(BE)/I_{TO}(FE)]$\approx 10^4$.[10] The corresponding luminescence

[10] It is worth noticing that while the free exciton line is, owing to the strict selection rules (Section 7.1.4), accompanied by emission of TO-phonons ($\hbar\omega_{TO} \approx$ 8meV) satisfying the quasi-momentum conservation law, the fine phonon structure at the iodine band points to participation of primarily LO-phonons since separation between the individual peaks in Fig. 1.2(b) is very close to the energy of $\hbar\omega_{LO} \cong 16$ meV. The reason is basically twofold: firstly, the free and bound

decay times found experimentally are $\tau_{\mathrm{Iod}} \approx 2 \times 10^{-5}$ s [43] and $\tau_{\mathrm{FE}} \approx 6 \times 10^{-8}$ s [44]. We write down simple relations for the intensities:

$$\mathrm{Iod}_{\mathrm{LO}}(\mathrm{BE}) = \frac{N_{\mathrm{Iod}}}{\tau_{\mathrm{r}}^{\mathrm{Iod}}},\ \mathrm{I}_{\mathrm{TO}}(\mathrm{FE}) = \frac{N_{\mathrm{FE}}}{\tau_{\mathrm{r}}^{\mathrm{FE}}}, \tag{7.25}$$

where $\tau_{\mathrm{r}}^{\mathrm{Iod}}$ and $\tau_{\mathrm{r}}^{\mathrm{FE}}$ are the radiative lifetimes and N_{Iod}, N_{FE} are the concentrations of 'excited iodine ions' and of free excitons at the given pumping level, respectively. By using relation (3.3) for the luminescence efficiencies η_{Iod}, η_{FE}, we obtain from (7.25)

$$\frac{\mathrm{Iod}_{\mathrm{LO}}(\mathrm{BE})}{\mathrm{I}_{\mathrm{TO}}(\mathrm{FE})} = \frac{N_{\mathrm{Iod}}}{N_{\mathrm{FE}}}\,\frac{\eta_{\mathrm{Iod}}}{\eta_{\mathrm{FE}}}\,\frac{\tau_{\mathrm{FE}}}{\tau_{\mathrm{Iod}}}. \tag{7.26}$$

Now, we introduce an effective enhancement factor $z_{\mathrm{ef}} = (N_{\mathrm{Iod}}/N_{\mathrm{FE}})(\eta_{\mathrm{Iod}}/\eta_{\mathrm{FE}})$ which comprises both the ratio of bound and free exciton concentrations (and thus takes into consideration the high effective cross-section of the free exciton trapping at impurity ions) and the ratio of quantum efficiencies of both the radiative processes. The factor z_{ef} then follows from relation (7.26) as

$$z_{\mathrm{ef}} = \frac{\mathrm{Iod}_{\mathrm{LO}}\ (\mathrm{BE})}{\mathrm{I}_{\mathrm{TO}}\ (\mathrm{FE})}\,\frac{\tau_{\mathrm{Iod}}}{\tau_{\mathrm{FE}}} \approx 10^4 \times \left(10^3/3\right) \cong 3 \times 10^6 (!).$$

This is an extremely high enhancement factor. The concentration of residual iodine in a nominally pure sample AgBr is $\sim 10^{16}\ \mathrm{cm}^{-3}$. Considering the host lattice contains $\sim 6 \times 10^{22}$ molecules of AgBr per cubic centimetre, there is a single iodine ion per approximately 10^7 ions of the host lattice and, at the same time, 'its' own luminescence is by many orders of magnitude more intense than the intrinsic free exciton luminescence characterizing the pure host lattice! How is this possible?

In addition to the efficient trapping of free excitons itself (which has already been discussed when introducing the bound excitons at the beginning of Section 7.2), two other factors contribute to the high efficiency of the radiative recombination of an already trapped exciton. The first one is the absence of Auger non-radiative recombination. An exciton bound to a neutral donor or acceptor represents three quasi-particles localized in a relatively small volume—either two electrons and one hole or two holes and one electron. This promotes a high probability of Auger recombination, therefore the luminescence of BE at neutral donors and acceptors is not particularly intense. On the contrary, the third quasi-particle is missing in the case of an exciton in an isoelectronic trap, therefore, Auger recombination does not take place at all.

A second factor, important mainly in indirect-bandgap semiconductors (both AgBr and GaP), is the already mentioned strong localization of one quasi-particle at the impurity atom. The Heisenberg uncertainty relations $\Delta x \Delta k \geq 1/2$ (where Δx is the uncertainty in the spatial coordinate and Δk that in the wavevector) imply, however, that a strong localization in real space ($\Delta x \to 0$)

excitons have different symmetry and, secondly, in radiative recombination of the impurity centre also local vibrations may participate. More detailed discussion of the phonon structure of the AgBr/I luminescence can be found in Czaja, W. and Baldereschi, A. (1979). *J. Phys. C: Solid State Phys.*, **12**, 405.

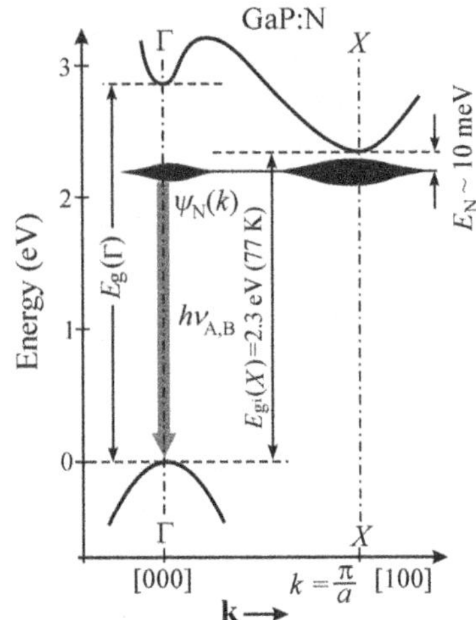

Fig. 7.27
Energy band structure of the GaP doped with nitrogen. Black filled areas denote the amplitude of the electron wavefunction $\psi_N(\mathbf{k})$. It is seen that $|\psi_N(\mathbf{k}=0)|^2 \neq 0$. Quasi-direct transitions at the Γ point lead to intense luminescence (the lines A and B in Fig. 7.26(a)). E_N stands for the binding energy of the electron at the isoelectronic trap N. After Holonyak *et al.* [45].

inevitably introduces a strong delocalization of the wavefunction in **k**-space. Figure 7.27 depicts the situation as can be visualized in the energy band scheme of GaP:N [45]. The uncertainty Δk in localization of the modulus squared of the wavefunction $|\psi_N(\mathbf{k})|^2$ of an electron trapped at the N atom, i.e. an electron occupying the energy level just below the conduction band minimum at $\mathbf{k} \cong \pi/a(100)$, extends up to the $\Gamma(\mathbf{k}=0)$ point. This makes *quasi-direct radiative recombination* of the electron with hole possible. The probability of such transitions—taking place without the need of phonon assistance—is naturally by several orders of magnitude higher than that of the indirect transitions. This leads therefore to an exceptionally intense luminescence even in indirect-bandgap materials.

It might be perhaps useful to emphasize once more the principal difference between the luminescence of shallow impurities (donors and acceptors) and the radiative decay of an exciton bound to an isoelectronic trap. At the shallow impurities, the electron or hole is attracted to a donor or acceptor, respectively, by the long-range Coulomb force. The emission spectrum of the exciton bound at a shallow impurity always bears the character of narrow lines. On the other hand, the isoelectronic trap captures the electron or hole through a short-range force and the wavefunction of the corresponding quasi-particle is strongly delocalized in momentum space. According to the type of the exciton–phonon interaction, the emission spectrum may take the form of either narrow lines (GaP, $S \cong 0$) or a phonon wing with a few phonon replicas in the case of medium exciton–phonon coupling (ZnTe/O, $S \cong 3$), or possibly a broad band in the case of very strong coupling (AgBr, $S \cong 9$).

The luminescence of the isoelectronic impurity N has found widespread use in the production of green-emitting electroluminescence diodes based on GaP/N. The emission spectrum of this kind of diode is shown in Fig. 7.28 [46]. In commercially produced diodes, ternary alloys $GaAs_{1-x}P_x$ ($x = 0.4$ to 1) are often used as the active medium which enables, apart from other things, the emission wavelength to be tuned. At high nitrogen doping, another radiative recombination channel due to the presence of coupled nitrogen atoms (NN-pair states) appears. A detailed discussion of the physics of the isoelectronic impurity N in $GaAs_{1-x}P_x$ can be found in the review article by Craford and Holonyak [47].

The question arises here as to whether it is possible to apply the principle of the isoelectronic trap also in homopolar semiconductors of group IV. It is a topical problem, in particular relevant to the search for light-emitting materials on a Si basis for future silicon photonics (Chapter 15). The doping of silicon with germanium or tin can hardly lead to any effect because the electronegativities of these chemical elements are almost the same ($X_{Si} = 1.9$; $X_{Ge} = 2.0$; $X_{Sn} = 2.0$). As for doping with carbon ($X_C = 2.6$), the desirable effect of strong luminescence is expected—according to what was explained above—to occur. Unfortunately, this is not the case. The carbon-related luminescence in crystalline silicon, represented by a narrow line at $\sim$0.97 eV, is very weak and, moreover, in order to appear, the material must first be activated by electron beam irradiation. The reasons for this are not fully clear. Both the small atomic radius of the carbon atom and the weak exciton–phonon interaction have a possible role to play. A closer analysis of issues of isoelectronic traps in silicon can be found in [48].

7.2.4 Self-trapped excitons

In polar semiconductors with sufficiently strong exciton–phonon interaction, free exciton trapping can occur also in a pure and unperturbed crystal lattice. This process is called self-trapping or auto-localization of the exciton; radiative recombination of such a self-trapped exciton (STE) then evidently represents an *intrinsic luminescence process*. One can infer—from the fact that the exciton–phonon interaction is strong—that the corresponding luminescence emission spectrum will be broad.

An intuitive illustrative idea of exciton self-trapping can be drawn in the following way: the translational movement of the exciton through the lattice is driven by its heavier constituent, i.e. the hole. If the polaron coupling constant of the hole α_h is high, the hole–polaron is heavy; its effective mass is described by relation (6.9). The movement of the heavy hole is then very slow, the hole thus strongly polarizes the surrounding lattice, and thereby its movement is further slowed down and, finally, a total localization of the hole close to some atom constituting the host lattice can occur owing to positive feedback. A light electron accompanying the hole will then describe its orbit in the close vicinity.

A more exact phenomenological description of the self-trapping of free quasi-particles as well as that of the entire exciton was put forward by Toyozawa [49]. In the case of an electrically neutral particle—an exciton—the interaction with the host lattice is, according to Toyozawa, caused not directly by the electric forces but mainly by the short-range (s) forces, and a redistribution of atoms (lattice relaxation—LR) occurs in the close vicinity of the exciton. This redistribution is within the configurational coordinate model accompanied by the lattice relaxation energy E_R given by relation (4.14). For the purpose of more exact identification, we denote here this relaxation energy as E^s_{LR}; thus $E^s_{LR} = S_0 \hbar\omega$, where S_0 is the Huang–Rhys factor. The short-range interaction is characterized by the parameter $g_s = E^s_{LR}/B$, where B is the total energy width of the exciton band.

The interaction of charged quasi-particles, electrons and holes, with the lattice is mediated by the Fröhlich mechanism employing LO-phonons. Consequently, long-range (ℓ) forces are involved and Toyozawa introduces the parameters E^ℓ_{LR} and $g_\ell = E^\ell_{LR}/B$ for their characterization. The Fröhlich interaction is naturally completely absent in homopolar semiconductors of group IV (Si, Ge), and therefore $g_\ell = 0$ there.

Toyozawa showed that excitons can occur really in two stable states that occupy local energy minima in the configurational coordinate model: free (F) and self-trapped (S) states. A self-trapped exciton gets localized in a potential well, created as a consequence of the strong lattice deformation over the exciton's surroundings. Toyozawa assigns this exciton to an effective mass $m' = m_{exc} \exp S_0$, where m_{exc} is the free exciton effective mass (thus the case of a strong exciton–phonon interaction, when $S_0 \gg 1$, entails $m' \to \infty$). The steady state switches rapidly from F to S as soon as the value of the coupling constant $g_s = E^s_{LR}/B$ exceeds unity.

The resulting phase diagram in the plane (g_s, g_ℓ) is shown in Fig. 7.29. The coupling of electrons, holes and excitons with phonons increases in the direction of rising g_s, g_ℓ. The F–S boundary is depicted by a solid line, and sections marked F(S) and S(F) indicate regions in which the self-trapped and free states

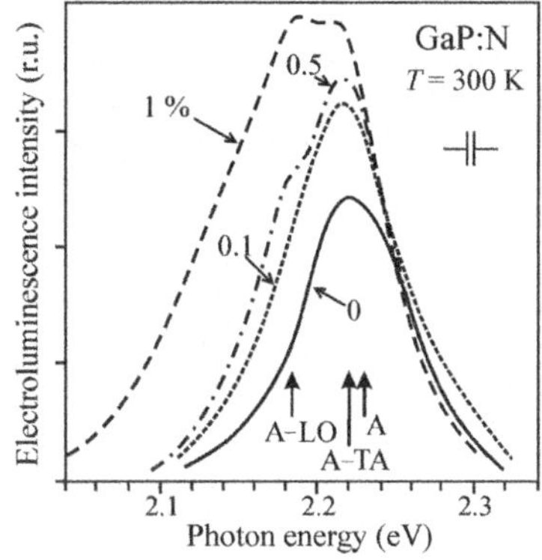

Fig. 7.28
Emission spectrum of the green electroluminescence from a GaP/N diode at room temperature. The GaN content in the melt is shown in percent at the individual curves. The positions of the individual lines from Fig. 7.26(a) are indicated; these lines merge into a single band at higher temperature. The red-shift of the lines in comparison with Fig. 7.26(a) is caused by the bandgap narrowing upon temperature increase 4 K → 300 K. After Vishnevskaya *et al.* [46].

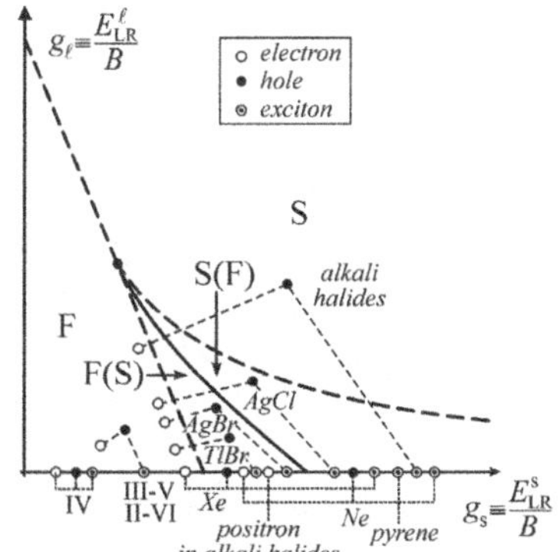

Fig. 7.29
Phase diagram in the (g_s, g_ℓ) plane for electrons, holes and excitons in a deformable lattice. The boundary F (free state)–S (self-trapped state) is depicted by a solid line. It is seen that self-trapping does not occur in IV, III-V and II-VI semiconductors. In AgCl and alkali halides, the hole and subsequently also the exciton become self-trapped. After Toyozawa [49].

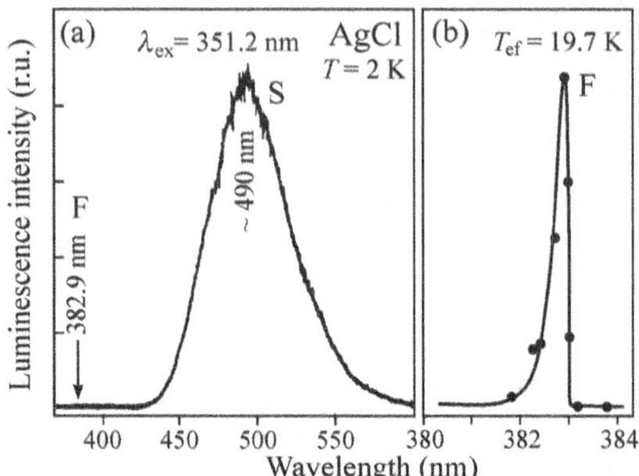

Fig. 7.30
(a) Emission spectrum of AgCl under pumping with laser pulses 351 nm (~ 1 ps) at T = 2 K. The position of the expected free exciton luminescence line is indicated by the arrow F. The broad band S at ~490 belongs to the self-trapped exciton. (b) Time integrated luminescence in the region of the expected free exciton emission. Its intensity is extremely weak (~ 70 counts/s) and the luminescence decays very quickly (~ 20 ps). This line has been attributed to the radiative decay of the free exciton. Points represent experiment; the full line is the Maxwell–Boltzmann shape (7.16) with an effective temperature of 19.7 K. After Kobayashi *et al.* [50].

can coexist (more details later). It can be clearly seen that a break occurs in silver halides—while both the carriers and the exciton are still free in AgBr (and also in TlBr), in AgCl the hole becomes self-trapped and, consequently, self-trapping of the exciton also occurs. Therefore, the luminescence of a (free) exciton in AgBr consists of a narrow resonant emission line (Fig. 7.15), while the emission spectrum of STE with $S_0 \geq 50 \gg 1$ in AgCl is represented by a broad band at ~490 nm (Fig. 7.30(a) [50] or Fig. 4.10(c)). Although AgBr and AgCl differ very little only in most physical, chemical and photochemical properties, the difference in their low-temperature luminescence is substantial!

Various microscopic mechanisms of hole self-trapping can exist. In AgCl the hole becomes self-trapped—even though it may seem strange at first sight—at an already positively charged cation Ag^+[51], thereby a 'molecular ion' $(Ag^+Cl_6^-)^{4-}$ is created, see Fig. 7.31. The reason for such hole localization lies, in the first place, in the fact that the valence band maximum of AgCl at L_3' comprises contributions both from 3p states of Cl^- and 4d states of Ag^+; therefore the self-trapped hole must be constructed from both the states of chlorine and silver. The Jahn–Teller effect leads to a lowered symmetry of the created complex, because of lifting the geometrical equivalence of the six chlorine ions—four of them move towards the central Ag^{2+} ion while the other two move in the opposite direction. In this way a prolonged axis of the complex in one of the equivalent [100] directions is created (Fig. 7.31).[11]

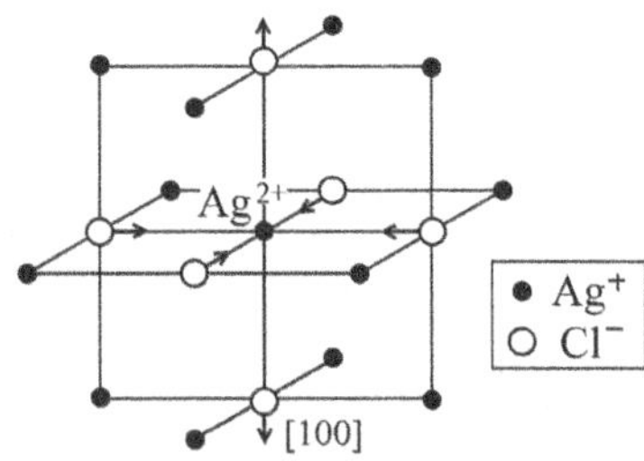

Fig. 7.31
Schematic of the microscopic structure of a self-trapped hole in AgCl depicting tetragonal distortion of the surrounding lattice. After von der Osten [51].

Free and self-trapped excitons may coexist. This holds true in segments marked as F(S) and S(F) in Fig. 7.29. One can imagine such coexistence in the framework of the configurational coordinate model as depicted in Fig. 7.32. In the particular case of AgCl, it is possible to interpret the configurational coordinate Q as the separation of Cl^- ligands from the central Ag^{2+} ion in the $(AgCl_6)^{4-}$ complex. The free exciton (minimum F in Fig. 7.32) is separated from the deeper minimum (S) of the self-trapped exciton by a potential barrier Δ. Free excitons created by light must overcome this barrier in order to reach the energetically favourable self-trapped state (by tunnelling at low temperatures); to do this, they need some time. Because the barrier is

[11] In alkali halides, on the other hand, the hole becomes self-trapped in the form of a molecular ion X_2^- (X = F, Cl, Br, I), which is the so-called V_K- centre oriented along the [110] direction.

low ($\Delta \cong 5$ meV) and thin in AgCl, it is easy to overcome. Therefore, the free exciton 'stays' in the F minimum only for a very short time and the probability of its radiative recombination there is negligible in comparison with penetrating the barrier and subsequently radiating an intense luminescence from the S minimum. In the standard steady-state luminescence experiment we thus record only a broad STE emission band at $\sim$ 490 nm (Fig. 7.30(a)).

A time-resolved luminescence experiment taking advantage of short enough excitation pulses ($\sim$ 1 ps) together with a very sensitive detection system, however, permits us to verify the presence of an extremely weak emission line due to the free exciton radiative annihilation at the F minimum. Besides the fast pumping, also a sufficiently low temperature ($T = 2$–4 K) is necessary because higher temperatures rapidly promote the probability of thermally overcoming the barrier Δ. This delicate experiment was performed by Kobayashi and co-workers [50] and the result agreed exactly with expectation—they succeeded in detecting a weak line with spectral profile corresponding to the Maxwell–Boltzmann distribution at a wavelength of about 383 nm (which corresponds to the indirect exciton gap 3.248 eV decreased by the energy $\hbar\omega_{TO} = 8$ meV of the quasi-momentum conserving phonon), see Fig. 7.30(b).

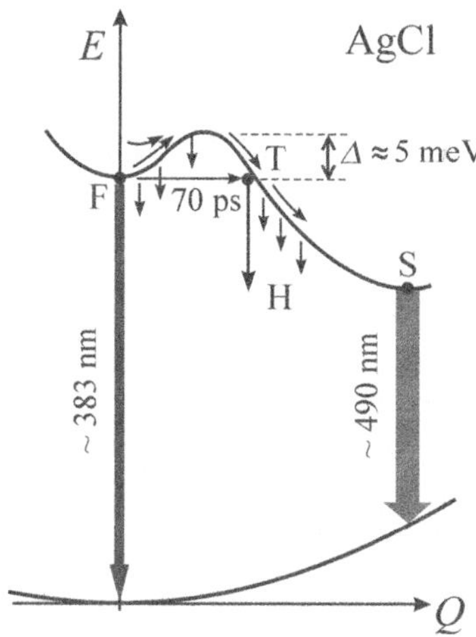

Fig. 7.32
Potential energy curve of a free (F) and self-trapped (S) exciton in the configurational coordinate model. Vertical lines depict luminescence processes, and Δ stands for the potential barrier against self-trapping. After Kobayashi *et al.* [50].

It is interesting to note that nanosecond pumping pulses are not sufficiently short for that purpose because the time required to penetrate the barrier appears to be very short, around 70 ps. However, also under pumping with relatively long ($\sim$10 ns) pulses time-resolved luminescence spectroscopy found a clear indication of short-living emission on the high-energy side of the STE emission band at $\sim$425 nm (Fig. 1.2(a)). This emission can be attributed to hot luminescence of STE (see the arrows H in Fig. 7.32) and to transient fusion of the free exciton with the self-trapped [52].

In general, the occurrence of exciton self-trapping in semiconductors is the exception rather than the rule. Experimental attention was, in addition to AgCl, paid to CdI_2 [53] and $ZnSe_{1-x}Te_x$ ($x \approx 0.01$) [54]. The latter case is, however, a ternary alloy containing defects of the disordered lattice that contribute significantly to the localization effect. Such a localization is no longer an intrinsic process, and is therefore called *extrinsic self-trapping* (the self-trapping occurs neither in pure ZnSe nor in pure ZnTe, in agreement with Fig. 7.29). Otherwise, it is possible to meet the intrinsic self-trapping of excitons in a variety of wide-bandgap non-semiconducting materials, both inorganic and organic. An overview of these phenomena is given in the monograph [55].

7.3 Problems

7/1: Show, employing the slope of the photon dispersion curve $\hbar c(\sqrt{\varepsilon_\infty})^{-1}$, that states corresponding to the polariton effects occupy within the first Brillouin zone only a very small volume around $\mathbf{k} = 0$. (This justifies why we consider $\hbar\omega(\mathbf{k} \cong 0)$ as the energy of phonons taking part in the thermalization and radiative recombination of the exciton–polariton, and it means also that the polariton effects play mostly a negligible role in the total emitted luminescence.)

7/2: Show that the maximum of the (X–LO) emission line is shifted from the energy of the free exciton ground state $n = 1$ towards lower energies not exactly by $\hbar\omega_0$, but by a temperature variable quantity $(\hbar\omega_0–(3/2)k_B T)$.

7/3: The quasi-momentum conservation law is formulated differently depending on whether the free exciton radiative decay is one- or two-phonon assisted. This leads not only to a slightly different shape of the corresponding emission lines (eqns (7.12) and (7.13)) but also to different temperature dependences of both the lines. Show that the ratio of the integral intensities $I_{sp}^{(1)}$ and $I_{sp}^{(2)}$ of the X–LO and X–2LO lines, respectively, can be expressed as $(I_{sp}^{(1)})/(I_{sp}^{(2)}) = \gamma T$, where γ is a constant. Proceed from relations (7.12) and (7.13).

7/4: *Free excitons and the configurational coordinate model.* A free exciton, considering its translational movement, surely does not represent a localized luminescence centre. Can we discuss its luminescence behaviour in the framework of the configurational coordinate model? According to what was formulated in Section 4.4 obviously not. However, the existence of the LO-phonon assisted radiative decay of free excitons points to the fact that an intrinsic exciton–phonon interaction occurs here. We can then consider the (X–LO), (X–2LO), etc. lines as being phonon satellites of the resonant exciton–polariton (no-phonon) line. Therefore, also here we are in principle allowed to use formally the configurational coordinate model, and the reason for this is the finite spatial extent of the exciton wavefunction. The rise and decay of the related electronic polarization then causes a moderate lattice deformation. The Huang–Rhys factor S is, naturally, very small ($S \doteq 0.1$). Deal with the Huang–Rhys factor of free excitons after Zhao and Kalt [56].

7/5: We have derived the symmetry-driven selection rules for indirect exciton transitions in AgBr by making use of (7.17). The applied approach was, however, somewhat simplified. That is, we tacitly assumed the vertical photon-related transitions to the nearby 'virtual' states (whose involvement results from second-order perturbation theory) to be dipole allowed, see transitions $L_1 - L_3'$ and Γ_1–Γ_{15} in Fig. 7.33. Show, using relation (7.14), that this assumption was justified. (You need not know either the tables of characters of point groups or to be proficient in their usage. You will find all the necessary relations throughout Subsections 7.1.3 and 7.1.4. We recall only that $\Gamma_1 \otimes \Gamma_X = \Gamma_X$ holds for an arbitrary representation Γ_X, because the Γ_1 representation is spherically symmetric.)

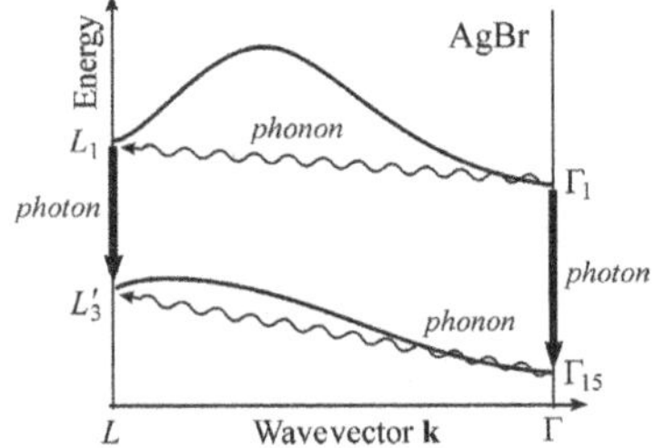

Fig. 7.33

References

1. Knox, R. S. (1963). *Theory of Excitons.* Academic Press, New York; Bassani, F. and Pastori Parravicini, G. (1975). *Electronic States and Optical Transitions in Solids.* Pergamon Press, Oxford.
2. Velický, B., Tauc, J., Trlifaj, M., and Prosser, V. (1965). *Interaction of electromagnetic radiation with solids* (In Czech: *Interakce elektromagnetického záení s pevnou látkou*). In *Solid State Theory* (In Czech: *Teorie pevných látek*), p. 229. NČSAV, Prague.

3. Klingshirn, C. F. (1995). *Semiconductor Optics*. Springer, Berlin.
4. Peyghambarian, N., Koch, S. W., and Mysyrowicz, A. (1993). *Introduction to Semiconductor Optics*. Prentice Hall, Englewood Cliffs, N.J.
5. Weisbuch, C. and Benisty, H. (2005). *phys. stat. sol. (b)*, **424**, 2345.
6. Joesten, B. L. and Brown, F. C. (1966). *Phys. Rev.*, **148**, 919.
7. Razeghi, M., Omnes, F., Nagle, J., Defour, M., Acher, O., and Bove, P. (1989). *Appl. Phys. Lett.*, **55**, 1677.
8. Koteles, E. S., Lee, J., Salerno, J. P., and Vassel, M. O. (1985). *Phys. Rev. Lett.*, **55**, 867.
9. Bose, S. S., Kim, M. H., and Lee, B. (1988). *J. Electr. Mater.*, **17**, S36.
10. Gross, E., Permogorov, S., and Razbirin, B. (1966). *J. Phys. Chem. Solids*, **27**, 1647.
11. Segall, B. and Mahan, G. D. (1968). *Phys. Rev.*, **171**, 935.
12. Gross, E. F., Permogorov, S. A., and Razbirin, B. S. (1971). *Sov. Phys. Usp.*, **14**, 104.
13. Bassani, F. and Pastori-Parravicini, G. (1975). *Electronic States and Optical Transitions in Solids*. Pergamon Press, Oxford.
14. Kunz, A. B. (1982). *Phys. Rev. B*, **26**, 2070.
15. von der Osten, W. and Weber, J. (1974). *Solid State Comm.*, **14**, 1133.
16. Thomas, G. A., Frova, A., Hensel, J. C., Miller, R. E., and Lee, P. A. (1976). *Phys. Rev. B*, **13**, 1692.
17. Hensel, J. C., Phillips, T. G., and Thomas, G. A. (1977). *The electron–hole liquid in semiconductors: experimental aspects*. In *The Electron–Hole Liquid in Semiconductors* (ed. H. Ehrenreich, F. Seitz and D. Turnbull), p. 88. *Solid State Physics* Vol. 32. Academic Press, New York.
18. Rashba, E. I. and Gurgenishvili, G. E. (1962). *Sov. Phys. – Solid State*, **4**, 759.
19. Haynes, J. R. (1960). *Phys Rev. Lett.*, **4**, 361.
20. Bebb, H. B. and Williams, E. W. (1972). *Photoluminescence I. Theory*. In *Semiconductors and Semimetals* (ed. R. K. Willardson and A. C. Beer), Vol. 8, p. 181. Academic Press, New York and London,
21. Bernussi, A. A., Barreto, C. L., Carvalho, M. M. G., and Motisuke, P. (1988). *J. Appl. Phys.*, **64**, 1358.
22. Hayes, W. and Stoneham, A. M. (1985). *Defects and Defect Processes in Nonmetallic Solids*. John Wiley, New York.
23. Hönerlage, B., Rössler, U., and Schröder, U. (1975). *Phys. Rev. B*, **12**, 2355.
24. Herbert, D. C. (1984). *J. Phys. C: Solid State Phys.*, **17**, L901.
25. Kaminskii, A. S. and Pokrovskii, Ya. E. (1970). *JETP Lett.*, **11**, 255.
26. Lampert, M. A. (1958). *Phys. Rev. Lett.*, **1**, 450.
27. Kirczenow, G. (1977). *Canad. J. Phys.*, **55**, 1787.
28. Tajima, M.(1982). *Quantitative impurity analysis in Si by the photoluminescence technique*. In *Semiconductor Technologies* (ed. J. Nishizawa), p. 1. North Holland, Amsterdam.
29. Pelant, I., Hála, J., Ambrož, M., Vácha, M., Valenta, J., Adamec, F., Kohlová, V., and Matoušková, J. (1990). *Impurity analysis in Si crystals by the photoluminescence method V* (In Czech: *Analýza příměsí v krystalech Si fotoluminiscenční metodou V*). Research Report for Tesla Rožnov, k.p., Charles University in Prague, Faculty of Mathematics & Physics, Prague.
30. Colley, P. McL. and Lightowlers, E. C.(1987). *Semicond. Sci. Technol.*, **2**, 157.
31. Pelant, I., Hála, J., Ambrož, M., Kohlová, V., and Vacek, K. (1989). *Czech. J. Phys A* (In Czech), **39**, 142.
32. Karaiskaj, D., Thewalt, M. L. W., Ruf, T., and Cardona, M. (2003). *phys. stat. sol. (b)*, **235**, 63.
33. Pelant, I., Hála, J., Ambrož, M., and Kohlová, V. (1988). *Photoluminescence of impurities and photoluminescence assessment of boron and phoshorus in silicon* (In Czech: *Fotoluminiscence příměsí a fotoluminiscenční stanovení boru a fosforu v křemíku*). In *Křemík '88*, part 2, p. 86. General Head Office Tesla–ES, Rožnov pod Radhoštěm.

34. Pelant, I., Hála, J., Ambrož, M., and Kohlová, V. (1988). *Quantitative analysis of low concentrations of boron and phosphorus in silicon by the laser luminescence spectroscopy—experimental aspects* (In Czech: *Kvantitativní analýza nízkých koncentrací boru a fosforu v křemíku laserovou luminiscenční spektroskopií – experimentální aspekty*). In *Lasers in Research and Industry* (6th Czechoslovak Conference), p. 252. Račkova dolina, SVŠT ČSSP Liptovský Mikuláš.
35. Kaminskii, A. S., Kolesnik, L. I., Leiferov, B. M., and Pokrovskii, Ya. E. (1982). *J. Appl. Spectros.*, **36**, 516.
36. Schumacher, K. L. and Whitney, R. L. (1989). *J. Electron. Mater.*, **18**, 681.
37. Broussell, I., Stolz, J. A. H., and Thewalt, M. L. W. (2002). *J. Appl. Phys.*, **92**, 5913.
38. Zimmermann, H., Boyn, R., Michel, C., and Rudolph, P. (1990). *phys. stat. sol. (a)*, **118**, 225.
39. Thewalt, M. L. W., Nissen, M. K., Beckett, D. J. S., and Lungren, K. R. (1989). *High-performance photoluminescence spectroscopy using Fourier-transform interferomety*, MRS Meeting Boston November/December, Symp. G. *MRS Symposium Proceedings*, **163**, (1990), 221.
40. Dekker, A. J. (1963). *Solid State Physics*. Prentice-Hall, Englewood Cliffs, N.J.; Curie, D. (1960). *Luminescence Cristalline*. Dunod, Paris.
41. Gershoni, D., Cohen, E., and Ron, A. (1985). *J. Luminescence*, **34**, 83.
42. Pelant, I. (1988). *Laser luminescence spectroscopy of some crystalline semiconductors and ionic crystals* (In Czech: *Laserová luminiscenční spektroskopie vybraných krystalických polovodičů a dielektrik*), DSc Thesis, Charles University in Prague, Faculty of Mathematics and Physics, Prague.
43. Tsukakoshi, M. and Kanzaki, H. (1971). *J. Phys. Soc. Japan*, **30**, 1423.
44. Stolz, H., von der Osten, W., and Weber, J. (1976). *Lifetime and time-resolved spectra of indirect excitons in AgBr*. In *Physics of Semiconductors* (ed. F. G. Fumi), p. 865. *Proceedings of the 13th International Conference on the Physics of Semiconductors*, Rome.
45. Holonyak, N. Jr., Campbell, J. C., Lee, M. H., Verdeyen, J. T., Johnson, W. L., Craford, M. G., and Finn, D. (1973). *J. Appl. Phys.*, **44**, 5517.
46. Vishnevskaya, B. I., Korneev, V. M., Kogan, L. M., and Yunovich, A. E. (1973). *Soviet Phys. Semicond.*, **6**, 1372.
47. Craford, M. G. and Holonyak, N. Jr. (1976). *The optical properties of the nitrogen isoelectronic trap in* $GaAs_{1-x}P_x$. In *Optical Properties of Solids. New Developments* (ed. B. O. Seraphin), p. 187. North Holland, Amsterdam.
48. Brown, T. G. and Hall, D. G. (1998). *Radiative isoelectronic impurities in silicon and silicon–germanium alloys*. In *Light Emission in Silicon: From Physics to Devices* (ed. D. J. Lockwood), *Semiconductors and Semimetals* Vol. 49, p. 77. Academic Press, San Diego.
49. Toyozawa, Y. (1970). *J. Luminescence*, **1**, 632; Toyozawa, Y. (1981). *J. Luminescence*, **24/25**, 23.
50. Kobayashi, M., Matsushima, Y., Nishi, O., Mizuno, K., and Matsui, A. H. (1995). *SPIE*, **2362** (*Exciton processes in condensed matter*), 225.
51. von der Osten, W. (1984). *Excitons and exciton relaxation in silver halides*. In *Polarons and Excitons in Polar Semiconductors and Ionic Crystals* (ed. J. T Devreese and F. Peeters), p. 293. Plenum Press, New York.
52. Pelant, I. and Hála, J. (1991). *Solid State Comm.*, **78**, 141.
53. Nakagawa, H., Muneda, Y., and Matsumoto, H. (1988). *J. Luminescence*, **40/41**, 485.
54. Lee, D., Mysyrowicz, A., Nurmikko, A. V., and Fitzpatrick, B. J. (1987). *Phys. Rev. Lett.*, **58**, 1475.
55. Song, K. S. and Williams, R. T. (1996). *Self-Trapped Excitons*. Springer Series in Solid State Sciences Vol. 105. Springer, Berlin.
56. Zhao, H. and Kalt, H. (2003). *Phys. Rev. B*, **68**, 12.

Highly excited semiconductors

8

The channels of radiative recombination, discussed in Chapters 5 and 7, are characteristic of conventional (weak) optical excitation when the excitation power densities on the sample surface amount to 0.01–$10\,\mathrm{W/cm^2}$. When the excitation intensity is increased to $\mathrm{kW/cm^2}$–$\mathrm{MW/cm^2}$ (which is attainable almost exclusively with the aid of pulsed lasers) the emission spectrum undergoes substantial modifications: New emission lines or bands appear. Figure 8.1 shows a typical example [1]. These lines or bands signify novel luminescence processes occurring in highly excited semiconductors, in particular: radiative decay of excitonic molecules (biexcitons), collisions between excitons, luminescence of the electron–hole liquid or electron–hole plasma and Bose–Einstein condensation of excitons or biexcitons. We are going to discuss these

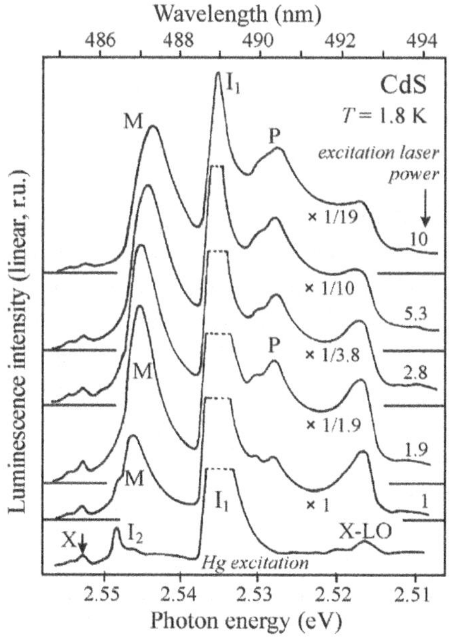

Fig. 8.1
Edge emission spectrum of CdS under various intensities of optical excitation (temperature $T = 1.8\,\mathrm{K}$). The lowermost spectrum was acquired at a conventional Hg-lamp excitation ($\lambda_{ex} = 365\,\mathrm{nm}$) and is dominated by I_1 and I_2 lines, which are due to excitons bound to residual impurities. The X-line belongs to the resonant emission of an exciton–polariton (compare with Fig. 7.11(a)). The remaining spectra were excited by a pulsed N_2-laser ($\lambda_{ex} = 337\,\mathrm{nm}$), relative excitation power densities being indicated to the right along with normalization factors. The figure illustrates that the I_1, I_2 lines gradually saturate with increasing excitation intensity while novel M and P lines appear. After Shionoya *et al.* [1].

phenomena in the present chapter. Optical properties of highly excited semiconductors are dealt with also, e.g. in special issues of *Physics Reports* [2, 3].

8.1 Experimental considerations

To understand the underlying physical nature of the novel luminescence processes occuring under high excitation, it appears advantageous to characterize the concentration level of created free excitons n_X. We can start with eqn (3.4) from which it follows that in steady-state conditions

$$n_X \cong I_{ex}\,\alpha\,\tau, \tag{8.1}$$

where I_{ex} stands for the excitation photon flux (photon/cm^{2}s) of energy $h\nu_{ex}$, $\alpha(\mathrm{cm}^{-1})$ denotes the absorption coefficient at energy $h\nu_{ex}$ and τ (s) means the exciton lifetime. For a typical case of band $\rightarrow$ band excitation we consider $\alpha \cong 10^5\mathrm{cm}^{-1}$, $\tau = 10^{-9}$s (direct semiconductor) and the excitation photon flux I_{ex} can be derived from the excitation intensity P_{ex}—usually expressed in watts per cm^2—using the relation $I_{ex} = (P_{ex}/h\nu_{ex})$. For $P_{ex} = 1\,\mathrm{W/cm^2}$ (conventional weak excitation) and $h\nu_{ex} = 3\,\mathrm{eV}$ we have $I_{ex} \cong 2 \times 10^{18}$ photon/cm^2 s and the density of photogenerated excitons (8.1) becomes $n_X \cong 2 \times 10^{14}\,\mathrm{cm}^{-3}$.

Given $\ell_X \cong \sqrt[3]{(1/n_X)}$) as the mean interexciton separation, we get in this case $\ell_X \cong 170\,\mathrm{nm}$. Comparing this value with a typical exciton Bohr radius $a_X \cong 5\,\mathrm{nm}$ (Table 7.1) implies $\ell_X \gg a_X$, and therefore free excitons are sufficiently separated from each other and do not enter into interactions. However, if we increase their density so that ℓ_X approaches the exciton diameter $2a_X$, the situation becomes completely different. The condition $\ell_X \cong 2a_X$ yields (keeping $a_X = 5\,\mathrm{nm}$) the exciton density $n_X \cong 10^{18}\,\mathrm{cm}^{-3}$ and the corresponding excitation photon flux is, using (8.1), equal to $I_{ex} \approx 10^{22}$ photon/cm^2 s or, equivalently, to the excitation intensity $P_{ex} = I_{ex}h\nu_{ex} \approx 5\,\mathrm{kW/cm^2}$.[1]

At such excitation intensities one can already expect the occurence of interactions between excitons. These, as we shall see, may dramatically modify the luminescence behaviour of the material under study. It is worth stressing, however, that relation (8.1) is not exact; the density of excitons or electron–hole pairs under standard one-photon excitation is considerably inhomogeneous. The concentration gradient pushes them from a thin subsurface layer with thickness $\alpha^{-1} < 1\,\mu\mathrm{m}$ deeper into the sample, thereby rapidly reducing their concentration. Besides, excitons close beneath the surface undergo efficient surface non-radiative recombination (Fig. 6.2). That is why the excitation intensities P_{ex} applied in experiments must in fact be higher than the above order-of-magnitude estimate—they usually range from $\sim 1\,\mathrm{kW/cm^2}$ up to 1–10 $\mathrm{MW/cm^2}$, where the upper bound is driven by the threshold of mechanical destruction of the sample by powerful laser pulses.

[1] For expressing the density of excitons or electron–hole pairs, one sometimes exploits the dimensionless parameter $r_s = (3/4\pi n_X)^{1/3}(1/a_X)$; high densities n_X are characterized by values $r_s < 1$.

Two remarks may be in order. The first one refers to most indirect-bandgap semiconductors (Si, Ge, GaP). Here, the exciton lifetime is much longer than 10^{-9} s and high exciton densities $r_s < 1$ can already be achieved using cw lasers or even conventional high-pressure lamps.

The second remark points to the principal superiority of two-photon excitation for studying luminescence of high-density excitonic systems. Owing to the very slow decrease of excitation intensity along the penetration depth inside the sample (Fig. 5.15(a)), relation (8.1) holds—contrary to the one-photon case—very accurately. We can thus achieve a pre-determined value of homogeneous exciton concentration over the whole volume of the sample, which is highly desirable for comparing the experimentally deduced density with various models of excitonic interactions discussed below. Of course, we have to replace the one-photon absorption coefficient α in (8.1) by its two-photon analogue $\alpha^{(2)}$, defined in the context of eqn (5.22) as

$$\alpha^{(2)} = aI_0 = aP_{\text{ex}}(\text{cm}^{-1}) \tag{8.2}$$

(that is, we denote the excitation power density as P_{ex} in this chapter). The reader is reminded that a is a material constant with a typical value of $a \cong 10^{-3}$ cm/MW.

However, there is one hitch in this, namely, if we realize how low the amount of energy deposited generally into the electronic system of the sample under two-photon excitation is. Introducing $\alpha^{(2)}$ from (8.2) to (8.1) and expressing P_{ex} as a function of the required exciton concentration, we get

$$P_{\text{ex}} = \sqrt{n_{\text{X}}\, h\, \nu_{\text{ex}}/\tau\, a}. \tag{8.3}$$

Therefore, if we wish to generate a homogeneous exciton concentration $n_{\text{X}} = 10^{17}\,\text{cm}^{-3}$, using typical values of $h\nu_{\text{ex}} = 3\,\text{eV} \approx 5 \times 10^{-19}$ J, $\tau = 10^{-9}$ s and $a = 10^{-3}$ cm/MW, it follows from (8.3) that the required excitation power density P_{ex} reaches $\sim 100\,\text{MW/cm}^2$! Such a high value of excitation power density is already comparable with the threshold of sample destruction. We have to carefully look up all possible material parameters (especially the value of a) in advance, as well as evaluate all the capabilities of our experimental set-up. Only then can we decide whether the possible advantage of achieving a homogeneous density of exciton gas would not be nullified by damaging the sample.

8.2 Excitonic molecule or biexciton

Referring to the similarity between a free exciton and a hydrogen atom, discussed in Section 7.1, we can expect that with increasing free exciton density a 'fusion' of two excitons into an excitonic molecule (EM) or biexciton will occur, in analogy with the fusion of two hydrogen atoms into a hydrogen molecule. An EM is a quasi-particle consisting of two electrons and two holes and propagating freely through the crystal lattice (more exactly, we should designate it as a *free biexciton*). It is obvious that the very existence of EMs in a given semiconductor will be critically dependent upon the ratio of the electron and hole effective masses $\sigma = m_{\text{e}}/m_{\text{h}}$, much like as in bound exciton

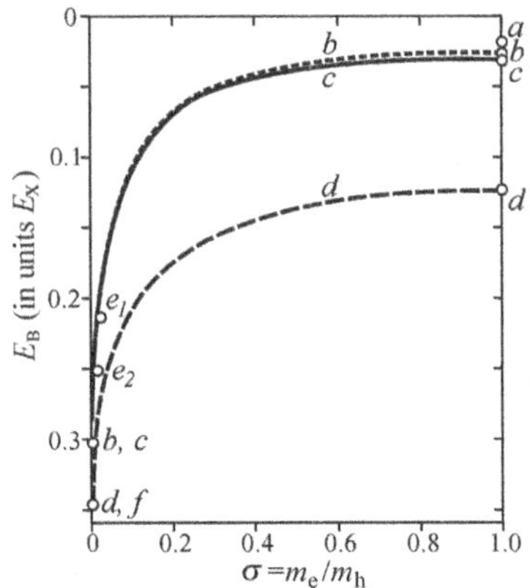

Fig. 8.2
Examples of theoretically calculated dependencies of the biexciton binding energy E_B upon the ratio of the electron and hole effective masses $\sigma = m_e/m_h$. E_B is expressed in units of the free exciton binding energy E_X. Small letters denote results obtained by different authors. Adapted from Quattropani and Forney [4], where the reader can find a more detailed discussion.

complexes discussed in Subsection 7.2.1. Both an electron and a hole are very light quasi-particles and thus the resulting large amplitude of biexciton zero-point vibrations will have a tendency to dissociate the excitonic molecule even at low temperatures. The measure of EM stability against dissociation into two free excitons is its binding energie E_B.

Theoretical calculations of E_B using diverse variation methods have been performed by many authors; the results, as compiled by Quattropani and Forney, are depicted in Fig. 8.2 [4]. It is observed that an EM is a stable entity ($E_M \geq 0$) for an arbitrary value of σ, even if, quite naturally, the binding energy strongly increases in the case of heavy holes.

It is important to note that the calculations predict, for typical values of σ in semiconductors ($\sigma \approx 0.1$–0.5), that the magnitude of the binding energy E_B will be approximately equal to one-tenth of the free exciton binding energy: $E_B \cong 0.1E_X$. A biexciton is therefore substantially less resistant to thermal dissociation than an exciton. An important message for experiments is that for studying biexcitons very low temperatures are required in most cases (about 1.3–50 K), accessible in liquid helium cryostats only.

Radiative recombination of EMs is a process in which one of the excitons forming the molecule undergoes a transition to the free exciton $n = 1$ state while the second one recombines, emitting a photon $h\nu_M$. This process gives rise to a novel emission line (e.g. the M-line in Fig. 8.1). In order to be able to assign reliably an experimentally observed line to excitonic molecule decay, we have to analyse its lineshape, intensity dependence, spectral position and temperature behaviour.

8.2.1 Identification of the EM emission line

The basic guideline for the indentification of an emission line as being due to radiative decay of an EM is its characteristic lineshape. It turns out to be suitable to discuss the EM emission lineshape separately in direct-bandgap and indirect-bandgap semiconductors.

Luminescence of excitonic molecule in a direct semiconductor

To begin, we determine the approximate spectral position of the EM emission line, neglecting the kinetic energies of all participating quasi-particles.

The formation of an EM through the 'fusion' of two excitons with energies (E_g–E_X), accompanied by a simultaneous release of binding energy E_B, leads to a biexciton in its ground state with an energy of $2(E_g - E_X) - E_B$ (see Fig. 8.3(a)). Radiative decay of the biexciton generates one photon, $h\nu_M$, and leaves behind one free exciton, therefore the total energy of the system reads $h\nu_M + (E_g - E_X)$. Comparing the above mentioned energies yields the overall energy balance

$$2(E_g - E_X) - E_B = h\nu_M + (E_g - E_X) \tag{8.4}$$

or

$$h\nu_M = (E_g - E_X) - E_B. \tag{8.5}$$

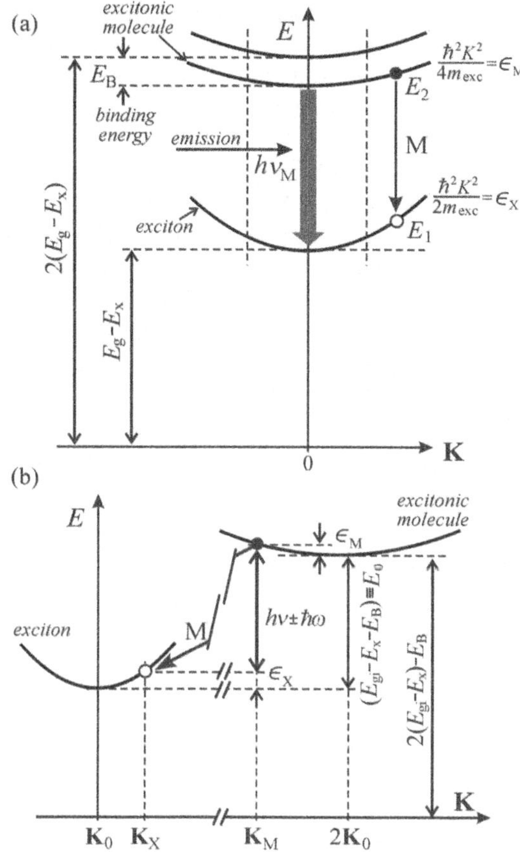

Fig. 8.3
Radiative decay of an excitonic molecule in an $E(\mathbf{K})$ diagram for (a) a direct-bandgap semiconductor and (b) an indirect-bandgap semiconductor with the indirect bandgap situated at $\mathbf{K}_0$. The scheme goes, of course, beyond the limit of the one-electron approximation (otherwise the energy level of the excitonic molecule would be 'submerged' high up into the conduction band).

Hence, the emission line of the EM is shifted with respect to the position of the free exciton absorption line (E_g–E_X) (or the luminescence line of an exciton–polariton) by the binding energy E_B towards low photon energies.

Now we take into consideration also the kinetic energies in the ensemble of excitonic molecules and free excitons. We shall arrive at important information on the characteristic biexciton spontaneous emission lineshape $I_{sp}^M(h\nu)$, namely, that it is of a mirror-like appearance with regard to the free exciton emission lineshape described by expressions (7.13) or (7.16). We speak of the inverse Maxwell–Boltzmann distribution (with effective temperature T_M of the biexciton gas), i.e.

$$I_{sp}^M(h\nu) \cong [(E_g - E_X - E_B) - h\nu]^{1/2}$$
$$\exp\{-[(E_g - E_X - E_B) - h\nu]/k_B T_M\}. \qquad (8.6)$$

Prior to deriving expression (8.6) rigorously, we shall attempt to understand this lineshape qualitatively employing Fig. 8.3(a). The biexciton dispersion curve $\epsilon_M(\mathbf{K}) = h^2K^2/4m_{exc}$ (where $m_{exc} = m_e + m_h$ is the exciton mass) is less curved compared with the exciton curve $\epsilon_X(\mathbf{K}) = h^2K^2/2m_{exc}$. The highest emitted photon energy thus belongs to the downward transition at $\mathbf{K} = |0\rangle$ and is, according to (8.5), equal to $(E_g$–$E_X)$–E_B; radiative decay of biexcitons with non-zero kinetic energy at $\mathbf{K} \neq |0\rangle$ results in luminescence photons with lower energy, which is depicted in Fig. 8.3(a) by the transition M between states E_2 and E_1. Therefore, the emission line will have an asymmetric shape characterized by a high-energy edge at $(E_g–E_X)–E_B$ followed by a low-energy tail.

To derive (8.6) we shall apply a procedure similar to that utilized in Subsection 5.2.1 for deducing the Maxwell–Boltzmann distribution (5.9a). Spontaneous emission of a luminescence photon pertaining to radiative decay of a biexciton $I_{\mathrm{sp}}^{\mathrm{M}}(h\nu)$ is driven by (i) the optical joint density of states $\rho_{\mathrm{M}}(h\nu)$, coming from the relevant densities of states in the exciton and biexciton bands,[2] (ii) the population fraction $f_{\mathrm{M}}(h\nu)$ of the biexciton states and (iii) the transition matrix element M_{M}:

$$I_{\mathrm{sp}}^{\mathrm{M}}(h\nu) \sim \rho_{\mathrm{M}}(h\nu) f_{\mathrm{M}}(h\nu) |M_{\mathrm{M}}|^2. \qquad (8.7)$$

The matrix element M_{M} is considered to be independent of $h\nu$. Referring to Fig. 8.3(a) we can write

$$E_2 = 2(E_g - E_X) - E_B + \hbar^2 K^2 / 4\, m_{\mathrm{exc}}, \qquad (8.8a)$$

$$E_1 = (E_g - E_X) + \hbar^2 K^2 / 2\, m_{\mathrm{exc}}, \qquad (8.8b)$$

which immediately implies

$$h\nu = E_2 - E_1 = (E_g - E_X - E_B) - \hbar^2 K^2 / 4\, m_{\mathrm{exc}}, \qquad (8.8c)$$

hence

$$K^2 = \frac{4\, m_{\mathrm{exc}}}{\hbar^2} [(E_g - E_X - E_B) - h\nu]. \qquad (8.9)$$

Inserting for K^2 from (8.9) to (8.8b) yields

$$E_1 = 2[(E_g - E_X - E_B) - h\nu] + (E_g - E_X). \qquad (8.10)$$

Let us suppose now that the density of states in the exciton band ρ_X has the common square-root form, fully analogous to the densities of states in the conduction and valence bands, expressed by formula (5.4):

$$\rho_X(E_1) = \frac{(2\, m_{\mathrm{exc}})^{2/3}}{2\, \pi^2 \hbar^3} \sqrt{E_1 - (E_g - E_X)}. \qquad (8.11)$$

By using the incremental relation

$$\rho_X(E_1)\, \mathrm{d}E_1 = -\rho_{\mathrm{M}}(h\nu)\, \mathrm{d}(h\nu)$$

(the '–' sign means here that the loss of one free exciton state is equivalent to the generation of one photon $h\nu$) along with (8.10) and (8.11) we find the optical joint density of states

$$\rho_{\mathrm{M}}(h\nu) = -\rho_X(E_1)\, \mathrm{d}E_1/\mathrm{d}(h\nu) = -\rho_X(E_1)(-2)$$
$$= \frac{\sqrt{2}(2m_{\mathrm{exc}})^{3/2}}{\pi^2 \hbar^3} \sqrt{(E_g - E_X - E_B) - h\nu}. \qquad (8.12)$$

[2] The reader is reminded that the possibilty of introducing the joint density of states follows from the verticality of the transitions depicted in Fig. 8.3(a).

Finally, the population factor $f_M(h\nu)$ in (8.7) can be determined quite easily: Since both biexcitons and excitons are composed of an even number of fermions, they have integer spin and are, therefore, bosons. Then, there is no need—contrary to direct interband *e–h* transitions—to formulate any mathematical condition concerning the occupancy of the exciton dispersion curve; the population fraction is driven exclusively by the occupation probability of the upper (biexciton) level. This is given by the Boltzmann factor

$$f_M(h\nu) \approx \exp(-\hbar^2 K^2/4\, m_{exc}\, k_B T_M)$$
$$= \exp\left[-(E_g - E_X - E_B - h\nu)/k_B T_M\right].$$

By combining this expression with (8.12) we obtain, according to (8.7), the emission lineshape of biexciton luminescence (8.6).

The reader will have certainly noticed that the above considerations were conducted under the tacit assumption of the validity of not only energy conservation, but also quasi-momentum conservation for the whole recombining biexciton population. In other words, all biexcitons are supposed to take part in the radiative decay, independently of the magnitude of their wavevector **K**. This represents a fundamental difference from the free exciton luminescence in direct-bandgap materials, where the radiative recombination can occur only for $\mathbf{K} \cong |0\rangle$ and when the polariton effects play an important role, as we have clarified in Subsection 7.1.2. When dealing with biexcitons we have not been speaking about polaritons at all. Is this standpoint correct? And if it is the case, then why?

Yes, this approach is basically correct for the following reason: The biexciton momentum $\hbar\mathbf{K}$ is carried away by the so-called recoil exciton, which is released during the luminescence process

$$\hbar\,\mathbf{K} = \hbar\,\mathbf{K}_{photon}(\cong 0) + \hbar\,\mathbf{K}_{exc}, \tag{8.13}$$

thereby ensuring the momentum conservation law for any value of **K**. For the same reason the influence of the polariton effects (being limited to a small region $|\mathbf{K}| \leq 10^5\,\mathrm{cm}^{-1}$) on the overall emission lineshape can be, in a good approximation, neglected. From what has just been said it follows immediately that biexciton luminescence can become a very *effective channel* of radiative recombination in highly excited semiconductors, and may even take part in stimulated emission.

The typical asymmetry of the free biexciton emission lineshape in direct-gap semiconductors is documented—in addition to Fig. 8.1—also in Fig. 8.4(a) [5]. The investigated material is crystalline copper chloride, CuCl; the solid curve denotes experimental data, the dashed curve represents the theoretical lineshape (8.6) with effective temperature $T_M = 18\,\mathrm{K}$ and binding energy $E_B = 28\,\mathrm{meV}$. The presence of two lines is due to transitions to the longitudinal and transverse exciton states, split by $\Delta_{LT} \approx 5\,\mathrm{meV}$. The relevant excitation photon ($h\nu_{ex} \cong 3.7\,\mathrm{eV}$) flux was $I_{ex} = 8 \times 10^{21}\,\mathrm{photon/cm^2\,s}$ ($P_{ex} \approx 4.8\,\mathrm{kW/cm^2}$). A further increase in excitation intensity leads to biexciton collisions accompanied by a deviation of the experimental lineshape from (8.6); the line becomes more symmetric. A temperature increase entails a similar effect. For the sake of comparison, Fig. 8.4(b) displays the biexciton emission line

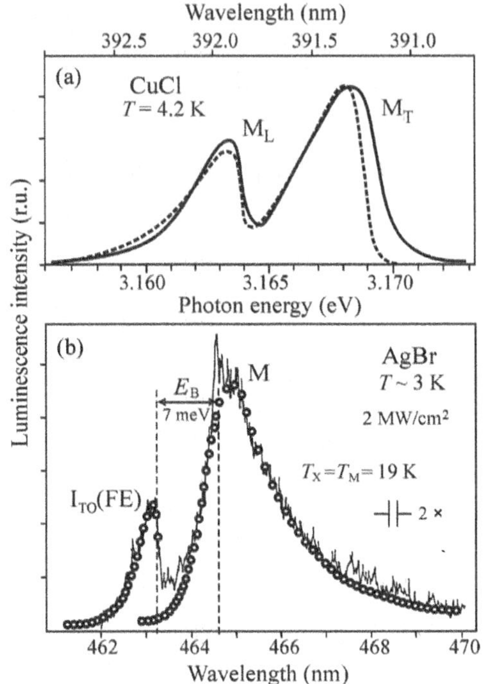

Fig. 8.4
(a) Biexciton emission spectrum in CuCl (direct-gap semiconductor) at a bath temperature of 4.2 K, excited by a pulsed N_2-laser (337 nm). Solid curve – experiment, dashed curve – theoretical lineshape after (8.6). The line $M_T(M_L)$ is due to a transition to the transversal (longitudinal) exciton level, respectively. After Grun *et al.* [5]. (b) Spectrum of edge emission of AgBr (indirect-gap semiconductor) at a bath temperature of 3 K under N_2-laser excitation, comprising a free exciton line I_{TO}(FE) and a biexciton line M. This figure demonstrates clearly the mirror symmetry of both lines. Solid curve – experiment, symbols – theoretical fit of the I_{TO}(FE) line employing (7.16) and of the M line according to (8.16), using the parameter $\Gamma = (\hbar^2/2a_M^2 m_{exc})/k_B T_M$ equal to 4. The effective temperatures of the exciton and biexciton gases turn out to be equal: $T_X = T_M = 19$ K. After Baba and Masumi [6].

in an indirect-gap semiconductor (AgBr [6]); the corresponding lineshape will now be discussed.

Luminescence of excitonic molecule in an indirect semiconductor

Given the exciton band minimum located at a wavevector $\mathbf{K}_0 \neq 0$, then biexcitons are characterized by wavectors around $2\mathbf{K}_0$. Consequently, phonons with momentum $\sim \hbar\mathbf{K}_0$ (and energy $\hbar\omega$) have to participate in the process of radiative biexciton decay. This is certainly an important difference with respect to the biexciton luminescence in direct-bandgap materials; nevertheless, we shall see that this fact has no principal impact upon the emission lineshape.

The line will be—at low temperatures—of a shape very similar to the inverse Maxwell–Boltzmann distribution. We can draw such a conclusion by looking at Fig. 8.3(b). Downward transitions from the minimum of the EM dispersion curve (where at $T \approx 0$ K the entire biexciton population occurs, having kinetic energy $\epsilon_M = 0$) give rise to photons with maximum available energy $h\nu_M = E_{gi} - E_X - E_B - h\omega$, provided the final state of the transition is located at the exciton dispersion minimum, i.e. at $\epsilon_X = 0$; all other transitions (to higher exciton states) may result in lower photon energies only. Consequently, the emission line will have a sharp high-energy threshold on the one side, and a low-energy tail on the other—as yet, everything is basically the same as in biexcitons in direct-gap semiconductors. Even so, with rising temperature now the high-energy edge will become broader, because biexcitons will gradually occupy also higher states $\epsilon_M > 0$ on their dispersion curve, and downward transitions from these states to the minimum of the exciton states $\epsilon_X = 0$ yield photon energies higher than $h\nu_M = E_{gi} - E_X - E_B - h\omega \equiv E_0 - h\omega$.

It is obvious that also in the indirect bandgap the entire population of biexcitons is participating in the luminescence process, because the law of momentum conservation is met again, this time owing to participating phonons $h\omega$.

Let us now attempt to indicate how one can formulate a mathematical description of the emission lineshape. The concept of joint density of states cannot be—unlike the direct bandgap—introduced here. We have to deal with the task in a slightly different way, namely, in analogy with expression (5.10) that determines the luminescence lineshape of recombining free electron–hole pairs in an indirect bandgap. This expression is sufficiently universal. We can thus, making use of (5.10), write down for the case of EM luminescence in the indirect bandgap

$$I_{\text{in}}^{\text{M}}(h\nu) \cong \int \rho_{\text{M}}(\epsilon_{\text{M}}) f_{\text{M}}(\epsilon_{\text{M}}) \rho_{\text{X}}(\epsilon_{\text{X}}) f_{\text{X}}(\epsilon_{\text{X}}) \left| M_{\text{in}}^{\text{M}} \right|^2 \mathrm{d}\epsilon_{\text{M}}, \tag{8.14}$$

where ϵ_M and ϵ_X stand for the biexciton and exciton kinetic energy, $\rho_M(\epsilon_M)$ and $\rho_X(\epsilon_X)$ denote the densities of states in the biexciton and exciton bands, respectively, and f_M and f_X mean the relevant occupation factors. Meanwhile, we keep the squared matrix element $| M_{\text{in}}^{\text{M}} |^2$ within the integral and we have not explicitly shown the integration limits. These can be found easily from Fig. 8.3(b). For a given constant value of $h\nu$ we have to perform the integration with respect to ϵ_M obviously from a certain lower limit to infinity; the lower limit can be derived from energy consideration (Fig. 8.3(b))

$$\epsilon_{\text{X}} + (h\nu + \hbar\omega) = \epsilon_{\text{M}} + (E_{\text{gi}} - E_{\text{X}} - E_{\text{B}}) \tag{8.15}$$

applied for $\epsilon_X = 0$. Accordingly, making use of (8.15) yields for the lower limit

$$\epsilon_{\text{M}}' = (h\nu + \hbar\omega) - (E_{\text{gi}} - E_{\text{X}} - E_{\text{B}}) = h\nu + \hbar\omega - E_0.$$

We denote this lower limit as $\overline{h\nu} = h\nu + \hbar\omega - E_0$.[3] In addition, with the aid of (8.15) one can write down

$$\epsilon_{\text{X}} = \epsilon_{\text{M}} - \overline{h\nu}$$

and for the densities of states in the parabolic approximation of $\epsilon(\mathbf{K})$ we obtain $\rho_M(\epsilon_M) \approx \sqrt{\epsilon_M}$, $\rho_X(\epsilon_X) \approx \sqrt{(\epsilon_M - \overline{h\nu})}$. Thereby, expression (8.14) becomes

$$I_{\text{in}}^{\text{M}}(h\nu) \approx \int_{\overline{h\nu}}^{\infty} \sqrt{\epsilon_{\text{M}}} \sqrt{\epsilon_{\text{M}} - \overline{h\nu}}\, f_{\text{M}}(\epsilon_{\text{M}}) f_{\text{X}}(\epsilon_{\text{X}}) \left| M_{\text{in}}^{\text{M}} \right|^2 \mathrm{d}\epsilon_{\text{M}}.$$

In the case of bosons occupying their ground state (i.e. excitons on their dispersion curve with a local minimum at $\mathbf{K}_0$) the occupation factor f_X does not need to be taken into account. The biexciton population factor referring to the

[3] It might be of interest to note that this lower limit $\overline{h\nu}$ may be considered an analogue of the upper limit in the integral (5.10). This 'exchange' of limits arises because here the parabolas of the excited (EM) and ground (FE) states are curved in the same sense.

upper dispersion curve reads $f_M(\epsilon_M) \approx \exp(-\epsilon_M/k_B T_M)$ and the biexciton emission line is therefore described by the integral

$$I_{in}^{M}(h\nu) \approx \int_{\overline{h\nu}}^{\infty} \sqrt{\epsilon_M}\sqrt{\epsilon_M - \overline{h\nu}}\,\exp\left(-\frac{\epsilon_M}{k_B T_M}\right)\left|M_{in}^{M}\right|^2 d\epsilon_M. \quad (8.16)$$

Until now we have always considered the matrix element M of the optical transition in integrals of the type (8.16) to be a constant, which can be put in front of the integral. However, this appears hardly acceptable here, for the simple reason: if $|M_{in}^{M}|^2 = \text{const}$, then (8.16) would result in a monotonic increase of the emission spectrum towards low-photon energies, which is physically impossible. Indeed, with decreasing $h\nu$ (and thus also $\overline{h\nu}$) the second term in the integrand (8.16) is increasing and, simultaneously, the integration range is getting broader while keeping all other parameters constant. Hence, various phenomenological models have been proposed to establish the dependence of M_{in}^{M} on ϵ_M or $\mathbf{K}$, i.e. $M_{in}^{M} = M_{in}^{M}(\epsilon_M) \approx M_{in}^{M}(\mathbf{K})$; this dependence should produce a rather sharp maximum of M_{in}^{M} at $\mathbf{K}_M = 2\mathbf{K}_X$. Cho [7] has proposed a matrix element of the form

$$\left|M_{in}^{M}\right|^2 = \left|\frac{C_0}{(\mathbf{K}_M/2 - \mathbf{K}_X)^2 + (1/a_M)^2}\right|^2, \quad (8.17)$$

where $C_0 = \text{const}$ and a_M stands for 'biexciton radius', or the average separation of the two holes in a biexciton. In other words, a_M is a quantity determining the spatial extent of the biexciton wavefunction; as such, it is of the same order of magnitude as the exciton Bohr radius a_X. The underlying physical background of expression (8.17) consists in the fact that the matrix element drops rapidly from its maximum value at $\mathbf{K}_M = 2\mathbf{K}_X$ down to almost zero just for $|\mathbf{K}_M/2 - \mathbf{K}_X| \approx a_M^{-1}$. This reflects the 'magnitude' of the biexciton in real space. The radius a_M thus becomes a crucial factor driving the emission lineshape; this factor is usually embodied in the dimensionless parameter $\Gamma = (\hbar^2/2a_M^2 m_{exc})/k_B T_M$. A detailed discussion of the emission lineshape (8.16) employing the matrix element (8.17) together with a hint for numerical integration can be found in Appendix C.

Figure 8.4(b) shows the emission spectrum of AgBr in the edge emission region, acquired under high photoexcitation with a pulsed nitrogen laser. The spectrum contains two lines—an I_{TO}(FE) line, ascribed to radiative decay of a free exciton accompanied with the emission of a TO-phonon, and an M-line being due to radiative decay of a biexciton. It is commonly supposed that biexciton annihilation is, apart from photon emission, accompanied with the emission of the same type of phonon as free exciton decay, i.e. a TO-phonon in this case. The biexciton line here therefore would have been correctly denoted as I_{TO}(M) or I_{TO}(EM) rather than simply M, if one insisted on strictly adhering to notation.[4] Theoretical fits (symbols) follow the experimental

[4] It can be shown that a biexciton has in most cases the same full symmetry as the crystal ground state (Γ_1), hence the selection rules for biexciton and exciton luminescence are similiar to each other.

data more than satisfactorily. It is seen that the typical M-line asymmetry—broadening towards the long-wavelength side, i.e. a resemblance to the inverse Maxwell–Boltzmann distribution—is indeed conserved even in indirect-gap semiconductors.

M-line intensity dependence

Apart from the lineshape analysis, the behaviour of the M-line as a function of the photoexcitation power density represents an important auxiliary guideline for firm identification of the experimentally observed biexciton emission line. Intuitively we should expect this dependence to be quadratic. Although experiment does confirm the dependence of the biexciton luminescence intensity upon the excitation intensity to be superlinear, $I^{\mathrm{M}} \sim I_{\mathrm{ex}}^{n}$, the exponent n amounts to only $n \cong 1.4$–1.6 in most cases. This rule holds independently of whether the forbidden gap is direct or indirect; $n = 2$ is seldom attained. An explanation has been put forward by the Knox *et al.* [8] through a kinetic theory, similar to the model discussed in Section 3.5, which describes the simultaneous occurrence of both monomolecular (here exciton) and bimolecular (here biexciton) luminescence. The authors [8] divide excitons photogenerated by band $\rightarrow$ band transitions into 'optical' and 'thermal' ones, supposing that biexcitons can be created by fusing thermal excitons only. They arrive at the M-line intensity dependence

$$I^{\mathrm{M}} \approx \mathrm{const}(\sqrt{1 + I_{\mathrm{ex}}/I_0} - 1)^2, \tag{8.18}$$

where I_0 stands for the characteristic material-dependent excitation intensity. Evidently, for $I_{\mathrm{ex}} \ll I_0$ we can approximate $\sqrt{1 + I_{\mathrm{ex}}/I_0} \approx 1 + I_{\mathrm{ex}}/2\,I_0$ and I^{M} scales with I_{ex} quadratically, while for $I_{\mathrm{ex}} \gg I_0$ the dependence (8.18) turns out to be linear: $I^{M} \sim I_{\mathrm{ex}}$. This corroborates qualitatively the experimental observations; the reader is reminded about the resemblance of (8.18) to relations (3.24) and (3.26). Simultaneously the model [8] predicts that the intensity dependence of free exciton luminescence should be linear, also in accordance with experiment.

Figure 8.5 shows the superlinear growth of biexciton luminescence in AgBr ($I^{\mathrm{M}} \sim I_{\mathrm{ex}}^{1.5}$). It may be of interest to specify the relevant experimental conditions [9]. The excitation was performed by ~ 2 ns pulses of a dye laser, the $\mathrm{I_{TO}}$(FE) and M lines were observable only in a temporal window coincident with the laser pulse; then the emission spectra and the intensity dependence were recorded. Further development of the spectra with increasing delay after the excitation pulse, recorded with the use of time-resolved spectroscopy, is displayed in Fig. 2.48. It can be seen how the fast edge emission $\mathrm{I_{TO}}$(FE) and M fades away, free excitons and biexcitons become gradually localized on the iodine ions and a slower emission band emerges at 500 nm, due to excitons localized on the isoelectronic impurity I^-.

To confirm further the interpretation of the emission M-line as being due to EM decay, it is highly recommended to analyse its spectral position and temperature behaviour. They are closely connected with the determination of two important EM parameters: the binding energy E_{B} and the radius a_{M}.

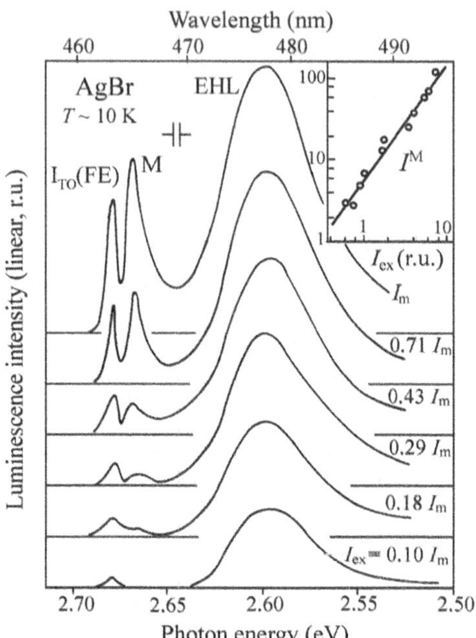

Fig. 8.5
Edge emission of AgBr under pulsed laser excitation at a bath temperature of $T \approx 10$ K. The nonlinear growth of the biexciton M-line with increasing excitation intensity I_{ex} (I_m corresponds to $\approx 0.7 \text{MW/cm}^2$) is clearly observable. The inset demonstrates the dependence of the integral M-line intensity $I^M \sim I_{ex}^{1.5}$. The I_{TO}(FE) line is due to free exciton decay, and the EHL line originates in the electron–hole liquid. After Pelant *et al.* [9].

8.2.2 Determination of biexciton parameters

The biexciton binding energy E_B in direct-bandgap semiconductors can be extracted readily and quite accurately from the emission spectrum. According to (8.5), E_B is equal to the energy separation between the emission line of a free exciton–polariton and that of a biexciton, or more precisely, to the separation between the centre of the polariton line and the high-energy edge of the biexciton line.

Less obvious is how to determine E_B in a semiconductor with an indirect bandgap. There, the high-energy edge of the M-line gets broader at $T > 0$ K, as discussed in the context of Fig. 8.3(b). Fortunately the model [7], making use of the matrix element (8.17), comprises in the theoretical curve not only the relevant lineshape but also an energy position pertinent to the minimum of the biexciton dispersion $\epsilon_M(\mathbf{K})$. This position is indicated in Fig. 8.4(b) by a vertical line at 2.669 eV ($\sim$ 464.5 nm). The determination of E_B is then the same as mentioned above: as the energy separation of this line from the energy coordinate pertinent to the minimum of the exciton dispersion $\epsilon_X(\mathbf{K})$, i.e. from the vertical line at 2.676 eV ($\sim$ 463.2 nm). This is how we get $E_B \approx 7$ meV in AgBr.

This mode of assessing the EM binding energy is designated as *spectroscopic*. An independent determination of E_B consists in investigating the intensity variations of the FE and M lines with varying temperature and can thus be called a *thermodynamic* method. It is applicable above all in indirect bandgap semiconductors where the free exciton emission line is sufficiently strong. The principle is as follows: In the course of sample heating, biexcitons undergo thermal dissociation more easily than free excitons do, because their binding energy is always smaller, $E_B < E_X$. Due to this, the M-line intensity drops substantially faster compared with the intensity fall of the FE-line. In thermodynamic equilibrium, and provided a quadratic proportion $I^{M} \sim I_{ex}^{2} \sim (I^{FE})^{2}$ is valid, we can anticipate that the intensity ratio $R = (I^{FE})^{2}/I^{M}$ will be apparently driven by a factor $\exp(-E_{B}/k_{B}T_{M})$. A more exact derivation leads to the even sharper dependence

$$R = \frac{(I^{FE})^{2}}{I^{M}} \approx T_{M}^{k} \exp\left(-\frac{E_{B}}{k_{B}T_{M}}\right), \tag{8.19}$$

where the value of the exponent k varies depending on whether the free movement of the quasi-particles FE, EM is not limited in any way ($k = 3/2$) or whether they are localized in a certain place of the sample, for instance in a potential well created by a local deformation of the sample ($k = 3$) [10]. A semi-logarithmic plot of $R^{*} = (I^{FE})^{2}/I^{M}T_{M}^{k}$ against $1/k_{B}T_{M}$ then yields E_{B} as the slope of the straight line.

Such a plot is shown in Fig. 8.6 for crystalline silicon at three different levels of laser excitation [10]. It can be seen that the slope, i.e. the binding energy E_{B}, is independent of the excitation intensity (in fact, there is no reason why it should be dependent); we get $E_{B} = 1.53$ meV, in good agreement with spectroscopic value $E_{B} = 1.46$ meV. In this case the exponent $k = 3$ was applied in (8.19), because the gas of free excitons and biexcitons was spatially localized in a potential well. This localization made it possible to attain sufficiently elevated concentrations of both FE and EM under not excessive optical pumping levels. As a consequence of this approach a purely quadratic dependence $I^{M} \sim I_{ex}^{2} \sim (I^{FE})^{2}$ was observed in experiment—since the condition $I_{ex} \ll I_{0}$ was fulfilled in (8.18)—justifying the use of eqn (8.19).

It is worth mentioning that experimentally observed binding energies E_{B} in various materials systematically exceed theoretical values. This is documented in Table 8.1 (the only exception to this rule being copper halides CuCl and CuBr). The reasons are not clear. Often one blames the theoretical approaches for not taking into account at least two factors: (i) the effect of lattice polarization in polar semiconductors, which may raise E_{B} substantially and (ii) the process of creating a biexciton from two excitons located at different valleys of the conduction (valence) band in indirect semiconductors. For in this way one can generate a two-electron triplet spin function connected with a lower

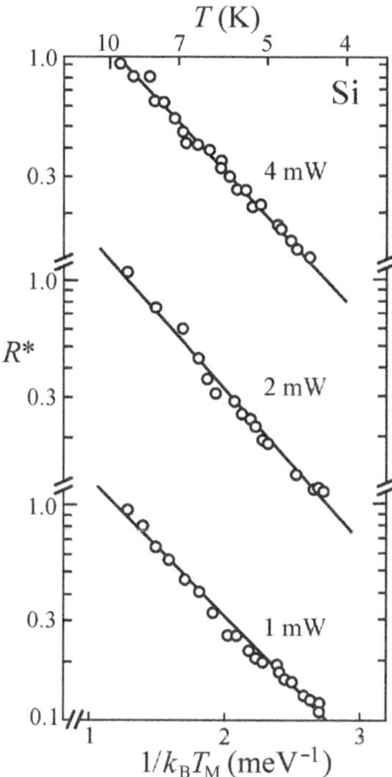

Fig. 8.6
Thermodynamic determination of the EM binding energy in silicon at three levels of laser excitation. Symbols—experiment, lines—fit by the quantity $R^{*} = R/T_{M}^{3}$ with the use of (8.19). Taking into consideration the relatively long lifetimes of participating quasi-particles along with not too high excitation, $T_{M} = T$ (bath temperature) holds. After Gourley and Wolfe [10].

Table 8.1 Comparison of the experimental and theoretical values of EM binding energy in several semiconductors (d – direct bandgap, i – indirect bandgap). The values of E_{B} are given in meV.

Semicondutor	E_{B} experimental	E_{B} calculated
CuCl (d)	28–34	41–44
AgBr (i)	~ 7	1.3–2.5
CdS (d)	~ 5	2–5
ZnO (d)	12–16	5–9
GaAs (d)	~ 0.7	~ 0.4
Si (i)	1.2–1.5	0.45–1.1
Ge (i)	~ 0.3	~ 0.1
HgI_{2} (d)	~ 6	~ 2.5

lying energy level compared with a singlet state of two excitons located in the same valley (there, conforming to the Pauli exclusion principle, two electrons must be of antiparallel spins, i.e. total spin $S = 0$). Thereby the EM dispersion curve in Fig. 8.3(b) may shift downwards, raising the binding energy E_B.

The above considerations are in qualitative agreement with Table 8.1. The most marked difference between the experimental and theoretical values of E_B occurs in AgBr, where—as in a polar, indirect-gap semiconductor—both factors enter the play. Any quantitative theoretical estimates are lacking, though.

The second biexciton parameter, the radius a_M, can be found relatively easily from luminescence measurements in indirect semiconductors. When fitting the experimental lineshape with relations (8.16) and (8.17), the variable parameters are $\Gamma = (\hbar^2/2a_M^2 m_{exc})/k_B T_M$ together with the reduced wavelength axis $\overline{h\nu}/k_B T_M = (h\nu + \hbar\omega - E_0)/k_B T_M$ [7]. This enables an independent determination of both T_M and a_M. The fit shown in Fig. 8.4(b) yields, for AgBr, $\Gamma = 4$, $T_M = 19\,\text{K}$ and $a_M \approx 2.6\,\text{nm}$. In direct-bandgap semiconductors, on the other hand, there is no similar straightforward and commonly accepted method to establish a_M experimentally.

The last remark in this section concerns *bound* (*localized*) biexcitons. It has been found that free biexcitons, similarly to free excitons, can get localized close to impurity atoms. Actually, this is not surprising if we recall bound multiexciton complexes (Subsection 7.2.1). A bound biexciton can be regarded as a limiting case of a bound exciton complex composed of two excitons, even though the process of creation (localization) may be slightly different. In the case of high excitation, biexcitons as a whole get localized preferentially. The relevant emission line, originating in radiative decay of bound biexcitons, is red-shifted with regard to $(E_{g(i)} - E_X - E_B)$ and features properties analogous to other extrinsic radiative channels, namely, its shape is symmetric, does not vary either with excitation intensity or temperature and the line exhibits saturation behaviour.

Bound biexcitons have been discovered via luminescence measurements in GaP, CuCl and AgBr.

8.3 Collisions of free excitons

Another type of interexciton interaction that can manifest itself in luminescence spectra at higher excitation power densities is inelastic collisions between free excitons (X–X collisions). Two FEs, instead of fusing and creating an EM, collide with one another while one of the two FEs dissociates into a free e–h pair and the remaining one recombines radiatively emitting a photon $h\nu_P$. This photon energy, however, is reduced compared to the $n = 1$ FE state by an amount passed over to the first exciton and making it dissociate—and this is the very exciton binding energy E_X.

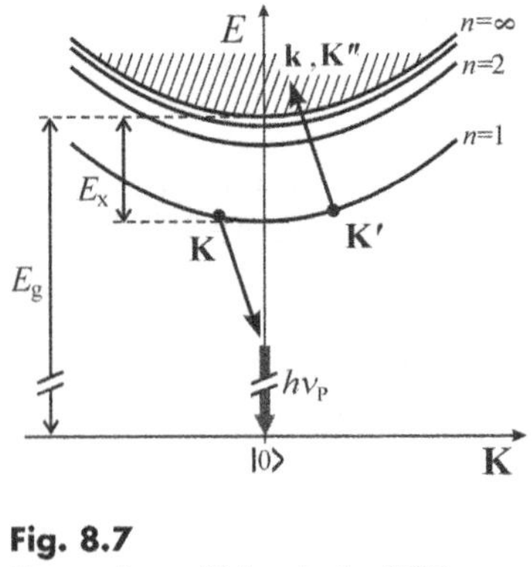

Let us consider a direct-bandgap semiconductor, Fig. 8.7. Energy balance then reads

$$2\text{ free excitons} \rightarrow \text{photon} + \text{free } e\text{–}h\text{ pair}$$

$$(E_g - E_X) + (E_g - E_X) \approx h\nu_P + (E_g + E_{e,h}^{kin}),$$

Fig. 8.7
Free exciton collisions in the $E(\mathbf{K})$ scheme. $\mathbf{K}$, $\mathbf{K}'$ denote the exciton wavevectors before the collision, $\mathbf{k}$ is pertinent to internal e–h pair movement $(h^2k^2/2m_r)$ and $\mathbf{K}''$ to its translation movement $(h^2K''^2/2(m_e + m_h))$ after the collision.

and thus

$$h\nu_{\rm P} \approx (E_{\rm g} - 2\,E_{\rm X}) - E_{\rm e,h}^{\rm kin}, \tag{8.20}$$

where $E_{\rm e,h}^{\rm kin} = \hbar^2 k^2/2\,m_{\rm r}+\hbar^2\,K''^2/2\,(m_{\rm e}+m_{\rm h})$ stands for the kinetic energy of the created electron–hole pair. A new emission P-line therefore appears, red-shifted from the spectral position of an exciton–polariton (which is $E_{\rm g}$–$E_{\rm X}$) roughly by $E_{\rm X}$(see Fig. 8.1 and Problem 8/1).

Experiment reveals that the final state in this collision process does not necessarily have to be always a free *e–h* pair. The exciton, acquiring energy at the collision, may happen to be excited only to some higher lying but still bounded exciton state ($n = 2, 3, 4, \ldots$) and the energy of the emitted photon is then

$$h\nu_{{\rm P}n} \approx (E_{\rm g} - E_{\rm X}) - E_{\rm X}(1 - 1/n^2), n = 2, 3, \ldots, \infty, \tag{8.21}$$

which is nothing but a generalization of (8.20) with simultaneous neglect of kinetic energy. In Fig. 8.8 there are emission spectra of a high-purity monocrystalline ZnO sheet under high excitation with an electron beam at $T = 10\,{\rm K}$ [11]. Owing to the extreme purity, the emission line of the exciton localized on a residual impurity (denoted I_9) is strongly suppressed, enabling the observation of two lines in the P-band, corresponding to photon energies $h\nu_{\rm P2}$ and $h\nu_{\rm P\infty}$ according to (8.21).

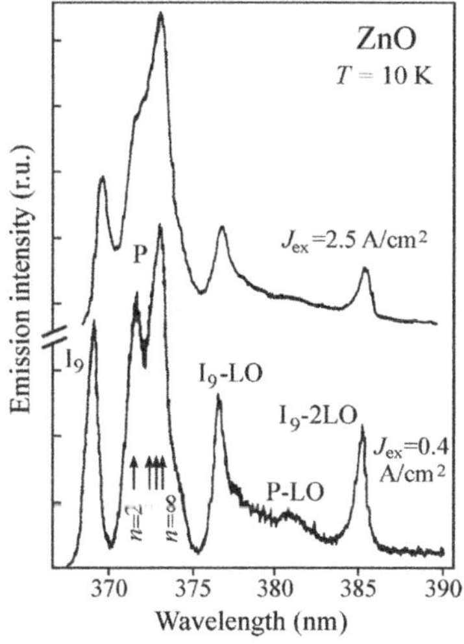

Fig. 8.8
Emission spectra of pure ZnO under electron beam excitation (40 kV, ~ 100 ns pulses) with two different current densities $J_{\rm ex}$. Apart from the P-band, the bound exciton line I_9 and its LO-phonon replicas can be seen. The P-band grows superlinearly with increasing excitation current density at the expense of other lines. Arrows indicate emission wavelengths calculated using (8.21) for $n = 2, 3, 4$, and ∞. $T = 10\,{\rm K}$. After Hvam [11].

The basic features of the spectral shape of the emission line accompanying the creation of a free *e–h* pair ($n = \infty$) can be readily guessed: The P_∞ line should have a high-energy threshold at $E_{\rm g}$–$2E_{\rm X}$ and is expected to be asymmetrically broadened towards the low-energy side, since for the *e–h* pairs there exists a continuum of energy states. The creation of an *e–h* pair with high kinetic energy $E_{\rm e,h}^{\rm kin}$ then leads, in compliance with (8.20), to the emission of lower-energy photons.

Moriya and Kushida studied this lineshape theoretically in the framework of perturbation theory and found that [12]

$$I_{\rm sp}^{\rm P\infty}(h\nu) \approx \left| M^{\rm P} \right|^2 h\,\nu^3\,n_{\rm X}^2 \int_0^\infty {\rm d}\xi \int_0^\infty {\rm d}t\,\frac{\xi^{1/2}}{(1+\xi/E_{\rm X})^4}$$
$$\times \exp\left[-t - \frac{1}{4\,t}\left(\frac{E_{\rm g} - 2\,E_{\rm X} - h\nu - \xi}{k_{\rm B}\,T}\right)^2\right] \tag{8.22}$$

($M^{\rm P}$ denotes the optical matrix element and $n_{\rm X}$ the concentration of free excitons). The physical meaning of the double integration in (8.22) consists in the fact that photons with a given energy $h\nu$ may originate from numerous combinations of initial ($\mathbf{K}, \mathbf{K}'$) and final ($\mathbf{k}, \mathbf{K}''$) wavevector pairs. In practise, however, this theoretical lineshape is seldom compared with experiment. There are several reasons for this. On the one hand, a background composed of many lines due to impurities (BE, e–A^0, h–D^0, etc.) often occurs in this spectral region, which makes the lineshape analysis of the P-line difficult; on the other hand, the very theoretical expression (8.22) is based on a simplified approximation of the interaction potential between the two excitons which is, in reality, rather complicated. It is worth noting that (8.22) contains a term n_X^2, therefore one can expect a quadratic dependence of I_{sp}^P upon the excitation intensity; Fig. 8.8 basically confirms this expectation.

The close resemblance of the M- and P-lines (superlinear growth, line asymmetry) invokes the following questions: When can we in an experiment expect the creation of excitonic molecules and when are inelastic exciton collisions more likely to occur? Or do both of these mechanisms take place always simultaneously, as Fig. 8.1 indicates?

The answer to these questions is neither simple nor unambiguous. There is no strict theoretical prediction as to when (at what excitation level) and where (in what material) we should expect this mechanism or that one to occur. Of course, in a qualitative manner we can come to the conclusion that the fusion of two excitons into an EM will prevail at relatively lower excitation power densities and provided E_B is sufficiently large. For if we create a very high concentration of free excitons possessing, moreover, a non-equilibrium excess kinetic energy ('optical excitons' introduced in Subsection 8.2.1), then the collision process will probably predominate. (The interacting excitons will 'have no time' to fuse; only a collision followed by one exciton annihilation will set in.) The observed predominance of exciton collisions under pumping with high current density electron beams seems to support this conjecture. However, there is certainly a relatively broad range of laser excitation power densities where both 'M' and 'P' actions occur simultaneously. Radiative decay of the EM as well as luminescence due to exciton collisions can easily result in stimulated emission, as we shall see later.

All considerations held in this section concerned direct-badgap semiconductors. Till now the P-line has not been observed in any indirect-bandgap material (Problem 8/3). On the contrary, in the next section we shall discuss a luminescence process taking place exclusively in indirect semiconductors, namely, radiative annihilation of an electron–hole liquid.

8.4 Electron–hole liquid (EHL)

Another qualitative alteration of luminescence (but not only luminescence) behaviour of a semiconductor, linked with high exciton densities, is exciton condensation into a new state of matter, named an *electron–hole liquid*, EHL. We have already introduced the concept of an exciton gas; the condensation of excitons represents a phase transition that can be regarded as a quantum-mechanical analogue of the well-known classical gas condensation into a liquid (water vapour molecules in the atmosphere condense into droplets constituting fog or rain). In the case of excitons one speaks of a Fermi liquid—a degenerate electron–hole system—composed of free fermions, electrons and holes. Similarly to the classical case, also particles of this particular liquid form spherical drops, inside which they are held together by means of internal forces and surface tension; hence also the term *electron–hole drops* (EHD) is sometimes used.

The typical diameter of these drops amounts to several μm. It should be stressed again that the drops do not contain excitons but free electrons and

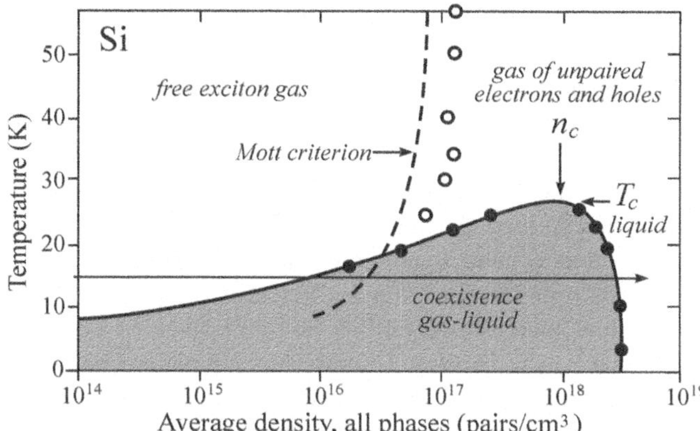

Fig. 8.9
The phase diagram of photoexcited electrons and holes in silicon. It can be deduced, e.g., that for an average *e–h* pair density of $\sim 10^{17}$ cm^{-3} and $T = 15$ K a saturated free exciton gas with a density about 10^{16} cm^{-3} coexists with drops of EHL; in each of these drops there is a concentration of *e–h* pairs $\sim 3 \times 10^{18}$ cm^{-3}. The dashed curve and open symbols denote theoretical and experimental values of the Mott insulator–metal transition (Section 8.5). After Shah *et al.* [13].

holes. The thing is that excitons in the condensate immediately 'break up', therefore the designation 'exciton drops', which is sometimes used, is not correct. The density of *e–h* pairs inside the drops may be higher by several orders of magnitude than the concentration of excitons remaining in the surrounding gas phase, as we shall see shortly.

To liquefy a classical gas, it is necessary to reach a critical pressure and to descend below a critical temperature, the whole process being advantageously sketched in a p, T (pressure, temperature) diagram. As for the exciton concentration, it appears more illustrative to apply an n, T (density, temperature) coordinate system and to speak about the critical density n_c of *e–h* pairs and the critical temperature T_c of the liquid. The phase diagram of an electron–hole system in silicon is shown in Fig. 8.9. The values of the critical temperature and the critical density are $T_c = 25$ K and $n_c = 1.2 \times 10^{18}$ cm^{-3}, respectively. An EHL cannot survive for $T > T_c$.

The *e–h* pair density in EHL at a given temperature is driven by the liquid interface of the two-phase coexistence region (shadowed area), where the *e–h* system separates into a liquid and the surrounding gas phase. It can be seen that this density is $n_0 \approx 3 \times 10^{18}$ cm^{-3} and is only weakly temperature dependent in the range 0–15 K. On the left and above the coexistence region only the gas of free excitons may subsist.

Similarly to the case of excitonic molecules, of fundamental importance is now to ask about the stability of the EHL, or to ask which parameter plays the role of the binding energy E_B here. Prior to establishing such a parameter, we define the ground state energy $E_G(n)$ of the liquid, allotted per one *e–h* pair. This energy is given by the sum of the kinetic, exchange and correlation energies and it exhibits, as a function of the *e–h* pair density, its minimal value at the equilibrium density n_0. Should the EHL be a stable and energetically favourable phase compared to the free exciton gas, this minimum must be located deeper than the ground exciton state level, referred to the bottom of the *e–h* continuum of states. In Fig. 8.10, this means that the relation $|E_G(n_0)| > E_X$ must hold. The difference $\varphi = |E_G(n_0)| - E_X$ then represents the (material dependent) EHL stability parameter we are looking for. φ is called the binding energy of the condensate with respect to free excitons (or, in analogy with thermionic electron emission, the *work function*). It is the energy

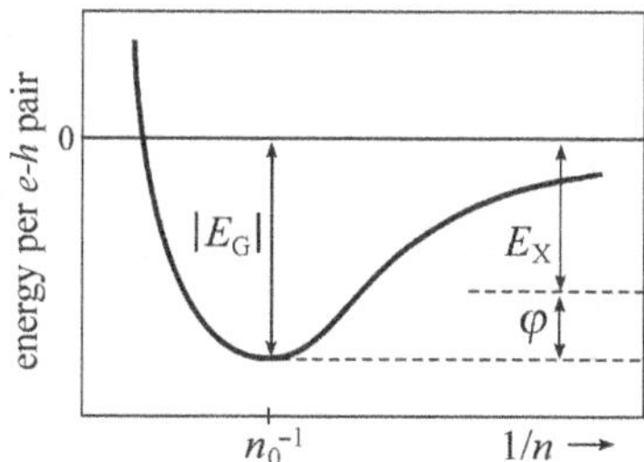

Fig. 8.10
Schematic drawing of the EHL ground state energy per electron–hole pair as a function of the reciprocal pair density.

required to 'evaporate' one e–h pair from the liquid.[5] The condensate is stable if and only if $\varphi > 0$.

Various theoretical approaches to calculate $E_G(n)$ or φ, along with obtained results, are summarized, e.g., in [14] and [15]. Theory reveals that the EHL stability may profit from specific features of the energy band structure of indirect-bandgap semiconductors. In particular, the long exciton lifetime and several equivalent conduction band minima ν_c, in, e.g., Si and Ge (or several equivalent valence band maxima ν_v in AgBr) promote the chance of photoelectrons (photoholes) to reduce efficiently their kinetic energy. This entails an important increase in stability and cohesiveness of the condensed phase. Quantitatively, the increased number of equivalent band extremes ν_c (ν_v) appears in the expression for the density of states in the bands. Instead of (5.4) we have to be more specific now and write down

$$\rho_c(E) = \nu_c \frac{(2\,m_{de})^{3/2}}{2\,\pi^2\hbar^3}\,E_e^{1/2},$$

$$\rho_v(E) = \nu_v \frac{(2\,m_{dh})^{3/2}}{2\,\pi^2\hbar^3}\,E_h^{1/2}. \tag{8.23}$$

The factors ν_c and ν_v describe the fact that the electrons and holes have at their disposal ν_c-times or ν_v-times more available states, respectively, in comparison with a semiconductor having a single band extreme. (In a semiconductor characterized by a simple band structure with a direct bandgap at the Γ point, $\nu_c = \nu_v = 1$ holds, while in silicon $\nu_c = 6$, $\nu_v = 2$.) Furthermore, instead of simple effective masses m_e, m_h, which belong to idealized parabolic bands, we have to consider now the so-called density-of-states effective masses m_{de}, m_{dh} that take into account the complex band structure of real semiconductors Si, Ge, AgBr, etc. Namely, they involve the anisotropy of the effective mass and the existence of heavy and light holes with markedly different masses [14].

The kinetic energy of photocarriers gets reduced because its mean value is equal to $3/5F_e$ and $3/5F_h(F_e,\ F_h$ being the Fermi energies of electrons and holes, respectively), see Problem 8/4. A pure mechanical analogue may be invoked: if we pour a liquid (here the photoexcited electron–hole fluid) uniformly into several vessels (here several valleys of the conduction or the valence band) instead of a single one, the liquid levels in the vessels become inversely proportional to the number of vessels used.

It has been demonstrated that further promotion of φ arises from the interaction of electrons and holes with LO-phonons. A very stable EHL with a high T_c therefore occurs in polar semiconductors with an indirect bandgap (AgBr, GaP). On the contrary, the short exciton lifetime along with the single conduction band valley in the Γ point of direct semiconductors do not allow the necessary conditions for exciton condensation to be reached, even though this issue had been intensively discussed throughout the 1970s and 1980s.

[5] Upon leaving the condensate, such a pair is immediately bound and creates a free exciton.

8.4.1 Luminescence determination of EHL parameters

The EHL represents an electronic excited state of a semiconductor, which fades away rapidly (10^{-9}–10^{-6} s) after the cessation of the excitation effect. However, the EHL emits its characteristic luminescence radiation that originates (even in the steady state) through the radiative recombination of electrons and holes constituting the condensate. The probability of radiative recombination in the EHL is even increased several times compared to a system of free carriers of the same density (due to the so-called enhancement factor) because of the correlation existing between the electrons and holes. The EHL luminescence is of intrinsic type and its emission spectrum features a relatively broad band rather than a single narrow line; this reflects the Pauli exclusion principle, according to which each level in a system of fermions can be occupied by two electrons (or holes) at the most (Fig. 8.11). Referring to what has been discussed above, this band appears in the emission spectrum as soon as the pumping level exceeds a certain threshold level, provided the condition $T < T_c$ holds at the same time.

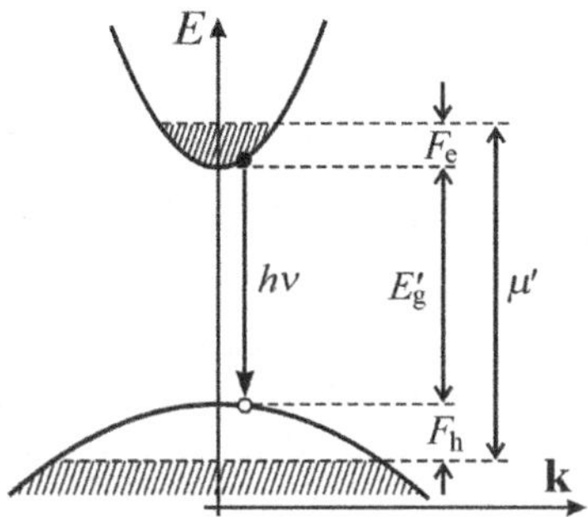

Fig. 8.11
Schematic of EHL recombination radiation. A photon $h\nu$ originates via the recombination of an electron with a hole from the corresponding Fermi seas. Electrons occupy the conduction band up to the electron Fermi level F_e; in a similar manner levels in the valence band are occupied by holes up to the hole Fermi level F_h. E'_g means a reduced bandgap ($< E_g$). For the sake of simplicity the transitions are drawn as direct transitons. $\mu' = E'_g + F_e + F_h$ stands for the EHL chemical potential.

Let us now discuss what the emission bandshape looks like; simultaneously we shall learn how we can, by analysing the emission spectrum, extract the two most important EHL parameters, namely, n_0 and φ. These are listed, for selected semiconductors, in Table 8.2.

The formation of EHL luminescence is evident from Fig. 8.11. The emission lineshape is defined through a combination of all the occupied electron states with all the occupied hole states, the momentum conservation rule for all recombining e–h pairs being ensured through the participation of a phonon $\hbar\omega$. Formally therefore we can start with the convolution integral (5.10) that has been derived for allowed indirect recombination processes

$$I^{\mathrm{EHL}}(h\nu) \approx |M|^2 \int_0^{\overline{h\nu}} \sqrt{E}\sqrt{\overline{h\nu} - E}\, f_c(E)\, f_v(\overline{h\nu} - E)\mathrm{d}E, \tag{8.24}$$

and where $\overline{h\nu} = h\nu + \hbar\omega - E_g$.

Now we perform in (8.24), compared to (5.10), two formal adjustments and then one essential physical modification.

Table 8.2 The basic EHL parameters of selected semiconductors (experimental values); n_0, φ and T_c denote the equilibrium density, the binding energy and the critical temperature, respectively.

Semiconductor	n_0 (cm^{-3})	φ (meV)	T_c (K)
Ge	2.4×10^{17}	1.5	6.5
Si	3.3×10^{18}	8	25
GaP	6×10^{18}	15	40
AgBr	8×10^{18}	55	~ 60 ($T_{ef} \approx 100$)$^{(1)}$
TlBr$^{(2)}$	1.4×10^{19}	–2.2	–

Note: $^{(1)}$ Owing to the short lifetime, the effective critical temperature T_{ef} is higher than the bath T_c.
$^{(2)}$ Theoretically calculated values.

Under formal adjustments we omit the phonon energy $\hbar\omega$ in (8.24) and replace E_g by the so-called reduced or renormalized bandgap $E'_g < E_g$; therefore we put $\overline{h\nu} = h\nu - E'_g$. The omission of $\hbar\omega$ serves only to simplify our reasoning; this step does not affect the emission spectral shape, shifting it only—as a whole—on the photon energy axis. Graphically, we can then illustrate the EHL recombination radiation as if it were due to direct transitions (Fig. 8.11); we shall see later that this simplification will turn into a certain benefit. Introducing the reduced bandgap E'_g has a deeper physical meaning: it tells us that the high density of *e–h* pairs inside the condensate droplets entails a reduction of the bandgap E_g (calculated within the one-electron approximation or found experimentally, e.g. with the aid of optical measurements using low-intensity light) down to a value $E'_g < E_g$. We shall investigate this effect more closely in Section 8.5. For luminescence measurements, this has a practical consequence, namely, the recombination radiation originating inside the drops is not absorbed in the surrounding lattice and can be easily detected.

The essential physical modification, indicated above, compared to the luminescence of an *e–h* system under weak excitation (Subsection 5.2.2) consists now in introducing into (8.24) for f_c and f_v the relevant Fermi–Dirac distribution functions in the conduction and the valence bands, respectively. Therefore, (8.24) becomes

$$I^{EHL}(h\nu) \approx \int_0^{\overline{h\nu}} \sqrt{E}\sqrt{\overline{h\nu} - E}\left[\exp\left(\frac{E - F_e}{k_B T}\right) + 1\right]^{-1} \left[\exp\left(\frac{\overline{h\nu} - E - F_h}{k_B T}\right) + 1\right]^{-1} dE. \qquad (8.25)$$

The integrand does not contain any singularities and thus to calculate (8.25) for fitting the experimental emission spectrum, any simple numerical method may be used.

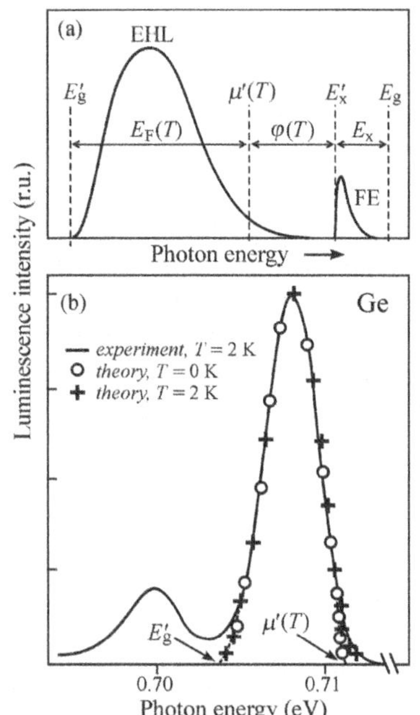

Fig. 8.12
(a) Schematic of the emission spectrum comprising the EHL band and a free exciton line FE, indicating at the same time how to determine the binding energy φ. (b) Experimental and theoretical (eqn 8.25) lineshapes of the EHL band (LA-phonon replica) in pure Ge. The FE line is not displayed. The less-intense band at ~ 0.7 eV is a TO-phonon replica of the EHL luminescence. After Benoit à la Guillaume and Voos [16].

Determination of the reduced bandgap E'_g and the potential μ'

Close to the temperature $T = 0$ K, the emission spectrum obviously starts on the short wavelength side at $h\nu = \mu'$ and terminates on the long-wavelength side at $h\nu = E'_g$ (let the reader compare Figs 8.11 and 8.12(a)). Photoluminescence therefore turns out to be a unique method for a direct determination of auxiliary EHL parameters such as the reduced bandgap E'_g and the chemical potential of the electron–hole liquid $\mu' = E'_g + F_e + F_h$.

The fit parameters[6] in (8.25) are F_e, F_h and T. They are not independent of each other, however, which somewhat complicates the fitting procedure based on eqn (8.25). This procedure is connected with the extraction of the density n_0 from the luminescence spectrum, a process that we are now going to describe.

[6] E'_g appears to be also a fit parameter, however, it does not affect the bandshape but shifts the overall spectrum on the wavelength axis only. It can be extracted from the final fitting of the calculated curve on the experimental spectrum.

Determination of the electron–hole density n_0

The simplest way to determine n_0 can be realized at very low temperatures. Let us consider the limit case $T = 0\,\mathrm{K}$. The EHL emission band has a full width (at its foot) $E_F = F_e + F_h = \mu' - E'_g$ (Figs 8.11 and 8.12). From here, n_0 will be extracted as follows: The electron density n_e (= the hole density n_h) is given by summing over all occupied states in unit volume, which is generally expressed as

$$n_0 = n_e = n_h = \int_0^\infty \rho_c(E) f_c(E, F_e)\,\mathrm{d}E = \int_0^\infty \rho_v(E) f_v(E, F_h)\,\mathrm{d}E$$

and, by considering explicitly the densities of states (8.23), the above expression transforms into

$$n_0 = n_e = n_h = \frac{\nu_{c(v)}(2\,m_{de(h)})^{3/2}}{2\pi^2\hbar^3} \int_0^\infty \sqrt{E}\left[\exp\left(\frac{E - F_{e(h)}}{k_B T}\right) + 1\right]^{-1} \mathrm{d}E. \tag{8.26}$$

For $T = 0\,\mathrm{K}$ (and solely in this case) the integration can be performed analytically which yields

$$n_0 = \frac{\nu_c}{3\pi^2\hbar^3}(2\,m_{de}\,F_e)^{3/2} = \frac{\nu_v}{3\pi^2\hbar^3}(2\,m_{dh}\,F_h)^{3/2} \tag{8.27a}$$

or

$$\nu_c^{2/3}\,m_{de} F_e = \nu_v^{2/3}\,m_{dh} F_h. \tag{8.27b}$$

Ultimately the relationship between the full width of the band E_F and the concentration n_0 follows from $E_F = F_e + F_h$, which combined with (8.27b) gives

$$F_e = \left[1 + \left(\frac{\nu_c}{\nu_v}\right)^{2/3} \frac{m_{de}}{m_{dh}}\right]^{-1} E_F. \tag{8.28}$$

Therefore, upon obtaining experimentally the full width E_F of the emission band we are able, making use of (8.27a), (8.28) and of the band structure parameters, to calculate n_0 immediately. It can be seen that $n_0 \sim E_F^{3/2}$; the wider the luminescence band, the denser the EHL.

For $T \neq 0\,\mathrm{K}$ the procedure is similar, the expression (8.26) being no longer computable analytically, though. The integration has to be performed numerically. The method of determining n_0 then cannot be separated from fitting the emission bandshape, as we have already mentioned.

The experimental EHL band is fitted with the aid of (8.25), letting the values of the couple F_e, F_h vary, but keeping their ratio fixed, as follows from the numerical integration of (8.26) or, in other words, from the condition of the electroneutrality of the condensate $n_e = n_h$. (This means that the *e–h* pair density $n_0 = n_e = n_h$ enters the spectral shape (8.25) implicitly.) The temperature T represents the next fit parameter, as already mentioned above; this effective temperature can again be higher than the lattice temperature. It should be noted, however, that at a sufficiently low bath temperature the simple determination of n_0 through a combination of (8.27) with an iterative analytical calculation of the bandshape (8.25) for $T = 0\,\mathrm{K}$ is often applicable. That is to say, the spectral shape undergoes only an insignificant modification around μ', as demonstrated in Fig. 8.12(b) [16].

Determination of the binding energy φ

From the above fitting procedure one gets the value of the reduced bandgap E'_g together with $F_e(T)$ and $F_h(T)$. Their sum defines the chemical potential: $\mu'(T) = E'_g + F_e(T) + F_h(T)$. We shall show before long that μ' is the *spectroscopic energy* characterizing the EHL ground state, i.e. it corresponds to the energy $E_G(n_0)$. Then it is readily seen in Figs 8.10 and 8.12(a) that the condensate binding energy φ (with respect to a free exciton) may be extracted very simply as the separation between μ' and the low-energy edge of the FE emission line, whose spectroscopic position is denoted E'_X in Fig. 8.12(a):

$$\varphi(T) = E'_X - \mu'(T).^7 \tag{8.29}$$

Why, then, does $\mu'(T)$ characterize the EHL ground state? We come to this conclusion by means of simplified reasoning, the origin of which is ascribed to Keldysh: the maximum EHL luminescence photon energy at $T = 0\,\mathrm{K}$ is determined by the radiative recombination of an electron with a hole which reside on their Fermi levels; in other words, such recombination leaves the EHL in its ground state (any other recombination event leaves behind the EHL in an excited state, because for instance an unoccupied state at energy lower than F_e appears in the electron Fermi sea, and a redistribution of the population on electron levels must follow to restore the ground state). Consequently, μ' is the spectroscopic energy relevant to the EHL ground state. This can also be shown computationally (Problem 8/5).

This determination of φ is essentially spectroscopic, much like the case of the determination of the binding energy E_B of an excitonic molecule. Similarly, also here the possibility exists to set φ thermodynamically. In this case φ is obtained as the activation energy from plotting, e.g., the exciton density on the gas–liquid boundary against temperature (Fig. 8.9). The excellent treatise of these methods as well as of other luminescence manifestations of the EHL can be found in the monograph [17].

8.4.2 Identification of the EHL emission band

Now we can summarize the typical spectral features that differentiate EHL luminescence from plenty of other lines or bands in the emission spectrum:

1. With increasing excitation intensity (even over several orders of magnitude) the EHL band firmly keeps its shape and position, i.e. neither band broadening nor a spectral shift occurs. This is due to the fact that the liquid density n_0 is constant at a given temperature (Fig. 8.9). An increase in the

[7] Also E'_g and E'_X are dependent on temperature but this dependence is weak in comparison with the temperature dependencies $F_e(T)$, $F_h(T)$ and $\mu'(T)$.

pump power density entails—no doubt—an increase in the total volume occupied by the drops and thus naturally also an increase in the emission intensity I^{EHL}, but the relevant spectral shape, driven by the density n_0, does not vary.

2. With increasing temperature two contradictory effects happen: the high-energy edge broadens (because more and more electrons and holes are excited above the respective Fermi levels), but, at the same time, the FWHM of the band gets slightly narrower (because the liquid density with increasing temperature drops slightly due to thermal expansion—see the liquid boundary of the phase diagram in Fig. 8.9). Consequently, the chemical potential $\mu'(T) = E'_{\mathrm{g}} + F_{\mathrm{e}}(T) + F_{\mathrm{h}}(T)$ is a decreasing function of T, or $\varphi(T)$ becomes, according to (8.29), an increasing function of temperature; it turns out that this function can be written as

$$\varphi(T) = \varphi(0) + a(k_{\mathrm{B}}T)^2,$$

where a is a constant of the order of 1 meV^{-1}.

3. The EHL phase diagram implies the existence of two distinct thresholds connected with EHL luminescence. The EHL band appears in the emission spectrum upon reaching an *intensity threshold*, which is nicely demonstrated in Fig. 8.13 [18]. On the other hand, the EHL band disappears

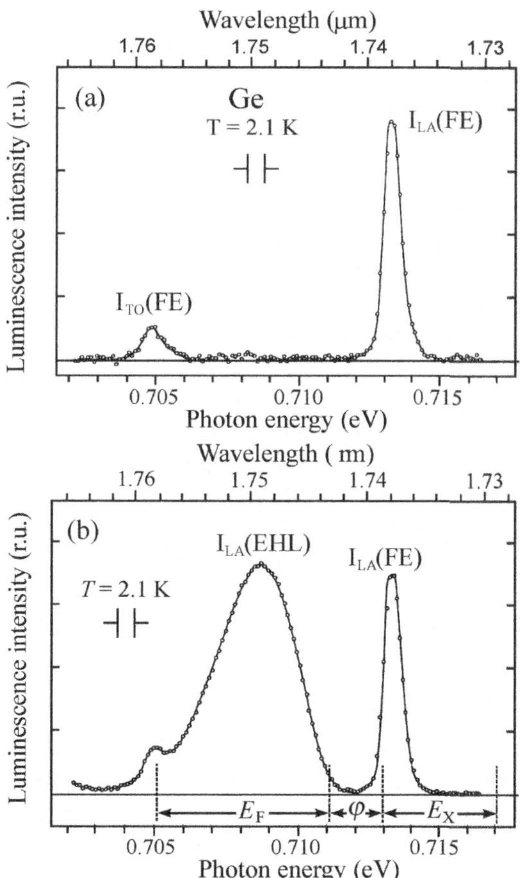

Fig. 8.13
Emission spectrum of ultrapure Ge at $T = 2.1$ K under excitation with a tungsten lamp. (a) Excitation intensity is below the threshold for exciton condensation into the EHL. Only LA- and TO- phonon replicas of the free exciton are present. (b) Excitation intensity has been 2.2× increased. Suddenly the EHL band (LA-replica) at ~ 708.5 meV appears. After Westervelt [18].

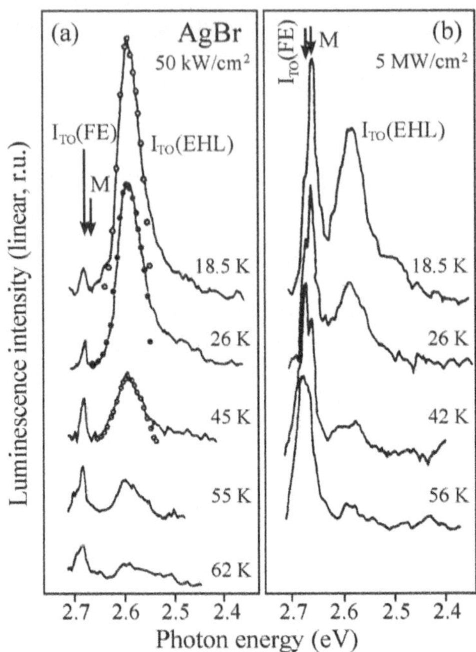

Fig. 8.14
Fast component of AgBr luminescence under pulsed laser excitation (380 nm, 6 ns). (a) Excitation intensity 50 kW/cm^2. Only the I_{TO}(EHL) band and a weak free exciton line I_{TO}(FE) occur. (b) Excitation intensity of 5 MW/cm^2. This high excitation induces, apart from the I_{TO}(EHL) and I_{TO}(FE) lines, also a very strong M-line due to excitonic molecules. Bath temperatures are indicated. The critical bath temperature is $T_c \sim 60$ K, which corresponds to the effective critical temperature of the *e–h* system of about 120 K. Symbols denote the theoretical shape (8.25) for $n_0 = 8 \times 10^{18}$ cm^{-3}. The intensity scales in (a) and (b) are not comparable. After Hulin *et al.* [19].

sharply upon reaching a *temperature threshold* (the critical temperature T_c), as can be seen in Fig. 8.14 [19]. The occurrence of both thresholds follows also from a simple kinetic model of nucleation and decay of electron–hole drops, see Appendix D.

The salient features (1.–3.) are very characteristic and enable one, as a rule, to make a facile identification of the EHL luminescence. Moreover, the considerable bandwidth of the EHL luminescence compared with, e.g., exciton lines may serve as primary guidance (see Fig. 1.1 where the replicas $I_{LO,TO}$(EHL) and I_{TA}(EHL) in silicon contrast with the narrow lines of free, FE, and bound, BE, excitons).

On the other hand, neither the analysis of the luminescence decay curves nor the shape of the intensity dependence of the EHL band is a reliable guide in this case. EHL luminescence decays faster than the free exciton emission (owing to Auger recombination in drops), but the shape of the decay curves as well as that of the intensity dependence may be strongly affected by the presence of parallel recombination processes (EM, BE, etc.).

8.4.3 Coexistence of excitonic molecules with electron–hole liquid

We have seen that under strong excitation various exciton interactions take place, among which especially EM and EHL affect significantly the low-temperature emission spectrum. Let us now ask whether in a semiconductor both the condensation of excitons into an EHL and the fusion of excitons into an excitonic molecule can occur simultaneously, or whether these are two 'disjunctive' processes in the sense that in a given material these may exist—depending upon the specific band structure—always just one of them.[8]

The attentive reader already knows the answer, of course. Comparison of Tables 8.1 and 8.2 shows that in AgBr, Si and Ge simultaneously $E_B > 0$ and $\varphi > 0$ hold, hence the gas of excitonic molecules can indeed coexist with the electron–hole liquid, In spite of this, there are certain points here—firstly, the remarkable historical context (the interpretation of the first luminescence experiments in highly excited Si and Ge crystals in the 1960s and 1970s oscillated between the EHL and EM models, because the correct answer to the above question was not known) and, secondly, interesting experimental aspects—which are worth mentioning.

Figure 8.14 compares AgBr emission spectra under a relatively low pulsed photoexcitation (50 kW/cm^2, panel (a)) with those taken with the excitation power density two orders of magnitude higher (5 MW/cm^2, panel (b)). Somewhat surprising is the occurrence of the M-line at the higher excitation only, while the EHL band is present in both these cases. It follows from these observations that, firstly, the equilibrium between the biexciton gas and the EHL (and, possibly, also the FE gas) may be limited to only a certain part of the phase diagram in the (n, T) plane, the so-called *biexciton pocket* [19]. Secondly, it turns out that the exciton condensation into an EHL is not entirely a typical 'high excitation effect'; a comparable, or perhaps an even more important role is played by a sufficiently low temperature. Condensation may often be achieved by making use of cw incandescent lamp excitation (see, e.g., Fig 8.13).

In Ge and Si the biexciton binding energy E_B is very small, not only in comparison with φ, but also absolutely. Consequently, at sufficiently low temperatures excitons very willingly condense into an EHL and thus the M-line cannot be observed at all, see Figs 1.1, 8.12 or 8.13. This is because the density of the surrounding free exciton gas, n_X, is low and biexcitons cannot form. If we simply decide to heat the sample in order to increase n_X and thus create biexcitons (which was realized in AgBr through heating the *e–h* system with powerful laser pulses, see Fig. 8.14(b)), we shall be faced with a problem. It is true that upon increasing the temperature by a couple of degrees we shall 'evaporate' more excitons, but at the same time we might already prevent biexciton formation because $k_B T \approx E_B$ will hold. Ultimately, at $T > T_c$ we shall dissolve the EHL entirely. Such a straightforward route cannot be applied, therefore the issue of the coexistence of EM with EHL remained unresolved for a long time.

We have to make the conditions for exciton condensation difficult, i.e. to make the EHL less stable, e.g. via stressing a Si sample along the [100] direction. In this way we lower the crystal symmetry and reduce the number of equivalent conduction band valleys ν_c from six to two. Therefore—according to what has been outlined above—we reduce the condensate binding energy φ. This means that the exciton gas surrounding the EHL drops gets denser, biexcitons can thus be created and, in the luminescence, the M-line appears. This is documented in Fig. 8.15 [20].

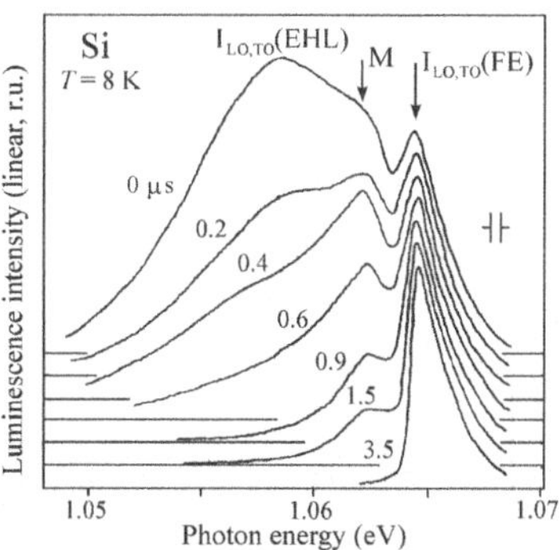

Fig. 8.15
Time-resolved emission spectra of a pure Si crystal, stressed along [100] at $T = 8$ K. Numbers at the curves indicate delay times (μs) after the excitation pulse. Besides the coexistence of all three phases—EHL, gas of FE and gas of EM—the figure also demonstrates the very rapid decay of the EHL luminescence. After Kulakovskii and Timofeev [20].

[8] With regard to what has been outlined in the introduction to Section 8.4, this inquiry concerns indirect-bandgap semiconductors.

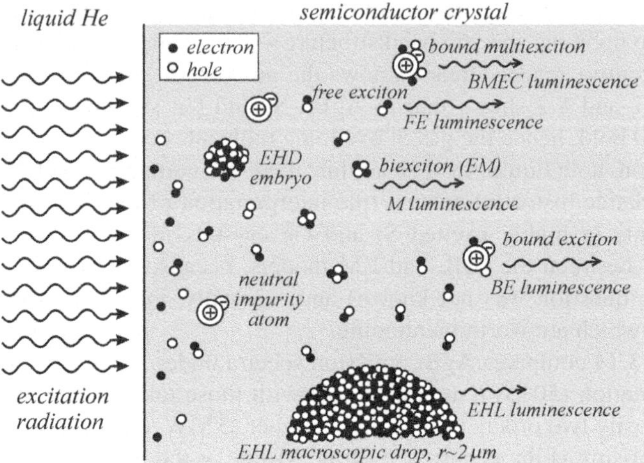

Fig. 8.16
Schematic cross-section of a (indirect-bandgap) semiconductor sample excited by light at low temperatures. Various luminescence processes are shown. After Westervelt [18].

An illustration of several parallel luminescence channels in semiconductors is given in Fig. 8.16. Let us add that, by using a special luminescence technique, the coexistence of an EHL with *free polyexcitons* (apart from biexcitons, with triexcitons and tetraexcitons) has been observed, particularly in silicon [21].

8.5 Electron–hole plasma (EHP)

8.5.1 Mott transition

Let us disregard for a moment the fact that the condensation of excitons can give rise to EHL drops. We thus have a gas of free excitons, the density of which, n_X, is continuously increased through the enhancement of excitation intensity. At the same time we also increase the number of free electron–hole pairs N, for a certain number of excitons always become ionized, even at low temperatures. In this way we reach a density N_M at which excitons as bound *e–h* pairs cease to exist. The reason for this resides in the effective screening of the Coulomb interaction between electrons and holes owing to the surrounding free carriers;[9] instead of (7.1), the interaction potential becomes

$$U_s(r) = -\frac{e^2}{4\pi\varepsilon_0\varepsilon r}\exp(-k_s r), \tag{8.30}$$

where $k_s = k_s(N) > 0$ is the so-called screening factor. Alternatively, $\lambda_s = k_s^{-1}$ is called the screening length. Relation (8.30) is known, e.g., from plasma physics.

The screening length $\lambda_s(N)$ gets shorter with increasing *e–h* pair density N. When λ_s approaches the free exciton radius a_X, excitons become unstable—the attraction between the electron and the hole is destroyed. As a first consequence of the screening (8.30), therefore, the transition of the exciton gas

[9] The proximity of other excitons operates in the same sense but it is commonly accepted that the main contribution to the screening comes from free electrons and holes.

(composed of bosons) through ionization to an *electron–hole plasma* (EHP) of a metallic character (i.e. driven by Fermi statistics) appears. This process represents one of a variety of possible realizations of the so-called *Mott insulator–metal transition*. The relevant density N_M is called the Mott density.

Bandgap renormalization arises as a second consequence. We have to realize that now we are far beyond the one-electron approximation. Due to Coulomb correlation and exchange effects, at high density of electron–hole pairs/excitons the semiconductor bandgap reduces to $E'_g(N) < E_g$. This can be demonstrated via a simple calculation, which compares the self-energy of a point charge in a many-electron system with that derived in the one-electron approximation [22]

$$\begin{aligned}\Delta E_g(N) &= E'_g(N) - E_g = \lim_{r\to 0}[U_s(r) - U(r)] \\ &= \frac{e^2}{4\pi\varepsilon_0\varepsilon}\lim_{r\to 0}\left[\frac{\exp(-k_s r) - 1}{r}\right] \\ &= \frac{e^2}{4\pi\varepsilon_0\varepsilon}\lim_{r\to 0}\left[-k_s\exp(-k_s r)\right] = -\frac{e^2}{4\pi\varepsilon_0\varepsilon}k_s(N) < 0, \quad (8.31)\end{aligned}$$

where we have used l'Hospital's rule. It follows immediately that $E'_g(N) < E_g$; the renormalization therefore entails a reduction of the bandgap ('gap shrinkage') in a dense electron–hole ensemble compared to the one-electron value E_g.

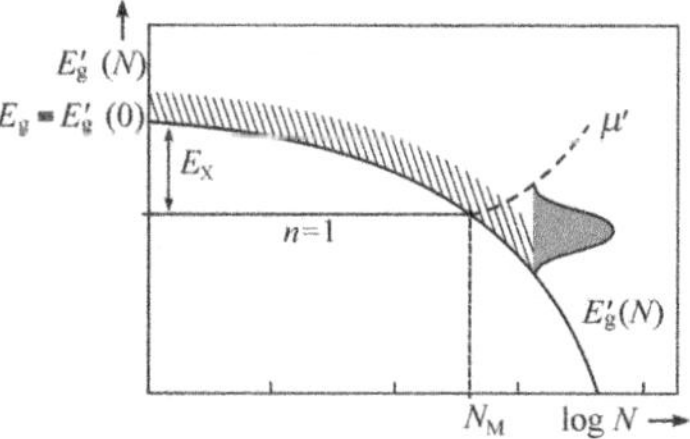

Fig. 8.17
The renormalized bandgap E'_g as a function of the *e–h* pair density N. The exciton binding energy of the $n = 1$ state is denoted E_X. The transition from the dielectric free exciton gas to ionized plasma EHP, proceeds smoothly. The EHP luminescence spectrum is depicted on the right; μ' stands for the plasma chemical potential, N_M is the Mott density.

The Mott transition in an excitonic system is displayed schematically in Fig. 8.17. Already at densities N smaller than N_M by two orders of magnitude the value E_g begins to drop; at the Mott density N_M the curves $E'_g(N)$ and the ground exciton level $n = 1$ cross each other, which means that $E'_g(N) < E_g(0) - E_X$ for $N > N_M$. In other words, for the *e–h* system it becomes energetically more favourable to be in the continuum of states of free *e–h* pairs than to be bound into excitons. It might seem strange that the exciton $n = 1$ level is independent of N. In fact, however, this indicates that the binding energy E_X also drops with increasing N (due to screening) and this drop in E_X is just compensated by the drop in $E'_g(N)$. This information is very important because it tells us that the Mott transition to EHP is—unlike EHL nucleation—gradual, i.e. excitons do not disappear suddenly upon reaching N_M, but their concentration decreases slowly as a function of the increasing excitation intensity.[10]

The Mott density N_M can be calculated within various approximations. The calculation is fairly easy to perform considering a non-degenerate EHP with relevant Fermi distribution functions approximated by Boltzmann tails. The screening length λ_s is then called the *Debye–Hückel* screening length and one gets

[10] Gap shrinkage and the corresponding loss of excitonic resonances entail important qualitative changes in the optical absorption spectrum. Their features are obviously dependent on the excitation intensity and thus these spectral modifications represent a typical example of nonlinear optical phenomena.

$$\lambda_s \sim N^{-1/2},\ N_M \approx \frac{k_B T}{8\pi a_X^3 E_X} = \frac{k_B T\, \varepsilon_0\, \varepsilon}{e^2 a_X^2}. \tag{8.32}$$

The Debye–Hückel model happens to be a viable approximation in semiconductors with large a_X, except at low temperatures. The temperature T in (8.32) may designate the effective temperature.

8.5.2 Luminescence of EHP

Electrons and holes in an EHP can recombine radiatively, of course. We are interested in what the EHP luminescence emission spectrum looks like. To begin, however, we have to answer an important question: What is the difference between an electron–hole plasma, EHP, and an electron–hole liquid, EHL, the existence of the latter having been—for the sake of simplicity—disregarded at the beginning of this section?

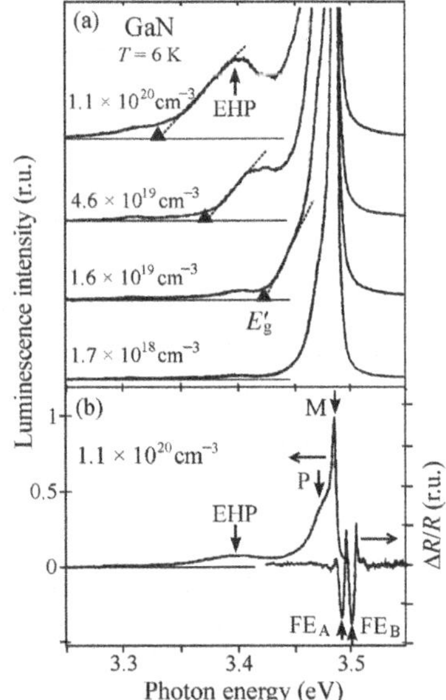

Fig. 8.18
(a). Onset of the EHP emission band in highly excited GaN at $T = 6$ K upon reaching the Mott density ($N_M \approx 10^{19}$ cm^{-3}). Triangles label the positions of the reduced bandgap E'_g at the indicated densities of e–h pairs. Panel (b) displays the overall shape of the emission spectrum. Apart from the EHP band, also the M-line due to biexciton luminescence and the P-line due to exciton collisions are present. The free exciton spectral position is given by the reflection spectrum $\Delta R/R$. After Nagai *et al.* [23].

EHL and EHP are two similar phases of the e–h system. However, while the condensation exciton gas $\rightarrow$ EHL starts abruptly and the liquid is concentrated into spatially clearly separated regions—spherical drops with a constant e–h pair density n_0, the transition exciton gas $\rightarrow$ EHP proceeds smoothly, continuously. Also, the plasma gradually fills the whole excited volume (there are no drops) and may diffuse both laterally and into the depth of the sample, due to the concentration gradient. The e–h pair density therefore is not, unlike EHL, spatially constant and, moreover, rises with increasing excitation intensity. And perhaps the most important: the EHP may exist—unlike EHL—at arbitrary temperatures, neither is there an EHP critical temperature nor binding energy.

The EHP is therefore a 'less ordered' relative of EHL. There is no radiative recombination enhancement factor in EHP, unlike EHL. It is worth stressing, however, that the concept of the renormalized gap, as introduced in Subsection 8.4.1, holds equally for both EHP and EHL! If condensation into an EHL happens, the e–h system simply 'jumps' somewhere to the right in Fig. 8.17, far beyond the N_M density, and remains there. Equally, we can imagine that Fig. 8.11 is valid for both EHL and EHP, but while $E'_g(n_0)$ is constant inside EHL drops, $E'_g(N)$ in EHP varies as a function of N (see Fig. 8.17).

The scenario under growing optical excitation intensity, e.g., in silicon at $T = 15$ K, can then be described as follows (see the horizontal arrow in Fig. 8.9): First, a gas of excitons with increasing density n_X is created, but before the Mott transition sets in, n_X reaches a critical value for exciton condensation into EHL drops with a very high equlibrilum e–h pair density $N = n_0 \approx 3 \times 10^{18}$ cm^{-3}. The density of the surrounding gas, n_X, henceforth remains constant[11] ($\sim 10^{16}$ cm^{-3}) and the gradual Mott transition will not occur. Generally it turns out that in indirect-bandgap materials, EHP and its luminescence reveal themselves only at temperatures $T = T_c$ or higher. To differentiate the contributions to the emission spectrum from EHL, EHP and possibly from other luminescence channels (biexcitons, etc.) under these

[11] The average density of the e–h system (all phases) will rise with the continuing growth of the excitation level, though, due to the growth the total volume of the EHL drops.

conditions is not easy. Some authors even take the view that EHP can also undergo its own condensation process, characterized by a second critical point in the phase diagram. Currently, there is no consensus on this issue and further discussion goes beyond the scope of this book.

In this respect, the situation is much simpler in direct-bandgap semiconductors, where EHL does not occur. The onset of EHP luminescence under sufficiently strong excitation is straightforward there, and according to the foregoing discussion this light emission has very specific spectral features:

1. A wide emission band (much like an EHL—radiative recombination in an ensemble of fermions), gradually emerging in the spectrum when the density of e–h pairs, N, approaches N_M.
2. The low-energy emission band boundary (determined by $E'_g(N)$) undergoes a red shift with increasing excitation intensity; for an EHL, this boundary does not shift.
3. With increasing excitation intensity the EHP emission band gets broader, because the plasma density increases; the EHL bandwidth remains fixed.

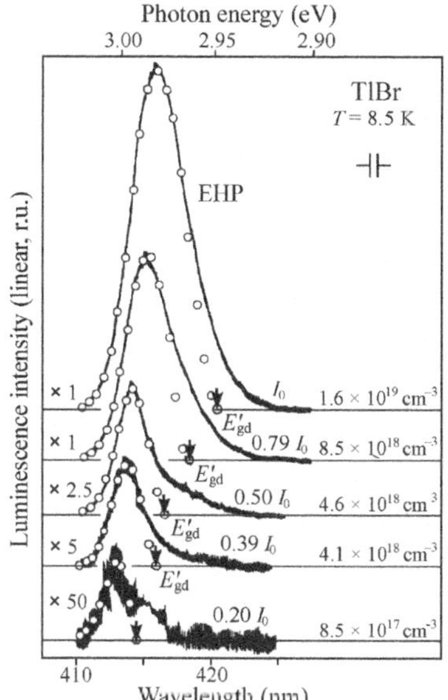

Fig. 8.19
EHP emission spectra in a highly excited TlBr crystal (a pulsed N_2-laser, 2 ns, 3.4 eV) at a bath temperature of 8.5 K. The excitation intensity I_0 corresponds to ~ 0.6 MW/cm^2. In TlBr, radiative recombination can occur both in indirect as well as in direct bandgaps; the displayed spectra originate in the direct one, whose reduced values E'_{gd} are labelled by arrows. After Kohlová *et al.* [24]. (Fig. 2.38(b) represents the EHP emission spectrum of TlBr at $T = 25$ K both in direct—A-line—and indirect—B-line—bandgaps. Arrows at ~ 3.02 eV and ~ 2.66 eV denote the spectral positions of relevant non-renormalized gaps corresponding to the one-electron approximation. The respective plasma densities and effective temperatures are also shown.)

Examples are shown in Figs 8.18 (GaN [23]) and 8.19 (TlBr [24]). The gradual increase of the EHP band around $N = N_M$—the salient feature (1)—can be nicely seen in Fig. 8.18, while Fig. 8.19 demonstrates features (2) and (3).[12] Open circles in Fig. 8.19 represent a theoretical fit from which the displayed EHP densities were derived.

It is important to stress that, although in TlBr the direct bandgap is concerned here, expression (8.25) together with the procedure described in Subsection 8.4.1 were applied for fitting. This is surprising because the convolution (8.25) was derived for an indirect bandgap. Here, strictly speaking, its application is not justified. Nevertheless, this phenomenological approach is currently being used because it is simple and provides us with both a reasonable reproduction of the emission bandshape and a realistic effective temperature and density of EHP. (This was one of the reasons to depict transitions in Fig. 8.11 as direct ones.) More elaborate routines for spectral modelling that comprise **k**-*conservation* have been proposed, of course, but they are rather sophisticated and the resulting spectral shape is almost the same, except for more accurate determination of E'_g [25].

Let us recall once more that the densities obtained through fitting are *average* EHP densities, unless we create in some way a spatially uniform plasma which, in fact, is highly desirable for the comparison of theory with experiment. To this end for example, two-photon excitation discussed in Section 5.5 can be used, as well as various spatial localizations of EHP in low-dimensional structures. Finally, let us note that radiative recombination of EHP in direct semiconductors may easily lead to stimulated emission (Chapter 10), which is of basic importance for light-emitting diodes and semiconductor lasers.

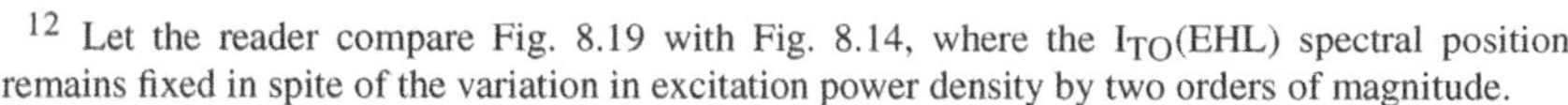
[12] Let the reader compare Fig. 8.19 with Fig. 8.14, where the I_{TO}(EHL) spectral position remains fixed in spite of the variation in excitation power density by two orders of magnitude.

8.6 Bose–Einstein condensation of excitons

An exciton as a quasi-particle consisting of two fermions (an electron and a hole) always has an integer total spin $S = 0$ or 1. An exciton is thus a boson and these can undergo, at high densities, so-called Bose–Einstein condensation. It is worth stressing that Bose–Einstein condensation represents a fundamentally different process with regard to exciton condensation into an EHL discussed above. In order to understand the fundamentals of Bose–Einstein condensation and to discover how this kind of exciton condensation can manifest itself in luminescence, we first recall the basic properties of Bose–Einstein statistics.

8.6.1 Properties of the Bose-Einstein distribution

The distribution or occupation function of ideal (non-interacting) bosons, determining the mean boson' number in a state with energy E, reads

$$f_{\mathrm{BE}}(E, T) = \frac{1}{\exp\left(\frac{E-\mu}{k_{\mathrm{B}}T}\right) - 1}, \tag{8.33}$$

where μ is the chemical potential. This function is called the Bose–Einstein distribution. Unlike the Fermi–Dirac distribution, where μ is more often known as the Fermi energy and has quite an illustrative meaning of an energy level with the occupation probability equal to 1/2, the chemical potential μ in (8.33) is less familiar. Its properties are:

1. $\mu < 0$ must be satisfied, otherwise the distribution function (8.33) in the ground state ($E = 0$) would acquire a negative value, which is not admissible.
2. $|\mu|$ is large at low boson densities, while with increasing density $|\mu|$ decreases and gradually approaches zero. This also follows directly from (8.33); a large value of $|\mu|$ implies a low mean number of bosons in a given state.

Let us now ask a question: Can the boson' number in the investigated system reach an arbitrary value? The total boson' number (per unit volume) n_{tot} can be found by summing over all occupied states and all possible energies:

$$n_{\mathrm{tot}} = \int_0^\infty \rho(E) f_{\mathrm{BE}}(E)\, \mathrm{d}E. \tag{8.34}$$

For bosons with a mass m and total spin $S = 0$ we shall consider the density of states of the form[13] $\rho(E) = (2m)^{3/2} E^{1/2} / 4\pi^2 \hbar^3$. The maximum available number of bosons, n_{CB}, can be obtained if we introduce into (8.34) a maximum value of f_{BE}, which, according to the foregoing discussion, occurs in the limiting case $\mu \to 0$. By performing the relevant calculation (Appendix E) we get

[13] The difference 1/2 compared to (5.4) is due to the difference in the spin degeneracy factor g between fermions ($g = 2$) and bosons with spin $S = 0 (g = 1)$.

$$n_{CB}(T) = \frac{2.31\,(2m)^{3/2}\,(k_B T)^{3/2}}{4\pi^2\hbar^3}$$
$$\cong 2.61\left(\frac{m\,k_B T}{2\pi\hbar^2}\right)^{3/2} \cong 6.4 \times 10^{15}\left(\frac{m}{m_0}\right)^{3/2} T^{3/2}(\mathrm{cm}^{-3}), \quad (8.35)$$

where m_0 stands for the free electron mass. Obviously, this is a finite number, thus a very important message from (8.35) is: an ensemble of bosons may always contain only a *finite* number of particles (dependent on temperature).

What happens now if we continuously add further particles into the ensemble? (In particular, if we keep creating a higher and higher exciton density through increasing the optical excitation?)

Einstein, shortly after the formulation of quantum statistics by the Indian physicist Bose, deduced: the excess particles must 'condense' to the ground state $E = 0$, which, in this way, somehow separates from other states. Relation (8.35) thus defines a *critical concentration* n_{CB} for the onset of this so-called Bose–Einstein condensation,[14] which occurs primarily in energy or momentum space (**k**-space), not in real space—unlike exciton condensation to an EHL.

Before we discuss manifestation of luminescence the of this peculiar condensation, we shall derive one simple mathematical consequence of (8.35). Upon reaching n_{CB} we can obviously express the boson density as

$$n_{tot} = n_0 + n_{CB}(T).$$

The ground state population n_0 will thus grow with a further increase of pumping, while n_{CB} at a given T remains constant. This relation can be rewritten as

$$\frac{n_0}{n_{tot}} = 1 - \frac{n_{CB}}{n_{tot}} = 1 - \frac{6.4 \times 10^{15}\,(m/m_0)^{3/2}\,T^{3/2}}{n_{tot}},$$

from which the *critical temperature* T_{CB} for the Bose–Einstein condensation can be defined via

$$\frac{1}{T_{CB}^{3/2}} = \frac{6.4 \times 10^{15}\,(m/m_0)^{3/2}}{n_{tot}}. \quad (8.36)$$

The announced mathematical follow-up of (8.35) then reads

$$\frac{n_0}{n_{tot}} = 1 - \left(\frac{T}{T_{CB}}\right)^{3/2}. \quad (8.37)$$

From (8.37) we can easily infer the meaning of the critical temperature:

For $T > T_{CB}$ .... (8.37) has no physical meaning,
$T = T_{CB}$ .... $n_0 = 0$ and the condensation just begins,

[14] At the same time a condition must be satisfied, namely, that the energy and the number of bosons can vary independently. This is possible, e.g., for atoms, excitons and biexcitons but not for photons and phonons; their chemical potential equals zero but, despite this fact, one cannot speak of a Bose–Einstein condensation. An equivalent statement reads that bosons undergoing Bose–Einstein condensation must have non-zero rest mass (see (8.35)), which holds neither for photons nor phonons.

$T < T_{CB}$ .... n_0 according (8.37) sharply increases,
$T = 0\,\text{K}$ .... $n_{tot} = n_0$.

On the basis of (8.36) and (8.37) we can quite clearly imagine what the Bose–Einstein condensation of excitons looks like in an experiment. Let us apply, e.g., a fixed excitation intensity, i.e. a fixed exciton density $n_X = n_{tot}$, and let us start the experiment at a relatively high temperature when, according to (8.35), $n_X < n_{CB}(T)$. Therefore there is no condensation. In the course of cooling we reduce $n_{CB}(T)$ till we drop to the critical temperature T_{CB} when $n_X = n_{CB}$. The condensation sets in and if we keep cooling, the concentration of 'condensed' excitons rises according to (8.37) up to the limit value $n_0 = n_X$ at $T = 0\,\text{K}$. Then, all of the photogenerated excitons occur in the condensate with energy $E = 0$.

8.6.2 Luminescence experiment: Bose–Einstein condensation yes or no?

The Bose–Einstein condensation represents the macroscopic population of a single state $E = 0$ or $\mathbf{k} = 0$. It is one of the fundamental consequences of quantum mechanics that leads—among others—to important, application-related effects like superfluidity of liquid He and superconductivity. The development of efficient laser and magnetic cooling of atoms has recently led to a fruitful investigation of experimental features of this condensation in an atomic gas, but at extremely low temperatures below 100 nK.

Semiconductor luminescence proposes unique conditions for a detailed study of Bose–Einstein condensation. If we adjust (8.36) to express explicitly the critical temperatures

$$T_{CB} \cong 2.9 \times 10^{-11} \left(\frac{m_0}{m}\right) n_{tot}^{2/3}\,(\text{K}), \tag{8.38}$$

it is observed that the critical temperature for an atomic gas ($m \gg m_0$) is by many orders lower compared with a gas of excitons ($m \approx m_0$) of the same density. Considering this density to be $n_{tot} = 10^{17}\,\text{cm}^{-3}$ (easily attainable under optical pumping), we obtain from (8.38) for $m = m_0$ a critical temperature $T_{CB} \approx 6.3\,\text{K}$. This is a 'reasonable' temperature that can be achieved in a current helium cryostat. Pumping of ^{4}He vapour enables going down to $T \approx 1.3\,\text{K}$ commonly. There are additional advantages: the exciton density and thus, according to (8.38), also T_{CB} can be varied readily via the variation of excitation intensity, or vice versa, at a given excitation power density we can finely tune the sample temperature in the environs of T_{CB}. And, most importantly, the exciton condensate reveals itself through a dramatic modification of the semiconductor emission spectrum. For all of these reasons the search for the Bose–Einstein condensation of excitons using luminescence methods has represented a worldwide challenge for more than 30 years.

How the condensation should manifest itself in the emission spectrum can be predicted first with the aid of simple reasoning. Since it arises from the accumulation of excitons in the $E = 0$ state, which corresponds to the bottom of the exciton band, we expect the recombining 'condensed' excitons to emit

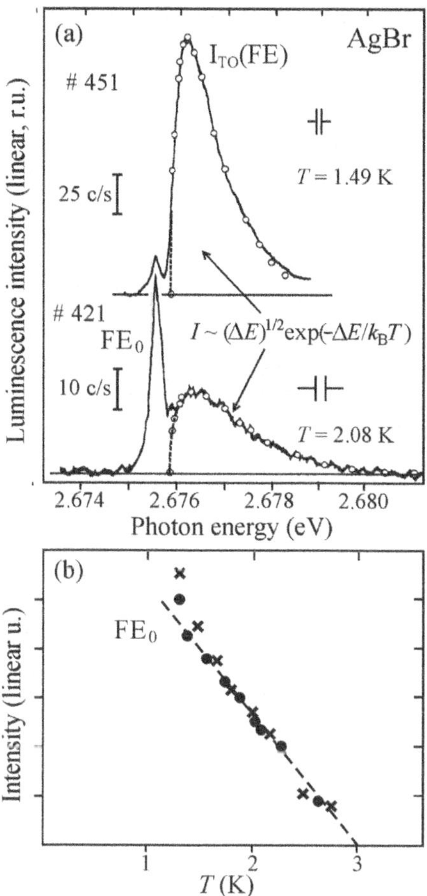

Fig. 8.20
(a) Low-temperature free exciton photoluminescence in AgBr accompanied with the emission of a TO(L)-phonon (i.e. I_{TO}(FE) line), in two samples #451 and #421. Open circles denote the theoretical lineshape fit based on (7.16). The very narrow peak labelled FE_0 has been identified as being due to the Bose–Einstein condensation of the free exciton gas. Panel (b) shows a sudden intensity increase of this peak in the course of cooling below 3 K. After Czaja and Schwerdtfeger [26]. Later experiments revealed that this interpretation was probably incorrect, though.

a very narrow peak on the low-energy edge of the free exciton line ($h\nu = E_{gi} - E_X - \hbar\omega$ in an indirect-bandgap semiconductor). This peak should appear in the spectrum suddenly in the course of cooling, upon reaching T_{CB}.

Exactly this kind of behaviour can be recognized in the spectra shown in Fig. 8.20(a), which come from two nominally pure AgBr crystals at $T = 2.08$ K and $T = 1.49$ K. The lower panel (b) shows that the intensity behaviour of the sharp FE_0 peak at ~ 2.6754 eV complies perfectly with (8.37), provided we put $T_{CB} = 3$ K. This represents one of the first luminescence observations interpreted in terms of the excitonic Bose–Einstein condensation [26]. Everything looks very convincing. However, a series of subsequent papers revealed this interpretation to be untenable, by finding the origin of the FE_0 peak in an exciton bound to a residual shallow impurity (Subsection 7.2.1).

This indicates that the experimental realization of the Bose–Einstein exciton condensation is not a simple task and that such a straightforward way as the one shown in Fig. 8.20 is hardly viable. A 'jump' from the Maxwell–Boltzmann distribution (circles on the experimental curves in Fig. 8.20(a)) immediately to the spectrally narrow condensate is unrealistic. The Maxwell–Boltzmann distribution is a classical limit, shared both by bosons and fermions. With

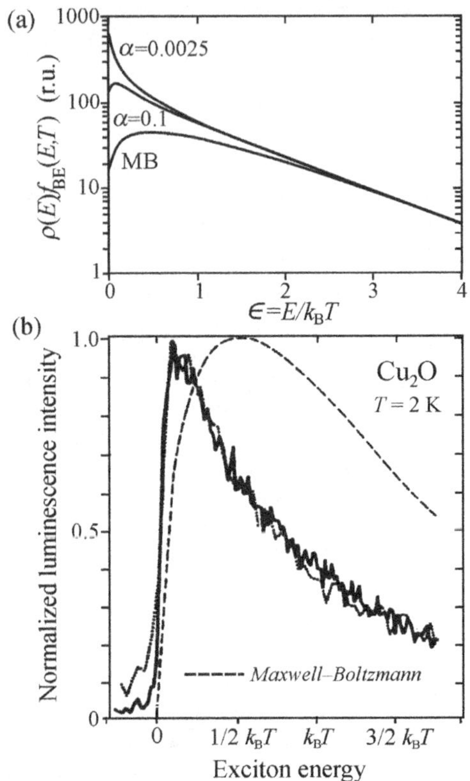

Fig. 8.21
(a) Product of the density of states $\rho(E) \approx \sqrt{E}$ with the Bose–Einstein distribution function (8.33) where $\alpha = -\mu/k_BT$. The lower curve (MB) is the Maxwell–Boltzmann approximation valid for low exciton densities ($\alpha > 4$). In the limit $\alpha \to 0$ a narrow peak remains only at $\epsilon = 0$. The curves are normalized at high values of ϵ. (b) Narrowing down of the exciton emission line (a phonon replica) in Cu_2O at high exciton density. The sample was immersed in liquid He ($T = 2$K) and excited by 10 ns pulses of an Ar^+-laser with excitation intensity of $\sim 40\,MW/cm^2$. The dashed curve (Maxwell–Boltzmann distribution) is considerably wider. After Snoke *et al.* [27].

increasing density n_X (therefore with decreasing $|\mu|$) the excitons should at first prove that they start to behave as genuine bosons. Consequently, the exciton statistics and also the emission lineshape should gradually transform to true Bose statistics. In this case, the lineshape width would undergo dramatic but gradual narrowing, which can be easily shown by calculating $\rho(E)f_{BE}(E)$ and which is displayed in Fig. 8.21(a). This is the so-called *Bose narrowing*. The reader is reminded that the FWHM of the Maxwell–Boltzmann distribution (7.16) is invariable, equal to about $1.8\,k_BT$.

This Bose narrowing has indeed been found in exciton luminescence of Cu_2O [27], as demonstrated in Fig. 8.21(b). Nevertheless, till now, no unambiguous experimental evidence of Bose–Einstein condensation has been achieved either in this material or in bulk semiconductors in general (at least unconditional evidence that would be accepted without taking exceptions to). Taking into consideration everything discussed above in this chapter, we understand why this is so. First of all, excitons have a finite lifetime. The ideal Bose–Einstein distribution (8.33) holds exactly for excitons as non-interacting quasiparticles with infinite lifetime in thermodynamic equilibrium only. Moreover, with increasing exciton density the important exciton–exciton interactions occur in most semiconductors (biexcitons, EHL, EHP). For instance, prior to reaching the critical concentration n_{CB}, the Coulomb e–h attraction may happen to be screened and the exciton gas transformed into EHP or EHL. (This is one of the reasons why especially Cu_2O appears to be a good candidate for the exciton Bose–Einstein condensation—owing to specific features of the Cu_2O band structure the interaction between excitons is repulsive, thus biexcitons do not originate here. Moreover, because the Bohr radius a_X is very small, see Table. 7.1, EHP does not occur in this material at high excitation.)

Another complication preventing excitons from Bose–Einstein condensation may be insufficient sample purity, as we have seen in the case of AgBr (Fig. 8.20)—the presence of bound excitons reduces n_X and, in the spectrum, may either mask a possible emission line arising from the Bose–Einstein condensate or cause an incorrect interpretation of the experimental observations. Finally, high optical excitation levels can both evoke spatial exciton diffusion and entail a high effective temperature of the exciton gas that hinders reaching T_{CB}. A more detailed discussion goes beyond scope of this book. It is worth mentioning though, that also biexcitons are bosons and one can expect their Bose–Einstein condensation together with its specific luminescence manifestation. The most investigated material in this regard is copper chloride, CuCl. The whole topic is covered by a special monograph [28].

It is worth noting that, strictly speaking, the above discussed 'classical' Bose–Einstein condensation may exist in infinite bulk materials only. In low-dimensional systems, a generalized definition of Bose–Einstein should be applied. In particular in two-dimensional structures the concept of condensation is linked to the phase-coherent superfluid movement of a bosonic system rather than to boson accumulation in the lowest energy level [29]. Recent results concerning Bose–Einstein condensation of exciton–polaritons in semiconductor low-dimensional microcavities will be briefly mentioned in Chapter 16.

8.7 Problems

8/1: Based on Fig. 8.1: (a) Estimate the binding energy E_B of an excitonic molecule in CdS. (b) Check that the spectral position of the P-band, which is due to inelastic exciton–exciton scattering, is in accord with the binding energy E_X in Table 7.1.

8/2: Luminescence of excitonic molecules in indirect-bandgap semiconductors, discussed in Subsection 8.2.1, has been described in the framework of allowed dipole transitions. Nakahara and Kobayashi [30] have generalized this approach for the case of *indirect forbidden* transitions, when the relevant transition matrix element is composed of two terms, the first of which is proportional to the magnitude of the biexciton wavevector $\mathbf{Q}$ and the second one to the magnitude of the wavevector $\mathbf{q}$ of the remaining exciton. Making use of the formalism applied in Appendix C, adjust the expression describing the emission lineshape to a form suitable for numerical integration (analogy with (C-15)) and show that the contribution of the second term is negligible.

8/3: Prove, using the $E(\mathbf{K})$ diagram (in analogy with Fig. 8.7), that in common semiconductors with indirect bandgap the luminescence channel due to inelastic free exciton collisions (P-line) cannot occur. Hint: consider that the indirect bandgap in common semiconductors is always defined by an extreme of the conduction or valence band situated at the boundary of the first Brillouin zone.

8/4: Show that the average kinetic energy (per electron) in a Fermi gas at $T = 0\,\mathrm{K}$ is equal to $(3/5)F_e (F_e =$ Fermi energy of electrons).

8/5: An electron–hole liquid (EHL) in the form of electron–hole drops (EHDs) represents an open system exchanging quasi-particles, i.e. the electron–hole pairs, with its surroundings. Open systems in thermodynamics are characterized, among others, by their chemical potential μ. Show that here μ is characteristic of the EHL ground state. Hint: start from the definition of the chemical potential as the energy required for passing from the ground state with N electron–hole pairs in a drop to the ground state with $(N+1)$ pairs in the drop. See also [31].

8/6: By using the relationship (8.32) and exploiting the parameters from Table. 7.1, calculate the Mott density N_M in selected semiconductors for $k_B T/E_X = 0.1$. Compare the obtained results with a rough estimate derived on the basis of the criterion $N_M \cong (4\pi a_X^3/3)^{-1}$. Discuss the illustrative meaning of this criterion.

References

1. Shionoya, S., Saito, H., Hanamura, E., and Akimoto, O. (1973), *Solid State Comm.*, **12**, 223.
2. Klingshirn, C. and Haug, H. (1981). *Phys. Rep.*, **70**, 315.
3. Hanamura, E., and Haug, H. (1977). *Phys. Rep., C*, **33**, 209.
4. Quattropani, A. and Forney, J. J. (1977). *Nuovo Cimento B*, **39**, 569.
5. Grun, J. B., Lévy, R., Ostertag, E., Vu Duy Phach, H., and Port, H. (1976). *Biexciton luminescence in CuCl and CuBr*. In: *Physics of Highly Excited States in Solids* (ed. M. Ueta and Y. Nishina), Lecture Notes in Physics Vol. 57, p. 49. Springer, Berlin.
6. Baba, T. and Masumi, T. (1977). *Nuovo Cimento B*, **39**, 609.
7. Cho, K. (1973). *Optics Comm.*, **8**, 412.
8. Knox, R. S., Nikitine, S., and Mysyrowicz, A. (1969). *Optics Comm.*, **1**, 19.
9. Pelant, I., Hála, J., Parma, L., and Vacek, K. (1980). *Solid State Comm.*, **36**, 729.
10. Gourley, P. L. and Wolfe, J. P. (1979). *Phys. Rev. B*, **20**, 3319.
11. Hvam, J. M. (1973). *Solid State Comm.*, **12**, 95.
12. Moriya, M. and Kushida, T. (1976). *J. Phys. Soc. Japan*, **40**, 1668.
13. Shah, J., Combescot, M., and Dayem, A. H. (1977). *Phys. Rev. Lett.*, **38**, 1497.
14. Rice, T. M. (1977). *The electron–hole liquid in semiconductors: theoretical aspects*. In: *The Electron–Hole Liquid in Semiconductors* (ed. H. Ehrenreich, F. Seitz and D. Turnbull), Solid State Physics Vol. 32, p. 1. Academic Press, New York.
15. Vashista, P., Kalia, R. K., and Singwi, K. S. (1983). *Electron–hole liquid: theory.* In: *Electron–Hole Droplets in Semiconductors* (ed. C. D. Jeffries and L. V. Keldysh), Modern Problems in Condensed Matter Sciences Vol. 6, p. 1. North Holland, Amsterdam.
16. Benoit à la Guillaume, C. and Voos, M. (1973). *Phys. Rev. B*, **7**, 1723.
17. Hensel, J. C., Phillips, T. G., and Thomas, G. A. (1977). *The electron–hole liquid in semiconductors: experimental aspects*. In: *The Electron–Hole Liquid in Semiconductors* (ed. H. Ehrenreich, F. Seitz and D. Turnbull), Solid State Physics Vol. 32, p. 88. Academic Press, New York.
18. Westervelt, R. M. (1983). *Kinetics of electron–hole drop formation and decay*. In: *Electron–Hole Droplets in Semiconductors* (ed. C. D. Jeffries and L. V. Keldysh),

Modern Problems in Condensed Matter Sciences Vol. 6, p. 187. North Holland, Amsterdam.
19. Hulin, D., Mysyrowicz, A., Combescot, M., Pelant, I., and Benoit à la Guillaume, C. (1977). *Phys. Rev. Lett.*, **39**, 1169.
20. Kulakovskij, V. D. and Timofeev, V. B. (1983). *Thermodynamics of electron-hole liquid in semiconductors.* In: *Electron–Hole Droplets in Semiconductors* (ed. C. D. Jeffries and L. V. Keldysh), Modern Problems in Condensed Matter Sciences Vol. 6, p. 95. North Holland, Amsterdam.
21. Steele, A. G., McMullan, W. G., and Thewalt, M. L. W. (1987). *Phys. Rev. Lett.*, **59**, 2899.
22. Haug, H. (1988). *Introduction.* In: *Optical Nonlinearities and Instabilities in Semiconductors* (ed. H. Haug), p. 1. Academic Press, New York.
23. Nagai, T., Inagaki, T. J., and Kanemitsu, Y. (2004). *Appl. Phys. Lett.*, **84**, 1284.
24. Kohlová, V., Pelant, I., Hála, J., Ambrož, M., and Vacek, K. (1987). *Solid State Comm.*, **62**, 105.
25. Capizzi, M., Modesti, S., Frova, A., Staehli, J. L., Guzzi, M., and Logan, R. A. (1984). *Phys. Rev. B*, **29**, 2028.
26. Czaja, W. and Schwerdtfeger, C. F. (1974). *Solid State Comm.*, **15**, 87.
27. Snoke, D. W., Wolfe, J. P., and Mysyrowicz, A. (1990). *Phys. Rev. B*, **41**, 11171.
28. Moskalenko, S. A. and Snoke, D. W. (2000). *Bose–Einstein Condensation of Excitons and Biexcitons*. Cambridge University Press, Cambridge.
29. Kavokin, A. V., Baumberg, J. J., Malpuech, G., and Laussy F. P. (2007). *Microcavities*. Oxford University Press, Oxford.
30. Nakahara, J. and Kobayashi, K. (1976). *J. Phys. Soc. Japan*, **40**, 189.
31. Combescot, M. and Noziéres, P. (1972). *J. Phys. C*, **5**, 2369.

Luminescence of disordered semiconductors

Disordered (amorphous) semiconductors are characterized by the absence of an ordered atomic arrangement over long distances from a chosen atom ('long-range order'). One is thus not allowed to employ the concepts familiar from the crystalline state such as the electron or hole wavevector **k**, the Bloch theorem or the energy band structure perceived as the energy dependence of the charge carrier on the wavevector **k**, i.e. $E = E(\mathbf{k})$. However, because the short range order (i.e. basically the periodic arrangement in the close surroundings of the given site) is kept, also the concepts of the conduction and valence bands, the forbidden gap and the density of states are basically conserved. These again are of decisive importance for understanding the luminescence processes, but the microscopic luminescence features in disordered solids differ quite considerably from the mechanisms that operate in their crystalline counterparts and that were discussed in Chapters 5, 7 and 8.

9.1 Densities of states in bands

The commonly accepted scheme of the electron and hole energy distribution in disordered semiconductors is depicted in Fig. 9.1(a): The electrons with energy higher than the so-called mobility edge E_{cm} are considered free and may take part in the charge transport; their movement is characterized by the so-called microscopic mobility μ_{m}. The electrons having energy $E < E_{\mathrm{cm}}$ are not free; they are spatially localized in traps whose microscopic origin consists in deflections of the bond lengths and angles from their fixed values in a regular crystalline lattice. Upon applying an electric field these carriers can participate in the charge transport only via the so-called hopping mechanism, when the electrons tunnel between nearby potential wells that act as charge traps, or by means of so-called multiple trapping. In that case the localized electrons are thermally excited above the mobility edge E_{cm}, migrate with the mobility μ_{m}, fall down into a distant trap, are thermally re-excited, etc. A mirror-like picture may be applied to the holes with relevant mobility edge E_{vm}.

The difference E_{cm}–E_{vm} roughly corresponds to the energy bandgap in crystalline semiconductors. The density of states of the electrons with energy $E > E_{\mathrm{cm}}$ is described by a square-root function $\rho(E) \simeq (E - E_{\mathrm{cm}})^{1/2}$, analogously to crystals. In an ideal crystalline semiconductor, however, the density of states within the bandgap is zero, while in disordered materials it remains non-zero even for $E < E_{cm}$. It has been established that the density of these states (originating, as already mentioned, owing to variations of the bonding parameters between atoms), $g(E)$, drops exponentially downwards from the mobility edge [1]

$$g_c(E) = g_{c0} \exp(-E/E_{c0}). \tag{9.1}$$

In this expression the energy E is referred to E_{cm} and considered positive downwards, see Fig. 9.1. A similar convention for the holes yields

$$g_v(E) = g_{v0} \exp(-E/E_{v0}). \tag{9.2}$$

The quantities E_{c0}, E_{v0} stand for the energy widths of the tail states (i.e. the states extending into the 'energy gap') of the conduction and valence bands, respectively. They amount typically to tens of meV. Nonequilibrium electron–hole pairs, created with the generation rate G, rapidly thermalize (within 10^{-13}–10^{-12} s) towards the corresponding mobility edges, then deeper into the tail states and eventually—on a much longer timescale—they may recombine radiatively via tunnelling between nearby sites with a recombination rate $1/\tau_r$ (Fig. 9.1(b)). This is generally believed to be the main radiative recombination channel in disordered semiconductors.

Within the bandgap of disordered semiconductors one can find other electron states. The deformed covalent bonds between neighbouring atoms are exposed to mechanical strain and may break. This rupture gives rise to so-called dangling or unsaturated bonds. They are very often passivated by hydrogen atoms (especially in hydrogenated amorphous silicon, a-Si:H), but some remain unpassivated, may capture an electron and thereby introduce deep localized electron states approximately in the middle of the bandgap. Important, from the luminescence standpoint, is that these deep defect levels act as efficient centres of *non-radiative recombination* (Fig. 9.1(b)), which is, after all, a general property of all deep defect states (see Chapter 6).

It should be noted also that the total density of the dangling bonds is not constant in time. It grows upon irradiating the sample with photons $h\nu \geq E_{cm} - E_{vm}$ that create free electron–hole pairs. Subsequent non-radiative recombination of these *e–h* pairs may release energy sufficient to break the weakened bonds, as explained already in Section 6.2. The process has been known for a long time as the Staebler–Wronski effect [2] but its microscopic mechanism has not been understood in all details till now. Recent results indicate that it is basically due to bimolecular non-radiative recombination [3].

The dangling bonds can be either empty (i.e. positively charged, D^+, because the negative electron charge is missing there), or occupied with one electron (D^0), or with two electrons of opposite spins (negatively charged dangling bond, D^-). The energy required to add the second electron to an already occupied dangling bond is called the correlation energy U, see Fig. 9.1(a).

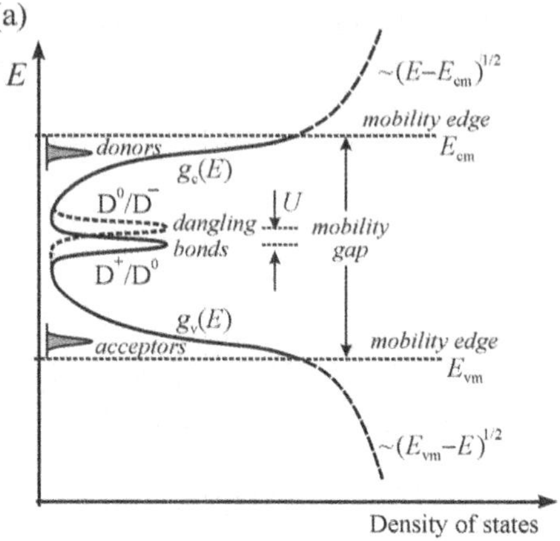

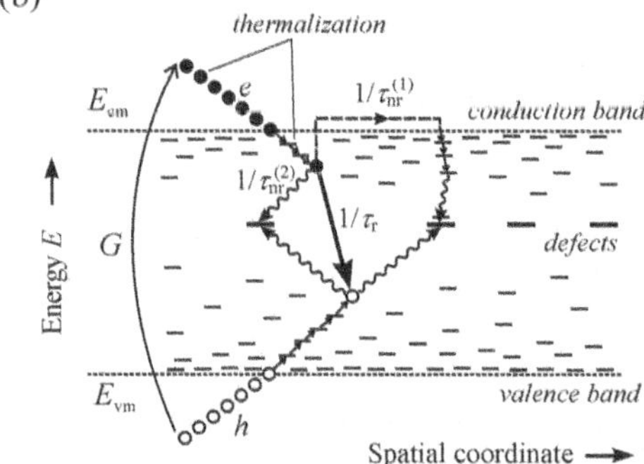

Fig. 9.1
(a) Schematic of the electron density of states in disordered semiconductors. Luminescence properties are mainly driven by occupation of the localized, so-called tail states $g_c(E)$, $g_v(E)$ that drop exponentially down from the relevant mobility edges into the 'forbidden' gap. The electrons above the mobility edge E_{cm} occupy free states, whose density is described approximately by a square-root shape (like the holes below the E_{vm} edge). Approximately in the middle of the bandgap there are levels D^+, D^0, D^- arising from the dangling bonds. These levels represent efficient non-radiative centres. (b) Schematic of thermalization and recombination of non-equilibrium electrons and holes in amorphous semiconductors. The radiative recombination of an e–h pair localized in the tail states $(1/\tau_r)$ competes with non-radiative tunnelling to a deep defect level $(1/\tau_{nr}^{(2)})$ or with non-radiative recombination due to thermal carrier excitation and subsequent diffusion towards the non-radiative centres $(1/\tau_{nr}^{(1)})$.

The last types of energy levels in disordered semiconductors are possible impurity levels in doped materials. These are located, similarly to crystalline counterparts, within the bandgap close below the relevant mobility edges (donors and acceptors in Fig. 9.1(a)).

9.2 Temperature dependence of luminescence

We have seen that the disorder does not introduce any new kinds of recombination processes fundamentally different from crystalline solids. However, the radiative recombination in the tail states exhibits a typical experimental dependence whose provenance can be simply derived. We start with the expression for luminescence quantum efficiency (3-2) by rewriting it as a relation suitable for the luminescence intensity as a function of temperature $I(T)$:

$$I(T) = I_0 \frac{p_r(T)}{p_r(T) + p_{nr}(T)} = I_0 \frac{1/\tau_r}{(1/\tau_r + 1/\tau_{nr})}. \tag{9.3}$$

Here p_r and p_{nr} stand for the probability of radiative and non-radiative recombination, respectively, and τ_r and τ_{nr} denote the corresponding recombination times. I_0 is a constant quantity whose meaning will be discussed below.

Expression (9.3) simply connects the luminescence intensity with its quantum efficiency η, namely, $I(T) = I_0\eta(T)$.

At a fixed temperature, the electron (or hole) in the tail state with energy E can, according to Fig. 9.1(b), either recombine radiatively $(1/\tau_r)$ or be excited thermally above the mobility edge and subsequently diffuse towards a non-radiative centre $(1/\tau_{nr}^{(1)} \equiv 1/\tau_{nr})$.[1] Thermal excitation causing luminescence quenching is described by the known relation (4.17) that now reads

$$1/\tau_{nr} = p_0 \exp(-E/k_B T)\,, \tag{9.4}$$

where $p_0 \approx 10^{12}\,s^{-1}$ is the so-called frequency factor. The electrons and holes in shallow states (E small) thus recombine mainly non-radiatively, while the photocarriers deep below the relevant mobility edge have no chance to 'escape non-radiatively' and rather they give rise to recombination luminescence radiation. It is thus possible to define a temperature dependent *demarcation energy* $E_D(T)$; carriers at this energy have the rate of radiative recombination just equal to the rate of non-radiative recombination given by (9.4):

$$1/\tau_r = p_0 \exp(-E_D(T)/k_B T)\,. \tag{9.5}$$

The significance of the demarcation energy therefore consists in separating light-emitting states from non-radiative ('dark') states.

Now we can express the quantum efficiency through the tail state densities as

$$\eta = \frac{p_r(T)}{p_r(T) + p_{nr}(T)} = \frac{\int_{E_D}^{\infty} g_v(E)\mathrm{d}E}{\int_0^{\infty} g_v(E)\mathrm{d}E} = \mathrm{e}^{-E_D(T)/E_{v0}}\,. \tag{9.6}$$

The integrals in (9.6) mean summing over all the tail states $(\int_0^{\infty} \mathrm{d}E)$ and over the states below the demarcation energy $(\int_{E_D}^{\infty} \mathrm{d}E)$. Note that we sum over the valence band states (9.2) only. This is justified by different widths of the tail states: The valence band tail states are usually wider than the conduction band ones $(E_{v0} > E_{c0})$. Consequently, even if both kinds of photocarriers located above the relevant demarcation energy may be thermally released, decisive for the efficiency of the non-radiative process is the release of deeper lying holes—the electrons liberated easily from their shallower states cannot recombine unless free holes, upon migrating through the sample, occupy the deep centres of non-radiative recombination.

Supposing further that η is very small compared with unity (i.e. $p_{nr}(T) \gg p_r(T)$, or $\eta = p_r/(p_r + p_{nr}) \approx p_r/p_{nr}$), expression (9.3) becomes

$$\frac{I_0}{I(T)} = \frac{I_0}{I_0 p_r/(p_r + p_{nr})} = \frac{p_r + p_{nr}}{p_r} = 1 + \frac{p_{nr}}{p_r} \cong 1 + \frac{1}{\eta}$$

[1] For the sake of simplicity we neglect here the (temperature independent) channel of direct non-radiative recombination $1/\tau_{nr}^{(2)}$.

or

$$\left(\frac{I_0}{I(T)} - 1\right) \cong \frac{1}{\eta},$$

which, with the aid of (9.6), yields the required expression for the characteristic temperature dependence of luminescence

$$\ln\left(\frac{I_0}{I(T)} - 1\right) = \frac{E_D(T)}{E_{v0}} \equiv \frac{T}{T_0}. \tag{9.7}$$

In (9.7) we have introduced, to a certain extent formally, the characteristic temperature T_0 via

$$E_{v0} = k_B T_0 \ln(p_0 \tau_r). \tag{9.8}$$

Such a definition of T_0 can be justified by a formal similarity of E_{v0} and $E_D(T)$, provided we express the latter quantity from eqn (9.5). The parameter T_0 characterizes the width of the tail states.

Therefore, plotting the experimentally acquired values of $\ln(I_0/I(T)-1)$ against T should yield—provided the above model is correct—a straight line with slope $1/T_0$. The curves in Fig. 9.2(a), measured in various samples of hydrogenated amorphous silicon, nicely demonstrate that this is the case [4]. The reciprocal slope value T_0, extracted from this figure, amounts to $T_0 = 22$–23 K. The parameter T_0 is very sensitive to the degree of disorder: In amorphous alloys, where, besides the topological disorder, also composition disorder occurs (e.g. in a-Si:H with high hydrogen content ≥ 10 at. %) the tail states are much wider. Figure 9.2(b) shows the dependence (9.7) for a-Si:H with 19 at. % of hydrogen [5]. The slope of the linear plot yields $T_0 = 62$ K. This means that, among others, the photocarriers after thermalization get localized in deep states and thus the thermal luminescence quenching is less effective. This corroborates Fig. 9.2(b)—it becomes evident that $I(T) \neq 0$ even at room temperature while in 'standard' a-Si:H with hydrogen content of about 5–10 at.% only, photoluminescence at $T = 295$ K in fact does not occur. The presence of room-temperature luminescence in amorphous alloys is supported probably also by a low carrier mobility, which makes the 'search' of released carriers for non-radiative centres more difficult.

Till now, we have not specified how to assess the constant I_0 and what is its meaning. This can be easily anticipated—since monotonic thermal luminescence quenching is involved with increasing temperature, as indicated, e.g. by the growth of the demarcation energy with temperature (9.5), I_0 should mean the luminescence intensity at the lowest temperature of measurement. (After all, it can be deduced from (9.7) that $I_0/I(T) > 1$ must always hold, for formal reasons.) Despite this, the assessment of I_0 may be ambiguous in some cases, because in disordered semiconductors the form of $I(T)$ at the lowest temperatures (about 4–70 K) is not always simple: Sometimes in this temperature range a 'plateau' is found; another time even a moderate increase of $I(T)$ with increasing temperature is observed. The question then arises which value of $I(T)$ to select as I_0. This selection may influence the left part of the curve $[I_0/I(T)-1]$, as becomes evident in Fig. 9.2(b) for the

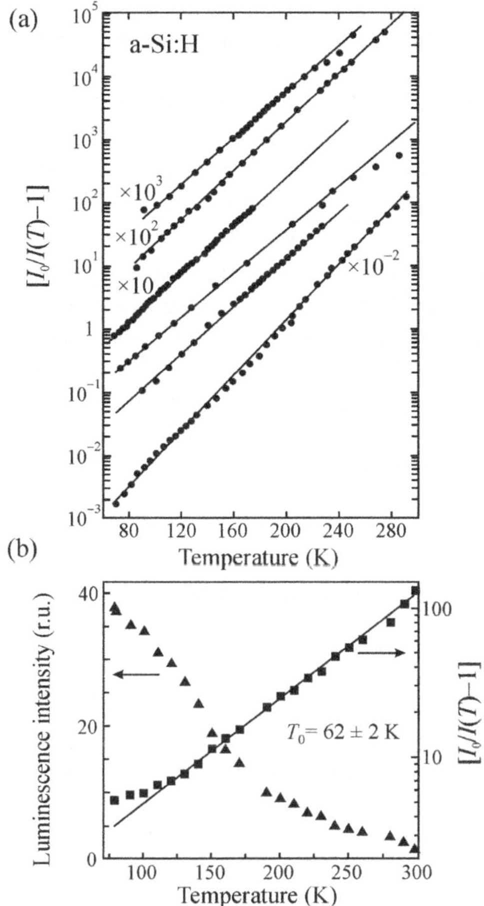

Fig. 9.2
(a) Plot of $[I_0/I(T)-1]$ against T in samples of a-Si:H prepared in various ways. After Collins *et al.* [4]. (b) The same plot (■) obtained in an a-Si:H alloy with high hydrogen content. Also the direct dependence of $I(T)$ on temperature is shown (▲). After Luterová *et al.* [5]. The symbols denote experimental data, and the lines are fits according to (9.7).

data at $T \leq 125$ K. For a more detailed discussion, the reader is referred to the literature [4]. Of course, I_0 is only a certain experimental parameter, not (unlike the fit parameter T_0) a characteristic material constant.

The model under discussion leads to the following consequence: The demarcation energy $E_D(T)$ increases with increasing temperature. Therefore, the luminescence emission spectrum as a whole (or its maximum) must shift to the red and, what is of the essence, this shift must exceed the temperature red-shift of the bandgap. Figure 9.3 shows that this is the case [6].

Let us now attempt to summarize all assumptions applied in deriving the temperature behaviour (9.7) and to determine the limits of its validity. First, as we have just stated, this model can be applied only in a temperature range where the luminescence intensity drops with increasing T. Besides, we have neglected the channel of direct non-radiative tunnelling ($1/\tau_{nr}^{(2)} = 0$). This, it is true, means implicitly that within the model the quantum efficiency η is equal to unity, provided the photocarriers thermalize into states lying below the demarcation energy. This may occur at sufficiently low temperatures when no carriers can be thermally excited to the free states, and thus all they can do

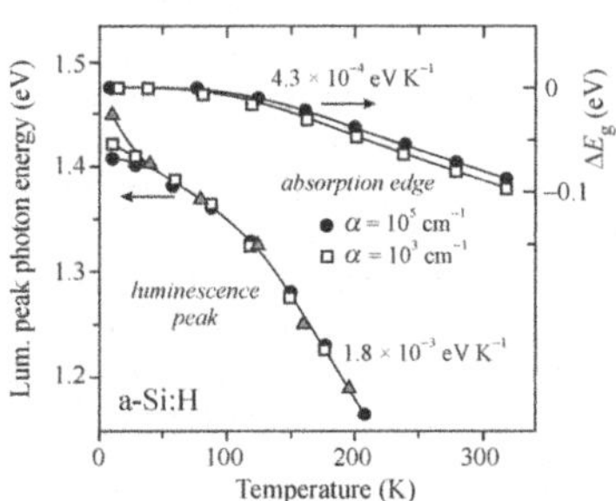

Fig. 9.3
Temperature shift of the emission band maximum in a-Si:H (with rate of 1.8×10^{-3} eV/K; different symbols denote different samples or different excitation modes) in comparison with the temperature shift of the absorption edge (4.3×10^{-4} eV/K). After Tsang and Street [6].

is to recombine radiatively. This fact, however, gets involved in a contradiction with the second assumption applied in deriving (9.7), namely, that $\eta \ll 1$. Is the model thus inconsistent?

It is better to say one should be cautious in making use of it. The mentioned contradiction may be removed if we admit $1/\tau_{nr}^{(2)} \neq 0$. The overall character of $I(T)$ remains unchanged, but the left-hand side of the definition (9.5) of the demarcation energy will comprise the low-temperature luminescence decay rate $1/\tau_L^0 = 1/\tau_r + 1/\tau_{nr}^{(2)}$ instead of the reciprocal radiative lifetime $1/\tau_r$. Similarly, τ_L^0 will substitute for τ_r in expression (9.8) for E_{v0}. We have to keep this in mind when deriving E_{v0} from the experimentally determined T_0, and also in a possible reciprocal procedure: if, for instance, we assess E_{v0} via optical absorption or photoelectric measurements (in addition to evaluating T_0 from luminescence experiments), we have to carefully analyse every time whether eqn (9.8) will yield τ_r or τ_L^0. This difference is not always sufficiently accentuated in the literature.

9.3 Distribution of luminescence lifetimes

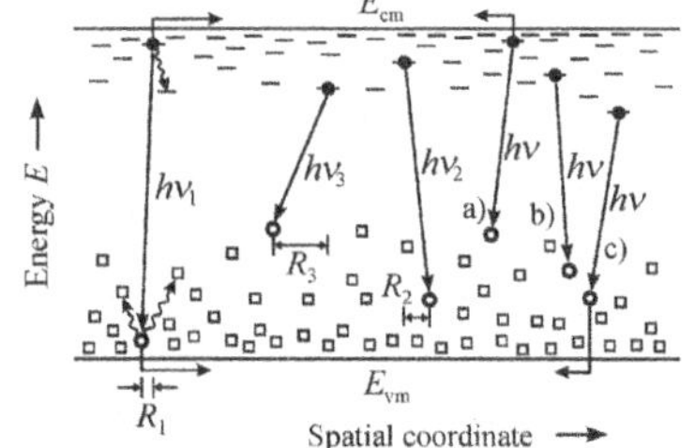

Fig. 9.4
Illustrating the distribution of luminescence decay times. The conduction band tail states are depicted by short horizontal lines, those of the valence band by squares (the tail state widths are exaggerated for the sake of clarity). Oblique arrows represent radiative recombination via tunnelling, horizontal arrows denote the escape of carriers towards non-radiative centres and the wavy lines indicate carrier thermalization.

In the preceding section we implemented for disordered semiconductors a single luminescence decay time τ_L, in full analogy with their crystalline counterparts. This is, however, an oversimplification that does not correspond to the facts. In disordered materials, generally speaking, a large range of luminescence decay times occurs, spanning from 10^{-9} s up to 10^{-1} s, in strong dependence upon temperature. The reason for this consists in the fact that the tail states, in which luminescence originates, are very broad and have a very variable spatial density.

The non-equlibrium electrons sitting in energy states close to the mobility edge E_{cm} can find in their close neighbourhood many 'deep' hole states (in proximity to E_{vm}) for tunnelling radiatively. That is, near the mobility edges the densities of states (9.1) and (9.2) acquire their maximal values per unit volume. Because the probability of radiative tunnelling can be expressed as $1/\tau_r \approx \exp(-2R/R_0)$, where R is the localized electron–localized hole separation and R_0 is a characteristic constant, this probability is large for small R. Therefore, high-energy luminescence photons $h\nu_1$, corresponding to a small distance R_1, will decay faster than photons $h\nu_2 < h\nu_1$ and $h\nu_3 < h\nu_2$ that belong to the tunnelling separations $R_2 > R_1$ and $R_3 > R_2$ (Fig. 9.4).

Furthermore, energetically higher lying carriers find more chance to thermalize (wavy lines) and also to escape non-radiatively into the free states (Fig. 9.4, left part). This, altogether, makes the high-photon-energy luminescence radiation decay faster that the low-energy radiation. Consequently, a continuous distribution of luminescence decay times $g(\tau_L)$ must exist; taking into account its considerable width, it is usually plotted as a function of the logarithm of the decay time $g(\log \tau_L)$.

The luminescence of every electron–hole pair localized in the tail states can certainly be regarded as a 'monomolecular' type of radiative recombination, i.e. according to (3.5) we shall consider the individual decay curves in the form

$$i_L(t) = i(0)e^{-t/\tau_L} = \frac{N_0\eta}{\tau_L}e^{-t/\tau_L}, \qquad (9.9)$$

where N_0 denotes the number of excited centres (*e–h* pairs with the decay time τ_L) and η stands for the luminescence efficiency. However, owing to the existence of $g(\log \tau_L)$, instead of (9.9) we now have to consider the sum over all possible τ_L:

$$i_L(t) = \eta N_0 \int_{-\infty}^{+\infty} d(\log \tau_L) g(\log \tau_L) \frac{\exp(-t/\tau_L)}{\tau_L}. \qquad (9.10)$$

An example of an experimentally acquired distribution $g(\log \tau_L)$ is shown in Fig. 9.5 [7].

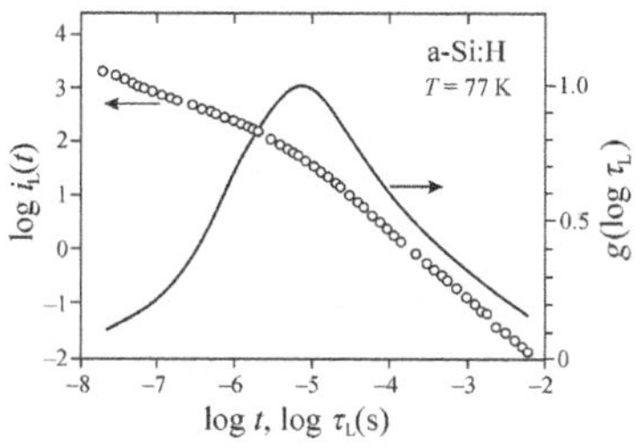

Fig. 9.5
Temporal shape of the decay of the spectrally integrated photoluminescence of a-Si:H at a temperature $T = 77$ K (symbols) and the relevant decay time distribution $g(\log \tau_L)$ (full curve) defined via (9.10). After Wang and Fritzsche [7].

Now we analyse how to determine the distribution function $g(\log \tau_L)$ experimentally. One of the possible ways is as follows. We have mentioned in Section 3.4 that the luminescence decay curve in disordered systems can be in most cases described phenomenologically by the temporal stretched exponential function

$$i_L(t) = i(0) \exp[-(t/\tau)^\delta], \qquad (9.9a)$$

where τ is called the decay time and $\delta \in (0, 1\rangle$ is the so-called dispersion factor. Now, let us recall the definition of the Laplace transform of a function $f(x)$

$$F(t) = \int_0^\infty f(x)e^{-tx}dx\,;$$

x is a real argument and $F(t)$ is called image of the Laplace transform. Substituting $x = 1/\tau$ yields

$$F(t) = \int_0^\infty f(1/\tau)\exp(-t/\tau)\frac{d\tau}{\tau^2}$$

and, if in (9.10) we write $d(\log \tau_L) \approx d\tau_L/\tau_L$ and replace formally $g(\log \tau_L) \rightarrow f(1/\tau_L)$, eqn (9.10) becomes

$$i_L(t) \approx \int_0^\infty f(1/\tau_L)\exp(-t/\tau_L)\frac{d\tau_L}{\tau_L^2}. \qquad (9.10a)$$

By comparing (9.9a) with (9.10a) it turns out that, under suitable normalization, one can write

$$\exp\left[-(t/\tau)^\delta\right] = \int_0^\infty f(1/\tau_L)\exp(-t/\tau_L)\frac{d\tau_L}{\tau_L^2}. \qquad (9.10b)$$

The last equation implies that, in order to reveal the distribution function $f(1/\tau_L)$, first of all we should determine τ, δ by virtue of fitting the experimental decay curve $i_L(t)$, and then we can calculate $f(1/\tau_L)$ as the inverse Laplace transform of the stretched exponential function.[2]

This procedure is applied whenever the distribution functions $g(\tau_L)$ or $f(1/\tau_L)$ are relatively narrow. An example will be given in Chapter 15 (Fig. 15.1). When the distribution function is sufficiently broad ($g(\log \tau_L)$), it can be determined by making use of a much simpler way, namely, directly from the measured decay curve $i_L(t)$ as $g(\log \tau_L) \simeq i_L(t)t$ (Problem 9/3).

The preceding discussion on the continuous character of $g(\log \tau_L)$ has not been fully exhausting. Even if we study the 'monochromatized' luminescence of disordered semiconductors, i.e. if we measure the decay time on a single selected wavelength, we obtain curves $g(\log \tau_L)$ similar to those displayed in Fig. 9.5, see Fig. 9.6(a) [8]. Referring to what has been said above this could hardly be expected (!) How, then, is it possible?

The answer is given in Fig. 9.4. The recombination of the *e–h* pairs (a) and (c) will be faster than that of the pair (b), because both the electron in case (a) and the hole in case (c) can be easily thermally excited into delocalized states and subsequently find a counterpart to recombine non-radiatively. At the same time the photon energy $h\nu$ may be in all three cases (a), (b) and (c) identical. Moreover, the random statistical distribution of the electron–hole separation R, occurring also for $h\nu = \text{const}$, will manifest itself.

In materials where the tail state widths E_{v0} and E_{c0} are large enough, the luminescence is present till room temperature. This is expected to be due to participation in the radiative recombination of predominantly those electrons and holes which are localized deepest in the bandgap. In this case we may expect a sizeable narrowing of $g(\log \tau_L)$. The inset in Fig. 9.6(b) [5] or its comparison with the curves shown in Fig. 9.6(a) demonstrate that this is indeed the case.

9.4 Spectral shape of the emission band

Figure 9.7 displays examples of photoluminescence emission spectra of several pure amorphous semiconductors [9–11]. All of these spectra exhibit a simple structureless band of considerable width (FWHM $\approx$ 200–400 meV). Its shape is approximately Gaussian—with a minor asymmetry towards low photon energies—and does not exhibit any indications of a potential fine structure.

To explain this spectral shape, firstly we shall attempt to apply some of the models discussed in previous chapters. However, because of the different microscopic essence and different density of the electronic states, participating in the radiative recombination in the amorphous and crystalline phases, it is not at all obvious whether we shall succeed. For instance, it stands to reason that the model of free *e–h* pairs (Section 5.2) is inapplicable, since now the light emission originates in *localized* states. Besides, the observed bandwidth is much larger than the corresponding Boltzmann factor $\sim k_B T$

[2] It is worth stressing that if the left-hand side of (9.10b) contains the stretched exponential function, the distribution $f(1/\tau_L)$ cannot be analytically expressed via simple functions such as the Gaussian distribution, etc.

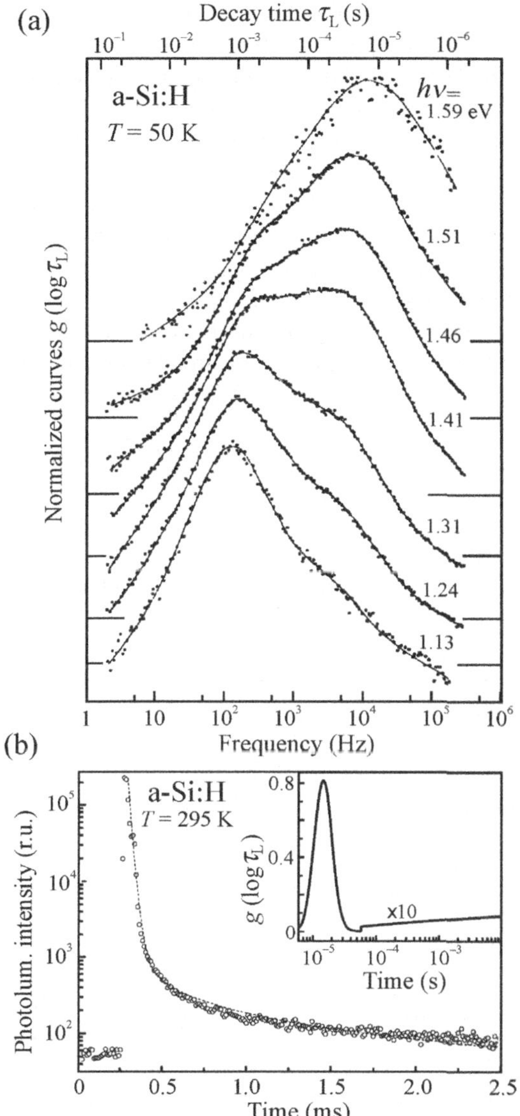

Fig. 9.6
(a) The distribution function $g(\log \tau_L)$ of photoluminescence of a-Si:H for various emitted photon energies at $T = 50$ K, after Oheda [8]. To acquire τ_L, the method of phase shift (Subsection 2.9.2) was used; the lower abscissa represents the modulation frequency of the excitation radiation. (b) The photoluminescence decay curve in an amorphous alloy a-Si:H with 19 at.% of hydrogen, room temperature. The inset demonstrates narrow distribution function $g(\log \tau_L)$. After Luterová *et al.* [5].

(≈ 25 meV at room temperature). For a similar reason we have to reject the radiative recombination of type e–A^0 or h–D^0 (Section 5.3). Excitons as the quasi-particles with a well-defined wavevector do not exist in amorphous materials, thus we are not allowed to apply any of the channels of the exciton luminescence, discussed in Chapter 7. The broad emission band could perhaps correspond to electron–hole liquid EHL (Section 8.4) or electron–hole plasma EHP luminescence (Section 8.5), but these are high-excitation effects while in disordered semiconductors this kind of light emission occurs independently of the excitation intensity and, moreover, it exhibits its specific temperature behaviour (Section 9.2) that is incompatible with the EHL or EHP models.

And yet, there remains one possibility. A broad structureless emission band can occur at the localized optical centre with a strong electron–phonon coupling (Section 4.5). In disordered semiconductors, the non-equlibrium carriers actually become localized before recombination and, to employ the configuration coordination concept, there is no need to meet the presumption about the long-term ordering (i.e. a regular crystalline lattice). This is the basic idea underlying the *phonon broadening* model, which has been submitted to interpret the broad emission band in a-Si:H by Street [12]. This concept therefore assumes that each *e–h* pair localized in the tail states has a strong phonon interaction with the surrounding matrix, which leads to a large Huang–Rhys factor $S \gg 1$ (or to a Debye–Waller factor $u_{DW} = 0$), to the absence of the no-phonon line and the appearance of a broad Gaussian emission band with a considerable Stokes shift. The bandshape is then described by (Appendix F)

$$I_{sp}^{a}(h\nu) \approx \text{const}\,\exp\left\{-\frac{[h\nu - (E_0 - E_R)]^2}{2\sigma^2}\right\}, \tag{9.11}$$

where E_0 is the transition no-phonon energy, $E_R = S\hbar\omega$ stands for the relaxation energy ($\hbar\omega$ being the phonon energy) and σ denotes the band halfwidth at the points of inflection.

Indeed, the emission band maximum in a-Si:H peaks at $\sim 1.4\,\text{eV}$ (Fig. 9.7), while the optical bandgap width amounts to $\sim 1.9\,\text{eV}$. The shift in the luminescence maximum from the absorption edge thus amounts to $2E_R \sim 0.5\,\text{eV}$. Enumerating the FWHM of the band, which is described by eqn (9.11), through relation (F-4), yields $\sigma_F = 4\sqrt{\ln 2 E_R \hbar\omega} \approx 370\,\text{meV}$ if we consider $E_R = 250\,\text{meV}$ and adopt $\hbar\omega = 50\,\text{meV}$ as a typical phonon energy. Everything

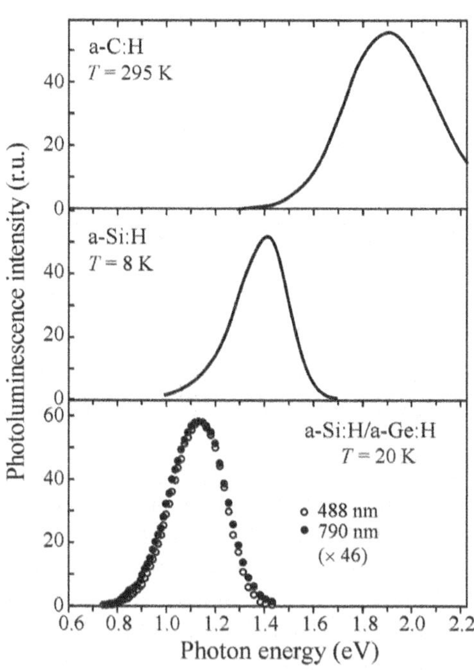

Fig. 9.7
Examples of the emission spectra of undoped hydrogenated amorphous semiconductors. From the top downwards: a-C:H (after Koós *et al.* [9]), a-Si:H (after Street and Biegelsen [10]) and multilayers a-Si:H (5 nm)/a-Ge:H (1 nm) (after Deki *et al.* [11]). The lowermost curve documents independency of the spectral shape of the excitation wavelength ($\lambda_{ex} = 488$ nm and $\lambda_{ex} = 790$ nm were applied). Notice also different measurement temperatures.

corresponds well to experimental data. The phonon broadening model thus reasonably describes the dominant spectral features, however, at the same time it involves implicitly the assumption about a fixed value of the no-phonon energy of all of the recombining *e–h* pairs (at most, the model can allow for a very narrow distribution of these energies only); otherwise an additional inhomogeneous broadening due to disorder would arise. This assumption is open to discussion and, in fact, denies to a large extent the effect of the finite width of the tail states. Moreover, the experimentally observed emission band is not fully symmetric but exhibits a slight low-energy tail.

Therefore, a diametrically opposite approach to the interpretation of the emission band—particularly in a-Si:H again—has been taken by Dunstan and Boulitrop [13, 14]. They have proposed a model of *disorder broadening* in which the tail states are supposed to be quite 'solid', i.e. without any electron–phonon interaction, and the sizeable emission bandwidth is interpreted as being due to a wide distribution of these states. Here, an idea might cross our mind, namely, that the authors copy the procedure we used previously to derive the spectral shape of the luminescence of free *e–h* pairs in indirect crystalline semiconductors (Subsection 5.2.2): the convolution of occupied electron and hole states, this time in the relevant tail states instead of in the conduction and valence bands.[3] It is thus tempting to adopt relation (5.10) in which we merely replace the square root functions by the exponential expressions (9.1) and (9.2).

However, we are not allowed to proceed in this way. Amorphous semiconductors, as already stressed several times, have their specific features. Above all, the tunnelling radiative lifetime τ_r is not constant but depends strongly on the electron–hole separation. The transition matrix element thus cannot be put outside the integral. In addition, in amorphous materials it is possible, no doubt, to introduce the Fermi–Dirac distribution function for electrons and holes, but not all the non-equilibrium *e–h* pairs are able to recombine radiatively (owing to the random spatial distribution of the tail states, see below). This is why the Fermi–Dirac function is not applicable to calculate the occupation of luminescence-active states. The crucial problem of the disorder broadening model is thus to determine the distribution function of carriers recombining radiatively in the tail states.

To this end, Boulitrop and Dunstan [13] suppose that a very fast thermalization of non-equilibrium carriers down to the lowest states within a critical volume (surroundings) V_c takes place before the recombination occurs. This volume is driven by the rate of hopping carrier diffusion, i.e. the rate of the non-radiative tunnelling transitions to lower unoccupied states. Let this neighbourhood contain N states (Fig. 9.8) and let us select one reference couple of *e–h* states. The probability p_n of another couple, next to the reference couple, having a lower energy (E_0-E) decreases exponentially with increasing

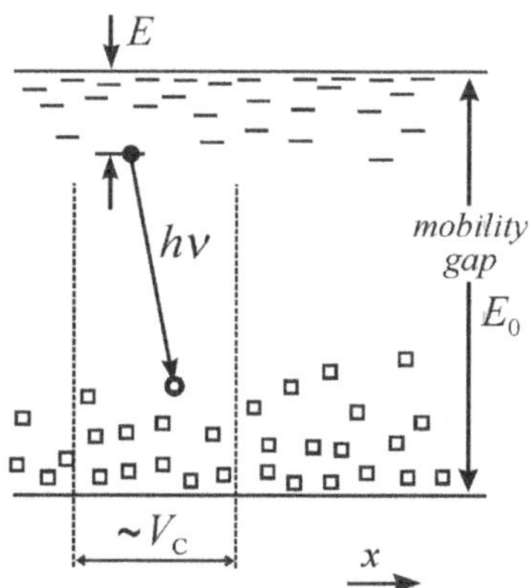

Fig. 9.8
Illustrating the calculation of the luminescence occupation probability of the tail states. In a volume V_c around an *e–h* pair with the lowest energy (i.e. with the maximum value of E) there are N other couples of states.

[3] Of course, now the wavevector concept and thus the notion of direct and indirect transitions are useless and, moreover, the recombining carriers are *localized*. Nevertheless, the *e–h* recombination is indirect in real space, because the electron and the hole cannot get localized on the same place; see Fig. 9.1(b) or Fig. 9.4.

E, i.e.

$$p_{\mathrm{n}} = \mathrm{e}^{-E/E_{\mathrm{c}0(\mathrm{v}0)}}.$$

This means that the probability of the occurrence of a neighbouring couple with a higher energy (E_0–E) is

$$p_{\mathrm{m}} = (1 - p_{\mathrm{n}}) = (1 - \mathrm{e}^{-E/E_{\mathrm{c}0(\mathrm{v}0)}}).$$

Should the number of such neighbours be N, then the occupancy probability of the relevant states will be

$$p_{\mathrm{ec(v)}}(E) = (1 - \mathrm{e}^{-E/E_{\mathrm{c}0(\mathrm{v}0)}})^N. \tag{9.12}$$

The density of occupied electron states $R_{\mathrm{c}}(E)$, as a product of the total density of states $g_{\mathrm{c}}(E)$ with the relevant occupation function p_{ec}, then reads

$$R_{\mathrm{c}}(E) = g_{\mathrm{c}}(E) p_{\mathrm{ec}}(E) \approx \mathrm{e}^{-E/E_{\mathrm{c}0}}(1 - \mathrm{e}^{-E/E_{\mathrm{c}0}})^N. \tag{9.13a}$$

Similarly, for the states occupied by holes we get

$$R_{\mathrm{v}}(E_0 - h\nu - E) \approx \mathrm{e}^{-(E_0 - h\nu - E)/E_{\mathrm{v}0}}[1 - \mathrm{e}^{-(E_0 - h\nu - E)/E_{\mathrm{v}0}}]^N \tag{9.13b}$$

and the shape of the emission spectrum is described, in analogy with (5.10), by the convolution

$$I_{\mathrm{sp}}^{\mathrm{a}}(h\nu) \approx \int_0^{E_0 - h\nu} R_{\mathrm{c}}(E) R_{\mathrm{v}}(E_0 - h\nu - E) \mathrm{d}E. \tag{9.14}$$

Let us note that expression (9.14) does not involve a distribution of radiative lifetimes; it holds for a single value of τ_{r}.

The material parameters of a-Si:H yield the value of the important parameter N to be $N \approx 50$ (see Problem 1/4). An arbitrary electron thus must not be combined with an arbitrary hole, unlike the recombination of free *e–h* pairs in an indirect bandgap crystalline semiconductor. The values of the occupied tail state densities, calculated with the aid of (9.13),[4] are shown in Fig. 9.9(a). The theoretical lineshape $I_{\mathrm{sp}}^{\mathrm{a}}(h\nu)$, obtained with the use of (9.14), is compared with two experimental spectra in Fig. 9.9(b). Obviously the calculated spectrum (curve c) follows well the experimental asymmetry (curves a, b) but is considerably narrower compared to the measured spectra. Taking into account the experimentally determined distribution of radiative lifetimes makes the calculation of $I_{\mathrm{sp}}^{\mathrm{a}}(h\nu)$ rather elaborate [14] and the result is represented by curve (d) in Fig. 9.9(b). This curve is wider than the theoretical spectrum (c), but still narrower than the experimental curves. Therefore, the 'disorder broadening' model does not do full justice to the microscopic nature of the luminescence band in disordered semiconductors, and clearly is to be combined with Street's phonon broadening model. A detailed analysis of the applicability of both models in the case of a-Si:H and related alloys can be found in [15].

Another way to improve the disorder broadening model consists in the following reasoning: The involved assumption about full thermalization of

[4] The approximation $[1 - \exp(-x)]^N \cong \exp[-N \exp(-x)]$ has been used.

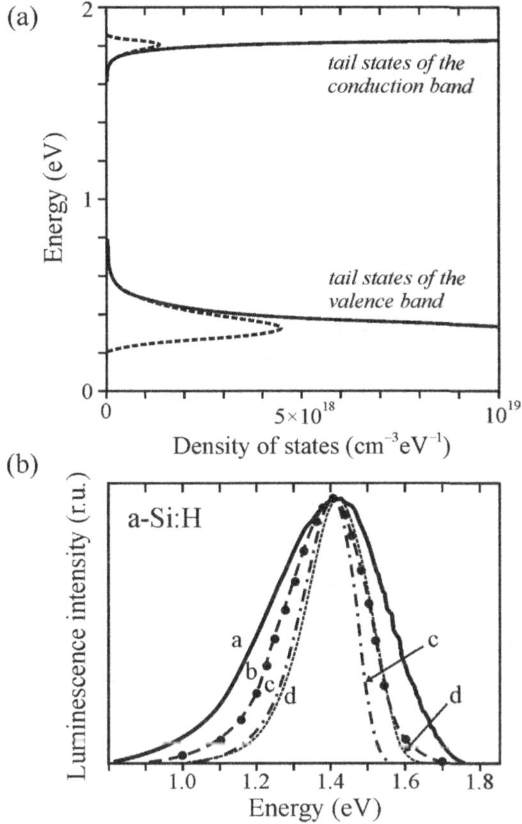

Fig. 9.9
(a) The densities of occupied tail states of the conduction and valence bands in a-Si:H (dashed lines) that participate in the radiative recombination. (b) Comparison of two experimental photoluminescence spectra of a-Si:H (curves a, b, $T = 8$ K) with the theoretical shape by Dunstan and Boulitrop, calculated both employing (9.14) for a single value of $\tau_r (= 10$ ms, curve c) and taking into account the experimental distribution function of τ_r with non-zero width (curve d). After Dunstan and Boulitrop [14].

carriers before recombining in fact discriminates the potential radiative recombination of 'hot carriers', represented by the left transition in Fig. 9.4 (and, to explain the broad distribution curve of lifetimes, the model offers merely a distribution of distances between the localized electrons and holes, like, after all, most other authors). If we admit the existence of the hot luminescence, then the theoretical curves in Fig. 9.9(b) will obviously broaden. The relevant mathematical description, however, would be rather complicated.

9.5 Some other properties of luminescence of disordered semiconductors

9.5.1 Correlation effects

Excitonic effects play an important role in the luminescence of crystalline solids. The Wannier exciton, a large-radius correlated electron–hole pair moving freely through the lattice, does not exist in disordered materials. However, one may think of an electron–hole correlation of another type. As the wavevector $\mathbf{k}$ of a non-equilibrium carrier is not a good quantum number, its uncertainty can be considered to be infinitely large, $\Delta\mathbf{k} \rightarrow \infty$, and the uncertainty relations then for the spatial coordinate $\mathbf{x}$ predict $\Delta\mathbf{x} \rightarrow 0$. One can thus, unlike the crystalline semiconductors, attribute to an electron and a hole a certain space coordinate. We have already met this concept in Figs 9.1 and 9.4 and when discussing the Boulitrop and Dunstan model. Many authors in this context consistently distinguish between recombination of the so-called *geminate* and *non-geminate e–h* pairs. Under the geminate pair one understands an electron–hole couple created through the absorption of a given photon in a given site (which means the recombining electron and hole are correlated to a certain extent). The non-geminate pair is represented by an electron and a hole that have originated in different parts of the sample as a consequence of two separated photon absorption events.

Because the photocarrier diffusion length in disordered semiconductors is short, obviously the recombination of geminate pairs dominates at low optical excitation levels when the created *e–h* pairs are sufficiently separated from each other, while at high excitation levels the *e–h* pairs may 'mix up' and the non-geminate recombination may no longer be negligible. One of the central issues of the luminescence of amorphous semiconductors has been for a long time the question of whether radiative recombination (i.e. the tunnelling in the tail states) takes place in arbitrary pairs or in geminate pairs only. Without going into details (we refer to the monograph by Street [1]), at present it is believed that it is the geminate pairs that dominate in luminescence while the annihilation of the non-geminate pairs represents a channel of non-radiative recombination and can be detected for instance via light-induced electron spin resonance.

It is worth noting, however, that Dunstan and Boulitrop in their model [14] do not explicitly differentiate between geminate and non-geminate recombination (even if it can be deduced that implicitly they had in mind the recombination of geminate pairs only). Besides, in the injection electroluminescence, as a matter of fact all *e–h* pairs ought to be considered non-geminate and, consequently, the electroluminescence of amorphous semiconductors would not occur at all. At the same time the very occurrence of electroluminescence in a-Si:H p-i-n structures has been established beyond any doubt for a long time [16] and, even if its intensity is comparatively low, it is applied for instance to studying the microscopic mechanisms of photocarrier recombination and their relations to charge transport processes [17]. Moreover, the electroluminescence of LED structures based on the p-i-n junction in amorphous hydrogenated silicon carbide a-SiC:H exhibits at room temperature a relatively high quantum efficiency ($\sim 10^{-3}\%$) and high brightness (of the order of $10\,\mathrm{cd/m^2}$). The emission is situated in the visible region and these devices have been considered potential candidates for optoelectronic light sources [18]. All these facts are hard to explain if we accept unreservedly the non-radiative recombination of non-geminate pairs.

9.5.2 Non-radiative recombination

Multiple passages in this chapter have mentioned possible paths of the non-radiative recombination of *e–h* pairs in amorphous semiconductors. These have been:

1. A recombination mediated by the dangling bonds, i.e. the Shockley–Read recombination at defect levels located approximately in the middle of the bandgap;
2. thermal quenching;
3. recombination of non-geminate *e–h* pairs.

Two additional mechanisms (occurring also in crystalline solids) should be attached, namely:

4. Auger recombination;
5. surface recombination.

Auger recombination gains importance at high excitation intensities (Subsection 6.1.2). Figure 9.10 [19] shows how the relative photoluminescence quantum efficiency η_L in a-Si:H and the relevant decay time (τ_L) distribution depend on the excitation photon flux ϕ; obviously for ϕ above a certain level a marked decrease in the luminescence quantum efficiency (not intensity!) occurs and, simultaneously, the decay time gets shorter. This undoubtedly indicates the opening of a new channel of non-radiative recombination, which may be plausibly identified with the Auger process (Problem 9/5): under comparatively high excitation, not only the deep tail states but also shallow localized states close to the mobility edge will be occupied. The latter are characterized by a high spatial density and are thus located close to each other. Under these circumstances the energy of such a recombining *e–h* pair may be transferred either to an electron or to a hole residing in the close neighbourhood. The rate of such an Auger process may even be increased in

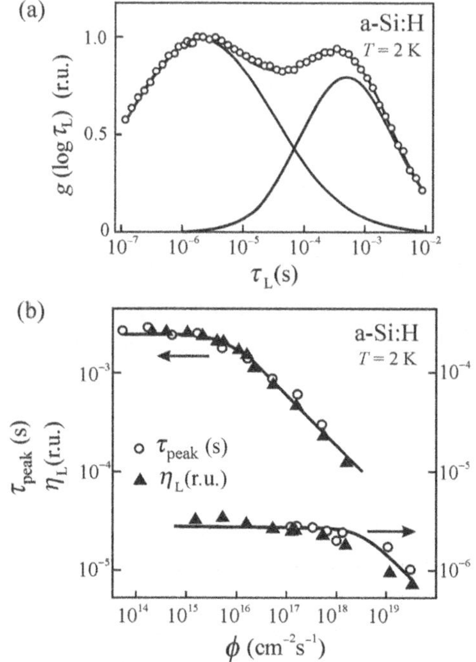

Fig. 9.10
(a) The distribution of photoluminescence decay times τ_L, in a-Si:H features, at a very low temperature (here $T = 2$ K), two maxima (see also Fig. 9.6(a)). (b) Plot of the temporal location of these maxima and of the relative luminescence quantum efficiency η_L versus the excitation photon flux ϕ. It can be seen how, under high excitation, τ_L gets shorter and η_L is reduced. After Stachowitz *et al.* [19].

comparison with crystalline counterparts because of the breakdown of the **k**-conservation rule.

There is, however, also an alternative point of view. Stachowitz and co-workers [19] have concluded, by analysing the experimental data in Fig. 9.10 that, instead of the Auger process, it is the non-radiative recombination of non-geminate pairs which is responsible for the observed drop of τ_L and η_L. They proceed even further, rejecting even the concept of geminate pair radiative recombination and proposing the only possible luminescence centre in a-Si:H to be *e–h* pairs with extremely small separation R between the electron and the hole ($R \sim 0.5$–1 nm), therefore a kind of 'small-radius localized exciton'. The occurrence of two maxima in the luminescence decay time distribution (Fig. 9.10(a)) they propose to interpret via recombination of singlet and triplet 'excitons' (whose existence is based on the exchange interaction between spins of closely spaced carriers in the 'exciton'). However, neither can this opinion be easily reconciled with the existence of electroluminescence mentioned in Subsection 9.5.1.

Surface recombination in a thin subsurface layer, owing to the presence of specific unsaturated bonds near the surface, was treated in Subsection 6.1.1. Microscopic features of this non-radiative recombination remain unchanged in disordered materials but, because the thickness of the luminescence 'dead' layer close to the surface depends on the photocarrier diffusion coefficient D which is small in disordered semiconductors, the thickness of this non-luminescent layer $L = \sqrt{D\tau}$ is also very small, typically $L \sim 0.01\ \mu\text{m}$ at low temperatures. Thus, although the concentration of the subsurface defects is large, the impact of the surface non-radiative recombination on the total luminescence intensity reduction is small ($\sim 10\%$).

Examining once more all of the five mechanisms of non-radiative recombination put forward above, we notice that from the microscopic point of view items (1), (2), (3) and perhaps even (5) can be reduced to a shared basis, which is the occurrence of dangling bonds: In thermal quenching, in surface recombination and perhaps even in non-geminate pair non-radiative recombination, the final dissipation stage of the electronic excitation energy proceeds via a deep defect level. Both kinds of photocarriers meet there in the end, upon accomplishing the diffusion motion, provided they did not find a way to recombine radiatively. A full consensus does not exist in this respect (an alternative chance for the non-geminate pairs is to recombine through direct non-radiative tunnelling between localized states), in any case, however, the role of the dangling non-saturated bonds in non-radiative recombination is substantial.

9.5.3 Luminescence of impurities and defects

Till now we have occupied ourselves with the luminescence of pure disordered semiconductors. It is legitimate to ask whether doping of amorphous semiconductors with donors and acceptors can induce novel luminescence processes, similarly to the crystalline analogues (e.g. e–A^0, h–D^0, etc.). Even though the donor and acceptor energy levels in amorphous materials are broadened due to

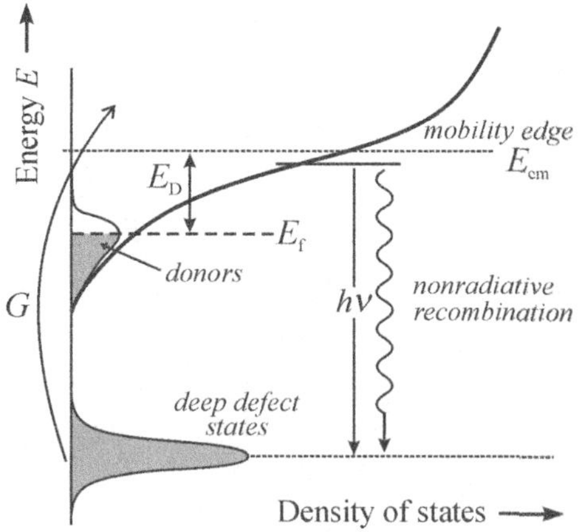

Fig. 9.11
Sketch of the density of the donor states (beneath the conduction band bottom, E_D stands for the donor binding energy) and of the deep defect states located approximately in the middle of the bandgap. One conceivable transition representing the defect-related luminescence ($h\nu \sim 0.8$–0.9 eV in a-Si:H) is shown. E_f denotes the Fermi level; G indicates the photogeneration rate of non-equilibrium electrons. The shadowed states are occupied in a non-excited semiconductor.

disorder (there is a distribution of the binding energies E_D, E_A), as indicated in Fig. 9.11, one can certainly imagine for instance a recombination of the type h–D^0, when an electron from some of the donor levels recombines radiatively with a hole residing in the valence band tail states.

It has been established experimentally that, e.g., a-Si:H doped with phosphorus or boron actually exhibits a new emission band in the near-infrared region (at ~ 0.8–0.9 eV against the 'intrinsic' emission peaked at ~ 1.4 eV), whose intensity is linked with the presence of the dopants, but the band is due to luminescence of deep defect levels rather than due to transitions e–A^0 or h–D^0. The introduction of dopants produces new dangling bonds and, accordingly, corresponding deep defect levels roughly in the middle of the band appear. These levels, as we have already stressed several times, mediate predominant non-radiative recombination, but if their concentration is sufficiently increased due to doping, they can manifest themselves through a weak but measurable luminescence (Fig. 9.11). This luminescence therefore originates owing to the radiative recombination of a photocarrier that is localized in the tail states with an oppositely charged photocarrier trapped at the defect level.[5] It should be stressed, of course, that the overall intensity of this long-wavelength emission amounts to a few percent of the principal luminescence band intensity only; the defect levels predominantly quench the 'intrinsic' luminescence and only then substitute it partially by a new channel of radiative recombination.

Figure 9.12(a) displays emission photoluminescence spectra of a-Si:H doped with phosphorus [20]. Obviously, the defect luminescence at ~ 0.85 eV can be excited, in line with Fig. 9.11, even by absorption of photons whose energy is considerably less compared to the bandgap width ($E_g \approx 1.5$ eV)). Since the band at ~ 0.85 eV is fairly large, the relevant transition has been suggested to be characterized by a strong electron–phonon interaction. This is qualitatively illustrated by the configurational coordinate model sketched

[5] The doping has to be relatively high ($> 10^{17}$ cm^{-3}). The Fermi level thereby shifts from the middle of the bandgap close to the relevant mobility edge, and charged dangling bonds along with their correlation energies enter the play. Details go beyond the scope of this book, see Street, R. A. (1991). *Hydrogenated Amorphous Silicon*. Cambridge University Press, Cambridge.

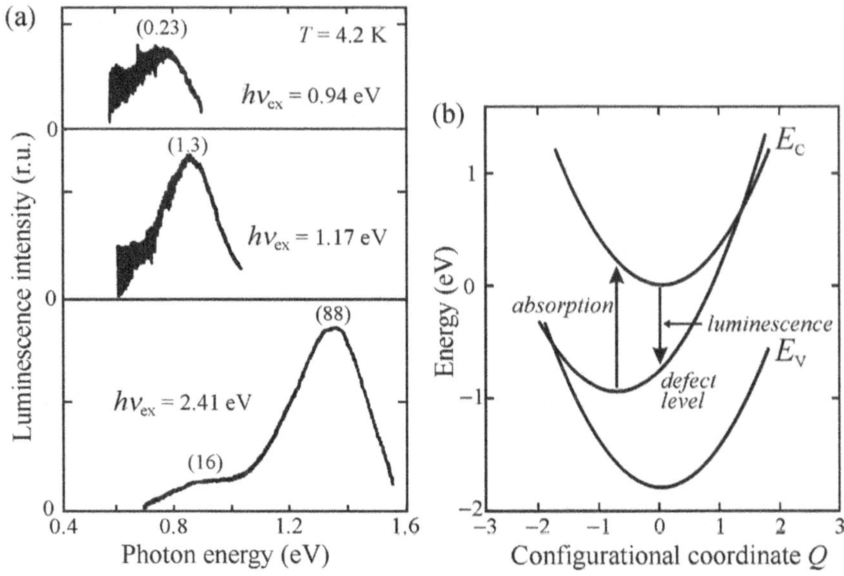

Fig. 9.12
(a) Low-temperature emission spectra of a-Si:H doped with phosphorus ($\sim 10^{18}$ cm^{-3}), under various excitation photons $h\nu_{ex}$. Numbers in parentheses denote relative intensities of the maxima. (b) Configurational coordinate diagram for the defect luminescence in a-Si:H:P. After Tajima *et al.* [20].

in Fig. 9.12(b). A note may be in order pointing to the resemblance with Fig. 6.1(c), which elucidates an analogous process in a localized centre of a crystalline phosphor.

However, the question of why radiative processes like (e–A^0) or (h–D^0) have not been observed in amorphous semiconductors is still to be answered here. The reason lies again in the unique feature of these materials—the occurrence of the tail states. From the luminescence point of view, the donor and acceptor states in an excited sample become in fact indistinguishable from the overlapping tail states, 'fusing' with them (see Fig. 9.1(a) or 9.11). This implies that the potential manifestation of their radiative recombination—if any—is included in the main emission band at $\sim$ 1.4 eV. Provided the concentration of donors or acceptors is very high ($\sim 10^{19}$–10^{20} cm^{-3}), comparable to or even higher than the total density of the tail states, one could perhaps expect some structure superimposed on this emission band. In this case, however, complete quenching of the $\sim$ 1.4 eV luminescence due to the high concentration of the defects mostly prevails.

9.5.4 Luminescence 'fatigue'

Photoluminescence in amorphous semiconductors may be accompanied by an undesirable adverse effect: gradual fading away of the main emission band, which is observable mainly under relatively high excitation. This is, as generally accepted, due to the Staebler–Wronski effect or the light-induced formation of the dangling bonds; these bonds represent, as already stressed, powerful quenching centres. An example is displayed in Fig. 9.13, which shows luminescence decay curves of a-Si:H, measured before and after illuminating the sample with an intense cw argon-ion laser (514 nm; 0.54 W; 45 min) [21].[6] A decrease of the luminescence intensity upon illuminating down to about 60 % of the initial value is evident, along with decay shortening. This corresponds to an increased rate of non-radiative transitions. It is worth mentioning that similar experiments performed in nominally pure a-Si:H revealed a moderate increase in intensity of the weak defect luminescence at ~ 0.85 eV [22], which fits well the concept of the Staebler–Wronski effect as a phenomenon caused by enhancement of the dangling bond density. Annealing of the sample at a temperature of 150–200°C for one hour usually leads to a complete recovery of the initial photoluminescence intensity.

It is well known that the application potential of amorphous semiconductors has overtaken—already a long time ago—our detailed microscopic understanding of their physical and chemical properties. The present knowledge of the luminescence processes confirms this state of the art—we have been emphasizing throughout this chapter that, as yet, no consensus has been achieved about many issues of both radiative and non-radiative recombination.

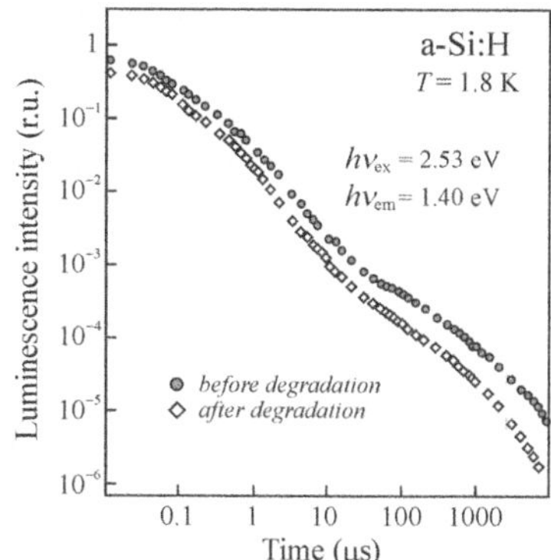

Fig. 9.13
Photoluminescence decay curves in a-Si:H (the emission band at ~ 1.4 eV, energy of excitation photons $h\nu_{ex} = 2.53$ eV, pulsed excitation 10 ns). Circles and diamonds denote measurements before and after light-induced degradation, respectively. After Hirabayashi *et al.* [21].

9.6 Problems

9/1: In expression (9.6) for η it has been tacitly supposed that all the tail states are occupied, otherwise in the integrals figuring in (9.6) an occupation function f would have appeared (here we apply $f = 1$). Prove that in case this supposition is not valid, one obtains the resulting temperature dependence formally identical with eqn (9.7), only instead of T a temperature $T' > T$ will appear. Determination of the slope $1/T_0$ therefore is not affected.

9/2: Assess the tail state width E_{v0} in hydrogen-rich a-Si:H, on the basis of the temperature dependence shown in Fig. 9.2(b). Values of the parameters p_0, τ_r can be found throughout the text of Chapter 9.

9/3: Show that (a) if the distribution of luminescence decay times $g(\log \tau_L)$ is very narrow, then expression (9.10) transforms into (9.9), while (b) if $g(\log \tau_L)$ is very broad, it can be approximated as

$$g(\log \tau_L) = g(\log t) \approx i_L(t)t,$$

or simply via multiplying the luminescence intensity $i_L(t)$ by the relevant time decay t related to the end of the excitation event. (For example the curve $g(\log \tau_L)$ in Fig. 9.5 has been determined in this way.) Start from eqn (9.10).

9/4: By making use of [13] explain how to estimate the value of the parameter N appearing in eqns (9.12)–(9.14), which represents the number of high-energy $(E_0–E)$ tail states in the neighbourhood of a particular lower-energy site with a trapped thermalized photocarrier.

[6] The photoluminiscence decay by itself was measured under excitation with a low-peak-power pulsed laser (0.8 kW) in order to avoid distortion of the results owing to the fatigue effect.

9/5: With the aid of the equation $dn/dt = G - n/\tau_r - \beta n^2$ comprising Auger non-radiative recombination prove that, at a very high photogeneration rate G, drops both in the luminescence decay time and quantum yield occur, driven by a proportion $\sim G^{-1/2}$. Comment: A specific 'extrinsic' Auger recombination of bimolecular type (see eqn (6.8)) in disordered semiconductors is considered.

References

1. Street, R. A. (1991). *Hydrogenated Amorphous Silicon*. Cambridge University Press, Cambridge.
2. Staebler, D. and Wronski, C. (1977). *Appl. Phys. Lett.*, **31**, 292.
3. Kudrna, J., Malý, P., Trojánek, F., Štěpánek, J., Lechner, T., Pelant, I., Meier, J., and Kroll, U. (2000). *Mater. Sci. Eng. B*, **69–70**, 238.
4. Collins, R. W., Paesler, M. A., and Paul, W. (1980). *Solid State Comm.*, **34**, 833.
5. Luterová, K., Pelant, I., Fojtík, P., Nikl, M., Gregora, I., Kočka, J., Dian, J., Valenta, J., Malý, P., Štěpánek, J., Poruba, A., and Horváth, P. (2000). *Phil. Mag. B*, **80**, 1811.
6. Tsang, C. and Street, R. A. (1979). *Phys. Rev. B*, **19**, 3027.
7. Wang, W. and Fritzsche, H. (1989). *Photoluminescence in a-Si:H films and multilayers*. In *Advances in Semiconductors*, Vol. 1, *Amorphous Silicon and Related Materials*, Vol. B (ed H. Fritzsche), p. 779. World Scientific, Singapore.
8. Oheda, H. (1993). *J. Non-Crystal. Solids*, **164–166**, 559.
9. Koós, M., Pócsik, I., and Tóth, L. (1993). *J. Non-Crystal. Solids*, **164–166**, 1151.
10. Street, R. A. and Biegelsen, D. K. (1980). *Solid State Comm.*, **33**, 1159.
11. Deki, H., Miyazaki, S., Ohmura, M., and Hirose, M. (1993). *J. Non-Crystal. Solids*, **164–166**, 841.
12. Street, R. A. (1978). *Phil. Mag. B*, **37**, 35.
13. Boulitrop, F. and Dunstan, D. J. (1983). *Phys. Rev. B*, **28**, 5923.
14. Dunstan, D. J. and Boulitrop, F. (1984). *Phys. Rev. B*, **30**, 5945.
15. Searle, T. M. and Jackson, W. A. (1989). *Phil. Mag. B*, **60**, 237.
16. Pankove, J. E. and Carson, D. E. (1976). *Appl. Phys. Lett.*, **29**, 620.
17. Wang, K., Han, D., Kemp, M., and Silver, M. (1993). *Appl. Phys. Lett.*, **62**, 157.
18. Kruangam, D. (1991). *Amorphous and microcrystalline silicon–carbide alloy light–emitting diodes: physics and properties*. In: *Amorphous & Microcrystalline Semiconductor Devices* (ed J. Kanicki), p. 195. Artech House, Boston.
19. Stachowitz, R., Schubert, M., and Fuhs, W. (1998). *J. Non-Crystal. Solids*, **227–230**, 190.
20. Tajima, M., Okushi, H., Yamasaki, S., and Tanaka, K. (1986). *Phys. Rev. B*, **33**, 8522.
21. Hirabayashi, I., Morigaki, K., and Nitta, S. (1980). *Jap. J. Appl. Phys.* **19**, L357.
22. Tajima, M., Okyama, H., Okushi, H., Yamasaki, S., and Tanaka, K. (1989). *Jap. J. Appl. Phys.*, **28**, L1086.

Stimulated emission

10

Stimulated emission can be viewed as a special case of luminescence. Its character conforms to the definition of luminescence (the excess electromagnetic radiation when compared to the equilibrium radiation as described by Planck's law, featuring simultaneously a 'sufficiently long' lifetime), although the thermodynamic conditions necessary for its occurrence differ fundamentally from the common occupancy of electronic states in spontaneous luminescence processes. A population inversion of energy levels, a state diametrically opposed to thermodynamic equilibrium, needs to be achieved. That is why, among other things, the spectral distribution of stimulated emission cannot be described using relations (5.9), (7.11) or (7.16), as these relations are valid provided that only minute variations from the equilibrium occupancy of states occur. It is thus obvious that the description of stimulated emission needs to be based on a completely different principle. Apart from the thermodynamic non-equilibrium, the nonlinear character of this optical process starts to be of particular importance.

In this chapter, we will first formulate the conditions necessary for the occurrence of stimulated emission in bulk semiconductors. Then, we will show how stimulated emission can originate in an electron–hole plasma (EHP) and discuss possible contributions of excitons and excitonic complexes to stimulated emission. Finally, the methods employed for the study of optical gain will be described.

10.1 Spontaneous versus stimulated emission. Optical gain

The phenomenon of stimulated emission is generally unconsciously perceived as the fundamental principle underlying the operation of all lasers. In luminescence spectroscopy, stimulated emission can be treated from two diametrically opposed points of view. If a particular phosphor is studied as a potential new active laser medium, any manifestation of stimulated emission is warmly welcome. If, on the other hand, importance is put on the basic research of the fundamentals of radiative recombination processes, should (unintentionally) stimulated emission arise, the emission spectrum or luminescence dynamics

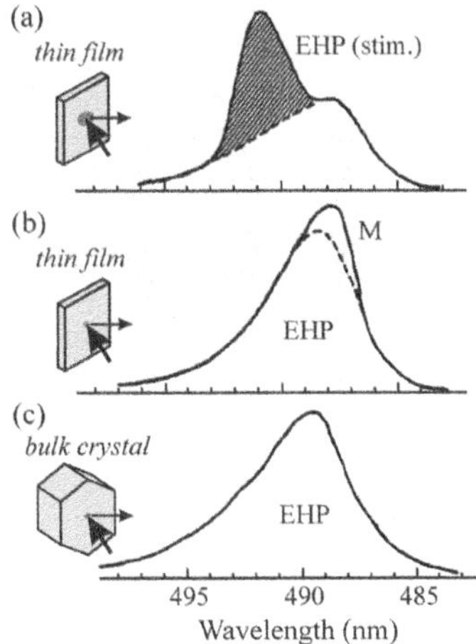

Fig. 10.1
Emission spectra of an electron–hole plasma (EHP) in various kinds of CdS crystals with different sizes of the excited area (bath temperature 4.2 K, excitation with a nanosecond dye laser 470 nm). (a) Thin film, 0.2 mm, excited area $0.8 \times 0.8\,\text{mm}^2$. (b) Thin film, 0.2 mm, excited area $0.1 \times 0.1\,\text{mm}^2$. (c) Bulk crystal with thickness of 5 mm, excited area $0.8 \times 0.8\,\text{mm}^2$. Excitation intensity was $5\,\text{MW/cm}^2$ in all cases. After Yoshida *et al.* [1].

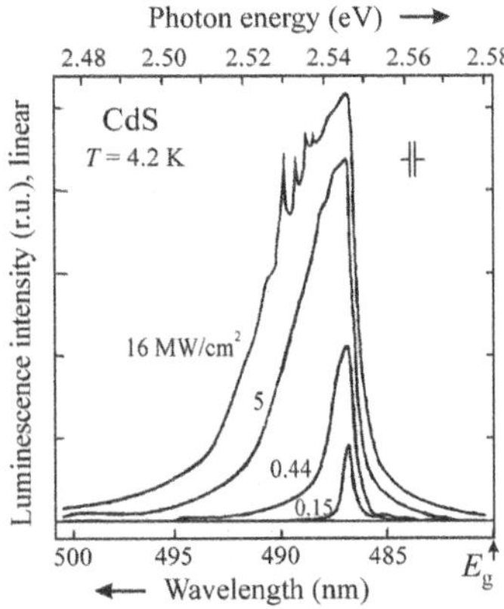

Fig. 10.2
EHP emission spectra of thin-film CdS. Excitation intensities of a pulsed nitrogen laser (337 nm, 10 ns) are noted at the corresponding curves. Although the excited area is extremely small, approximately $10 \times 10\,\mu\text{m}^2$, an indication of several lasing modes appears at $16\,\text{MW/cm}^2$. After Lysenko *et al.* [2].

can be influenced in such a way that they give rise to the formulation of an incorrect model of the luminescence centre and to erroneous conclusions. This can easily happen when measurements are carried out at low temperatures and in direct-bandgap semiconductors, which typically exhibit high radiative recombination rates. Consequently, it is indispensable to know, in the first place, what experimental conditions are favourable for making the occurrence of stimulated emission in experiment highly probable, and thus what measures can be taken to prevent it if necessary.

An immensely important role in this context is played even by the shape of the sample and the applied geometry of the experiment (Fig. 10.1) [1]. If the excited spot is a relatively large (diameter ≈ 1 mm) area of a sample with a surface of high optical quality (e.g. a semiconductor thin film, either free-standing or on a polished substrate), the spontaneous luminescence, travelling in the plane of the sample within the excitation spot, will very likely stimulate the emission of other photons. This situation is illustrated in Fig. 10.1(a), where the hatched area demonstrates the contribution of stimulated emission from an electron–hole plasma (EHP) in a thin CdS slab with the excited area of $0.8 \times 0.8\,\text{mm}^2$ at $T = 4.2$ K. The spontaneous EHP emission itself is represented by a broad background band. The contribution of stimulated emission can even be enhanced if the detection system also collects the photons emitted from the edge of the sample (the waveguiding effect in the 'reflection' geometry at 90° angle mentioned in Section 2.1).

Reducing the excitation area to $0.1 \times 0.1\,\text{mm}^2$ (Fig. 10.1(b)) obviously results in the suppression of EHP stimulated emission, however, a new M line appears due to the radiative decay of excitonic molecules. On the other hand, if a bulk crystal of relatively large dimensions with irregular edges is used, pure EHP spontaneous emission can be detected (Fig. 10.1(c)).

It is important to point out the fact that at very high excitation fluences even a radical decrease in size of the excited area to the order of $10 \times 10\,\mu\text{m}^2$ does not necessarily prevent the onset of stimulated emission, as is demonstrated in the top curve in Fig. 10.2; this graph again displays low-temperature light emission of a CdS thin film under pulsed laser excitation [2].

Let us come back for a short while to Fig. 10.1. The identification of the hatched area in the spectrum with actual stimulated emission is definitely not straightforward; the assignment might be made clearer, however, by employing, e.g., time-resolved luminescence spectroscopy. The results of such measurements, carried out on an EHP in a material related to cadmium sulphide, namely in cadmium selenide CdSe, are shown in Fig. 10.3. Whereas the time-integrated spectrum (top curve) comprises both spectral components (i.e. those originating in both spontaneous and stimulated emission), spectra measured at different time delays after the 20-ps excitation pulse confirm that the long-wavelength component decays much faster, which is a strong indication of stimulated emission [3].

Our goal for now will be to formulate a relation describing the spectral lineshape or, more specifically, the shape of the emission band corresponding to stimulated emission in semiconductors. First, we will make a short digression into atomic spectroscopy to recall the basic concepts of spontaneous and stimulated transitions in the optical region. Based on an analogy with a two-level atomic system, the concept of optical gain will be introduced.

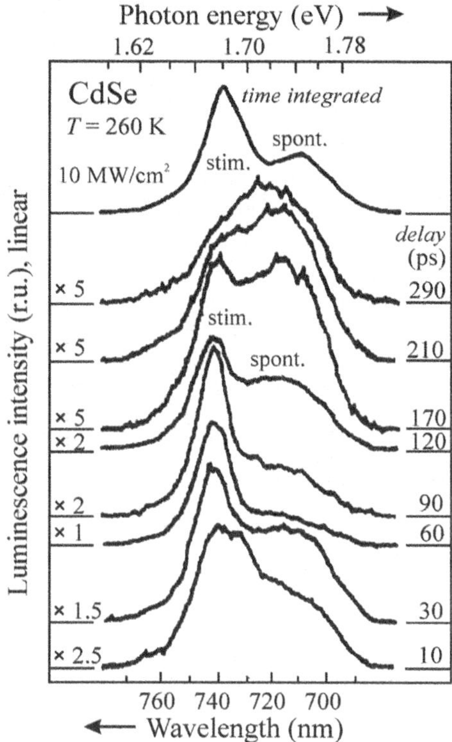

Fig. 10.3
Time-resolved emission spectra of EHP in a thin-film CdSe slab (0.3 mm). Excitation was performed using a pulsed Nd:YAG laser system (532 nm, 20 ps, diameter of the beam 2–3 mm); excitation intensity was 10 MW/cm^2. A relatively high bath temperature of 260 K was chosen so that the contributions of spontaneous and stimulated emission were more or less comparable; at higher temperatures, the conditions necessary to achieve stimulated emission can be met only with difficulty, while at sufficiently low temperatures the stimulated-emission contribution dominates. After Yoshikuni *et al.* [3].

This concept, along with the relevant emission spectra, is commonly exploited in spectroscopy for characterizing active laser media; in many cases, especially in semiconductor lasers, gain is used even more frequently than the spectrum. To a large extent, we will follow the reasoning used in [4] and [5].

In a two-level atomic system (Fig. 10.4(a)), the rate of transitions from level $|2\rangle$ to level $|1\rangle$, accompanied by spontaneous emission of a photon $h\nu$ and being independent of external conditions, is equal to the Einstein coefficient of spontaneous emission $A_{21} = 1/\tau_{sp}$ (s^{-1}), where τ_{sp} is the spontaneous lifetime of the level $|2\rangle$.[1] Stimulated emission between levels $|2\rangle$ and $|1\rangle$, as well as the inverse transition $|1\rangle \rightarrow |2\rangle$ (i.e. the absorption of a photon $h\nu$), are conditioned by the presence of a resonant radiation with the spectral energy density $u(\nu)(\mathrm{J/cm^3 s^{-1}})$. The corresponding rates of optical transitions can then be expressed as

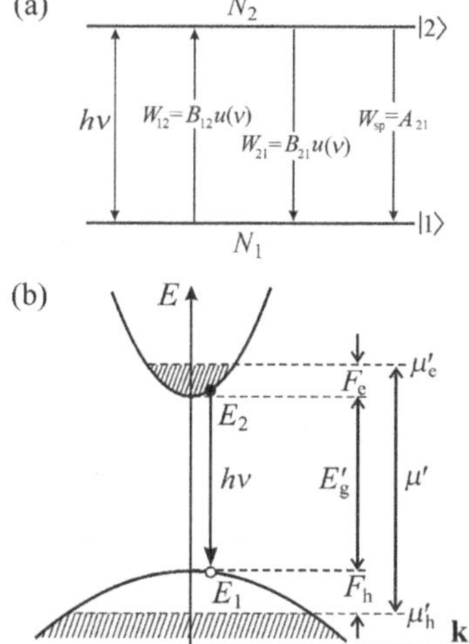

Fig. 10.4
Schematic of possible optical transitions (a) in a two-level atomic system and (b) in a direct-bandgap semiconductor close to the absorption edge (in the absence of excitonic effects).

$$W_{21} = B_{21}u(\nu) \quad (s^{-1}) \qquad \text{(stimulated emission)}$$

and

$$W_{12} = B_{12}u(\nu) \quad (s^{-1}) \qquad \text{(absorption)},$$

[1] In a single atom, this lifetime is equivalent to the 'radiative' lifetime, because energy dissipation (through its transformation into heat) is not possible in this case.

where B_{12} is the Einstein coefficient of absorption and B_{21} the Einstein coefficient of stimulated emission. The relation connecting the coefficients reads

$$\frac{A_{21}}{B_{21}} = \frac{8\pi h\,\nu^3}{c^3} \tag{10.1}$$

(c being the velocity of light in vacuum) and in the simplest case of non-degenerate levels also $B_{12} = B_{21}$ holds. The densities of corresponding electron transitions then depend on the populations of the individual levels. If N_1 and N_2 denote the population densities of levels $|1\rangle$ and $|2\rangle$, then the numbers of atoms participating in the corresponding transitions per unit volume and unit time are

$$p_{sp} = A_{21}N_2\ (\mathrm{cm^{-3}s^{-1}}) \qquad \text{spontaneous emission,} \tag{10.2a}$$

$$p_{21} = W_{21}N_2 = B_{21}u(\nu)N_2\ (\mathrm{cm^{-3}s^{-1}}) \qquad \text{stimulated emission,} \tag{10.2b}$$

$$p_{12} = W_{12}N_1 = B_{12}u(\nu)N_1\ (\mathrm{cm^{-3}s^{-1}}) \qquad \text{absorption.} \tag{10.2c}$$

In atomic spectroscopy, the description of these processes involves the introduction of a normalized emission or absorption lineshape function $\gamma(\nu{-}\nu_0)$ ($1/\mathrm{Hz} = \mathrm{s}$), which takes into account the narrow, though finite linewidth around the ν_0 maximum.[2] Subsequently, the energy spectral density can be expressed as $u(\nu) = u_\nu\gamma(\nu{-}\nu_0)$, where u_ν stands for the energy per unit volume, and formulas (10.2) take on the form

$$p_{sp} = A_{21}N_2, \tag{10.3a}$$

$$p_{21} = W_{21}N_2 = B_{21}N_2u_\nu\gamma(\nu{-}\nu_0), \tag{10.3b}$$

$$p_{12} = W_{12}N_1 = B_{12}N_1u_\nu\gamma(\nu{-}\nu_0). \tag{10.3c}$$

Applying these equations, we can easily derive a formula for the absorption coefficient $\alpha(\nu)$ in the Lambert–Beer law or, as the case may be, for the so-called *optical gain coefficient* $g(\nu)$. Let a (probe) parallel beam of photons $h\nu$, whose power flux density, or intensity, is given by $I_\nu = cu_\nu(\mathrm{W/cm^2})$, impinge on our atomic system in the z- direction from an external source (such as a spectral lamp). The variation in the intensity on the path element $\mathrm{d}z$ can obviously be written as

$$\mathrm{d}I_\nu = h\nu(A_{21}N_2 + W_{21}N_2 - W_{12}N_1)\mathrm{d}z.$$

Usually, the contribution of isotropic spontaneous emission in the given direction can be assumed to be much smaller than the contribution of stimulated emission, i.e. $A_{21}N_2 \ll W_{21}N_2$ and thus

$$\mathrm{d}I_\nu = h\nu(W_{21}N_2 - W_{12}N_1)\mathrm{d}z. \tag{10.4a}$$

It follows from $B_{12} = B_{21}$ and (10.1) that

$$\frac{\mathrm{d}I_\nu}{\mathrm{d}z} = (N_2 - N_1)\frac{c^2}{8\pi\nu^2 t_{sp}}\gamma(\nu - \nu_0)I_\nu. \tag{10.4b}$$

[2] Besides, it is of importance in many cases whether $\gamma(\nu{-}\nu_0)$ describes a homogeneously or inhomogeneously broadened spectral line.

From the phenomenological point of view, the assumption about the validity of the Lambert–Beer law $I_\nu(z) = I_\nu(0)\exp(-\alpha(\nu)z)$ is fully justified, immediately leading to

$$\frac{dI_\nu}{dz} = -I_\nu\alpha(\nu). \tag{10.5}$$

The comparison of the microscopic (10.4b) and macroscopic (10.5) approaches finally yields the formula for the absorption coefficient

$$\alpha(\nu) = (N_1 - N_2)\frac{c^2}{8\pi\nu^2 t_{sp}}\gamma(\nu - \nu_0)\ (\mathrm{cm}^{-1}). \tag{10.6}$$

If the levels $|1\rangle$ and $|2\rangle$ are occupied in a standard way, i.e. $N_1 > N_2$, the probe beam is attenuated since $\alpha(\nu) > 0$. If, on the other hand, population inversion, i.e. $N_2 > N_1$, is achieved, it follows from (10.5) that the intensity I_ν can rise exponentially along the z-direction, since $\alpha(\nu) < 0$. Under such circumstances, the quantity $g(\nu) = -\alpha(\nu) > 0$ is referred to as the optical gain. This term will be specified shortly. Please note here that in the atomic case the spectral lineshape of the optical gain is determined by the shape of the corresponding (spontaneous) emission line.

10.2 Optical gain in semiconductors

In many of the preceding chapters, the theoretical spontaneous emission lineshape for various mechanisms of radiative recombination was discussed. As regards stimulated emission, which is the case being analysed right now, neither a general theoretical description of the lineshape nor its comparison with experiment can be applied in most cases. In principle, two underlying reasons can be found. First of all, the strongly nonlinear character of stimulated emission makes the process, and naturally also its emission spectral lineshape, extremely dependent on the intensity of the stimulating light. Moreover, as was shown in the preceding section, spontaneous and stimulated emission in semiconductors often occur simultaneously in luminescence experiments (Figs 10.1–10.3), which brings even more difficulties in attempts at spectral separation of stimulated emission and its comparison with a theoretical lineshape. Nevertheless, a theoretical guideline exists even in this case; it should not be based on the emission spectrum itself but, instead, on the spectral distribution of the optical gain, which can often be both theoretically well described and—as will be shown in Section 10.6—in many cases reliably experimentally determined.[3] Our goal for now will be the formulation of a general formula for the spectral shape of the optical gain in semiconductors.

We proceed from the mechanism of EHP radiative recombination, a process which finds application in the operation of semiconductor injection lasers. The issue of how to mathematically formulate the optical gain in semiconductors

[3] Even if the spectral distribution of the optical gain depends also on the excitation intensity, this dependence can be quantified, and, moreover, the optical gain does not contain any ingredient of spontaneous emission.

dates back to the beginning of the 1960s [6, 7], which is the time when the very first lasers of this type were built [8, 9].

One can easily find that, unlike the atomic case, stimulated transitions between the quasi-continua of electron and hole states (Fig. 10.4(b))—not between discrete energy levels—need to be taken into account. Factors playing important roles will thus be the densities of states in the conduction $\rho_c(E_2)$ and valence $\rho_v(E_1)$ bands, respectively. For the case of a direct-bandgap semiconductor, in Section 5.2 we replaced these densities with the joint density of electron and hole states in the bands $\rho(\nu)$. Let us recall that under low-fluence excitation (i.e. without EHP) the formula for the probability density of spontaneous band-to-band emission reads

$$I_{sp}(h\nu) = \frac{1}{\tau_r}\rho(\nu) f_e(\nu), \tag{10.7a}$$

where $f_e(\nu)$ stands for the probability of 'suitable occupancy of states' in the conduction and valence bands. The role of the spontaneous lifetime t_{sp} is, in semiconductors, taken on by the radiative lifetime of electron–hole recombination τ_r. When deriving $f_e(\nu)$, we considered only a small deviation from thermodynamic equilibrium (due to the low-fluence excitation), a situation that could be described satisfactorily by a single Fermi level E_f pinned in the middle of the bandgap, as described by the well-known Fermi–Dirac distribution function

$$f(E) = \frac{1}{\exp(E - E_f/k_B T) + 1}. \tag{10.8}$$

The probability $f_e(\nu)$ of (or the condition for the occupancy of levels for) emission of a photon was then given as the product $f_e(\nu) = f_c(E_2)[1 - f_v(E_1)]$, standing for the co-occurrence of an electron in the conduction band and a hole (with the same wavevectors) in the valence band. In other words, relation (10.7a) was rearranged into the form

$$I_{sp}(h\nu) = \frac{1}{\tau_r}\rho(\nu) f_c(E_2)\left[1 - f_v(E_1)\right], \tag{10.7b}$$

resulting then in the Maxwell–Boltzmann lineshape of spontaneous emission.

Let us now try to find a formula analogous to eqn (10.7a) or (10.7b), but suitable for stimulated emission under high-fluence excitation. We proceed from relations (10.2) or (10.3). Before starting we have to realize that this time the concept of the volume density of atoms N_2 cannot be applied; instead, a term proportional to the product $\rho(\nu) f_e(\nu)$ and characterizing the 'occupied joint states per unit volume (1 cm^3) which are about to emit a photon' will take on its role. Please note also that the quantities W_{21} and W_{12} in (10.2) represent the rates of optical transitions in an isolated atom or molecule, being thus in s^{-1} units. In a semiconductor, however, the closely packed atoms in a solid need to be taken into account, which results in the quasi-continuous distribution of energy levels, and the possibility of emission or absorption of a photon has to be related to some frequency interval $(\nu, \nu + d\nu)$; therefore, we can specify that the role of N_2 will be taken on by the factor $\rho(\nu) f_e(\nu)d\nu$, having the same units as N_2 (cm^{-3}). Finally, it follows also that it will not be possible to study the optical gain (or absorption) employing quasi-monochromatic light with a single frequency ν_0 and a spectral density of energy $u_\nu\,\gamma(\nu-\nu_0)$, but it will be necessary to employ, in general, light with a broad spectral distribution $u(\nu)$.

It follows from what has been said above that the probability density of stimulated emission $r'_{\mathrm{stim}}(h\nu)$, often referred to as the stimulated emission rate (i.e. the number of photons $h\nu = E_2 - E_1$ emitted per second per 1 Hz and per 1 cm^3 of the semiconductor [10]), can be expressed, in analogy with (10.3b), as

$$r'_{\mathrm{stim}}(h\nu_0)\,\mathrm{d}\nu_0 = B_{21}\,u(\nu_0)\,\gamma(\nu-\nu_0)\,\rho(\nu_0)\,f_{\mathrm{e}}(\nu_0)\,\mathrm{d}\nu_0 \ (\mathrm{cm}^{-3}).$$

The overall stimulated emission rate can then be obtained by integrating over all the allowed frequencies ν_0, or

$$r_{\mathrm{stim}}(h\nu) = \int r'_{\mathrm{stim}}(h\nu_0)\,\mathrm{d}\nu_0 = \int B_{21}\,u(\nu_0)\rho(\nu_0)f_{\mathrm{e}}(\nu_0)\,\gamma(\nu-\nu_0)\,\mathrm{d}\nu_0$$

and, as the linewidth of the individual energy levels from the quasi-continuum of electronic states in semiconductors is much smaller than the frequency band of stimulated emission, the approximation $\gamma(\nu-\nu_0) \to \delta(\nu-\nu_0)$ can be exploited, yielding

$$r_{\mathrm{stim}}(h\nu) = B_{21}\,u(\nu)\,\rho(\nu)\,f_{\mathrm{e}}(\nu). \tag{10.9a}$$

Taking into account the 'reciprocity' of stimulated emission and absorption transitions, the photon absorption rate can be obtained by replacing the probability $f_{\mathrm{e}}(\nu)$ in (10.9a) by the occupancy condition for absorption $f_{\mathrm{a}}(\nu) = f_{\mathrm{v}}(E_1)[1-f_{\mathrm{c}}(E_2)]$:

$$r_{\mathrm{abs}}(h\nu) = B_{12}\,u(\nu)\,\rho(\nu)\,f_{\mathrm{a}}(\nu). \tag{10.9b}$$

Consequently, the resulting rate of observable stimulated emission (taking into account the absorption-related losses) can be derived from (10.9a) and (10.9b), considering the equality $B_{12} = B_{21} = B$, as

$$R_{\mathrm{stim}}(h\nu) = r_{\mathrm{stim}}(h\nu) - r_{\mathrm{abs}}(h\nu) = B\,u\,(\nu)\rho(\nu)\,\left[f_{\mathrm{c}}(E_2) - f_{\mathrm{v}}(E_1)\right]. \tag{10.10}$$

Finally, an equation analogous to (10.4a) will now be of the form

$$\mathrm{d}I(\nu) = h\nu\,[r_{\mathrm{stim}}(h\nu) - r_{\mathrm{abs}}(h\nu)]\,\mathrm{d}z = h\nu\,R_{\mathrm{stim}}(h\nu)\,\mathrm{d}z, \tag{10.11}$$

where $I(\nu) = (c/n)u(\nu)$ is the spectral distribution of intensity, i.e. the energy of optical radiation per 1 Hz per area of 1 cm^2 per second. Taking into account the fact that $I(\nu)$ again satisfies the Lambert–Beer law (10.5)

$$\mathrm{d}I(\nu) = -\alpha(\nu)\,I(\nu)\,\mathrm{d}z,$$

we find, making use of the comparison of the last two formulas, that the absorption coefficient in this case reads

$$\alpha(\nu) = -\frac{h\nu\,R_{\mathrm{stim}}(h\nu)}{I(\nu)},$$

which can, using (10.10), (10.1) and $A_{21} = 1/\tau_r$, be easily transformed to express the optical gain as

$$g(\nu) = -\alpha(\nu) = \frac{h\nu}{I(\nu)} B\, u(\nu)\rho(\nu)\,[f_c(E_2) - f_v(E_1)]$$
$$= \frac{1}{\tau_r}\frac{c^2}{8\pi\nu^2 n^2}\rho(\nu)\,[f_c(E_2) - f_v(E_1)] \qquad (10.12)$$
$$= \frac{1}{\tau_r}\frac{\lambda_0^2}{8\pi n^2}\rho(\nu)\,[f_c(E_2) - f_v(E_1)]\,;$$

λ_0 is the wavelength of light in vacuum and n stands for the refractive index of the semiconductor. Thus, we have succeeded in finding a formula for stimulated emission, which is analogous to eqn (10.7b) describing the spectral shape of spontaneous emission at low excitation fluences.[4] Obviously, the spectral shape of the gain is influenced largely by the factor in square brackets (sometimes also referred to as the Fermi inversion factor), i.e. by the occupancy of the electron states in the conduction band and the hole states in the valence band, in other words, by the pump intensity.

Before we discuss the spectral shape of $g(\nu)$ on the basis of (10.12), let us stop for a while to examine the time characteristics of stimulated emission. By formally comparing (10.10) with (10.7b), we can easily see that instead of the reciprocal spontaneous radiative lifetime τ_r^{-1} we can write for the case of stimulated emission

$$\frac{1}{\tau_{\text{stim}}} = B\, u(\nu),$$

which can, using (10.1), be rearranged to

$$\frac{1}{\tau_{\text{stim}}} = \frac{1}{\tau_r}\frac{c^2}{8\pi n^2\nu^2}\frac{I(\nu)}{h\nu} = \frac{1}{\tau_r}\frac{\lambda_0^2\phi(\nu)}{8\pi n^2}, \qquad (10.13)$$

when we have applied the substitution $t_{\text{sp}} \to \tau_r$ and defined the mean spectral density of photons $\phi(\nu) = I(\nu)/h\nu(\text{cm}^{-2})$ in units of the number of photons per second per cm^2 per 1 Hz.

Let us for now imagine an experiment in which the stimulated emission transitions are just setting in, causing the simultaneous occurrence of both spontaneous and stimulated transitions in the detected emission spectrum. The overall radiative lifetime τ_{ov} at the chosen wavelength can then be expected to behave as

$$\frac{1}{\tau_{\text{ov}}} = \frac{1}{\tau_r} + \xi\frac{1}{\tau_{\text{stim}}} = \frac{1}{\tau_r}\left(1 + \xi\frac{\lambda_0^2\phi(\nu)}{8\pi n^2}\right) > \frac{1}{\tau_r}. \qquad (10.14)$$

[4] The role of the probe beam is, in luminescence experiments, taken on by the spontaneous emission in a given direction $I_{\text{sp}}(h\nu)$. By solving an equation similar to (10.11) it is possible to derive a formula for the amplification of stimulated emission in the given direction $I_{\text{stim}}(h\nu, z) \sim I_{\text{sp}}(h\nu)\exp[g(\nu)z]$, which then in fact describes the spectral shape of the stimulated emission band. This illustrates, among other things, the above-mentioned nonlinearity of the process. This problem will be treated in more detail in Section 10.6.

The factor $0 < \xi(t) < 1$ here stands for the relative part of the excited volume of material that takes part in the stimulated emission along the direction under investigation. Consequently, the onset of stimulated emission effectively shortens the radiative lifetime and thus also the experimentally measured luminescence lifetime. This shortening is a feature typical of the onset of stimulated emission (see Fig. 10.3) and can become quite significant, indeed, as a result of the positive feedback implicit in eqn (10.14). Nevertheless, stimulated emission typically depletes the population inversion very quickly, fading itself away as a result, and the whole process of luminescence decay finishes through the slower spontaneous emission.

We should point out, however, that the above discussion is valid only near the very threshold of the onset of stimulated emission, whereas in the case of real laser action inside a resonator it no longer serves any useful purpose. In the latter case, the rapid 'emptying' of electron states in the conduction band, rather than being directly linked to τ_r, is in a decisive way influenced by the quality and construction of the resonator, along with the level of pumping.

10.3 Spectral shape of the optical gain

Let us now treat in more detail the spectral shape of function $g(\nu)$ defined in eqn (10.12). In this case, obviously, the quasi-equilibrium state characterized by a single Fermi energy level cannot be applied, but we need to define (as indicated also in Fig. 10.4(b)) separate (quasi-)Fermi levels for electrons F_e, or μ'_e and for holes F_h, or μ'_h. We thus write

$$f_c(E_2) = (\exp\left(\frac{E_2 - \mu'_e}{k_B T}\right) + 1)^{-1} \tag{10.15}$$

$$f_v(E_1) = (\exp\left(\frac{E_1 - \mu'_h}{k_B T}\right) + 1)^{-1}.$$

In order to be more specific, let us add that the energies E_1, E_2, μ'_e and μ'_h share a common system of coordinates whereas F_e and F_h are values relative to the corresponding band extremes. At the same time, the Fermi energies are linked through the relation $E'_g + F_e + F_h = \mu'_e - \mu'_h (= \mu'$, the plasma chemical potential). An easy derivation of the Fermi inversion factor $\Gamma(\nu) = f_c(E_2) - f_v(E_1)$ by making use of (10.15) shows that $\Gamma(\nu)$ can be expressed as a fraction with a positive denominator, possessing a numerator in the form of

$$1 - \exp[(h\nu - E'_g - F_e - F_h)/k_B T]. \tag{10.16}$$

In order to achieve positive optical gain (i.e. to describe the amplification of light by stimulated emission, and not attenuation through absorption), the factor $\Gamma(\nu) = f_c(E_2) - f_v(E_1)$ has to be positive according to (10.12), which, based on (10.16), implies

$$h\nu < E'_g + F_e + F_h.$$

Since the lowest photon energy $h\nu$ which can be amplified in a given case is $h\nu = E'_g$, we obtain the following formulation of the condition for the occurrence of optical gain in semiconductors

$$E'_g < h\nu < E'_g + F_e + F_h = \mu'_e - \mu'_h. \tag{10.17}$$

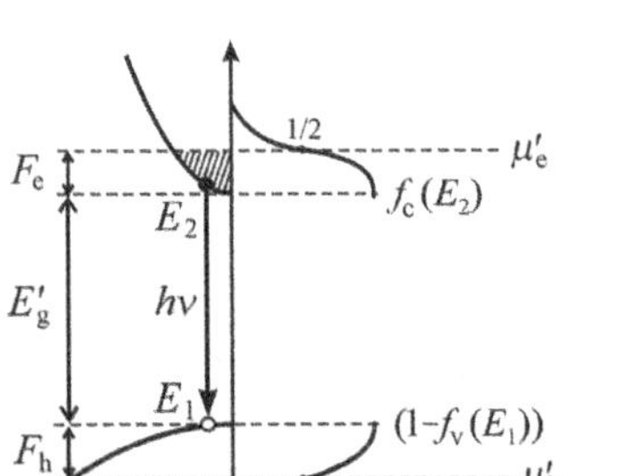

Fig. 10.5
Schematic sketch of the Bernard-Duraffourg condition for the inverse occupancy of energy states. Amplification of radiation through stimulated emission can occur only between electron states with occupancy higher than 1/2 ($f_c(E_2) > 1/2$), and an analogous condition applies also for holes ($1 - f_v(E_1 > 1/2)$. States denoted by hatched areas are occupied largely by electrons.

This is the well-known Bernard–Duraffourg condition for inverse population of states and amplification of radiation through stimulated emission in a degenerate semiconductor [6]. It is valid for any temperature.

This condition can be reformulated to better illustrate the bottom line: The amplification of light through stimulated emission in semiconductors can occur only for photons which arise from radiative transitions between states in the conduction band with the probability of being occupied by electrons larger than 1/2, and states in the valence band with probability of being occupied by electrons lower than 1/2. The corresponding states in the valence band are thus largely occupied by holes, as depicted in Fig. 10.5. This case is in fact an analogy to the case of inversion population in discrete atomic levels shown in Fig. 10.4(a) ($N_2 > N_1$).

One can easily get the basic idea about the spectral dependence of the optical gain by looking at formula (10.12). This dependence is determined in a crucial way by the spectral behaviour of the factor $\Gamma(\nu) = f_c(E_2) - f_v(E_1)$ and of the joint density of states $\rho(\nu)$; the parameter λ_0^2, or ν^{-2}, on the other hand, does not influence the spectral shape of $g(\nu)$, as will be shown shortly. Based on (10.15), it is easy to show that $\Gamma(\nu)$ monotonically decreases with ν (i.e. $\mathrm{d}\Gamma(\nu)/\mathrm{d}\nu < 0$, see also Problem 10/3) and intersects zero at $h\nu = E'_g + F_e + F_h$, which is shown in Fig. 10.6(a). The very same figure also illustrates the well-known square-root shape $\rho(\nu) \sim \sqrt{h\nu - E'_g}$ and the resulting curve $\rho(\nu)\Gamma(\nu)$ (solid circles). The shape of the curve $\rho(\nu)\Gamma(\nu)\nu^{-2}$ is also shown for comparison by open circles; this curve denotes the accurate shape of the optical gain $g(\nu)$ according to (10.12). The comparison proves that the factor ν^{-2} in fact does not influence the shape of the product $\rho(\nu)\Gamma(\nu)$ (for $h\nu$ of the order of 1 eV and for typical widths of gain curves of the order of tens of meV); it can thus be considered constant and $g(\nu) \cong \rho(\nu)\Gamma(\nu)$ can be used as a good approximation.

Now, however, we encounter a problem: Experimentally obtained spectral gain shapes in most cases do not copy the symbols-denoted curve in Fig. 10.6(a). Rather, the experimental data can be better fitted using the curve shown as circles in Fig. 10.6(b), which starts with a slow increase of $g(\nu)$ near the threshold value $h\nu = E'_g$. We generated this curve by replacing *ad hoc* the square-root dependence $\rho(\nu) \sim \sqrt{h\nu - E'_g}$ by a quadratic dependence $\rho(\nu) \sim (h\nu - E'_g)^2$. We have already met this factor in the previous chapters in the text on band-to-band recombination of electron–hole pairs in an indirect semiconductor, i.e. in the case when the recombination of any electron–hole pair is allowed irrespective of the corresponding wavevectors. This implies that during EHP stimulated emission in a direct semiconductor the quasi-momentum conservation law probably is not satisfied. We have already come across a similar effect—we mentioned in Subsection 8.5.2 that the emission

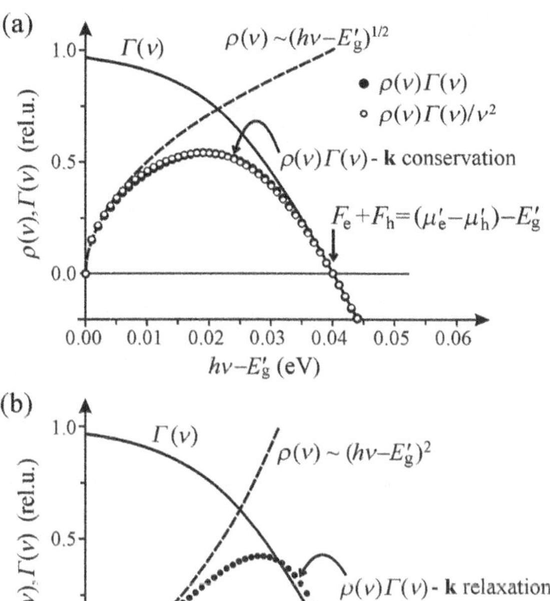

Fig. 10.6
(a) Spectral dependence of the Fermi inversion factor $\Gamma(\nu)$, the joint density of states $\rho(\nu) \sim \sqrt{h\nu - E_g'}$, and the products $\rho(\nu)\Gamma(\nu)$ and $\rho(\nu)\Gamma(\nu)/\nu^2$. Curves correspond to direct transitions with the quasi-momentum conservation law 'on'. (b) The same dependencies as in case (a) but considering the density of states in the form $\rho(\nu) \sim (h\nu - E_g')^2$. Parameters of the calculation: $E_g' = 1\,\text{eV}$, $m_e = m_h$, $F_e + F_h = 20 + 20 = 40\,\text{meV}$, $T \approx 60\,\text{K}$.

spectrum of spontaneous EHP luminescence in a direct semiconductor can be well fitted with a mathematical model ignoring the quasi-momentum conservation law. Now, we came to a similar conclusion also for the EHP stimulated emission. If this is really the case, however, we are not allowed to describe the shape of $g(\nu)$ simply by substituting $\rho(\nu) \sim (h\nu - E_g')^2$ in (10.12). The proper treatment in this case is to integrate over all suitable pairs of states in the conduction and valence bands. We can exploit formula (8.24) or (8.25), in which, however, the spontaneous-emission occupancy condition $f_c(E) f_v(\overline{h\nu} - E)$ will be replaced by the condition for stimulated emission $[f_c(E) - f_v(\overline{h\nu} - E)]$:

$$g(\nu) \simeq \int_0^{\overline{h\nu} = h\nu - E_g'} \sqrt{E}\sqrt{\overline{h\nu} - E}\left[f_c(E) - f_v(\overline{h\nu} - E)\right] \mathrm{d}E \qquad (10.18)$$

$$= \int_0^{\overline{h\nu} = h\nu - E_g'} \sqrt{E}\sqrt{\overline{h\nu} - E}\ \left[f_c(E) + f_h(\overline{h\nu} - E) - 1\right]\ \mathrm{d}E.$$

In this relation, the occupancy probability of the corresponding state by a hole is denoted as $f_h = 1 - f_v$.

Figure 10.7 demonstrates the results of a calculation carried out according to (10.18). Panel (a) shows one of the first detailed theoretical calculations of

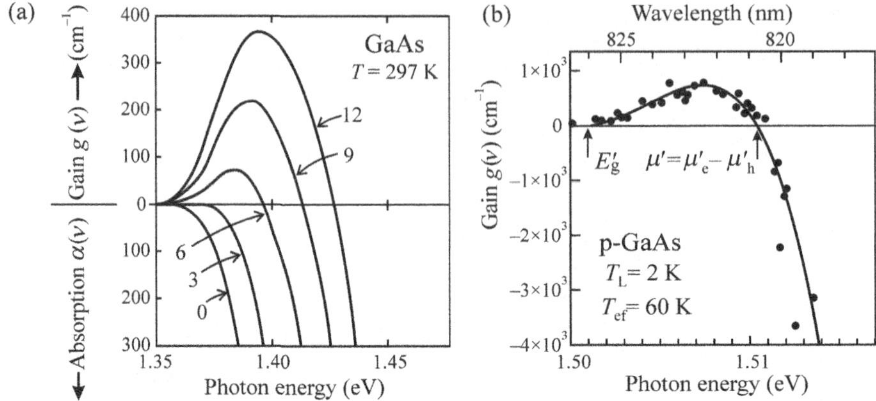

Fig. 10.7
(a) Optical gain $g(\nu)$, or absorption coefficient $\alpha(\nu)$, calculated for GaAs within the model of the relaxation of the **k**-vector, qualitatively described by the relation (10.18). Parameters shown next to the curves stand for the relative level of electrical pumping. Curves are calculated for room temperature. After Stern [11]. (b) EHP optical gain in GaAs (measured using the technique of thin-film optical transmission: laser pump and probe) at a bath temperature of $T_L = 2$ K. Points denote experimental data, the solid curve is a theoretical fit considering **k**-vector relaxation according to (10.18). $\mu' = \mu'_e - \mu'_h$ is the plasma chemical potential. After Hildebrand and Göbel [12].

gain in GaAs by Stern [11]. It is obvious that the overall shape of the curves is in very good accord with the dependence $g(\nu) \sim \rho(\nu)\Gamma(\nu)$ in Fig. 10.6(b). In addition, we can easily see that in the absence of pumping, i.e. when the conduction band is empty ($f_c(E) = 0$), the valence band is completely filled ($f_v(\overline{h\nu} - E) = 1$) and $E'_g = E_g$, the gain attains negative values, which means that by applying this approach, we can obtain the onset of the absorption coefficient $\alpha(\nu) = -g(\nu)$ near the absorption edge. In panel (b) of Fig. 10.7 the theory is compared to experiment, showing nice agreement [12]; see also Appendix G.

Before moving on it is useful to take note of one more thing. Figure 10.7(a) depicts not only the shape of the gain curves but also the numerical values of g(ν), which are of the order of hundreds of cm^{-1}. Up to now in luminescence spectroscopy, we have come across only the shape of spectra in arbitrary units; the optical gain, however, in fact a counterpart of the optical absorption coefficient, can be quantified and is, in this sense, an exception for us.

Nevertheless, in order to carry out a correct calculation of $g(\nu)$ values it would be indispensable firstly to add the squared modulus of the transition matrix element $|M|^2$ to (10.18) and, secondly, to insert a numerical factor in front of the integral, containing, apart from universal constants, also the effective masses of the carriers. This means that in the case of a particular material it would be necessary to take into account the existence of several valence bands (light and heavy holes) and obtain the resulting values of $g(\nu)$ via summing over all bands. At the same time, a sufficiently accurate calculation of $|M|^2$ for a given band structure represents a problem on its own. Such procedures are, however, beyond the scope of the present textbook and anyone interested in this problem should refer to the specialized literature [13, 14], containing further references.

A simple estimate of at least an order of magnitude of $g(\nu)$ according to (10.12) is, on the other hand, within our powers. Considering that the typical densities of states at the band edges (usually expressed in units of $\mathrm{cm^{-3}eV^{-1}}$) attain values of $10^{17}\,\mathrm{cm^{-3}eV^{-1}}$, we can approximate the joint density of states to be of the same order $\rho(h\nu) \approx 10^{17}\,\mathrm{cm^{-3}\,eV^{-1}}$. This implies that $\rho(\nu) = \rho(h\nu)\mathrm{d}(h\nu)/\mathrm{d}\nu = h\rho(h\nu) \approx 4 \times 10^{2}\,\mathrm{cm^{-3}/s^{-1}}$. The optical frequencies are of the order of $\nu \approx 10^{14}$ Hz and in a direct-bandgap semiconductor $\tau_r \approx 1$ ns. As the Fermi inversion factor $\Gamma(\nu) = f_c(E_2) - f_v(E_1)$ near the extreme of the gain-curve maximum is of the order of unity (see Fig. 10.6), we can approximate $\Gamma(\nu) = 1$, and relation (10.12) finally yields $g(\nu) \approx 200\,\mathrm{cm^{-1}}$. This confirms the values of the vertical axis in Fig. 10.7(a), i.e. it shows that the peak values of the optical gain in semiconductors are of the order of 10^2–$10^3\,\mathrm{cm^{-1}}$.

Now, let us come back to the model (10.18) ignoring the quasi-momentum conservation law, or, put more simply, the **k** conservation law. Although, in most cases, the model describes well the optical-gain spectrum, the physical reason underlying its application is not clear. What may cause the **k** conservation law to be violated? The answer can lie, e.g., in the large number of inhomogeneously distributed electron–hole pairs in an EHP, which distorts the ideal translational symmetry of the lattice and thus violates the **k** conservation law. A similar effect arises from, as we already know, the presence of impurity atoms, which might in fact be a description closer to reality: semiconductor lasers are based on a p-n junction, which inevitably requires doping. Other explanations include the opinion that, during radiative EHP recombination, the **k** conservation law is actually being satisfied locally, through multiple collisions between electrons and holes. These—or similar—arguments, however, always have their weaknesses.

If, for example, the high concentration of electron–hole pairs were the most important reason underlying the relaxation of the **k** conservation law, the shape of the gain curve would naturally have to depend on the intensity of pumping and, at the transition to the negative values of $g(\nu)$, it would have to change to a square-root dependence $-g(\nu) \simeq \sqrt{h\nu - E_g}$, which is not backed up by experimental observations. If, on the other hand, the doping of the semiconductor is the cause, the mentioned 'smearing' of the absorption edge can be explained by the existence of tail states inside the bandgap, but the characteristic shape of the gain curve described by (10.18) should not be observed under optical pumping of high-purity samples, which actually does happen.

The applicability of the **k**-vector relaxation model has thus been a hotly debated topic ever since the beginnings of semiconductor-laser research, see e.g. [15]. Over time, even models respecting the **k** conservation law were proposed as an explanation of the shape of the gain curve [16]. The currently accepted opinion is that the **k** conservation law is satisfied during EHP radiative recombination, in accordance with the basic laws of quantum physics; the gentle increase of $g(\nu)$ close to E_g' instead of a square-root dependence and the overall asymmetry of the gain curve are believed to be due to two 'correction' factors. The first one is the so-called *final state damping* in the EHP while the second one is referred to as *excitonic enhancement*. The final state damping can be understood as a phenomenon when, once the recombination of an electron with a hole in the EHP is accomplished, one unoccupied state remains among the conduction-band electrons and an electron with higher energy quickly relaxes towards it (while, at the same time, among the holes close to the top of the valence band one state filled with an electron appears, which leads to an analogous relaxation via rearrangement of holes). Such a relaxation is very fast and, as a result of the uncertainty principle, manifests itself as a spectral broadening of the gain curve. This, however, would lead to a more or less systematic smearing of the curve shown in Fig. 10.6(a). The observed shift of the centre of gravity of the curve towards $(\mu'_e - \mu'_h) - E'_g$ is then the result of excitonic enhancement. Roughly speaking, the excitonic enhancement means that electron–hole pairs with high energy, i.e. close to $\mu' = \mu'_e - \mu'_h$ where filled and empty states are in close neighbourhood, are—at least partially—describing relative circular orbits, which evokes memories of a bound excitonic state [17]. This effect entails an enhancement in EHP luminescence and the relevant gain curves on their high-energy edge.

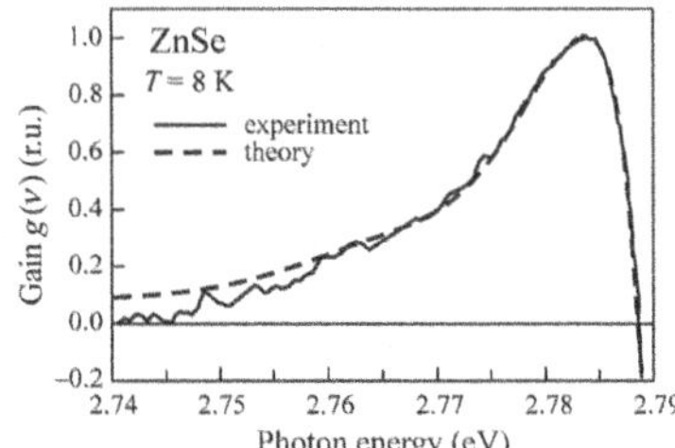

Fig. 10.8
EHP optical gain in thin-film (60 μm) ZnSe under pulsed laser pumping ($I_{ex} = 2\,\mathrm{MW/cm^2}$), measured by a pump and probe method. The dashed curve represents a theoretical fit, taking into account the **k**-vector conservation law and including both the final state damping and excitonic enhancement. $T_L = 8$ K, effective EHP temperature $T_{ef} = 20$ K. After Kunz *et al.* [18].

Figure 10.8 then confirms that the above ideas are correct, as it represents an experimentally obtained spectrum of EHP optical gain in ZnSe as well as a theoretical fit taking into account the **k** conservation law, final state damping and excitonic enhancement [18]. The agreement between experiment and theory is once again good; obviously, both the models, i.e. both the **k**-vector relaxation and the **k** conservation law, are able to model the experimental curves of gain more than satisfactorily. In this 'spectral' sense, the argument about which of the models is the correct one can be viewed more as an academic debate, because the values of the basic fitting parameters of the theoretical gain curves (the density of the plasma entering the fit through μ' and the effective plasma temperature) turned out to be practically identical in both approaches.

Finally, let us address three more issues: the spectral position of the gain with respect to the maximum of the spontaneous emission band, the behaviour of the gain with varying temperature and—once again—the ν^{-2} factor in eqn (10.12).

As for the spectral position of gain: looking at Fig. 10.5 or even better at Fig. 10.9(a) we can immediately see that the stimulated emission in the EHP is always situated on the longer-wavelength part of the spontaneous emission band, because stimulated emission can occur only for photon energies $h\nu < \mu'_e - \mu'_h$, whereas regarding spontaneous luminescence, even electron–hole pairs with energy higher than $\mu' = \mu'_e - \mu'_h$ contribute to the signal. Moreover, the possible onset of optical absorption, e.g. from the absorption-band tail, can negatively influence the higher-energy side, competing with the stimulated recombination transitions. Corresponding spectra of spontaneous emission and gain are schematically sketched in 10.9(b) and, in addition, the whole situation is very well documented by a series of EHP emission spectra in Fig. 10.3. A word of caution, however, is that in the case of low-dimensional semiconductor structures, the onset of stimulated emission on the longer-wavelength side of the spontaneous emission band is not necessarily the rule.

Now, let us move on to the influence of temperature on optical gain. Most experimental curves shown in this section correspond to low temperatures

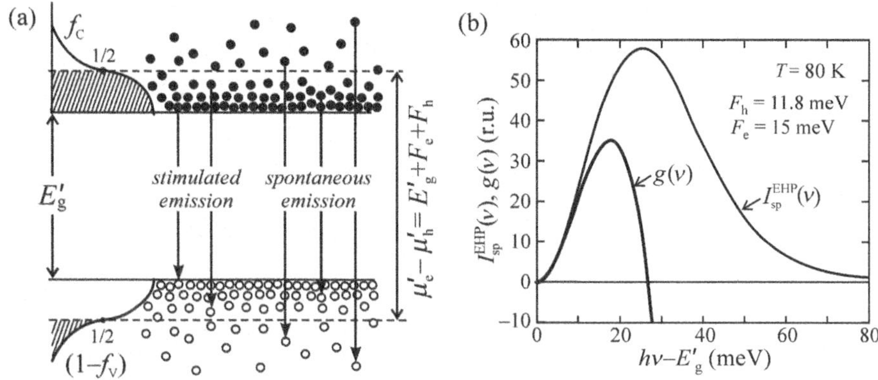

Fig. 10.9
(a) EHP radiative recombination with highlighted spontaneous and stimulated transitions, which are separated by the Bernard–Duraffourg condition. (b) Calculated spectra of the optical gain $g(\nu)$ and spontaneous emission $I_{sp}^{EHP}(\nu)$ for a semiconductor with parabolic energy bands, Fermi energies of $F_e = 15$ meV, $F_h = 11.8$ meV and EHP temperature equal to 80 K. The calculation was performed in the same units for both curves in the approximation of **k**-vector relaxation (10.18) or (8.25). After Lasher and Stern [7].

(2–80 K). As the most important applications of stimulated emission in semiconductors are injection lasers functioning at $T = 295$ K, the question arises of how the conditions for achieving stimulated emission vary with increasing temperature. The answer to this question is as easy as pie: the conditions worsen. Let us recall that the inevitable condition for light amplification by stimulated emission in EHP is a degenerate electron–hole plasma, i.e. a plasma obeying Fermi–Dirac statistics. Speaking more accurately, the plasma chemical potential $\mu' = \mu'_e - \mu'_h$ needs to be, according to condition (10.17), larger than the reduced bandwidth, $\mu' > E'_g$.[5] The higher the temperature, however, the more difficult it is to fulfil this condition, because electrons and holes are thermally excited to higher states, the occupancy of lower states decreases and, consequently, the Fermi levels μ'_e, μ'_h shift towards the middle of the bandgap and their difference is reduced. What is more, thermal quenching of luminescence sets in.

Let us formulate a simple mathematical condition for EHP degeneration. The Fermi–Dirac distribution functions can be written in the coordinate systems of the corresponding bands as

$$f_c = \{\exp\left[(E_2 - E_c - F_e)/k_B T_{ef}\right] + 1\}^{-1},$$
$$f_v = \{\exp\left[(E_v - E_1 - F_h)/k_B T_{ef}\right] + 1\}^{-1}.$$

When the temperature T_{ef} is high enough so that both the Fermi energies happen to be situated within the bandgap (i.e. F_e, $F_h < 0$), and, at the same time, not extremely high so that $\exp[(\Delta E_{e(h)} - F_{e(h)})/k_B T_{ef}] \gg 1$ applies,

[5] This, however, does not necessarily mean that both the Fermi levels need to be situated within the corresponding bands, as is the case, e.g., in Figs 10.4 or 10.5—just a single Fermi level suffices to be immersed within a band. This is a common situation for achieving optical gain in GaAs: μ'_e is situated in the conduction band ($F_e > 0$) while μ'_h lies inside the bandgap ($F_h < 0$). The underlying reason for this situation is that in GaAs $m_h(\sim 0.52 m_0) \gg m_e(\sim 0.067 m_0)$ holds. See also Problem 10/4.

Fermi–Dirac statistics are transformed into Maxwell–Boltzmann statistics and the Bernard–Duraffourg condition does not make sense any more; it cannot be employed ($\Delta E_e = E_2 - E_c$, $\Delta E_h = E_v - E_1$ are the energies of carriers inside the bands). Since the energies ΔE_e, ΔE_h are of the order of $k_B T_{ef}$, the condition for a transition of the EHP from quantum statistics (degenerate gas) to classical statistics (non-degenerate gas) reads

$$\exp\left(-F_{e\,(h)}/k_B T_{ef}\right) \gg 1. \tag{10.19}$$

Finally, as regards the ν^{-2} factor in eqn (10.12), although it does not influence the shape of the $g(\nu)$ curve, as was shown in Fig. 10.6, it definitely enhances its magnitude. This effect is more pronounced at low frequencies, which can be considered one of the reasons why the very first laboratory semiconductor lasers generated long-wavelength radiation in the near-infrared spectral region (> 800 nm) and, moreover in accordance with the preceding discussion, solely at low temperatures (~ 80 K). It was not until the concept of semiconductor heterostructures that the problem of laser action at room temperature was successfully solved.

10.4 Stimulated emission in an indirect-bandgap semiconductor

Up to now, we have been discussing optical gain and stimulated emission in direct-bandgap semiconductors only. Can we, however, eliminate *a priori* indirect-bandgap semiconductors from this discussion? Naturally, we know that their radiative lifetime τ_r is by several orders of magnitude longer than in direct-bandgap materials and, accordingly, we would expect the optical gain to scale to lower values, in conformity with relation (10.12). This does not necessarily imply, however, that a (at least very low) positive optical gain coefficient cannot in principle occur in indirect-bandgap semiconductors. By the way, the peak value of the optical gain in ruby (optical transitions in a Cr^{3+} ion) is very low, about 0.2 cm^{-1} only at the typical concentration of Cr^{3+} ions of 10^{19} cm^{-3}. In spite of this, the ruby laser was the very first realized laser type and in some laboratories is still being used.

In indirect-bandgap semiconductors, or at least in the most common materials such as silicon and germanium, there is another obstacle that needs to be overcome in order to achieve stimulated emission, namely, free carrier absorption (FCA). Before we start a detailed discussion of stimulated emission in an indirect-bandgap semiconductor, let us concisely recall the basics of FCA.

In an ideal pure semiconductor at $T = 0$ K the conduction band is empty and the valence band is filled with electrons. If, however, free electrons are introduced into the minimum of the conduction band by any means, photons of electromagnetic radiation can be absorbed and the excess energy can be exploited for transferring these electrons to allowed higher lying states in the very same band (this type of absorption is thus not restricted to energies given by the well-known condition $h\nu > E_g$; it can occur even at lower photon energies).

A quantitative description of free carrier absorption can be most easily given in the framework of the classical Drude model. If a free electron gas with the free electron concentration N_0, mass m_0 and a relaxation time τ (in the simplest solid-state approximation it can be the mean time between collisions of the electron with either the lattice, i.e. phonons, or impurities) is inserted into a field of optical radiation at the frequency ν, the imaginary part of the dielectric constant ε_2 is described by the relation [19]

$$\varepsilon_2(\nu) \approx \frac{2\pi\varepsilon_\infty \nu_\mathrm{p}^2}{\nu\tau(4\pi^2\nu^2 + 1/\tau^2)}, \tag{10.20}$$

where ε_∞ is the high-frequency (real) dielectric constant and $\nu_\mathrm{p}^2 = N_0 e^2/4\pi^2\varepsilon_0\varepsilon_\infty m_0$ is the plasma frequency ν_p squared. In the visible or near-infrared spectral regions, for semiconductors $\nu \gg 1/2\pi\tau$ holds and (10.20) is reduced to $\varepsilon_2(\nu) \approx \varepsilon_\infty \nu_\mathrm{p}^2/2\pi\nu^3\tau$. The absorption coefficient $\alpha_\mathrm{FCA}(\nu)$ as a function of frequency ν then reads

$$\alpha_\mathrm{FCA}(\nu) = \frac{2\pi\nu}{c\,n(\nu)}\varepsilon_2(\nu) \approx \frac{N_0}{\nu^2} \approx N_0\lambda^2. \tag{10.21}$$

This implies that free carrier absorption is of a non-resonant character and rises monotonically with wavelength λ (in (10.21) c stands for the light velocity in vacuum and $n(\nu)$ is the refractive index, whose dependence on frequency is quite weak). Possible deviations in (10.21) in wavelength power from the value of two, which are sometimes observed in certain materials, can be due to variations of τ as a function of ν. The absorption cross-section of free carriers σ_FCA, a material constant, is then introduced by the relation $\alpha_\mathrm{FCA} = \sigma_\mathrm{FCA}\, N_0$.[6] It is of the order of $\sigma_\mathrm{FCA} \doteq 10^{-18} - 10^{-17}\,\mathrm{cm}^2$.

The absorption of radiation in a system of free electrons residing in the conduction band minimum is schematically sketched in Fig. 10.10(a); an analogous mirror scheme applies for holes in the valence band. The fact that one deals with an intraband type of absorption is evident. An important piece of information stemming from Fig. 10.10(a) is the non-verticality of the transition, which signifies that a phonon $\hbar\omega$ has to take part in the process apart from a photon $h\nu$ (and an electron or a hole, of course) in order to compensate for the change in quasi-momentum $\hbar\Delta\mathbf{k}$ (alternatively, relaxation of the quasi-momentum conservation law might occur as a result of the presence of impurities).[7] From the point of view of perturbation theory, these processes are of second order, which will shortly turn out to be very important.

The importance of free carrier absorption for lasing in semiconductors becomes clearer if we realize that the absorption transitions in Fig. 10.10(a) coexist with the radiative recombination of electrons and holes across the bandgap. This scenario is shown in Fig. 10.10(b). We can immediately notice that generated photons $h\nu = E_2 - E_1$ can be in no time (re)absorbed by free electrons and holes in the corresponding bands without ever leaving the

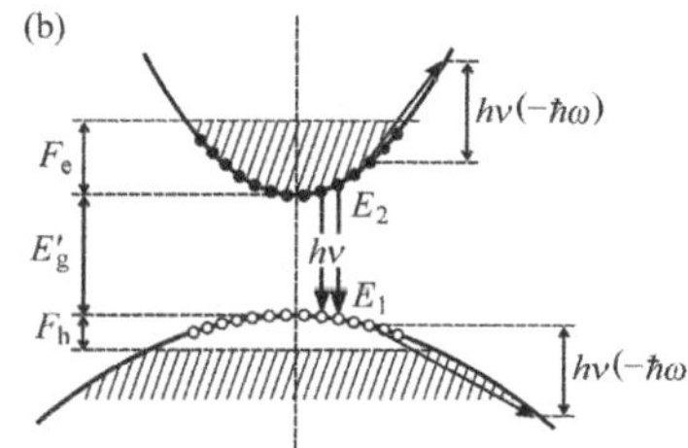

Fig. 10.10
(a) Illustrating the absorption of photons $h\nu$ on free EHP electrons in the minimum of the conduction band; the electrons are raised to free states above the Fermi level. A change in quasi-momentum $\hbar\Delta\mathbf{k}$ needs to be compensated for. (b) A process when radiative EHP recombination across the bandgap takes place and resulting luminescence photons are reabsorbed by free carriers. Only low temperatures are considered, with the optical transition being accompanied by the emission of a phonon $\hbar\omega$.

[6] When an electron–hole plasma is present in the semiconductor, N_0 denotes the concentration of electron–hole pairs.

[7] This implies that free carrier absorption is not allowed in vacuum.

material. The corresponding absorption coefficient $\alpha_{FCA}(\nu)$ will then compete with the optical gain $g(\nu)$ and as long as the former is high enough, the difference

$$G(\nu) = g(\nu) - \alpha_{FCA}(\nu) \tag{10.22}$$

becomes negative. According to the Lambert–Beer law, the generated luminescence will thus be attenuated instead of being enhanced via stimulated emission. This process will prevent the material from becoming an active medium for a semiconductor laser. The quantity $G(\nu)$ defined by relation (10.22) can be referred to as the 'net' optical gain; additional subtle distinctions will be mentioned in Subsection 10.6.1.

To be more specific, the situation in two typical examples of direct (GaAs) and indirect (Si) semiconductors is compared in Fig. 10.11. The radiative recombination in GaAs (even though some doubts were discussed in the preceding section) occurs via direct transitions, whereas the free carrier absorption is due to, as was emphasized above, indirect transitions, i.e. due a second-order process. In Si, on the other hand, both these transitions are indirect, conditioned by phonon participation, and thus their rates will be comparable. This fundamental difference when it comes to laser action was pointed out for the first time by Dumke in 1962 [20] and the values of $g(\nu)$ and $\alpha_{FCA}(\nu)$, calculated based on his theory, are quoted in Fig. 10.11 for $T = 4$ K. In GaAs at photon energies $h\nu \approx E_g \approx 1.5$ eV we get $g(\nu) \gg \alpha_{FCA}(\nu)$ and the influence of free carrier absorption is thus negligible, resulting in $G(\nu) \approx g(\nu) > 0$. On the other hand, in Si for $h\nu \approx E_g \approx 1.1$ eV one gets $g(\nu)$ smaller than $\alpha_{FCA}(\nu)$ by about an order of magnitude, i.e. $G(\nu) < 0$. Let us point out that this fact is not obvious *a priori*; it is determined by the band structure of a particular semiconductor and only a detailed calculation or an experiment can decide.[8] Similarly, the calculation yields $G < 0$ for germanium, which implies that neither Si nor Ge can be exploited for the construction of a semiconductor

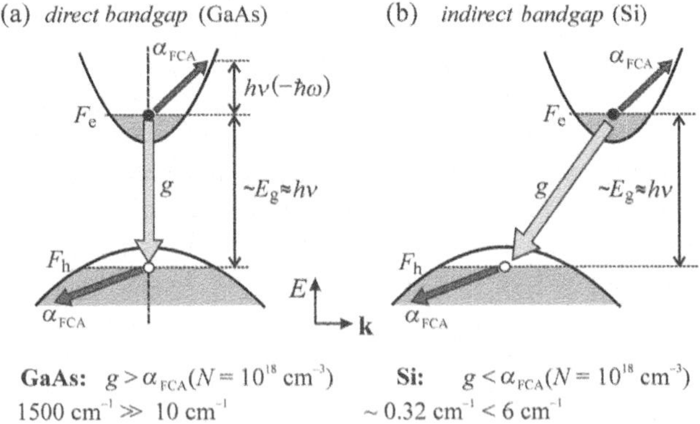

Fig. 10.11
Competition between the optical gain g and free carrier absorption, characterized by the absorption coefficient α_{FCA}, for (a) GaAs and (b) Si. The coefficients g and α_{FCA} were calculated according to theoretical work [20] for pumping levels giving rise to an EHP with the density of e–h pairs $N = 10^{18}$ cm^{-3} and for $T \approx 4$ K.

[8] For example in $Al_xGa_{1-x}As$, which is an indirect-bandgap semiconductor for $x \geq 0.44$, the occurrence of stimulated emission was proved experimentally by Kalt, H., Smirl, A. L., and Boggess, T. F. (1989). *J. Appl. Phys.*, **65**, 294.

laser operating via electronic transitions across the bandgap at low temperatures. This signifies, however, that laser action is even less probable at room temperature, when the conditions for achieving optical gain, as we already know, generally worsen and, besides, the luminescence quantum yield in both Si and Ge is extremely low, $\eta \approx 10^{-6}$. Up to now, no experiment has cast doubts on the results of these calculations.

This conclusion about silicon not being suitable as a laser medium puts severe restrictions on the application potential of this material, which otherwise represents the cornerstone of nearly all present-day microelectronics. A large number of authors therefore tried to circumvent this situation by applying more or less sophisticated approaches. For example Rediker [21] filed a patent for an electrically pumped injection laser based on a p-n junction in an indirect semiconductor, with the bottom line being the introduction of a pulsed phonon 'avalanche' into the junction just at the moment when sufficiently high population inversion is reached. These phonons should have caused a fast radiative electron 'sweeping down' across the bandgap, presumably accompanied with emission of a laser pulse (the principle was thus very like the generation of giant pulses, commonly exploited in optically pumped lasers). The idea was never experimentally realized, though. Recently Trupke *et al.* [22] theoretically analysed Dumke's calculations and concluded that in the case of indirect-bandgap semiconductors Dumke [20] underestimated the possibility of radiative recombination accompanied with the absorption of phonons, which in principle could help to overcome the losses caused by free carrier absorption. Up to now, however, the idea has not met with a positive response, and all functional injection lasers are still based on III-V direct-bandgap semiconductors such as GaAs. The issue of a silicon laser still remains open and is hotly debated, though, mostly due to the intense development of so-called silicon photonics, which will be treated in more detail in Chapter 15.

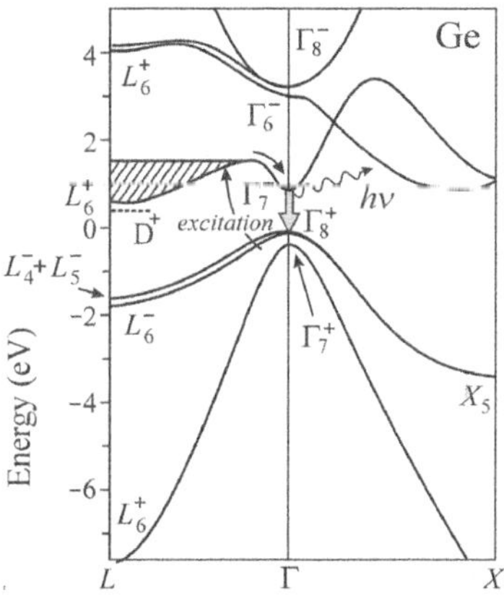

Fig. 10.12
Part of the band structure of germanium, demonstrating how to realize direct radiative recombination and lasing. (The scheme is an oversimplified visual demonstration and should not be taken literally. In fact it is the levelling out of Fermi levels in L_6^+ and Γ_7^- valleys which leads to preferential $\Gamma_7^- \rightarrow \Gamma_8^+$ radiative recombination.)

Recently, a very interesting idea of how to attain direct radiative recombination in germanium has been proposed [23]. It is based on the fact that the absolute minimum of the conduction band at L_6^+ is not excessively deep (Fig. 10.12). If this minimum were fully filled with electrons from ionized donors, under optical or electrical pumping the excess non-equilibrium electrons in the conduction band would prefer fast scattering to the Γ_7^- minimum, and radiative recombination with a hole via $\Gamma_7^- \rightarrow \Gamma_8^+$ direct transitions, or even laser action, would follow. Obviously, a high level of doping with donors is required. If, at the same time, a suitable mechanical deformation is applied in order to reduce the number of equivalent minima at the L point, estimates show that the necessary doping level would still be realistic—around 8×10^{19} cm^{-3}. The question of whether an analogous approach is feasible in silicon is left for the reader to answer (Problem 10/5).[9]

[9] Optically pumped lasing in germanium at room temperature, based on this scheme, has recently been achieved (Liu, J., Sun, X., Camacho-Aguilera, R., Kimerling, L. C., and Michel, J. (2010). *Optics Letters*, **35**, 679).

10.5 Participation of excitons in stimulated emission

So far, stimulated emission has been discussed with regard to the high-density system of free electrons and holes, namely, the electron–hole plasma EHP. In the previous chapters of this book, however, the luminescence of bound electron–hole pairs, excitons, has been widely covered. It is thus only logical to ask if and how stimulated emission can occur in various radiative recombination processes involving excitons, or their complexes. We will focus here on semiconductors with direct bandgap and, most importantly, we will show that luminescence arising from the recombination of a system containing solely free excitons (i.e. without the participation of other quasi-particles) cannot be amplified through stimulated emission; if LO-phonons (radiative A–LO or X–LO recombination) are present, however, stimulated emission can be achieved. In addition, we will also explain that stimulated emission can occur in the luminescence accompanying collisions of free excitons (X–X collisions) and during the radiative decay of excitonic molecules (EMs). These effects, however, are usually related to low temperatures without a direct application potential for optoelectronic components.

In order to simply but adequately describe stimulated emission in a system of free excitons ($h\nu \approx E_g - E_X$, Fig 10.13(a)) we will apply, like in the previous chapters of this book, kinetic equations [24]. For this purpose, they need to contain explicitly the intensity of generated radiation; this can be achieved

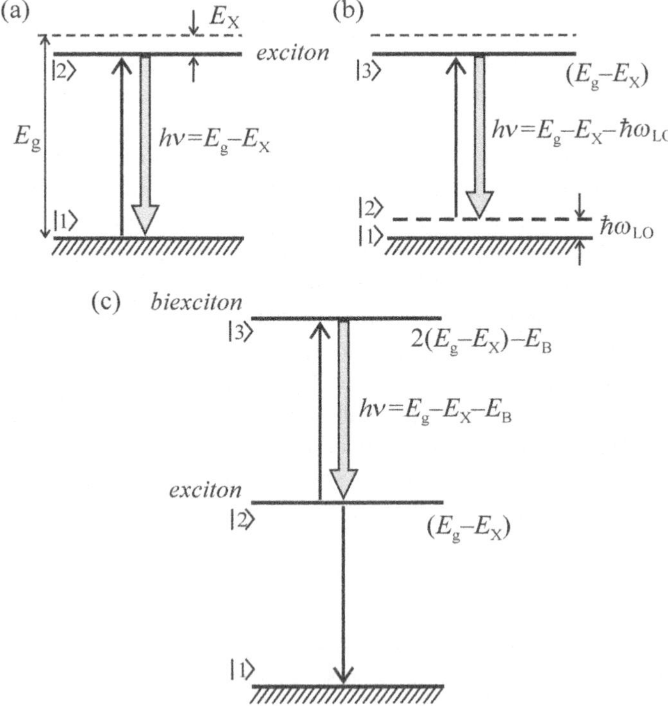

Fig. 10.13
Schematic of radiative decay of (a) a free exciton–polariton, (b) a free exciton with the emission of an LO-phonon, and (c) a biexciton or an excitonic molecule. Arrows directed upwards illustrate reabsorption. In contrast to (a), in (b) and (c) amplification by stimulated emission can occur. The similarity of (a) to a two-level atomic system, or (b) and (c) to a three-level atomic system is conspicuous. It is well known that whereas the three-level system (e.g. Cr^{3+} in sapphire, i.e. ruby) makes lasing possible, a two-level system does not. Such an analogy, however, has to be viewed with caution, because the lower level |1⟩ denotes the ground state of the entire crystal, which makes it difficult to define the concept of population inversion.

by employing the number of photons $n_{\rm f}$. The overall number of excitons (generated in the system under investigation e.g. via optical pumping), which are allowed to recombine radiatively, satisfying quasi-momentum conservation and emitting a photon to a given mode of the electromagnetic field, will be denoted as $N_{\rm X}$. If $\tau_{\rm r}$ stands for the exciton radiative lifetime, the rate of increase in the number of photons in the given electromagnetic mode through spontaneous emission is equal to $N_{\rm X}/\tau_{\rm r}$. Stimulated emission, taking place in parallel, contributes to the increase in the number of photons by $n_{\rm f}N_{\rm X}/\tau_{\rm r}$ and the overall rate of photon generation will then read

$$\frac{N_{\rm X}}{\tau_{\rm r}}(n_{\rm f}+1). \tag{10.23a}$$

These photons, however, undergo resonant reabsorption, as was mentioned in Subsection 7.1.2. The important thing now is that both excitons and photons are bosons and the roles they play can be considered symmetric (i.e. we are not restricted by the Pauli exclusion principle imposing constraints on the occupancy of levels, see e.g. the occupancy condition in (10.9) or (10.10)). This fact can also be rephrased as 'the absorption of a photon is accompanied by both stimulated and spontaneous emission of an exciton' and the corresponding term in the kinetic equations can be derived from (10.23a) by interchanging $N_{\rm X}$ and $n_{\rm f}$:

$$-\frac{n_{\rm f}}{\tau_{\rm r}}(N_{\rm X}+1). \tag{10.23b}$$

Finally, some photons have to leave the investigated sample (or laser resonator) so that we could obtain a measurable luminescence signal, which can be described with a loss coefficient γ and a term

$$-\gamma\, n_{\rm f}. \tag{10.23c}$$

All in all, the kinetic equation for photons then contains all three terms (10.23a), (10.23b), and (10.23c):

$$\begin{aligned}\frac{{\rm d}n_{\rm f}}{{\rm d}t} &= \frac{1}{\tau_{\rm r}}N_{\rm X}(n_{\rm f}+1) - \frac{1}{\tau_{\rm r}}n_{\rm f}(N_{\rm X}+1) - \gamma\, n_{\rm f} \\ &= \frac{1}{\tau_{\rm r}}(N_{\rm X}-n_{\rm f}) - \gamma\, n_{\rm f}. \end{aligned} \tag{10.24}$$

This brings us to an important conclusion: as the number of photons on the right-hand side of (10.24) stands with a negative sign, the number of photons in a given mode cannot grow in time, i.e. amplification of free exciton luminescence via stimulated emission cannot occur. This fact was recognized both experimentally and theoretically in the early 1960s in connection with the very first luminescence experiments in semiconductors under high pumping. This conclusion is in fact not very surprising if we recall the previous discussion on polaritons in Chapter 7; in this sense, excitons and photons can also be seen as a system of bound oscillators.

In other types of radiative decay of excitons—i.e. those involving another quasi-particle—stimulated emission, on the other hand, can occur. Let us consider the radiative annihilation of an exciton accompanied by the emission of an LO-phonon (X–LO, see Subsection 7.1.3). The approximate energy of the emitted photons derives from the energy conservation law and amounts to

$$h\nu \cong E_g - E_X - \hbar\omega_{LO},$$

where E_X is the exciton binding energy and $\hbar\omega_{LO}$ is the LO-phonon energy, see also Fig. 10.13(b). The key here is that the energy of the luminescence photon is now lower by $\hbar\omega_{LO}$ with respect to the exciton energy, which prevents reabsorption-related losses (at low temperatures, when the phonon population is negligible).

Let us keep denoting by n_f the number of photons. Then $n_\mathbf{q}$ will stand for the number of phonons with wavevector $\mathbf{q}$, and $N_\mathbf{q}$ for the number of excitons with the same wavevector (or, more accurately speaking, these are always mean occupation numbers in the given state or mode). We know that excitons with any value of $\mathbf{q}$ can participate in the process, since their quasi-momentum $\hbar\mathbf{q}$ is taken on by phonons. Therefore, the overall number of initial states contributing to the radiative process is much larger than in the preceding case of simple excitonic recombination, which is an additional factor facilitating the onset of stimulated emission. Let the overall number of excitons be N_X, implying $N_X = \sum\limits_{\mathbf{q}} N_\mathbf{q}$.

As photons, phonons and excitons are bosons, the rate of their generation is proportional to their numbers increased by unity: $(n_f + 1)$, $(n_\mathbf{q} + 1)$ and $(N_\mathbf{q} + 1)$. The rate of annihilation, on the other hand, is proportional only to their numbers: n_f, $n_\mathbf{q}$ and $N_\mathbf{q}$. The contribution of relevant terms to the kinetic equation will thus read:

- The rate of generation of photons (emitted together with phonons)

$$C_\mathbf{q} N_\mathbf{q}(n_f + 1)(n_\mathbf{q} + 1),$$

  where $C_\mathbf{q} > 0$ is the corresponding rate constant.
- The rate of losses due to an inverse process

$$C_\mathbf{q}(N_\mathbf{q} + 1)n_f n_\mathbf{q}.$$

- Losses caused by the escape of photons from the sample (resonator)

$$-\gamma_{LO} n_f.$$

Hence, the kinetic equation for the number of photons reads

$$\frac{dn_f}{dt} = -\gamma_{LO} n_f + \sum_\mathbf{q} C_\mathbf{q}\left[N_\mathbf{q}(n_f + n_\mathbf{q} + 1) - n_f n_\mathbf{q}\right]. \tag{10.25}$$

This equation can be simplified provided the excitonic gas is in thermal equilibrium and can be described by a thermal distribution

$$N_\mathbf{q} = N_X f_\mathbf{q}(T);$$ [10]

[10] If we assume, for the sake of simplicity, the effective gas temperature to be equal to the lattice temperature T, it follows for low temperatures and not too high densities of exci-

further, at low temperatures we can approximate $\exp(\hbar\omega_{\mathrm{LO}}/k_{\mathrm{B}}T) \gg 1$ and thus it follows from (4.6) that

$$n_{\mathbf{q}} \approx \exp(-\hbar\omega_{\mathrm{LO}}/k_{\mathrm{B}}T) \ll 1,$$
$$n_{\mathbf{q}} \ll N_{\mathbf{q}}.$$

Equation (10.25) then takes on the form

$$\frac{\mathrm{d}n_{\mathrm{f}}}{\mathrm{d}t} = -\gamma_{\mathrm{LO}}n_{\mathrm{f}} + \sum_{\mathbf{q}} C_{\mathbf{q}}\left[N_{\mathrm{X}} f_{\mathbf{q}}(n_{\mathrm{f}}+1) - n_{\mathrm{f}}n_{\mathbf{q}}\right]$$
$$\approx -\gamma_{\mathrm{LO}}n_{\mathrm{f}} + \beta N_{\mathrm{X}}(n_{\mathrm{f}}+1), \qquad (10.26)$$

where $\beta = \sum_{\mathbf{q}} C_{\mathbf{q}} f_{\mathbf{q}}$. Now it is obvious that amplification of luminescence by stimulated emission is possible. Whether it occurs or not depends on which one of the factors on the right-hand side of eqn (10.26) prevails. Conditions supporting stimulated emission may be achieved through maximizing the positive-sign term comprising the number of photons, i.e. through sufficiently high excitation (maximizing N_{X}) and under fast radiative recombination (maximizing β).

A similar approach can confirm that stimulated emission can influence a number of other luminescence processes: Luminescence of free excitons with emission of multiple phonons, collisions of free excitons (X–X), collisions of free exciton–electrons or free exciton–holes, annihilation of free excitons accompanied by emission of plasmons and, most of all, the radiative decay of an excitonic molecule (EM) into a photon and a free exciton, see Fig. 10.13(c). In all these cases, the decrease in energy of a luminescence photon by the energy of a participating quasi-particle (phonon, free carriers, plasmon, recoil exciton) prevents photons from being reabsorbed and, moreover, makes it possible for the whole excitonic population to take part in the radiative recombination.

Let us now attempt to estimate the spectral shape of the optical-gain curve $G^{\mathrm{LO}}(h\nu)$ in the process of X–LO luminescence. In analogy between the system of free excitons and an atomic gas this process can be viewed (Fig. 10.13(b)) as a (quasi-)atomic process, in which the shape of the gain $g^{\mathrm{LO}}(h\nu)$ is determined by the shape of the corresponding (spontaneous) spectral line. In the case of X–LO luminescence, the spontaneous emission lineshape is described by relation (7.12)

$$I_{\mathrm{sp}}^{\mathrm{LO}}(h\nu) \cong g^{\mathrm{LO}}(h\nu) \cong N_{\mathrm{X}}(h\nu - E_0)^{3/2}\exp\left(-\frac{h\nu - E_0}{k_{\mathrm{B}}T}\right), \qquad (10.27)$$

where $E_0 = (E_{\mathrm{g}} - E_{\mathrm{X}}) - \hbar\omega_{\mathrm{LO}}$ and, moreover, now the overall number of excitons N_{X} has been added as a coefficient. The reason is that this expression needs to be compared to a competing reabsorption process, which increases proportionally to the occupancy number of phonons $n_{\mathbf{q}} \sim \exp(-\hbar\omega_{\mathrm{LO}}/k_{\mathrm{B}}T)$

tons that $f_{\mathbf{q}}(T) \simeq (k_{\mathrm{B}}T)^{-3/2}\exp(-\hbar^2q^2/2m_{\mathrm{exc}}k_{\mathrm{B}}T)$, i.e. the Boltzmann distribution can be employed.

at the level $|2\rangle$, Fig. 10.13(b). The rate of these absorption processes will, in addition to that, be proportional to the relevant joint density of states, which can be considered here to be represented by the term $(h\nu - E_0)^{3/2}$, and the expression for the absorption coefficient thus reads

$$\alpha^{\mathrm{LO}}(h\nu) \approx (h\nu - E_0)^{3/2} n_{\mathbf{q}}. \tag{10.28}$$

The net gain coefficient can then be obtained as the difference between (10.27) and (10.28), yielding

$$\begin{aligned} G^{\mathrm{LO}}(h\nu) &= g^{\mathrm{LO}}(h\nu) - \alpha^{\mathrm{LO}}(h\nu) \\ &\approx AN_{\mathrm{X}}(h\nu - E_0)^{3/2} \exp\left(-\frac{h\nu - E_0}{k_{\mathrm{B}}T}\right) - B(h\nu - E_0)^{3/2} n_{\mathbf{q}} \\ &= AN_{\mathrm{X}}(h\nu - E_0)^{3/2} \exp\left(-\frac{h\nu - E_0}{k_{\mathrm{B}}T}\right)\left(1 - \eta \exp\left(\frac{h\nu - E_0}{k_{\mathrm{B}}T}\right)\right), \end{aligned} \tag{10.29}$$

where A, B are constants independent of photon energy $h\nu$ and $\eta = Bn_{\mathbf{q}}/AN_{\mathrm{X}}$. A more theoretically detailed calculation [25] yields a very similar formula. One can easily see from (10.29) that at low temperatures, when $n_{\mathbf{q}} \to 0$, the gain coefficient will be positive $G^{\mathrm{LO}}(h\nu) > 0$; with rising temperature, however, η rapidly grows and the term in the last bracket will quickly acquire a negative value.

The reader can show, by employing analogous reasoning, that the gain coefficient in the process of EM radiative decay can be written down in the form of (Problem 10/6)

$$G^{\mathrm{EM}}(h\nu) \approx A' N_{\mathrm{XX}}(E_0' - h\nu)^{1/2} \exp\left(-\frac{E_0' - h\nu}{k_{\mathrm{B}}T}\right)\left(1 - \eta' \exp\left(-\frac{E_0' - h\nu}{k_{\mathrm{B}}T}\right)\right), \tag{10.30}$$

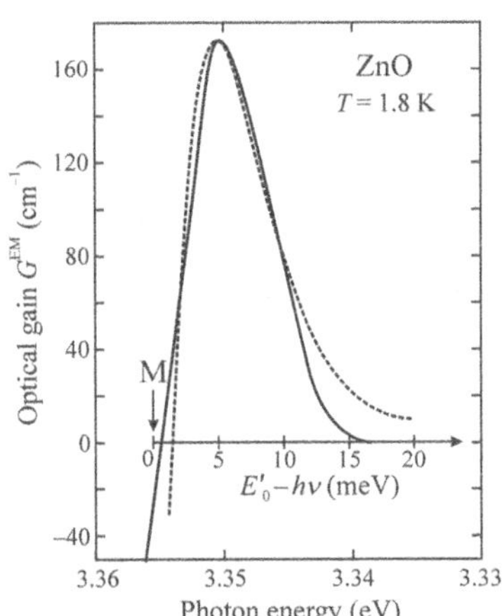

Fig. 10.14
Optical gain in EM luminescence: experimental (solid line) and theoretical (dashed line) gain curves, the latter being calculated using relation (10-30), in thin-film ZnO. The letter M denotes the position of the high-energy edge of the spontaneous emission line $E_0' = E_{\mathrm{g}} - E_{\mathrm{X}} - E_{\mathrm{B}}$. Effective temperature of the gas of excitons and excitonic molecules as one of the fitting parameters yields $T_{\mathrm{ef}} = 39$ K, bath temperature was $T = 1.8$ K, excitation was carried out with a pulsed N_2 laser (337 nm). After Hvam [28].

where A' and B' are again constants, N_{XX} is the number of excitonic molecules, $E_0' = E_{\mathrm{g}} - E_{\mathrm{X}} - E_{\mathrm{B}}$ and $\eta' = (B'/A')(N_{\mathrm{X}}/N_{\mathrm{XX}})$; E_{X} and E_{B} stand for the binding energy of an exciton and an excitonic molecule, respectively (Subsection 8.2.1). Anyone interested in more details about relation (10.30) is referred to the literature [26, 27]. One more thing to bear in mind is that, even though for the sake of simplicity thermal equilibrium of all the quasi-particles with the lattice was assumed, in reality T in (10.29) and (10.30) denotes the effective temperature of the excitonic or biexcitonic gas, which is always higher than the lattice temperature.

Figures 10.14 and 10.15 demonstrate the above discussion. Figure 10.14 shows both the experimentally obtained and theoretically calculated (using eqn (10.30)) shape of the excitonic-molecule gain in ZnO [28]. It is obvious that the gain shape keeps the typical asymmetry of the spontaneous EM emission line (Subsection 8.2.1): a tail extended towards lower photon energies. The high-energy edge of the gain curve is simultaneously shifted towards lower energies with respect to the edge of spontaneous emission $E_0' = E_{\mathrm{g}} - E_{\mathrm{X}} - E_{\mathrm{B}}$.

This shift is due to a process in which high-energy photons are those which are most strongly reabsorbed; this is quantitatively described by the coefficient η' in (10.30). Similarly to EHP stimulated emission, the gain curve is thus situated on the long-wavelength side of the spontaneous emission line.

This red-shift is easily discernible in Fig. 10.15, where one can immediately compare the spectrum of spontaneous emission (a) with the gain curve (shaded area in (b)) in thin-film ZnS [29]. While the spontaneous EM luminescence in (a) peaks at ~ 3.785 eV, the corresponding gain attains its maximum value at ~ 3.781 eV. Also, there is one more feature which makes this figure interesting: it contains optical-gain features due to two other mechanisms of radiative exciton recombination. The first one is gain at the line of free exciton collisions X–X, which was already mentioned in this section, and in addition to that, gain at the line of a bound exciton D^0–X is present. The latter mechanism of optical gain has not yet been mentioned; its existence is not at all surprising, though. An exciton bound at an impurity atom or ion is, as we already know, a synonym for an excited electronic state of such an atom. In principle, a similar laser mechanism is encountered, e.g., in a ruby or neodymium laser (even if the differences, e.g. in the concentration of impurities or the strength of their interaction with the lattice–matrix, may be substantial).

Now, let us move on to experimental techniques which yield the spectra of optical gain, shown, e.g., in Figs 10.14, 10.15 and other in Sections 10.3 and 10.5.

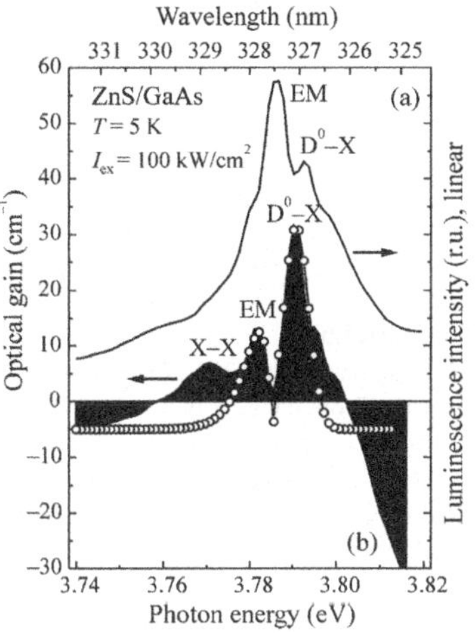

Fig. 10.15
(a) Spontaneous emission and (b) spectral shape of net optical gain in thin-film epitaxially grown ZnS on a GaAs substrate. The optical gain contains three contributions: from the luminescence of a bound exciton (D^0–X), excitonic molecule (EM) and free exciton collisions (X–X). Circles represent the theoretical shape of the gain, calculated for EM according to relation (10.30). Effective temperature of the gas of excitons and excitonic molecules was $T_{ef} = 36$ K, bath temperature $T \approx 5$ K, excitation was carried out with a pulsed laser (308 nm; 4.02 eV). After Valenta *et al.* [29].

10.6 Experimental techniques for measuring the optical gain

At the beginning of this chapter, the experimental manifestation of the optical gain in luminescence experiments was discussed mostly as a phenomenon slightly difficult to control, which can (but not necessarily) appear accidentally under conditions of strong excitation. The availability of an experimental method which is focused intentionally on confirming the presence of stimulated emission and enabling the value of the optical gain to be quantified is certainly desirable. We will now focus on two such methods: the so-called variable stripe length (VSL) method and a pump and probe (P&P) method.

Both of these techniques are based on a direct measurement of the amplification of a weak probe beam propagating through an optically pumped sample of the material under study. The most significant difference between these two methods lies in the origin of the probe beam.

10.6.1 Variable stripe length (VSL) technique

This method is in principle a photoluminescence method: the spontaneous emission itself, coming from an optically excited area of the sample, is exploited as the probe beam; the excited spot has the shape of a thin stripe of small width t. Such a shape can be obtained by focusing a pumping laser beam with a cylindrical lens. This method was introduced in the early 1970s by Shaklee *et al.* [30, 31] and its principle is depicted in Fig. 10.16. The length of

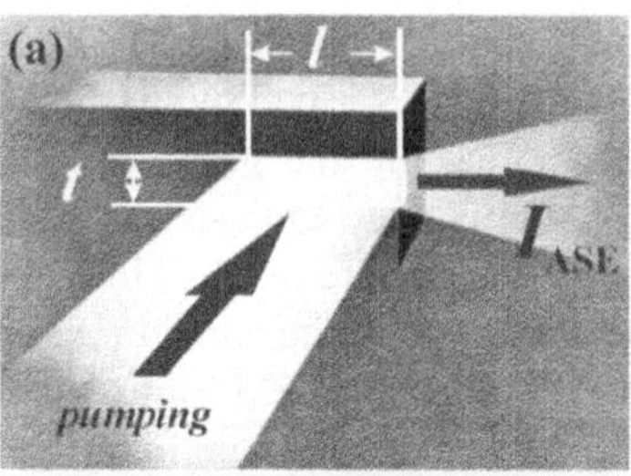

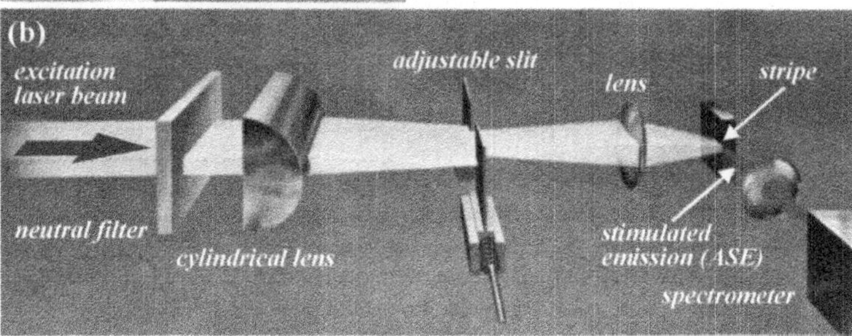

Fig. 10.16
(a) Schematic of the VSL method for measuring optical gain in a stripe characterized by length l and width t. (b) Typical experimental setup. After Shaklee *et al.* [31], reproduced courtesy of Elsevier.

the excitation stripe l varies in a defined way during the measurement (utilizing an adjustable slit or a shifting razor blade) and, at the same time, the radiation emanating from the edge of the sample in the direction of the axis of the stripe is detected, i.e. the measured signal corresponds to amplified spontaneous emission (ASE). Provided its intensity scales with increasing stripe length l superlinearly, the amplification of the (spontaneous) emission actually occurs through stimulated emission, i.e. net optical gain is present.[11]

Before we point out the characteristic features (and also some drawbacks) of this method, the whole process will be described quantitatively provided the stripe can be considered infinitely narrow ($l \gg t$) and thus treated as a one-dimensional object in the direction of the x axis. Spontaneous luminescence radiation arising in the stripe is emitted randomly in all directions and the spontaneous photons can initiate acts of stimulated emission along the x axis, which is then monitored by recording the output intensity of radiation I_{ASE} from the edge of the sample. The excited stripe thus serves (provided that sufficiently high population inversion was reached) as a single-passage amplifier[12] for photons travelling along the x axis. Consequently, the net gain $G(\nu)$ can be defined as a relative change in the intensity of radiation $I(x, \nu)$ on an infinitesimal path element $\mathrm{d}x$:

$$G(\nu) = [g(\nu) - K(\nu)] = \frac{\mathrm{d}I(x, \nu)}{\mathrm{d}x} \frac{1}{I(x, \nu)}. \tag{10.31}$$

[11] Referring to Section 2.1, this experimental set-up can be viewed as a special case of a 'reflection' arrangement under an angle of 90°. The sample needs to have both its excitation area together with its output edge polished, or it should be prepared as a thin film on a similarly treated substrate.

[12] Generally speaking, such a device is not a common laser (laser oscillator), because the positive feedback, in a laser realized by inserting the active medium into a resonator, is intentionally left out. The amplification of spontaneous emission is also often classified as lasing without mirrors.

The coefficient $K(\nu)$ now describes all the losses that in a real material act against amplification, i.e. not only the free carrier absorption as in (10.22), but also the absorption on residual impurities, light scattering on possible inhomogeneities of the sample, etc.

The overall increase in the detected intensity $\mathrm{d}I_{\mathrm{ASE}}(x,\nu)$, related to the increase in the stripe length $\mathrm{d}x$, is then

$$\mathrm{d}I_{\mathrm{ASE}}(x,\nu) = G(\nu)I_{\mathrm{ASE}}(x,\nu)\,\mathrm{d}x + I_{\mathrm{sp}}(\nu)\,\mathrm{d}x; \tag{10.32}$$

as compared with relation (10.31), the term $I_{\mathrm{sp}}(\nu)\mathrm{d}x$, originating from (non-amplified) spontaneous emission, has been added to this equation. The quantity $I_{\mathrm{sp}}(\nu)$ thus stands for the intensity of spontaneous emission coming from unit length of the stripe and being independent of x; the latter is also presumed to be a property of the gain coefficient $G(\nu)$.

By solving (10.32)—an inhomogeneous linear differential equation—we obtain the intensity $I_{\mathrm{ASE}}(l,\nu)$ coming out of the stripe. The general solution of the corresponding homogeneous equation (when putting $I_{\mathrm{sp}}(\nu) = 0$) is simple and yields $I_{\mathrm{ASE}}(l,\nu) = \text{const} \times \exp(Gl)$, the application of the initial condition $I_{\mathrm{ASE}}(l = 0,\nu) = 0$ implying const $= 0$. Thus, in order to solve the equation, one has no choice but to find the particular solution with a 'non-zero right-hand side' $I_{\mathrm{sp}}(\nu) \neq 0$ and the variation of constants method [32] leads to the solution

$$\begin{aligned} I_{\mathrm{ASE}}(l,\nu) &= \left[\int_0^l I_{\mathrm{sp}}(\nu)\exp\left(\int -G(\nu)\mathrm{d}x\right)\mathrm{d}x\right]\exp\int_0^l G(\nu)\mathrm{d}x \\ &= \frac{I_{\mathrm{sp}}(\nu)}{G(\nu)}\left(\mathrm{e}^{G(\nu)l} - 1\right). \end{aligned} \tag{10.33}$$

Consequently, by measuring the dependence of I_{ASE} on l and by subsequently fitting this dependence using relation (10.33) we obtain values of the gain coefficient $G(\nu)$ as a fitting parameter for chosen wavelengths $\lambda(= c/\nu)$. It is also possible to retrieve the overall spectral profile of the gain by measuring, for two particular stripe lengths l_1 and $l_2 = 2l_1$, relevant emission spectra and dividing them. For this case, we can get from (10.33)

$$G(\nu) = \frac{1}{l_1}\ln\left(\frac{I_{\mathrm{ASE}}(l_2,\nu)}{I_{\mathrm{ASE}}(l_1,\nu)} - 1\right). \tag{10.34}$$

From the terminological point of view, it is useful to comment on the net gain coefficient here. A waveguiding effect can be present in the sample as a result of either intentional preparation of the sample in the form of a planar (or any other) waveguide, or because of an increase in the refractive index within the area of the stripe due to the high concentration of photoexcited carriers that may occur. Under such circumstances, it is necessary to take into account the mode confinement factor Γ_{M}, defined as the fraction of the radiant power confined in the core of the waveguide ($0 < \Gamma_{\mathrm{M}} < 1$, $\Gamma_{\mathrm{M}} = 0.8$–$0.9$ in high-quality waveguides). Instead of the term 'net gain', $G(\nu)$ is then referred to as the *modal* gain, $g(\nu)$ is called the *material* gain and one can write down

$$G(\nu) \equiv G_{\mathrm{modal}} = \Gamma_{\mathrm{M}}g(\nu) - K(\nu). \tag{10.35}$$

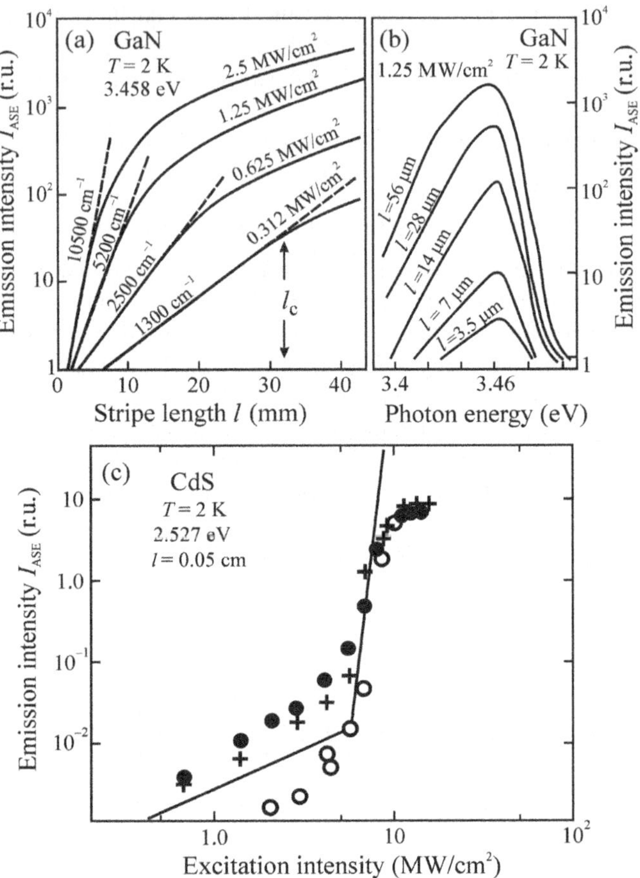

Fig. 10.17
(a) GaN at $T = 2$ K: dependence of $I_{ASE}(h\nu = 3.458\,\text{eV})$ on the stripe length l in a log–lin scale. The gain $G(\nu)$ can be determined from the slope of the linear parts of the curves. The values of $G(\nu)$ and of the excitation intensity (a pulsed N_2-laser) are shown at each curve. (b) GaN at $T = 2$ K: development of the emission spectrum $I_{ASE}(l, \nu)$ with increasing stripe length at an excitation intensity of $1.25\,\text{MW/cm}^2$. The emission originates in radiative EHP recombination. After Dingle *et al.* [33]. (c) CdS at $T = 2$ K: intensity dependence of $I_{ASE}(h\nu = 2.527\,\text{eV})$ at a constant stripe length $l = 500$ μm. Different symbols denote different samples. In this case, gain occurs at the P-line (X–X collisions). After Shaklee and Leheny [30].

One of the very first applications of the VSL method is depicted in Fig 10.17 [30, 33]. There is one more aspect which makes this figure noteworthy: it dates back to as long ago as 1971 when it demonstrated an exceptionally high optical gain in GaN. Although this demonstration singled out GaN as an ideal material to serve as an active medium in an injection laser, it took 30 more years before an InGaN-based laser was put on the market. The journey from basic research to commercially available application output is not necessarily straightforward and various obstacles of technological problems might be encountered on the way. The story of the 'blue' GaN-based laser is summarized in a monograph [34].

In addition to that, one more thing catches the eye in Fig. 10.17(a). The fitted curves determined by (10.33)—dashed lines—systematically deviate from experimental dependencies (solid lines) upon exceeding a certain critical stripe length l_c, which gets shorter with increasing excitation intensity. This effect is due to gain saturation, a phenomenon well-known from laser physics: at high photon densities, the rapid onset of stimulated emission is quickly depleted, which, in this case, leads to only a linear increase in I_{ASE} with increasing l

for $l > l_c$.[13] The saturation effect is not included in the assumptions applied when deriving (10.33) and the occurrence of the saturation deviation in the VSL curve such as that in Fig. 10.17(a) can be considered another proof supporting the onset of stimulated emission. Frequently, an abrupt increase in the slope of the intensity dependence of $I_{ASE}(l, \nu)$ at a constant stripe length l (Fig. 10.17(c)), and a corresponding narrowing of the emission spectrum, are also viewed as additional supporting arguments. As for the narrowing, however, it does not have to be necessarily evident in every sample (we should not forget that resonator feedback is not employed); it can be missing, e.g. if the luminescence originates in the EHP or if several parallel emission mechanisms take place. Moreover, what definitely cannot be left out and what is of fundamental importance in evaluating spectral narrowing here is a careful determination of the zero-signal level.

Nevertheless, it is obvious this method works flawlessly for high values of the gain $G(\nu)$. If $G(\nu)l \geq 2.5$, then $\exp(G(\nu)l) \gg 1$, the I_{ASE} intensity rises purely exponentially with increasing l and the occurrence of stimulated emission cannot be doubted.

The reliability of VSL, however, gets worse at low net gain values. This can be simply deduced from the fact that at $|G|l \ll 1$ the exponential factor can be approximated as $\exp(Gl) \approx 1 + Gl$ and from (10.33) it follows that $I_{ASE}(l, \nu) \approx I_{sp}(\nu)l$. The emission intensity scales only linearly with increasing l and, consequently, the method is unable to distinguish small positive gain ($G > 0$) from weak attenuation ($G < 0$). Generally, already on condition that $Gl \approx 1$, a large number of experimental parameters can negatively influence the measurement, which did not have to be considered up to now:

(a) Finite stripe width t; if it is too large, the one-dimensional approximation (10.33) is no longer valid. (Widths below approximately 10–20 μm cannot commonly be achieved, without employing specialized equipment, however.)
(b) Constant pump intensity along the whole stripe length. From this point of view, what appears to be ideal in the VSL method is the application of lasers or excitation sources with a rectangular beam profile and uniformly distributed pump energy density over the entire beam cross-section (N_2-laser 337 nm, XeCl excimer laser 308 nm). Such a beam is transformed into a stripe of constant t and constant excitation intensity over the whole length l via focusing by means of a cylindrical lens (Fig. 10.16(b)). If other types of lasers are exploited (with a Gaussian beam profile), only a small central part of the beam needs to be selected, using, e.g., an expander.
(c) Constant efficiency of coupling of the I_{ASE} emission to the monochromator, or the detector, from any part of the stripe.
(d) Diffraction of the pump beam on the slit or a razor blade, limiting the utilizable stripe length.

[13] For $l \leq l_c$, the term 'small-signal gain coefficient' is sometimes used.

Unsuitably chosen parameters (a) through (c) or diffraction effects can lead to experimental artefacts (artificial nonlinear I_{ASE} growth), which can be mistakenly interpreted as the manifestation of stimulated emission.

Can this be prevented? Yes, to a certain extent, and the corresponding instructions on how to do it can be found in [35, 36]. Let us start with a short overview of diffraction effects, which are inevitably present, but whose influence can be minimized by choosing a suitable imaging system between the cylindrical lens and the sample. As a rule, the imaging system is usually chosen so that it geometrically reduces the length of the stripe 2–3 times with regard to its length in the focal plane of the cylindrical lens, where the slit is inserted (Fig. 10.16(b)). This reduces correspondingly the overall length of the diffraction area as against the case when the slit is positioned directly on the sample surface. In addition, the diffraction pattern can be easily detected directly during the experiment by collecting, together with the I_{ASE} emission, the scattered pump radiation for every fixed stripe length l. The relevant part at the beginning of the stripe with a distinctive diffraction pattern will then be excluded from analysis of the experiment, see Fig. 10.18 [37].

Now, let us focus on item (c) from the list, i.e. the coupling of the emission from the stripe to the detector. In principle, two undesirable effects can cause the signal to be collected preferentially from a certain spot or area of the excited stripe. They are the waveguiding effect and the confocal effect. Firstly, the *waveguiding effect* can occur in samples in which the luminescence stripe is excited in the core of a planar, or any other type of waveguide, having a primarily positive influence on the amplification of spontaneous emission. On the other hand, both guided and substrate modes can leave the edge of the sample under particular angles, see Fig. 10.19(a). These angles can be different for different stripe lengths l, being larger for short stripes near the edge of the sample. Consequently, part of the radiation leaving the stripe might happen

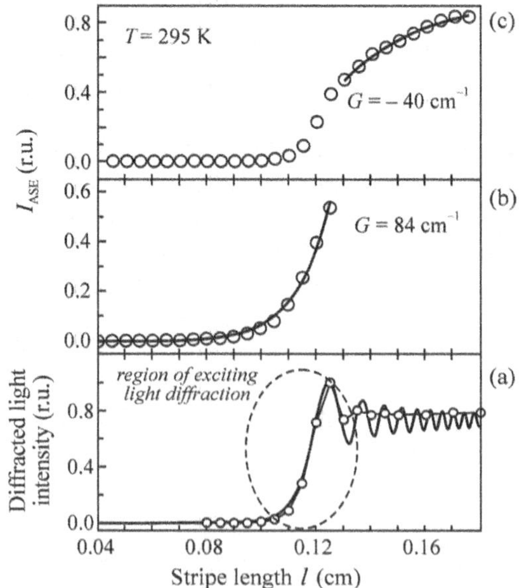

Fig. 10.18
Influence of pump light diffraction in the VSL method. The stripe begins at the coordinate $l_0 \approx 0.11$ cm. (a) Diffraction pattern of the pump radiation (Ar^+ laser 365 nm) at the beginning of the stripe obtained experimentally (symbols) as well as computationally (solid curve), (b) VSL measurement (symbols) and fit according to eqn (10.33)—solid curve. The exponential growth is an artefact and with fitting parameter $G = 84\ \mathrm{cm}^{-1}$ means apparent gain only. (c) Theoretical VSL fit outside the diffraction area yields $G = -40\ \mathrm{cm}^{-1}$, i.e. in reality the sample exhibits only optical losses. Silicon nanocrystals are the material under investigation. After Dal Negro *et al.* [37].

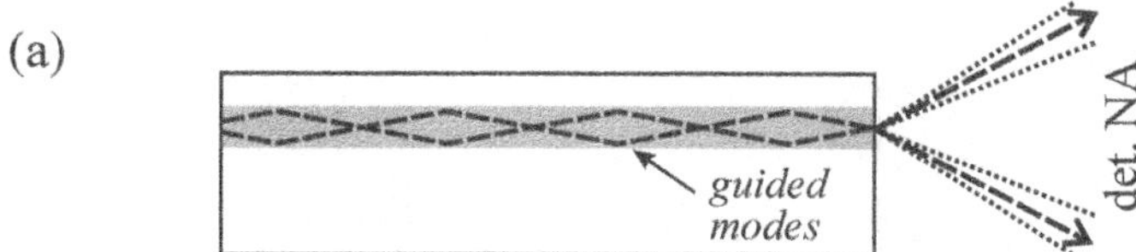

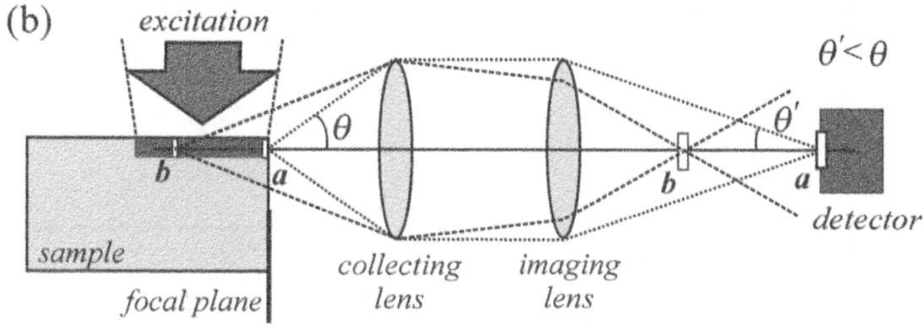

Fig. 10.19
(a) Waveguiding effect in the VSL method. Guided modes are highlighted in the cross-section of a planar waveguide; emission leaves the waveguide towards a detection system characterized by a numerical aperture NA. (b) Confocal effect in the VSL method.

not to be collected by the numerical aperture (NA) of the collecting lens, while all the radiation is collected for longer stripe lengths l. Therefore, the $I_{ASE}(l)$ dependence can appear to contain a nonlinear increase in its initial parts, posing as false gain.

Secondly, the *confocal effect* is depicted in Fig. 10.19(b). It consists in the fact that the very edge of the sample (spot a in Fig. 10.19(b)) is usually the area which is imaged by the optical system onto the detector or the entrance slit of the spectrometer. Commonly, a combination of a large NA and a short focal length of the collecting lens is applied in order to maximize the collecting system's efficiency while a longer focal length of the imaging lens is often preferred to fit the NA of the spectrometer (Section 2.5). The image is thus magnified in the detection plane, as also indicated in Fig. 10.19(b). During the measurements of small gains, a long stripe needs to be excited, leading to a situation when the image of emitting spots distant from the output edge (spot b in Fig. 10.19(b)) is formed in front of the detector or slit. If the detector area is small or if the entrance slit is narrow, part of the emitted radiation will be lost for detection. The larger the NA of the collecting lens (i.e. the smaller the depth of focus) and the larger the magnification of the optical system, the more significant influence this effect has. In principle, the very same effect is successfully exploited in confocal microscopes to gain depth resolution; in the VSL method, however, this is an undesirable phenomenon. In particular, if the focal point of the collecting lens is not located exactly at the edge of the sample but lies a bit deeper inside, artefacts posing as false growth of $I_{ASE}(l)$ with increasing l can appear.

What would help to prevent this problem is to use a collecting lens with small NA, and therefore with long depth of focus; however, such a collecting system significantly reduces the collection efficiency of I_{ASE}. Fortunately, both the confocal and waveguiding effects can be to a large extent corrected for by combining the VSL method with a method called *shifting excitation spot* (SES) [38].

The SES approach represents in fact a modification of the VSL method. It is based on employment of the same experimental setup with a solitary

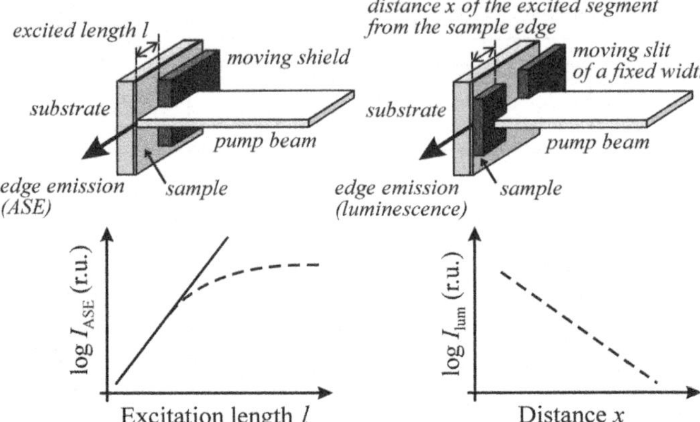

Fig. 10.20
Comparison of the VSL (left) and SES (right) methods.

change: only a very short part (a segment) of the stripe is excited instead of exciting the whole length of the stripe, and this segment is shifted along the axis of the stripe, see Fig. 10.20. At the same time, the radiation leaving the edge of the sample is detected; this radiation, however, certainly remains non-amplified by stimulated emission, because it propagates only through the non-excited part of the sample. On the contrary, it can be attenuated via various loss mechanisms (residual absorption, scattering) on its way through the sample, as demonstrated on the right-hand side of Fig. 10.20. Let the intensity of this emission be $I_{SES}(x, \nu)$; by summing up the contributions of all the segments over the path l we obtain

$$I_{SES}(l, \nu) = \int_0^l I_{SES}(x, \nu)\mathrm{d}x \qquad (10.36)$$

and then we can compare this curve (as a function of l) with the $I_{ASE}(l, \nu)$ curve described by relation (10.33).

If we find that both the curves coincide along the whole length of the stripe (including the possible false exponential part), i.e. $I_{SES}(l, \nu) \equiv I_{ASE}(l, \nu)$, the investigated material obviously does not exhibit optical gain. If, on the other hand, a deviation of the $I_{ASE}(l, \nu)$ curve from the $I_{SES}(l, \nu)$ curve towards higher intensity values is observed, especially for longer stripe lengths, this can be considered proof of the occurrence of optical gain. Nevertheless, if the curves are to be compared, they need to be normalized 'in a suitable way', the intensity dependence of emission needs to be taken into account and the influence of possible absorption bleaching remains to be clarified thoroughly. Several examples will be given in Chapter 15 and the reader is referred to [36] and [38] for a more detailed discussion.

Finally, we have to mention one more, equally important application of the SES method. It can be easily employed to determine the loss coefficient K of the material under study. If a 'point-like' segment is excited with a weak excitation beam and if the spontaneous-emission signal leaving the edge of the sample is detected at the same time, the intensity of this emission is attenuated as the photons propagate a non-excited medium, according to the well-known relation exp $(-Kx)$, where x stands for the distance of the segment from the edge. Thus K can be directly obtained by fitting the decrease of the spontaneous emission (measured for different values of x) with the above-mentioned exponential function.

10.6.2 Pump and probe (P&P) method

The P&P method exploits, like the VSL method, optical pumping of a certain part of the sample with a strong laser beam. It does not, however, rely on the occurrence of spontaneous photoluminescence, but monitors the processes in the excited area using a weak external probe beam (Fig. 10.21). Changes in the absorption coefficient of the probe beam (or the sample optical transmittance) are then analysed. Let I_0 be the intensity of the incident probe beam. The intensity of this beam after passing through the sample (no pumping is applied) will be denoted as I_{OFF}. In the presence of the pump beam let the intensity of the probe behind the sample be I_{ON}. Furthermore, let $\alpha_0(\nu_s)$ denote the absorption coefficient at the frequency ν_s of the probe beam in the absence of pumping (also referred to as the linear absorption coefficient), and $\alpha_0(\nu_s) + \Delta\alpha(\nu_s)$ will denote this absorption coefficient altered by $\Delta\alpha(\nu_s)$ as a result of nonlinear processes induced by the pump beam.

In most cases, the so-called *differential transmittance* is then the investigated quantity. It is defined as $T_D = (I_{ON} - I_{OFF})/I_{OFF}$, which can be rewritten using the Lambert–Beer law in the form

$$
\begin{aligned}
&T_D(\nu_s) \\
&= \frac{(1-R)I_0(\nu_s)\exp[-(\alpha_0(\nu_s)+\Delta\alpha(\nu_s))d] - (1-R)I_0(\nu_s)\exp(-\alpha_0(\nu_s)d)}{(1-R)I_0(\nu_s)\exp(-\alpha_0(\nu_s)d)} \\
&= \exp(-\Delta\alpha(\nu_s)d) - 1, \qquad (10.37)
\end{aligned}
$$

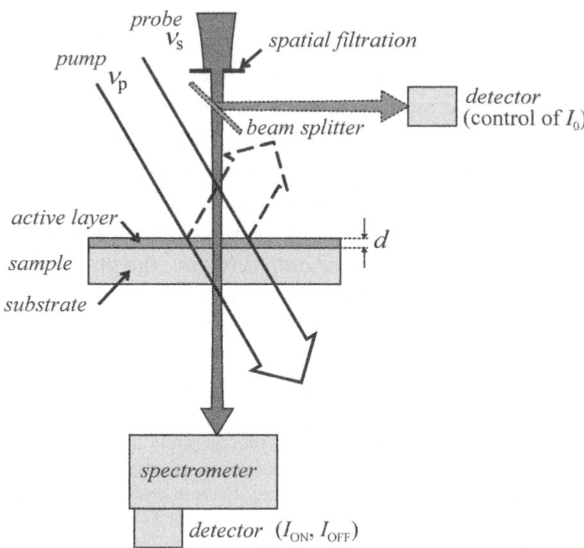

Fig. 10.21
Typical schematic of the pump and probe method for the measurement of optical gain.

where d is the thickness of the sample and R its reflectivity. The possible change of R as a result of sample irradiation with the pump beam was not taken into account in (10.37); such an approximation is, however, usually considered reasonable.

If experiment reveals $T_D > 0$, thus $\Delta\alpha < 0$, then—at first glance—we could deduce that amplification of the intensity of the probe beam by stimulated emission was achieved in the sample, which implies that positive optical gain was measured. In such a case, however, we should exercise prudence.

Two different scenarios are possible:

(1) $|\Delta\alpha| \leq \alpha_0$. Therefore, only a nonlinear decrease in absorption at the frequency ν_s is present as a result of bleaching.
(2) $|\Delta\alpha| > \alpha_0$. Only now has the gain overcome the losses and real amplification by stimulated emission takes place. The difference $G = |\Delta\alpha| - \alpha_0$ is then called the net gain G.[14]

Thus, it is necessary to compare quantitatively the linear absorption α_0 with the differential absorption $\Delta\alpha$ in order to confirm confidently the presence of net optical gain. (If we are sure that $|\Delta\alpha| \gg \Delta\alpha_0$ we naturally do not have to subtract the 'background' due to linear absorption.) Figure 10.22 [39] serves as an example of a P&P measurement. We would just like to remind the reader that the gain spectra displayed in Figs 10.7(b) and 10.8 were also obtained using the P&P method.

Now, let us focus on the experimental details. The pump beam usually comes from a high-power pulsed laser and it is advantageous to send it to the sample at an inclined angle, so that the transmitted and scattered pump radiation does not hit the detector monitoring the intensity of the probe behind the sample. The diameter of the excited spot should be several times larger than that of the probe beam on the sample (e.g. 500 μm/50 μm) in order to detect only in the area of homogeneous excitation within the pumped spot. To ensure spatial overlap of the two beams, a spatial filter (pinhole) can be applied, or the area can be checked using a camera or a microscope with a long working distance. The probe beam intensity needs to be at least one order of magnitude weaker than that of the pump beam not to cause nonlinear optical effects on its own.

Both the faces of the sample should be of high optical quality and the sample should have the form of either a self-standing membrane or a thin film on a transparent substrate. The thickness of the sample d is limited by both the necessity to ensure perfect overlap of the pump and probe beams within the whole excited volume, and by the requirement imposed by the pump frequency ν_p on the sample to appear *optically thin*: the product $\alpha_0(\nu_p)d$ has to be sufficiently small to allow the pump beam to penetrate the entire depth of the

[14] The P&P method yields in most cases *material* gain, because the beam is not sensitive to waveguiding effects during its passage through a thin film. In correct terminology, we should therefore write $g = |\Delta\alpha| - \alpha_0$. In principle, it is thus possible to encounter a situation when $g > 0$, with low optical quality of the film (waveguide), causing the modal gain G to be negative, preventing the onset of stimulated emission or lasing. Even in custom-made laser structures, the material gain can be much higher than the modal gain (Chapter 14). For the sake of simplicity, in the following discussion we shall not turn our attention to the difference between g and G.

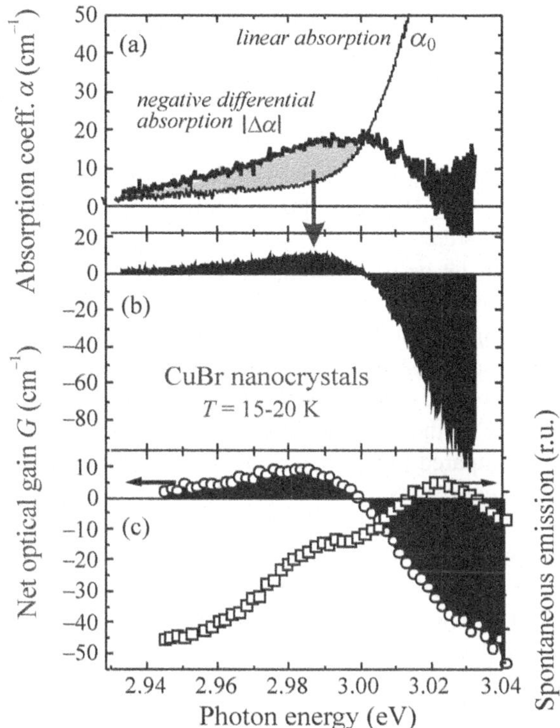

Fig. 10.22
Comparison of the P&P and VSL methods. Measurements were carried out on CuBr nanocrystals embedded in a glass matrix under an excimer 308-nm laser at $T = 15$–$20\,\text{K}$. (a) Linear absorption coefficient α_0 in the exciton region and negative differential absorption $-\Delta\alpha = |\Delta\alpha|$. (b) Net optical gain $G(\nu)$ given by the difference $G = |\Delta\alpha| - \alpha_0$. (c) Net gain spectrum $G(\nu)$ (circles) and spontaneous emission spectrum $I_{sp}(\nu)$ (squares), determined by the VSL method. Both methods yield nearly the same magnitude and spectral dependence of $G(\nu)$. After Valenta *et al.* [39].

sample and thus to generate a homogeneous population of photocarriers also in the direction normal to the surface. For example, by allowing the maximum acceptable decrease of the pumping intensity due to interaction with the sample to be 10%, we obtain $\ln(I/I_0) \approx -\alpha_0(\nu_p)d$; since $\ln(0.9) \approx -0.105$, then the upper bound of the thickness is $d \approx 1\ \mu\text{m}$ for $\alpha_0(\nu_p) = 10^3\ \text{cm}^{-1}$.

Now, we wish to estimate the sensitivity of this method. First, we realize that the lowest detectable change in the absorption coefficient $\Delta\alpha$ is determined mostly by the dynamic range of the detection system. The dynamic range as such is defined as the ratio of the maximum detectable signal S_{max} to the noise level of the detector, i.e. to the noise equivalent power NEP (Section 2.2). Besides, there is a limitation due to shot noise, which scales with the measured signal like $\sqrt{S}$. In other words, the induced relative change in transmission T_D has to be larger than the reciprocal value of the *effective dynamic range* $r = S_{max}/(\text{NEP} + \sqrt{S})$. It thus follows from (10.37) that

$$\frac{I_0 \exp[-(\alpha_0 + \Delta\alpha)d] - I_0 \exp(-\alpha_0 d)}{I_0 \exp(-\alpha_0 d)} \geq \frac{\text{NEP} + \sqrt{S}}{S_{max}} = r^{-1},$$

or

$$\exp(-\Delta\alpha d) - 1 \geq r^{-1}. \tag{10.38}$$

To simplify the problem, let us consider $|\Delta\alpha| \gg \alpha_0$ and therefore $G = -\Delta\alpha(> 0)$. This converts (10.38) to

$$Gd \geq \ln(1 + r^{-1}),$$

which for $r \gg 1$ enables us to apply the well-known approximation $\ln(1 + 1/r) \approx r^{-1}$, yielding

$$Gd \geq r^{-1}.$$

By inserting a typical value $r = 10^3$ we obtain the sensitivity limit of the P&P method in the form $Gd \geq 10^{-3}$ and if $d \approx 1$ μm, the minimum detectable gain amounts to $G_{\min} \approx 10\,\mathrm{cm}^{-1}$.

Last but not least, let us try to compare the P&P and VSL/SES methods. From the experimental point of view, it is easier to employ the VSL method as it requires a single (pump) laser beam and, besides, in principle it is applicable even to samples prepared on non-transparent substrates. Because of the possible occurrence of artefacts, discussed in Subsection 10.6.1, the threshold sensitivity of the VSL method—expressed in terms of the product of Gl—is worse compared to the case of the P&P method; it can be estimated to be $Gl \approx 1$. However, since usually the relation $l \gg d$ applies, l can reach values as high as $l \approx 1$ cm if the sample is large enough and of sufficiently high quality, implying the lower limit of detectable gain is roughly about $G_{\min} \approx 1\,\mathrm{cm}^{-1}$. Nevertheless, the VSL method features also an upper bound of applicability, stemming from two phenomena: firstly, saturation of the I_{ASE} signal limits the method to $Gl_c \leq 5$ and, secondly, diffraction effects on the slit (razor blade) make it impossible to measure gain reliably if the stripe is shorter than about 10 μm. Thus, the maximum detectable net gain value (producing saturation of the I_{ASE} signal for stripes as short as $l = 10$ μm) amounts to $G_{\max} \approx 5 \times 10^3\,\mathrm{cm}^{-1}$. This upper limit does not, however, cause problems in real-life measurements, because gains reaching higher values are in reality rarely encountered, except perhaps for, e.g., GaN (see Fig. 10.17). All these considerations are summed up in Table 10.1.

To close, the P&P method is more suitable for materials with higher optical gain, while the VSL method (combined with the SES approach) is more advantageous if the values of G are lower. Obviously, a relatively wide interval of the values of G in which both methods are applicable exists, as demonstrated in Fig. 10.22. The handicap of lower sensitivity of the P&P method can be overcome if the sample is prepared in a special way, i.e. in the form of a channel waveguide. The thickness of such a sample, or—more precisely in this particular case—the length of the waveguide d, can reach values from several millimetres to centimetres (Fig. 10.23). The preparation of these special samples (of sufficiently high quality), however, is not at all simple, which is probably why such samples are, for the time being, only scarcely used. They can be exploited for measuring modal gain by the P&P method.

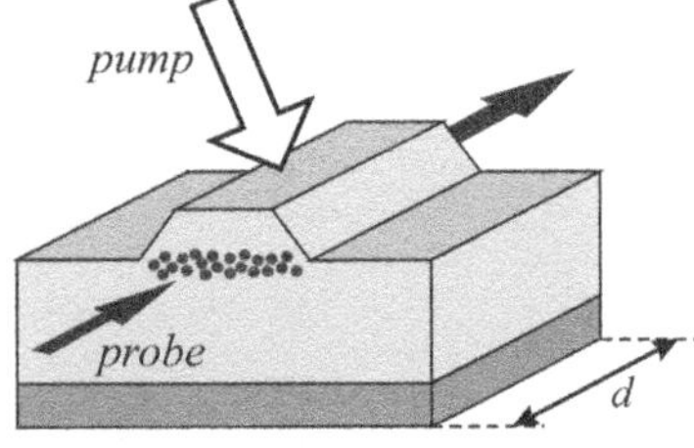

Fig. 10.23
Schematic of the pump and probe method applied to a thick sample in the form of a channel waveguide.

Table 10.1 Limits of applicability of the P&P and VSL methods for measurements of optical gain.

Method	Lower limit	Upper limit	Note
P&P	$Gd > 10^{-3}$	–	$\alpha(\nu_p)d \leq 0.1,\ d \leq 1$ μm
VSL	$Gl \geq 1$	$G \leq 5 \times 10^3\,\mathrm{cm}^{-1}$	

10.7 Problems

10/1: Try to find the reasons for the M line being present in Fig. 10.1(b), in contrast to the spectra shown in Figs 10.1(a) and 10.1(c).

10/2: Show that if electron levels $|1\rangle$ and $|2\rangle$ are degenerate (the corresponding degeneracy factors being g_1 and g_2), the relation $B_{12} = B_{21}(g_2/g_1)$ applies for the Einstein coefficients. This implies that the higher the degeneracy of the upper level $|2\rangle$, the harder it is to achieve stimulated emission.

10/3: Using relation (10.14) and Fig. 10.4(b), give a derivation of the Bernard–Duraffourg condition (10.16) by calculating the Fermi inversion factor $\Gamma(\nu)$. Next, show that $\mathrm{d}\Gamma(\nu)/d\nu < 0$. For the sake of simplicity, assume equality of the effective masses $m_e = m_h$.

10/4: Show that degeneracy of the electron or hole gas in a semiconductor can result from either (a) light masses of quasi-particles or (b) low (effective) temperature of the gas or (c) high density of the gas; it can evidently be due to a combination of all three factors. (Hint: express the density of particles as a function of the Fermi energy F_e or F_h and apply relation (10.19).)

10/5: Making use of the energy band structure of silicon (e.g. Fig. 7.16) estimate the concentration of donors necessary for filling all the equivalent minima of the conduction band near the X point. Would it then be possible for direct *e–h* recombination at the Γ point to occur? Apply the parabolic approximation of the conduction band and use $m_{de} = 0.322m_0$ for the electron density-of-states effective mass.

10/6: Consider the luminescence decay of an excitonic molecule to a free exciton FE and a photon: $\mathrm{EM}(\mathbf{K}) \rightarrow \mathrm{FE}(\mathbf{K}) + h\nu$. The inverse process of EM generation via reabsorption of a photon by an exciton ($\mathrm{FE}(\mathbf{K}) + h\nu \rightarrow \mathrm{EM}(\mathbf{K})$, i.e. optical absorption induced by a strong excitation), constitutes losses during amplification of spontaneous luminescence by stimulated emission. Show that the corresponding optical gain coefficient arising from the EM line can be described by relation (10.30). Hint: start with the inverse Maxwell–Boltzmann shape of the spontaneous emission line (8.6) and, in addition to that, consider that the thermal distribution of excitons on level $|2\rangle$ in Fig. 10.13(c) is proportional to $\exp(-\hbar^2 K^2/2m_{exc}k_B T)$.

10/7: Answer qualitatively the following questions:

(1) Why is the gain curve arising from the EHP luminescence line considerably broader (full width at half maximum is the order of tens of meV) than in exciton-based gain mechanisms (the order of meV)?

(2) Why are high-energy photons those being mostly reabsorbed in the loss process ($\mathrm{FE} + h\nu \rightarrow \mathrm{EM}$) that reduces gain on the line of an excitonic molecule?

(3) Is it in principle possible to achieve positive optical gain on the line of the radiative decay of an excitonic molecule in an indirect-bandgap semiconductor?

References

1. Yoshida, H., Saito, H., Shionoya, S., and Timofeev, V. B. (1980). *Solid State Comm.*, **33**, 161.
2. Lysenko, V. G., Revenko, V. I., Tratas, T. G., and Timofeev, V. B. (1975). *JETP*, **68**, 335.
3. Yoshikuni, Y., Saito, H., and Shionoya, S. (1979). *Solid State Comm.*, **32**, 665.
4. Liu, J.-M. (2005). *Photonic Devices*. Cambridge University Press, Cambridge.
5. Yariv, A. (1967). *Quantum Electronics*. John Wiley, New York.
6. Bernard, M. G. A. and Duraffourg, G. (1961). *phys. stat. sol.*, **1**, 699.
7. Lasher, G. and Stern, F. (1964). *Phys. Rev. A*, **133**, 553.
8. Hall, R. N., Fenner, G. E., Kingsley, J. D., Soltys, T. J., and Carlson, R. O. (1962). *Phys. Rev. Lett.*, **9**, 366.
9. Nathan, M. I., Dumke, W. P., Burns, G., Hill, F. D., and Lasher, G. J. (1962). *Appl. Phys. Lett.*, **1**, 62.
10. Saleh, B. E. A. and Teich, M. C. (1991). *Fundamentals of Photonics*. John Wiley, New York.
11. Stern, F. (1976). *J. Appl. Phys.*, **47**, 5382.
12. Hildebrand, O. and Göbel, E. (1976). *Investigations on the ground state energy of the electron–hole plasma in GaAs*. In: *Physics of Semiconductors* (ed. F. G. Fumi), p. 942. *Proceedings of the 13th international conference on the physics of Semiconduct*, Rome.
13. Cassey, H. C. Jr. and Panish, M. B. (1978). *Heterostructure Lasers. Part A: Fundamental Principles*. Academic Press, New York.
14. Thompson, G. H. B. (1980). *Physics of Semiconductor Laser Devices*. John Wiley, Chichester.
15. Göbel, E. (1974). *Appl. Phys. Lett.*, **24**, 492.
16. Majumder, F. A., Swoboda, H.–E., Kempf, K., and Klingshirn, C. (1985). *Phys. Rev. B*, **32**, 2407.
17. Klingshirn, C. (2005). *Semiconductor Optics*. Springer, Berlin.
18. Kunz, M., Pier, T., Bhargava, R. N., Reznitsky, A., Kozlovskii, V. I., Müller-Vogt, G., Pfister, J. C., Pautrat, J. L., and Klingshirn, C. (1990). *J. Crystal Growth*, **101**, 734.
19. Peyghambarian, N., Koch, S. W., and Mysyrowicz, A. (1993). *Introduction to Semiconductor Optics*. Prentice-Hall, Englewood Cliffs, N.Y.; Yu, P. Y. and Cardona, M. (1996). *Fundamentals of Semiconductors*. Springer, Berlin.
20. Dumke, W. P. (1962). *Phys. Rev.*, **127**, 1559.
21. Rediker, R. H. (1972). United States Patent No 3 636 471.
22. Trupke, T., Green, M. A., and Vürfel, P. (2003). *J. Appl. Phys.*, **93**, 9058.
23. Michel, J., Liu, J., Sun, X., Bernardis, S., Hong, C.-Y., Beals, M. A., and Kimerling, L. C. (2008). *Advanced Ge devices for electronic-photonic integration*. E-MRS Spring Meeting, Strasbourg.
24. Lévy, R., Bivas, A., Grun, J. B., and Nikitine, S. (1975). *Interaction between excitons at high concentrations*. In *Excitons at High Density* (ed. H. Haken and S. Nikitine), Springer Tracts in Modern Physics, Vol. 73, p. 171. Springer, Berlin.
25. Haug, H. (1968). *J. Appl. Phys.*, **39**, 4687.
26. Kushida, T. and Moriya, T. (1975). *phys. stat. sol (b)*, **72**, 385.
27. Saito, H. and Göbel, E. O. (1985). *Phys. Rev. B*, **31**, 2360.
28. Hvam, J. M. (1978). *Solid State Comm.*, **26**, 987.
29. Valenta, J., Guennani, D., Manar, A., Hönerlage, B., Cloitre, T., and Aulombard, R. L. (1996). *Solid State Comm.*, **98**, 695.
30. Shaklee, K. L. and Leheny, R. F. (1971). *Appl. Phys. Lett.*, **18**, 475.
31. Shaklee, K. L., Nahory, R. E., and Leheny, R. F. (1973). *J. Luminescence*, **7**, 284.
32. Arfken, G. B. and Weber, H. J. (1995). *Mathematical Methods for Physicists*, 4th edn. Academic Press, San Diego.
33. Dingle, R., Shaklee, K. L., Leheny, R. F., and Zetterstrom, R. B. (1971). *Appl. Phys. Lett.*, **19**, 5.

34. Nakamura, S., Pearton, S., and Fasol, G. (2000). *The Blue Laser Diode. The Complete Story*. Springer, Berlin.
35. Valenta, J., Luterová, K., Tomašiunas, R., Dohnalová, K., Hönerlage, B., and Pelant, I. (2003). *Optical gain measurements with variable stripe length technique*. In *Towards the First Silicon Laser* (ed. L. Pavesi, S. Gaponenko and L. Dal Negro), NATO Science Series, Vol. 93, p. 223. Kluwer Academic Publishers, Dordrecht.
36. Dal Negro, L., Bettotti, P., Cazzanelli, M., Pacifici, D., and Pavesi, L. (2004). *Optics Commun.*, **229**, 337.
37. Dal Negro, L., Cazzanelli, M., Daldosso, N., Gaburro, Z., Pavesi, L., Priolo, F., Pacifici, D., Franzò, G., and Iacona, F. (2003). *Physica E*, **16**, 297.
38. Valenta, J., Pelant, I., and Linnros, J. (2002). *Appl. Phys. Lett.*, **81**, 1396.
39. Valenta, J., Dian, J., Gilliot, P., and Hönerlage, B. (2001). *phys. stat. sol. (b)*, **224**, 313.

11 Electroluminescence

Most of the previous chapters dealt with luminescence processes excited by optical means, i.e. with photoluminescence. This prevalence closely correlates with the fact that the number of scientific publications related to photoluminescence problems highly outnumbers that focusing on the rest of the luminescence processes, i.e. cathodoluminescence, thermoluminescence, chemiluminescence, etc. It mainly stems from the relative simplicity of the photoluminescence technique and the minimal requirements on the shape, surface treatment and, among other things, also the electrical conductivity of the studied samples.

This approach fully meets the requirements of basic research. However, from the point of view of the present-day applications of luminescence processes in electronics, optoelectronics or our everyday life, the privileged status is given to electroluminescence as a technique of a direct, i.e. non-thermal transformation of electric energy into light. Let us devote this chapter both to the explanation of the basic terms of electroluminescence processes and to the corresponding experimental techniques. We will not, however, go into details of commercial displays, electroluminescence or LED diodes,[1] let alone semiconductor injection lasers. In this context, the reader is referred, e.g. to publications [1–4].

11.1 Historical notes

Electroluminescence radiation arises from the application of an electric field to a luminescent material.[2] If we adhere to historical terminology, we distinguish two categories of electroluminescence in semiconductors: the so-called

[1] We use the term 'LED' diode despite its being an evident pleonasm (LED = light-emitting diode). It is, however, quite a common collocation, which helps to distinguish the light-emitting diode from other diodes. Besides, the term 'LED' is sometimes used as an abbreviation for a 'light-emitting device' and, in this sense, can also refer to other types of light sources.

[2] The term 'electroluminescence' does not include gas discharge. The term *photoelectroluminescence* stands for a process when a phosphor, excited by means of an electric field, is simultaneously irradiated by light. *Electrophotoluminescence* then stands for luminescence excited by light, while its intensity is increased or decreased by an additional electric field.

Destriau effect and *injection electroluminescence*. Firstly, the Destriau effect, also referred to as *high-field electroluminescence*, is based on the excitation of certain luminescence centres by free (majority) carriers accelerated in a high electric field (of the order of up to MV/cm); both direct (DC) and alternating fields (AC) are possible. This process was first observed by Destriau in the 1930s during the application of an electric field to a suspension of zinc sulphide (ZnS) luminescent particles in an insulator [5]. These materials, in particular ZnS:Mn together with rare earth chalcogenides (SrS, SrSe, CaS), remain at the focus of electroluminescence research up to now. The application stimulus behind this research lies in a search for new materials for special types of flat displays, which have now been commercially available for about 20 years.

The second type of electroluminescence, namely injection electroluminescence, is characterized by the radiative interband recombination of majority and minority carriers injected into a semiconductor p-n junction. Although the first observation of a similar effect dates back to as early as 1907 (in a Schottky rectifying junction in silicon carbide, SiC), it is widely accepted that a thorough investigation of this effect had to await the birth of the p-n junction and transistor between 1947 and 1948, and the correct explanation was provided as late as four years later, in the beginning of the 1950s in the USA [6, 7]. Until then, light emitted by a rectifying metal–semiconductor contact had been perceived only as an unimportant curiosity and had not been well-known in the wider scientific community. However, the remarkable contribution of the almost forgotten Soviet physicist Losev is definitely noteworthy.

Although he never formally graduated from a university, during the 1920s and 1930s he published (just by himself, without any co-author) a series of publications on the observation of light emission in ZnO and SiC rectifying diodes used, in those days, in radio receivers [8]. Moreover, he correctly explained the nature of this effect as the 'inversion photoelectric effect'. As is indicated in a recent memorial article [9], it is probable that even signs of stimulated emission can be traced in his work. Losev starved to death as a technician in Leningrad Medical Institute in 1942 during the Leningrad blockade and the effect of injection electroluminescence had to await its rediscovery for another ten years [6, 7].[3]

Nevertheless, although the origin and main features of injection electroluminescence were determined reliably in the 1950s, its application potential was not fully appreciated at first. The renowned textbook of *Solid State Physics* by Dekker in its seventh edition from 1963 [10] devotes only a short closing paragraph in the chapter on Luminescence to this effect: *When a p-n junction of germanium or silicon is biased in the forward direction, electrons from the n-region penetrate into the p-region and holes flow from p to n. The minority*

[3] A similarity with the story of Nakamura immediately comes to mind. Nakamura, as an unknown scientist in the laboratories of the Nichia Corporation, developed just by himself an LED and injection laser based on GaN (Nakamura, S., Pearton, S., and Fasol, G. (2000). *The Blue Laser Diode. The Complete Story*. Springer, Berlin). After all, similar features can also be traced back in the story of Maiman who, in 1960 as a young researcher at the Hughes Aircraft Laboratories, almost alone fabricated the very first functioning laser (Maiman, T. (1960). *Nature*, **187,** 493) in spite of being disdained and underestimated by world-leading laboratories aiming at the same target. The question arises if such scenarios are the rule rather than a rare exception.

carriers so injected will recombine with their counterparts and one might expect emission of photons. This has indeed been observed by Haynes and Briggs. The emitted radiation has a wavelength which agrees well with the optical absorption associated with band-to-band transitions. For Ge and Si the radiation lies in the infrared ($\lambda = 1.77\,\mu m$ and $1.12\,\mu m$, respectively). The emission is localized in the junction region.

The physical phenomenon which by means of the discoveries of LED diodes and injection lasers radically contributed to a revolutionary change not only in telecommunications (optoelectronic light sources for optical communications) but also in many aspects of our everyday life (CD-ROM disks, bar code readers, traffic lights, etc.) had thus to await the full exploitation of its potential for another decade. The observation of the high quantum yield of electroluminescence emitted from a p-n junction in GaAs upon injection of minority carriers, published by Keyes and Quist in 1962 [11], is often viewed as an important milestone here. Those interested in further reading are referred to the review articles [12, 13].

11.2 High-field electroluminescence

High-field electroluminescence can be divided into three subcategories:

- electroluminescence of powder phosphors (the original discovery by Destriau falls into this category—the phosphor particles were dispersed in a dielectric);
- electroluminescence of thin films (a homogeneous thin layer of a phosphor filling up the space between capacitor plates—electrodes);
- electroluminescence of a reverse-biased p-n junction.

Firstly, we will examine several experimental questions, such as the sample structure and properties, and we will also mention some particularities of electrical excitation. Then, we will explain the principles of the excitation mechanism of luminescence centres in high fields; the explanation of electroluminescence of the reverse-biased p-n junction is left to Section 11.4.

11.2.1 Experimental considerations

The collection of electroluminescence radiation, its dispersion into a spectrum and subsequent detection proceeds wholly analogously to the photoluminescence case, and we can therefore refer the reader to Chapter 2. The fundamental difference here, however, is the excitation technique itself, which is reflected primarily in the sample shape and structure. They can no longer be as simple as in the case of photoluminescence.

A typical cross-section of a sample for a high-field electroluminescence measurement is depicted in Fig. 11.1 (which describes mainly the case of powder phosphors and thin films). The substrate can be a non-conductive solid slab covered with a metal electrode, as in Fig. 11.1(a). However, a semiconductor (silicon) wafer with adequate electrical conductivity substituting simultaneously the bottom electrode is often exploited. Thin insulator layers are

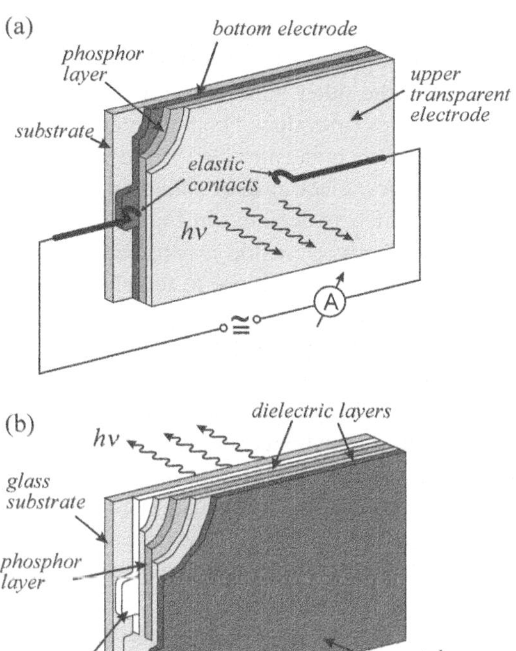

Fig. 11.1
(a) Principal schematic of a sample for high-field electroluminescence studies. (b) Arrangement of the electroluminescent panel for excitation by an alternating field. The phosphor is completely encapsulated in a dielectric, which stabilizes its properties and prevents electrical breakdown. The phosphor layer thickness ranges from hundreds of nanometres to several micrometres in both cases.

sometimes inserted in between the phosphor and the electrodes. Then, strictly capacitance coupling, not a galvanic connection, arises (Fig. 11.1(b)), and it is thus necessary to apply AC or pulsed voltage. The active layer thickness is typically hundreds of nm to several micrometres. Tin oxide, SnO_2, sometimes with indium alloy (indium-tin-oxide, ITO) usually serves as a transparent electrode. It is also possible to use a thin film of many other metals (Au, Ag, Al, ...); in the form of a thin film, they are transparent to visible light, however, they are soft and can be damaged easily. Naturally, to analyse the emission spectrum in the correct way, it is necessary to take into account, besides the spectral sensitivities of the dispersion element and the detector, also the spectral dependence of the optical transmission of the transparent electrode.

As for the preparation of these types of samples, technical equipment for thin film deposition (evaporation, sputtering, plasma-enhanced chemical vapour deposition (PECVD), etc.) has to be available. It is also possible to apply, in particular for the preparation of the active phosphor layer, a wide range of other techniques, for example spin-coating or deposition from a liquid phase.

As was mentioned earlier, both AC and DC fields can be applied for the excitation of electroluminescence; it is related, to a considerable extent, to the phosphor structure and the microscopic excitation mechanism (details will be specified in Subsection 11.2.2). It is often advantageous to work with a pulsed excitation since long-term loading with a high DC field can lead, in particular in novel and unknown materials, to fast degradation of the sample. The pulsed excitation, however, can also cause difficulties in many cases. We will mention qualitatively one of them, namely, a consequence of polarization effects.

The pulsed electric field perceives most phosphor samples as if they were capacitors because the bandgap width of typical phosphors (ZnS, etc.) is close to that of dielectrics. On the other hand, these dielectrics are far from being ideal insulators as impurities (introduced both unintentionally or intentionally) or defects cause a non-zero concentration of free carriers. As a result of polarization by free carriers, electric-field shielding occurs upon application of an external voltage, which is depicted in Fig. 11.2 [14]. The field is then attenuated in the sample body or inside a particular grain of the phosphor, whereas the field becomes enhanced close to the electrodes or near the grain boundaries (Fig. 11.2(b)). On switching the external field off (Fig. 11.2(c)), a field of opposite polarity suddenly appears, due to the polarized free carriers. Therefore, even if we apply for example an ideal rectangular voltage pulse with very fast rise time (which is easily done with present-day pulse generators) and with repetition rate $f = 1/T$, the field exciting the luminescence centres inside the sample may have a considerably different shape—see Figs 11.2(d) and (e).

This electric field behaviour obviously manifests itself also in the electroluminescence kinetics (rise and decay). The rate of the changes shown in Figs 11.2(d), (e) is driven by the time lag required to establish the equilibrium between free carrier drift and diffusion. This time lag τ_M, referred to as the *dielectric relaxation time* (or Maxwell relaxation time, respectively),

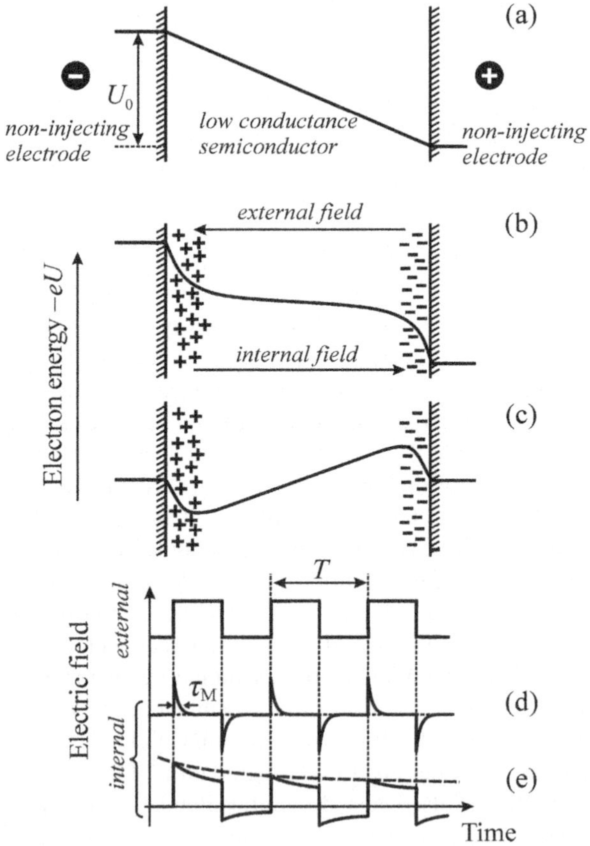

Fig. 11.2
Impact on the electric field distribution of semiconducting phosphor polarization by free carriers in the case of non-injecting electrodes or grain boundaries (capacitance coupling). (a) Spatial distribution of the electron potential energy at the moment of application of the external voltage U_0. Polarization has not yet set in; there is still a homogeneous electric field distribution in the sample. (b) The voltage is switched on; the phosphor becomes polarized after an elapsed time τ_M due to the free carrier displacement and the formation of the space charge near electrodes. (c) Moment of switching off the external voltage. A transient field of opposite polarity decaying with time constant τ_M emerges in the phosphor bulk. (d) Time dependence of the internal electric field under a periodic external pulsed voltage (with a period T), fulfilling the condition $\tau_M \ll T$. (e) Same as (d) but at $\tau_M \gg T$. After Henisch [14].

can be defined also as the mean time necessary for an extra volume charge in an insulator or semiconductor to disperse—due to the DC electric conductivity σ_{DC}, not as a result of recombination. It can be written in the form $\tau_M = \varepsilon_\infty \varepsilon_0 / \sigma_{DC}$, where ε_∞ is the relative high-frequency dielectric constant of the material [15]. Whereas ε_∞ of electroluminescent materials lies in a narrow range of values from 2 to 10, σ_{DC} can vary by many orders of magnitude. For $\varepsilon_\infty = 10$ and in relatively strongly conductive materials, where $\sigma_{DC} = 1\Omega^{-1}\,\mathrm{cm}^{-1} = 10^2\,\mathrm{Sm}^{-1}$, we therefore obtain (considering $\varepsilon_0 = 8.85 \times 10^{-12}\,\mathrm{C}^2/\mathrm{Nm}^2$) $\tau_M \simeq 9 \times 10^{-13}\mathrm{s}$, therefore τ_M is in the order of picoseconds. In this case, complete shielding of the electric field inside the phosphor will probably occur (high free carrier concentration); the time dependence of the internal field is depicted in Fig. 11.2(d). For less conductive samples, which is more typical, though, τ_M can be by many orders of magnitude longer (microseconds to milliseconds) and the field shielding will be weaker; the internal field time relaxation for $\tau_M \gg T$ is shown in Fig. 11.2(e).

How these effects manifest themselves in the electroluminescence intensity and temporal behaviour will be discussed in Subsection 11.2.3. However, it becomes evident that the sample preparation, the mode of high-field electroluminescence excitation and the subsequent interpretation of results are, as a consequence of a combination of both electrical and optical effects, considerably more complicated in comparison with photoluminescence. In fact, a wide range of other phenomena can moreover play an important role here, e.g. the very high electric fields at grain boundaries and in the vicinity of the electrodes (see Fig. 11.2(b)) and the presence of hot electrons in these regions; the processes of the recombination of electrons with holes and of surface recombination, which can lead to deviations from the equilibrium state; inhomogeneities of the phosphor and the presence of potential barriers, etc. Obviously, a sufficiently general analysis of the conditions of the excitation by an electric field is in fact impossible.

What is also worth mentioning is the question of the measurement temperature. We remember that the rate of non-radiative processes in semiconductors commonly decreases rapidly with decreasing temperature and therefore the luminescence intensity increases (Section 4.6). From this point of view, it appears favourable for basic research to perform experiments at low temperatures. On the other hand, the electrical conductivity $\sigma_{DC} = en\mu_n + ep\mu_p$ decreases with decreasing temperature. The reason for this lies in the decrease of both the free electron n and hole p concentrations and (at very low temperatures ≤ 100 K) also in the decrease of the corresponding mobilities μ_n and μ_p. This may adversely affect the efficiency of excitation of luminescence centres (Subsection 11.2.2). For this reason, studies of electroluminescence at low temperatures are not nearly as common as photoluminescence studies. Besides, the importance of applications of electroluminescence requires that electroluminescence processes are investigated at room temperature or at least in its close vicinity. In other words, if we want to investigate the microscopic nature of a luminescence process, the almost exclusive approach is by means of photoluminescence; electroluminescence and its investigation, on the other hand, have significance as a tool of targeted or applied optoelectronic research.

From this point of view, the electroluminescence efficiency is undoubtedly one of the important parameters. It is also referred to as the quantum yield in the case of photoluminescence (Section 3.1); for electroluminescence there is no point in using this term (although it is still sometimes used anyway). The electroluminescence efficiency is defined in a different way and in the case of high-field electroluminescence even several different efficiencies are introduced [16]:

1. Excitation efficiency

$$\eta_{\mathrm{ex}} = \frac{\text{Number of excited luminescence centres (cm}^{-3})}{Q/e\ (\mathrm{cm}^{-2})}$$

where Q is the total transferred electric charge per unit area and e is the elementary electron charge.

2. Radiative efficiency

$$\eta_{\mathrm{rad}} = \frac{\text{Number of centres decaying radiatively (cm}^{-3})}{\text{Number of excited centres (cm}^{-3})}$$

3. Outcoupling (external) efficiency

$$\eta_{\mathrm{out}} = \frac{\text{Number of photons emitted through the surface (cm}^{-2})}{\text{Generated photons (cm}^{-3})}$$

The number of photons emitted from the surface is limited mainly by the optical transmission of the semitransparent output electrode, total internal reflection and by the surface morphology.

4. The product of the above-mentioned partial efficiencies determines the total electroluminescence efficiency

$$\eta_{\mathrm{tot}} = \eta_{\mathrm{ex}}\eta_{\mathrm{rad}}\eta_{\mathrm{out}} = \frac{\text{Number of photons emitted through the surface (cm}^{-2})}{Q/e\ (\mathrm{cm}^{-2})}$$

Although a little in advance, we can mention here also an expression for the total efficiency of the injection electroluminescence:

$$\eta_{\mathrm{tot}} = \frac{\text{Number of photons emitted through the surface (cm}^{-2})}{\text{Number of injected electron–hole pairs (cm}^{-2})}$$

11.2.2 Mechanisms of high-field electroluminescence

To begin with, it may be useful to reflect briefly on which of the numerous mechanisms of radiative recombination discussed in the previous chapters can apply in electroluminescence. As mentioned earlier, the study of low-temperature electroluminescence is rather exceptional; hence we can exclude all luminescence processes occurring only at low temperatures. This, to be specific, stands for all the channels of radiative recombination of free and most bound excitons (Chapter 7) and luminescence of the electron–hole liquid, EHL (Section 8.4). Next, the effects of high excitation are also excluded (Chapter 8) because it is hardly possible to attain high current densities without damaging the phosphor in high-field electroluminescence.[4] Hence, there is not much left—in fact, only three potential radiative recombination channels remain:

(1) Extrinsic luminescence of localized centres, primarily ions of the transient metals and rare-earth ions (Section 5.6).
(2) Recombination processes involving the participation of neutral donors and acceptors, in particular the luminescence of donor–acceptor pairs (Section 5.4). The prerequisite here, however, is that the ionization energy of donors and acceptors is sufficiently high for the impurities to exist in their neutral state also at room temperature.
(3) Intrinsic recombination of free electron–hole pairs. The probability of such a process is, however, in comparison with processes (1) and (2) very low at room temperature.

A condition necessary for high-field electroluminescence is then the presence of *free carriers*. They are accelerated in an electric field and due to collisions they can transfer a portion of their energy to the above-mentioned luminescence centres, exciting them in this way. Possibly also free electron–hole pairs, which subsequently recombine radiatively or pass their energy to the above-mentioned extrinsic centres, can be generated during this process. In the case of a sufficiently conductive phosphor and galvanic coupling, carriers can be introduced into the luminescent material via injection from electrodes; in the capacitively coupled thin films of phosphors, on the other hand, carrier injection can be realized, e.g. from surface states located in the vicinity of the electrodes or by thermal release (with the assistance of the electric field) from deep traps in the material volume, which is the so-called Poole–Frenkel effect.

Let us leave these considerations for now and assume that free electrons or holes are present in the luminescent layer. We will now explain in more detail the essence of the excitation mechanism. Let us emphasize here that, unlike photoluminescence experiments and theoretical work which—thanks to development of laser technology in the 1960s and 1970s—boosted the discovery and interpretation of a series of new recombination processes in semiconductors, the past 40 or 50 years did not bring any fundamentally new results in electroluminescence. Strictly speaking, we will recapitulate the classical knowledge summarized comprehensively, e.g., in the review chapters of monographs [16] and [17].

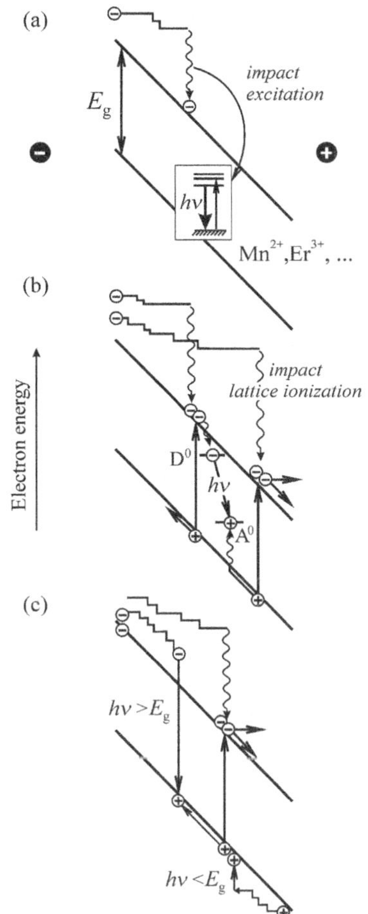

Fig. 11.3
Possible electroluminescence processes in high fields. (a) Impact excitation of an extrinsic luminescence centre (the transition metals or rare earth type of ions) by an accelerated electron: the most important electroluminescence process in ZnS:Mn. (b) Excitation of a donor–acceptor pair via impact ionization of the lattice. (c) Intrinsic electroluminescence process: Impact ionization of the pure semiconductor lattice. With a certain probability, it is possible to detect luminescence photons with energy both much higher and lower than E_g.

Impact ionization and excitation

Typical luminescence centres behind the luminescence of thin films of ZnS are Mn^{2+} ions, where photons are emitted owing to the electron transitions between levels of the not fully occupied 3d shell. A similar role can also be played by di- and trivalent rare earth ions. Processes of excitation of such centres occurring in high electric fields are depicted in Fig. 11.3. The electric field accelerates the free carriers propagating in the ZnS lattice; we will specifically

[4] The maximum attainable steady-state drift velocity of carriers in most semiconductors (in a field intensity $\geq 100\,\mathrm{kV/cm}$) is of the order of $v = 10^7$ cm/s and typical current densities are $j = 4\,\mathrm{A/cm^2}$, the relevant carrier density being $n = j/ev \approx 2 \times 10^{12}\,\mathrm{cm^{-3}}$ (e is the electron charge). The high excitation effects set in at carrier densities of at least three or four orders higher.

speak of electrons because they are the majority carriers in semiconductors of this type. The electrons transfer their energy to the luminescence centres by inelastic collisions and relax to the bottom of the conduction band (so-called impact excitation, Fig. 11.3(a)). This is the excitation mechanism that occurs in a 'standard' electroluminescence material ZnS:Mn^{2+}. For completeness, two additional types of excitation processes which can occur in high fields in other semiconductors are shown in Figs 11.3(b) and (c), namely, impact lattice ionization leading to the luminescence of donor–acceptor pairs or to a specific intrinsic emission.

It is apparent that the minimum energy of an accelerated electron in ZnS:Mn^{2+} must amount to at least $\sim 2\,\text{eV}$, which corresponds to the wavelength of about 600 nm of the radiation emitted by the Mn^{2+} centre. Electrons with such a high energy (measured relative to the bottom of the conduction band) do not occur at standard conditions of charge transport in semiconductors. The temperature T_{ef}, participating in the energy distribution function of the electrons, is then much higher than the lattice temperature T. We speak of so-called *hot electrons* (or holes).[5] In order to better understand their role in electroluminescence processes, we have to discuss them in more detail. Most convenient will be to begin with classical electrical conduction in semiconductors in weak electric fields ($\leq 10^3\,\text{V/cm}$).

In weak fields F, as is well-known, the statistical energy distribution of the electrons at the bottom of the conduction band is described by the (non-degenerate) Boltzmann distribution. Under these conditions (sufficiently low concentration, sufficiently high temperatures), electrons travel for the mean free path

$$\langle l \rangle = \langle v_{\text{T}} \rangle \tau_{\text{m}} \tag{11.1}$$

isotropically in all directions between collisions; here, τ_{m} is the mean time between collisions due to electron scattering and $\langle v_{\text{T}} \rangle$ is the mean velocity of the thermal motion. The mean velocity follows from the thermal equilibrium of the electron gas with the lattice as descibed by the equipartition theorem $m_{\text{e}}\langle v_{\text{T}} \rangle^2/2 = (3/2)k_{\text{B}}T$, or

$$\langle v_{\text{T}} \rangle = \sqrt{\frac{3k_{\text{B}}T}{m_{\text{e}}}}. \tag{11.2}$$

The electric force $-eF$ affects the equilibrium distribution of velocities only as a small perturbation, the drift velocity of electrons v_{d} in the field direction being much smaller than $\langle v_{\text{T}} \rangle$ and Ohm's law holding true. We may write down this law in the form $j = \sigma F$, where

$$\sigma = en\mu_{\text{n}}$$

[5] Hot carriers can be generated equivalently both by accelerating them in high electric fields or exciting by light highly above the gap (even if the detailed mechanisms through which the carriers acquire energy may be different). In Section 5.3 we put the accent on the concept of the *effective temperature* T_{ef} of the exciton or photocarrier gas rather than on that of hot carriers. Such terminology was employed because the term 'hot carriers' is used more commonly in the narrower sense of the word just for the case of high-field charge transport in semiconductors.

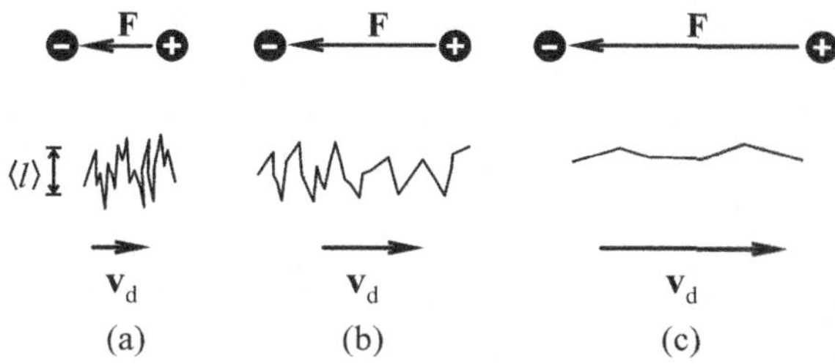

Fig. 11.4
Path evolution of an electron that is 'heated up' in a semiconductor in an electric field with increasing intensity F. Beginning with low fields and increasing F from (a) via (b) to (c), the drift velocity v_d increases, however, the mobility decreases (see text).

is the electrical conductivity, n stands for the electron concentration and

$$\mu_n = \frac{e\tau_m}{m_e} = \frac{e\langle l\rangle}{m_e\langle v_T\rangle} \tag{11.3}$$

is their mobility, representing a proportion between the drift velocity and the electric field intensity

$$v_d = \mu_n F. \tag{11.4}$$

The transport of electrons in low fields is shown schematically in Fig. 11.4(a). The drift velocity v_d is much lower than the mean thermal velocity $\langle v_T\rangle$.

Now, let us have a look at what happens when increasing the field intensity F. This represents a rather sophisticated complex of processes, in which the decisive role is played by scattering mechanisms (collisions) in which electrons change the direction of their movement and lose part of their energy [15, 18, 19]. Starting from a certain field intensity F_0, the electrons begin to 'feel' the accelerating pull of the electric field along their free path $\langle l\rangle$, and the mean velocity between collisions $\langle v'_T\rangle$ begins to rise (unlike the thermal velocity (11.2)), which means that the electron system heats up. The electron drift velocity increases; however, as long as one may still regard the field as not too high, electrons do not gain too much energy and the decisive collision mechanism in that case remains the scattering on acoustic phonons. The mean free path $\langle l\rangle$ remains constant (does not increase with increasing electron energy) and we may deduce from (11.1) and (11.3) that a decrease in τ_m and thus also a decrease in the mobility of electrons μ_n occurs for $\langle v'_T\rangle > \langle v_T\rangle$. Mobility is thus no longer a constant, but starts to be a function of F:$\mu_n = \mu_n(F)$.

The increase in the drift velocity is depicted in Fig. 11.4(b), and the accompanying decrease in mobility μ_n, which, at a first glance, might seem somewhat paradoxical, is demonstrated in Fig. 11.5 via decreasing the slope of the v_d (F) plot in segment (b).

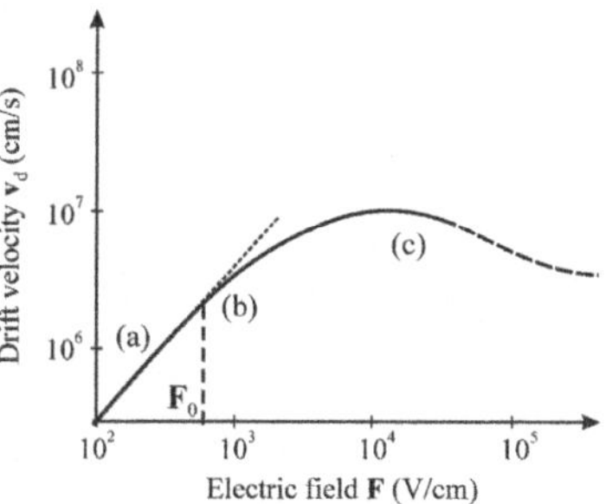

Fig. 11.5
Approximate sketch of the dependence of the electron drift velocity on the electric field intensity in a semiconductor (at room temperature). The segments (a), (b), (c) correspond to the labelling in Fig. 11.4.

A further increase in the field intensity leads to a further, although slower, increase in the free electron energy. Once this energy becomes comparable with the energy of the optical lattice vibrations $\hbar\omega_0$, the generation of optical phonons becomes the prevailing mechanism of electron scattering. Since the energy $\hbar\omega_0$ is rather high, an electron loses almost all its energy during such a 'collision' or inelastic scattering. Therefore, a further increase in v_d with increasing F is now slow and approaches saturation characterized by a further decrease in mobility (Fig. 11.5, segment (c)). Only a very small number of hot electrons manage to retain an energy exceeding $\hbar\omega_0$ during the given time interval. The magnitude of the saturated drift velocity is, in most semiconductors, of the order of 10^7 cm/s (Problem 11/2) and, at such a high value, not only the impact excitation of luminescence centres but also the lattice ionization accompanied by the generation of free electron–hole pairs can occur. These effects were, by the way, already demonstrated in Fig. 11.3. High-field electroluminescence processes therefore generally lie at the edge of semiconductor avalanche breakdown. A full development of the electric breakdown is, however, prevented by both the presence of active luminescence impurities and also of a suitable energy band structure (see below). Moreover, a critical issue here is the high homogeneity of the material. In the standard ZnS:Mn^{2+} material, however, lattice ionization is very improbable, as we will see shortly.

Let us keep following with our eyes the plot $\mu_n = \mu_n(F)$ in Fig. 11.5. In some semiconductors with a suitable band structure, even a decrease in the drift velocity can occur under the highest fields (dashed curve in Fig. 11.5). This is a rather astonishing effect, characterized also by the term *negative differential conductivity*, arising from transitions of the accelerated electrons from the absolute minimum of the conduction band at the Γ point ($\mathbf{k} = (0, 0, 0)$) to a higher-situated minimum at some other points of the first Brillouin zone. Such a transition between band minima (intervalley scattering) is not, of course, possible at thermal equilibrium of electrons with the lattice and in low fields. Besides the presence of a high field, another condition must therefore be fulfilled, namely, that the energy separation between the minima is much higher than $k_B T$ and, at the same time, smaller than the bandgap width E_g, otherwise electron–hole pair generation by impact ionization across the forbidden gap prevails.

The lateral energy minima usually exhibit smaller band curvature and therefore higher effective mass m_e. This, according to (11.3), means yet another decrease in electron mobility and, consequently, a lower drift velocity v_d. It was believed for rather a long time that the transfer of hot electrons between valleys of the conduction band occurs solely in the case of GaAs and InP, where the so-called Gunn effect takes place based on this principle, facilitating the production of tiny microwave generators. However, the ZnS band structure, depicted in Fig. 11.6 [20], shows that ZnS also fulfils the above-stated criteria for electron transitions from the central Γ minimum to side valleys, namely to X ($\mathbf{k} = (1, 0, 0)$) and L ($\mathbf{k} = (1, 1, 1)$). At the same time, it can be clearly seen that these valleys are in fact much shallower than the valley at the Γ point, and that hot electrons appreciably decrease their drift velocity there. An important question is, then, whether the transfer of hot electrons $\Gamma \rightarrow X$, L, which is now believed to exist in ZnS, is of any importance for electroluminescence processes.

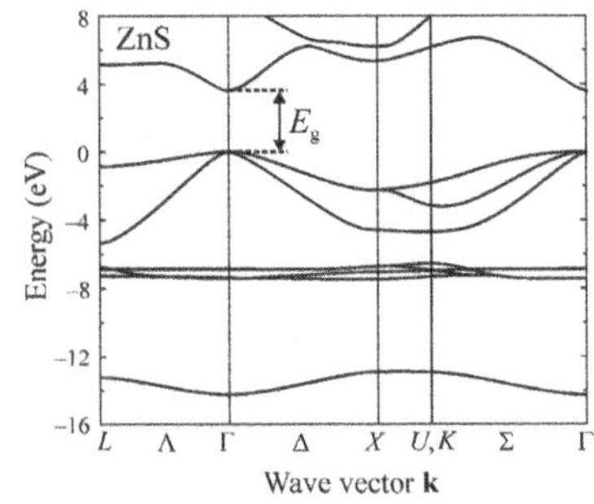

Fig. 11.6
Energy band structure of ZnS. After Luo *et al.* [20].

It actually turns out to be of substantial importance. That is to say, it increases the efficiency of impact excitation of the Mn^{2+} centres. This efficiency is quantitatively described by a quantity referred to as the impact excitation cross-section σ_{im}. Recent calculations [21] have shown that there is a wide range of parameters which the impact excitation cross-section depends on; in particlar σ_{im} scales with the free electron effective mass m_e and its drift velocity v_d as

$$\sigma_{\rm im} \approx \frac{m_{\rm e}}{E_{\rm ge}^2 v_{\rm d}}; \tag{11.5}$$

E_{ge} stands for the energy loss of electrons corresponding to the excitation of the Mn^{2+} centre from its ground to the excited state. From relation (11.5), it is immediately clear that a heavy mass m_e and a low velocity v_d in the side valleys lead to the increase in σ_{im}.

For many years, another supporting factor to which $ZnS:Mn^{2+}$ and related semiconductors owe their leading role among inorganic electroluminescent materials—as the ideal combination of host lattice and extrinsic luminescence centre—is found in the suitable energy separation of the X and L valleys from the absolute minimum at the Γ point: this separation is much lower than E_g (≈ 3.7 eV at room temperature). The energy distribution of hot electrons then, as calculations show, just matches the absorption spectrum of the Mn^{2+} ions in ZnS and, consequently, only a tiny fraction of electrons can reach energies high enough to be able to ionize the ZnS lattice. Therefore, there is no immediate danger of avalanche breakdown of the material.

It is necessary to stress one more fact: Although the decrease in mobility of hot electrons with increasing F turns out to be a prerequisite for efficient and safe electroluminescence, we have to realize that the initial high mobility μ_n in low fields (the high initial slope (a) in Fig. 11.5) is no less important. It helps to attain the desired energy of hot electrons in fields of a reasonable intensity F.

Now, let us move on to the question of free carrier generation in high-field electroluminescence.

Generation of free carriers

Several mechanisms of free carrier generation for impact excitation of luminescence exist:

(1) injection of majority carriers from electrodes;
(2) release of carriers from deep traps localized in the phosphor, due to both high electric field and tunnelling;
(3) release of carriers from traps on the boundary between a dielectric and the phosphor.

The mechanism (1) is active at the direct galvanic contact of a conductive electrode with the semiconductor-phosphor, while the carriers release (3) will apparently apply in an electroluminescence device with capacitance coupling and AC excitation. Carrier generation (2) can occur for both galvanic and capacitance coupling, with very variable importance with respect to the two above-mentioned possibilities.

We will now explain the physics underlying these mechanisms.

(1) *Injection of majority carriers from electrodes.* One deals essentially with a process occurring in the well-known rectifying metal–semiconductor Schottky junction. However, in this case, the junction is exploited in a specific way.

Firstly, let us recapitulate what is to be understood under the concept of 'charge carrier injection'. This is a disturbance of the equilibrium space distribution of carriers by a local increase of their concentration beneath the surface, or at an interface with another material or, in particular, near a contact with a metal. The depth into which this disturbance penetrates depends on the magnitude of the injection current and carrier lifetime (at low current and low lifetime, injected charges will be localized only in the close vicinity of the injecting contact, and in a negligible concentration).

Now, it is important to distinguish between the injection of minority and majority carriers. It usually goes without saying that the term 'injection' refers to the injection of minority carriers, because injecting minority carriers is much simpler than injecting majority carriers. This stems from the following reasons: In both the cases, an additional local electric field and consequently also the drift currents are generated by the injection, introducing a compensating charge of opposite sign into the area of injection. At the same time, the diffusion current of the injected carriers flows in the opposite direction. Specifically, in an n-type semiconductor (where $n_0 \gg p_0$ holds true for the carrier concentrations), we generate a local increase in hole concentration Δp during minority carrier injection; these holes will tend to diffuse into the material volume. Simultaneously, as a result of the relation $n_0 \gg p_0$, the drift current of majority carriers—electrons—will be comparatively higher. Therefore, if we keep the Δp concentration constant with the aid of an external activity, other electrons are attracted to the area of injection and the excess concentration of minority carriers—holes—remains relatively stable.

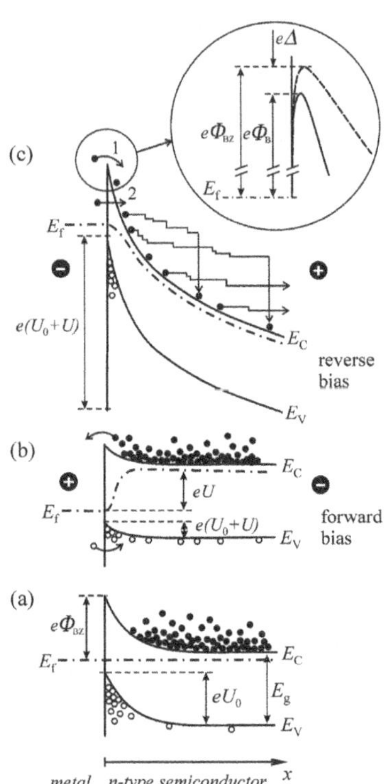

Fig. 11.7
Energy band scheme of a metal–semiconductor junction (Schottky barrier) for an n-type semiconductor. (a) Steady state without external bias, characterized by a built-in potential barrier Φ_{BZ}. (b) Bias U is applied in the forward direction. Minority carrier (hole) injection into the semiconductor takes place. (c) Bias U applied in the reverse direction. Majority carrier (electron) injection into the semiconductor takes place either thermally across the lowered barrier (transition 1) or via tunnelling (transition 2). Simultaneous lowering and narrowing of the barrier is shown in the enlarged inset.

If, on the contrary, in an n-type semiconductor we increase locally via charge injection the electron concentration by Δn, then the diffusion electron current flowing all around into the surrounding lattice overwhelms (again) the current caused by attracted holes; thus compensation of the negative space charge is realized mostly via spreading of the electron cloud. Therefore, the excess concentration Δn tends to vanish very rapidly and it is indeed very difficult to keep it constant. In order to do so, a very high rate of the electrons is indispensable.

In order to achieve electroluminescence in high fields, however, it is just the injection of majority carriers that is the crucial process. In the typical case of $ZnS:Mn^{2+}$, as we mentioned earlier, it is the injection of electrons because ZnS is inherently an n-type semiconductor. Let us now have a look at Fig. 11.7, which shows the carrier energy scheme in the vicinity of a contact between a metal and an n-type semiconductor. Steady state is depicted in panel (a). In the forward direction (panel (b), negative bias U applied to the semiconductor), injection of the holes, i.e. minority carriers occurs. Unfortunately, such an injection is not suitable for excitation of electroluminescence. Consequently, it is necessary to apply reverse bias (panel (c)) when, due to an electric field F, the built-in potential barrier for electrons is decreased from its initial value Φ_{BZ} by the quantity $\Delta = (eF/4\pi\varepsilon_0\varepsilon_s)^{1/2}$ [22, 23] and electron injection occurs (this effect is sometimes referred to as the *Schottky effect*). The effect of the high electric field F is twofold: firstly, this field keeps the injected electron current at a magnitude sufficient to sustain the stability the electron concentration (as explained above), in spite of the very small number of injected electrons (reverse bias!). Secondly, this field has a major role to play also in the impact excitation of luminescence centres by accelerated electrons, as is also indicated in Fig. 11.7(c).

The electron injection current density in this case reads [22, 23]

$$\begin{aligned} j &\cong T^2 \exp\left(\frac{-e(\Phi_{\mathrm{BZ}}-\Delta)}{k_{\mathrm{B}}T}\right) \\ &= T^2 \exp\left[-\frac{e(\Phi_{\mathrm{BZ}}-(eF/4\pi\varepsilon_0\varepsilon_{\mathrm{s}})^{1/2})}{k_{\mathrm{B}}T}\right] \\ &= T^2 \exp\left(\frac{aU^{1/2}}{T}-\frac{e\Phi_{\mathrm{BZ}}}{k_{\mathrm{B}}T}\right), \end{aligned} \tag{11.6}$$

where ε_s stands for the static dielectric constant of the semiconductor. Therefore, the current density increases exponentially with the square root of the applied bias $U = Fd$.

Let us emphasize that (11.6) describes the injection of thermal electrons across the lowered potential barrier (arrow 1 in Fig. 11.7(c)). Under the application of a very high field, however, the injection of electrons can be enhanced due to their tunnelling through the barrier (arrow 2 in Fig. 11.7(c)). That is to say, the barrier not only gets lower but also narrower with increasing field F, as depicted in the enlarged inset in Fig. 11.7(c). Thus, the barrier penetrability for electrons quickly increases (so-called *Fowler–Nordheim tunnelling*). The effect sets in at a barrier width below about 0.01 μm.

(2) *Release of carriers from deep traps*. Lattice defects in semiconductors act as deep traps (potential wells) for charge carriers. The concentration of these defects is relatively high mainly in wide-bandgap materials (ZnS, CaS, SrS), which have appropriate band-structure parameters for electroluminescence in high fields. Especially electron traps, as we know, are essential in this context. Similarly to the above-discussed case of injection from contacts, upon the application of an electric field, a decrease in the potential barrier of the trap occurs as is shown in Fig. 11.8. This leads to the thermal release of trapped electrons into the conduction band, which is known as the *Poole–Frenkel effect*. An expression for the electron current density is very similar to

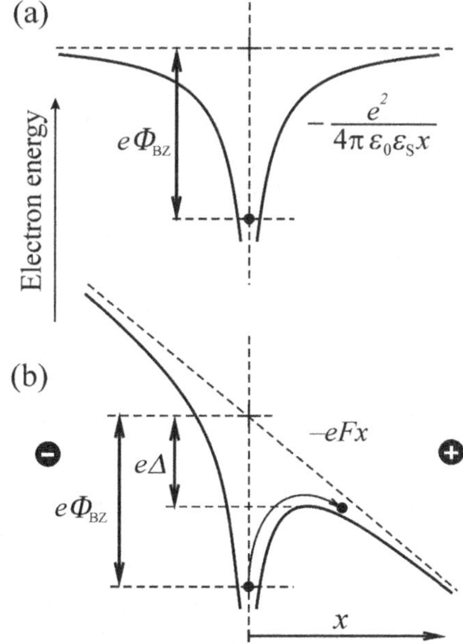

Fig. 11.8
Energy band scheme of an electron in a deep trap described by the potential $-e/4\pi\varepsilon_0\varepsilon_s x$. (a) Without external bias. (b) A high external field of intensity F is switched on. A decrease of the potential barrier height by $\Delta = (eF/\pi\varepsilon_0\varepsilon_s)^{1/2}$ and field-assisted thermal emission of the electron from the trap occur (Poole–Frenkel effect.)

eqn (11.6) and has the form of [22, 23]

$$j \cong F \exp\left[-\frac{e\left(\Phi_{\mathrm{BZ}}-(eF/\pi\varepsilon_0\varepsilon_s)^{1/2}\right)}{k_B T}\right] \approx U \exp\left(\frac{bU^{1/2}}{T}\right), \tag{11.7}$$

where U is the voltage across the luminescent layer.

Moreover, also in this case the electrons can be injected from the trap into the conduction band via direct tunnelling through the narrowed barrier at fields F of the order of 10^6 Vcm^{-1}. Under such conditions, the current density naturally does not depend on temperature and is given by [22–24]

$$j \cong U^2 \exp\left(-\frac{b'}{U}\right), \tag{11.8}$$

where b' is a constant. It is widely believed that this type of tunnelling constitutes the dominant contribution to the electron emission into the conduction band in very high fields.

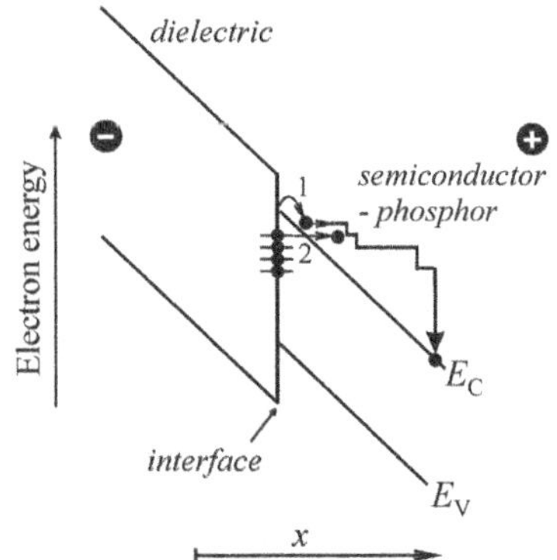

Fig. 11.9
Band scheme of the interface between a dielectric and a luminescent layer. Electron injection can happen both thermally (1) and via tunnelling (2).

(3) *Release of carriers from traps on the boundary between a dielectric and a phosphor.* If a semiconductive phosphor is in contact with an insulator (which is typically the case of AC powered electroluminescent components where the phosphor is sandwiched between two insulating layers—see the scheme in Fig. 11.1(b)) – lattice defects are formed at the interface. The corresponding electronic states are localized within the forbidden band, as is indicated in Fig. 11.9. Under application of an electric field, electrons captured in these traps can be injected into the conduction band, again either thermally owing to the Poole–Frenkel effect or by tunnelling.

11.2.3 Intensity, spectral and temporal characteristics

Throughout this subsection we give several examples of typical experimental results, also paying attention to the question of whether and how it is possible to use the experimental demonstration of high-field electroluminescence to identify the relevant excitation mechanism, and potentially also to help uncover the luminescence centre. Concentrating here on the basic physics—we are to investigate a sample of a new material exhibiting electroluminescence—we leave the purely application-related aspects aside.

The dependence of the electroluminescence intensity (often referred to as the electroluminescence film brightness B) on the applied voltage U is evidently an easily measurable characteristic. If we make the assumption that the brightness B is proportional to the injection current density j, it springs to mind to inspect the differences in the dependencies (11.6), (11.7) and (11.8) as functions of the applied voltage U. These curves are (in a convenient normalization and suitable choice of the constants a, b and b') depicted in Fig. 11.10. It is evident that the differences between the overall curve shapes are not too important. Thus, if we plot B directly as a function of U (both quantities in linear scale), it is very difficult to determine the injection mechanism, especially if these mechanisms happen to occur simultaneously. An

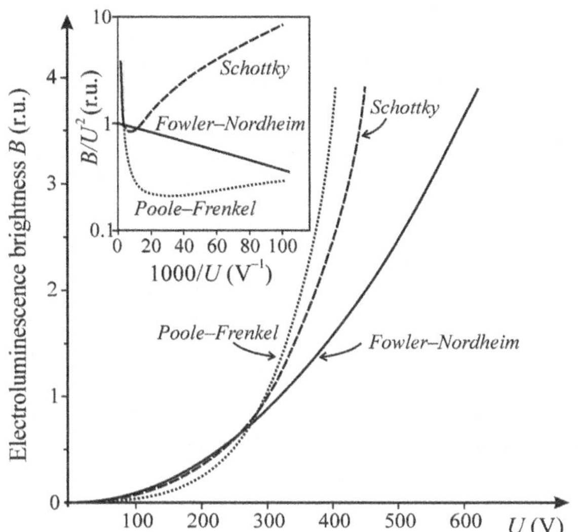

Fig. 11.10
Dependence of the high-field electroluminescence brightness B on the applied voltage U for three different mechanisms of free carrier generation. The inset shows the same curves, just plotted in a different way. The calculation parameters, i.e. the values of the constants a, b and b' in eqns (11.6), (11.7) and (11.8), were chosen as close to a typical experiment as possible, and room temperature was considered.

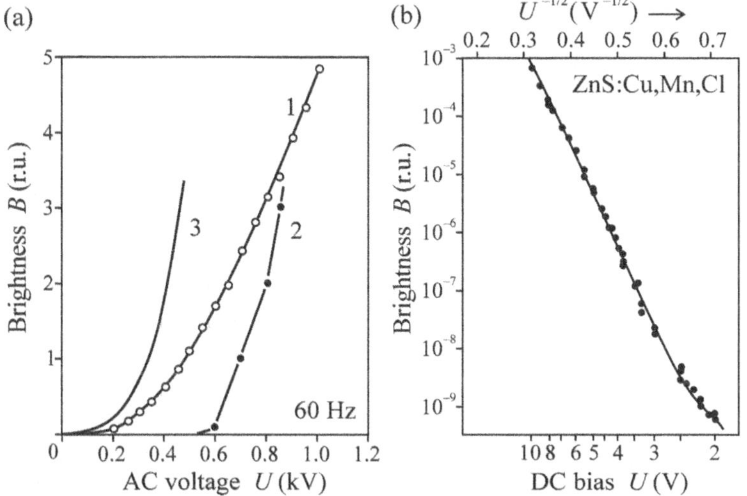

Fig. 11.11
Examples of experimental curves of the electroluminescence brightness B against the applied voltage U (at room temperature). (a) In linear scale (different curves correspond to different powder materials dispersed in a dielectric matrix, AC voltage was applied. After Henisch [14]). (b) In the form of $\log B = f(1/\sqrt{U})$ for a ZnS film of thickness of $\sim$ 1 μm doped with Cu, Mn, Cl; DC bias, after Thornton [26].

example of several experimental curves $B = B(U)$ in a linear scale is given in Fig. 11.11(a) [14].

We might seem closer to solving the problem, however, if we plot in a logarithmic scale $B/U^2 \sim j/U^2$ as a function of $1/U$ according to (11.8): see the inset in Fig. 11.10. This way we obtain the so-called Fowler–Nordheim straight line for the case of tunnelling and if the experimental data follow such a dependence, some authors conclude that Fowler–Nordheim tunnelling is the primary source of free carriers. However, a very careful examination of data is indispensible and the measurements should be performed, if possible, in a sufficiently wide interval of voltages.

Quite frequently, in particular in older works, one can meet the dependence

$$B \approx \exp(-b''/\sqrt{U}). \tag{11.9}$$

This relation was derived by Zalm in his doctoral thesis [25] devoted to zinc sulphide, ZnS, doped with different 'activators'. Exploiting the specific conditions of this semiconductor, he assumed that under the negative electrode near the ZnS surface a potential barrier (depletion layer) of the Schottky type (Subsection 11.2.2) emerges from ionized donors. Despite the barrier being relatively thick ($\sim 100\,\mathrm{nm}$), the author considers that tunnelling of electrons into ZnS under the influence of external bias from the cathode occurs, followed by their acceleration in the high-field region of the barrier and subsequent impact excitation or ionization of the luminescence centres, rather than thermal overcoming of the barrier by the electrons. From our modern terminology, this is a specific case of Fowler–Nordheim tunnelling. For ZnS, the relation (11.9) holds very precisely over several orders of magnitude (Fig. 11.11 (b) [26]), both for DC and AC excitation. The given assumptions, however, considerably limit the general validity of this model, which was relatively widely discussed from its early days. In addition, breaks on the curve sometimes occur, which may indicate either a non-exact model or the participation of more injection mechanisms. It is therefore evident that the study of the $B = B(U)$ dependence on its own can be a sufficiently reliable approach for the determination of the transport or injection mechanisms in electroluminescence in exceptional cases only, rather than as a routine tool.[6]

It might have crossed the reader's mind that the temperature behaviour of the brightness B can be considered an important additional piece of information because relations (11.6)–(11.8) show different temperature behaviour. However, the corresponding luminescence experiment cannot differentiate the temperature dependence of the excitation current density from that of the radiative process itself. Parallel measurements of the current density j and brightness B under varying temperature can of course be performed, if we have at our disposal an instrument sufficiently sensitive for measuring small currents. However, such an experiment is already more complex and an unambiguous result (i.e. separation of the temperature dependence of the charge transport from the temperature dependence of the radiative recombination) is, remembering the above mentioned possible superposition of more transport processes and the occurrence of polarization effects, by no means guaranteed.

As for the spectral composition of electroluminescence, the situation simplifies substantially. If we measure (at the same temperature) the emission spectrum of a given electroluminescence sample under both electrical and optical pumping (the latter realized, e.g. by illuminating the transparent electrode with ultraviolet radiation) and both the spectra look the same in their salient features, we can quite reliably conclude that in both cases the same luminescence centre is active. A wide range of methods offered by photoluminescence spectroscopy, which were discussed in detail in the previous chapters, can then be employed. This approach has been adopted from the very beginning of

[6] It is also necessary to be aware of the simplifying assumptions. Firstly, is the assumption about linear scaling of the electroluminescence brightness with current density; here, however, it is not easy to determine the effective area of the injecting electrodes and, in addition, we have to rely on homogeneity of the flowing current. Further, the probability of the excitation or ionization process itself may depend on the applied voltage, etc.

electroluminescence research, as is indicated by the examples in Fig. 11.12, showing a comparison of a photoluminescence and electroluminescence spectrum of ZnS/CdS powder from 1940 in panel (a) [27] and analogous curves used during a contemporary study of the properties of silicon nanocrystals in panel (b) [28]. Generally speaking, we cannot expect complete identity of both spectra because, e.g., the photoluminescence emission spectrum can slightly depend on the excitation wavelength, and, similarly, the electroluminescence spectrum can depend on the applied voltage, frequency of the AC field, etc. Neither can the influence of the semitransparent electrode be neglected. By the way, these effects are also demonstrated in Fig. 11.12. Attention must be paid—we cannot help but emphasize again—to the main features of the spectra, and to the occurrence of several similar bands, even though their relative intensities may differ. It is also necessary to keep in mind that the emission spectrum of a thin layer can have an interference structure modulated on it.

An example of the temporal characteristics of electroluminescence in high fields is given in Fig. 11.13 [29]. It refers to thin films of the so-called IIa–VIb compounds, doped with rare earth ions: blue-green emitting $SrS:Ce^{3+}$, and $CaS:Eu^{2+}$ luminescing in the red.[7] A double-isolating sandwich sample

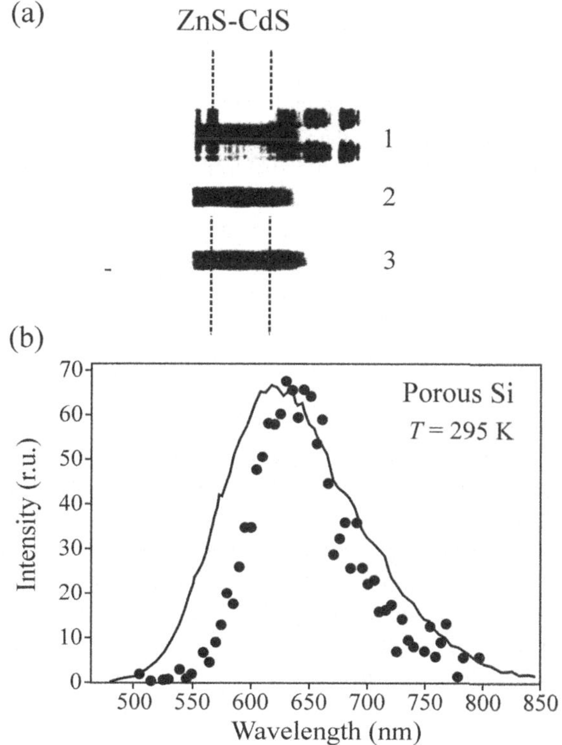

Fig. 11.12
(a) Comparison of a photographic record of the photoluminescence emission spectrum of mixed ZnS/CdS crystals (1, 2) with their electroluminescence spectrum (3) excited by an AC field of amplitude 5×10^5 V/cm. Spectrum 1 is sandwiched between reference spectra of argon lines. Wavelength increases from right to left, measured at room temperature, after Destriau and Loudette [27], dated 1940. Reproduced with kind permission of EDP Sciences. (b) Electroluminescence (points) and photoluminescence (line) emission spectra of an electroluminescence diode manufactured from porous silicon. Room temperature, after Linnros and Lalic [28], dated 1995. Among other things, this figure represents an interesting comparison of the level of presentation of results in scientific journals 70 years ago and nowadays.

[7] Such phosphors are sometimes called Lenard phosphors and have been the subject of study for a long time. With regard to their hygroscopicity, it is necessary to passivate the electroluminescence films of these materials well to prevent direct contact with air. These semiconductors are attractive primarily because of their ability to accept a wide range of dopants including rare earth

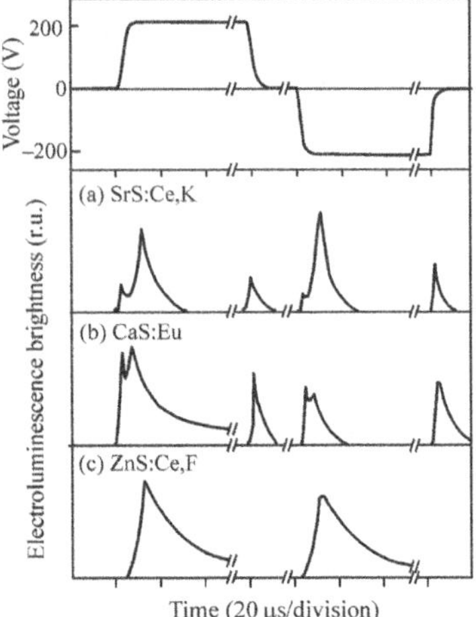

Fig. 11.13
Time behaviour of electroluminescence of thin films (a) SrS:Ce,K, (b) CaS:Eu and (c) ZnS:Ce,F. The top panel depicts the shape of the applied voltage pulses, measurements performed at room temperature. After Tanaka [29].

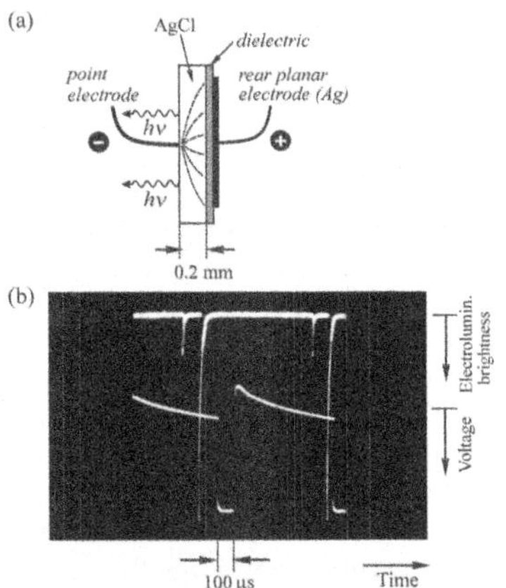

Fig. 11.14
(a) Cross-section of an AgCl thin sheet fabricated for electroluminescence measurements and (b) oscillographic record of the temporal behaviour of the electroluminescence (upper trace) during excitation by electric pulses (lower trace) with a duration of 100 μs. For the sake of transparency, both waveforms are mutually phase shifted. Measurement temperature $T = 77$ K (sample in liquid nitrogen). After Vacek [31], reproduced with kind permission of Wiley-VCH Verlag GmbH&Co. KGaA.

structure, similar to that in Fig. 11.1(b), is employed. The characteristic feature in this particular case is that the luminescence, pumped with electrical pulses, has the form of short flashes both during switching on and off the voltage. This means that the second flash appears in a field of opposite polarity, as a result of the polarization effects occurring near the interface between the phosphor and the insulating layers, as depicted in Fig. 11.2. From this one can infer the excitation mechanism: the first thing which most probably occurs is the impact excitation of the Ce^{3+} and Eu^{2+} centres by the electrons accelerated in the applied pulsed field, then their ionization in high electric field takes place, thereby creating Ce^{4+} and Eu^{3+} ions. Electrons released into the conduction band in this way get subsequently localized in traps. At the moment when the pulsed voltage is switched off, these electrons are again released into the conduction band, captured back by ionized rare earth centres, and finally, the electroluminescence photon is emitted due to the radiative transition $4f^{n-1}5d \rightarrow 4f^{n}$ inside the luminescence centre.

Such temporal behaviour of electroluminescence (when the frequency of the luminescence flashes is doubled with respect to the frequency of the pumping pulses) is not of course observed in all materials, as is demonstrated in panel (c) in Fig. 11.13, representing the electroluminescence of Ce^{3+} centres in ZnS. Here, just a single luminescence pulse corresponds to a given polarity of the electric pulses. This means that no ionization of the Ce^{3+} centres occurs, but, like the 'classical' case of $ZnS:Mn^{2+}$, impact excitation by the accelerated electrons should be considered instead. This substantial difference between electroluminescence of ZnS and CaS or SrS is caused by both the differences in the energy band structure and the differences between ionic radii of the lattice cation constituents and their ratio to the ionic radii of dopants–activators (Ce^{3+}, Eu^{2+}). Details can be found in [29, 30].

In the previous chapters we quite often mentioned crystals of silver halides as suitable model materials. We shall close this subsection with a brief remark on the electroluminescence of AgCl. The temporal behaviour of the electroluminescence of a sheet of pure silver chloride (excited at a temperature of liquid nitrogen $T = 77$ K by electric pulses of duration 100 μs) is, together with the sample cross-section, shown in Fig. 11.14 [31]. It is obvious that the luminescence brightness exhibits two short flashes, which are in phase with the leading and trailing edges of the electric pulse (the luminescence and voltage traces are intentionally shifted with respect to one another in Fig. 11.14); the first flash is substantially weaker than the second one. The applied voltage has a negative polarity on the point electrode, which is in direct contact with the sample surface. On the basis of the contemporary understanding of mechanism of low-temperature AgCl photoluminescence, consisting in the radiative decay of the self-trapped exciton (see Subsection 7.2.4), we may qualitatively interpret the observed kinetics of AgCl electroluminescence from Fig. 11.14(b) in the following way.

ions and thus cover with their luminescence the whole visible spectrum, which is ideal for colour thin film electroluminescent displays.

The negative point electrode injects electrons into the sample. They are immediately strongly accelerated because the field under the point electrode is very high, as is indicated by the field lines (dashed curves) in Fig. 11.14(a). These electrons can therefore ionize the lattice, thereby generating free electron–hole pairs.

The holes in AgCl immediately become self-trapped and the electrons soon become localized at the traps, which occur in sufficient concentration in AgCl (e.g. interstitial ions, Ag^+). At the same time, a certain number of secondary electrons may bind with the holes to create self-trapped excitons, which subsequently recombine radiatively with a typical decay time of $\sim 10\,\mu s$. This is how the first weak electroluminescence flash arises.

During the application of the external voltage, however, an internal field appears at the same time, due to both the generation of a space charge of free carriers and to ionic lattice polarization. Therefore, upon switching off the voltage pulse, a field of opposite polarity appears inside the sample for a moment (Fig. 11.2(c)). This field may subsequently release the localized electrons via the Poole–Frenkel mechanism; the electrons then come back to the point electrode. They are captured owing to Coulomb forces by the holes previously self-trapped beneath the surface, thereby creating self-trapped excitons. Their radiative decay then gives rise to the second, more intense electroluminescence flash.

This result can serve as a rather atypical example of low temperature intrinsic electroluminescence. The kinetics itself depicted in Fig. 11.14 would not of course suffice for an unambiguous identification of the origin of this electroluminescence; we should at least add that it is observed that the emission spectra of photoluminescence and electroluminescence are almost identical in AgCl.

11.3 Injection electroluminescence

What is meant by the term 'injection electroluminescence' is in fact the radiative recombination induced by the injection of minority carriers into a semiconductor p-n junction. Since the description of the rectifying mechanism of a p-n junction and its electrical properties are included in every basic textbook on solid-state physics, we will keep our exposition on this topic to a minimum. In order to get a more detailed description of the effects connected with charge transport and light emission in a semiconductor p-n junction, the reader is referred, e.g. to textbooks [1–4].

The energy band scheme of a semiconductor p-n junction is depicted in Fig. 11.15. If the p and n semiconductors are in contact, the diffusion of electrons from the n region to the p region as well as the diffusion of holes from the p region to the n region occur—without any application of an external bias. The n region thus gets charged with a positive space charge; the p region, on the other hand, charges negatively. An electric field, preventing further diffusion of carriers, therefore arises in the vicinity of the interface; this field has the direction from the n towards the p region. Thus, a steady state is reached afterwards, Fermi levels are equalized and a potential barrier U_D (also

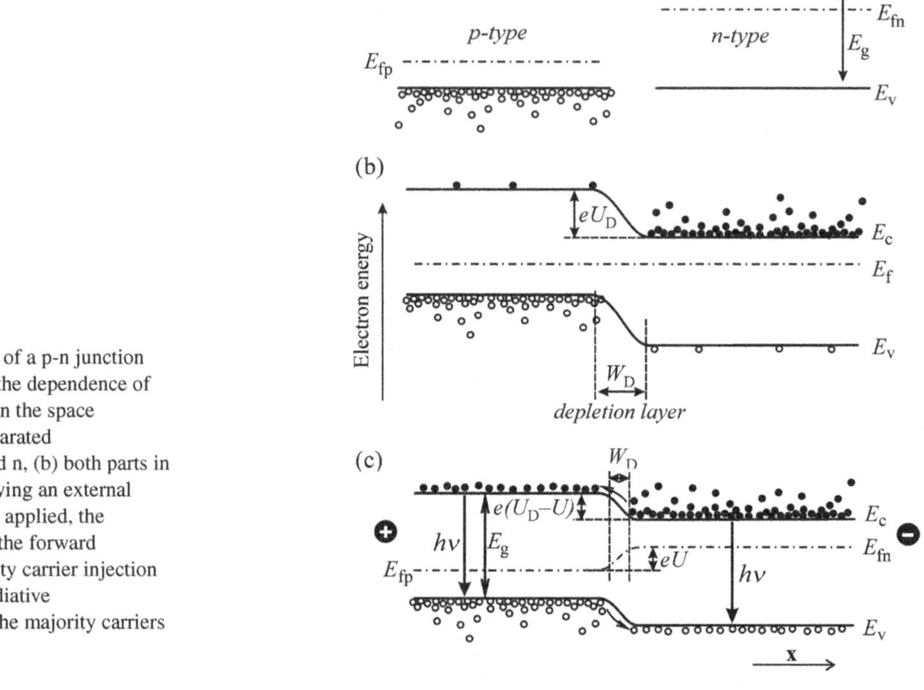

Fig. 11.15
Energy band scheme of a p-n junction (homojunction), i.e. the dependence of the electron energy on the space coordinate x. (a) Separated semiconductors p and n, (b) both parts in contact without applying an external bias, (c) voltage U is applied, the junction is biased in the forward direction, and minority carrier injection and their possible radiative recombination with the majority carriers occur.

called the diffusion potential) sets in, see Fig. 11.15(b). Its electric field is localized in a narrow region near the interface, from where all free carriers disappeared because of diffusion. This region is therefore called a *depletion layer* and its electrical conductivity is substantially lower than the conductivity of unaffected p and n regions. The width of the depletion layer will be denoted as W_D.

Upon applying a suitable external voltage U (positive polarity on the p region, i.e. the forward direction), the coincidence of the Fermi levels is perturbed, the internal potential barrier decreases to a value (U_D-U) and excess electrons are injected from the n region to the p region. Symmetrically, excess holes are injected into the n region. This 'mixing' of high concentrations of minority and majority carriers leads to their recombination and if such recombination is radiative, it gives rise to injection electroluminescence. Before we focus on its closer description, we will sum up the electrical properties of a p-n junction.

11.3.1 Electrical properties of a p-n junction

The potential barrier height U_D, its width W_D and their variations as a function of the applied bias are the determining factors of the behaviour of the p-n junction. For the sake of simplicity, we will assume (as is common in this context) that all dopants, i.e. donors of concentration N_D in the n region and acceptors of concentration N_A in the p region, are fully ionized, thus

$n = N_D$, $p = N_A$ holds for the free electron (n) and hole (p) concentrations, respectively. We can then expect that the diffusion voltage U_D in the depletion layer will be in some way proportional to the concentration of both donors and acceptors; by solving the equations describing the equality of the drift and diffusion currents of electrons and holes (e.g. [24]) one can obtain

$$U_D = \frac{k_B T}{e} \ln \frac{N_A N_D}{n_i^2}, \tag{11.10}$$

where n_i is the intrinsic carrier concentration in the semiconductor under study.

The width of the depletion layer W_D depends also on the concentrations N_A and N_D. The higher these concentrations are, the narrower the region of the semiconductor that has to be depleted of free carriers, originating from ionized dopants, in order to form the given diffusion potential U_D. Therefore, W_D decreases with increasing N_A and N_D. By applying an external bias in the forward direction, we actually 'return' the electrons to the n region and the holes to the p region; consequently, we can expect that W_D will decrease with increasing U. The resulting expression has the form

$$W_D = \sqrt{\frac{2\varepsilon_0\varepsilon_r}{e}(U_D - U)\left(\frac{1}{N_A} + \frac{1}{N_D}\right)}. \tag{11.11}$$

From this expression it is, among other things, evident that in the case of a different level of doping of the n and p regions the space charge will be concentrated predominantly in the region with a lower concentration of impurities. Naturally, application of the reverse bias (positive polarity on the n region) will have the opposite effect and the depletion layer's width will increase on increasing (absolute value of) the reverse bias as $W_D \approx \sqrt{(U_D + U)}$.

As the depletion region has a high electrical resistance, the applied bias easily decreases or increases the potential barrier height. As mentioned earlier, under forward bias the electrons and holes are injected as minority carriers into the regions of opposite conductivity type, which makes a current flow across the p-n junction; this current increases with increasing bias. By solving the continuity equation—and taking into account the generation and recombination mechanisms of carriers [24]—it is possible to obtain an expression for the current density (sometimes referred to as the *Shockley equation*)

$$\begin{aligned} j &= en_i^2 \left(\sqrt{D_p/\tau_p}\frac{1}{N_D} + \sqrt{D_n/\tau_n}\frac{1}{N_A} \right) \exp(eU/k_B T - 1) \\ &= j_s \exp(eU/k_B T - 1). \end{aligned} \tag{11.12}$$

Here, D_n or D_p stand for the diffusion coefficient of electrons or holes, respectively; τ_n or τ_p denote the lifetime of electrons or holes as minority carriers, respectively.

If a reverse bias ($U' = -U < 0$) is applied, the approximation $\exp(eU'/k_B T) \ll 1$ can be written for a sufficiently high absolute value of this bias and eqn (11.12) yields, for this case, a current flowing in the opposite direction, whose magnitude no longer depends on the bias:

$$j = en_i^2\left(\sqrt{D_p/\tau_p}\frac{1}{N_D} + \sqrt{D_n/\tau_n}\frac{1}{N_A}\right) = -js. \qquad (11.13)$$

The current of density j_s is called the *saturation current*; it is usually much lower than the forward current. A p-n junction therefore has, as is of course well-known, rectifying properties and is thus the cornerstone of semiconductor diodes. A typical current–voltage characteristic (11.12) in the vicinity of the origin of the coordinate system is shown in Fig. 11.16(a).

Equation (11.13) can be rewritten in a slightly different way which very well illustrates the physical meaning of the saturation current. Let us first consider that the carrier diffusion length is determined by the relation $L_{n,p} = \sqrt{D_{n,p}\tau_{n,p}}$. Then, the left term in parentheses of eqn (11.13) can be expressed as $\sqrt{D_p/\tau_p}(1/N_D) = L_p/(\tau_p N_D)$. Also, it is known that the square of the intrinsic concentration of carriers in semiconductors is given by the product of the free electron and free hole concentrations (*law of mass action*), for the n region of the investigated p-n junction we can therefore write $n_i^2 = n_0^n p_0^n$; in this formula, n_0^n and p_0^n stand for the equilibrium densities of electrons or holes in this region, respectively.[8]

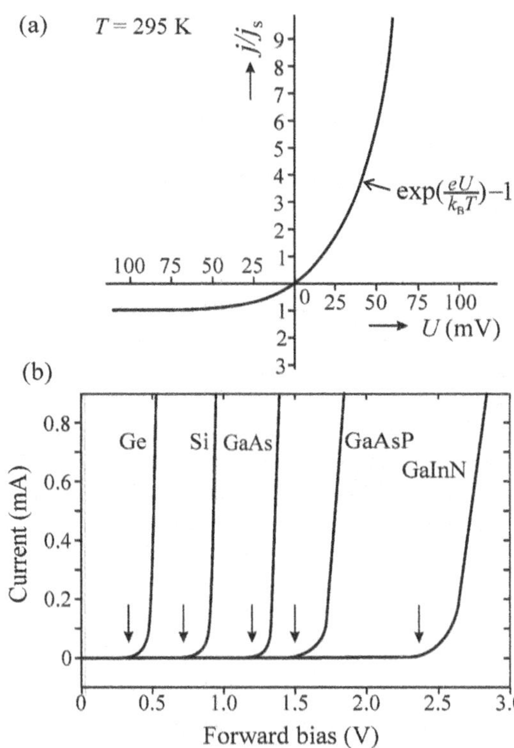

Fig. 11.16
(a) Current–voltage characteristic of a p-n junction calculated in the vicinity of the origin of the coordinate system using expression (11.12). (b) Current–voltage characteristic in the forward direction of p-n junctions manufactured from different semiconductors. Room temperature, arrows mark the corresponding threshold biases. After Schubert [2].

[8] We consider the reverse direction. In the forward direction, the injection of minority carriers occurs in the vicinity of the depletion layer, which is a non-equilibrium state and the law of mass action thus cannot be applied.

However, in this region we assumed that $n_0^{\rm n} = N_{\rm D}$ and thus the hole component of the saturation current density will—considering what was just said—be equal to

$$j_{\rm s(p)} = en_{\rm i}^2\sqrt{D_{\rm p}/\tau_{\rm p}}\frac{1}{N_{\rm D}} = \frac{ep_0^{\rm n}L_{\rm p}}{\tau_{\rm p}}. \tag{11.14}$$

From this, a microscopic interpretation of the saturation current follows: $p_0^{\rm n}/\tau_{\rm p}$ stands for the number of holes recombining per second; therefore, in steady state it means also the rate of thermal generation of holes. Holes generated in this way diffuse from their place of birth and recombine after travelling a mean distance $L_{\rm p}$. Consequently, under reverse bias, the electrons in the n region do not contribute to the current at all (do not overcome the potential hill $U_{\rm D} + U$); as for the holes, only those generated in a distance shorter than $L_{\rm p}$ from the depletion layer contribute to the current. Similar reasoning can also be applied to the electrons in the p region. The magnitude of the saturation current is therefore driven by the diffusion length and lifetime of minority carriers and the external bias has no impact on it.

In the special case of sufficiently high forward bias $U \gg k_{\rm B}T/e$ it is possible to further modify (11.12) to

$$j = j_{\rm s}\exp(eU/k_{\rm B}T),$$

which, using (11.12) and (11.10) and considering the fact that the voltages $U_{\rm D}$ and U have opposite polarities, gives

$$j = e(\sqrt{D_{\rm p}/\tau_{\rm p}}N_{\rm A} + \sqrt{D_{\rm n}/\tau_{\rm n}}N_{\rm D})\exp((U - U_{\rm D})/k_{\rm B}T). \tag{11.15}$$

This expression shows that a significant increase of current density in the forward bias does not occur until $U \approx U_{\rm D}$. This bias is often referred to as the *threshold bias* $U_{\rm th}(\approx U_{\rm D})$. Looking at Fig. 11.15(c), one may guess that this increase in current can be expected the moment that the edges of the conduction and valence bands begin to equalize on both sides of the junction, so that the potential barrier almost disappears and the minority carrier injection increases rapidly. Using the same figure, we can easily see that

$$E_{\rm g} = e(U_{\rm D} - U) + (E_{\rm c} - E_{\rm fn}) + eU + (E_{\rm fp} - E_{\rm v}).$$

In the common case of highly doped semiconductors, the differences $(E_{\rm c} - E_{\rm fn})$ and $(E_{\rm fp} - E_{\rm v})$ are negligible with respect to the bandgap width $E_{\rm g}$. Using the previous equation, we obtain

$$U_{\rm D} \approx U_{\rm th} \approx E_{\rm g}/e. \tag{11.16}$$

The threshold bias $U_{\rm th}$ (and also the energy of emitted photons) thus substantially depends on the material from which the p-n junction is manufactured; see Fig. 11.16(b) and also Problem 11/4.

To end this subsection we wish to add two notes. The first one is related to the Shockley equation (11.12); this equation describes the ideal theoretical current–voltage characteristics of a p-n junction. In order to characterize a real experimental curve, a dimensionaless *ideality factor* $n_{\rm ideal} \geq 1$ is introduced and relation (11.12) gains the form $j = j_{\rm s}\exp(eU/n_{\rm ideal}k_{\rm B}T - 1)$. For a 'theoretical' p-n junction, $n_{ideal} = 1$ holds true; in real diodes n_{ideal} can reach values up to $n_{ideal} = 6$.

The second note is to point out the facts that commercial electroluminescence (LED) diodes are not based on a simple p-n (homo)junction, which we analysed hitherto and a schematic of which is depicted in Fig. 11.15, but they use the principle of a *heterojunction*. In heterojunctions, two different semiconductors are in contact: an active (luminescent) region with narrower bandgap (e.g. GaAs) and a barrier region with wider bandgap ($Al_xGa_{1-x}As$). Such a structure commonly consists of two highly doped barriers—therefore two parts with wider bandgaps—surrounding the central active narrower-gap region, which is usually doped lightly or left completely undoped, see Fig. 11.17. Such a structure is known as a *double heterostructure*. The double heterostructure introduces two factors that significantly improve the luminescence performance of p-n junctions. First, is the spatial localization of the injected electrons and holes in a relatively narrow active region ($W_{DH} \approx 0.1$–$1\ \mu m$, see Fig. 11.17(b)). The material volume in which the radiative recombination occurs is no longer determined by the diffusion length of minority carriers, which can reach values of up to 10–15 μm. An increased concentration of electrons and holes in active region then manifests itself in increased electroluminescence efficiency.

A second factor is then a waveguiding effect in the active region, which is due to the higher refractive index of semiconductors possessing narrower bandgaps [32]; the whole structure then represents a thin planar waveguide with efficient spatial confinement of the radiant power. As a result, double heterostructures represent a technological base not only for LED diodes but also for semiconductor injection lasers. The optical transparency of the barrier layers—owing to the wider bandgap—for luminescence radiation generated in the active region (i.e. much smaller absorption losses in comparison with homojunctions) can also be put on the list of the benefits of heterojuntions.

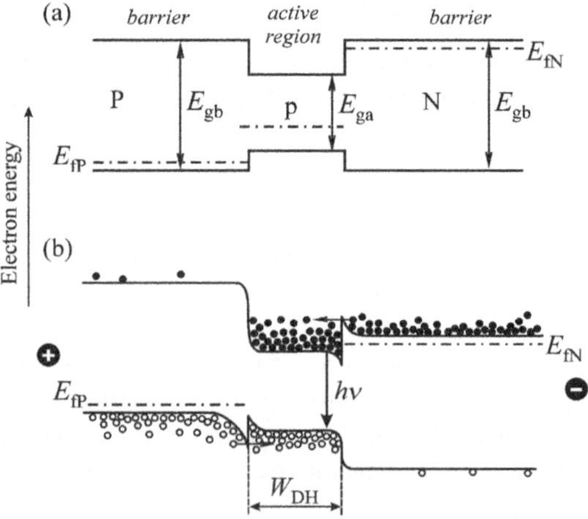

Fig. 11.17
Double heterostructure of P-p-N type. (a) Band scheme of the non-interacting components, (b) heterostructure biased in the forward direction.

11.3.2 Intensity, spectral and temporal characteristics of LEDs

Nowadays, injection electroluminescence is not (and in fact has never been) a tool for basic research in the sense of obtaining basic information about the mechanisms of radiative recombination in a luminescent material.[9] Thus, a treatise on injection electroluminescence goes, to a large degree, beyond the scope of this book, namely teaching the reader to decode the information carried by luminescence radiation (mainly by its spectral composition) about the place and the mechanism of its origin. On the other hand, injection luminescence presents, as we have already pointed out several times, an important application output from the basic research of the optical properties of solids towards electronic devices achieving a direct transformation of electric energy into light. The main application domains of LED diodes emitting in the visible spectral region are light indicators of various colours, and recently also energy-saving interior lighting or high-brightness traffic lights. Another wide field of application of LED diodes, this time those emitting in the infrared region, is represented by optical telecommunications systems. Let us focus in this subsection on several illustrative examples of the emission characteristics of LED diodes biased in the forward direction. Only data obtained at room temperature will be presented.

Intensity characteristics

An example of the dependence of the electroluminescence intensity on injection current is shown in Fig. 11.18 [33]. The onset of photon emission is conditioned by reaching the threshold bias $\sim E_g/e$ (although weak luminescence is already observable at lower bias, see also Problem 11/5), and afterwards, a linear increase of the electroluminescence intensity with increasing injection current follows in most cases. An indication of saturation often occurs at higher current values. Saturation can generally be caused by several factors; firstly, it can indicate an extrinsic origin of the emission: the luminescence labelled as 'blue' in Fig. 11.18 is thought to be due to the recombination of donor–acceptor pairs (Section 5.4) in the active InGaN layer doped with silicon and zinc as a substitutional donor and acceptor, respectively. On the other hand, the 'violet' emission with maximum at 385 nm originates from band-to-band intrinsic bimolecular recombination in the active layer free from any donor–acceptor pairs and does not exhibit saturation.[10] However, saturation can also be caused for example by so-called carrier overflow, i.e. leakage of injected carriers from the fully occupied potential well represented by the active layer in the heterostructure.

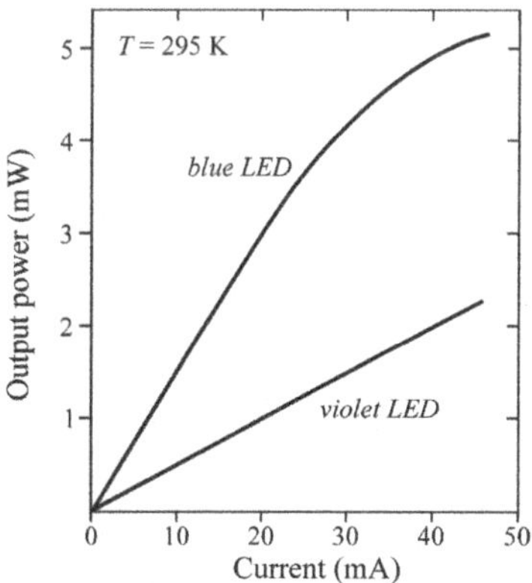

Fig. 11.18
Output light power of double heterostructure InGaN/AlGaN LED diodes as a function of the forward current. After Nakamura *et al.* [33].

If a markedly superlinear increase of the output power with increasing current occurs when investigating light emission from a laboratory sample of a p-n junction, this probably indicates that an excitation mechanism other than

[9] Among many other reasons which we will emphasize soon, this is so because the preparation of high-quality p-n junctions or heterostructures with both sufficient luminescence intensity and durability often goes beyond the capabilities of basic research.

[10] The occurrence of injection electroluminescence of various origins at room temperature is, first of all, due to the efficient spatial localization of carriers in the heterostructure.

minority carrier injection (e.g. impact ionization in a high electric field, see also Section 11.4) is taking place.

Spectral characteristics

This is the right place to point out that the material basis for the fabrication of commercial LED diodes has undergone a certain evolution since the 1960s. From the beginning of the 1960s, parallel research was conducted by different laboratories, both on light diodes for the visible region prepared from gallium phosphide, GaP (red emission when doped with donor–acceptor pairs, O–Zn, and yellow or green emission when doped with nitrogen, N) and diodes on the basis of GaAs for the near-infrared region (this research was given a significant boost by the observation of stimulated emission in a GaAs p-n junction as early as in 1962). These devices were put on the market in the late 1960s. Ternary alloys of $GaAs_{1-x}P_x$ appeared shortly afterwards. In the 1980s, AlGaAs alloys were developed for high-brightness red diodes. None of these semiconductor materials, however, was able to cover the blue spectral region. Blue-emitting diodes had been manufactured from silicon carbide, SiC, for some time but their brightness had been low and, moreover, the high temperatures necessary for the synthesis of this material had posed a problem. Then, for a long time hopes were pinned on wide-bandgap II-VI semiconductors, in the first place on ZnSe. Although the fundamental problem of suitable doping for reaching p-type conductivity in this material was successfully solved and laboratory samples of light-emitting diodes on the basis of a p-n junction were prepared, fast degradation of these devices as a result of non-radiative recombination on lattice defects prevented their commercial implementation. In the end, mastering the technology of producing high-quality layers of gallium nitride, GaN, and of related ternary compounds (InGaN, AlGaN) for diodes emitting in both the blue and near-ultraviolet region turned out to be a revolutionary solution in the beginning of the 1990s. Consequently, nowadays III-V semiconductors have thus taken over the market with electroluminescence optoelectronic light sources and their research keeps continuing, mainly towards the infrared region to fit the needs of optical communication systems.

Figure 11.19 presents examples of normalized emission spectra of LED diodes by different manufacturers. As we mentioned previously, the minimum wavelength λ_m of the emitted radiation is determined by the bandgap width as $\lambda_m = hc/E_g$, however, the injected electrons and holes in most cases do

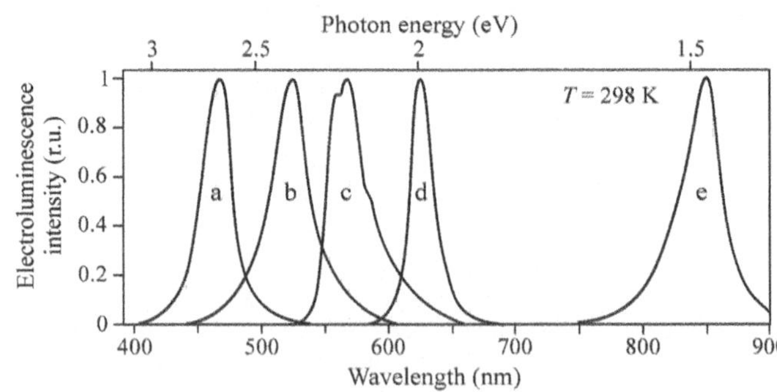

Fig. 11.19
Normalized emission spectra of LEDs by different manufacturers. Active layers and the corresponding colour perceptions are: (a) GaInN (blue 470 nm), (b) GaInN (green 525 nm), (c) GaP:N (green 565 nm), (d) AlGaInP (red 625 nm) and (e) AlGaAs (infrared 850 nm).

not recombine radiatively via a bimolecular mechanism across the bandgap, but extrinsic luminescence due to donor–acceptor pairs or an exciton bound to an intentionally introduced impurity occurs instead. The maximum of the emission band is then shifted towards longer wavelengths in comparison with λ_m. Moreover, the emission spectra of LED diodes may be influenced by light absorption and scattering at impurities and in the materials surrounding the active region, including the plastic encapsulation. The spectral shape is also influenced (in ternary and quaternary compounds) by broadening of the emission band owing to random fluctuations in the chemical composition of the active layer (so-called alloy broadening).

A typical spectral width is 50–100 meV and this increases slightly with increasing current.[11] In most cases, the spectra are smooth curves, however, e.g. curve (c) in Fig. 11.19 corresponding to GaP:N is an exception from the rule and demonstrates the fact that gallium phosphate is an indirect-bandgap semiconductor: the short-wavelength peak is often attributed to the emission of the X–TO type and the neighbouring peak to the X–LO process. An indication of structure on the long-wavelength wing corresponds to the so-called NN luminescence due to the recombination of an excited pair of nearest neighour nitrogen atoms.

From the physical point of view, the process of how the isoelectronic nitrogen impurity in GaP and $GaAs_{1-x}P_x$ radically influences the luminescence properties of these materials is very interesting (as discussed already in Subsection 7.2.3) and, what is no less important, how substantial the impact of this phenomenon on applications in optoelectronics is. The N impurity together with the variable composition of the ternary $GaAs_{1-x}P_x$ compounds contributes to the possibility of continuous tuning of the emission wavelength

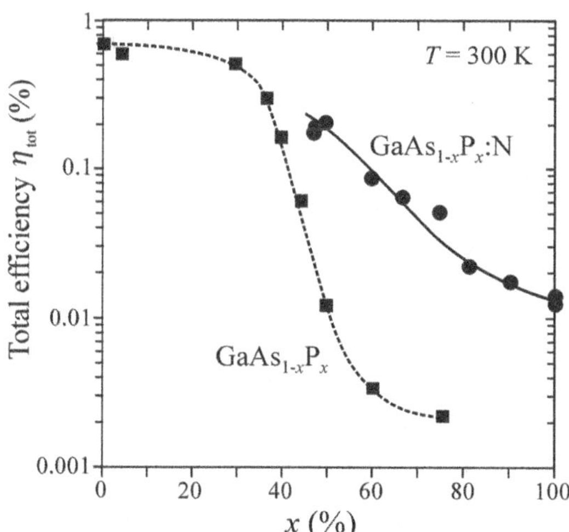

Fig. 11.20
Total (external) efficiency of injection electroluminescence in $GaAs_{1-x}P_x$ as a function of the phosphorus molar fraction x. Symbols stand for experimental data. A sharp decrease of the efficiency in the vicinity of $x = 50\%$ in $GaAs_{1-x}P_x$ without nitrogen is due to the direct → indirect bandgap transformation. Doping with nitrogen compensates to a large degree for this inconvenient effect. After Campbell [34].

[11] Typical values of the total efficiency of LED diodes η_{tot} (here also referred to as the external efficiency) are $\eta_{tot} = 1$–10% and the luminous efficiency (i.e. the emitted light flux at an input power of 1 W) is of the order of 10–100 lm/W. Further technical parameters can be found, e.g. in [2].

in the range from about 550 to 690 nm. That is, the bandgap E_g and thus the energy of the emitted photon varies with the molar fraction x, this energy being lowered with respect to E_g by the localization energy of an exciton at the isoelectronic impurity N. The reabsorption of the emitted radiation is thus significantly reduced. Most importantly, however, the presence of the N atoms induces quasi-direct radiative transitions in $GaAs_{1-x}P_x$ alloys for $x \geq 0.5$, which is a boundary for transformation from direct to indirect bandgaps. These effects are demonstrated in Fig. 11.20 [34]. This is, in fact, one of the first practical applications of the Heisenberg uncertainty relations.

It might also be of interest to mention also white-emitting LED diodes, which are nowadays a commonly available commodity. White emission can be obtained by means of so-called *luminescence converters*, i.e. phosphors covering the active p-n junction of a blue LED diode in the form of a thin layer. Blue injection electroluminescence, originating e.g. in an InGaN chip, excites in an appropriately chosen converter photoluminescence emission spectrally located in the green, yellow or red regions. Part of the blue electroluminescence which was not absorbed in the converter leaves the diode and additive colour mixing then invokes a white perception in the human eye. The principle is thus very similar to a classical fluorescent tube containing mercury vapour and phosphor deposited on the inner surface of the tube.

A commonly used converter is yttrium-aluminium garnet $(YGd)_3(AlGa)_5O_{12}$ doped with cerium ($YAG:Ce^{3+}$), which has yellow luminescence. An example of the emission spectrum of such an LED diode is shown in Fig. 11.21. A certain drawback of the $YAG:Ce^{3+}$ converter is the absence of a red spectral component, resulting in a 'cold' white spectrum. In order to achieve 'warm' white light (with a yellowish tint), a combination of two converters—$YAG:Ce^{3+}$ and $CaS:Eu^{2+}$—is used. New types of converters on the basis of (oxy)nitrides doped with Ce^{3+} and Eu^{2+}, exhibiting almost 100% photoluminescence quantum efficiency, are being studied nowadays.

Temporal characteristics

Upon opening or closing the electrical circuit with an LED diode, transient electronic processes manifest themselves in different temporal behaviours of the current passing through the diode, the diode voltage and the rise or decay of the luminescence emission. These processes are influenced by a wide range

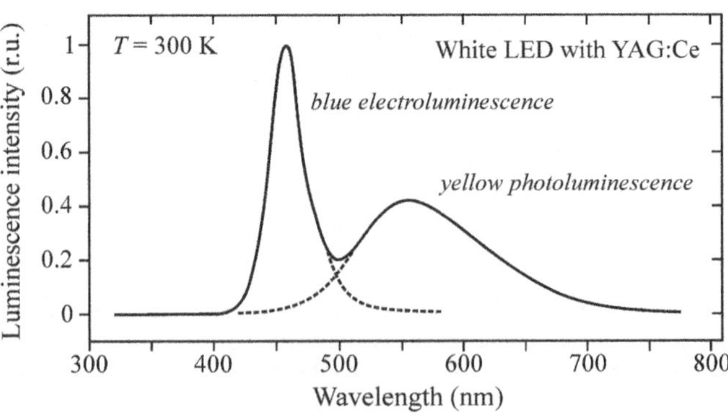

Fig. 11.21
Emission spectrum of a white LED diode on the basis of GaInN with a $YAG:Ce^{3+}$ converter. Produced by Nichia Chemical Industries Corporation. After Schubert [2]. This diode already contains a low-dimensional element—one quantum well (Chapter 12).

of factors; apart from the rates of radiative and non-radiative recombinations themselves, electrical parameters and the geometry of the p-n junction as well as of the external circuit have an important role to play. An example is shown in Fig. 11.22 [35]. One can see that, quite logically, the emission of light sets in with a certain delay after the current (about 1.5 μs here): minority carrier injection across the depletion region, possible filling of the electron and hole traps and carrier capture at active impurities—all these effects take some time. On the other hand, the fact that also the voltage across the diode is delayed after the current can be rather surprising; however, this suggests that a p-n junction represents, to a large extent, a capacitance load for the external source. This capacitance is, moreover, owing to the variable depletion layer width (11.11), dependent on the bias and thus, because of the transient effects, also on time. A simple theoretical description of the curves shown in Fig. 11.22 is then almost impossible. The only thing that can be said here is that the rise of luminescence begins roughly when the voltage approaches its saturated value and that this delay of the light-emission rise time does not depend on the resistance of the external circuit.

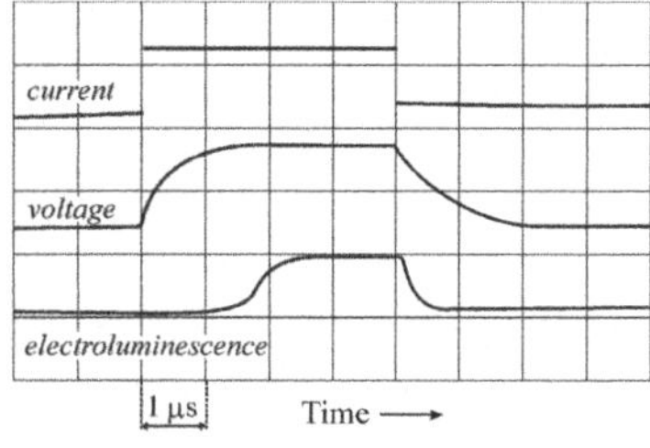

Fig. 11.22
Example of temporal behaviour of the characteristic quantities of an LED diode: current, voltage across the diode and the electroluminescence intensity. After Kogan [35].

The speeds of the rise and decay of electroluminescence are of course very important parameters for assessing the suitability of LED diodes to be used in optical communications, when their light emission is directly modulated by the flowing current. In technical practice, the rise in time of the light output under an electric pulsed excitation may be formally described by the relation (see Fig. 11.23(a))

$$I_{\mathrm{r}}(t) = I_0(1 - \exp(-t/\tau_1))$$

and its decay by

$$I_{\mathrm{d}}(t) = I_0 \exp(-t/\tau_2).$$

Then, it is common to introduce the rise time τ_{r} and the decay time τ_{d} as time intervals between the points corresponding to the 90% and 10% levels of the

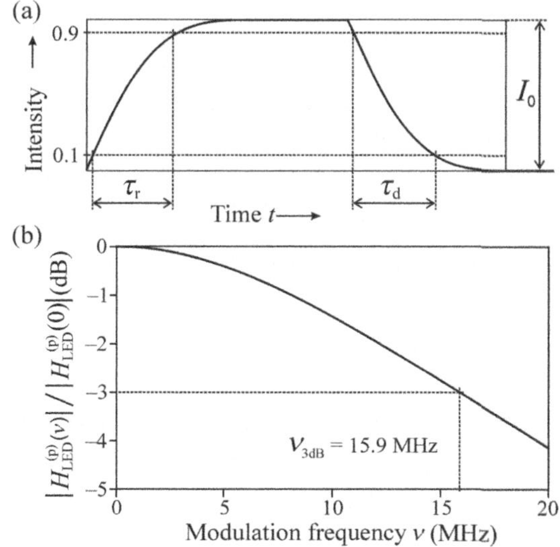

Fig. 11.23
(a) Definition of the rise time τ_{r} and decay time τ_{d} of LED-diode emission for the determination of its frequency bandwidth. (b) An example of the frequency response of an LED diode. After Liu [1].

signal, as indicated in Fig. 11.23(a). The bandwidth of the LED frequency response is then defined by means of these quantities.

The determination of τ_r and τ_d is very simple. Using the definition of τ_r we write

$$I_0 \times 0.1 = I_0(1 - \exp(-t_{10}/\tau_1)), \tag{11.17a}$$

$$I_0 \times 0.9 = I_0(1 - \exp(-t_{90}/\tau_1)) \tag{11.17b}$$

and from eqn (11.17a) we obtain by elementary adjustments

$$t_{10} = -\tau_1 \ln 0.9;$$

similarly, it follows from (11.17b) that

$$t_{90} = -\tau_1 \ln 0.1$$

and thus

$$\tau_r = t_{90} - t_{10} = \tau_1 \ln 9 \cong 2.2\,\tau_1.$$

In a similar way one can get

$$\tau_d = \tau_2 \ln 9 \cong 2.2\,\tau_2. \tag{11.18}$$

If we look on the LED diode as a linear system (which is naturally a simplification), a *transfer function* $H(\nu)$ can be assigned to the diode; this function is, generally speaking, complex and is equal to the ratio of the output to the input signal as a function of frequency ν. (The function $H(\nu)$ is determined via Fourier transformation of the impulse response function.) Because an LED diode is a system characterized by an exponential impulse response $\exp(-t/\tau_2)$, its power transfer function reads [36]

$$H_{\text{LED}}^{(\text{p})}(\nu) = \frac{1}{1 + \mathrm{i}2\pi\nu\tau_2}.$$

The upper bound of the diode frequency band is then determined as the frequency at which the modulus of the transfer function $|H_{\text{LED}}^{(\text{p})}|$ drops to half of its low-frequency value (the signal level decreases by 3 decibels (dB) because $\log(1/2) = -0.3\,\text{B} = -3\,\text{dB}$). This frequency limit (cut-off) $\nu_{3\text{dB}}$ is thus determined from the condition

$$\left| \frac{1}{1 + \mathrm{i}2\pi\nu_{3\text{dB}}\tau_2} \right| = \frac{1}{2},$$

from which we immediately obtain

$$\nu_{3\text{dB}} = \frac{\sqrt{3}}{2\pi\tau_2},$$

which, using (11.18), yields $\nu_{3\text{dB}} = \sqrt{3}\ln 9/2\pi\tau_d$. If in addition we consider the rising edge with the time constant τ_r, we obtain

$$\nu_{3\text{dB}} = \frac{\sqrt{3}\ln 9}{\pi(\tau_d + \tau_r)} \cong \frac{1.2}{(\tau_r + \tau_d)}.$$

For example, if $\tau_r = \tau_d = 2\,\text{ns}$, we get $\nu_{3\text{dB}} \cong 300\,\text{MHz}$. The cut-off frequencies of LED diodes range from 10 MHz to 1 GHz, see Fig. 11.23(b); higher frequencies of the modulating current cannot be transferred by the diode without substantial distortion.

11.4 Electroluminescence of a p-n junction biased in the reverse direction

As early as in the 1950s, a surprising fact was experimentally discovered, namely, that a p-n junction biased even in the reverse direction emits electroluminescence radiation. This kind of behaviour can *a priori* hardly be expected; after all, the diode is a rectifying element, which almost does not transmit electric current under reverse bias! The relevant emission spectrum was found to be quite broad and located—again unexpectedly—at photon energies substantially exceeding the bandgap width. This was a very unusual observation and if the electroluminescence exhibits any new, specific mechanism of radiative recombination in comparison with photoluminescence, it is just this emission. Let us devote this short independent section to this amazing effect.

The emission electroluminescence spectrum of a reverse-biased p-n junction in silicon and germanium extends to the visible region (Fig. 11.24 [37]). At the same time, it was discovered that this radiation originates first in the isolated spots in the close vicinity of the junction area, and with increasing reverse bias these spots merge into a single homogeneous emitting area. The explanation of this effect is based on the close link between electroluminescence and the electrical (pre-)breakdown of a reverse-biased p-n junction, as depicted in Fig. 11.25(a). At reverse bias, as we know, the 'correct injection' of carriers, i.e. the injection of electrons from the n region to the p region and of holes from the p region to the n region, cannot be achieved. The saturation current across the junction arises from thermally generated minority carriers on both sides of the junction. They at first diffuse towards the depletion region, are dragged in its high electric field and finally are accelerated in this field. Then, processes emerge there which we met earlier in avalanche diodes (Section 2.2) or when

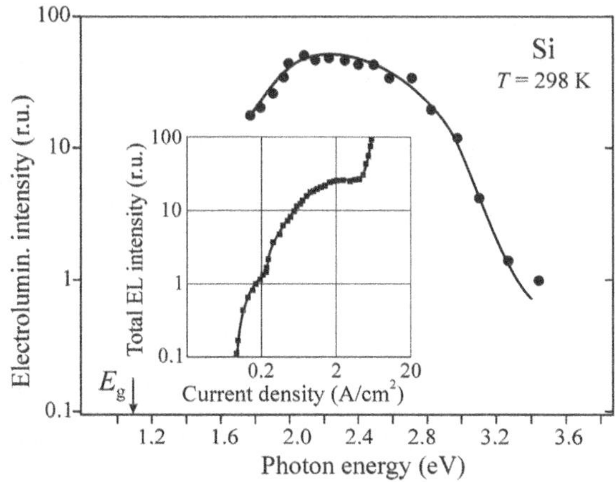

Fig. 11.24
Electroluminescence spectrum of a silicon p-n junction, biased in the reverse direction at room temperature. The spectrum is corrected for reabsorption in the material. The position of the bandgap E_g is marked. The inset shows an example of the dependence of the overall intensity of this light emission on the current density across the junction. After Newman [37].

discussing the mechanisms of high-field electroluminescence: an accelerated electron or hole ionizes the lattice, a free *e–h* pair is created and an avalanche-like increase in the number of free carriers evolves. The current is increasing substantially; however—as the total number of the carriers taking part in this process is low—the total electric breakdown and device destruction do not occur, provided the bias does not exceed a critical value.

Electron–hole pairs with substantial excess energy gained by the acceleration can, naturally, recombine radiatively. Several mechanisms of this recombination have been proposed in the course of time (band-to-band recombination of hot electrons and holes, bremsstrahlung of the electrons corresponding to transitions within the conduction band, recombination of holes within the valence band). Nowadays, it is widely believed that the luminescence photons are created via the band-to-band recombination of hot *e–h* pairs (Fig. 11.25(b)); various estimates of the effective carrier temperature in Si yield values between 3800 and 7500 K. Such a high effective temperature is exactly what underlies the specificity of this recombination mechanism, because the recombination of such carriers does not commonly occur in the steady-state electroluminescence (let alone photoluminescence) regime. In the case of photoluminescence excited high up into the bands it is the high rate of thermalization of hot carriers which prevents such recombination from taking place. In high-field electroluminescence, conditioned by impact excitation of the impurity ions in phosphors of $ZnS:Mn^{2+}$ type, the energy of the accelerated electrons is effectively lost in collisions with the active impurity ions and thus photons with energy higher than E_g do not appear in the emission spectra either.[12]

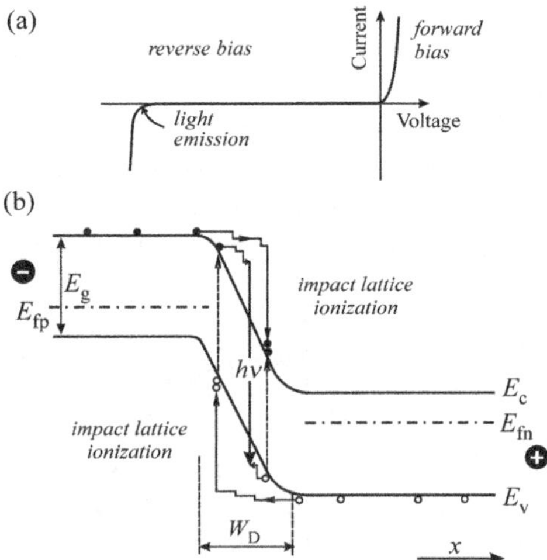

Fig. 11.25
Light emission in a reverse biased p-n junction. (a) The current–voltage characteristic; (b) energy band scheme which demonstrates current amplification due to impact ionization and depicts radiative recombination of an *e–h* pair giving $h\nu > E_g$.

[12] Calculations and various experiments show that the threshold kinetic energy E_{th} for the generation of a free *e–h* pair by impact excitation across the bandgap is in common semiconductors appreciably higher than E_g. Estimates give $E_{th} \approx 1.2$ eV for silicon. This means that the emission spectrum should begin to decrease rapidly for the energy of emitted photons $h\nu = (E_{th} + E_g) \approx 2.3$ eV. This is in agreement with Fig. 11.24. In ZnS, field-accelerated electrons can hardly achieve the high threshold energy $E_{th} > E_g \approx 3.7$ eV, as we already mentioned in Subsection 11.2.2.

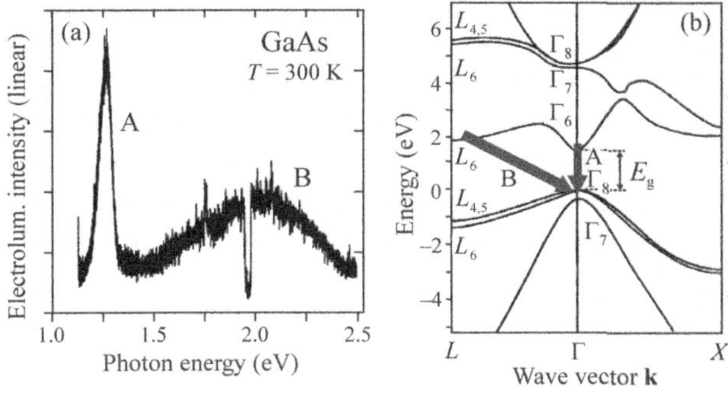

Fig. 11.26
Electroluminescence spectrum of a commercial GaAs LED diode biased in the reverse direction. The dip at $\sim$ 1.9 eV is an experimental artefact. After Lahbabi *et al.* [38]. (b) Corresponding interband recombination of hot carriers in the band scheme of GaAs. The emission band denoted A arises from direct transitions at the Γ point (Γ_6—Γ_8), and the B band is due to the indirect recombination of electrons from the local minimum of the conduction band L_6 with holes in Γ_8.

In silicon, indirect band-to-band recombination of the *e–h* pairs (with phonon assistence) occurs, otherwise the spectrum would have to contain photons with energy higher than the direct forbidden gap $E_{gd}(\Gamma_{15} - \Gamma'_{25}) \approx 3\,\text{eV}$. In p-n junctions made of GaAs, the emission spectrum contains a component due to both direct and indirect interband transitions of hot electrons and holes (Fig. 11.26), which is the result of the specific GaAs band structure [38].

The random and highly nonlinear character of avalanche ionization is reflected also in the intensity dependence of electroluminescence, which can, in a log-log scale, show high values of the slope as well as breaks on the curve (inset in Fig. 11.24). The message for basic research is then: if a markedly nonlinear intensity dependence of electroluminescence occurs in an experiment performed on a sample of a new material (whose thin layer is surrounded by—not very well characterized—oxide barriers, metal electrodes, a covering layer, etc.), it is always necessary to think of the occurrence of avalanche (pre-) breakdown in one of the involved reverse biased depletion layers.

Although the total efficiency of this electroluminescence is very low ($\sim 10^{-6}$ %), it does not pose any problem for contemporary highly sensitive photodetectors and even various potential applications of this electroluminescence have been proposed. Most important is the luminescence emission microscopy of these devices, i.e. the connection of the microscopic monitoring of semiconductor devices (with high spatial resolution) with the spectral investigation of electroluminescence arising in different zones of both discrete elements and silicon CMOS integrated circuits. That is, the discussed light emission is not restricted, as it could perhaps seem from the previous explanation, only to standard electric or LED diodes, but it occurs also in other p-n barriers and similar structures, whose miniature dimensions themselves naturally lead to high electric fields: in memory integrated circuits, the conductive channel of MOSFET transistors, etc. In this way, the phenomenon makes it possible for us to monitor optically the homogeneity of details, prebreakdown states in the diodes and transistors and to analyse their causes, to test integrated circuits, etc. More information can be obtained in the review articles [39, 40]. Impact electroluminescence in the region of avalanche breakdown, for instance, contributed to a large extent to the discovery of rapid crystallization of hydrogen-rich a-Si:H, induced by local electric breakdown in p-i-n structures [41].

Finally, let us briefly mention one more thing, namely the dynamics of this impact electroluminescence. Obviously, it must be a very fast luminescence process (even in silicon) because it is the hot carriers in highly non-equilibrium states that are recombining. This opens up the possibility of manufacturing a fast pulsed LED diode, destined, e.g. for the above-mentioned diagnostic purposes [40]. Systematic steps in this direction have been missing for a long time. Recently, however, a report on the laboratory realization of such a diode has appeared [42].

11.5 Problems

11/1: Unlike semiconductors, hot electrons do not occur in metals. Try to explain qualitatively why this is so.

11/2: Show that the saturation drift velocity of hot electrons is $v_{sd} = (\hbar\omega_0/m_e)^{1/2}$, where $\hbar\omega_0$ is the optical phonon energy and m_e is the effective electron mass. Using typical values of $\hbar\omega_0 = 40\,\text{meV}$ and $m_e = 0.1m_0$, one obtains $v_{sd} \cong 2 \times 10^7\,\text{cm/s}$. Hint: assume that in high electric fields F, the rate of electron energy increase is fully compensated by the energy loss rate owing to the emission of optical phonons, thus $\mathrm{d}\langle E\rangle/\mathrm{d}t = eFv_s - \hbar\omega_0/\tau_{en}$, where τ_{en} is the energy relaxation time. Write an analogous equation also for the electron quasi-momentum $m_e v_s$. See also [19].

11/3: By virtue of the energy band structure of Si and Ge (e.g. Fig. 7.16), discuss the suitability of these semiconductors for possible infrared luminescence panels based on high-field electroluminescence (impact excitation and ionization).

11/4: One might conclude from Fig. 11.15(c) that total equalization of the band edges occurs upon increasing the forward bias up to the value $U = U_D$, the depletion layer then completely disappearing ($W_D = 0$, see relation (11.11)). In other words, this would mean that the full voltage equal to E_g/e will be applied to the junction at that moment. This situation, however, cannot in fact be achieved; considering the widths of the forbidden gaps of the most common semiconductors at room temperature to be $E_g(\text{Ge}) \approx 0.67\,\text{eV}$, $E_g(\text{Si}) \approx 1.12\,\text{eV}$ and $E_g(\text{GaAs}) \approx 1.425\,\text{eV}$, this fact is illustratively depicted in Fig. 11.16(b). Explain qualitatively the reason for this.

11/5: Electroluminescence in a forward biased p-n junction emerges already at very low currents when the bias applied to the junction is lower than the threshold voltage $\approx E_g/e$. This means that the barrier of the junction is overcome by a small fraction of electrons (or holes) with energy high above the bottom of the conduction band or below the top of the valence band, respectively (Fig. 11.27). These charge carriers gain their energy from the lattice. The release of this energy in the form of photons then, of course, leads to inevitable cooling of the close vicinity of the junction. Discuss this effect after [43] or [44].

11/6: Figure 11.26 shows electroluminescence spectra of a reverse biased LED diode manufactured from GaAs. Some of the recombining hot

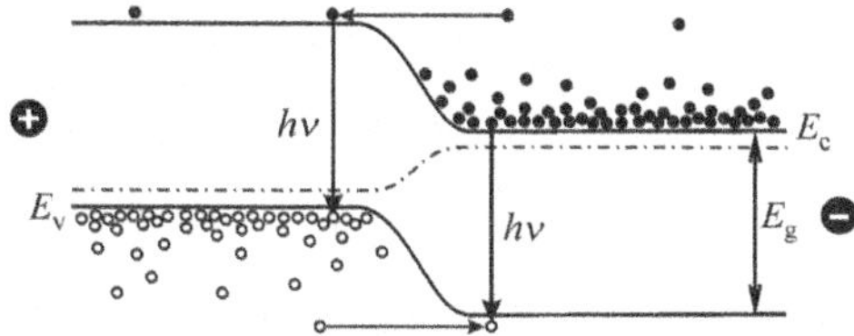

Fig. 11.27

electrons originating during the avalanche diode breakdown are localized at the absolute minimum of the conduction band at the Γ point (A emission band) and, at the same time, others can be found in the local minimum at the L point (B emission band). Theoretical modelling of the experimental emission spectra yields the effective temperature $T_{\mathrm{ef}}^{\mathrm{A}}$ of electrons at the Γ minimum to be much lower than the temperature $T_{\mathrm{ef}}^{\mathrm{B}}$ of the electrons at the L minimum [38]. This contradicts the intuitive concept, namely, that the Γ electrons should be hotter in the electric field owing to their lower effective mass ($m_\Gamma = 0.064\, m_0$, $m_{\mathrm{L}} \approx 0.22\, m_0$) and thus their higher mobility. Explain this apparent contradiction.

References

1. Liu, J.-M. (2005). *Photonic Devices.* Cambridge University Press, Cambridge.
2. Schubert, E. F. (2003). *Light-Emitting Diodes.* Cambridge University Press, Cambridge.
3. Moss, T. S., Burrell, G. J., and Ellis, B. (1973). *Semiconductor Opto-Electronics.* Butterworth, London.
4. Carroll, J., Whiteaway, J., and Plumb, D. (1998). *Distributed Feedback Semiconductor Lasers.* IEE Circuits, Devices & Systems series, Vol. 10. The Institution of Electrical Engineers/SPIE Optical Engineering Press, London.
5. Destriau, G. (1936). *J. de Chim. Phys.*, **33**, 597.
6. Lehovec, K., Accardo, C. A., and Jamgochian, E. (1951). *Phys. Rev.*, **83**, 603.
7. Haynes, J. R. and Briggs, H. B. (1952). *Phys. Rev.*, **86**, 647.
8. Losev, O. (1928). *Phil. Mag.*, **6**, 1024.
9. Zheludev, N. (2007). *Nature Photonics*, **1**, 189.
10. Dekker, A. J. (1963). *Solid State Physics.* Prentice-Hall, Englewood Cliffs, N.J.
11. Keyes, R. J. and Quist, T. M. (1962). *Proc. IRE*, **50**, 1822.
12. Rediker, R. H. (2000). *IEEE J. Selected Topics Quant. Electron.*, **6**, 1355.
13. Grimmeiss, H. G. and Allen, J. W. (2006). *J. Non-Crystal. Solids*, **352**, 871.
14. Henisch, H. K. (1962). *Electroluminescence.* Pergamon Press, Oxford.
15. Sah, C.-T. (1991). *Fundamentals of Solid-State Electronics*, World Scientific, Singapore.
16. Gumlich, H.-E., Zeinert, A., and Mauch, R. (1998). *Electroluminescence.* In *Luminescence of Solids* (ed D. R. Vij), p. 221. Plenum Press, New York.
17. Tanaka, S., Kobayashi, H., and Sasakura, H. (1999). *Fundamentals of luminescence.* In *Phosphor Handbook* (ed. S. Shionoya and W. M. Yen), p. 123. CRC Press, Boca Raton.
18. Blakemore, J. S. (1985). *Solid State Physics.* Cambridge University Press, Cambridge.
19. Yu, P. Y. and Cardona, M. (1996). *Fundamentals of Semiconductors.* Springer, Berlin.
20. Luo, W., Ismail-Beigi, S., Cohen, M. L., and Louie, S. G. (2002). *Phys. Rev. B*, **66**, 195215.

21. Allen, J. W. (1989). *Developments in the theory of electroluminescence mechanisms*. In *Electroluminescence. Proceedings of the Fourth International Workshop* (ed. S. Shionoya and H. Kobayashi), Springer Proceedings in Physics Vol. 38, p. 10. Springer, Berlin.
22. Sze, S. M. (1981). *Physics of Semiconductor Devices.* 2nd edn. John Wiley, New York.
23. Kasap, S. and Capper, P., eds. (2006). *Springer Handbook of Electronic and Photonic Materials,* Chap. 29. Springer Science and Business Media, New York.
24. Böer, K. W. (1992). *Survey of Semiconductor Physics. Volume II: Barriers, Junctions, Surfaces, and Devices*. Van Nostrand Reinhold, New York.
25. Zalm, P. (1956). *Philips Res. Rep.*, **11**, Part I p. 353, Part II p. 417.
26. Thornton, W. A. (1961). *Phys. Rev.*, **122**, 58.
27. Destriau, G. and Loudette, P. (1940). *J. Phys. Rad.*, **1**, 51.
28. Linnros, J. and Lalic, N. (1995). *Appl. Phys. Lett.*, **66**, 3048.
29. Tanaka, S. (1988). *J. Luminescence*, **40&41**, 20.
30. Tanaka, S. (1990). *J. Crystal Growth*, **101**, 958.
31. Vacek, K. (1967). *phys. stat. sol.*, **23**, 105.
32. Moss, T. S. (1985). *phys. stat. sol. (b)*, **131**, 415.
33. Nakamura, S., Pearton, S., and Fasol, G. (2000). *The Blue Laser Diode. The Complete Story.* Springer, Berlin.
34. Campbell, J. C. (1974). *J. Appl. Phys.*, **45**, 4543.
35. Kogan, L. M. (1983). *Semiconductor Light Emitting Diodes* (in Russian: *Poluprovodnikovyje svetoizluajušije diody*). Energoatomizdat, Moscow.
36. Saleh, B. E. A. and Teich, M. C. (1991). *Fundamentals of Photonics.* John Wiley, New York.
37. Newman, R. (1955). *Phys. Rev.*, **100**, 700.
38. Lahbabi, M., Ahaitouf, A., Fliyou, M., Abarkan, E., Charles, J.-P., Bath, A., Hoffmann, A., Kerns, S. E., and Kerns, D. V., Jr. (2004). *J. Appl. Phys.*, **95**, 1822.
39. Deboy, G. and Kölzer, J. (1993). *Semicond. Sci. Technol.*, **9**, 1017.
40. Kramer, J., Seitz, P., Steigmeier, E. F., Auderset, H., Delley, B., and Baltes, H. (1993). *Sensors and Actuators A*, **37–38**, 527.
41. Luterová, K., Pelant, I., Fojtík, P., Nikl, M., Gregora, I., Kočka, J., Dian, J., Malý, P., Kudrna, J., Štěpánek, J., Poruba, A., and Horváth P. (2000). *Phil. Mag. B*, **80**, 1811.
42. Kuai, S. and Meldrum, A. (2009). *Physica E*, **41**, 916.
43. Tauc, J. (1957). *Čs. čas. fyz.* (in Czech), **7**, 246.
44. Dousmanis, G. C., Mueller, C. W., Nelson, H., and Petzinger, K. G. (1964). *Phys. Rev.*, **133**, A316.

Electronic structure and luminescence of low-dimensional semiconductors

12

All the previous chapters described in detail the luminescence properties of the so-called 'bulk' semiconductors, either crystalline or amorphous. This term applies to materials which are in principle homogenous and whose dimensions (in all directions) are substantially larger than the exciton Bohr radius. The properties of such bulk semiconductor samples are determined by their chemical composition, crystalline structure, and certainly also by structural defects and impurities. The number of stable, 'natural' semiconducting materials, which can be grown, investigated, and prospectively exploited in electronics and optoelectronics is somewhat limited to, let us say, 30 or so.

In addition to that, however, a tempting approach to change the discrete spectrum of available semiconductor properties into a nearly continuous spectrum exists. To start with, the de Broglie wavelength of electrons λ_e and holes λ_h in semiconductors as well as the Bohr radius of Wannier excitons a_X (typically 1–10 nm) are fortunately many times larger than the lattice constant of the corresponding semiconductor crystal (typically 0.5–0.6 nm). Therefore, it is in principle possible to decrease the size of the crystal (in one or more directions) down to a size comparable to (or even smaller than) the Bohr radius of an exciton while keeping the basic structure of the semiconductor crystal unaltered. In such a spatially confined crystal, its 'chemically determined' electronic states become dependent on its size—the so-called *quantum confinement effect* sets in. The principle of quantum confinement lies in the quantization of the kinetic energy of a quasi-particle whose motion is restricted to a region comparable with its de Broglie wavelength.

Thanks to the advanced semiconductor technology and the highly developed theoretical models in the semiconductor field, it is now possible to fabricate structures with custom-designed properties tailored to specific applications. Such procedures are sometimes referred to as *quantum engineering* or *band engineering*. Starting from the end of the 1980s, semiconductor research gradually diverted away from bulk semiconductors, which are almost perfectly understood, to the investigation of low-dimensional semiconductor structures.

As this chapter is of fundamental importance for the further comprehension of luminescence properties and the application potential of semiconductor nanostructures, which are the subject of the remainder of the book, we shall have to delve deeper into the theory of the electronic energy structure and optical properties of low-dimensional semiconductors. However, we will put aside other aspects, such as the important field of electrical transport properties. The reader can find more details in a number of monographs and textbooks which are currently available and cover the whole field. We can recommend the textbook by Davies as an excellent introductory treatise [1].

12.1 Basic types of low-dimensional semiconductors

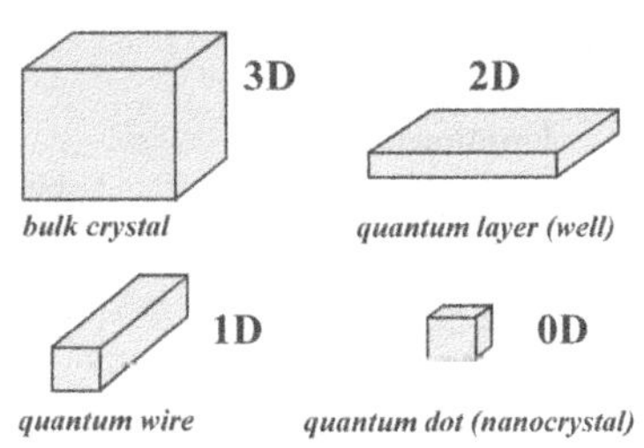

Fig. 12.1
Basic types of low-dimensional semiconductor nanostructures characterized by the number of dimensions which are restricted to the size where quantum confinement effects start to play an important role.

Primarily, low-dimensional semiconductors can be divided into groups according to the number of dimensions in which the characteristic size is small enough to fit into the quantum confinement limit. The structures confined in one, two or three dimensions are called *quantum layers* (or *quantum wells*), *quantum wires*, and *quantum dots* or *nanocrystals*, see Fig. 12.1. Alternatively, we can speak of 'quasi-' *two-dimensional* (2D), *one-dimensional* (1D) and *zero-dimensional* (0D) structures, in contrast to a bulk, three-dimensional (3D) material.

As mentioned above, the quantum-confinement effect sets in if the structures under consideration have dimensions on the order of tens of nanometres or less, which means that the fabrication technology for preparing such materials belongs to the area of nanotechnologies. These technologies are based on the application of two approaches: a synthesis of materials in the 'bottom-up' direction and the patterning of materials in the 'top-down' direction.

Bottom-up growth is a chemistry-based procedure of the synthesis of the desired material from appropriate precursors. In contrast to conventional technologies, the size of the growing structure needs to be controlled so that the growth can be terminated in due time.

The top-down disintegration process again proceeds chemically, when chunks of the original material are transformed from the initial solid structure to a mobile phase (gas or liquid) and removed. Most frequently, various types of lithographic techniques are applied; in these procedures, a mask typically protects parts of the surface from erosion, giving rise to nano-patterned structures. In fact, lithography is the basic method for the fabrication of semiconductor integrated circuits. However, in the case of nanostructures, the masks have to be accurate with nanometric accuracy, indicating that conventional optical lithography cannot be employed (as its wavelength is of the order of hundreds of nanometres) but electron-beam or ion-beam lithography (or perhaps a scanning tunnelling microscope) needs to be exploited.

12.1.1 Semiconductor heterostructures

Low-dimensional semiconductor structures usually do not contain a single type of material confined into small dimensions but they tend to combine several types of semiconductors (or even metals and insulators) of various composition. The reasons for this are multiple: (i) the need for passivation and protection of the nanostructured surface by another material; (ii) embedding of a nanostructure into a macrostructure, which is, in contrast to a nanostructure, suitable for manipulation; and (iii) the necessity to fabricate the appropriate energy structure, which meets the functionality requirements put on the low-dimensional structure. Such compound materials are commonly referred to as *heterostructures*. The simplest example of a low-dimensional heterostructure is a *quantum well*—a small segment of a semiconductor (thin layer, spherical nanocrystal, etc.) surrounded by a different semiconductor with wider bandgap. Excited electrons and holes (excitons) are localized in the corresponding extremes of the conduction or valence band, i.e. inside the quantum well, which is constituted by a semiconductor with a narrower bandgap. The surrounding material with wider bandgap acts as a potential barrier.

The basic technologies for the fabrication of such semiconductor heterostructures are epitaxial methods for thin-film depositions. During epitaxial deposition, one type of semiconductor is deposited onto a (typically) crystalline substrate of the same (homoepitaxy) or a different (heteroepitaxy) semiconductor. The growing semiconductor chemically bonds to atoms of the substrate, thus accepting its crystalline structure. Consequently, heteroepitaxy is well suited only to the combinations of semiconductors with the same crystalline structure, appropriate lattice match and similar properties of chemical bonds. If a semiconductor is forced to grow on a structure much different from its natural form, mechanical strain is introduced into the growing layer. When the layer exceeds a certain critical thickness, the strain is relaxed through the formation of structural defects (see Fig. 12.2 [2]). Strain in a heterostructure induces changes in the semiconductor band structure, which can sometimes be constructively exploited for proper engineering of the energy structure of the device.

Other subtypes of low-dimensional heterostructures include the so-called *multiple* quantum wells, i.e. several quantum wells side by side. If the number of neighbouring quantum wells is very large (the order of hundreds) and the barriers in between them are very thin, so-called *superlattices* arise. Due to the thin potential barriers, the wavefunctions of carriers in neighbouring quantum

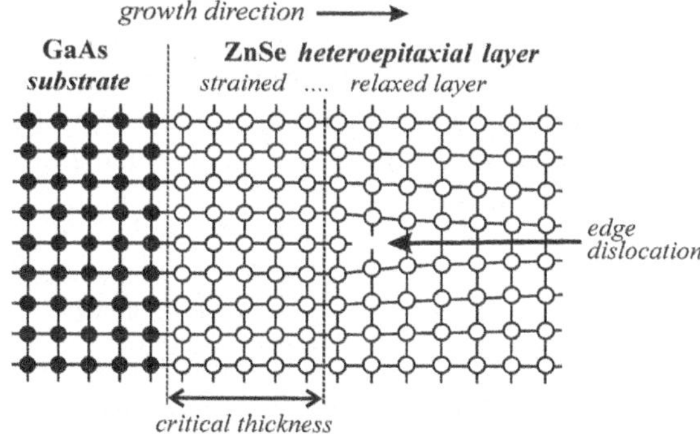

Fig. 12.2
Simplified scheme of epitaxial growth of a ZnSe layer on a GaAs substrate. The epitaxial layer is strained due to the lattice mismatch (~ 0.27%) and the difference in thermal expansion coefficients of the two materials. When the layer becomes thicker than a certain critical thickness (around 150 nm in this case), the strain is released as a result of forming dislocations; a layer with a thickness of about 1 μm is already fully relaxed. Adapted from Yao *et al.* [2].

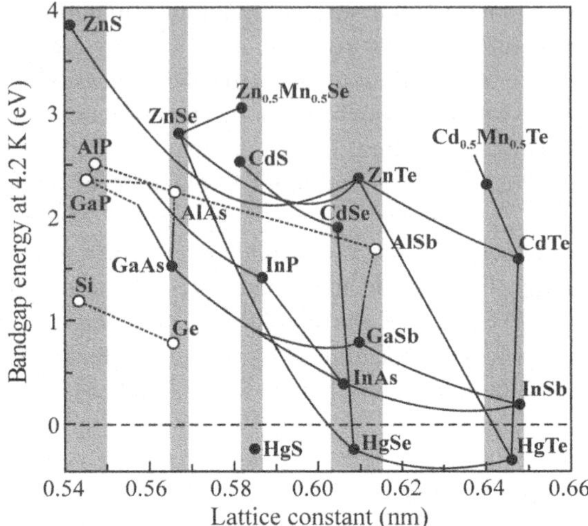

Fig. 12.3
Overview of semiconductor materials sorted by their low-temperature lattice constants and bandgap values. The materials forming stable alloys are connected by lines. White circles and dashed lines indicate indirect-bandgap materials, while black circles and solid lines denote direct-bandgap materials. (The negative bandgap values in HgS, HgSe, and HgTe signify that these materials are near the border between metals and semiconductors.) The grey vertical rectangles highlight groups of materials with close lattice constants, suitable for the fabrication of heterostructures. Adapted from Davies [1] and Yu and Cardona [4].

wells significantly overlap, thus allowing for tunnelling between wells. This means that a brand new energetic band structure, together with specific electric transport properties, emerges.

The most important epitaxial techniques are molecular beam epitaxy (MBE), liquid phase epitaxy (LPE), metal-organic chemical vapour deposition (MOCVD), and metal-organic vapour-phase epitaxy (MOVPE). A description of these techniques is beyond the scope of this book, and the reader is referred to, for example, the monograph by Kelly [3].

What kind of materials can be combined in heterostructures and how broad is the range? The range is indeed plentiful, as is illustrated in Fig. 12.3, which displays the lattice constants and bandgaps of the most common semiconductors and their alloys [1, 4]. The archetypal example of a 'traditional' hetero-system is definitely GaAs–$Al_xGa_{1-x}As$, which has negligible lattice mismatch and which was the cornerstone of the first operating devices with heterojunctions.[1]

12.1.2 Basic types of quantum-well heterostructures

Now, we will inspect the electronic energy levels in heterostructures in more detail. The essential factor which determines the properties of a heterostructure is the relative shift between the conduction and the valence bands, the so-called *band offset*. A first estimate of band offsets can be obtained by applying the *Anderson rule*, which states that the vacuum levels of both materials connected

[1] Interestingly, AlAs is an indirect semiconductor, chemically unstable and reacts with air moisture. As a result of lucky chance, it was found that the $Al_xGa_{1-x}As$ alloy is stable enough and, moreover, it has a direct bandgap for $x < 0.45$. This discovery opened the door to heterostructure devices towards the end of the 1960s. Alferov and Kroemer were awarded half the Nobel Prize in physics in 2000 for the development of semiconductor heterostructures.

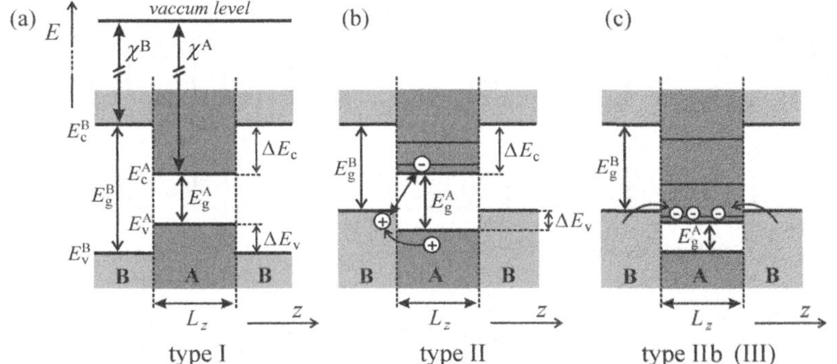

Fig. 12.4
Basic types of quantum wells and illustration of the band-offset calculation from electron affinities—the Anderson rule: (a) type I, (b) type II, and (c) type IIb, sometimes called type III.

in a heterojunction must be aligned, as illustrated by Fig. 12.4(a). Let the two interconnected semiconductors A and B have electron affinities (i.e. the energy required to release an electron from the bottom of the conduction band into vacuum—out of the crystal) χ^A, χ^B and bandgap widths E_g^A, E_g^B. Then, the conduction and valence band offsets ΔE_c and ΔE_v are given by

$$\Delta E_c = \chi^A - \chi^B,$$
$$\Delta E_v = E_g^B - E_g^A - \Delta E_c = \Delta E_g - \Delta E_c.$$

For example, GaAs and $Al_{0.3}Ga_{0.7}As$ have electron affinities 4.07 and 3.74 eV, respectively, and the difference in bandgap widths is $\Delta E_g = 0.37$ eV. Consequently, the offsets are $\Delta E_c = 0.33$ eV for the conduction bands and only $\Delta E_v = 0.04$ eV for the valence bands. Sometimes, a relative offset of the conduction band with respect to the bandgap is introduced as the factor $Q = \Delta E_c/\Delta E_g$. For the above example, we obtain $Q \simeq 0.89$. However, in real heterostructures, the situation tends to be more complicated; experiments revealed that $Q = 0.62$ for $GaAs/Al_xGa_{1-x}As$ with $x < 0.45$. The exact determination of band offsets in a heterostructure is quite difficult, in contrast to the measurement of bandgap widths (the values of which can be obtained for example from optical absorption data).

Quantum wells may be divided into two basic groups, according to their band offsets. If the minimum of the conduction band and the maximum of the valence band are both situated in the same layer, we speak of a *quantum well of the first type* (Fig. 12.4(a)). If, on the other hand, these extremes lie in different (neighbouring) layers (materials), it is a so-called *quantum well of the second type* (Fig. 12.4(b)), whose energy structure reminds as of that of an indirect-bandgap semiconductor. However, here the band extremes are not misaligned in **k**-space, but in real space, implying that electrons and holes are spatially separated.

A special case of the type II structure arises when the conduction and valence bands of neighbouring materials partially overlap (Fig. 12.4(c)). Then, the neighbouring materials can freely exchange electrons and holes until equilibrium due to induced local fields is reached (similarly to p-n junctions). Such structures are sometimes referred to as *quantum wells of the third type*.

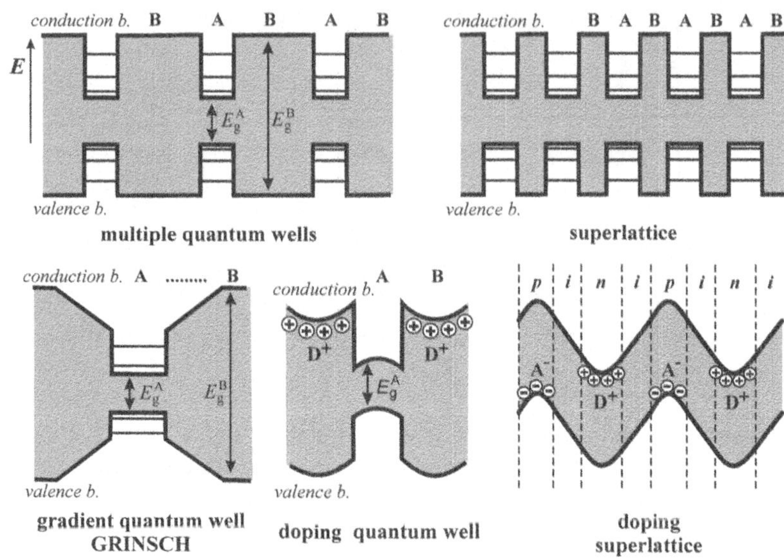

Fig. 12.5
Selected types of heterostructures with quantum wells: The upper row shows multiple (non-interacting) quantum wells and a superlattice. The lower row illustrates a quantum well with a gradual transition between two materials, the so-called GRINSCH structure (graded index separate confinement heterostructure), a modulation doping quantum well and a doping superlattice (or nipi structure) formed within one semiconductor via alternating the type of doping (D^+ and A^- indicate ionized donors and acceptors, respectively). Adapted from Klingshirn [6].

A wide variety of quantum well structures have been studied and described in the literature, some of them being illustrated in Fig. 12.5. Quantum wells are often prepared as multiple wells, meaning that the same sequence of layers is repeated several times. If such wells are separated to such an extent that the wavefunctions of the quasi-particles in the neighbouring wells do not overlap, the wells behave as independent structures. On the other hand, when they are so close that quasi-particles localized in the neighbouring wells can interact, the energy levels split (due to the Pauli exclusion principle) and tunnelling from well to well becomes possible. A structure consisting of a large number of interacting wells is called a *superlattice*, as mentioned above. A superlattice can be fabricated not only by a periodic alternation of layers of two different semiconductors but also via the alternation of heavy doping by donors and acceptors within a single type of semiconductor; such a material is known as a *doping superlattice* or a *nipi structure*.

12.2 Density of states in low-dimensional semiconductors

Distinctive features of low-dimensional semiconductor structures can be illustrated by a simple application of the effective mass approximation. Let us take the quasi-particle wavefunction in the form of a plane wave $\Phi(\mathbf{r}) = K\exp(\mathbf{ik}\cdot\mathbf{r})$. As usual, the product of a wavefunction with its complex conjugate $\Phi(\mathbf{r})\Phi^*(\mathbf{r})\mathrm{d}V = w(\mathbf{r})\mathrm{d}V$ corresponds to the probability of finding a particle inside an elementary volume $\mathrm{d}V = \mathrm{d}x\mathrm{d}y\mathrm{d}z$ at the position described by vector $\mathbf{r}$. The integration of the probability density over the whole system (with a volume V_{syst}), in which the particle is located, must be equal to unity.

This normalization condition allows us to find the value of the coefficient K in the wavefunction:

$$1 = \int_{\text{syst}} w(\mathbf{r})\mathrm{d}V = \int_{\text{syst}} \Phi(\mathbf{r})\Phi^*(\mathbf{r})\mathrm{d}V = K^2 \int_{\text{syst}} \mathrm{e}^{\mathrm{i}\mathbf{k}\cdot\mathbf{r}}\mathrm{e}^{-\mathrm{i}\mathbf{k}\cdot\mathbf{r}}\mathrm{d}V$$

$$= K^2 V_{\text{syst}} \Rightarrow K = \frac{1}{\sqrt{V_{\text{syst}}}}.$$

Obviously, for an infinite system $V_{\text{syst}} \to \infty$ and K becomes infinitely small, $K \to 0$. In the case of 3D, 2D, and 1D systems with finite volumes of L^3, L^2, and L, K is equal to $L^{-3/2}$, L^{-1}, and $L^{-1/2}$.

Let the motion of a 'quasi-free' particle (described by a plane wavefunction) be restricted in the direction of the coordinate x to a segment of length L and let the barrier limiting its motion be impenetrable, i.e. the probability of finding the particle outside this region is zero (actually, this problem, sometimes referred to as a 'particle-in-a-box', was an elementary example treated in a variety of textbooks long before the advent of nanophysics, e.g. [5]). Then, the boundary condition requires that the wavefunction reaches zero at the walls of the box (see Fig. 12.6); these conditions are satisfied only by standing waves (formed by a superposition of waves with inverse wavevectors $\mathbf{k}$ and $-\mathbf{k}$) whose integer multiples of half-wavelength equal the segment length L. In other words, $|\mathbf{k}| = n(\pi/L)$, where $n = 1, 2, 3, \ldots$ This condition holds true for every dimension in which the particle's motion is restricted.

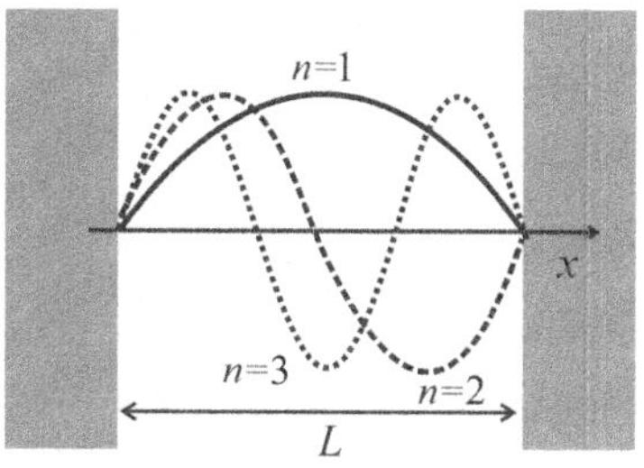

Fig. 12.6
A particle in a box of length L: Possible wavefunctions of a particle confined within impenetrable barriers are harmonic functions with nodes at the barrier walls and periods equal to an integer multiple of π/L.

From the point of view of **k**-space, the allowed states are distributed along the k_x, k_y, k_z axes with a period of π/L. Consequently, one state occupies a volume of $(\pi/L)^d$ in **k**-space (here, d stands for the dimension of the space, see Fig. 12.7).

Now, we can calculate the density of states $\rho(k)$ in **k**-space as the number of available states whose wavevector modulus $|\mathbf{k}| = k$ falls within the interval $(k + \mathrm{d}k)$. This calculation is illustrated in Fig. 12.7 [6] for the case of two-dimensional space. In one dimension, we need to count the number of states within the segments $(k, k + \mathrm{d}k)$ and $(-k, -(k + \mathrm{d}k))$.

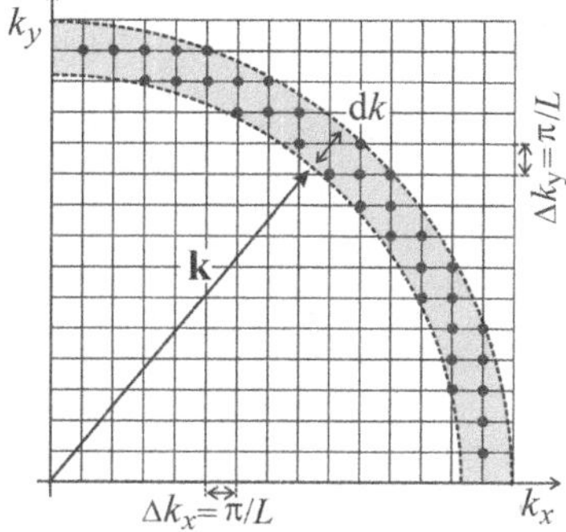

Fig. 12.7
Illustration to the calculation of the density of states in two-dimensional **k**-space. One state occupies a volume of $(\pi/L)^2$. After Klingshirn [6].

In two dimensions, the number of states is determined within an annulus and in three dimensions within a shell confined between radii of $(k, k + \mathrm{d}k)$. Henceforth, we will consider only wavevectors with positive values of k_x, k_y and k_z because one state is formed by a superposition of waves with opposite wavevectors, as mentioned above.

Thus, the density of states $\rho(k)$ is given by

$$\rho(k)\mathrm{d}k = g_S \mathrm{d}V_k \frac{1}{(\pi/L)^d}, \tag{12.1}$$

where g_S is the degeneracy factor (equal to 2 for fermions) and $\mathrm{d}V_k$ denotes the volume of **k**-space, in which the wavevector projections k_x, k_y and k_z are positive and their amplitude k lies within the range of $(k, k + \mathrm{d}k)$. For the 1D, 2D, and 3D cases, the relevant elemental volumes read

$$
\begin{aligned}
d &= 1, \quad \mathrm{d}V_k = \frac{1}{2}\mathrm{d}k, \\
d &= 2, \quad \mathrm{d}V_k = \frac{\pi k}{2}\mathrm{d}k, \qquad (12.2)\\
d &= 3, \quad \mathrm{d}V_k = \frac{\pi k^2}{2}\mathrm{d}k.
\end{aligned}
$$

After substituting $\mathrm{d}V_k$ from (12.2) into (12.1), we obtain the density of states $\rho(k)$ in each single dimension:

$$
\begin{aligned}
d &= 1, \quad \rho(k)\mathrm{d}k = gs\frac{L}{2\pi}\mathrm{d}k, \\
d &= 2, \quad \rho(k)\mathrm{d}k = gs\frac{L^2}{2\pi}k\mathrm{d}k, \qquad (12.3)\\
d &- 3, \quad \rho(k)\mathrm{d}k = gs\frac{L^3}{2\pi^2}k^2\mathrm{d}k.
\end{aligned}
$$

Having divided these quantities by the volume L^d (normalization) and neglecting the constant terms, we obtain a general relation for the density of states in **k**-space in the form of $\rho(k) \simeq k^{d-1}$. In other words, the density of states in one, two, and three dimensions is expressed through a constant, linear, and quadratic function of k, respectively.

Now, we can proceed from the density of states in **k**-space $\rho(k)$ to the density of states $\rho(E)$ in 'energy space', i.e. the number of states per unit volume and per unit energy. In order to do so, we may employ the incremental relation

$$
\rho(E)\mathrm{d}E = \rho(k(E))\frac{\mathrm{d}k}{\mathrm{d}E}\mathrm{d}E \qquad (12.4)
$$

or, in a more general form (in the case of anisotropy),

$$
\rho(E)\mathrm{d}E = \rho(k(E))\frac{1}{|\mathrm{grad}_{\mathbf{k}} E(\mathbf{k})|}\mathrm{d}E.
$$

Let us perform the calculation of $\rho(E)$ for the case of a simple parabolic shape of the conduction and valence bands in the effective mass approximation

$$
E(k_\mathrm{h}) = -\frac{\hbar^2 k_\mathrm{h}^2}{2m_\mathrm{h}}, \quad E(k_\mathrm{e}) = E_\mathrm{g} + \frac{\hbar^2 k_\mathrm{e}^2}{2m_\mathrm{e}}.
$$

For an electron in the conduction band, we can easily derive the modulus of the wavevector and its derivative with respect to E (for the sake of simplicity, we omit the subscript e):

$$
k = \frac{\sqrt{2m}}{\hbar}\sqrt{E - E_\mathrm{g}}, \quad \frac{\mathrm{d}k}{\mathrm{d}E} = \frac{\sqrt{2m}}{2\hbar}\frac{1}{\sqrt{E - E_\mathrm{g}}}. \qquad (12.5)
$$

The density of states of electrons in the conduction band (per unit volume $L = 1$) is obtained by substituting eqns (12.3) and (12.5) into (12.4):

$$d = 1, \quad \rho(E) = gs\frac{\sqrt{2m}}{4\pi\hbar}\frac{1}{\sqrt{E - E_g}}, \tag{12.6a}$$

$$d = 2, \quad \rho(E) = gs\frac{m}{2\pi\hbar^2}, \tag{12.6b}$$

$$d = 3, \quad \rho(E) = gs\frac{(2m)^{3/2}}{4\pi^2\hbar^3}\sqrt{E - E_g}. \tag{12.6c}$$

Generally, the density of states of an electron in the conduction band and a hole in the valence band can be written down as

$$\rho_c(E) \sim (E - E_g)^{d/2-1}, \quad E > E_g$$
$$\rho_v(E) \sim |E|^{d/2-1}, \quad E < 0.$$

The corresponding shapes of the densities of states for an electron are illustrated in Fig. 12.8.

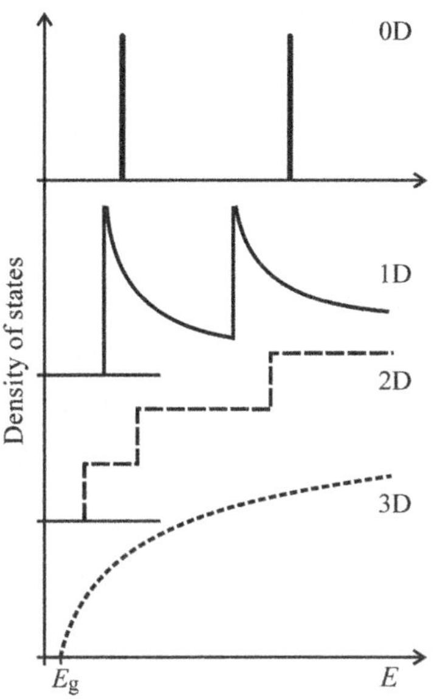

Fig. 12.8
Sketch of the density of states in the conduction band of three−, two−, one−, and zero-dimensional semiconductor structures. In the case of reduced dimensions (0D, 1D, 2D), we assume the existence of several conduction bands as explained below in this chapter.

In brief, the case of a 3D system is described by the density of states which takes on the well-known square-root form. The 2D system is then characterized by a constant density of states for each parabolic conduction band. From the point of view of optical transitions, the key feature here is the finite value of the density of states at the bottom of the conduction band, in contrast to 3D semiconductors, where the density of states approaches zero at the bottom of the band. The 1D system has (in this simplest description) a singularity at the bottom of the band and then the density of states decreases as $1/\sqrt{E}$ with increasing energy. For a (quasi-) 0D system, one can expect the density of states to have the form of discrete delta functions, like in atoms.

12.3 Quantum wells (layers)—two-dimensional semiconductors

The first type of low-dimensional semiconductor structure that was fabricated in reasonable quality and investigated thoroughly from the beginning of the 1970s was the *quantum layer* or *quantum well*. These materials emerged from the previous development of single and double heterojunctions.

12.3.1 Single quantum well with infinite barriers

The simplest 2D system to be treated in more detail now is a quantum well with infinite barriers. Let the layer be oriented in parallel to the x,y axes and let its thickness be L_z. If the well is placed symmetrically with respect to the origin of the coordinate system, the potential energy of a quasi-particle $V(z)$ within the well is described by the relations

$$V(z) \begin{cases} = 0 & \text{for} \quad -(L_z/2) < z < (L_z/2) \\ = \infty & \text{for} \quad |z| > (L_z/2). \end{cases}$$

The stationary Schrödinger equation for a quasi-particle confined in the quantum well $V(z)$ has the form

$$\left[-\frac{\hbar^2}{2m}\nabla^2 + V(z)\right]\psi(x,y,z) = E\psi(x,y,z). \tag{12.7a}$$

In order to solve this equation, it is convenient to separate it into the parts describing the motion in the $\{x, y\}$ plane, where the quasi-particle moves freely, and the motion in the direction parallel to the z axis, where the confinement takes effect. In order to do so, we separate the Laplace operator, wavefunction, and energy eigenvalues:

$$\nabla^2 = \frac{\partial^2}{\partial x^2} + \frac{\partial^2}{\partial y^2} + \frac{\partial^2}{\partial z^2} = \nabla_\perp^2 + \frac{\partial^2}{\partial z^2},$$

$$\psi(x,y,z) = \Phi(x,y)\zeta(z), \quad E = E_\perp + E_z,$$

where the $\{x, y\}$ plane was labelled by $\perp$, denoting the direction perpendicular to the confinement direction. Thus, instead of the original wave equation (12.7a), we solve two separate equations:

$$\begin{gathered} -\frac{\hbar^2}{2m}\nabla_\perp^2\Phi(x,y) = E_\perp\Phi(x,y), \\ \left[-\frac{\hbar^2}{2m}\frac{\partial^2}{\partial z^2} + V(z)\right]\zeta(z) = E_z\zeta(z). \end{gathered} \tag{12.7b}$$

The first equation describes the free motion of a quasi-particle within the $\{x, y\}$ plane. Consequently, its eigenfunctions are plane waves and the dispersion relation has the well-known form of parabolic bands

$$E_\perp = \frac{\hbar^2}{2m}\left(k_x^2 + k_y^2\right) = \frac{\hbar^2}{2m}k_\perp^2.$$

Thus, the electrons and holes are described by

$$E_{\mathrm{e}}(k) = E_{\mathrm{g}} + \frac{\hbar^2}{2m_{\mathrm{e}}}\left(k_x^2 + k_y^2\right), \qquad E_{\mathrm{h}}(k) = -\frac{\hbar^2}{2m_{\mathrm{h}}}\left(k_x^2 + k_y^2\right).$$

The remaining equation corresponds to the motion of a quasi-particle in the direction perpendicular to the layer

$$\left[-\frac{\hbar^2}{2m}\frac{\partial^2}{\partial z^2} + V(z)\right]\zeta(z) = E_z\zeta(z). \tag{12.8}$$

As the particle cannot, due to the infinite height of the barriers, escape outside the well, we will look for the wavefunction only in the interior of the well where $V(z) = 0$. Hence, we obtain the differential equation

$$-\frac{\hbar^2}{2m}\frac{\partial^2}{\partial z^2}\zeta(z) = E_z\zeta(z) \quad \Rightarrow \quad \frac{\partial^2\zeta(z)}{\partial z^2} + k_z^2\zeta(z) = 0. \tag{12.9}$$

This equation resembles the equation of motion of a harmonic oscillator. Therefore, we can look for the solution in the form of a combination of sines and cosines

$$\zeta(z) = A\sin(k_z z) + B\cos(k_z z). \quad (12.10)$$

These functions have to satisfy the continuity boundary condition, implying that they have to approach zero at the walls of the well since the wavefunction outside the well is zero.

Considering the symmetry of the well, the solution can be either even or odd, i.e. in the form of $\zeta^+(z) = B\cos(k_z z)$ or $\zeta^-(z) = A\sin(k_z z)$. The normalization condition (the overall probability of finding the particle is equal to unity) allows the coefficients A and B to be specified

$$\langle\zeta(z)|\zeta(z)\rangle = 1 \quad \Rightarrow \quad A = B = \sqrt{\frac{2}{L_z}}$$

and the boundary conditions determine the allowed values of the wavevectors

$$\sqrt{\frac{2}{L_z}}\cos\left(\frac{k_z^+ L_z}{2}\right) = 0 \Rightarrow \frac{k_z^+ L_z}{2} = \left(j_e - \frac{1}{2}\right)\pi, \quad j_e = 1, 2, 3, \ldots,$$

$$\sqrt{\frac{2}{L_z}}\sin\left(\frac{k_z^- L_z}{2}\right) = 0 \Rightarrow \frac{k_z^- L_z}{2} = j_o\pi, \quad j_o = 1, 2, 3, \ldots$$

Therefore, the resulting wavefunctions and the corresponding energies, determined on the basis of eqn (12.7b), have the forms

$$\zeta^+(z) = \sqrt{\frac{2}{L_z}}\cos\left(\frac{2\pi(j_e - 1/2)}{L_z}z\right),$$

$$E_z^+ = \frac{4(j_e - 1/2)^2\hbar^2\pi^2}{2mL_z^2}, \quad j_e = 1, 2, 3, \ldots, \quad (12.11)$$

$$\zeta^-(z) = \sqrt{\frac{2}{L_z}}\sin\left(\frac{2\pi j_o}{L_z}z\right), \quad E_z^- = \frac{4j_o^2\hbar^2\pi^2}{2mL_z^2}, \quad j_o = 1, 2, 3, \ldots \quad (12.12)$$

The lowest energy of a quasi-particle in a well, that is to say the ground state, corresponds to the first even solution. The first higher (excited) state is then described by the first odd solution, followed by the second even solution, the second odd solution, the third even solution, etc. Let us write down at least the four lowest states and let us plot their wavefunctions (see Fig. 12.9):

The ground state

$$j_e = 1, \quad E_z^{1+} = \frac{\pi^2\hbar^2}{2mL_z^2}, \qquad \zeta^+(z) = \sqrt{\frac{2}{L_z}}\cos\left(\frac{\pi z}{L_z}\right). \quad (12.13)$$

The first excited state

$$j_o = 1, \quad E_z^{2-} = 4\frac{\pi^2\hbar^2}{2mL_z^2}, \quad \zeta^-(z) = \sqrt{\frac{2}{L_z}}\sin\left(\frac{2\pi z}{L_z}\right). \quad (12.14)$$

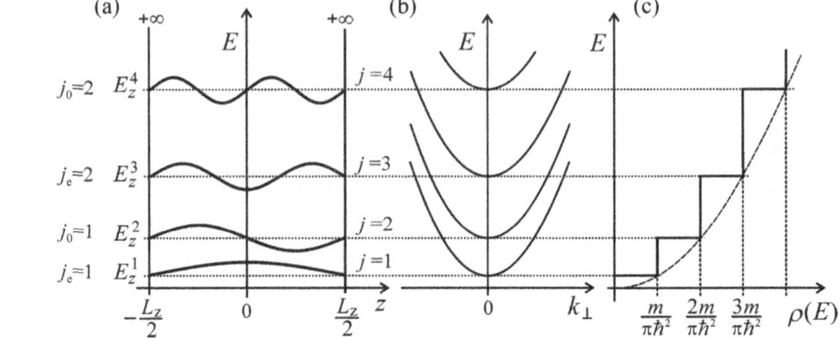

Fig. 12.9
Two-dimensional quantum well with infinite barriers: (a) wavefunctions, (b) dispersion relation, and (c) density of states of the four lowest quantum states.

The second excited state

$$j_e = 2, \quad E_z^{3+} = 9\frac{\pi^2\hbar^2}{2mL_z^2}, \quad \zeta^+(z) = \sqrt{\frac{2}{L_z}}\cos\left(\frac{3\pi z}{L_z}\right). \tag{12.15}$$

The third excited state

$$j_o = 2, \quad E_z^{4-} = 16\frac{\pi^2\hbar^2}{2mL_z^2}, \quad \zeta^-(z) = \sqrt{\frac{2}{L_z}}\sin\left(\frac{4\pi z}{L_z}\right). \tag{12.16}$$

The overall energy of a quasi-particle in an infinitely deep quantum well is then obtained by summing both partial energies and by introducing a quantum number j

$$E = E_z + E_\perp = \frac{\hbar^2}{2m}\left(\frac{j^2\pi^2}{L_z^2} + k_\perp^2\right), \quad j = 1, 2, 3, \ldots \tag{12.17}$$

By applying eqn (12.6b), the two-dimensional density of states takes on the form

$$\rho^{(2)}(E) = g_S\frac{m}{2\pi\hbar^2}\sum_j H(E - E_z^j) = \frac{m}{\pi\hbar^2}\sum_j H(E - E_z^j),$$

where $E_z^j = (\hbar^2/2m)(j\pi/L_z)^2$ are the allowed values of the kinetic energy of the quasi-particle motion in the direction perpendicular to the layer and H is the Heaviside step function, i.e. $H(E) = 0$ for $E < E_z^j$ and $H(E) = 1$ for $E \geq E_z^j$. Figure 12.9 illustrates the fact that, for every state characterized by the quantum number j, one parabolic band $k_\perp$ exists. This band contributes with a constant density-of-states value of $m/\pi\hbar^2$.

The two-dimensional joint density of states $\rho_s^{(2)}$, which drives optical absorption, is then derived by combining the density of states of electrons and holes (Subsection 5.2.1), or

$$\rho_s^{(2)}(E) = \frac{m_r}{\pi\hbar^2}\sum_{i,j} H\left(E - E_g - E_i^e - E_j^h\right), \tag{12.18}$$

where $m_r = (m_e^{-1} + m_h^{-1})^{-1}$ is the reduced mass of the electron and hole.

12.3.2 Quantum well with finite barriers

In a real quantum well, the barriers are of finite height only. The principal difference from the previous case of an infinitely deep well consists in the fact that there is a non-zero probability of finding a particle outside the finite-depth well. While the solution to the wave equation in the interior of such a quantum well remains basically unaltered, we will look for the solution outside the well in the form of an evanescent (exponentially decaying) wave. Obviously, we have to ensure the wavefunctions are continuous and continuously differentiable on the well boundary.

Let the well be described by

$$V(z) \begin{cases} = 0 & \text{for} \quad -(L_z/2) < z < (L_z/2) \\ = V_0 & \text{for} \quad |z| > (L_z/2). \end{cases}$$

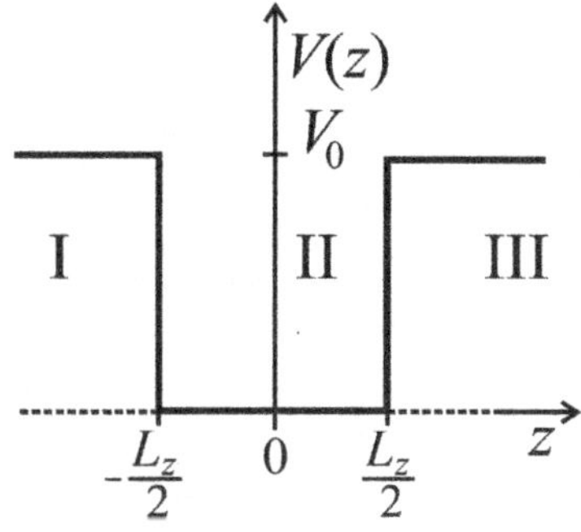

Fig. 12.10
The shape of the potential of a quantum well with finite barriers.

According to Fig. 12.10, the regions on the left of, inside, and right of the well will be labelled as I, II, and III, respectively. Within the regions I and III, we have to solve the following wave equation

$$\frac{\partial^2 \zeta(z)}{\partial z^2} - \frac{2m}{\hbar^2}(V_0 - E_\lambda)\zeta(z) = 0.$$

Let us denote $(2m/\hbar^2)(V_0 - E_\lambda) = K_z^2$ and look for a solution to the wave equation in the form of an evanescent wave exponentially decaying from a value of $E < V_0$ on the well wall (as we are interested in the states localized inside the well) to zero with increasing distance from the well (since the resulting wavefunction must be normalizable). More specifically,

$$\text{region I}: \quad \zeta(z) = C\exp(+K_z z), \qquad \text{region III}: \quad \zeta(z) = D\exp(-K_z z).$$

Now, the evanescent waves must match the internal wavefunctions (region II) described by eqn (12.10). For the even states, this condition obviously yields $A = 0$ and $C = D$, while for the odd states it follows that $B = 0$ and $C = -D$. In addition, the boundary conditions require continuity of the wavefunctions and their first derivatives on the well walls. The boundary conditions for the even states take on the form of

$$B\cos\left(k_z^+ \frac{L_z}{2}\right) = C\exp\left(-K_z^+ \frac{L_z}{2}\right)$$
$$-k_z^+ B\sin\left(k_z^+ \frac{L_z}{2}\right) = -K_z^+ C\exp\left(-K_z^+ \frac{L_z}{2}\right).$$

Dividing the second equation by the first and multiplying the result by –1, we obtain the equation

$$k_z^+ \mathrm{tg}\left(k_z^+ \frac{L_z}{2}\right) = K_z^+,$$

which can be transformed by substituting $k_z^+ = \sqrt{(2mE_z^+/\hbar^2)}$ and $K_z^+ = \sqrt{(2m(V_0 - E_z^+)/\hbar^2)}$ into the transcendental equation

$$\sqrt{E_z^+}\tan\left(\sqrt{\frac{mE_z^+}{2\hbar^2}}L_z\right) = \sqrt{V_0 - E_z^+} \tag{12.19a}$$

without an analytical solution.

Certainly, it can be solved by numerical methods but a graphical solution is very instructive [1]. In order to be clear and concise, we introduce a dimensionless variable θ and a dimensionless parameter θ_0

$$\theta = \sqrt{\frac{mE_z^+}{2\hbar^2}}L_z, \qquad \theta_0 = \sqrt{\frac{mV_0}{2\hbar^2}}L_z.$$

Thus, eqn (12.19a) can be written down in the simple form

$$\tan(\theta) = \sqrt{\frac{\theta_0^2}{\theta^2} - 1}. \tag{12.19b}$$

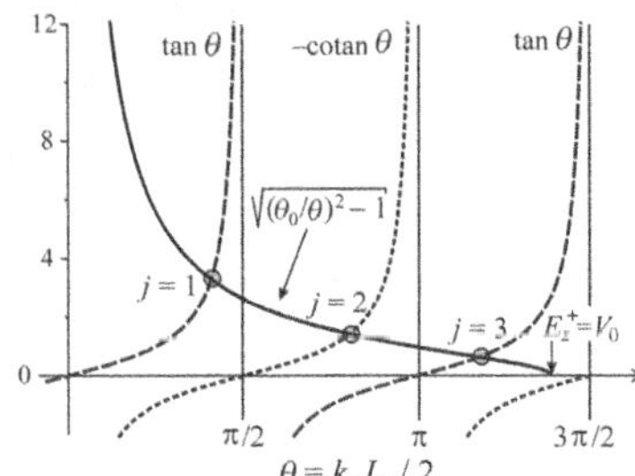

Fig. 12.11
Graphical solution of the transcendental equations (12.19b) and (12.19c) determining the energy values of a particle inside a finite-barrier quantum well. The parameter $\theta_0^2 = 19$ corresponds to the case of electrons in a 12-nm-wide GaAs well with a barrier of 0.3 eV. After Davies [1].

Now, in Fig. 12.11, we plot the left as well as the right side of eqn (12.19b) as functions of the variable θ. The value of the dimensionless parameter $\theta_0^2 = 19$ was calculated for the case of electrons in a $L_z = 12$-nm-wide GaAs quantum well with a barrier height of $V_0 = 0.3$ eV, where the electron effective mass is $m = m_e = 0.067\,m_0$. The solutions to the transcendental equation are given by the points of intersection of the left- and right-side functions. It is evident that the number of solutions (possible localized states) depends on the barrier height V_0 because the function $\sqrt{\theta_0^2/\theta^2 - 1}$ decreases monotonically with increasing θ down to the value $\theta = \theta_0$, that is to say $E_z^+ = V_0$. Interestingly, even for an arbitrarily shallow well $V_0 > 0$ always at least one localized state—namely, the first even solution $E_z{}^+$—exists. Naturally, the deeper the well, the larger the number of localized states that can be reached.

Odd solutions then lead to a transcendental equation similar to (12.19b)

$$-\sqrt{E_z^-}\,\text{cotan}\left(\sqrt{\frac{mE_z^-}{2\hbar^2}}L_z\right) = \sqrt{V_0 - E_z^-} \quad \Rightarrow \quad -\text{cotan}(\theta) = \sqrt{\frac{\theta_0^2}{\theta^2} - 1}. \tag{12.19c}$$

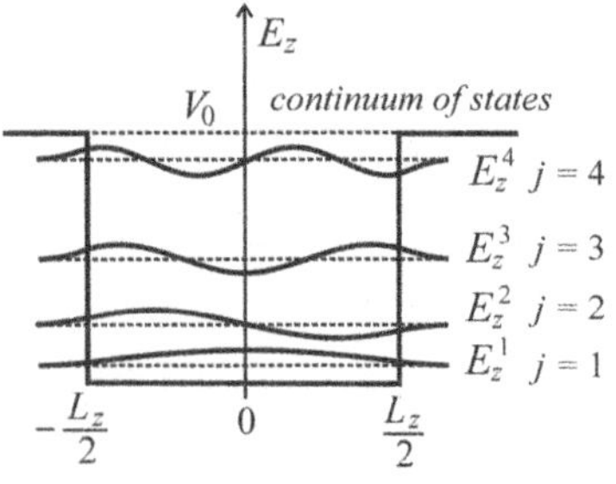

Fig. 12.12
Wavefunctions of the first four states localized in a finite-barrier quantum well. Note the increasing extension of the evanescent waves into the surrounding barriers at higher states.

Its graphical solution can be obtained by supplying the function $-\text{cotan}(\theta)$ to Fig. 12.11. Now, an odd localized state exists only if $\theta_0 > (\pi/2)$. In other words, the well depth must be equal to or larger than

$$V_0 \geq \frac{\pi^2\hbar^2}{2mL_z^2}.$$

The overall picture of wavefunctions (including the evanescent tails) for the first four localized states of a particle in a finite-depth quantum well are plotted in Fig. 12.12.

12.3.3 Excitons in a quantum well

The previous description was based on the effective mass approximation, and quasi-particle interactions were intentionally omitted. Thus, to treat a quasi-particle in a potential well, we have been dealing with single-particle wave equations, not comprising the Coulomb interaction. Such a simple description enabled us to understand the basic feature of low-dimensional semiconductor structures, namely, the quantum confinement phenomenon—the quantization of the quasi-particle kinetic energy in a narrow potential well.

When we wish to take into account the Coulomb interaction between an electron and a hole, we shall work with a two-particle Hamiltonian containing the kinetic energies of the electron and hole, the quantum well confinement potential V_{conf} and the potential of the Coulomb interaction V_{Coulomb}:

$$H = -\frac{\hbar^2}{2m_{\mathrm{e}}}\nabla_{\mathrm{e}}^2 - \frac{\hbar^2}{2m_{\mathrm{h}}}\nabla_{\mathrm{h}}{}^2 + V_{\mathrm{conf}} + V_{\mathrm{Coulomb}}.$$

The confinement potential has to describe the depth of the well separately for electrons and for holes as these values may significantly differ from each other (depending on the particular semiconductor materials forming the well and the barriers, see Fig. 12.4):

$$V_{\mathrm{conf}} = \Delta E_{\mathrm{c}} + \Delta E_{\mathrm{v}} = V_0(z_{\mathrm{e,h}}).$$

A common approach to treating the Hamiltonian

$$H = -\frac{\hbar^2}{2m_{\mathrm{e}}}\left(\frac{\partial^2}{\partial x_{\mathrm{e}}^2} + \frac{\partial^2}{\partial y_{\mathrm{e}}^2} + \frac{\partial^2}{\partial z_{\mathrm{e}}^2}\right) - \frac{\hbar^2}{2m_{\mathrm{h}}}\left(\frac{\partial^2}{\partial x_{\mathrm{h}}^2} + \frac{\partial^2}{\partial y_{\mathrm{h}}^2} + \frac{\partial^2}{\partial z_{\mathrm{h}}^2}\right)$$
$$+ V_0(z_{\mathrm{e,h}}) - \frac{e^2}{4\pi\varepsilon_0\varepsilon r},$$

where $r = |\mathbf{r}_{\mathrm{e}} - \mathbf{r}_{\mathrm{h}}|$, relies on the formulation of the electron and hole motion in the $\{x, y\}$ plane by means of the exciton centre of mass. Considerable simplification is then achieved by exploiting the so-called *strong quantum confinement limit*, in which the effect of the Coulomb interaction on the exciton energy in the z direction is assumed to be negligible in comparison with the manifestation of quantum confinement. Thus, the Hamiltonian takes on the form

$$H = -\frac{\hbar^2}{2m_{\mathrm{e}}}\frac{\partial^2}{\partial z_{\mathrm{e}}^2} - \frac{\hbar^2}{2m_{\mathrm{h}}}\frac{\partial^2}{\partial z_{\mathrm{h}}^2} - \frac{\hbar^2}{2M_{xy}}\left(\frac{\partial^2}{\partial X^2} + \frac{\partial^2}{\partial Y^2}\right)$$
$$- \frac{\hbar^2}{2m_{xy}}\left(\frac{\partial^2}{\partial x^2} + \frac{\partial^2}{\partial y^2}\right) + V_0(z_{\mathrm{e,h}}) - \frac{e^2}{4\pi\varepsilon\varepsilon_0 r},$$

where $M_{xy} = (m_{\mathrm{e}} + m_{\mathrm{h}}) = m_{\mathrm{exc}}$ is the total and $m_{xy}(= m_{\mathrm{r}})$ the reduced exciton mass. The subscript xy emphasizes that the excitonic effects manifest themselves only in the $\{x, y\}$ plane. The coordinates of the exciton centre of mass and the relative coordinates of the electron and the hole are indicated by the capital letters (X, Y) and the lower-case letters (x, y), respectively.

The steady-state Schrödinger equation with the above Hamiltonian can then be divided into two parts describing separately the motion in the $\{x, y\}$ plane and in the z direction. The wavefunction is thus separated into three parts

$$\psi(\mathbf{r}) = \Phi_n^{xy}(r_{xy})\zeta_{ei}(z_e)\zeta_{hj}(z_h),$$

where the second and third terms describe the motion of the electron and of the hole in the z direction. The first term, on the other hand, is the combination of the Bloch functions u_{c0} and u_{v0} with an 'envelope' $\phi_n^{xy}(r_{xy})$, describing the (quasi-)free motion of the exciton with principal quantum number n and wavevector $\mathbf{K}_{xy}$ in the $\{x, y\}$ plane:

$$\Phi_n^{xy}(r_{xy}) = u_{c0}u_{v0}\phi_n^{xy}(r_{xy})e^{i\mathbf{K}_{xy}\cdot\mathbf{R}_{xy}}.$$

The resulting energy levels of excitons in a two-dimensional quantum well finally read [7]

$$E_n^{2D} = E_g - \frac{E_X}{(n_j - 1/2)^2} + \frac{\hbar^2\pi^2 j^2}{2m_r L_z^2}, \quad n_j = 1, 2, 3, \ldots, \tag{12.20}$$

where E_X is the free exciton binding energy in the bulk semiconductor.[2,3]

Let us recall that the series of excitonic energies in the centre of the Brillouin zone of a 3D crystal is given by (see Section 7.1)

$$E_n = E_g - \frac{E_X}{n^2}.$$

We can now compare the energies of excitonic levels in 3D and 2D structures of the same material (referred to the bottom of the conduction band):

$$E_X^{3D} = \frac{E_X}{n^2}, \quad E_X^{2D} = \frac{E_X}{(n - 1/2)^2} \Rightarrow \frac{E_X^{2D}}{E_X^{3D}} = \frac{n^2}{(n - 1/2)^2} > 1.$$

One can see that in 2D structures the excitonic energies are always deeper in the bandgap, implying that the 2D exciton binding energy is increased. The largest difference occurs for the excitonic ground state $n = 1$, for which the 2D value is four times larger than in the 3D case. Because the squared exciton Bohr radius a_X^2 scales with the exciton binding energy E_X like $a_X^2 \sim 1/E_X$, the exciton Bohr radius in 2D structures decreases as compared to 3D crystals:

$$E_X^{2D} = \frac{\hbar^2}{2m_r(a_X^{2D})^2}, \quad E_X^{3D} = \frac{\hbar^2}{2m_r(a_X^{3D})^2} \Rightarrow \frac{a_X^{2D}}{a_X^{3D}} = \frac{(n - 1/2)}{n} < 1.$$

In a simplified picture, an electron and hole are 'squeezed' closer together in a relatively narrow well than the corresponding equilibrium distance a_X^{3D}, therefore the exciton radius is decreased and the electron–hole binding interaction is increased.

[2] The binding energy of an exciton in a bulk material is $E_X = e^4 m_r/2(4\pi\varepsilon\varepsilon_0)^2\hbar^2$.

[3] Each localized state of the electron–hole pair in a quantum well, described by a quantum number j, has a series of excitonic states $n_j = 1, 2, 3, \ldots$ For the sake of simplicity, we shall omit the subscript j in the following description. In wide quantum wells, the total exciton mass m_{exc} substitutes for its reduced mass m_r in eqn (12.20) and one then deals with the quantized movement of the exciton centre of mass (weak quantum confinement regime).

The third characteristic of the excitonic transition is its oscillator strength f (see Appendix I). This scales reciprocally with the third power of the quantum number n, or

$$f_{\mathrm{X}}^{3\mathrm{D}} \approx n^{-3}, \quad f_{\mathrm{X}}^{2\mathrm{D}} \approx (n-1/2)^{-3} \Rightarrow \frac{f_{\mathrm{X}}^{2\mathrm{D}}}{f_{\mathrm{X}}^{3\mathrm{D}}} = \frac{n^3}{(n-1/2)^3} > 1.$$

Thus, the ground state of an exciton $n = 1$ has eightfold increased oscillator strength in a 2D well. An important consequence here is also the fact that the oscillator strength of higher excitonic states decreases faster compared to a 3D crystal (the values of the increase in the oscillator strength are 8, ~ 2.4, ~ 1.7, etc. when going from the lowest to higher excitonic states). Therefore, we can expect relatively weaker optical manifestation of the higher excitonic states in comparison with the significantly strengthened ground state.

In the limit of an infinitely narrow quantum well, the excitonic energy has to approach the values characteristic of the barrier material. Figure 12.13 gives the theoretical dependence of the exciton binding energy in a GaAs quantum well with an $Al_{0.3}Ga_{0.7}As$ barrier [8], for the so-called light (lh) and heavy (hh) holes; these terms will be explained in the next subsection. It is clearly seen that in the case of real wells, one can find an optimum thickness value at which the exciton stability is maximized.

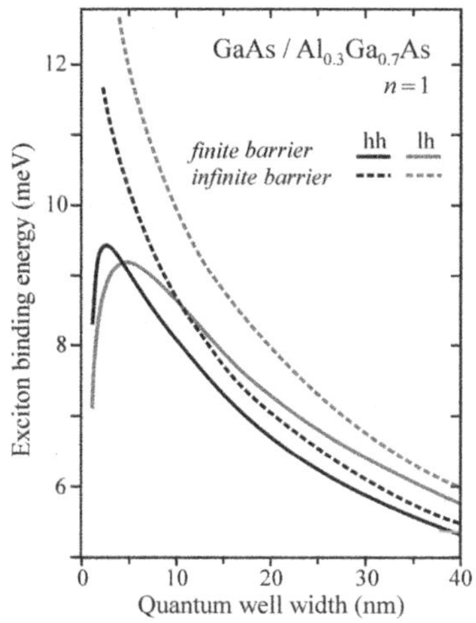

Fig. 12.13
Theoretical values of the exciton binding energy in a GaAs quantum layer with either infinite (dashed curves) or finite $Al_{0.3}Ga_{0.7}As$ barriers (solid lines). After Greene *et al.* [8].

12.3.4 Optical transitions in a quantum well

As discussed in Chapter 5, the probability of an optical transition is proportional to the product of the squared modulus of the transition matrix element (in the dipole approximation) and the joint density of states. In the case of a symmetrical quantum well, we showed the wavefunctions to be either even or odd (symmetrical or antisymmetrical).

In a bulk semiconductor, optical transitions are governed by the following selection rule: the matrix element is non-zero (thus the transitions are allowed) only between states with *different parity* [9]. Therefore, in the case of quantum wells one could deduce, at first glance judging from Fig. 12.9, that the selection rule for interband transitions reads $\Delta j = \pm 1, \pm 3, \pm 5, \ldots$ In fact, however, the correct selection rule for the change of quantum number is

$$\Delta j = \pm 0, \pm 2, \pm 4, \ldots \tag{12.21}$$

What is the reason for this discrepancy? The answer is that we considered only one part of the problem. The non-zero value of the matrix element is primarily given by the symmetry of the wavefunctions in the extremes of bulk semiconductor bands, where parity does apply as a quantum number. Therefore, the strength of optical transitions in a material is primarily determined by its bulk properties. In addition to that, however, other integrals, such as

$$\int \zeta_{\mathrm{e}i}(z)\zeta_{\mathrm{h}j}(z)\mathrm{d}z$$

(with wavefunctions determined by eqns (12.13)–(12.16)), come into play in the formula for the absorption coefficient in a quantum well. In a well with infinite walls, the integration is carried out from $-L_z/2$ to $+L_z/2$ and a non-zero value is ensured only if $i = j$. Consequently, the selection rule reads

$$\Delta j = 0. \tag{12.22}$$

In other words, the transitions are allowed only between quantized electron and hole sub-bands with the same quantum number j. In the case of wells with finite barriers, the strict selection rule (12.22) is somewhat relaxed into the form of relation (12.21), but the probability of a transition between states with equal j is still substantially higher than for other 'allowed' transitions.

Experimentally obtained absorption spectra of quantum wells[4] confirm the validity of the $\Delta j = 0$ selection rule, see Fig. 12.14(a) and (b) [10]. In these spectra, the absorption band undergoes a blue-shift with decreasing well width; in addition, strengthening of excitonic effects and a step-like increase in the absorption background obviously take place in the well. As a result of the high binding energies of excitons in 2D structures, excitonic resonances are often observed at room temperature. Moreover, some splitting of the transitions may

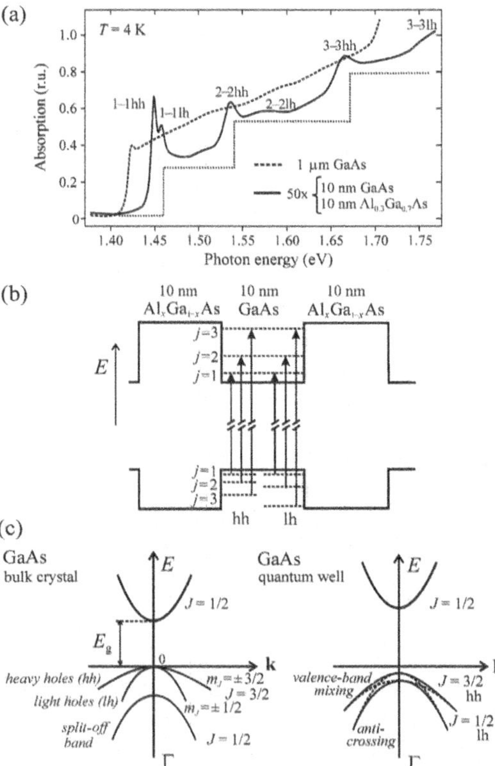

Fig. 12.14
(a) Absorption of a multiple (50×) quantum well 10 nm GaAs/10 nm $Al_{0.3}Ga_{0.7}As$ (solid line) compared to the absorption of a 1-μm-thick layer of GaAs (dashed line). The dotted line indicates an approximate shape of the joint density of states. After Schmitt-Rink *et al.* [10]. The interpretation of peaks is given in panel (b): the transitions take place between electron and hole states with identical quantum number j. The splitting of transitions is due to complex intermixing and splitting of the degenerate states of the uppermost valence band, as shown in panel (c).

[4] The absorption spectrum of a single quantum well cannot be measured by conventional methods as the absorption coefficient of 2D excitons is only of the order of 10^4 cm^{-1}, which, for a well thickness of 10 nm, yields a value of absorbance of only 10^{-2}. Therefore, the transmittance is usually measured in samples containing multiple quantum wells.

be observed; this splitting results from lifting the degeneracy of the valence bands and their intermixing. We will give a brief explanation of this effect, as it is often encountered in the study of optical properties of low-dimensional semiconductors.

The bulk band structure close to the valence band maximum at $\mathbf{k} = 0$ is usually quite complicated in comparison with a single parabolic minimum of the conduction band. As an example, we can recall the band structure of hexagonal CdS (Fig. 7.12), where the maximum consists of three bands A, B, and C. In III-V cubic semiconductors such as GaAs (or the cubic modification of CdS—the zincblende structure), the situation is similar but due to higher symmetry of the lattice the two uppermost valence bands A and B are degenerate at the Γ point. This situation is illustrated in the left panel in Fig. 12.14(c).[5] These bands are the so-called heavy-hole (hh) and the light-hole (lh) bands, referring to their curvature.

Yet, in quantum wells, the confinement potential lifts this degeneracy as a result of symmetry breaking. The hh and lh bands are therefore split, which is the phenomenon displayed in Fig. 12.14(a)–(c) along with the scheme of the corresponding optical absorption transitions. This splitting is increased with decreasing well width L_z. More details can be found in [1, 4, 9].

The situation is usually simpler in the case of luminescence transitions as the energy spacing between the hh and lh bands is often so large (> 10 meV) that after the thermalization process of photoexcited holes is completed only the higher band (hh) is occupied. Thus, the light holes usually do not participate in the luminescence of quantum wells. The role of light holes is also suppressed due to the very low offset of the lh band in many materials. Now, we will deal with the luminescence of quantum wells in more detail.

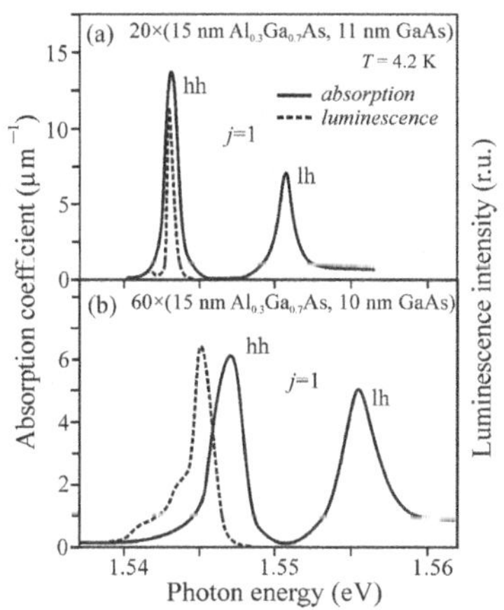

Fig. 12.15
Comparison of absorption (solid lines) and luminescence (dashed lines) spectra of GaAs/AlGaAs multiple quantum wells. Unlike absorption, the luminescence spectra contain only one emission band corresponding to the lowest transition. There is usually also a significant Stokes shift of this band against the corresponding absorption peak. The lower panel (b) shows spectra from a sample with the number of wells three times as high, revealing broader spectral bands due to inhomogeneous broadening (well thickness fluctuations). Adapted from Klingshirn [6].

12.3.5 Luminescence of quantum wells

Luminescence spectra typically differ from absorption spectra in several features:

(a) Only a single emission band, corresponding to the lowest transition, is usually observed (see Fig. 12.15 [6]). This follows from the efficient relaxation of higher excited states into the lowest state, from which radiative recombination takes place. This was already mentioned in the previous subsection.

(b) The luminescence peak is generally shifted to lower energies compared to the corresponding absorption band (the so-called Stokes shift, see Fig. 12.15). The main reason for such a shift lies in the energy relaxation within an inhomogeneously broadened band. That is, the width of a real quantum well is not perfectly constant, but has a certain distribution. Thus, the local quantum confinement acts with different strength and the local energy levels vary accordingly, resulting in the 'smearing' of energy levels, usually referred to as inhomogeneous broadening due to *interface*

[5] The hole wavefunction is of p-type ($L = 1$), which in combination with spin gives $J = 3/2$ or $J = 1/2$. The lower lying $J = 1/2$ valence band is referred to as the split-off band.

roughness. Obviously, the inhomogeneous broadening is expected to be more pronounced in a larger ensemble of quantum wells, as documented by the difference between the spectra in Figs 12.15(a) and (b). (To some extent, the width of luminescence bands can thus be used to characterize the quality and homogeneity of the fabricated structure.) Another reason for the observed Stokes shift can be, e.g., the interaction with phonons (see Chapter 4).

The above-mentioned pieces of knowledge imply the following most important differences in luminescence properties of a 'free' exciton in a bulk semiconductor and in a 2D quantum well. (Most of these features actually manifest themselves even more distinctly in 1D and 0D structures.)

(1) The wavelength of excitonic luminescence from a 2D well, in contrast to 3D crystals, depends on the size of the structure (well thickness). The emission shifts rapidly to shorter wavelengths (higher photon energies) with decreasing size L_z. Conversely, if the well thickness L_z increases, the luminescence peak approaches the limit of a bulk crystal (this limit is about $L_z \sim 10\,\text{nm}$ or several tens of nm, depending on the composition of the heterostructure). Therefore, were the heterostructure properties well-controlled and well-characterized, in principle a simple measurement of the exciton luminescence spectral position from a quantum well could be used as a quick method to extract the value of the well thickness, without the necessity of resorting to sophisticated electron-microscopy techniques.
(2) The 'free-exciton' emission linewidth should be more than twice as narrow in a 2D structure compared to a 3D crystal (the full width at half maximum, FWHM, is about $0.7\,k_BT$ as against $1.8\,k_BT$, see Fig. 12.16). This phenomenon results from the differing shape of the density of states, which is constant in a 2D well while it has a square-root form in 3D crystals. The emission line narrowing is important, e.g. for the application of nanostructures in laser diodes, as it makes it easier for the population inversion to be achieved, thus lowering the threshold current density for lasing.
(3) Excitonic phenomena are more pronounced and may be observable at higher temperatures (as compared to 3D crystals). The exciton binding energy is increased in low-dimensional structures and, in addition to that, the probability of excitonic transitions is enhanced due to the larger overlap of the electron and hole wavefunctions confined in a quantum well. Consequently, excitonic features in some structures can be observed even at room temperature in contrast to bulk crystals, where room-temperature exciton luminescence is excluded.

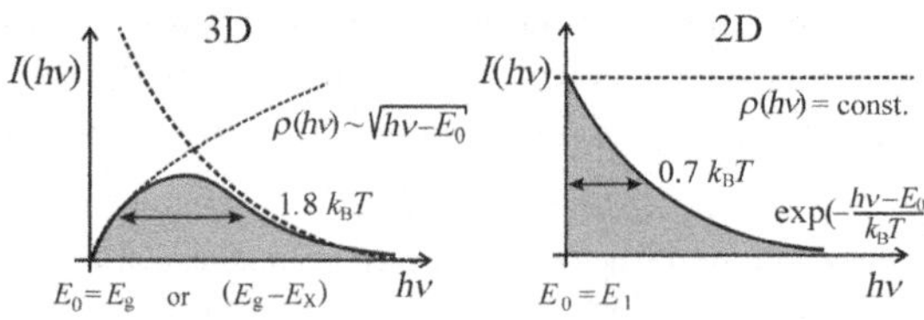

Fig. 12.16
Narrowing the 'free' exciton luminescence line when going from a 3D to a 2D structure, as a consequence of the modified shape of the density of states $\rho(E) = \rho(h\nu)$.

(4) Nonlinear optical phenomena tend to be much stronger and take place for much lower excitation power densities. This is a consequence of the confinement of the excited electron–hole pairs, which, as a result, cannot migrate freely in a crystal. Thus, the effective population density can reach very high values even for moderate excitation power densities (provided that a fast depopulation mechanism does not appear at the same time).

In the foregoing discussion, the attribute 'free' used in connection with an exciton was always enclosed in quotation marks, indicating that excitons cannot be fully free (delocalized) in low-dimensional structures, in contrast with bulk crystals. We have already mentioned that in quantum wells local fluctuations in the well width and possibly also in its chemical composition inevitably exist, especially in the case of ternary and quaternary alloys. Moreover, the wells may contain intentionally introduced impurity atoms. All these spots act as possible localization traps for excitons. As the total amount of material in a quantum well (or another low-dimensional structure) is small, local potential fluctuations such as localization centres are far more important for optical properties than in 3D semiconductors. Therefore, the observed excitonic emission band should be considered as inhomogeneously broadened due to the superposition of the radiative recombination of 'free' excitons and excitons localized in the tail states, the latter being similar to those occurring in amorphous semiconductors (Chapter 9). In brief, in low-dimensional structures one can encounter almost exclusively *quasi-free excitons*, having smaller or higher degree of localization depending on the type of structure and its quality. Only in exceptional-high-quality and relatively wide quantum wells [11] or quantum wires [12] has true free exciton photoluminescence succeeded in being observed.

Direct evidence of the inhomogeneous broadening of exciton emission lines has been supplied by the micro-photoluminescence technique with high spatial resolution. Figure 12.17 shows emission spectra of a 1D structure—an array of $GaAs/Al_{0.3}Ga_{0.7}As$ quantum wires [13]. The uppermost curve was detected in a standard photoluminescence arrangement, i.e. with an excitation laser spot of relatively large diameter of 35 μm. The remaining spectra were excited and detected through small openings fabricated by a lithographic technique in a thin aluminium layer deposited on the sample surface. Clearly, the ensemble smooth spectrum falls into a group of narrow emission lines whose number decreases with reduced size of the opening (down to submicron diameter). These lines originate from individual excitons localized in different spots of the nanowire as seen through the opening. This topic brings us to quantum wires, which are the subject of the following section.

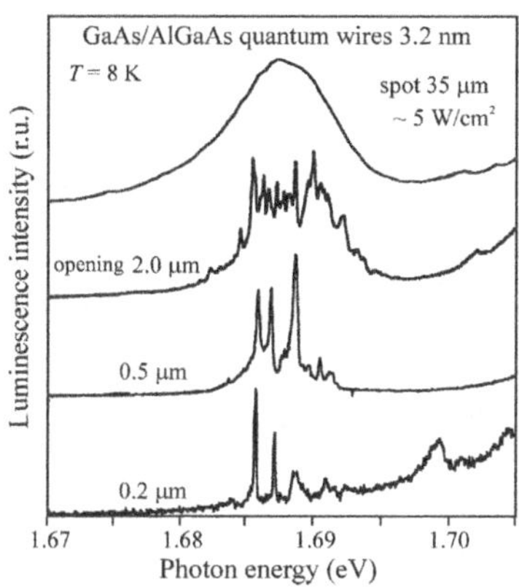

Fig. 12.17
Photoluminescence spectra of GaAs/AlGaAs quantum wires with a diameter of 3.2 nm, detected at $T = 8$ K through small openings of various sizes etched in a non-transparent mask. Excitation was provided by an Ar^+-laser at 488 nm with a power density of 240 W/cm^2. The upper spectrum was detected without masking with a laser spot size of 35 μm. Adapted from Vouilloz *et al.* [13].

12.4 Quantum wires

The term *quantum wires* refers to structures in which the motion of electrons and holes is confined in two directions, i.e. to quasi-one-dimensional structures with lateral dimensions of the order of several nm or tens of nm. Such structures are fabricated, e.g., by etching of narrow stripes in 2D quantum wells or by the bottom-up growth of a nanowire on a metallic nanodot. Similarly to 2D quantum wells, the unique electronic and optical properties of nanowires arise from a specific density of states; in what follows we indicate how its peculiar shape can be described [9].

Let the long axis of a quantum wire be parallel to the y axis and let L_x, L_z stand for its lateral dimensions. The motion of electrons and holes in then quantized in the x and z directions. In close analogy to eqns (12.7a, b), we can put together the Schrödinger equation, for example, for an electron and subsequently separate the wavefunction into the electron motion along the x, y and z axes. The resulting energy levels in the x and z directions then take on the familiar quantized form of (12.17)

$$E_{j_x} = \frac{\hbar^2}{2m_e}\left(\frac{j_x\pi}{L_x}\right)^2, \quad E_{j_z} = \frac{\hbar^2}{2m_e}\left(\frac{j_z\pi}{L_z}\right)^2, \qquad j_x, j_z = 1, 2, 3\ldots$$

while in the y direction the kinetic energy remains a continuous parabolic function of the wavevector $\mathbf{k}_y$. Consequently, the overall energy reads

$$E = \frac{\hbar^2}{2m_e}\left[\left(\frac{j_x\pi}{L_x}\right)^2 + \left(\frac{j_z\pi}{L_z}\right)^2 + k_y^2\right]. \tag{12.23}$$

The one-dimensional density of states is easily derived from an equation analogous to (12.4): The derivative (dk_y/dE) is obtained easily from (12.23) and by recalling that $\rho^{(1)}(k) = 1/\pi$ (because of spin degeneracy, $g_S = 2$) we find the solution

$$\rho^{(1)}(E) = \frac{1}{\pi}\left(\frac{m_e}{2\hbar^2}\right)^{1/2}\frac{1}{\sqrt{E - E_{j_x} - E_{j_z}}}.$$

The joint density of states (describing the difference energy band or, in other words, determining the number of electronic states in the conduction and valence bands that are separated by a given photon energy), related closely to optical transitions, is derived by substituting the electron–hole (exciton) reduced mass m_r for the electron mass m_e and by summing over all allowed pairs of quantum numbers j_x and j_z:

$$\rho_s^{(1)}(E) = \frac{1}{\pi}\left(\frac{m_r}{2\hbar^2}\right)^{1/2}\sum_{j_x, j_z, j'_x, j'_z}\frac{1}{\sqrt{E - E_g - E^e_{j_x} - E^e_{j_z} - E^h_{j'_x} - E^h_{j'_z}}}. \tag{12.24}$$

It might be of interest to compare eqn (12.24) with the more general form (12.6a), derived before introducing the quantum confinement effect—a phenomenon that changes the continuous energy bands into discrete levels.

The shape of the 1D joint density of states scales as an inverse square root $1/\sqrt{x-a}$: this function of x 'starts' with a singularity at an abscissa $x = a$ and rapidly falls off to zero for $x > a$. Figure 12.18 compares the joint density of states in quantum wells and quantum wires.

Optical absorption and emission of quantum wires have many features in common with quantum wells. Also in quantum wires we can imagine an

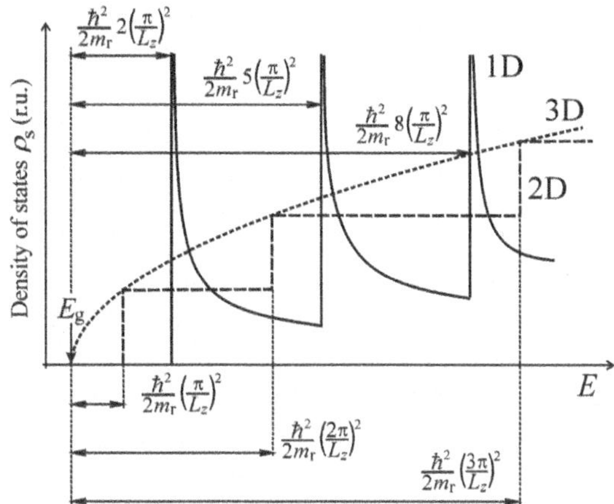

Fig. 12.18
Theoretical shape of the joint density of states in a 2D (12.18) and 1D (12.24) semiconductor (for the special case $L_x = L_z$). The dotted curve shows the density of states in a 3D crystal and E_g is the corresponding bandgap. (This scheme refers only to a relative comparison of ρ_s shapes using the appropriate normalization; a direct comparison is not possible as ρ_s have different dimensions in quantum structures of different dimensionalities.)

exciton moving 'freely' along one direction (it has one degree of freedom). The quantum confinement effect enhances the exciton binding energy with decreasing lateral dimensions of the wire. Optical absorption is then characterized by excitonic resonance; the singularities in the density of states of free carriers (12.24) do not manifest themselves particularly in the observed absorption spectra. The oscillator strength of the lowest transition prevails over the higher transitions, like in 2D structures. Furthermore, the splitting of light- and heavy-hole states also manifests itself. A blue-shift of the absorption due to quantum confinement is partially compensated by the increasing exciton binding energy, which causes a shift in the opposite direction. In realistic samples, we once again encounter excitons which are not completely free but 'quasi-free', as discussed at the end of Subsection 12.3.5.

Figure 12.19 [14] illustrates a basic comparison of photoluminescence excitation spectra (which are connected to optical absorption) and emission spectra of a (5.5-nm-thick) GaAs quantum well sandwiched between $Al_{0.5}Ga_{0.5}As$ barriers and a GaAs wire ('cut' from the GaAs layer with an ion-beam technique) with a cross-section of 5.5 nm × 100 nm. In the case of the wire, both

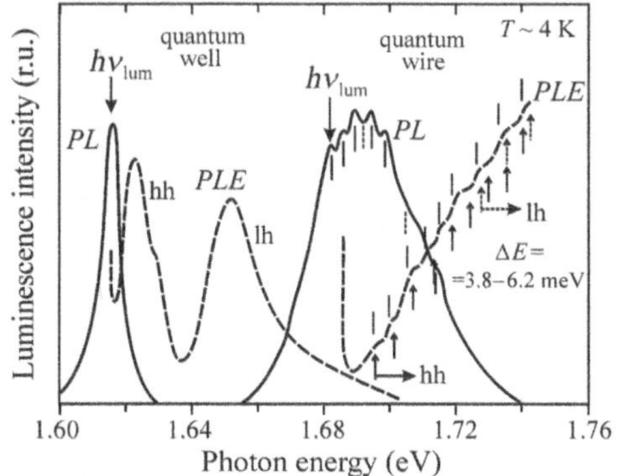

Fig. 12.19
Comparison of photoluminescence excitation (PLE) and emission (PL) spectra of a quantum well and a quantum wire (having a relatively weak lateral confinement in one dimension). Adapted from Klingshirn [6].

spectra are (under otherwise the same experimental conditions) blue-shifted when compared to the quantum well. This blue-shift is, in principle, in accord with Fig. 12.18. Futhermore, the difference in the shape of the density of states between a well and a wire (i.e. a plateau versus peaks) also manifests itself in Fig. 12.19, namely as the 'fine structure' modulating the emission and excitation spectra of the wire. The 'dense' arrangement of the small maxima in the spectrum reflects a relatively weak quantum confinement effect in one of the lateral dimensions of the wire ($L_x \gg L_z$), giving rise to closely spaced 1D singularities. In both structures, the splitting of light- (lh) and heavy-hole (hh) bands is clearly observed.

Finally, we shall demonstrate explicitly how the decreasing lateral dimension of quantum wires influences their luminescence spectra. Figure 12.20(a) [15] shows luminescence spectra of a series of InGaAs(wire)/InP(barrier) quantum wires with one fixed lateral dimension $L_z = 5\,\text{nm}$ and the second dimension L_x varying between 46 and 10 nm. The emission band is significantly blue-shifted with decreasing width L_x by up to 70 meV. Increasing pump power (Fig. 12.20(b)) induces profound spectral broadening on the high-energy side. This is a clear demonstration of nothing less than the Pauli exclusion principle, with which the quasi-particles must comply and thus with increasing excitation occupy higher levels resulting from the lateral confinement. Due to the overall low number of available 1D states, such a situation can be reached even at relatively low excitation levels ($\sim 0.5\,\text{kW/cm}^2$). These effects are treated in more detail in Chapter 13. A noteworthy analogy to be pointed out here and now, however, is the similarity between Fig. 12.20(b) (a quantum wire) and Fig. 13.7(a) (a quantum well).

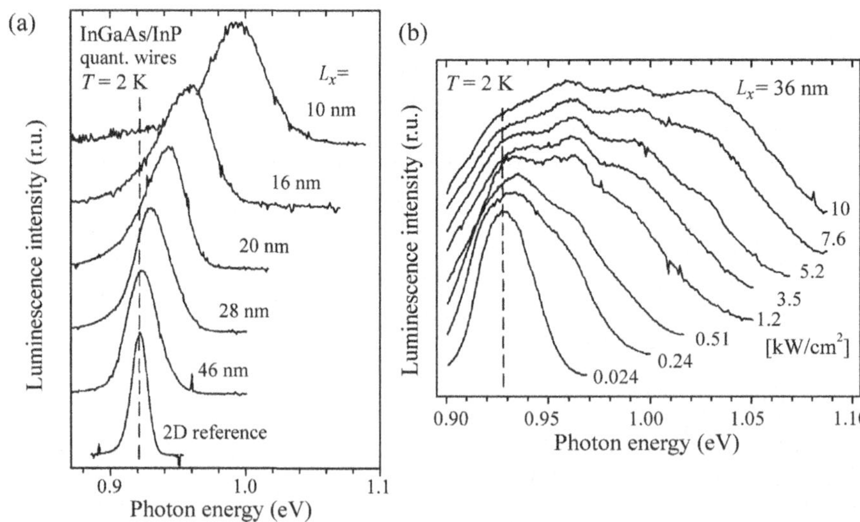

Fig. 12.20
(a) Luminescence emission spectra of InGaAs/InP quantum wires with different lateral dimensions L_x (sharing the second dimension $L_z = 5\,\text{nm}$). The excitation power density from a continuous Ar^+-laser was $40\,\text{W/cm}^2$. (b) Luminescence spectra of a quantum wire ($L_x = 36\,\text{nm}$) under various excitation power densities with a pulsed Ti-sapphire laser. $T = 2\,\text{K}$, adapted from Forchel *et al.* [15].

12.5 Quantum dots—nanocrystals

Now, we will move on to structures confined in all dimensions. Such quasi-zero-dimensional structures are commonly referred to as *quantum dots* or even as *artificial atoms* (as a consequence of some similarities in the energy structure). The term denoting 'real-life' quantum dots is often *nanocrystals*. Nanocrystals are the subject of countless reviews, published both as books and in peer-reviewed journals; probably the easiest one to understand is the textbook by Gaponenko [16].

Before we start describing the zero-dimensional structures, it is useful to make a terminological digression here: In the literature, the terms 'quantum dots' and 'nanocrystals' are frequently not considered equivalents. Zero-dimensional objects of pyramidal or hemispherical shape self-assembled on a planar substrate tend to be referred to as quantum dots (see Subsection 14.3.2). On the other hand, nanocrystals are often of a more or less spherical shape and are homogeneously dispersed in a liquid or solid matrix. The basic physics underlying the behaviour of both types of 0D objects is, nevertheless, the same and, therefore, later on we will use both of these terms as if they were synonyms.

12.5.1 Quantum dot with spherically symmetric potential

A quasi-zero-dimensional quantum dot with radius R, exhibiting the quantum-confinement effect in all directions, can be most easily described using a spherically symmetric potential well, defined by

$$V(r) \begin{cases} = 0 & \text{for } r \leq R, \\ = V_0 & \text{for } r > R, \end{cases} \tag{12.25}$$

where $r^2 = x^2 + y^2 + z^2$. The treatment of such a spherically symmetric well with finite depth can be carried out using separation of variables and finding an analytical solution; more details can be found, e.g., in [17].

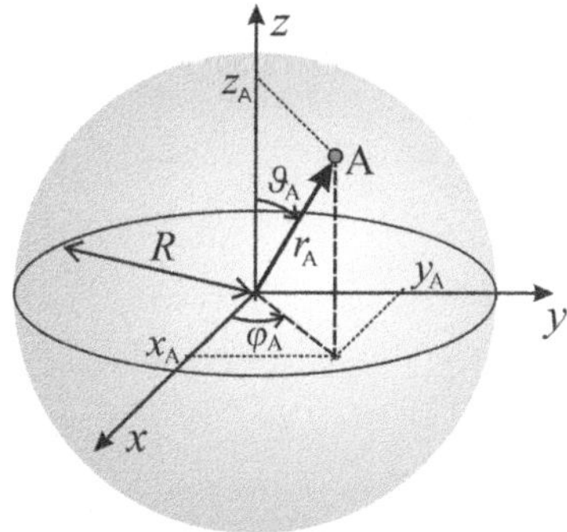

Fig. 12.21
Illustration of a spherically symmetric potential well and a description of the location of the point A using Cartesian (x, y, z) and spherical (r, ϑ, φ) coordinates.

In view of the symmetry of the problem, it is useful to employ the transformation of the Cartesian coordinate system into spherical coordinates (see Fig. 12.21):

$$x = r \sin\vartheta \cos\varphi, \quad y = r \sin\vartheta \sin\varphi, \quad z = r \cos\vartheta.$$

The transformation changes the relevant Hamiltonian into

$$H = -\frac{\hbar^2}{2m}\nabla^2 + V(r) \quad \Rightarrow \quad -\frac{\hbar^2}{2mr^2}\frac{\partial}{\partial r}\left(r^2\frac{\partial}{\partial r}\right) - \frac{\hbar^2\Lambda}{2mr^2} + V(r),$$

where the operator Λ stands for

$$\Lambda = \frac{1}{\sin\vartheta}\left[\frac{\partial}{\partial\vartheta}\left(\sin\vartheta\frac{\partial}{\partial\vartheta}\right) + \frac{1}{\sin\vartheta}\frac{\partial^2}{\partial\varphi^2}\right].$$

The wavefunction can be separated into the functions of the individual variables $\psi = R(r)\,\Theta(\vartheta)\Phi(\varphi)$. It is often written down as

$$\psi_{n,l,m}(r,\vartheta,\varphi) = \frac{u_{nl}(r)}{r} Y_{lm}(\vartheta,\varphi), \tag{12.26}$$

where Y_{lm} denote the *Laplace spherical harmonics* and the radial part $u(r)$ is the solution of

$$-\frac{\hbar^2}{2m}\frac{\mathrm{d}^2 u}{\mathrm{d}r^2} + \left[V(r) + \frac{\hbar^2}{2mr^2} l(l+1)\right] u(r) = Eu(r), \qquad l = 0, 1, 2, \ldots \tag{12.27}$$

This equation enables us to find the eigenvalues of the energy for a particle in a spherical potential well.[6]

Up to now, the derivations we have made in this subsection were *de facto* identical to the analysis of the motion of a particle under a central force (i.e. in a field of spherical symmetry), treated in countless textbooks on quantum physics, e.g. [17, 18]. In the present case, however, unlike the Coulomb potential term $V(r) = -e^2/4\pi\varepsilon_0 r$ applied traditionally in the well-known problem of a hydrogen atom, $V(r)$ is described by (12.25) and the solution to (12.27) will thus differ from the hydrogen problem.

As long as the well is infinitely deep ($V_0 \to \infty$), the radial part of the wave function $u(r)$ will equal zero for $r = R$ and the energy levels corresponding to the localized states will be described by the simple equation [7, 9]

$$E_{nl} = \frac{\hbar^2}{2m}\left(\frac{\chi_{nl}}{R}\right)^2, \tag{12.28}$$

where χ_{nl} stands for the nth root of the lth-order spherical Bessel function. The values of the roots are listed in Table 12.1. We can immediately see that if $l = 0$, the roots are equal to $\pi, 2\pi, 3\pi, \ldots$ It is definitely of interest to compare this result with the one-dimensional 'quantum box': if $l = 0$, the wave equation (12.27) transforms into the equation of a one-dimensional well

Table 12.1 Roots of the Bessel functions.

l	$n = 1$	$n = 2$	$n = 3$
0	3.142 (π)	6.283 (2π)	9.425 (3π)
1	4.493	7.725	10.904
2	5.764	9.095	12.323
3	6.988	10.417	
4	8.183	11.705	
5	9.356		
6	10.513		
7	11.657		

[6] For the sake of consistency, we probably should, in analogy with the above-treated cases of 2D and 1D structures, use the letter j to stand for the principal quantum number. However, as in this case the principal quantum number cannot be mistaken for the principal quantum number of an exciton, henceforth the letter n will be used in accordance with common usage in the literature.

(12.8) and the solution must thus correspond to eqns (12.11) and (12.12), i.e. $E_n = (n\pi\hbar)^2/2mL_z^2$, $n = 1, 2, 3, \ldots$

However, there are also important differences as against the one-dimensional 'box' case. First of all, the radial function $u(r)$ must converge to zero for $r \to 0$ (otherwise, the value of $u(r)/r$ would diverge and the wavefunction (12.26) would not be normalizable). Therefore, the solution in the form of the even functions $u(r)$ cannot be applied here, the ground state is described by an odd wavefunction $u(r) \sim \sin(\pi r/R)$ and by quantum numbers $n = 1, l = 0$ (at the same time, $n = j/2$, where j is the quantum number of the one-dimensional 'box'). As a result, when the problem of a finitely deep well is treated (when the wavefunctions 'leak' through the boundaries determined by the radius R, like in Fig. 12.12), no bound state will exist in a 'shallow' quantum dot, i.e. when $V_0 < (\pi\hbar)^2/2m(2R)^2$; see also Problem 12/5. This is a crucial piece of information that needs to be kept in mind when preparing quantum dots and choosing the barrier materials. The finite well depth in real-life nanocrystals usually stems from their encapsulation with a material with wider bandgap, or embedment into a similar matrix, or from surface capping by various passivation molecules.

We have introduced two quantum numbers n, l in order to describe the eigenstates in a spherical well; in addition to these, however, a third quantum number m is usually used. These quantum numbers are the so-called *principal, orbital* and *magnetic* quantum numbers. The energy of the state depends only on the n and l quantum numbers (the magnetic quantum number m can acquire $(2l + 1)$ values of $0, \pm 1, \ldots, \pm l$, or, in other words, every nl state is $(2l + 1)$–times degenerate and this degeneracy can be lifted in a magnetic field). The states corresponding to the orbital quantum number values of $l = 0, 1, 2, 3, 4$ are denoted by letters s, p, d, f, g, ...(and so on in alphabetical order).

Figure 12.22 displays the energy levels of an infinitely deep spherical well. The energy is expressed in units $\hbar^2/2mR^2$, that is to say the squares of the roots of the Bessel functions χ_{nl} are plotted there.

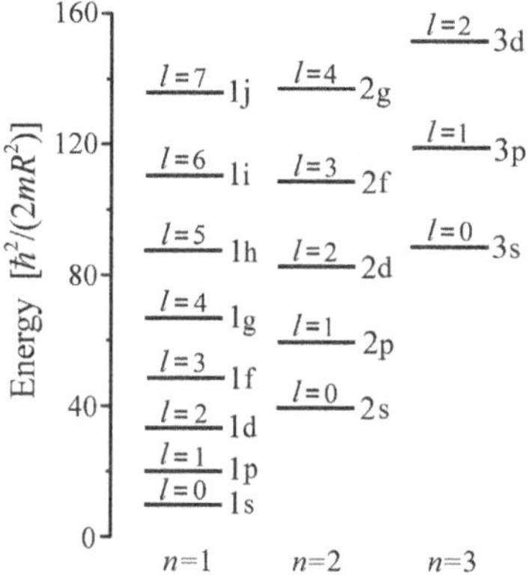

Fig. 12.22
Energy levels of an infinitely deep spherical well. Energy is given in units $\hbar^2/2mR^2$, that is the values correspond to the squares of the roots of the Bessel functions χ_{nl}.

12.5.2 Types of quantum dots according to the strength of the quantum confinement effect

If we base our description on the approximation of effective masses (and assume that the band structure of a semiconductor and the effective masses of the quasi-particles keep the characteristics of a bulk crystal), the energy levels in the quantum dot will depend on the strength of the quantum confinement effect and the Coulomb interaction between electrons and holes. Depending on the relative importance of these two effects, two extreme situations can occur; these two cases are commonly referred to as the *strong* and *weak quantum confinement regimes*, with the middle ground of the *intermediate quantum confinement regime* in between them. If the size of a nanocrystal is smaller than approximately 1–1.5 nm when the nanocrystals consist of only a few hundreds of atoms or less, the approximation of the effective mass can hardly be applied any more. Such small particles no longer possess a periodic crystal structure

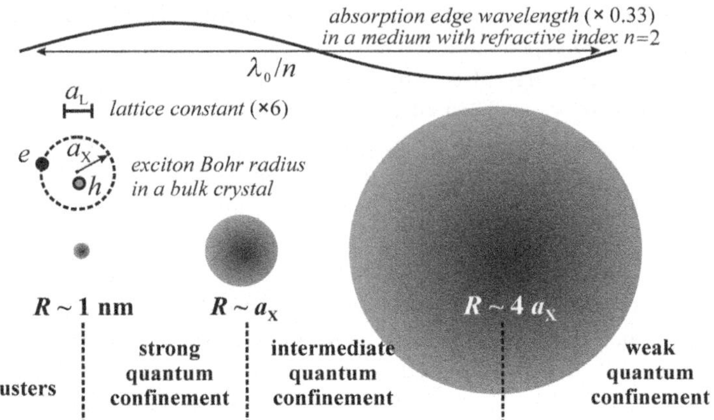

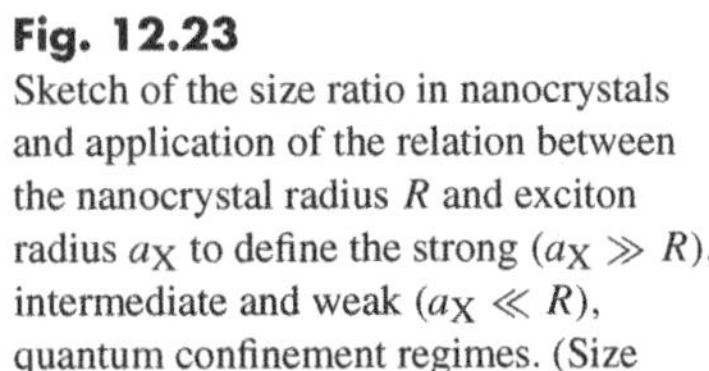

Fig. 12.23
Sketch of the size ratio in nanocrystals and application of the relation between the nanocrystal radius R and exciton radius a_X to define the strong ($a_X \gg R$), intermediate and weak ($a_X \ll R$), quantum confinement regimes. (Size ratios shown hold for the case of ZnSe.)

and as such have to be treated as *clusters*, which can only be described by quantum-chemical approaches (Fig. 12.23).

Strong quantum confinement regime

We continue to consider a spherical well with infinite potential barrier. Let the radius R of the quantum dot be much smaller than the exciton Bohr radius a_X in the material under consideration; an additional condition to be satisfied here should be the requirement that R is much larger than the crystal lattice constant a_L, otherwise we could not—even approximately—assume a periodical crystal, but we would have to treat the particle as a cluster instead. Or, mathematically,

$$a_L \ll R \ll a_X.$$

Under these conditions, the influence of the Coulomb interaction is considered smaller when compared to that of quantum confinement and, as a crude approximation, the interaction between the electron and the hole can be neglected. Thus, the states of each quasi-particle are described by (12.28). Selection rules allow transitions only between the electron and hole states with the same principal and orbital quantum numbers.

Thus, the absorption spectrum consists of discrete peaks at energies

$$E_{nl} = E_g + \frac{\hbar^2 \chi_{nl}^2}{2m_e R^2} + \frac{\hbar^2 \chi_{nl}^2}{2m_h R^2} = E_g + \frac{\hbar^2 \chi_{nl}^2}{2m_r R^2}.$$

Nevertheless, we have to keep in mind that the increased overlap of the electron and hole wavefunctions, as a result of the spatial localization of the quasi-particles in a small volume, inevitably enhances the Coulomb interaction. Therefore, this interaction cannot be completely disregarded even in the strong quantum confinement regime. Applying the calculus of variations to the problem in which both the quantum confinement and the Coulomb interaction are included results in a formula for the ground state of an electron–hole pair in the form

$$E_{1s,1s} = E_g + \frac{\pi^2 \hbar^2}{2m_r R^2} - 1.786 \frac{e^2}{2(4\pi\varepsilon_0)\varepsilon R}, \tag{12.29}$$

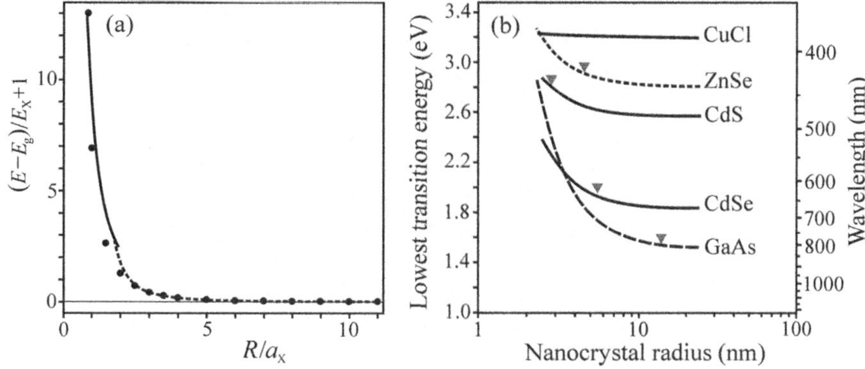

Fig. 12.24
(a) Energy of the lowest allowed optical transition $E_{1s,1s}$ in an ideal nanocrystal (with strong and intermediate quantum confinement) as a function of its radius, given in units of the exciton Bohr radius a_X. The energy shift with respect to the bulk exciton ground state is expressed in units of the exciton binding energy E_X. (b) Energy of the principal optical transition in several common semiconductors. Triangles denote the exciton Bohr radius in corresponding bulk crystals (in CuCl, $a_X = 0.7$ nm holds, outside the range of the plot). After Gaponenko [16].

where the subscript 1s,1s stands for the $n = 1, l = 0$ states of both the electron and the hole, respectively, and the last term describes the effective Coulomb interaction of an e–h pair. This term is substantially larger in comparison with the bulk exciton binding energy $e^2/2(4\pi\varepsilon_0)\varepsilon a_X$, because $a_X \gg R$. This is a significant difference when compared to 2D and 1D structures, where the contribution of the Coulomb interaction of a free e–h pair can be very close to zero in the strong quantum confinement regime.

A more general and more accurate form of eqn (12.29) for the exciton ground state reads

$$E_{1s,1s} = E_g + \pi^2 \left(\frac{a_X}{R}\right)^2 E_X - 1.786 \left(\frac{a_X}{R}\right) E_X - 0.248 E_X, \qquad (12.30)$$

where the last term represents the correlation energy. The energy versus size relationship (12.30) is plotted in Fig. 12.24(a) by the solid line, as a universal material-independent relationship valid for all semiconductors. Figure 12.24(b) depicts this dependence as calculated using (12.30) and additional approximations for selected examples of particular semiconductors [16].

There is a special case when the solution of the Schrödinger equation can be expressed analytically. This happens when the effective mass of the hole is much larger than that of an electron, $m_h \gg m_e$, which means the reduced mass can be approximated by the electron effective mass, $m_r \approx m_e$. Then, if we introduce the Bohr radius separately for electrons and for holes

$$a_e = \frac{(4\pi\varepsilon_0)\varepsilon\hbar^2}{m_e e^2}, \qquad a_h = \frac{(4\pi\varepsilon_0)\varepsilon\hbar^2}{m_h e^2}, \qquad a_X = a_e + a_h,$$

the inequality $a_h \ll R \ll a_e, a_X$ applies. The hole can be treated as motionless, localized at the centre of the nanocrystal. The situation under consideration is the so-called donor-like exciton, a parallel to the Born–Oppenheimer approximation (a heavy nucleus whose motion can be ignored as against that of the electrons). The energy levels and the absorption spectrum are then determined mostly by the quantization of the electron motion; each of these levels, however, is split into several sublevels due to the Coulomb interaction of the electron and hole.

Yet another phenomenon influencing the exciton energy levels is the surface polarization of the nanocrystal in question. This is a consequence of the dielectric medium by which the nanocrystal is surrounded; the dielectric constant of this medium is smaller than the dielectric constant of the nanocrystal itself. This difference might result in self-trapping of charge carriers on the surface and subsequent localization of both the wavefunction and the charge density close to the nanoparticle surface instead of in its core. This effect plays a more important role for the heavier quasi-particle, i.e. it is more important for the hole.

Weak quantum-confinement regime

The so-called weak quantum–confinement regime refers to the situation when the radius of the quantum dot is several times as large as the exciton Bohr radius. Under these circumstances, the excitonic effects dominate and the quantum-confinement effect only causes the kinetic energy of the exciton centre-of-mass motion to become quantized. Here, we can start with the formula for the bulk exciton dispersion (7.6) and substitute the quantized energy levels of a particle in a spherical quantum well for the free exciton kinetic energy, or

$$E_{n_{\mathrm{exc}}nl} = E_{\mathrm{g}} - \frac{E_{\mathrm{X}}}{n_{\mathrm{exc}}^2} + \frac{\hbar^2 \chi_{nl}^2}{2m_{\mathrm{exc}}R^2},$$

where n_{exc} stands for the principal quantum number of the internal excitonic states governed by the $e-h$ interaction (they can be denoted here in analogy to the hydrogen atom as 1S, 2S, 2P, 3S, 3P, ...) and n, l correspond to the quantum numbers of the quantization of the exciton centre-of-mass motion, perceiving the exciton as a whole, i.e. as a quasi-particle with total mass m_{exc} confined in a spherically symmetric external potential well (1s, 1p, 1d, ..., 2s, 2p, 2d, ..., see Fig. 12.22). To distinguish between the internal and external states, capitals are being used for the former and lower case for the latter states [16].

The lowest-lying state is characterized by $n_{\mathrm{exc}} = 1, n = 1, l = 0$, thus

$$E_{1\mathrm{S}1\mathrm{s}} = E_{\mathrm{g}} - E_{\mathrm{X}} + \frac{\hbar^2\pi^2}{2m_{\mathrm{exc}}R^2} = E_g - E_{\mathrm{X}}\left[1 - \frac{m_{\mathrm{r}}}{m_{\mathrm{exc}}}\left(\frac{\pi a_{\mathrm{X}}}{R}\right)^2\right].$$

The above relation clearly shows that the exciton ground state is shifted by $(m_{\mathrm{r}}/m_{\mathrm{exc}})(\pi a_{\mathrm{X}}/R)^2$ towards higher energy with respect to a bulk crystal. However, this shift is smaller than the exciton binding energy as long as the condition $R \gg a_{\mathrm{X}}$ is satisfied. Hence the term ‘weak confinement’.

12.5.3 Luminescence of quantum dots

The luminescence of an ensemble of quantum dots, either in a colloidal dispersion or embedded in a solid matrix, is one of the most-studied phenomena in the last 20 years or so. Since a large variety of preparation methods exist and each one has its own characteristic features, pros and cons, every single sample is in fact an original entity with its own distinctive luminescence properties. This is why it is not an easy task to find generally valid, unifying features. However, it is possible to list several common characteristics as follows:

(1) The emission spectrum of a (sufficiently small) nanocrystal is blue-shifted with respect to its 3D counterpart and shifts further towards shorter wavelengths with decreasing size of the nanocrystal.
(2) Edge emission is substantially enhanced compared to the wide extrinsic emission bands exhibited by the 3D counterparts.
(3) Quite commonly, luminescence can be observed even at elevated temperatures, frequently up to room temperature.
(4) Both the emission and excitation spectra are inhomogeneously broadened as a result of the slightly differing shape and finite distribution of sizes of the nanocrystals in the ensemble under study. Up to now, unfortunately, no preparation method capable of facing up to this drawback and producing an ensemble of nanocrystals with exactly predefined shape and uniform distribution has been found; consequently, the broadening hinders the fundamental research study of properties of nanocrystals as well as the prospective applications. An example of a relatively narrow size distribution of nanocrystals is shown in Fig. 12.25 [19]. As a result, the emission spectra one is likely to meet in practise exhibit features characteristic of inhomogeneous broadening: under excitation into the emission band, it is line narrowing and a shift of the luminescence maximum; in addition, a distribution of lifetimes occurs and, under strong pulsed monochromatic laser excitation, also spectral hole-burning.

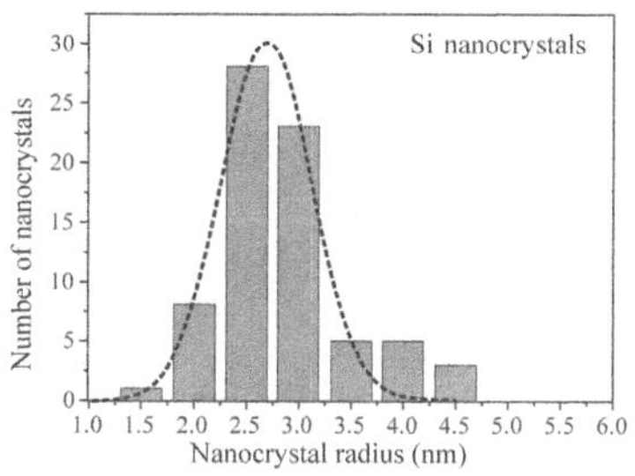

Fig. 12.25
Histogram of the distribution of diameters of Si nanoparticles prepared by electrochemical etching. After Dohnalová *et al.* [19].

These features will now be illustrated with as broad a spectrum of semiconductor materials as possible serving as examples.

In the previous chapters, we referred to silver halides as a suitable model material for a large variety of luminescence phenomena. Let us start also now with low-temperature photoluminescence of silver bromide AgBr. The experimental results presented in Fig. 12.26 [20] can serve as a perfect illustration of the first two properties on the list. The bottom spectrum corresponds to a bulk-crystal reference, the shape of the spectrum being almost identical with that in Fig. 7.26(b), including a very weak free-exciton emission I_{TO}(FE) at ~ 463 nm (~ 2.68 eV). As the diameter of the nanocrystal decreases (middle and top spectrum), an extraordinary increase in edge emission intensity at 450–470 nm (by four orders of magnitude!) as well as its blue-shift occur. At the same time, the broad impurity-related band of residual iodine Iod_{LO}(BE) at 500 nm completely fades away. The reason behind such a stark change is simple: the concentration of residual iodine is low, at the ~ 1 ppm level, and, therefore, in an AgBr nanocrystal of a typical radius of $R = 5$ nm consisting of about 20,000 ions, the probability of finding an impurity I^- ion is negligible. This naturally results in the increased edge emission of free[7] and weakly bound

[7] In analogy to 2D and 1D systems, in nanocrystals no truly 'free' or delocalized excitons exist; we can only speak in terms of 'quasi-free' excitons, all the more so when the free motion of quasi-particles is restricted in all three dimensions.

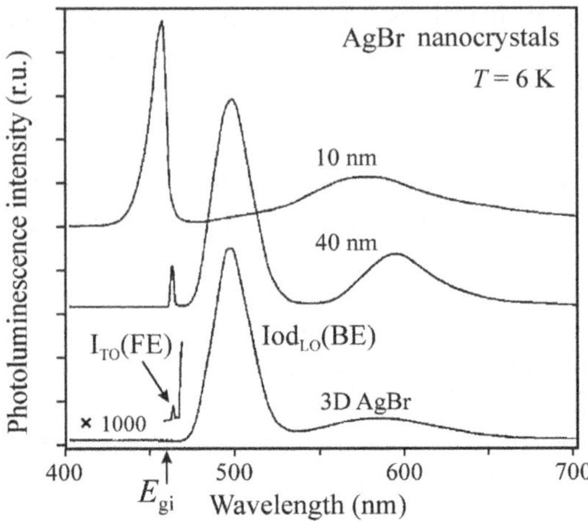

Fig. 12.26
Comparison of low-temperature emission spectra of variously sized AgBr nanoparticles. The bottom spectrum corresponds to bulk AgBr, and the arrow denotes the bandgap position. Note the increased edge-emission intensity in the uppermost spectrum, its blue-shift with respect to the bottom spectrum and the absence of the Iod_{LO}(BE) band at 500 nm. Cw excitation 325 nm with a HeCd laser, $T = 6$ K. After Freedhoff *et al.* [20].

excitons, as these were deprived of the chance to localize themselves in the deep I^- ion traps.

For the sake of interest, let us add that in low-temperature photoluminescence of silver chloride, AgCl, nanocrystals no edge emission occurs; only a significantly Stokes-shifted broad emission band at ~ 510 nm (~ 2.4 eV) emerges, being a result of the radiative recombination of self-trapped excitons (STEs). Consequently, the emission closely resembles to that of bulk AgCl samples (see Fig. 7.30(a)). Such behaviour confirms the well-known fact that this band is of intrinsic character, not impurity related. The absence of edge emission then reflects the fact that the strong exciton–phonon interaction in AgCl, leading to exciton self-trapping, exists also in nanocrystals and, very probably, is not influenced by their size too much.

Room-temperature photoluminescence of nanocrystals is shown in Fig. 12.27. These spectra of a colloidal dispersion of CdS nanocrystals in water are of interest for several reasons. First of all, this is one result of the pioneering works of the Berlin school [21], coming from the study of semiconductor quantum dots at the beginning of the 1980s (at that time, quantum dots were often called Q-particles). These measurements clearly confirmed the occurrence of quantum confinement through its optical manifestations and became one of the driving forces for the development of the nanophotonics field. Furthermore, Fig. 12.27 demonstrates a transition from the strong quantum confinement regime (top panel) to the intermediate (in the middle) and the weak (bottom panel) regimes. Freshly prepared nanocrystals (top spectrum) were heated in the dispersion, which in turn caused their growth: follow the top-down direction in Fig. 12.27. The original small nanocrystals, or perhaps correctly speaking clusters, with quantum states markedly spaced in energy, exhibit a structured absorption edge as well as a broad emission band of a rather molecular type. Gradually, however, as the diameter of the nanocrystals grows the absorption edge red-shifts and becomes much smoother, which is a characteristic of a solid. In addition, as the Coulomb interaction starts to prevail over quantum confinement, the emission concentrates into a narrow excitonic maximum at ~ 450 nm.

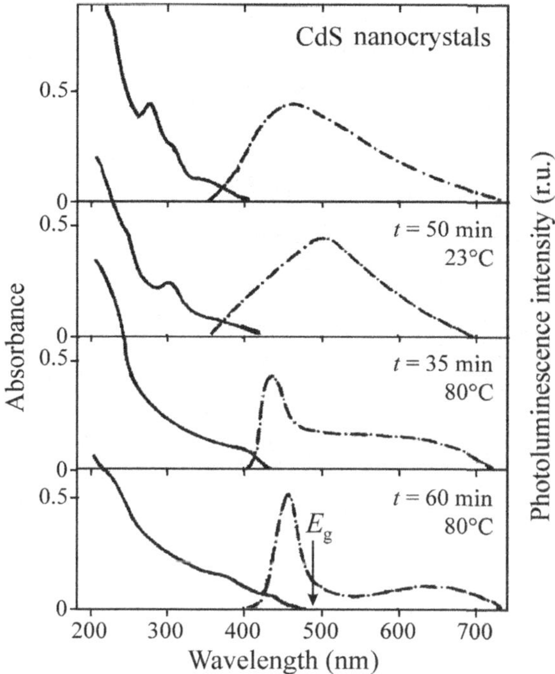

Fig. 12.27
Absorption (solid curves) and emission (dash-dot curves) spectra of a colloidal dispersion of CdS nanocrystals with increased diameter (as realized through heating; top-down sequence of the panels). The arrow denotes the bandgap of bulk CdS. The spectra were taken at room temperature. After Fojtík *et al.* [21].

One of the manifestations of an inhomogeneously broadened photoluminescence spectrum is shown in Fig. 12.28, using the particular case of CdSe nanocrystals with a mean radius of $R = 2.2\,\mathrm{nm}$ embedded in a glass matrix [22]. In their low-temperature emission spectrum (bottom panel), a distinctive pattern composed of the A_F line and its LO-phonon replica appears; a mirror image of this pattern arises also in the excitation spectrum (A_T, LO, top panel). The authors interpret this pattern as a splitting of the ground energy level of an *e–h* pair, due to the exchange interaction and crystal field, to the A_F, A_T doublet spaced in energy by several meV (inset in Fig. 12.28). It can be nicely seen that the pattern shifts as a whole with varying excitation photon energy, according to the subset of nanocrystals which is resonantly excited.

Finally, an example of the distribution of luminescence lifetimes in an ensemble of (Si) nanocrystals is given in Chapter 15 (Fig 15.1(c)).

So much for the luminescence characteristics of a macroscopic ensemble of nanocrystals; special properties of semiconductor quantum dots will be dealt with further through Chapters 13–16. Chapter 17 is then devoted to the experimentally challenging technique of the luminescence spectroscopy of single nanocrystals.

12.6 Exciton–phonon interaction. Phonon bottleneck

The exciton–phonon interaction plays an important role in luminescence processes (Chapter 4). When the dimensionality is reduced, what can happen

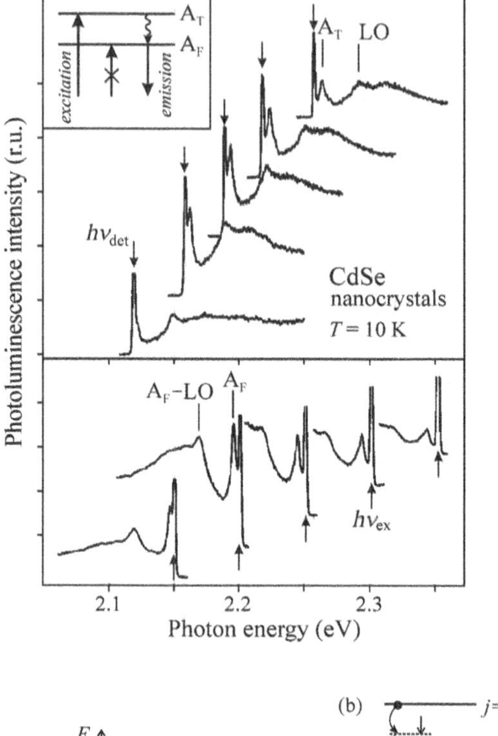

Fig. 12.28
Photoluminescence of CdSe nanocrystals, an example of inhomogeneously broadened emission (bottom panel) and excitation (top panel) spectra. Detection and excitation photon energies are marked by arrows. The inset depicts a sketch of optical transitions to the split lowest excited level; direct excitation to the A_F state is forbidden in the dipole approximation. $T = 10\,\mathrm{K}$, after Woggon *et al.* [22].

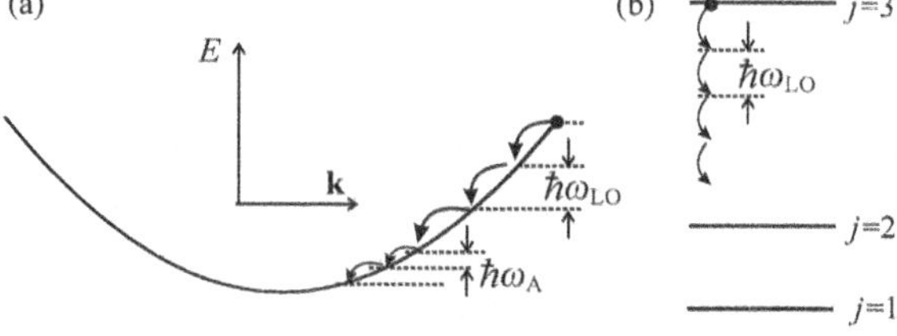

Fig. 12.29
(a) Relaxation of an excited electron state via the emission of phonons in a 3D semiconductor (a) and a semiconductor nanocrystal (b).

among other things is a change in the probability of the emission of a phonon by the excited electronic system; this will, consequently, influence the relaxation rates of hot photocarriers and excitons.

The traditionally applied sketch of carrier relaxation, in which the dispersion curve $E(\mathbf{k})$ is employed, is shown in Fig. 12.29(a). The important piece of information here is the fact that the energy spectrum of carriers is (quasi-) continuous and thus a cascade emission of LO-phonons can take place. LO-phonons, as we know, are characterized by a large energy $\hbar\omega_{\mathrm{LO}}$ and in polar semiconductors have the strongest interaction with carriers. A system of hot photocarriers can thus rapidly cool down and reach equilibrium with the lattice. Emission of acoustic phonons with a low energy $\hbar\omega_{\mathrm{A}}$ is much less important in this context and comes in to play only when the remaining excess energy of carriers is lower than $\hbar\omega_{\mathrm{LO}}$ and the emission of the next LO-phonon is thus impossible. The radiative *e*–*h* recombination itself takes place in the very end of this process; here again, the carrier interaction with phonons can manifest itself, this time via emission lines of X–LO type.

These processes are applied if a quasi-continuous $E(\mathbf{k})$ curve exists at least for one dimension of **k**-space, i.e. in 3D, 2D, and 1D semiconductors.

In sufficiently small nanocrystals, however, this concept fails and the photocarriers can relax only by releasing energy equal to the energy spacing between the discrete levels, which may be significantly larger than $\hbar\omega_{LO}$ (Fig. 12.29(b)). Relaxation can then take place only through a combined multi-phonon LO + LA/TA emission, any real intermediate states not being available, though. Therefore, what can be expected is that the relaxation rate between the excited states of a nanocrystal is dramatically reduced. This effect is commonly referred to as the *phonon bottleneck*.

This qualitative reasoning, supported also by quantum-mechanical calculations, has been more or less straightforward so far. In reality, however, the situation gets somewhat more complicated. Although some authors have experimentally observed the phonon bottleneck, it has not been confirmed by the others. For example, Fig. 12.30 [23] suggests that the phonon bottleneck clearly occurs. Its experimental manifestation in this case consists in the fact that the decrease in size of the nanocrystals leads to a rapid fading off of lines denoted as S–1LO, S–2LO and S–3LO, which are nothing but phonon replicas of free-exciton (X) and donor-bound-exciton (D^0–X) emissions. Such a decrease in intensity implies that the strength of the exciton–phonon interaction decreases with increasing exciton binding energy in smaller nanocrystals. Qualitatively speaking, this effect can also be explained via the smaller electric dipole moment of an exciton in a nanocrystal (because the magnitude of the dipole moment is proportional to the distance between an electron and a hole), which in turn results in weaker interaction of the exciton with the electric field induced by LO-vibrations of the polar lattice, or, in other words, in weakening the Fröhlich interaction (Section 4.2).

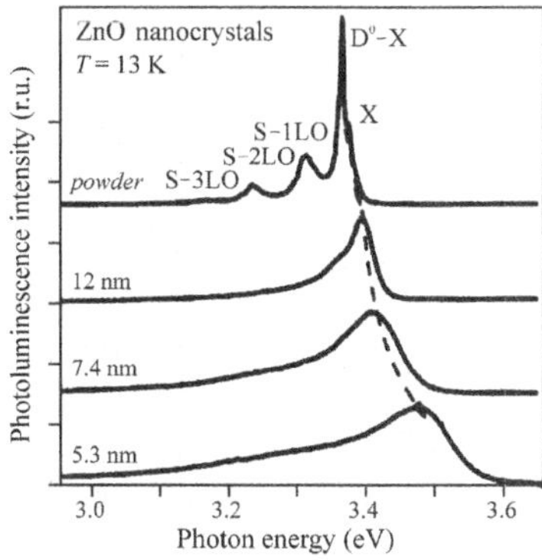

Fig. 12.30
Photoluminescence spectra of ZnO nanocrystals with variable size at $T = 13$ K. The dashed curve indicates the blue-shift of the 'free' exciton energy as a result of the quantum confinement effect. After Hsu *et al.* [23].

On the other hand, completely opposite results were obtained, e.g. by Cooney *et al.* [24]. These authors showed, using the pump and probe method, that hole relaxation in CdSe quantum dots proceeds very fast (< 0.5 ps). Figure 12.31 illustrates the growing rate of energy loss of holes with decreasing radius of nanoparticles, which directly contradicts the predictions of the phonon bottleneck concept. In order to explain this experimental observation, the authors proposed a new relaxation mechanism mediated by a surface passivating shell (vibrations of the surface of the quantum dot), the so-called adiabatic transitions. Another mechanism which can break and counterbalance the phonon bottleneck, proposed earlier particularly for electrons, is a special type of Auger recombination in which the Coulomb electron–hole interaction applies. This interaction is important especially in small quantum dots. By

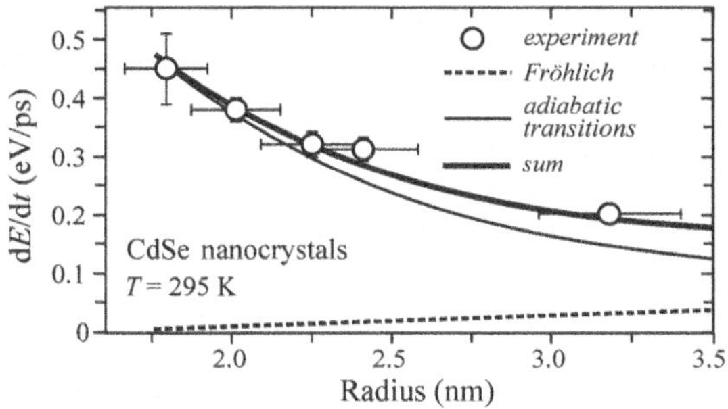

Fig. 12.31
Experimentally observed rate of energy loss (cooling) of holes as a function of the radius of colloidal CdSe quantum dots (circles). The contribution of the Fröhlich interaction with LO-phonons corresponds to the dashed curve, and the adiabatic relaxation channel via surface vibrations is shown as the thin solid curve. The sum of both these contributions, given as the thick solid curve, matches the observed dependence very well. Room temperature, after Cooney *et al.* [24].

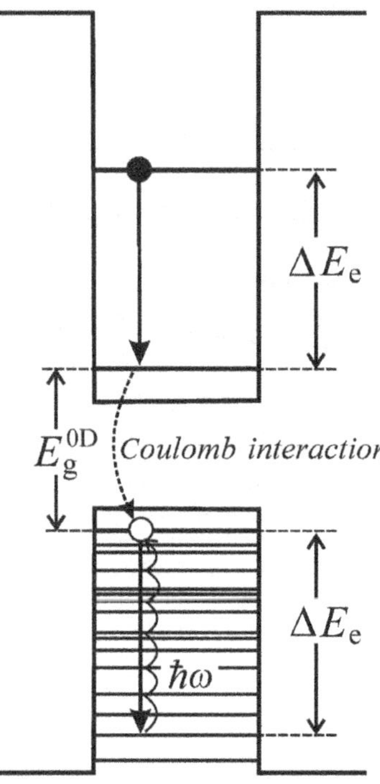

Fig. 12.32
Sketch of Auger recombination, capable of overcoming the restriction on the thermalization rate (phonon bottleneck) of electrons: the excess energy of an electron ΔE_e is transferred to a hole via Coulomb interaction.

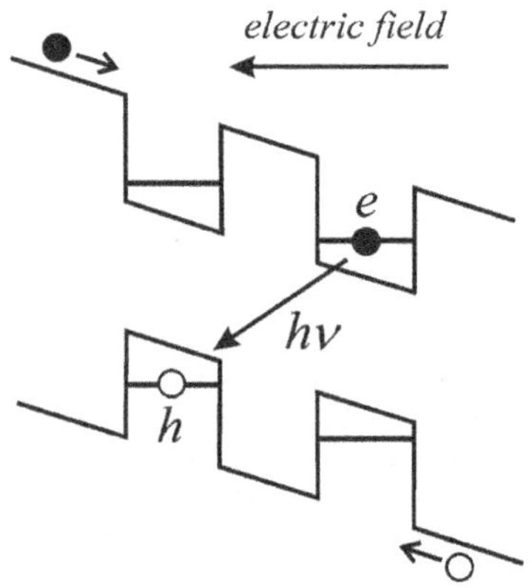

Fig. 12.33
Spatial separation of electrons and holes in a double quantum well under an external electric field.

means of the Coulomb interaction, an electron can transfer its energy to a hole and rapidly relax to the bottom of the conduction band via an interband transition (Fig. 12.32), while the hole compensates this energy difference by jumping to an excited level in the valence band. Afterwards, the hole can thermalize rapidly by emitting phonons, taking advantage of the hole energy spacing in quantum dots being generally smaller than that of electrons. This stems from a large effective hole mass and the degeneracy of the valence band. Obviously, this mechanism must be very sensitive to the parameters of the band structure.

Presumably, an important role in overcoming the phonon bottleneck in quantum dots may be played also by the shape of the quantum dot. However, a large number of combinations of various mechanisms and different aspects of the presence or absence of the phonon bottleneck are yet to be clarified, the literature still containing contradictory results.

12.7 Some special phenomena

The reduced dimensionality of semiconductor structures entails the emergence of a wide range of new phenomena having no parallels in bulk crystals. They are not necessarily and exclusively due to quantum confinement. Luminescence is no exception to the rule. In the following discussion, we shall mention four examples of such unusual luminescence phenomena, namely, the radiative transitions of excitons which are 'indirect' in real space, enhancement of luminescence intensity in some thin films of semiconductor nanocrystals during irradiation, enhancement of luminescence by metal nanoparticles, and carrier (exciton) multiplication.

Excitons indirect in real space

Excitons indirect in space can be quite easily realized utilizing a double quantum well, as shown in Fig. 12.33. If an external electric field is applied to the double quantum well, the energy band structure 'is inclined', causing the electrons and holes (injected optically or electrically) to be spatially separated in individual quantum wells. Such an *e–h* pair can be thought of as an exciton which is 'indirect in real space'. Since the electron–hole wavefunction overlap is reduced, the spontaneous emission rate due to radiative electron–hole recombination becomes substantially slower. Thus, the energy of the excitons as well as their recombination rate can be easily manipulated by means of an external field; this principle serves as a physical basis for the realization of various photonic integrated circuits and allows for new methods of the study of collective electronic excitations.

Photoluminescence spectra of a series of three double GaAs quantum wells (each pair of wells containing a thin AlAs barrier) are depicted in Fig. 12.34 [25]. Obviously, when zero external field is applied, the DX line, arising from the recombination of excitons direct in real space, dominates the spectrum (i.e. the electron and the hole are both localized in the same well), however, at an external field as low as 1 V most of the emission intensity jumps to lower photon energies, characterizing the luminescence of indirect-real-space excitons (the IX line). For voltages within the interval $-3\text{V} < U < -1\text{V}$, the DX line is still present, but its intensity is almost negligible; the IX line, on the other hand, exhibits a noticeable red-shift, which conforms to the increasing inclination of the energy bands. Evidently, this shift later saturates as a result of shielding of the external field by a sufficient concentration of the generated indirect excitons.

Such an 'exciton-based' photonic integrated circuit-to-be is in principle capable of simple operations such as fast switching and controlled transport of electrons in the plane of the well over relatively long distances due to prolonged lifetimes. Also, one can imagine a Bose–Einstein condensation of such long-lived indirect excitons because the long lifetime makes only a relatively weak excitation source sufficient for the generation of a critical concentration of excitons. Yet another effect which could find its use in the construction of an excitonic integrated circuit is the effect of a magnetic field. Theoretical calculations suggest that if an external magnetic field with induction **B** in the plane of the well is applied, excitons become indirect simultaneously even in reciprocal space. This, in principle, allows further modifications of the radiative recombination rate. So far, however, such possibilities lie within the framework of fundamental research or theoretical considerations.

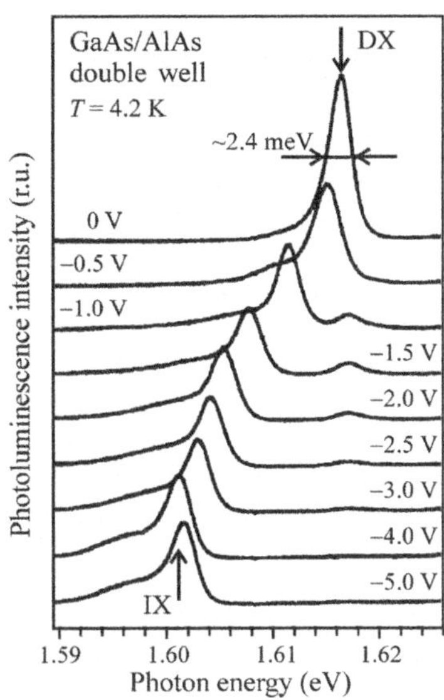

Fig. 12.34
Photoluminescence spectra of double quantum wells GaAs/AlAs for varying external voltage (corresponding values are shown at the curves). DX stands for a direct exciton, IX for an indirect exciton (in real space). $T = 4.2$ K, after Orlita *et al.* [25].

Enhancement of photoluminescence intensity during measurement

Enhancement of photoluminescence intensity in the course of measurements in thin film semiconductor nanocrystals is a very interesting, though somewhat mysterious phenomenon. It was observed most importantly in CdSe nanocrystals and corroborated by a large number of independent experiments at different laboratories. An example of such behaviour can be seen in Fig. 12.35 [26]. During the measurement of emission spectra of CdSe nanocrystals in a thin film (layer thickness around 200 nm) prepared on a silicon substrate, the detected photoluminescence intensity exhibited as much as twenty-fold

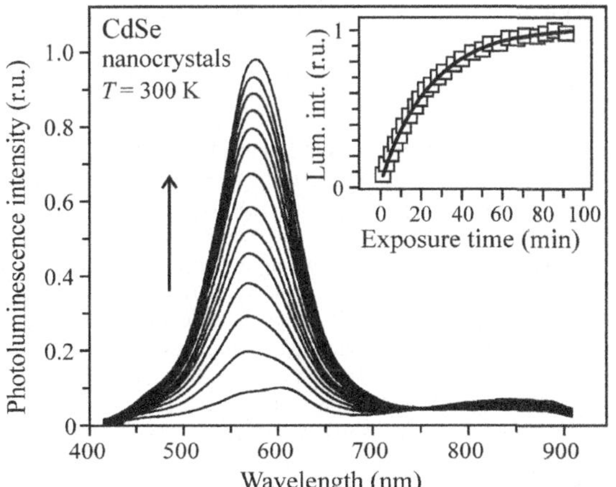

Fig. 12.35
Time development of photoluminescence spectra of CdSe nanocrystals under excitation with a Ti:sapphire laser (second harmonics, 390 nm, 90 fs, 200 MW/cm^2) at room temperature. After Šimurda *et al.* [26].

increase within the time span of 90 minutes.[8] The occurrence of the effect, however, strongly depends on external parameters, most notably on the material of the substrate, the thickness of the layer and the surrounding atmosphere. Results by different authors vary in this context, as well as do the proposed mechanisms explaining the observed enhancement. One of the possible mechanisms here is a photo-induced change in the composition of the surface of the nanocrystals. In particular, the passivation of surface defects, dangling bonds, acting as efficient non-radiative recombination centres, has been proposed. Water molecules could act as viable passivating species, however, the effect has been observed also in vacuum. Enhancement of photoluminescence intensity, which is definitely important from the application point of view, thus still awaits its detailed understanding and prospective applications in real-life devices.

Enhancement of luminescence intensity by metal nanoparticles

A similar phenomenon, namely, the enhancement of photoluminescence of various light-emitting nano-objects (molecules, semiconductor quantum dots), has been experimentally observed in the last 10 years also when the nano-objects under investigation are located close to metal nanoparticles. A similar phenomenon is well-known also from Raman scattering measured on molecules adsorbed on a metal surface (surface-enhanced Raman scattering—SERS). The physics underlying this phenomenon lies firstly in the interaction of the light-emitting objects with a strong electric field arising due to surface plasmons[9] generated in the metal nanoparticles. Secondly, also the so-called *Purcell effect* has an important role to play. In order to explain it in more detail, we have to go back to as long ago as 1946, when Purcell and his co-workers were the first to report [27] that the spontaneous emission act is not just an intrinsic property of the emitter itself, but depends also on the surrounding medium, more specifically, it is influenced by the strength of the interaction between the emitter and the surrounding electromagnetic field. If no suitable configuration (mode) of the field is available around the emitter, photon emission is not possible; on the contrary, the enhanced photon local density of states in close proximity to the emitter can lead to an increase in the radiative decay rate.

[8] From the experimental point of view, this phenomenon is really exceptional. In most cases, a small decrease of luminescence intensity is observed with time instead; this decrease can be, e.g., due to heating of the sample or the fatigue of the excitation source.

[9] A surface plasmon in an electromagnetic wave (i.e. oscillations of surface electrons and the corresponding electromagnetic field) propagating along the interface between a metal and air or another surrounding insulator. If the interface is planar, surface plasmons can occur but cannot be generated by means of optical excitation. The optical excitation is possible, on the other hand, when the interface is, e.g., periodically modulated. In the case of spherical metal nanoparticles with diameter much smaller than the wavelength of light, the quasi-static approximation can be applied. This says that the electrical field is constant over a distance equal to the particle diameter and it induces vibrations of electrons inside the nanoparticle. The electrons oscillate in phase and thus generate around the particle a locally enhanced electric field, very similar to the field from an electric dipole.

Naturally, this effect cannot be experimentally observed in bulk 3D emitters—phosphors, whose dimensions exceed many times the wavelength of light. The recombining charge carriers 'feel' all around a continuous density of available modes of electromagnetic waves (or, equivalently, the photon density of states) $\rho(\nu) \approx \nu^2$. The situation, however, is dramatically altered in the case of nanoparticles. In photoluminescence measurements, the two enhancement mechanisms invoked above can influence the emission. One of them acts during the photoexcitation process and the latter during photon emission [28]. If no metal nanodot is located nearby, the excitation radiation is absorbed directly by the luminescent nanocrystal, while if a metal nanoparticle is located in the close proximity of the luminescent nanophosphor, the excitation beam at first produces plasmonic oscillations within the metal nanodot. This surface plasmon concentrates the optical field into the luminescent nanocrystal, thereby increasing the excitation rate. Subsequently, surface plasmons can also cause an increase in the emission efficiency, via the Purcell effect (the so-called 'hot spots' in plasmonic nanostructures [29]).

The described enhancement of photoluminescence is influenced by a wide range of factors. First of all, the frequency of both the optical excitation and emission must be close to the plasmon frequency (in the visible region, gold or silver nanodots with diameters of the order of 10 nm are usually employed). Secondly, the whole process strongly depends on the radius of the metal nanodot and, moreover, the distance between the metal surface and the nanocrystal also plays an important role: if the distance is too short, quenching of luminescence takes place instead of its enhancement. In the case of a large ensemble of metal nanoparticles, the interaction of the metal nanodots with one another is yet another factor which comes into play. So far, experimental observations by different laboratories do not result in a coherent description of the whole effect; it is, however, obvious that—if fully understood—this effect has significant application potential in optoelectronics and optical sensors.

Carrier (exciton) multiplication

For a very long time, the fact that the absorption of a single photon with energy $h\nu_{ex} > E_g$ produces a single *e–h* pair or exciton has been generally accepted as an indisputable concept in semiconductor physics. The excess energy $h\nu_{ex}-E_g$, no matter how large, is transferred to the lattice in the form of phonons; it is thus lost for possible exploitation by the luminescence process. This causes the quantum efficiency to be always ≤ 1 : one absorbed photon produces one emitted photon in the best-case scenario. (This 'one-to-one photon' bound relates also to the above plasmonic photoluminescence enhancement.)

Such a restriction, however, is not supported by any basic physical law. From the point of view of energy conservation, there is no fundamental ban on, e.g., the process of generation of two *e–h* pairs by a photon with energy $h\nu_{ex} = 2E_g$. In the subsequent radiative recombination, two photons $h\nu_{lum} \leq E_g$ could arise, accounting for the photoluminescence quantum yield to be equal to 2. In reality, however, such a process is hindered by the fact that the relaxation and thermalization processes in the conduction and valence bands are so fast that the energy $h\nu_{ex}-E_g \cong E_g$ transforms into phonons before the second *e–h* pair can be formed. Eventual radiative recombination can thus produce only a single photon $h\nu_{\text{lum}} \le E_g$.

Nevertheless, provided the charge carrier interaction prevails over the electron–phonon interaction for some reason, the generation of two luminescence photons could take place. The best host medium for such a process to occur, as we saw in Section 12.5, would be inside nanocrystals, where the localization of the *e–h* pair in a very small volume enhances the interaction between charge carriers. Nowadays, a sufficient number of experimental observations have already confirmed that the multiplication of photocarriers can indeed occur under certain circumstances. The microscopic mechanism responsible is most probably the so-called inverse Auger effect, in which one hot carrier from the primary pair, e.g. an electron, falls to the bottom of the conduction band and the released excess energy is utilized for the generation of a second *e–h* pair via the Coulomb interaction. Evidently, both the overall energy and the quasi-momentum are conserved in the process, or, in sufficiently small nanocrystals, the quasi-momentum conservation law can relax.

Of course, the energy band structure has to comply with some prerequisites. The most favourable band structure for the multiplication of carriers seems to be that in which the band edges are almost parallel, implying a high density of states. At the same time, it is not completely clear how the two generated *e–h* pairs cope with efficient non-radiative Auger recombination. Obviously, apart from direct luminescence applications (lasers or LED diodes with increased external quantum efficiency), this effect could be beneficial also in photovoltaic cells, by substantially boosting the conversion efficiency of short-wavelength solar radiation to electricity [30]. In photovoltaics, the benefits can be even more straightforward than in luminescence because the *e–h* pairs rapidly dissociate in the strong electric field of the cell and the carriers are extracted to an external circuit, which prevents their recombination. Research on this topic, however, is still in its infancy. The review article by Klimov [31] is highly recommended to readers as a reference.

12.8 Problems

12/1: (a) Check, by making a straightforward calculation, the validity of the selection rule $\Delta j = 0$ (12.22) for optical transitions in a quantum well. (b) Explain qualitatively why the weaker variant $\Delta j = \pm 0, \pm 2, \pm 4, \ldots$ (12.21) applies in the case of a finite-barrier quantum well.

12/2: Determine the number of localized states for electrons in an AgBr quantum well of width $L_z = 7$ nm and depth $V_0 = 0.2$ eV. The effective electron mass in AgBr is $m_e = 0.22\, m_0$.

12/3: In eqn (12.20), one term is missing if it is to describe the overall exciton energy correctly. Write down this term, and explain its physical relevance as well as the reason(s) why it is usually left out.

12/4: (a) What is the general relation between the quantum number j in eqn (12.17) and the quantum numbers j_e and j_o appearing in formulas preceding this equation? (b) In Fig. 12.22, the energy levels of a quantum dot are denoted as 1p, 1d, etc. This notation is not used in a hydrogen atom or, more generally, in atomic spectroscopy (because $l = 0, 1, \ldots, (n-1)$ holds there). Why is this notation permissible in connection with quantum dots?

12/5: (a) Sketch the wavefunctions $u(r)$ and the corresponding energy levels in a spherically symmetric quantum dot (in the $l = 0$ state) with infinitely high barriers and compare them with a one-dimensional quantum well. (b) Using Fig. 12.11 explain why a quantum dot, in order to contain at least one localized electronic state, must have a barrier exceeding a minimum threshold value; determine this value.

References

1. Davies, J. H. (1998). *The Physics of Low-Dimensional Semiconductors: An Introduction*. Cambridge University Press, Cambridge.
2. Yao, T., Okada, Y., Matsui, S., Ishida, K., and Fujimoto, I. (1987). *J. Crystal Growth* **81,** 518.
3. Kelly, M. J. (1995). *Low-Dimensional Semiconductors: Materials, Physics, Technology, Devices*. Oxford University Press, Oxford.
4. Yu, P. Y. and Cardona, M. (1999). *Fundamentals of Semiconductors: Physics and Materials Properties*. Springer, Berlin.
5. Dekker, A. J. (1963). *Solid State Physics*. Prentice-Hall, Englewood Cliffs, N.J.
6. Klingshirn, C. (2005). *Semiconductor Optics*. Springer, Berlin.
7. Haug, H. and Koch, S. W. (1998). *Quantum Theory of the Optical and Electronic Properties of Semiconductors*. World Scientific, Singapore.
8. Greene, R. L., Bajaj, K. K., and Phelps, D. E. (1984). *Phys. Rev. B*, **29**, 1807.
9. Peyghambarian, N., Koch, S. W., and Mysyrowicz, A. (1993). *Introduction to Semiconductor Optics*. Prentice Hall, Englewood Cliffs, N.J.
10. Schmitt-Rink, S., Chemla, D. S., and Miller, D. A. B. (1989). *Adv. Phys.*, **38,** 89.
11. Deveaud, B., Clérot, F., Roy, N., Satzke, K., Sermage, B., and Katzer, D. S. (1991). *Phys. Rev. Lett.*, **67**, 2355.
12. Hoang, T. B., Moses, A. F., Zhou, H. L., Dheeraj, D. L., Fimland, B. O., and Weman, H. (2009). *Appl. Phys. Lett.*, **94**, 133105.
13. Vouilloz, F., Wiesendanger, S., Oberli, D. Y., Dwir, B., Reinhardt, F., and Kapon, E. (1998). *Physica E*, **2**, 862.
14. Hirayama, Y., Tarucha, S., Suzuki, Y., and Okamoto, H. (1988). *Phys. Rev. B*, **37**, 2774.
15. Forchel, A., Kieseling, F., Braun, W., Ils, P., and Wang, K. H. (1995). *phys. stat. sol. (b)*, **188**, 229.
16. Gaponenko, S. V. (1998). *Optical Properties of Semiconductor Nanocrystals*. Cambridge University Press, Cambridge.
17. Landau, L. D. and Lifshitz, E. M. (1991). *Quantum Mechanics. Non-Relativistic Theory. Course of Theoretical Physics*, Vol. 3. Butterworth–Heinemann, Oxford.
18. Beiser, A. (1969). *Perspectives of Modern Physics*. McGraw Hill, New York.
19. Dohnalová, K., Pelant, I., Kůsová, K., Gilliot, P., Gallart, M., Crégut, O., Rehspringer, J.-L., Hönerlage, B., Ostatnický, T., and Bakardjieva, S. (2008). *New J. Phys.*, **10**, 063014.
20. Freedhoff, M. I., Marchetti, A. P., and McLendon, G. L. (1996). *J. Luminescence*, **70**, 400.
21. Fojtík, A., Weller, H., Koch, U., and Henglein, A. (1984). *Berichte der Bunsen-Gesellschaft, Phys. Chem. Chem. Phys.*, **88**, 969.
22. Woggon, U., Gindele, F., Wind, O., and Klingshirn, C. (1996-I). *Phys. Rev. B*, **54**, 1506.
23. Hsu, W.-T., Lin, K.-F., and Hsieh, W.-F. (2007). *Appl. Phys. Lett.*, **91**, 181913.

24. Cooney, R. R., Sewall, S. L., Anderson, K. E. H., Dias, E. A., and Kambhampati, P. (2007). *Phys. Rev. Lett.*, **98**, 177403.
25. Orlita, M., Grill, R., Zvára, M., Döhler, G. H., Malzer, S., Byszewski, M., and Soubusta, J. (2004). *Phys. Rev. B*, **70**, 075309.
26. Šimurda, M., Němec, P., Trojánek, F., and Malý, P. (2004). *Thin Solid Films*, **453–454**, 300.
27. Purcell, E. M., Torrey, H. C., and Pound, R. V. (1946). *Phys. Rev.*, **69**, 37.
28. Sun, G., Khurgin, J. B., and Soref, R. A. (2009). *Appl. Phys. Lett.*, **94**, 101103.
29. Gaponenko, S. V. (2010). *Introduction to Nanophotonics*. Cambridge University Press, Cambridge.
30. Timmerman, D., Valenta, J., Dohnalová, K., de Boer, W. D. A. M., and Gregorkiewicz, T. (2011). *Nature Nanotech.*, **6**, 710.
31. Klimov, V. I. (2007). *Ann. Rev. Phys. Chem.*, **58**, 635.

Effects of high excitation in low-dimensional structures

13

In Chapter 8 we described manifestations of luminescence of various interactions between excitons (or free carriers) in the case of highly excited bulk, 3D semiconductors. It is reasonable to ask whether similar interactions can also occur in low-dimensional structures and if so, whether we can recognize them in luminescence spectra. We will first deal with 2D semiconductors, i.e. quantum wells and superlattices. Here, we can identify a wide range of high excitation effects analogous to their 3D counterparts. It is even possible to say that quantum wells and superlattices are, in a way, ideal systems for the studying luminescence phenomena at various levels of excitation. This is due to a variety of reasons—firstly, using cutting-edge technology, it is possible to prepare samples with parameters tailored to meet various requirements; secondly, existing laser excitation systems enable us to tune excitation wavelengths for the required excitation mode (either resonant or to a continuum of states); and, finally, a 2D increase in the exciton binding energy results in the amplification of certain features of the emission spectra. These issues are discussed in, e.g., the review paper [1].

The regimes for the occurrence of various luminescence processes in 2D structures are mostly characterized by area (surface) densities of photocarriers or excitons $n_X^{(2)}(\mathrm{cm}^{-2})$.[1] Thus, in the low-density mode, $n_X^{(2)} < 10^{10}\mathrm{cm}^{-2}$, we observe the radiative decay of bosons, i.e. free or bound excitons. The medium-fluence excitation mode, $n_X^{(2)} \approx 10^{11}\mathrm{cm}^{-2}$, is where the luminescence originating in trions and biexcitons, and luminescence due to inelastic exciton collisions comes into play. The high-density mode ($n_X^{(2)} \geq 10^{12}\mathrm{cm}^{-2}$) leads to the ionization of excitons and bandgap renormalization; under these conditions, luminescence arises from the radiative decay of a system of fermions—free carriers—in a 2D electron–hole plasma (EHP) or an electron–hole liquid

[1] We can get an estimate of this density if we multiply the volume concentration of excitons $n_X^{(3)}(\mathrm{cm}^{-3})$ determined, for example, from the excitation photon flux using relation (8.1), with the exciton diameter of $2a_X$: $n_X^{(2)} = n_X^{(3)} 2a_X$. Here, we assume the width of the well to be comparable to $2a_X$.

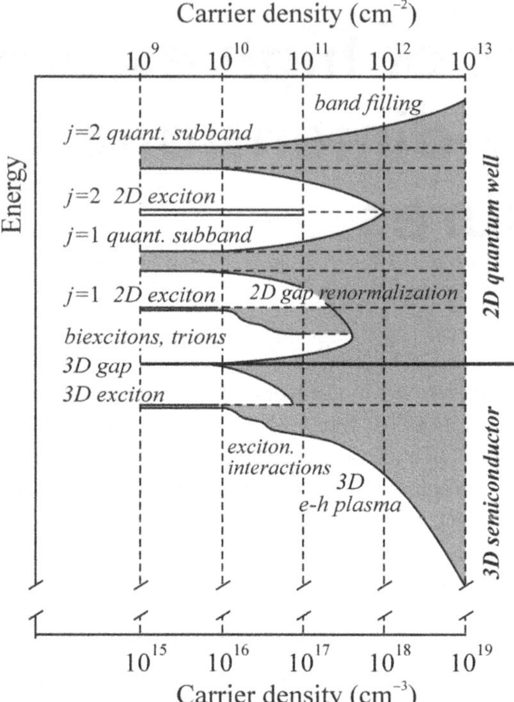

Fig. 13.1
Comparison of radiative recombination processes in 3D and 2D semiconductors as a function of the density of generated excitons or photocarriers. After Cingolani and Ploog [1].

(EHL). Figure 13.1 [1] presents a comparison of the energy spectrum of radiative recombination processes in 3D semiconductors and 2D quantum structures.

13.1 Excitonic molecule (biexciton) in a quantum well

In principle, a two-dimensional gas of excitons and free carriers allows for more favourable conditions for the formation of an excitonic molecule than bulk samples. Firstly, two degrees of freedom of motion within the plane of the well still provide sufficient room for the free motion of excitons, which enables them to take part in interactions with one another and, above all, the higher exciton binding energy in a quantum well (Subsection 12.3.3) also implies a higher binding energy and thus higher stability of a biexciton. For example, while the biexciton binding energy E_{B} in 3D gallium arsenide, GaAs, amounts to only about $E_{\mathrm{B}} = 0.4 - 0.7$ meV, in a quantum well of the same material, it is, as we shall see shortly, $E_{\mathrm{B}}^{(2)} \approx 1.1$ meV. Therefore, in this material, biexcitons can be easily observed at temperatures up to about 10 K. Moreover, the density of electron states in low-dimensional systems is generally reduced (Section 12.2) and thus lower excitation intensities suffice to reach the threshold densities of excitons for the appearance of the effects of high excitation; a continuous gas laser is often satisfactory.

The mechanism of 'fusion' of two free excitons into a biexciton is analogous to the 3D semiconductor case. However, the necessary prerequisite is to minimize the random fluctuations in the well width and in alloy composition so that excitons remain mostly delocalized and can participate in interactions. (The same applies to the generation of trions explained in the following section.) The radiative decay of a biexciton then shows as a new line in the emission spectrum, shifted with respect to the free exciton emission by $E_B^{(2)}$ towards lower energies. Judging by the 'step-like' shape of the density of states in a 2D structure, we can quite easily infer the shape of the biexciton emission line if we proceed from an analogy to the 3D case. In the case of a direct semiconductor, we derived expression (8-6) that reads

$$I_{sp}^{M}(h\nu) \simeq |M_M|^2 \left[(E_g - E_X - E_B) - h\nu\right]^{1/2} \exp\left\{-\left[(E_g - E_X - E_B) - h\nu\right]/k_B T_M\right\}, \tag{13.1}$$

where E_X is the binding energy of a free exciton and T_M denotes the effective temperature of the biexciton gas. In this case, the transition matrix element M_M is a constant, independent of $h\nu$. The second term on the right hand side of (13.1) represents the optical joint density of states derived from the square-root shaped densities of states in the excitonic and biexcitonic bands. If we now take into account the fact that both the exciton and biexciton densities of states in a 2D system are equal to zero for photon energies $h\nu < E_g^{(2)} - E_X^{(2)} - E_B^{(2)}$ while being a (non-zero) constant for $h\nu > E_g^{(2)} - E_X^{(2)} - E_B^{(2)}$, we simply replace the second term in (13.1) by a constant as well and we have

$$I_{sp}^{M(2)}(h\nu) \simeq \exp\left\{-\left[(E_g^{(2)} - E_X^{(2)} - E_B^{(2)}) - h\nu\right]/k_B T_M\right\}. \tag{13.2}$$

This, however, means that the emission line should begin with a steep onset at the threshold energy $h\nu_0 = E_g^{(2)} - E_X^{(2)} - E_B^{(2)}$ and fall exponentially towards lower photon energies. However, such a spectrum is not observed experimentally; it is believed that, in fact, the values $E_X^{(2)}$ and $E_B^{(2)}$ are dispersed due to the insufficiently planar interface in a real quantum well, which results in a line broadening. To describe the broadening, it has been suggested to introduce a distribution function of the biexciton binding energy $g(E_B^{(2)})$ [2], thus transforming (13.2) into

$$I_{sp}^{M(2)}(h\nu) \simeq \int g\left(E_B^{(2)}\right) H(E_g^{(2)} - E_X^{(2)} - E_B^{(2)} - h\nu) \times \exp\left[-(E_g^{(2)} - E_X^{(2)} - E_B^{(2)} - h\nu)/k_B T_M\right] \mathrm{d}E_B^{(2)}, \tag{13.3}$$

where H denotes the Heaviside step function.

Figure 13.2(a) gives a comparison with experiment in the case of multiple GaAs/AlGaAs quantum wells [2]. We can see that both the M and X lines are broadened into almost symmetric shapes. (A Gaussian form for the function $g(E_B^{(2)})$ was assumed, see Problem 13/1; next the fitting parameters were: $E_g^{(2)} - E_X^{(2)} = 1551.1$ meV, the mean value of the binding energy $\bar{E}_B^{(2)} = 1.1$ meV, $T_M = 10$ K for bath temperature $T = 1.6$ K.) The intensity dependence

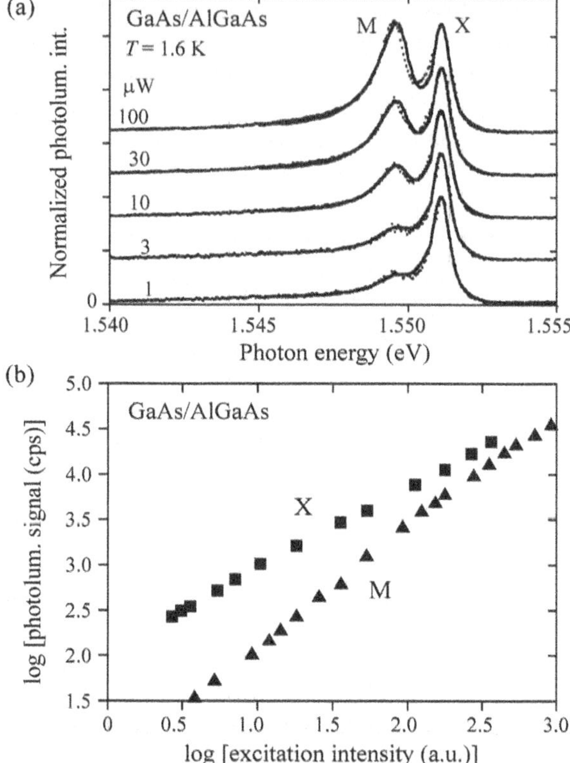

Fig. 13.2
A quantum well in GaAs/AlGaAs. (a) Gradual development of a biexciton emission line (M) with increasing excitation intensity. X denotes the normalized emission line of a free exciton. Points—experiment, solid lines—theory, using relation (13.3). (b) Exciton and biexciton luminescence intensity as a function of optical excitation power density (a continuous, tunable laser). Bath temperature was 1.6 K, after Phillips *et al.* [2].

of the X and M lines is shown in Fig. 13.2(b). We can see that the biexciton emission increases faster with increasing excitation (expressed as a function of the free exciton emission intensity I_{exc}, one finds $I_{\mathrm{sp}}^{\mathrm{M}(2)} \sim I_{\mathrm{exc}}^{1.6}$, which is very similar to 3D semiconductors).

13.2 Trions in a quantum well

Trions are quasi-particles consisting of three photocarriers—two electrons and a hole (denoted X^-) or two holes and one electron (X^+). They are sometimes called charged excitons. Their existence or non-existence had been discussed for a long time from as early as from the beginning of the 1970s. However, since we may expect their binding energy (i.e. the energy required to separate an electron or a hole and re-create a free exciton) and thus also their stability to be very low, conditions more favourable for their study appeared only after 2D structures had started to be studied. In these structures, as we have already mentioned several times, the binding energies of excitons and their complexes are increased, which applies to trions as well.

Trion luminescence is caused by the radiative recombination of an *e–h* pair while releasing the remaining quasi-particle, an electron or a hole. Trions manifest themselves via a new emission line in the luminescence spectrum, appearing, as the excitation power density increases, on the low-energy side of the exciton emission. Based on reasoning analogous to the biexciton case, we can expect the trion emission line to be asymmetric with a tail towards lower energies [3]. Indeed, it is not difficult to show that the emission line of a trion in a quantum well will have a theoretical shape (Problem 13/2)

$$I_{\mathrm{sp}}^{\mathrm{T}(2)}(h\nu) \simeq |M_{\mathrm{T}}|^2 \exp\left\{-\frac{m_{\mathrm{e}}}{(m_{\mathrm{e}}+m_{\mathrm{h}})}\left[(E_{\mathrm{g}}^{(2)} - E_{\mathrm{X}}^{(2)} - E_{\mathrm{BT}}^{(2)}) - h\nu\right]/k_{\mathrm{B}}T_{\mathrm{T}}\right\}, \tag{13.4}$$

where $E_{\mathrm{BT}}^{(2)}$ is the trion binding energy and T_{T} is the effective temperature of the trion gas. The formal similarity to relation (13.2) is obvious. Therefore, from this point of view, the trion line also features a steep onset at a certain threshold energy $h\nu_0 = E_{\mathrm{g}}^{(2)} - E_{\mathrm{X}}^{(2)} - E_{\mathrm{BT}}^{(2)}$ and offers a chance to easily estimate the magnitude of $E_{\mathrm{BT}}^{(2)}$ as the energy difference between $h\nu_0$ and the exciton emission line. Since the trion luminescence line is red-shifted with respect to the exciton, it is not easy to distinguish possible trion luminescence from biexciton luminescence. The fact that the emission lineshapes in real 2D systems lose, to a considerable extent, their characteristic asymmetries (as, after all, already suggested by Fig. 13.2) further contributes to the complexity of the problem and, as we saw in the previous section, we need to convolute the simple theoretical shape with a certain distribution function to allow for a comparison with theory. Besides, the trion luminescence line also increases superlinearly with excitation intensity, similarly to the biexciton case.

The X^- as well as X^+ (according to the doping type or excess concentration of one kind of carrier) trion luminescence has been observed in quantum wells of both III-V semiconductors (GaAs/AlAs) and II-VI semiconductors (e.g. $CdTe/Cd_{1-x-y}Mg_xZn_yTe$). An example is given in Fig. 13.3; here, the generation of *negative* trions X^- was ensured by accumulating a 'background' 2D electron gas via a flow of photoelectrons (i.e. their spatial separation from the photoholes) from a narrower to a wider quantum GaAs well separated by a barrier of AlAs [4]. The authors assume that elastic collisions of trions with free electrons play an important role in their radiative recombination mechanism since both kinds of quasi-particles are charged. These collisions may lead to broadening of the X^- line with an increase in excitation intensity.

However, Combescot *et al.* have recently objected to the above simple trion model [5]. In principle, they argue that trions are fermions and thus they cannot occupy a single energy level *en masse* to give rise to a narrow optical line (as opposed to excitons and biexcitons). This is also related to the fact that (in order to achieve good agreement between theory and experiment) one must consider the matrix element M_{T} in eqn (13.4) not to be a constant, as e.g. for a biexciton, but rather a quantity depending strongly on $h\nu$ [3]. Likewise, the broadening of the trion line with an increase in excitation intensity may be a sign of the fermionic nature of the trion gas. The authors of [5] approach the problem in a way that is completely different from the above 'standard' model and ascribe trion behaviour to multi-particle effects due to Coulomb and Pauli interactions in the exciton-carrier system. Thus, no unambiguous

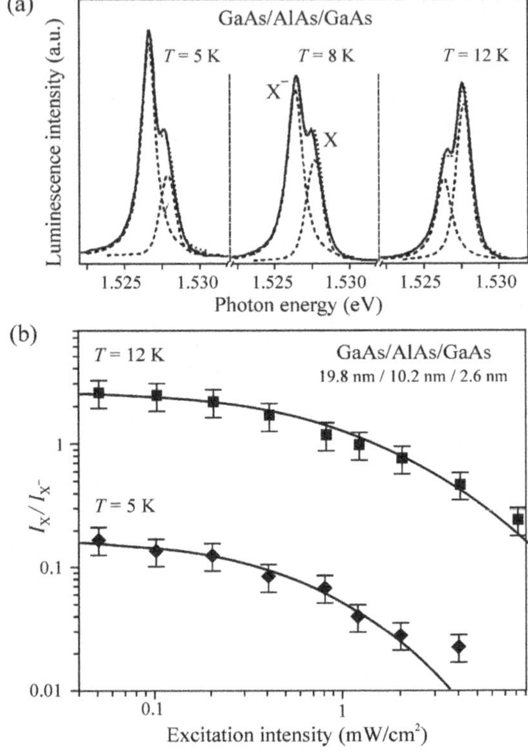

Fig. 13.3
(a) Photoluminescence spectra of GaAs/AlAs quantum wells. Each experimental emission spectrum (solid curves) splits into two lines: X^- (trion) and X (exciton). With increasing temperature, the line X^- disappears very quickly. (b) The ratio of the X line intensity to the X^- intensity as a function of excitation intensity demonstrating a nonlinear increase of the X^- line. Ti:sapphire laser excitation. After Manassen *et al.* [4].

consensus has been reached so far so as to the existence of trions and their optical signature.

13.3 Collisions of free excitons in a quantum well

Analogous to the 3D case discussed in Section 8.3, the collision mechanism of luminescence (inelastic collisions of free excitons, X–X collisions) occurs also in two-dimensional semiconductor structures. It is characterized by new luminescence lines P_n, red-shifted with respect to free exciton luminescence roughly by the amount of energy $E_X^{(2)}$. To be more specific, the photon energy satisfies

$$h\nu_{Pn} = (E_g^{(2)} - E_X^{(2)}) - E_X^{(2)}(1 - 1/n^2) - (3/2)k_B T_{ef}, \tag{13.5}$$

where $n(= 2, 3, 4, \ldots, \infty)$ is the main quantum number of the remaining exciton and $(3/2)k_B T_{ef}$ represents its kinetic energy (or the kinetic energy of the generated free electron–hole pair).

Figure 13.4 represents an example of this emission, detected in multiple quantum wells of ZnO with barriers consisting of BeZnO layers. In Fig. 13.4(a), one can see photoluminescence emission spectra measured at various temperatures and excited by a cw HeCd laser [6]. In addition to the free exciton line (X), a very distinct line denoted P_2 appears. Figure 13.4(b) then gives normalized room-temperature emission spectra for both a (relatively weak) continuous excitation and for strong excitation with a pulsed excimer ArF laser (0.42MW/cm^2). Under pulsed excitation, the spectrum is dominated by the P_∞ line.

The emission lines in Fig. 13.4 were assigned to processes P_2 and P_∞ in accordance with eqn (13.5), using fitting parameters as follows: $(E_g^{(2)} - E_X^{(2)}) \approx 3.55$ eV, $E_X^{(2)} = 263$ meV, P_2–$P_\infty \simeq 66$ meV. Therefore, the quantum-confinement effect results here in an enormous increase in the exciton binding energy from a value of $\sim$ 60 meV to 263 meV. Thus, exciton lines are the salient feature in the spectrum even at room temperature. The observation that under pulsed excitation the P_2 line transforms into the P_∞ line can be interpreted as being due to strong excitation into energy levels high up in the respective bands ($h\nu_{ex} = 6.42$ eV), which makes both the density and kinetic energy of excitons very high before the collision occurs, and one of the excitons then becomes fully ionized during the collision ($n \to \infty$).

As the excitation power density increases further, P_∞-line stimulated emission sets in.

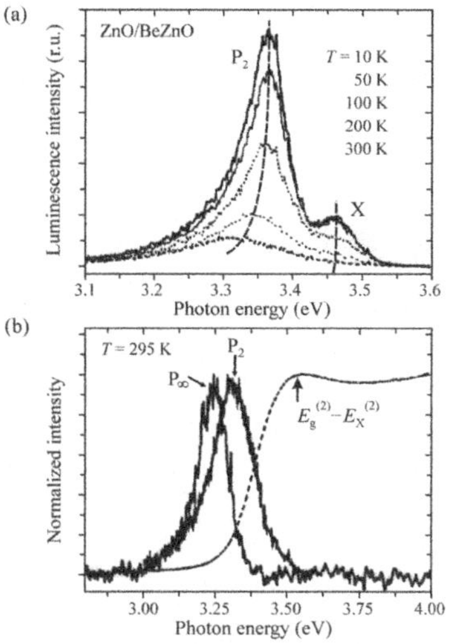

Fig. 13.4
(a) Photoluminescence spectra of multiple quantum wells ZnO/BeZnO at various temperatures. Continuous excitation at 325 nm. (b) Normalized photoluminescence spectra of the same sample at room temperature under continuous excitation (curve P_2) and strong pulsed excitation by a 193-nm ArF laser (curve P_∞). The absorption spectrum measured at the same temperature is also shown (dashed). After Ryu *et al.* [6].

13.4 Electron–hole plasma (EHP) and electron–hole liquid (EHL) in 2D structures

Like the case of bulk semiconductors, at high rates of carrier photogeneration, the excitons in 2D semiconductor structures are screened off and a dense (two-dimensional) electron–hole plasma is generated, i.e. the Mott transition occurs. This is accompanied by a renormalization of the bandgap, band filling by free carriers, and conditions for the onset of stimulated emission may be met, too.

All this is reflected in distinct alterations in the luminescence emission spectrum. Before dealing with them, however, we need to point out that certain differences do exist in the mechanism and efficiency of screening of the Coulomb interaction between the 2D and 3D cases. Screening in bulk materials is almost exclusively due to the surrounding free carriers and excitons, while in 2D structures another factor comes into play—the so-called *phase space filling*. This term refers to the consequences of a reduced density of states in low-dimensional systems. This is because each exciton, albeit a boson, consists of electrons and holes which are fermions. With increasing excitation intensity, these gradually fill all available states in the given volume of **k**-space (and thus also of **r**-space), which limits the highest attainable exciton concentration.[2] This phenomenon, which makes the screening less efficient, can be qualitatively understood by using the concept of field lines. We may imagine that the Coulomb interaction between an electron and a hole bound in an exciton (as represented by field lines) is homogeneously screened in 3D space all around the *e–h* pair by the dielectric constant of the material ε (Figure 13.5(a)). However, for an exciton in a quantum well, field lines mainly penetrate through

[2] In principle, this mechanism is independent of the semiconductor dimension but it manifests itself more distinctly in 2D systems than in bulk samples.

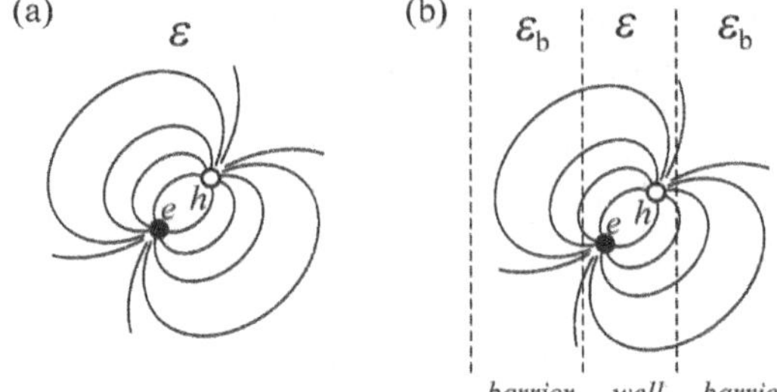

Fig. 13.5
Coulomb interaction screening. (a) A 3D semiconductor, (b) a quantum well.

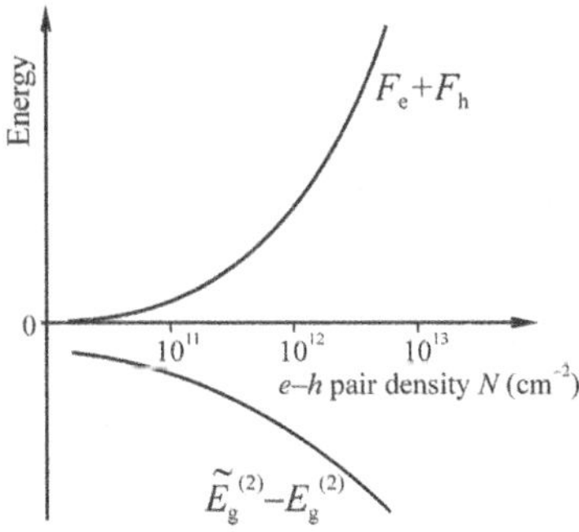

Fig. 13.6
A qualitative sketch of the trend of $(F_e + F_h)$ development and reduction in the bandgap width $(\tilde{E}_g^{(2)} - E_g^{(2)})$ in a quantum well, as a function of the $e-h$ pair density.

a barrier characterized by a dielectric constant $\varepsilon_b < \varepsilon$ (Fig. 13.5(b)), that is, Coulomb interaction screening is weaker. This phenomenon, apart from other factors, adds to the increase in the exciton binding energy in low-dimensional structures since an electron and a hole 'feel' each other more strongly. In the literature, this is often called *dielectric confinement* in analogy to quantum confinement.

Thus, when proceeding from a 3D to a 2D electron–hole system, the screening efficiency decreases. And yet, it is still qualitatively true that the width of the reduced bandgap $\tilde{E}_g^{(2)}$ decreases with increasing electron–hole pair density while the sum of Fermi energies $(F_e + F_h)$ increases (Fig. 13.6).[3] To obtain an *a priori* estimate of the Mott density $N_M^{(2)}$ based on the known two-dimensional Bohr exciton radius $a_X^{(2)}$, we can use a criterion analogous to 3D semiconductors, which now has the form $N_M^{(2)} \cong [\pi(a_X^{(2)})^2]^{-1}$. Figure 13.7(a) will serve as an example of a luminescence experiment allowing for the quantitative determination of carrier density in an EHP as well as of the renormalized bandgap based on analysis of the spontaneous photoluminescence emission lineshape. These emission spectra originate in a simple In_xGa_{1-x} As quantum well with barriers of InP [7]. The emission spectra graphically illustrate that with increasing excitation intensity, subbands gradually fill up up to the quantum number $j = 3$: once the luminescence intensity in the neighbourhood of the energy $E_{11(hh)}$ becomes saturated, which corresponds to the filling of the first subband $j = 1$, a steep increase appears at the energy $E_{22(hh)} = 0.9$ eV and the subband $j = 2$ starts filling up. A complete filling of this subband corresponds to saturation at the excitation intensity of ~ 20 kW/cm^2. At the highest excitation intensities ≥ 30 kW/cm^2, one can see signs of the onset of a new band at $h\nu \geq 1$eV when the third subband $j = 3$ starts filling up. Further development of the spectrum is not possible because by this time electrons pass into the conduction band of the barriers consisting of indium phosphide (see Fig. 13.7(b)).

Therefore, a qualitative view of the spectrum confirms that it really originates from the radiative recombination of the EHP: the spectrum is substantially broadened, mainly on the high-energy side, which corresponds to an increase in the value of $(F_e + F_h)$. The low-energy side broadens, too, as a result of bandgap renormalization. However, in order to determine

[3] Due to technical reasons, hereafter we denote the reduced bandgap by $\tilde{E}_g$ instead of E_g' used before.

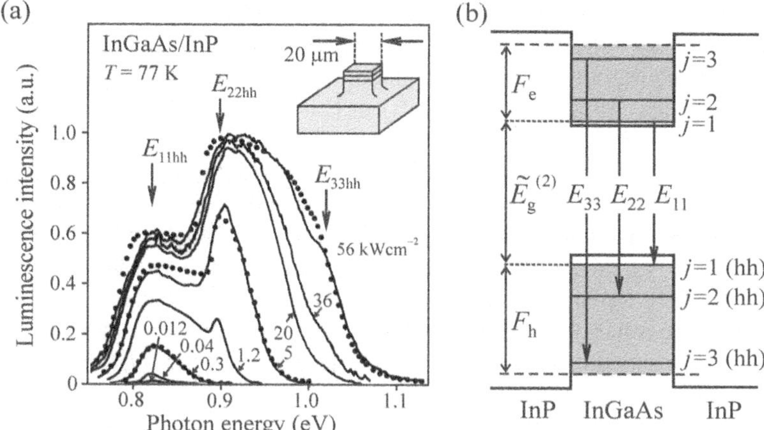

Fig. 13.7
(a) Experimental emission spectra (solid curves) and relevant theoretical lineshape (points) in a simple InGaAs quantum well 15 nm wide. Parameters standing by the curves represent the excitation power density from a continuous Kr^+-laser (647 nm); the temperature was 77 K. The inset depicts the geometrical shape of the sample. After Kulakovskii *et al.* [7]. (b) The corresponding recombination transitions in the well. Only the states of heavy holes (hh) apply.

quantitatively the carrier density or the corresponding bandgap renormalization, a theoretical shape has to be found to fit the experimental spectrum. The shape was described by eqn (8.24) in the case of 3D semiconductors; now we need to replace the square-root density of states in the convolution in (8.24) by a two-dimensional step-like density of states $\rho_c^{(2)}(E,\gamma)$, $\rho_v^{(2)}(\overline{h\nu}-E,\gamma)$ and to sum over all the occupied electron and hole subbands:

$$I_{\mathrm{sp}}^{\mathrm{EHP}(2)}(h\nu) \simeq \sum_{\mathrm{c,v}} \int_0^{\overline{h\nu}=h\nu-\tilde{E}_g^{(2)}} \rho_c^{(2)}(E,\gamma)\rho_v^{(2)}(\overline{h\nu}-E,\gamma) f_c(E) f_v(\overline{h\nu}-E)\,\mathrm{d}E. \tag{13.6}$$

In this expression, γ denotes a phenomenological broadening factor, introduced primarily to describe the low-energy wing of the spectrum.[4] The 'broadened' density of states can then be approximated by

$$\rho_c^{(2)}(E,\gamma) \simeq \frac{1}{1+\exp\left[-(E-\tilde{E}_c/\gamma)\right]}, \tag{13.7}$$

where $\tilde{E}_c$ is the edge of the corresponding subband. The functions f_c and f_v are the Fermi–Dirac distribution functions in the corresponding bands; fit parameters include, besides the *e–h* pair density N and the value of $\tilde{E}_g^{(2)}$, also the effective temperature.

As we know, however, relation (13.6) does not take into account the conservation of quasi-momentum during radiative recombination. More sophisticated models do, and yet the resulting fits do not differ much. The fits shown in Fig. 13.7 by solid points were obtained utilizing a theoretical analysis that does respect the conservation law [8]. The resulting plasma densities range from $2\times10^{11}\,\mathrm{cm}^{-2}$ to $7\times10^{12}\,\mathrm{cm}^{-2}$ and the effective temperatures are between 80

[4] Otherwise, the theoretical fit would begin with a steep onset here. However, the physical meaning of the parameter γ is not quite clear.

and 140 K (at a bath temperature of 77 K). We can see the calculated spectral shapes reproduce the experimental curves very well.

What remains to be mentioned is an important experimental aspect that supported this successful theoretical analysis. This is the specific shape of the sample to be measured, the so-called mesa structure, depicted in Figure 13.7(a). It resulted from etching off the substrate in the neighbourhood of a rectangular area of dimensions approximately $20 \times 20\ \mu m^2$ and it prevents lateral plasma diffusion within the quantum well plane, otherwise the EHP generated by photoexcitation is spatially inhomogeneous and quickly expands. Mesa structures keep the EHP density constant in a small volume[5] and enable the plasma density to be reliably determined through luminescence experiments; without them, the fit usually only yields a plasma density averaged over the inspected volume. We actually already met this problem in Subsection 8.5.2.

It will now be quite interesting to discuss the issue of the existence or non-existence of an electron–hole liquid (EHL) in 2D semiconductor structures. The properties of the EHL in 3D indirect-bandgap semiconductors, where free excitons condense below a critical temperature to form drop-shaped aggregates consisting of free charge carriers, are described in detail in Section 8.4. The EHL is gradually formed as *e–h* pairs coalesce in nucleation centres (impurity atoms, sample surface, spontaneous fluctuations of the exciton gas density) and, upon reaching a critical size, the seed starts growing up to its final diameter (typically micrometres to tens of micrometres). We know the critical radius of the nucleation centre is about 10 nm, which is comparable to the well width in many experiments with quantum wells. Therefore, *a priori* it is not obvious whether an EHL can even form in quantum wells although some experimental results on luminescence in special types of superlattices have already been interpreted as a signature of the radiative decay of the EHL.

The situation looks somewhat less ambiguous in wider 2D structures that, however, are not quantum wells in the true sense of the word. Silicon layers about 100 nm thick grown on a thin layer of SiO_2 covering a silicon substrate (silicon on insulator, SOI) represent a nanostructure suitable for the relevant luminescence experiment. Figure 13.8 gives photoluminescence spectra obtained on Si layers 50 nm and 340 nm thick under increasing intensity of excitation with an Ar^+-laser [9]. Because the thickness of the Si layers is much larger than the bulk Si exciton radius (~ 4.9 nm), the spectral position of emission lines in the exciton region is not affected by quantum confinement. Therefore, in this figure, we can see a $I_{LO,TO}$(FE) line at $h\nu \simeq 1.098$ eV ($\sim$ 1129 nm) and a wider $I_{LO,TO}$(EHL) band at 1.080 eV ($\sim$ 1148 nm).[6] The $I_{LO,TO}$(EHL) band shows an evident intensity threshold. Thus, the EHL obviously does develop in these samples. The important point here is that the EHL band is far more intense—under the same excitation intensity—in the thinner layer (50 nm) or, to be more exact, the EHL band *de facto* does not occur

[5] The laser-excited trace had dimensions corresponding to the etched island and, in fact, there is no inhomogeneity in the normal direction due to the very small thickness of the well.

[6] This can be compared to Fig. 1.1, which represents bulk silicon; the minute differences in spectral positions are due to different measurement temperatures or possibly due to mechanical stress inside the thin Si layer.

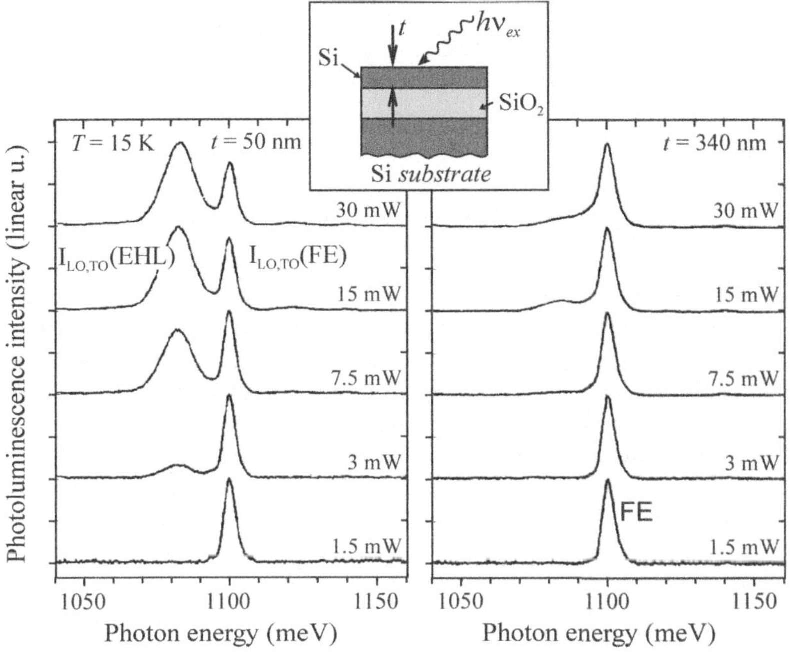

Fig. 13.8
Emission spectra of two thin Si layers on SiO_2 for various intensities of excitation with a continuous Ar^+-laser. The inset represents the sample shape and structure. To achieve a sufficient signal-to-noise ratio, excitation radiation (an Ar^+-laser 488 nm) was strongly focused by an objective to give a small-diameter spot of $\sim 1.5\ \mu m$. Additionally, the luminescence radiation was collected by a lens of large numerical aperture. $T = 15$ K; after Nihonyanagi and Kanemitsu [9].

in the wider layer, even at a relatively low temperature of 15 K. Therefore, the thin layer has more favourable nucleation conditions. And because the impurity concentration is the same in both samples we can conclude that the interface between Si and SiO_2 or between Si and air plays the role of nucleation centres. (The surface-to-volume ratio in a thin layer is much greater.) The authors assume that hemispherical *e–h* droplets form at both interfaces in these structures.

At the same time, the experiment reveals that a lower bound of the well width for the co-existence of the EHL and exciton gas exists; it amounts to about 15 nm. Below this limit, the emission band of the *e–h* system is considerably blue-shifted (the result of a combination of the quantum-confinement effect with bandgap renormalization) but, above all, the system loses its EHL character and transforms into an EHP of varying density [10]. This, however, is in good agreement with the above concept of the critical radius of nucleation centres in the EHL. Estimating the equilibrium EHL density in a 2D silicon structure from the shape of the emission spectrum using relation (13.6) yields $\sim 2 \times 10^{12} \text{cm}^{-2}$.

The last topic we shall briefly mention in this section is the existence of a single-component plasma in quantum wells; such a many-particle system has very specific properties. This kind of plasma can be created via so-called *modulation doping*: as the sample grows one inserts one or several monolayers consisting of donor or acceptor atoms into the barrier in the vicinity of the well, see Fig. 13.9(a). If the atoms (e.g. donors) become ionized, the released electrons flow into the well since it is an energetically more favourable state. This causes band bending (Fig. 13.9(b)), and in the vicinity of the interface

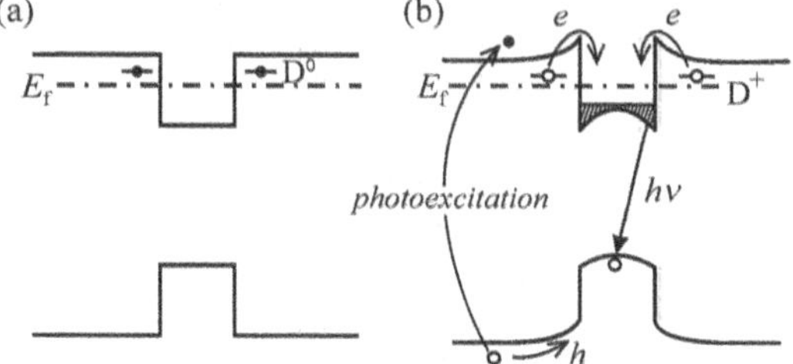

Fig. 13.9
Schematic of luminescence from a modulation doped quantum well with (a) a non-interacting system, and (b) with donor ionization and simultaneous photoexcitation.

a 'gas' consisting of free electrons appears in the well—an electron plasma with controlled density. This plasma is spatially separated from the ionized donors, which opens up very important applications in, e.g., the production of high-mobility transistors. (We have already seen a somewhat analogous effect applied to X^- trion formation in Section 13.2).

These materials show unique optical features such as the absence of the exciton peak in absorption (even in standard absorption measurements utilizing a weak optical beam) and a blue-shifted absorption edge. Luminescence spectroscopy provides an auxiliary way of studying the behaviour of a single-component plasma. However, it is necessary to introduce a low concentration of holes into the well through (weak) photoexcitation so that the *e–h* pairs could recombine radiatively. All emanating luminescence results from the recombination of free carriers since excitons cannot exist in the well. This enables us, among other options, to study the properties of a single-component plasma as a function of its density determined by the doping level, not by strong optical injection of electron–hole pairs. One can even describe the emission spectrum using an explicit expression differing from both relation (13.6) and the classical Maxwell–Boltzmann distribution, see Problem 13/4. Besides, another specific luminescence feature can appear here, the so-called Fermi level singularity. All these interesting phenomena are treated in more detail in [1].

If, however, photoexcitation becomes stronger and the hole concentration is no longer negligible compared to that of electrons, the system gradually turns into a two-component electron–hole plasma characterized by the properties discussed above.

13.5 Biexcitons, EHP, and EHL in quantum wires

In principle, one may expect the occurrence of high excitation effects analogous to 2D semiconductor systems also in 1D structures—in quantum wires. This stems from the fact that a continuous dispersion relation $E(\mathbf{k})$ still holds here albeit in a single dimension. Yet, in this regard, experimental results have been rather scarce so far. One of the reasons is the fact that it is quite difficult to prepare high-quality 1D samples for optical experiments and so most of the spectroscopic photoluminescence studies in low-dimensional semiconductors deal with 2D structures.

What is more, theoretical papers yield contradictory results here—from predicting that a Mott transition and optical gain in 1D systems cannot be

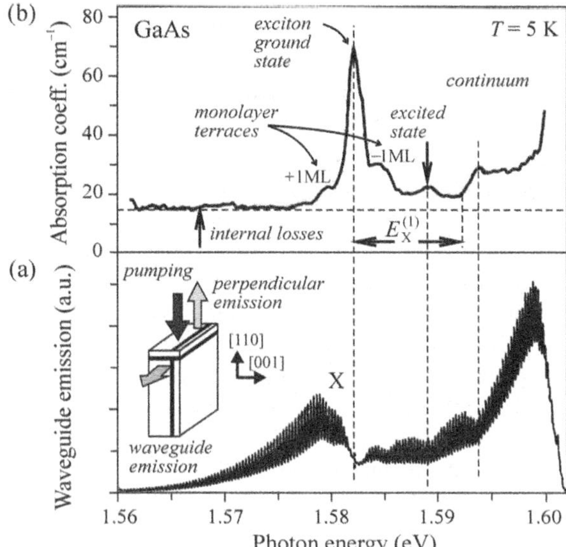

Fig. 13.10
(a) Spontaneous photoluminescence spectrum from a GaAs quantum wire recorded in a waveguide configuration but point-excited into a small spot of diameter $\sim 1\,\mu$m. The 'noise' on the curve is due to interference inside the sample. (b) Absorption spectrum derived from panel (a). $T = 5$ K; after Hayamizu *et al.* [11].

achieved even for the highest exciton gas density[7] (since screening of the Coulomb interaction, leading to EHP formation, is even weaker here than in 2D systems) to predicting the onset of optical gain in a one-dimensional *e–h* plasma already at densities of $n_X^{(1)} = 2 - 3 \times 10^5 \text{cm}^{-1}$.

We shall only give a few examples here. A recent experimental paper [11] established that the basic sequence of phenomena under increasing excitation intensity is indeed analogous to 2D and 3D materials: first exciton luminescence, then biexciton (trion) luminescence, and lastly a Mott transition to an EHP.[8] In the case of optical experiments, quantum wires are often prepared as 'intersections' of planar 2D formations (quantum wells about 10 nm wide) grown by molecular beam epitaxy. Their cross-section is thus V- or T-shaped. This was the case here, too: one dealt with a T-type intersection of a (001) $Al_{0.07}Ga_{0.93}As$ quantum well with a (110) GaAs well ($14 \times 6 \text{ nm}^2$). The inset in Fig. 13.10(a) represents the sample structure together with an excitation scheme (either point-like excitation, or stripe-like excitation along the wire using a continuous Ti-sapphire laser) and photoluminescence recording (perpendicular to the structure axis or from the sample edge—waveguide emission). The particular spectrum in this figure belongs to low-temperature spontaneous emission. It contains a free exciton band (X) at the energy $h\nu \approx 1.58$ eV and a band of carrier recombination across the bandgap at $h\nu \approx 1.60$ eV. Figure 13.10(b) shows the corresponding absorption spectrum (from which one can obtain easily, for example, the binding energy of an

[7] We can estimate the density of excitons or *e–h* pairs in 1D structures, $n_X^{(1)}(\text{cm}^{-1})$, if we, e.g., multiply the corresponding volume concentration of excitons $n_X^{(3)}(\text{cm}^{-3})$ by the area of the wire cross-section, provided it is comparable with the exciton cross-section: $n_X^{(1)} \approx n_X^{(3)}(2a_X)^2$.

[8] Here, due to stronger spatial localization, the biexciton binding energy can be higher than in a quantum well by up to tens of percent. Regarding trions in 1D structures, several theoretical papers were published and below we shall present also one of their experimental verifications.

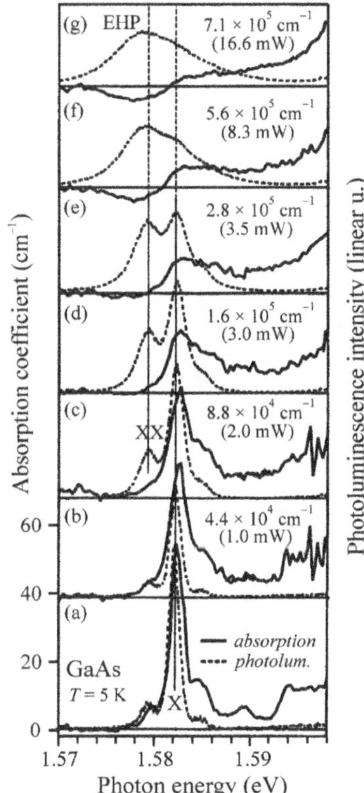

Fig. 13.11
Absorption (solid curves) and emission (dashed curves) spectra in a T- type GaAs quantum wire at various excitation power densities. The corresponding power of continuous excitation by a Ti-sapphire laser (mW) and the estimated densities of $e-h$ pairs (cm^{-1}) are shown by each pair of curves. $T = 5\text{K}$; after Hayamizu *et al.* [11].

exciton in the wire $E_X^{(1)} \approx 12$ meV as the separation between the ground exciton state and the onset of interband absorption; in bulk GaAs, E_X is only equal to 4.2 meV). It is also interesting to notice the little steps at the foot of the exciton absorption line—they originate from those areas of the wire the width of which is changed by $\pm$ one monolayer (± 1 ML).

The effects occurring in both absorption and emission as the excitation density increases are shown in Fig. 13.11. Luminescence is now—as opposed to Fig. 13.10—due to a homogeneous stripe-like excitation. For the lowest excitation density (laser power of 1 mW), the spectra from Fig. 13.10 (an exciton peak in absorption and, with a small Stokes shift, in emission, too) are reproduced, perhaps with a slight modification. As the excitation intensity increases a new peak appears in the luminescence; this is the XX peak about 3 meV away from the X peak towards lower photon energies. The new peak grows quadratically as a function of intensity of the X line. Both peaks then become broader and blend into a single, wide band (Fig. 13.11(f), (g)). In absorption, this is accompanied by quenching of the exciton absorption line at the *e–h* pair density of about $1 - 2 \times 10^5 \text{cm}^{-1}$ and, under the highest excitation, even by a broad region of negative absorption on the low-energy side.

If we sum up all these features, it is not hard to infer that the line denoted XX presents itself to be interpreted as the radiative decay of a biexciton with a binding energy $E_B^{(1)} \approx 3$ meV, followed by a gradual Mott transition to an ionized EHP. In Fig. 13.11, we can also clearly see the gradual bandgap renormalization and even the presence of positive optical gain, in the biexciton line at first and then, for the highest excitation, also in the EHP. These experimental results establish the possibility of all three basic phases—excitons, biexcitons, and EHP—coexisting together in 1D semiconductor systems.

The data from Fig. 13.11 imply the absence of an EHL since with increasing excitation intensity, the density of the *e–h* system keeps growing. We would expect something like that to occur in a direct GaAs semiconductor. This makes the recent results indicating, in contrast, the *existence* of an electron–hole liquid in quantum wires made of another direct semiconductor, InAs [12], even more surprising(!). With increasing excitation, the low-temperature photoluminescence spectra of these wires with an average diameter of 20 nm show a new, broad band with a well-defined intensity threshold, appearing on the low-energy side of the negative trion X^- emission line (Fig. 13.12(a)). The general behaviour of the band, particularly its invariable halfwidth and distinct temperature threshold, hints at EHL luminescence with a critical temperature $T_c = 35$ K and a critical density $n_c \approx 7.5 \times 10^4 \text{cm}^{-1}$, which is depicted in the phase diagram in Fig. 13.12(b). Comparing this figure to the phase diagram of the EHL in bulk silicon (Fig. 8.9) is definitely worth the reader's effort. Experiments also revealed intriguing, never before observed behaviour of the trion X^- line, occurring around the onset of EHL emission: the asymmetry of the line becomes clearly emphasized, as can be nicely seen in the upper spectra in Fig. 13.12(a). The explanation consists of the idea that, following the initial nucleation of the *e–h* liquid, the liquid volume increases rapidly, which leads to a 'suppression' of the trion gas in a relatively narrow wire. Enhanced trion localization (narrowing of the trion wavefunction in real space) then results in a substantial broadening of the X^- line towards lower photon energies. This is an explanation that has some logic in it; at the same time however, one can formulate the following objection against the model: The 'compression' of the trion should also lead to an increase in its binding energy and thus to a red-shifted X^- line (farther away from the free exciton energy). Yet, Fig. 13.12(a) reveals nothing like this. It appears to be an issue open to discussion that will require further experimental as well as theoretical investigation.

Now, however, a natural question arises immediately: what do we know about the electron–hole liquid in 1D *indirect-gap* semiconductor structures? The fact that an EHL exists in direct-gap wires makes its occurrence in indirect-bandgap wires all the more likely. Too narrow a wire probably does not provide conditions suitable for the formation of stable nucleation centres. On the other hand, if one somehow succeeded in 'pouring' the EHL into a relatively wide quantum wire made of an indirect semiconductor, we should intuitively expect the presence and perhaps even higher stability (that is, also a higher critical temperature limiting its existence) of the EHL due to the generally increased stability of excitons and their complexes in low dimensions. Therefore, the idea of studying somewhat 'wider' silicon or germanium wires immediately comes to mind. The concept has been realized quite recently: photoluminescence originating in an *e–h* system of freely moving carriers in silicon wires of diameter 50–300 nm was observed at a temperature of 10 K [13]. The established density of *e–h* pairs was close to the equilibrium density of the EHL in bulk silicon, however, unambiguous interpretation of this luminescence as a manifestation of the presence of an EHL in Si quantum wires remains open to question for the time being.

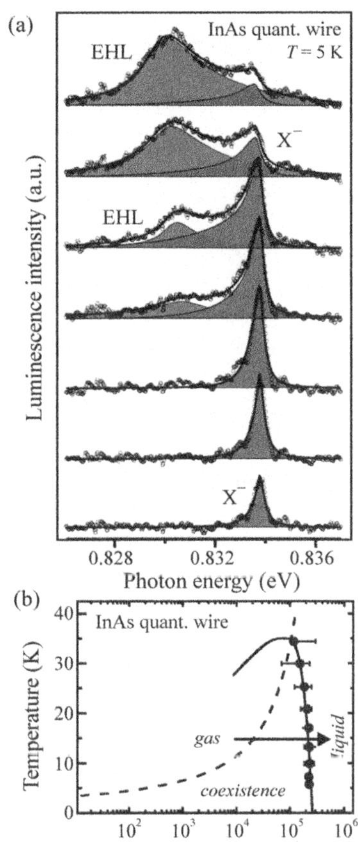

Fig. 13.12
(a) Development of photoluminescence emission spectra in InAs quantum wires with increasing excitation power (from bottom to top) at a temperature $T \approx 5$ K. (b) The phase diagram of the corresponding electron–hole system. The diagram indicates the presence of an EHL with a constant concentration of electron–hole pairs $\sim 2 \times 10^5$ cm^{-1} and with a critical temperature $T_c \approx 35$ K; circles denote experimental data. The dashed curve represents an estimate of the saturated density of exciton or trion gas. After Alén *et al.* [12].

13.6 Effects of high excitation in quantum dots (nanocrystals)

Nanocrystals or also 0D semiconductors go beyond the universal pattern of high-density luminescence phenomena, which was discussed in the preceding sections. This is due to the complete absence of freely moving excitons[9] or photocarriers in at least one dimension. The excitons are strongly localized within a tiny volume given by the diameter of the nanocrystal, being commonly just a few nanometres. Potential barriers or surface states surrounding the nanocrystal from all quarters prevent excitons from entering into immediate interactions with their colleagues in the neighbouring nanocrystals and thus, *a priori*, we cannot really expect collective interactions—that is, collisions of free excitons, EHP or even EHL—to occur in an ensemble of nanocrystals,

[9] The term 'exciton' as used in relation to nanocrystals is rather different from the strict meaning of the word exciton as introduced in 3D semiconductors for denoting the lowest electronic excitation of the entire crystal that is allowed for free diffusion motion on macroscopic scales. In the case of nanocrystals, the word exciton describes a correlated electron–hole pair (a boson) localized within the nanocrystal volume. This is related to the fact we mentioned in Chapter 12, i.e. that the Coulomb interaction between an electron and a hole never completely vanishes in small nanocrystals as a result of their overlapping wavefunctions.

although here we can give some thought to this idea to see whether this is not just a semantic problem. Let us consider a common, spherical nanocrystal of diameter 4 nm, the volume of which is about $3.2 \times 10^{-20}\text{cm}^3$. If it comprises of single excited *e–h* pair, one finds the corresponding density of *e–h* pairs inside the nanocrystal to be $\sim 3.2 \times 10^{19}\text{cm}^{-3}$, which is the typical EHP density in bulk semiconductors! If there are even more *e–h* pairs in the nanocrystal (under stronger excitation) then the Coulomb interaction can become screened, etc. Nevertheless, the mutual isolation of individual nanocrystals regarded like 'plasmatic elements' is strong enough and thus there is a general consensus about the fact that the effects typical for collective excitations, such as EHP or EHL, do not dominate the optical properties of an ensemble of nanocrystals.

Anyway, the fundamental parameter determining the luminescence behaviour of electronic excitations in quantum dots is the mean number of excitons or *e–h* pairs $\langle N \rangle$ created in the given nanocrystal by an external optical or electric excitation. To ascertain this number, it is important to know the absorption cross-section of the nanocrystal σ_{abs}, which is of the order of 10^{-17}–10^{-14}cm^2. With optical excitation, one can distinguish two special cases:

(a) A very thin layer of nanocrystals and a very short excitation pulse (typically ≤ 100 fs, i.e. substantially shorter than the lifetime of the *e–h* pair τ in a nanocrystal) carrying the energy E at an optical frequency ν when the total number of photons in the pulse is $N_{\text{f}} = E/h\nu$. Therefore, if focused on an area S containing the nanocrystals, the density of excitation photons is $n_{\text{f}} = N_{\text{f}}/S = E/h\nu S$ (cm^{-2}) and the mean number of *e–h* pairs per nanocrystal immediately after the excitation pulse will be

$$\langle N \rangle = n_{\text{f}}\sigma_{\text{abs}}. \tag{13.8}$$

If the layer is thicker than roughly several tens of nanometres and σ_{abs} is large enough, the lower-lying nanocrystals will be excited significantly less and (13.8) then represents an estimate of the upper bound of $\langle N \rangle$ in the given sample.

(b) A layer of densely packed nanocrystals with total thickness $l \gg$ diameter of the nanocrystal, characterized by a *filling factor* ξ (which is the relative part of the sample volume filled by the nanocrystals) and excited by a continuous optical beam; under these circumstances, the lifetime τ will already display its role. It is appropriate to introduce an auxiliary quantity—the absorption coefficient of an ensemble of nanocrystals

$$\alpha = \sigma_{\text{abs}} N_{\text{nc}} \quad (\text{cm}^{-1}),$$

where N_{nc} is the volume concentration of nanocrystals in the sample. The filling factor ξ can then be defined using the mean volume of one nanocrystal V_{nc} as

$$\xi = V_{\text{nc}} N_{\text{nc}} \ (< 1).$$

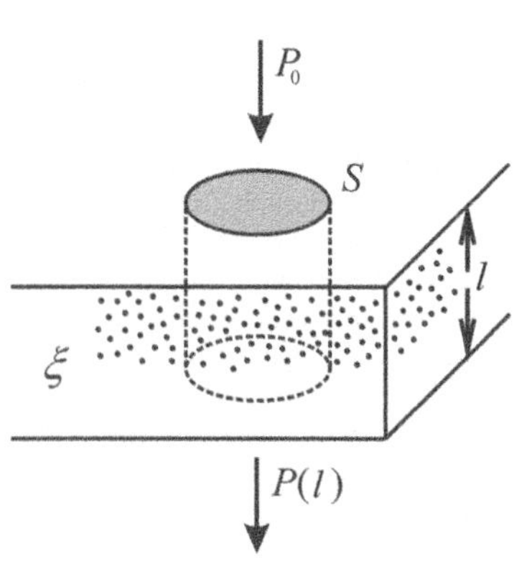

Fig. 13.13
Illustrating the determination of the mean number of $e-h$ pairs in a nanocrystal. P_0 is the excitation power incident on the sample, and $P(l)$ is the optical power after passing through the sample.

Let the excited area of the sample be S again and the incident exciting power be P_0 (Fig. 13.13). The intensity of the excitation beam then satisfies (assuming the matrix does not absorb the excitation radiation)

$$\frac{P(l)}{S} = \frac{P_0(1-R)}{S}\exp(-\alpha l) = \frac{P_0(1-R)}{S}\exp(-\sigma_{abs}N_{nc}l),$$

where R denotes the reflectivity. Thus, the absorbed power incident on the area S, fully exploited for the generation of e–h pairs, reads

$$P_{ex} = P_0(1-R)-P(l) = P_0(1-R)\,(1-\exp(-\sigma_{abs}N_{nc}l)) \approx P_0(1-R)\frac{\sigma_{abs}l\xi}{V_{nc}}, \tag{13.9}$$

after taking into consideration that the absolute majority of thin layers containing nanocrystals obeys $\sigma_{abs}N_{nc}l \ll 1$.

This power is absorbed in a volume Sl, within which there are $\xi(Sl/V_{nc})$ nanocrystals. Therefore, on average, $P_{ex}/h\nu(\xi Sl/V_{nc})$ e–h pairs are created per second and per nanocrystal and the steady-state mean number of e–h pairs in one nanocrystal finally reads

$$\langle N\rangle = \frac{P_{ex}\tau}{h\nu(\xi Sl/V_{nc})},$$

which, considering (13.9), yields

$$\langle N\rangle \approx \frac{P_0(1-R)}{S}\frac{\sigma_{abs}\tau}{h\nu}. \tag{13.10}$$

As could have been expected, neither the filling factor ξ nor the layer thickness l appear in the final expression (13.10) for $\langle N\rangle$.

Yet, we need to realize that in both the above cases, these are only estimates relying mainly upon our knowledge of the lifetime τ of 'excitons' in the nanocrystal. However, the lifetime is not inevitably constant; in general, it is a function of the excitation intensity. This is because with increasing excitation intensity when $\langle N\rangle \geq 2$, Auger non-radiative recombination of the $e-h$ pairs can manifest itself very efficiently, since the pairs are strongly localized in a small volume of the nanocrystal with no chance to escape.[10] This leads to appreciable shortening of τ and quenching of luminescence.

The mutual competition between the non-radiative and radiative lifetimes of $e-h$ pairs then drives, to a large extent, the specific optical properties of nanocrystals under strong excitation. It has been experimentally established that the optical spectra of nanocrystals can feature—as stable entities—not only excitons (in Si) but also correlated exciton pairs, such as biexcitons (in CdS) or even triexcitons (in CuCl). The above sequence replicates the increasing binding energy of excitons and also the decreasing radiative lifetime τ_r with increasing lattice bond polarity Si $\rightarrow$ CdS $\rightarrow$ CuCl. It thus turns out that the effects of Auger recombination are comparatively least important in I-VII nanocrystals while in silicon nanocrystals, it is presumed that luminescence originating from multiexciton complexes is probably not observed at all (the rapid non-radiative recombination of several $e-h$ pairs prevents their

[10] For a high value of N_{nc}, nanocrystals can be so densely packed that excitons diffuse, mainly from small nanocrystals into larger ones, which have a narrower bandgap. The statistical distribution of excited $e-h$ pairs also plays a role here. However, an approximate description of such a system is only possible using numerical modelling.

formation).[11] These effects are of particular importance in stimulated emission generated in a system of densely packed nanocrystals. The issues of stimulated emission and lasing in low-dimensional semiconductors have great application impact and, thus, the entire following chapter is devoted to them.

13.7 Problems

13/1: The emission lineshape expressed by relation (13.3) represents a convolution integral again. Write down its explicit form so that the convolution becomes evident. Assuming the function $g(E_B^{(2)})$ has a Gaussian form, try to estimate, based on Fig. 13.2(a), its parameter $2\sigma_B$ (i.e. the full width of the Gaussian curve between its inflexion points). Make use of, e.g., the analogy with Fig. A.2. What integration limits will appear in eqn (13.3)?

13/2: *Radiative decay of a trion.* For the case of a trion, sketch an energy diagram analogous to Fig. 8.3(a). Based on the diagram and using a procedure similar to the biexciton case (8.7 through 8.12), derive the theoretical expression for the emission lineshape of a trion in a quantum well (13.4), i.e.

$$I_{sp}^{T(2)}(h\nu) \simeq |M_T|^2 \exp\left\{-\frac{m_e}{(m_e+m_h)}\left[(E_g^{(2)}-E_X^{(2)}-E_{BT}^{(2)})-h\nu\right]/k_B T_T\right\}.$$

13/3: Show that at low temperatures the sum of the electron and hole Fermi energies in an electron–hole plasma (or liquid) in a given subband of a 2D semiconductor structure scales linearly with the density of electron–hole pairs, N, i.e. $(F_e + F_h) = \text{const}\ N$ (as against the 3D case where $(F_e + F_h) \sim N^{2/3}$ holds).

13/4: Derive an expression for the spectral shape of an emission line arising from recombination of free carriers in modulation-doped quantum wells and take into account conservation of the **k**-vector (a direct semiconductor) [14]:

$$I_{sp}^{MD}(h\nu) \simeq \rho_s^{(2)}(h\nu,\gamma)\,\frac{\exp\left(-\frac{m_e}{(m_e+m_h)}\frac{(h\nu-\tilde{E}_g^{(2)})}{k_B T}\right)}{\exp\left[\left(\frac{m_h}{(m_e+m_h)}(h\nu-\tilde{E}_g^{(2)})-F_e\right)/k_B T\right]+1}.$$

Here, $\rho_s^{(2)}(h\nu,\gamma) \simeq \{1+\exp[-(h\nu-\tilde{E}_g^{(2)})/\gamma]\}^{-1}$ denotes the joint density of states and γ is a phenomenological broadening factor. Hint: Proceed in analogy to Subsection 5.2.1.

13/5: Explain the origin of the differences between the luminescence spectra in Figs 13.10(a) and 13.11(a), i.e. the blue-shift of the exciton line and the quenching of the interband *e–h* emission under changeover from photoexcitation of a small spot ($\sim 1\,\mu$m) to excitation into a stripe of length 500 μm. Further, explain why the XX emission band in

[11] Nevertheless, some recent results indicate that radiative recombination of trions may appear, under specific circumstances, to be an efficient luminescence channel in silicon nanocrystals.

Fig. 13.11 cannot be interpreted as a result of inelastic exciton–exciton collisions.

13/6: (a) A thin layer of CdS nanocrystals, passivated by a surface ZnS shell and embedded in a SiO_2 matrix is excited by 100 fs pulses of wavelength 400 nm. The excitation energy density in the pulse (already allowing for reflection on the sample) is 200 $\mu J/cm^2$. Considering the absorption cross-section of the nanocrystals $\sigma_{abs} = 5.28 \times 10^{-15} cm^2$, estimate the mean number of $e-h$ pairs in the nanocrystals immediately after the excitation pulse is turned off. (b) A layer of silicon nanocrystals (with an average diameter of 3 nm) of thickness $l = 250$ nm is embedded in a SiO_2 matrix. It is excited by a 488 nm line of an Ar^+-laser with a cw power density of $50\,mW/cm^2$. Considering the lifetime of an $e-h$ pair in a silicon nanocrystal $\tau = 10\,\mu s$, refractive index of the layer $n \approx 1.75$, and absorption cross-section $\sigma_{abs}(422$ nm$) \approx 2 \times 10^{-17} cm^2$, determine the mean number of $e-h$ pairs per nanocrystal.

References

1. Cingolani, R. and Ploog, K. (1991). *Adv. Phys.* **40**, 535.
2. Phillips, R. T., Lovering, D. J., Denton, G. J., and Smith, G. W. (1992). *Phys. Rev. B*, **45**, 4308.
3. Esser, A., Runge, E., Zimmermann, R., and Langbein, W. (2000). *Phys. Rev. B*, **62**, 8232.
4. Manassen, A., Cohen, E., Ron, A., Linder, E., and Pfeiffer, L. N. (1996). *Phys. Rev. B*, **54**, 10609.
5. Combescot, M., Tribollet, J., Karczewski, G., Bernardot, F., Testelin, C., and Chamarro, M. (2005). *Europhys. Lett.*, **71**, 431.
6. Ryu, Y. R., Lubguban, J. A., Lee, T. S., White, H. W., Jeong, T. S., Youn, C. J., and Kim, B. J. (2007). *Appl. Phys. Lett.*, **90**, 131115.
7. Kulakovskii, V. D., Lach, E., Fochel, A., and Grützmacher, D. (1989). *Phys. Rev. B*, **40**, 8087.
8. Tränkle, G., Leier, H., Forchel, A., Haug, H., Ell, C., and Weimann, G. (1987). *Phys. Rev. Lett.*, **58**, 419.
9. Nihonyanagi, S. and Kanemitsu, Y. (2004). *Appl. Phys. Lett.*, **85**, 5721.
10. Pauc, N., Calvo, V., Eymery, J., Fournel, F., and Magnea, N. (2004). *Phys. Rev. Lett.*, **92**, 236802.
11. Hayamizu, Y., Yoshita, M., Takahashi, Y., Akiyama, H., Ning, C. Z., Pfeiffer, L. N., and West, K. W. (2007). *Phys. Rev. Lett.*, **99**, 167403.
12. Alén, B., Fuster, D., Muoz-Matutano, G., Martínez-Pastor, J., Gonzáles, Y., Canet-Ferrer, J., and Gonzáles, L. (2008). *Phys. Rev. Lett.*, **101**, 067405.
13. Demichel, O., Oehler, F., Noé, P., Calvo, V., Pauc, N., Gentile, P., Baron, T., Peyrade, D., and Magnea, N. (2008). *Appl. Phys. Lett.*, **93**, 213104.
14. Cingolani, R., Stolz, W., and Ploog, K. (1989). *Phys. Rev. B*, **40**, 2950.

Stimulated emission and lasing in low-dimensional structures

Low-dimensional semiconductors have several features favourable in view of the onset of stimulated emission, namely, tunability of the emission wavelength, increased probability of radiative recombination (or suppression of non-radiative transitions), more pronounced exciton luminescence processes, and reduced density of electronic states, entailing in principle easier achievement of population inversion. Therefore, the optical properties of quantum wells and superlattices have been studied extensively since, roughly, the mid-1980s. Shortly afterward, both optically and electrically pumped lasers on 2D structures were demonstrated in the laboratory; such lasers are now standard commercial components. In this chapter, we shall first discuss the physical mechanisms leading to optical gain in quantum wells, then the effects of stimulated emission in quantum wires and nanocrystals, and finally, we shall address the issues of so-called random lasing.

14.1 Stimulated emission in quantum wells

Extensive experimental research has established that in quantum wells multiple and diverse microscopic mechanisms of optical gain occur; which one of them will prevail in the given case then depends on the experimental conditions (sample material and type, measurement temperature, intensity and mode of excitation). We shall illustrate these mechanisms mostly using the examples of quantum wells in II-VI semiconductors because this is where the widest range of gain mechanisms can be found. On the other hand, the 'ordinary' 2D lasers based on III-V semiconductors work predominantly in a single 'standard' optical-gain regime of dense EHP plasma, which can be generated there even under relatively weak excitation; therefore, III-V semiconductors will be mentioned only briefly. In II-VI semiconductors, the situation is different as their exciton radius is smaller (because of the smaller dielectric constant and larger effective carrier masses) and thus the Mott density for exciton screening is much higher. Therefore, exciton transitions play an important role in the lasing processes, too. The reader is referred to, e.g., [1] and [2] for an overview of the physical properties of LED diodes and lasers based on II-VI semiconductors emitting in the blue-green spectral region.

14.1.1 Localized excitons

Ding *et al.* [3] established the mechanism of optical gain in $Zn_xCd_{1-x}Se$ quantum wells with ZnSe barriers at low and medium temperatures (≤ 200 K) and medium excitation level, namely, as being due to population inversion in a system of *localized* (bound) exciton states.[1] The threshold density of photoexcited *e–h* pairs for the onset of stimulated emission is $n_T \approx 7 \times 10^{11}$ cm^{-2} which is almost an order of magnitude lower than the estimated density for the Mott transition to EHP in zinc selenide ZnSe ($N_M^{(2)} \approx [\pi(a_X^{(2)})^2]^{-1} \approx 5.1 \times 10^{12}$ cm^{-2} for the two-dimensional exciton Bohr radius $a_X^{(2)} \approx 2.5$ nm in ZnSe).

The optical or possibly electrical excitation results in the formation of free excitons; the fundamental question of where to look for the origin of the localization traps for excitons was discussed in Chapter 12. The traps are created as a result of fluctuations—firstly, of local alloy fluctuations in the composition of $Zn_xCd_{1-x}Se$ (fluctuation in x) and, secondly, of fluctuations in the quantum well width. The luminescence line is thus inhomogeneously broadened. The localization energy of excitons in these traps amounts to about 15 meV and if the traps are sufficiently filled by excitons, population inversion with respect to the ground state of the well arises.

A schematic of a phenomenological model of exciton gain is shown in Fig. 14.1. Free excitons $|X\rangle$, generated by the excitation, become very quickly localized in the traps characterized by a Gaussian profile of the inhomogeneous broadening $D_i(E)$, which gives rise to localized exciton states $|X'\rangle$. With increasing pump intensity, these excitons, albeit bosons, gradually occupy ever higher states within the $D_i(E)$ profile—this is a sign that the states of the phase space of an inhomogeneous system with a reduced dimension are being filled (Section 13.4). Occupation of the localized exciton states can then be described in full analogy with an electron or hole system using the conventional Fermi–Dirac function

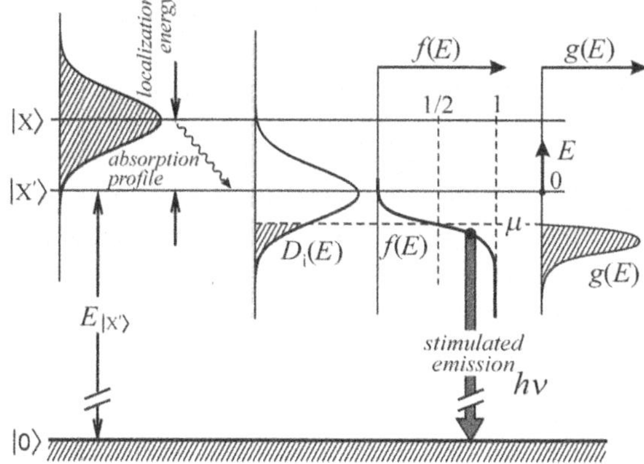

Fig. 14.1
The model of optical gain in an inhomogeneously broadened system of the tail states of localized excitons. $|X\rangle$ denotes the $n = 1$ state of a free exciton with a heavy hole (hh), $|X'\rangle$ is the $n = 1$ state of a localized exciton, $|0\rangle$ is the ground state. In this three-level system, population inversion occurs between $|X'\rangle$ and $|0\rangle$. For optical gain, one can write $g(E) \simeq D_i(E)[2f(E) - 1]$ (see text).

[1] Let us recall that in a system of free excitons, laser action is not possible without assistance of another quasi-particle.

$$f(E) = \frac{1}{\exp[(E-\mu)/k_B T_{ef}] + 1},$$

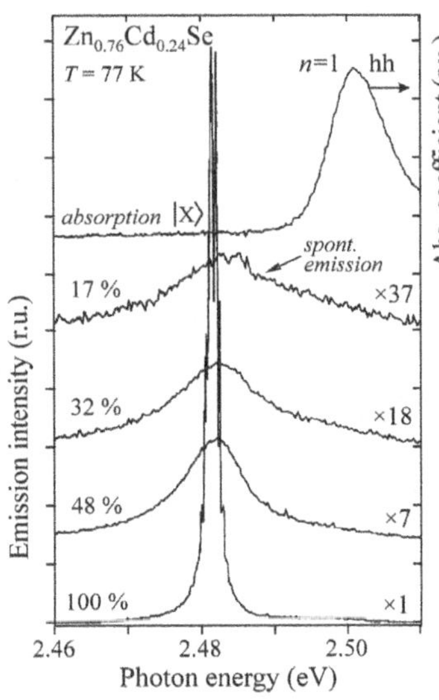

Fig. 14.2
Laser emission from a sixfold quantum well of $Zn_{0.76}Cd_{0.24}Se$ in the exciton spectral region at $T = 77$ K. A reference absorption spectrum is also shown. The laser line (lower spectrum) has a distinct modal structure since the sample has the form of a resonator. A pulsed laser (5 ps) excitation to the $|X\rangle$ state; the pump power density corresponding to 100% excitation was 180 kW/cm^2. After Ding *et al.* [3].

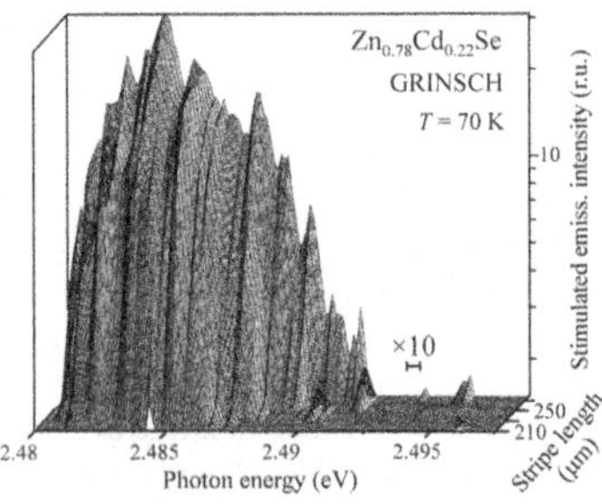

Fig. 14.3
Stimulated emission spectra of a simple quantum well of $Zn_{0.78}Cd_{0.22}Se$ of the GRINSCH type at $T = 70$ K. Excited by a XeCl pulsed laser (15 ns, 150 kW/cm^2), measured by the variable stripe length (VSL) method. After Tomašiūnas *et al.* [4], reproduced courtesy of the American Physical Society.

where E is referred to the centre of the exciton resonance (Fig. 14.1). Taking into consideration that in this case the occupation condition of population inversion for stimulated emission ($f_c + f_h - 1$), standing in eqn (10.18), transforms into $(2f - 1) > 0$, we can write on the basis of (10.18) a simplified expression for the spectral shape of the optical gain (see Problem 14/1)

$$g(E) \approx D_i(E)\,[2\,f(E) - 1] = D_i(E)\left[\frac{2}{\exp[(E-\mu)/k_B T_{ef}] + 1} - 1\right]. \tag{14.1}$$

We can easily transform this relation to the photon energy scale using the relation $h\nu = E_{|X'\rangle} + E$.

The optical gain mechanism we have just explained is not common. Now let us have a look at the supporting experimental data.

(a) Stimulated emission emerges directly from the profile of spontaneous emission around the $n = 1$ (hh) exciton resonance, i.e. without the significant Stokes shift typical for EHP luminescence (Fig. 14.2) [3].
(b) The spectral position of spontaneous emission remains constant over four orders of magnitude of the excitation intensity.
(c) A 'fine structure' modulated on the stimulated emission spectrum has been directly observed (Fig. 14.3). This structure, evolving randomly with increasing stripe length when applying the VSL method, represents the individual contributions of the excitons localized in (spatially) inhomogeneously distributed traps [4].

We shall stay for a while at the spectra displayed in Fig. 14.3 or, better to say, at the type of the investigated sample. This was a simple quantum well of $Zn_{0.78}Cd_{0.22}Se$ where the barriers consisted of a $(Zn_xCd_{1-x})Se$ alloy with high zinc content and a linear gradient of x from 0.95 to 1.00 (Fig. 14.4(a)). Such a structure is called a graded index separate confinement heterostructure, GRINSCH; see Fig. 12.5. An example of an application of the VSL method to studying the optical gain in this particular quantum well is given in Fig. 14.4(b). As the values of the net gain G are quite high here, they can be easily determined from the slope of relationship (10.33) if plotted using a logarithmic scale. It might be of interest now to put the GRINSCH type of samples in a historical context and thus also answer the question of what the gradient barriers are good for.

They were designed to increase the efficiency and decrease the threshold current of semiconductor injection lasers with quantum wells. The threshold current depends on the volume of the material where population inversion is to be achieved. From this perspective, one well is better than multiple wells; if, however, electrons and holes are to be efficiently localized inside and proper functioning of the quantum–confinement effect is to be ensured, the well width cannot exceed approximately 10 nm. Yet, such a narrow well then can no longer properly localize electromagnetic radiation because it is too narrow to do that (its width is much smaller than the wavelength of light). The

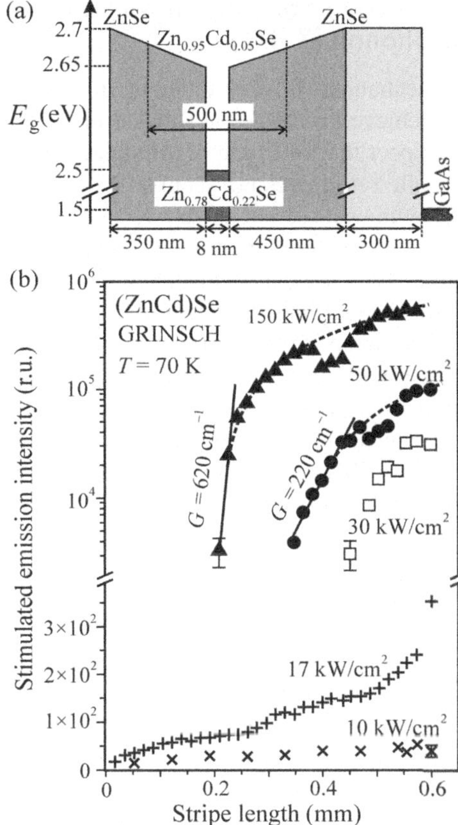

Fig. 14.4
(a) Schematic of the GRINSCH quantum well emitting stimulated radiation the spectra of which are shown in Fig. 14.3. (b) The corresponding amplified spontaneous emission (spectrally integrated) as a function of the stripe length using the VSL method. The values of the net optical gain are shown at two curves obtained under the highest optical excitation (a XeCl pulsed laser). $T = 70$ K, after Tomašiūnas *et al.* [4].

GRINSCH structure is thus designed in such a way that its width at the end of the funnel shape is comparable to the wavelength and it is there where the photons are guided, while the narrow well itself serves to localize electrons and holes. The inclined walls are designed to help in attaining efficient collection of the injected charge carriers in a p-i-n laser diode.

However, in the case of II-VI semiconductors, an additional aspect arises. It took quite a long time to accomplish acceptor doping which resulted in a p-type material because spontaneously only n-type samples of II-VI grow. Therefore, crystal growers failed to prepare p-n transitions, which are a necessary prerequisite for an injection laser. Thus, as an alternative approach, in the beginning of the 1990s it was suggested that the GRINSCH structures should be tried out instead; the intention here was to ensure an efficient flow of charge carriers into the well solely by means of the drift in the electric field of the gradient barriers without the need for a p-n junction. Shortly afterward, however, the GRINSCH concept in II-VI compounds was basically abandoned when InGaN materials of III-V type made an unexpected breakthrough in the blue part of the spectrum.

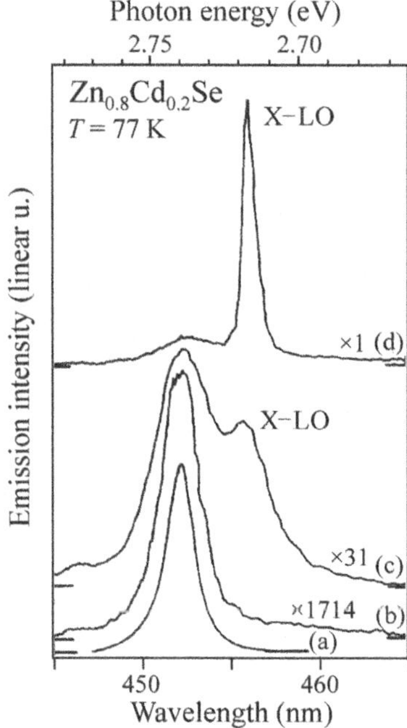

Fig. 14.5
Laser emission spectra from a fifteen-fold MBE quantum well of $Zn_{0.80}Cd_{0.20}Se/ZnSe$. The bottom spectrum (a) was excited by a weak continuous beam while the other spectra were excited by a blue pulsed dye layer (5 ns) with a pumping power density (b) $0.079\,kW/cm^2$, (c) $6.3\,kW/cm^2$, and (d) $12.5\,kW/cm^2$. $T = 77$ K; after Kawakami *et al.* [5].

14.1.2 Radiative decay of an exciton with emission of an LO-phonon (X–LO)

This luminescence mechanism is very efficient in 3D semiconductors and allows one to easily achieve lasing, as we showed in Section 10.5. Thus, one might unthinkingly expect that this type of stimulated emission asserts itself in 2D structures as well. Yet, it does not seem so certain on closer inspection. As we saw in the preceding subsection and as we shall see in addition later on, most authors interpret their experimental results using other mechanisms. For example, based on Fig. 14.2, we can immediately exclude the X–LO process from our considerations regarding the particular sample, as the stimulated emission line is approximately 18 meV only away from the maximum of the exciton absorption line while the LO-phonon energy, determined by Raman spectroscopy, amounts to $\hbar\omega_{LO} \cong 30\,meV$.

Nevertheless, e.g. Kawakami *et al.* interpret their experimental data on stimulated emission from $Zn_{0.80}Cd_{0.20}Se/ZnSe$ quantum wells, prepared by the MBE method, within the framework of the X–LO model [5]. In Fig. 14.5 we show an example representing the emission spectra originating from 15 quantum wells, each 1.5 nm wide, at $T = 77$ K. With increasing level of optical pumping, around a threshold value of $\sim 1\,kW/cm^2$ the intensity of the narrow peak denoted X–LO grows superlinearly (spectra (c) and (d)); the peak is ~ 27 meV away from the maximum of the bottom spectrum of spontaneous exciton emission (a). It is definitely noteworthy that the sample was prepared in the form of a Fabry–Perot resonator formed by the intrinsic cleavage planes and pumped using a cylindrical lens—thus, this is genuine lasing and not only amplified spontaneous emission (ASE). As $\hbar\omega_{LO} \cong 30\,meV$, the energy conservation law $h\nu \cong$ (exciton energy $- \hbar\omega_{LO}$) is well satisfied and it is thus justified to denote the peak as X–LO.

Additionally, the constant shift of ~ 30 meV has also been observed in quantum wells of the same origin and analogous composition but of different width (3 nm, 5 nm, 12 nm) as shown in Fig. 14.6. This confirms that the dominant lasing mechanism here is derived from luminescence of an exciton assisted with an LO-phonon.

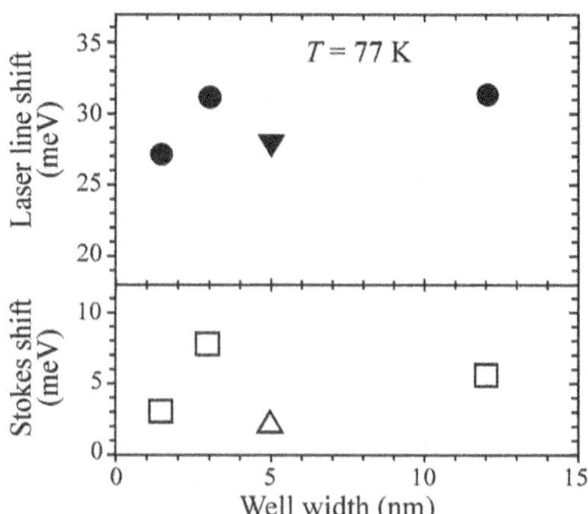

Fig. 14.6
Laser line shift (top panel) and Stokes shift (the difference between absorption and emission energies of an exciton, bottom panel) in MBE multiple quantum wells of $Zn_{0.80}Cd_{0.20}Se/ZnSe$(●, □) and $Zn_{0.85}Cd_{0.15}Se/ZnS_{0.08}Se_{0.92}$(▼, △). $T = 77$ K; after Kawakami *et al.* [5].

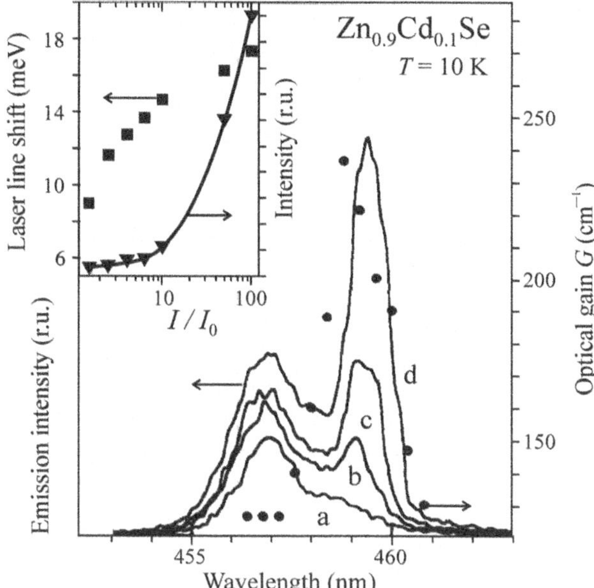

Fig. 14.7
Spectra of stimulated emission from $Zn_{0.90}Cd_{0.10}Se/ZnSe$ multiple quantum wells under increasing excitation intensity: (a) 69, (b) 114, (c) 183, and (d) 287 kW/cm^2. Solid circles represent the optical gain spectrum; the inset shows the change in slope in the intensity dependence upon reaching the lasing threshold. Well width 2.8 nm, excited by an N_2-laser, $T = 10$ K. After Cingolani *et al.* [6].

It is obvious that there is a fundamental difference between the stimulated emission mechanisms in samples of quantum wells of varying origin but of otherwise very similar structure. The situation is further complicated by the fact that if the optical excitation intensity is increased high above the lasing threshold in samples with the X–LO lasing mechanism active, the characteristic features of a third type of stimulated emission, exciton–exciton collisions (X–X), emerge. It means that the laser line gradually moves away from the spontaneous exciton emission, increasing the original separation of about 30 meV up to ~ 41 meV, which corresponds to the estimated binding energy of an exciton in the several-nanometre-wide $Zn_{0.80}Cd_{0.20}Se/ZnSe$ wells ($E_X^{(2)} \approx 37 - 45$ meV). In this respect, we can also refer to Fig. 13.4(b) where, as the pumping intensity in multiple quantum wells of ZnO/BeZnO increases, a sharp P_∞ peak at ~ 3.21 eV gradually develops from the P line and then progressively suppresses other features in the spectrum.

Before attempting to find a pattern in the otherwise somewhat chaotic experimental facts, we shall mention the last type of lasing occurring in quantum wells, namely, the optical transitions in a pool of free carriers in an electron–hole plasma.

14.1.3 Stimulated emission in electron–hole plasma (EHP)

We shall keep paying attention to quantum wells of the ZnCdSe type. Cingolani *et al.* [6] studied the spectra of both stimulated emission and optical gain in multiple wells of $Zn_{0.90}Cd_{0.10}Se/ZnSe$ using the VSL method with strong excitation by sub-nanosecond pulses. The evolution of the emission spectrum from the edge of the sample is shown in Fig. 14.7. Here, the salient features are:

(a) Stimulated emission does not originate directly from the maximum of the spontaneous band but it is red-shifted instead.
(b) This shift ($\leq 17\,\text{meV}$) is small in comparison with the energy of an LO-phonon as well as in comparison with the binding energy of an exciton in the well.
(c) The width of the optical gain curve is relatively large (more than 20 meV at the foot).
(d) There is a radical change in slope of the intensity dependence of the emission (inset in Fig. 14.7).

Firstly, item (a) rules out localized excitons from possible interpretation of the results and item (b) does the same with the X–LO and X–X mechanisms. Secondly, item (c) is then a strong indication that the stimulated emission originates in the EHP where the gain profile width at low temperatures is given by the sum of Fermi energies ($F_e + F_h$). Last, item (d) is an important sign of the onset of stimulated emission.

To confirm whether the transitions in a system of localized excitons or of free carriers (EHP) are the dominant mechanism of radiative recombination, one can experimentally monitor the shift of the emission line in an external magnetic field. A free exciton in a magnetic field behaves roughly as an atom—it shows the so-called Langevin diamagnetism when the susceptibility is proportional to the number of electrons and to the squared atom radius. In an exciton, this is a weak effect, which becomes even weaker as the exciton gets localized in the minimum of a random fluctuation of the quantum well potential (the electron and the hole are very close to one another; some researchers even speak of a 'quantum dot' localized in a quantum well). Neither spontaneous nor stimulated emission from an ensemble of localized excitons is thus, *de facto*, influenced by the magnetic field. In contrast, radiative recombination of free electrons and holes in a magnetic field shows a small nevertheless measurable shift, the so-called diamagnetic shift (of the order of 1 meV/T), which originates from Landau diamagnetism: free carriers describe quantized circular orbits in the plane of the well, perpendicular to the applied magnetic field. This spectral shift occurs towards higher photon energies since the system of free carriers gains energy as we can conclude from, e.g., Lenz's rule. Indeed, relevant experiments confirmed that stimulated emission documented by Fig. 14.7 originates from free carriers [6].

Thus, we can now attempt to put forth a certain systemization of the signatures of stimulated emission in 2D structures of II-VI semiconductors. First of all, let us state that no correlation between the method of preparation of the samples and the specific type of stimulated emission has been established.[2] What we see is rather the effects of pump intensity and alloy composition (well depth). Generally, the lowest threshold pump intensities ($\sim 1\,\text{kW/cm}^2$) were established for lasing of the X–LO type; upon exceeding this level several

[2] Therefore, one cannot just simply say that the samples grown using the cutting-edge method of molecular beam epitaxy (MBE) are always structurally more perfect than 2D wells prepared via metalorganic vapour phase epitaxy (MOVPE) and thus they do not feature lasing in a system of localized excitons. One may encounter this stimulated emission channel in both sample types.

times, the X–X collision mechanism begins to dominate stimulated emission. With further increase in the excitation level to about 10–100 kW/cm^2, lasing from an ensemble of localized excitons reveals itself; being a collective excitation effect (occupation of the low-energy tail of the exciton absorption band), it has a higher threshold. At the highest pumping level enabling the system to reach the Mott density, stimulated emission from an ensemble of free carriers in the EHP sets in.

The effect of the depth of the well shows itself by the fact that the exciton gain is preferred in sufficiently deep wells and, in addition—when lasing comes from an ensemble of localized excitons—if the excitons are sufficiently localized in sites of potential fluctuations. In shallow quantum wells, in contrast, excitons are less stable and, likewise, compositional disorder has a lesser impact; stimulated emission will thus originate from an ensemble of free carriers rather than from localized excitations. The $Zn_xCd_{1-x}Se/ZnSe$ wells are shallow if x is not too different from unity. This is exactly the case shown in Fig. 14.7. Naturally, there is a rather broad range of temperatures, excitation intensities, and material composition where optical gain mechanisms may overlap.

Now, we shall briefly consider III-V quantum wells of the type GaAs(well)/$Al_xGa_{1-x}As$(barrier), which stood in the very beginning of interest in 2D semiconductor structures in the 1980s. Here, the lasing mechanism consists in radiative recombination of the EHP. We shall illustrate this using an example of an electrically pumped injection laser with a simple quantum well of $GaAs/Al_xGa_{1-x}As$. With increasing pump current, there is a sudden step-like increase in gain accompanied by a similarly abrupt shortening of the output wavelength. The only logical explanation for such behaviour is the emergence of population inversion in a higher subband $j = 2$ in the EHP (just as Fig. 13.7 shows the filling of higher subbands during spontaneous emission). In this case, we may substitute a step-like shape density of electron states D_e for the Gaussian profile of localized excitons D_i in the expression for the spectral shape of the optical gain (14.1), yielding

$$G(h\nu) = D_e(h\nu)[2 f_c(h\nu) - 1] = C \sum_{j=1}^{2} H[(h\nu - E_g)S - E_j][2 f_c(h\nu) - 1], \tag{14.2}$$

where H stands for the Heaviside step function and C represents the maximum achievable modal gain.[3] The quantity S denotes the relative offset of the conduction band of the well, i.e. $S = \Delta E_c/\Delta E_g$ and f_c is the Fermi distribution function of electrons. The situation is schematically elucidated in Fig. 14.8(a) which also displays the gain spectra calculated using (14.2) for a high (top) and a very high (bottom) pumping level [7]. One can clearly see the step in the magnitude of the gain at the energy corresponding to the onset of the second subband when a simultaneous steep increase in the photon energy $h\nu$ occurs. Figure 14.8(b) represents a comparison of experimental gain values $G(h\nu)$

[3] Due to the fact that holes are heavier than electrons it is sufficient to consider only the first step of their density of states, which does not influence the form of expression (14.2).

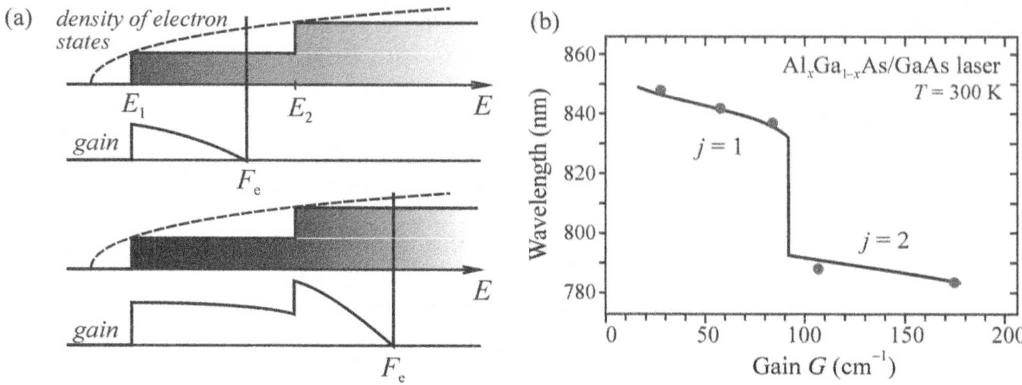

Fig. 14.8
(a) Filling of the electron subbands in an injection laser with a simple, 10 nm wide quantum well of $GaAs/Al_xGa_{1-x}As$ along with optical gain spectra calculated using (14.2) at room temperature. The bottom spectrum is calculated for a higher excitation level than the top one; the positions of the Fermi energy of electrons F_e are shown. (b) Comparison of experimental values of laser wavelengths and magnitudes of modal gain (symbols) with predictions by a model of lasing in the EHP (curve), expressed by eqn (14.2). After Mittelstein *et al.* [7].

and lasing wavelengths with the above model (modified further by introducing a phenomenological broadening factor). There is very good agreement, indeed.

14.2 Stimulated emission in quantum wires

We happened to meet some indications of stimulated emission in quantum wires in Section 13.5. Interest in the potential use of stimulated emission from 1D systems in miniature photonic components on the nanometre scale lies in (in addition to the facts we have already mentioned several times, such as the increased probability of radiative recombination and an easier-to-reach population inversion) the attractive geometric shape of the wires: just by themselves, they can represent a cylindrical optical cavity where the axial modes of the Fabry–Perot resonator can play their natural role in lasing. However, activities in this field have so far been restricted by technological limitations—preparation of quantum wires of the required shape, length, crystal quality, and suitable reflectivity on their ends is not a simple task. Up to now, scientific attention has been focused mainly on two materials—zinc oxide, ZnO, and gallium nitride, GaN; they both exhibit luminescence in the near-ultraviolet region. It turns out that the diameter of the wire is another limiting factor here: in nanowires of a diameter comparable or even smaller than the wavelength of the generated light, radiant energy escapes from the core via diffraction losses and thus lasing can hardly be achieved. This is because the reflectivity coefficient at the ends of the wire decreases strongly as the wire gets thinner since only that part of the mode energy contained within the wire gets reflected. The effective resonator quality thus drops.

Figure 14.9 displays an example of laser oscillations in an individual ZnO quantum wire under a pulsed laser pumping at room temperature [8]. Panel (a) demonstrates how the lasing oscillation threshold is achieved with increasing

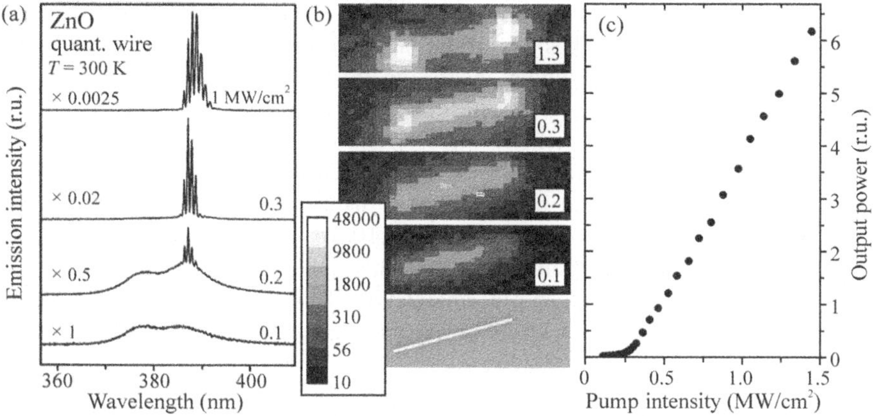

Fig. 14.9
A quantum wire of ZnO, 12 μm long and 150 nm across, as a miniature 'laser rod'. (a) Output spectra as functions of the pump intensity (given in units of MW/cm^2), (b) the wire scanned by a CCD camera, and (c) the total output power as a function of the pump intensity. Excitation by laser pulses from a Nd:YAG laser (355 nm, 6 ns), room temperature. After Zimmler *et al.* [8]; reproduced courtesy of the American Institute of Physics.

excitation level; the threshold clearly manifests itself in the emission spectrum as a transition from a broad spontaneous emission band to lasing in narrow, longitudinal (axial) resonator modes. Panel (b) then shows how the intensity of radiation originally emitted isotropically along the entire wire gradually concentrates at its ends, and panel (c) represents the intensity dependence of the total emission detected at one end of the nanowire. The break around $0.25\,MW/cm^2$ confirms that the lasing threshold has been reached. Wires of ZnO of radii smaller than 150 nm did not show lasing.

Analogous effects, observed in a single GaN quantum wire, are shown in Fig. 14.10 [9]. Also here, the diameter of the investigated wires ranged around 100 nm.

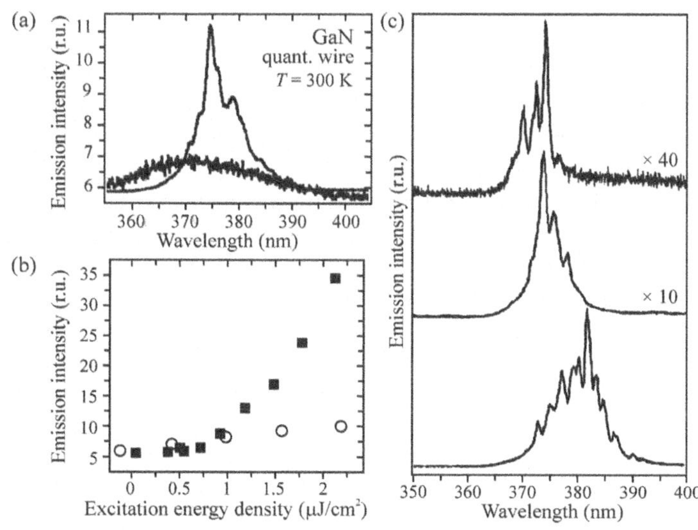

Fig. 14.10
A GaN quantum wire: (a) emission spectrum of spontaneous photoluminescence (noisy curve) and of the lasing output from the end of the wire, (b) intensity dependence of laser emission (solid symbols) and that of photoluminescence, the latter being recorded from a non-lasing area (circles), and (c) evolution of the laser spectrum with increasing excitation energy density (0.7, 1.2, and 2 $\mu J/cm^2$). Pumping: laser pulses of the fourth harmonic of a Ti:sapphire laser coupled with a parametric oscillator (~ 100 fs), room temperature. After Johnson *et al.* [9].

The lasing mechanism has not been studied in details so far. However, we get the basic information by analysing the behaviour of the stimulated emission spectra under increasing pump intensity. In ZnO, the stimulated emission emerges from the maximum of the luminescence band and with increasing pump intensity there is almost no spectral shift (Fig. 14.9(a)), which is an indication of lasing in a system of localized excitons (Subsection 14.1.1). In GaN, on the contrary, a distinct red-shift with increasing excitation power density appears (Fig. 14.10(a)), indicating that the dominant mechanism of optical gain here is probably radiative recombination of the EHP accompanied by bandgap renormalization. This is in agreement with the fact that compared to ZnO the binding energy of an exciton in GaN is less than half of that in ZnO.

14.3 Stimulated emission in nanocrystals

Nanocrystals can be prepared in a relatively simple way using a variety of methods (direct chemical synthesis, epitaxial growth, sol-gel method, etc.) with dimensions as little as several nanometres. In small nanocrystals with a distinct quantum-confinement effect, the energy separations between electronic states may be much larger than the thermal energy $k_B T$ and the strength of interaction between excitons and LO-phonons may be reduced. This makes thermal quenching of the excited levels more difficult and one can thus expect an easy-to-reach and almost temperature-independent lasing threshold. This together with an emission tunable within a wide range of wavelengths and a reduced density of states makes the prospect for commercially produced nanocrystal lasers very attractive. Therefore, researchers dedicated a large number of papers to the effects of stimulated emission in semiconductor nanocrystals and also nowadays there is intense on-going research. We shall attempt to comprehend the state of the art and to address open issues.

14.3.1 Nanocrystals dispersed in a matrix

In the case of 'zero-dimensional' nanocrystals, we obviously cannot—unlike the 1D quantum wires—entertain simply the idea of realizing a separate 'nanolaser' but, instead, stimulated emission can only be achieved in a macroscopic ensemble of nanocrystals similarly to, e.g., impurity centres of the Cr^{3+} atomic type in a ruby laser. In principle, we can handle nanocrystals like atoms or molecules, which means that we can embed them in a transparent glass, semiconductor or polymer matrix, in a waveguide or even make a colloidal dispersion of them in a suitable chemically inert solvent. This is important for lasing because we feel intuitively that in order to reach the onset of the collective stimulated emission, we need to achieve certain critical concentration of active centres—in this case of nanocrystals. Yet another important criterion is to have a narrow distribution of nanocrystal sizes. If this criterion is not met, the pumping energy spreads over a relatively broad range of wavelengths during the emission process and its efficiency in achieving population inversion is reduced. Semiconductor nanoparticles can be, in principle (although not in all materials), prepared with a relatively narrow size distribution ($\leq 5\%$). One can

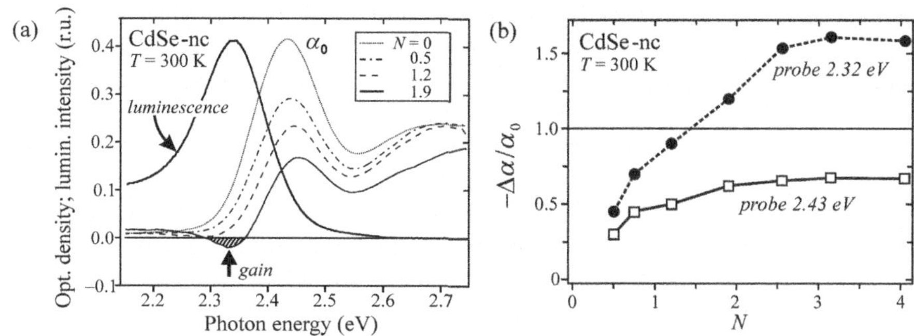

Fig. 14.11
(a) Evolution of the optical gain in CdSe nanocrystals ($R = 1.3$ nm), established by measurement of transient absorption in the exciton spectral region using the pump and probe method with a high temporal resolution. (b) Dependence of induced transient absorption on the number of *e–h* pairs, N, in a nanocrystal. Room temperature, after Klimov *et al.* [10].

also achieve a high quantum yield of photoluminescence η at room temperature here ($\eta > 50\%$), which is another favourable factor. In spite of that, unfortunately, an adverse competing process acting against the stimulated emission, namely, intrinsic Auger non-radiative recombination, can occur frequently.

Auger recombination is strongly concentration dependent (Subsection 6.1.2). Its rate $1/\tau^{\mathrm{A}}$ increases with decreasing volume in which photoelectrons and photoholes are localized—in the case of nanocrystals, therefore, with decreasing radius of the nanocrystal, R; consequently, $1/\tau^{\mathrm{A}} \simeq R^{-3}$ holds true. The simplest imaginable concept of a quantum dot as a two-level system with double spin degeneracy tells us that positive optical gain can be achieved already with the average number of *e–h* pairs in a nanocrystal equal to one, $N = 1$. With $N = 2$, we then have gain saturation (i.e. total population inversion). For $N = 2$, at the same time, very efficient Auger recombination can take place since two electrons and two holes in a nanocrystal of radius R of several nanometres imply a high carrier concentration of the order of 10^{19}–$10^{20}\mathrm{cm}^{-3}$(!). The crucial criterion determining the occurrence of optical gain is then that the stimulated emission lifetime τ_{stim} be shorter than τ^{A}. High luminescence yield η does not represent an absolute priority in this context, because we know one can achieve $\eta \to 100\%$ while the radiative lifetime τ_{r} (controlling essentially also the time τ_{stim}) is relatively very long. Thus, the direct-bandgap semiconductors with $\tau_{\mathrm{r}} = 0.1 - 1$ ns are clearly the front-runner here.

Figure 14.11 serves as an example of how the optical gain can behave as a function of N. Using the optical pump and probe method with femtosecond pulses [10] applied to a thin layer of CdSe nanocrystals, Klimov *et al.* established that with increasing intensity of the pump beam, absorption of the first exciton band at ~ 2.43 eV is bleached out gradually (Fig. 14.11(a)). The variation in absorption $\Delta\alpha$ is thus negative but its magnitude is smaller than the original value of the absorption coefficient α_0 for quite a while, i.e. $|\Delta\alpha| < \alpha_0$. Only under the highest pump intensity, absorption at $h\nu \approx 2.32$ eV 'flips' to negative values and the (material) optical gain $g = |\Delta\alpha| - \alpha_0 > 0$ appears; this can be quantified using the ratio $-\Delta\alpha/\alpha_0 > 1$. After rescaling the pump

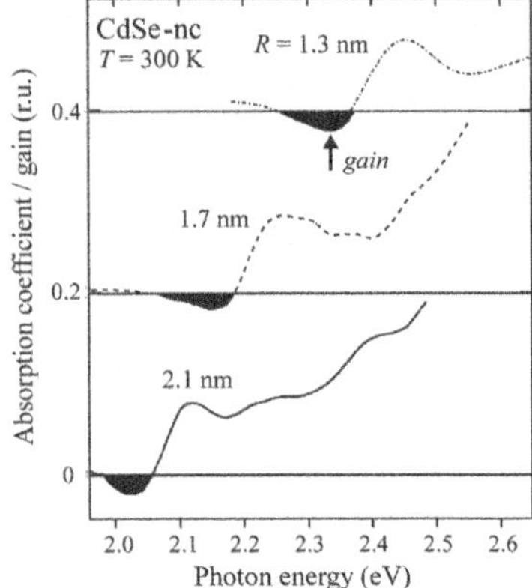

Fig. 14.12
Optical gain spectra of three thin films containing nanocrystals of CdSe of various size. After Klimov *et al.* [10].

intensity to the average number N of e–h pairs in a nanocrystal the plot of $-\Delta\alpha/\alpha_0$ against N is shown in Fig. 14.11(b). One can see that the optical gain sets in at the average value $N \simeq 1.4$, that is, near $N \simeq 1$ in accordance with the above considerations. Figure 14.12 represents an analogous result for CdSe nanocrystals of various sizes. Among other points, this figure also demonstrates the spectral tuning capabilities of lasers with quantum dots.

The reader perhaps remembers that in Chapter 10 we postulated the impossibility of achieving population inversion and optical gain in a system of free 3D excitons on their own. This is a result of the fact that absorption and stimulated emission of free excitons are perfectly balanced. Likewise, in quantum wells, we did not associate exciton lasing mechanisms with a pool of non-interacting free excitons but, instead, stimulated emission also involved either phonons, or collisions of two free excitons or exciton localization. In nanocrystals, where excitons actually do not have translational symmetry, excitons often become quickly localized, usually close to the surface. This leads to a shift in the absorption line with respect to the luminescence line, which disturbs the balance between absorption and stimulated emission and permits 'single-exciton' optical gain. The effect may be enhanced by appropriate surface passivation decreasing the overlap of electron and hole wavefunctions and thus also leading to the suppression of Auger recombination [11].

In general, however, the above-mentioned Stokes shift between exciton absorption and emission in nanocrystals is not a constant quantity but depends strongly on the quality of the samples (particularly on the surface of the nanocrystals and, as has already been stressed, on its passivation). Therefore, the occurrence of gain in various samples has been random to some extent so far and a range of unsuccessful experimental results was obtained when searching for it. For the same reason, one cannot predict exactly the wavelength on which the narrow line of stimulated emission starts rising up from the spontaneous luminescence band in luminescence experiments. This usually happens on the long-wavelength side, as shown in Fig. 14.13, but it can also be in the neighbourhood of the maximum.

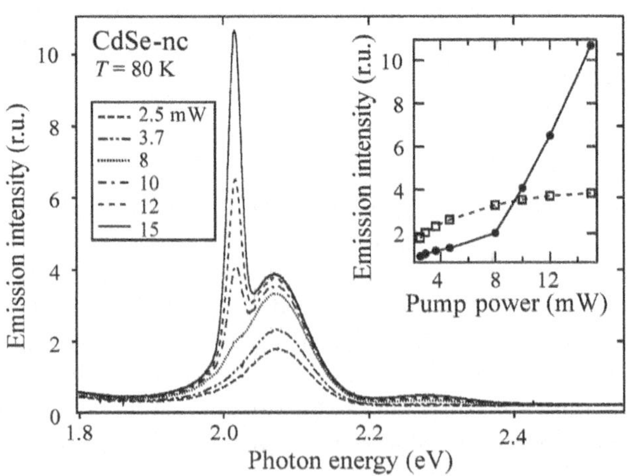

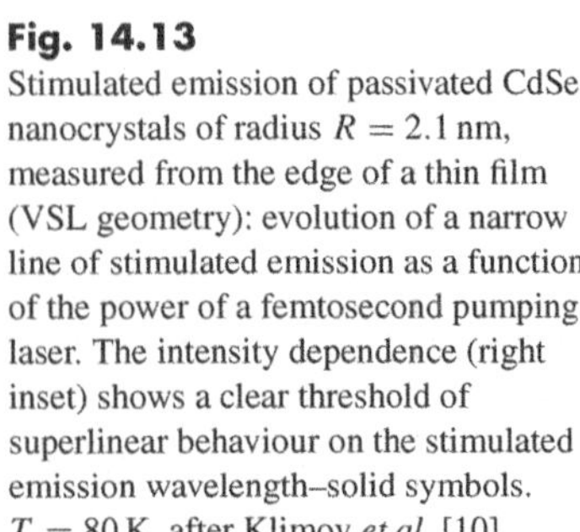

Fig. 14.13
Stimulated emission of passivated CdSe nanocrystals of radius $R = 2.1$ nm, measured from the edge of a thin film (VSL geometry): evolution of a narrow line of stimulated emission as a function of the power of a femtosecond pumping laser. The intensity dependence (right inset) shows a clear threshold of superlinear behaviour on the stimulated emission wavelength–solid symbols. $T = 80$ K, after Klimov *et al.* [10].

We have mentioned the crucial criterion for the occurrence of optical gain that reads $\tau_{\text{stim}} < \tau^{\text{A}}$. We shall now show that this criterion implies an important condition for the density of nanocrystals in a matrix (the so-called filling factor). The time τ_{stim} will surely become shorter with increasing the magnitude of the gain g; one can show (Problem 14/3) that the relation is a simple reciprocal

$$\tau_{\text{stim}} = \frac{n}{cg}, \tag{14.3}$$

where c is the velocity of light in vacuum and n is the refractive index. Let n_0 (cm^{-3}) be the density of nanocrystals in a film. We can then introduce the gain cross-section σ_{g} (cm^2) using the relation $g = \rho n_0 \sigma_{\text{g}}$ where $0 \le \rho \le 1$ denotes the fraction of nanocrystals in a state where population inversion is higher than the threshold value. We define the filling factor $\xi(\le 1)$, i.e. the volume fraction filled by nanocrystals, as

$$\xi = (\text{nanocrystal volume}) \times n_0 = 4\pi n_0 R^3/3. \tag{14.4}$$

Substituting (14.4) into (14.3) gives

$$\tau_{\text{stim}} = \frac{4\pi R^3}{3} \frac{n}{\rho \xi \sigma_{\text{g}} c}.$$

If we also express the above-mentioned empirical dependence of the rapid increase in Auger lifetime τ^{A} with radius R of the nanocrystal via

$$\tau^{\text{A}} = \beta R^3,$$

then, comparing the last two equations, the condition $\tau_{\text{stim}} < \tau^{\text{A}}$ leads to the important inequality

$$\xi > \frac{4\pi n}{3c\rho\beta\sigma_{\text{g}}} = \xi_{\text{m}}. \tag{14.5}$$

This relation says that in order to achieve optical gain in a thin layer of nanocrystals dispersed in a transparent matrix, a certain minimum filling factor ξ_{m} has to be attained. The smaller the gain cross-section σ_{g}, the more densely the nanocrystals must be arranged in the matrix. Therefore, stimulated emission occurs in thin films with a high concentration of quantum dots (typically $n_0 \approx 10^{18}$ cm^{-3}, i.e. $\xi_{\text{m}} \approx 1 - 10\%$) and under strong pumping ($\rho \to 1$). An obvious prerequisite here is sufficient optical quality of the film and, of course, as narrow nanocrystal size distribution as possible.

Nevertheless, the requirement for closely packed quantum dots often directly collides with the requirement of high optical quality of the sample—nanocrystals tend to cluster into larger systems, which can lead to considerable losses through scattering of light or, possibly, the sample can even incur mechanical micro-fissures. In that case, the gain will no longer overcome the losses and stimulated emission will not appear. Some unsuccessful experiments designed for the search for optical gain are probably also related to this effect.

Let us put the concept of stimulated emission mechanisms in quantum dots in more precise terms. We have stated above that, basically, stimulated

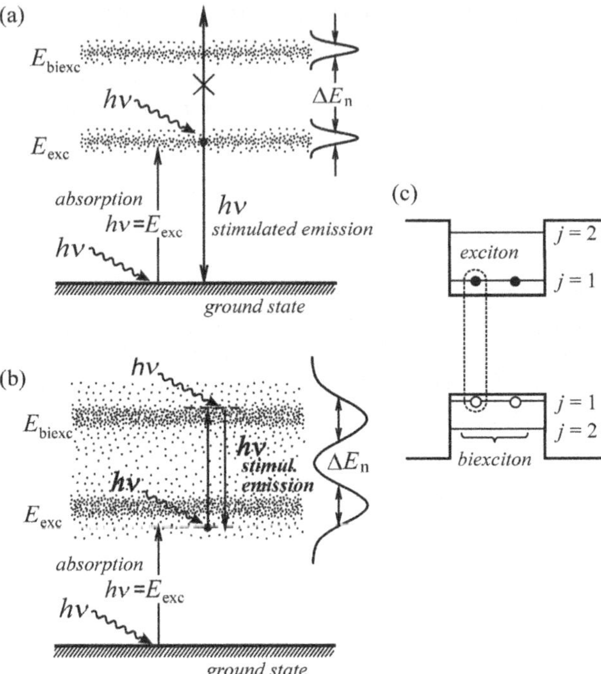

Fig. 14.14
Mechanisms of stimulated emission in nanocrystals, resulting from inhomogeneous broadening of levels ΔE_n. (a) Exciton mechanism at $\Delta E_n < (E_{biexc}-E_{exc})$, (b) biexciton mechanism at $\Delta E_n > (E_{biexc}-E_{exc})$ and strong pumping. Scheme (c) shows the biexciton as a doubly excited quantum dot in one-particle approximation.

emission in nanocrystals with doubly degenerate levels can occur both with $N = 1$ and with $N = 2$; we speak of the *exciton* and *biexciton* mechanisms. What is crucial here is the ratio between the inhomogeneous broadening of the exciton and biexciton levels ΔE_n and their energy separation $(E_{biexc} - E_{exc})$. If $\Delta E_n < (E_{biexc} - E_{exc})$, see Fig. 14.14(a), the quantum dot already containing a single exciton cannot absorb another photon $h\nu = E_{exc}$ and stimulated emission occurs through the exciton mechanism. This corresponds to the situation in Fig. 14.11. However, if $\Delta E_n > (E_{biexc} - E_{exc})$ holds, in a quantum dot containing one exciton, as the pump intensity increases at first enhanced absorption to the biexciton state can be observed but, in the end, this absorption can turn into biexciton gain (see Fig. 14.14(b)). The curves of gain are symmetric in both cases.

The biexciton gain in II-VI quantum dots has a characteristic signature if we apply the VSL method, namely, a specific shape of amplified spontaneous emission I_{ASE} as a function of the stripe length l. The onset of amplification only appears in the experimental curve $I_{ASE} = I_{ASE}(l)$ above a certain *threshold length* of the stripe l_{th}. This can be clearly seen in Fig. 14.15(a) where a thin layer of CdSe nanocrystals shows $l_{th} \approx 0.06\,\text{cm}$ [12]. This information is complemented by Fig. 14.15(b) demonstrating that a narrow band of amplified luminescence at $\sim 635\,\text{nm}$ only starts to emerge from a broad spontaneous emission spectrum for a stripe length exceeding $\sim 0.06\,\text{cm}$. This behaviour stems from the fact that spontaneous and stimulated emissions are of different microscopic origins here. Originally, the VSL method implicitly assumed that spontaneous and stimulated emissions result from the same recombination

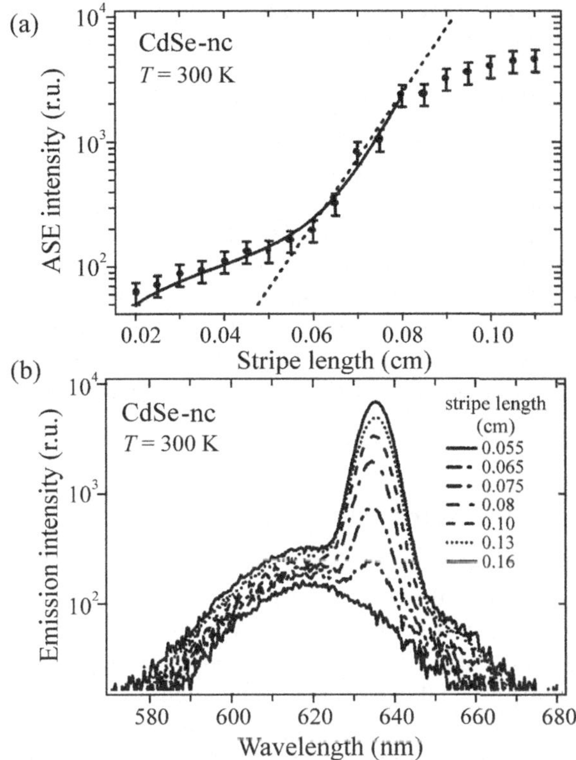

Fig. 14.15
(a) Biexciton gain signature: The ASE intensity at 635 nm as a function of the stripe length (CdSe nanocrystals, $R = 2.5$ nm). Points represent the experimental data; the solid curve is a fit using eqn (14.6); the dashed curve gives the standard fit using relation $I_{\text{ASE}}(l) \simeq I_{\text{sp}}(\mathrm{e}^{Gl} - 1)$. (b) The corresponding emission spectra for various stripe lengths l. Excited by pulses of 3 eV, 100 fs at room temperature. After Malko *et al.* [12].

process and, therefore, we now need to modify expression (10.33) to give it the form

$$I_{\text{ASE}}(l) \simeq I_{\text{X}}l + \frac{I_{\text{XX}}}{G_{\text{XX}}}(\mathrm{e}^{G_{\text{XX}}l} - 1), \tag{14.6}$$

where I_{X} and I_{XX} are constants proportional to the spontaneous emission of excitons and biexcitons and G_{XX} denotes the biexciton modal gain. Because biexcitons have a very short lifetime (a consequence of Auger recombination) their contribution to spontaneous emission is negligible ($I_{\text{XX}}/I_{\text{X}} \ll 1$) and the exponential increase in intensity only reveals itself for longer stripes when the product $G_{\text{XX}}l$ is sufficiently large. An equation similar to relation (14.6) may be used also in the general case when two different gain mechanisms are active.

Our considerations in this section so far have applied to materials with a simple electron band structure—a direct bandgap at the Γ point and only a double spin degeneracy. For semiconductors with multiple degeneracy of the lowest excited state in a quantum dot, the conditions for optical gain will probably be harder since, to achieve population inversion of their levels, one needs to create a higher number of *e–h* pairs which, however, sharply increases the probability of the competing Auger recombination at the same time. Nevertheless, experiments have shown that the situation is far from being hopeless. Lead selenide, PbSe, is a semiconductor with a narrow direct bandgap $E_{\text{g}} \approx 0.28$ eV, located at the L point of the first Brillouin zone. There are four equivalent L points in the first Brillouin zone of the cubic structure which, taking spin into consideration, leads to an eightfold degeneracy of the lowest excited state in a PbSe nanocrystal. Furthermore, the rate of non-radiative recombination—at least in 3D materials—grows with decreasing width of the gap (Section 6.1); therefore, the prospects of using PbSe nanocrystals for lasing seem negligible. In spite of this, amplified spontaneous emission with a high cross-section of optical gain on a wavelength of $\sim 1.5\,\mu$m has been observed in high-quality TiO_2 thin films containing densely packed (filling factor $\xi > 15\,\%$) and suitably surface-passivated nanocrystals of PbSe [13]. The established threshold value of *e–h* pairs of $N \approx 8$ was only twice as high as the theoretically estimated threshold. This firstly indicates the application potential of these nanocrystals for optical communications in the near-infrared region and, at the same time, it also gives good prospects for other materials with higher bandgap degeneracy.

14.3.2 Heterostructures with ordered quantum dots

Quantum wells embedded with self-organized nanocrystals of III-V semiconductors have a special place in the field of lasing in 1D systems. These are optical nanostructures connected with Alferov's Saint Petersburg school; these nanostructures take advantage of the spontaneous growth properties of some quantum dots when the molecular beam epitaxy (MBE) or metalorganic vapour phase epitaxy (MOVPE) methods are applied. This is the so-called heteroepitaxial growth via the Stranski–Krastanow (or similar) method. If, for example, indium arsenide, InAs, is being deposited on an atomic plane of gallium arsenide, GaAs, or of a ternary AlGaAs compound using atomic beams, a thin layer of InAs starts developing (the so-called wetting layer). Taking into account that indium arsenide has a quite different lattice constant (see Fig. 12.3), the lattice mismatch and resulting mechanical stress effects will make the layer form miniature 3D islets. An ensemble of quantum dots of InAs thus grows spontaneously, the dots usually being in the form of a pyramid or a dome.[4] One lets a layer of GaAs overgrow these and repeats the whole procedure. This generates nanocrystals of InAs again, this time positioned over the lower pyramids. Repeating the entire operation multiple times thus yields a 'column' of nested quantum dots of InAs embedded in quantum wells of GaAs. All this is schematically sketched in Fig. 14.16 [14]; a real-life illustration is shown in Fig. 14.17 [15]. Over the sophisticated lithographic growth technique, the method of growing self-organized quantum dots has a clear advantage in its simplicity, its drawback being a broader distribution of quantum dot sizes.

The growth of self-organized nanostructures can be considered as an example of macroscopic order forming spontaneously from an originally random, disordered state. Indium arsenide has a narrower bandgap than GaAs or InGaAs, however the difference in bandgaps between the dots and the well is relatively slight; attention must thus be paid to the dots not being too small,

[4] It is interesting that, originally, crystal growers considered this effect undesirable.

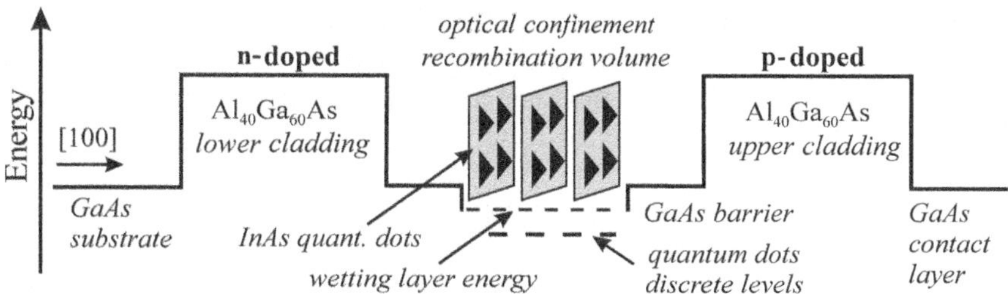

Fig. 14.16
Diagram of a low-dimensional structure with self-organized quantum dots arranged into a 3D chain pointing along the growth direction. The typical surface density of the dots is of the order of $10^{10}cm^{-2}$, and the characteristic lateral dimension of the dots is about 10 nanometres, similar to the separation between the planes on which they reside. The quantum dots are embedded into a quantum well and all this is designed as a p–n junction enabling electrical pumping of stimulated emission. After Bimberg *et al.* [14].

otherwise even a single electron level could not exist there (Section 12.5); self-organized dots, however, mostly do not involve this risk, usually growing to larger dimensions (the minimum diameter of the dots ranges around 5 nm).

As a whole, such a structure involves a range of favourable features that predestine it for lasing and, indeed, it represented a breakthrough in optoelectronics at the end of the last century. First of all, quantum dots prepared in this way are well surface passivated and, thus, potential non-radiative transitions are suppressed. In addition, the quantum-confinement effect here is in principle doubled, which leads to high material optical gain of the order of up to 10^5cm^{-1} and high temperature stability. Next, the relative arrangement of the dots further results in their high effective density which, as we know, is a necessary prerequisite for the occurrence of stimulated emission. And, finally, the semiconductor character of the matrix ensures (unlike nanocrystals randomly dispersed in a dielectric matrix) suitable electrical conductivity of the device, enabling electrical control. In this way, temperature-stable infrared injection lasers with long lifetime and low threshold density of the order of only tens of A/cm^2 have been realized. This is immense progress as compared to the first double heterostructure lasers in the 1970s, in which the threshold current density amounted to up to tens of thousands of A/cm^2. The first electrically driven laser with ordered quantum dots was described in 1994 and the overall development is shown in Fig. 14.18 [16]. Similar heterostructures with quantum dots based on other materials, e.g. (In,Al)GaN–GaN semiconductors, lase at the blue spectral region.

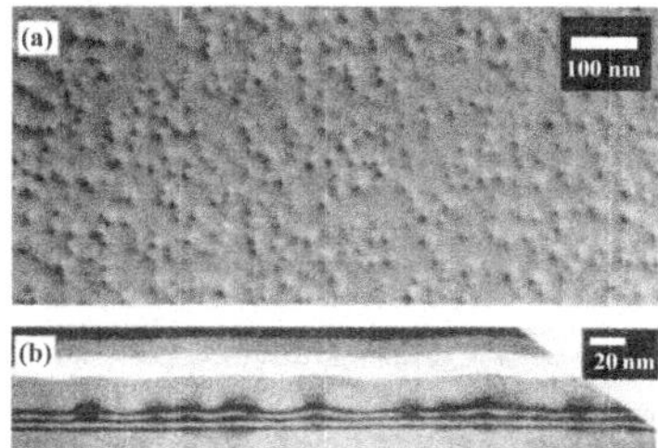

Fig. 14.17
Three layers of self-organized quantum dots of InAs in a GaAs matrix: (a) a surface scan by an atomic force microscope (AFM), and (b) a cross-section of the structure as seen by a transmission electron microscope (TEM). The sample was grown using the MOVPE method at the Institute of Physics of the Academy of Sciences of the Czech Republic in Prague.

The relevant stimulated emission mechanism is not different from the mechanisms discussed in the preceding subsection because the electronic band structure of the III-V semiconductors used here matches up with the concept of a quantum dot as a two-level system with double spin degeneracy; that is, it can involve either an exciton or a biexciton process. Figure 14.19 [17] shows the results of one of the first detailed experimental and theoretical studies of optical gain in an injection laser with quantum dots of InAs in a GaAs quantum well. In order to enable direct comparison of the results with a similar laser structure composed of only a simple quantum well, just a single thin layer of InAs quantum dots was embedded in the studied sample device. Measurements were performed at liquid nitrogen temperature. Panel (a) represents electroluminescence spectra both below and above the laser threshold. Panel (b) shows the results of direct measurements of the modal gain using the VSL method under optical pumping (right scale) along with the material gain determined from the estimated mode confinement factor Γ_M (which is very small here, $\Gamma_M \approx 1.2 \times 10^{-4}$, because the width of the well containing the InAs quantum dots is much smaller than the emission wavelength). The gain curve is quite broad indicating that the stimulated emission process also involves higher excited states of the quantum dots. Finally, panel (c) displays material gain spectra in a simple quantum well without quantum dots but otherwise of comparable parameters. One can see that the material gain at $T = 77$ K is by one order of magnitude lower here as compared to the structure involving quantum dots of InAs (panel (b)), which is a direct consequence of a more efficient quantum-confinement effect in the dots and the lower number of energy states that need to be filled to achieve a given value of the optical gain.

In the monograph [18], the reader can learn further details about heterostructures with quantum dots, e.g. how to prepare and model them, and how to study their optical and laser properties experimentally.

14.4 Random lasing

Usually, a laser consists of two basic elements: an active medium providing stimulated emission and an optical resonator determining the direction of the output radiation, its wavelength, and modal structure. Random lasing is based on an analogous principle yet the modes are not determined by the resonator but by multiple scattering of light instead. Laser emission is then spectrally narrow and coherent but it can generally be directed into a full solid angle of 4π [19].

Multiple scattering is a well-known phenomenon occurring in nature as well as in artificially prepared optical materials that appear opaque or turbid (such as fog, clouds, murky solutions, and the like). Light rays entering such a medium undergo thousands of scattering processes into random directions. This is a

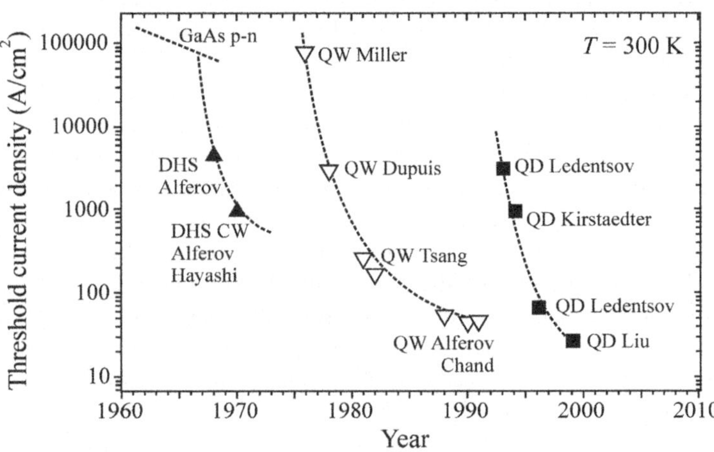

Fig. 14.18
Development of the threshold current density of heterostructure semiconductor lasers since 1960 (DHS – double heterostructure, QW – quantum wells, QD – quantum dots). After Ledentsov *et al.* [16].

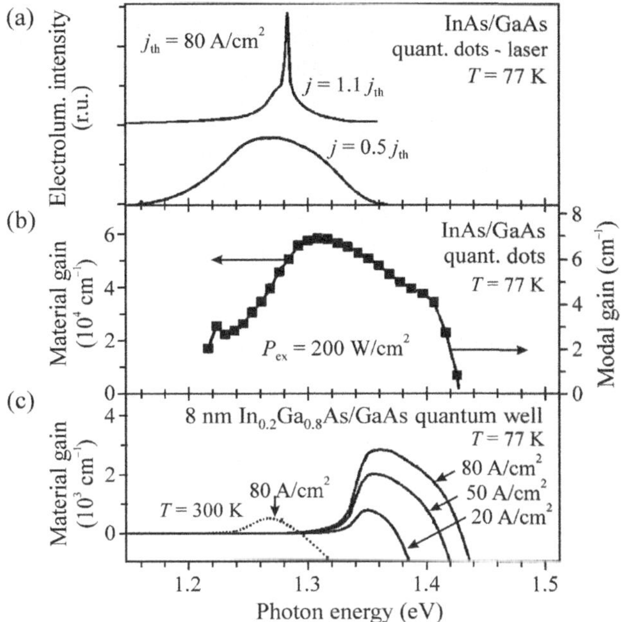

Fig. 14.19
(a) Laser emission and (b) optical gain spectrum in an injection laser with a single thin layer of InAs quantum dots in a GaAs quantum well. Temperature $T = 77$ K. For the sake of comparison, panel (c) shows the optical gain spectra in a separate, simple $In_{0.2}Ga_{0.8}As$ quantum well with GaAs barriers. After Kirstaedter *et al.* [17].

type of 'random walk' used to simulate, for example, the electrical transport properties of disordered materials or the Brownian motion of particles in a liquid. The important point is that light scattering in optical materials is fully coherent. This means that the phase of each optical wave is strictly defined and interference may occur. A phenomenon that we all know and that supports this—although we mostly do not realize it—is the laser speckle: the speckled image we perceive when illuminating a sheet of paper or a wall by a He-Ne laser or a laser pointer.[5]

So far, no sound theory that would describe random lasing in detail exists. It turns out that gain saturation plays an important role here while coherent feedback—represented by a resonator in most cases—is not indispensable. Random lasing has been observed in a multitude of different materials containing a large number of tiny scattering (and, at the same time, amplifying) centres, as schematically shown in Fig. 14.20; these were fine powders of laser crystals or semiconductors, light-scattering luminescent particles in a fluid, and the like. A wide range of prospective low-dimensional semiconductor structures are of interest here. We shall demonstrate this using an example of nanorods or nanoneedles of ZnO [20]. The ZnO nanoneedles arranged in parallel to each other and grown perpendicular to a planar substrate (Fig. 14.21(a)) represent—together with their luminescence emission at 390 nm—a suitable system for multiple scattering.

Fig. 14.20
Multiple light scattering in a medium with optical gain. Spherical luminescent microparticles are excited by an external light source to a state of population inversion; arrows represent scattering and amplification of the emitted radiation. After Wiersma [19], reproduced courtesy of Macmillan Publishers Ltd..

As the ZnO nanocolumns are parallel to each other, one can expect the occurrence of a closed feedback loop of the scattered radiation within the

[5] The common concept of (diffusive) light scattering is a simplified term denoting multiple scattering if we neglect interference phenomena.

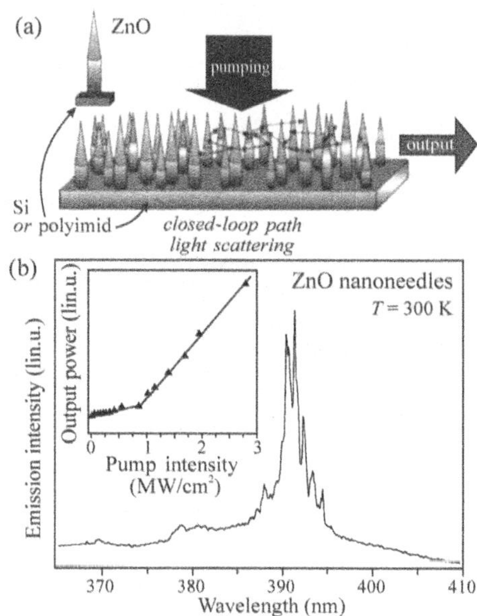

Fig. 14.21
(a) Schematic of an ensemble of ZnO nanoneedles designed to study random lasing. The nanoneedles were grown on a flexible, plastic substrate; their mean diameter is $\sim 100\,\mathrm{nm}$. (b) The emission spectrum excited by nanosecond pulses of 355 nm above the lasing threshold (the threshold pump intensity is $\sim 1\,\mathrm{MW/cm^2}$, see inset). One can clearly recognize the mode structure. Room temperature; after Lau *et al.* [20]. Reproduced courtesy of the American Institute of Physics.

sample plane as is indicated in Fig. 14.21(a). This is thus a special case of random lasing when coherent feedback is employed and the laser emission is restricted to the sample plane.

The emission, the spectrum of which is shown in Fig. 14.21(b), was excited in a fashion similar to the VSL method, that is, by laser pulses focused to a thin stripe by a cylindrical lens. Therefore, Fig. 14.21(b) will also help us to explain the difference between the 'normal' amplified spontaneous emission (ASE) and random lasing. ASE results from the amplification of luminescence radiation by means of stimulated transitions and its spectrum is, in principle, determined by the gain curve of the active material although it naturally gets narrowed at the output, see, for example, Figs 14.13 and 14.15. Notice that the spectra never involve mode structure since the system lacks any resonator. In contrast, with random lasing, the multiple scattering events define optical modes of certain frequencies and spectral profiles, which are very easily discernible in Fig. 14.21(b).

Easy and cheap production ranks among the main benefits of this kind of laser. The idea of an active lasing material, e.g. in the form of a suspension which can be applied as a coating emitting coherent laser radiation with a wide angular distribution, evokes many potential applications not only for illumination purposes but also in medical diagnostics or in military applications. However, the entire field is only in the initial phase of research.

14.5 Problems

14/1. Based on the article [3], discuss the assumptions necessary to proceed from eqn (10.18) to eqn (14.1). Also, discover a serious misprint in the article [3].

14/2. List the experimental criteria enabling us to distinguish the X–LO lasing mechanism (LO-phonon assisted radiative recombination of an exciton) from the X–X mechanism (collisions of excitons) in quantum wells. Proceed from the paper [5].

14/3. Show that, provided a positive optical gain $g(\nu)$ happens in a system of nanocrystals, the expression for the rate of stimulated emission (10.13) reduces to $\tau_{\mathrm{stim}}^{-1} = g(\nu)(c/n)$, where c is the velocity of light in vacuum and n denotes the refractive index. Hint: quantum dots are characterized by a discrete structure of electron energies; thus, modify (10.13) using the atomic relation (10.6) and introduce $N_0 = N_2 - N_1$ as the density of nanocrystals in the state of population inversion.

14/4. Calculate the minimum concentration of nanocrystals (with a radius $R = 2\,\mathrm{nm}$) in a matrix required to achieve stimulated emission if the gain cross-section is $\sigma_{\mathrm{g}} = 10^{-17}\,\mathrm{cm}^2$ and the coefficient β that links the Auger recombination time τ^{A} with R^3 via $\tau^{\mathrm{A}} = \beta R^3$ has a magnitude of $\beta = 5\,\mathrm{ps/nm}^3$. Assume that all the nanocrystals contribute to the gain.

14/5. When studying the optical gain, G, instead of using relation $I_{\mathrm{ASE}}(l) \simeq \exp(Gl)-1$ (see (10.33)), the amplified spontaneous emission I_{ASE} as a function of the stripe length l can be modelled by a somewhat more sophisticated expression

$$I_{\mathrm{ASE}}(l) \simeq \exp[Gl_{\mathrm{a}}(1 - \mathrm{e}^{-l/l_{\mathrm{a}}})] - 1, \tag{14.7}$$

where l_{a} stands for the product of the stimulated emission lifetime with the velocity of light in the amplifying medium [21]. Try to explain the physical meaning of eqn (14.7) and determine which part of the experimental dependence $I_{\mathrm{ASE}}(l)$ is better fitted by (14.7) than by the simple expression (10.33).

14/6. Among a variety of stimulated emission channels discovered in quantum wells, laser action of trions was also observed experimentally. Discuss a detailed mechanism of the stimulated emission $\mathrm{X}^- \to h\nu + \text{electron}$ according to [22], taking notice of similarities (as well as differences) with the process of the excitonic molecule stimulated decay $\mathrm{EM} \to h\nu + \mathrm{FE}$, see Section 10.5.

References

1. Nurmikko, A. and Gunshor, R. L. (1997). *Semicond. Sci. Technol.*, **12**, 1337.
2. Kalt, H. (2004). *(Cd,Zn)Se quantum wells*. In: *Landolt-Börnstein New Series–Group III Condensed Matter, Semiconductor Quantum Structures. Optical Properties* (ed. C. Klingshirn), Vol. 34C2, p. 90. Springer, Berlin.
3. Ding, J., Jeon, H., Ishihara, T., Hagerott, M., Nurmikko, A. V., Luo, H., Samarth, N., and Furdyna, J. (1992). *Phys. Rev. Lett.*, **69**, 1707.
4. Tomašiūnas, R., Pelant, I., Hönerlage, B., Lévy, R., Cloitre, T., and Aulombard, R. L. (1998). *Phys. Rev. B*, **57**, 13077.
5. Kawakami, Y., Hauksson, I., Stewart, H., Simpson, J., Galbraith, I., Prior, K. A., and Cavenett, B. C. (1993). *Phys. Rev. B*, **48**, 11994.
6. Cingolani, R., Rinaldi, R., Calganile, L., Prete, P., Sciacovelli, P., Tapfer, L., Vanzetti, L., Mula, G., Bassani, F., Sorba, L., and Franciosi, A. (1994). *Phys. Rev. B*, **49**, 16769.

7. Mittelstein, M., Arakawa, Y., Larsson, A., and Yariv, A. (1986). *Appl. Phys. Lett.*, **49**, 1689.
8. Zimmler, M. A., Bao, J., Capasso, F., Müller, S., and Ronning, C. (2008). *Appl. Phys. Lett.*, **93**, 051101.
9. Johnson, J. C., Choi, H.-J., Knutsen, K. P., Schaller, R. D., Yang, P., and Saykally, R. J. (2002). *Nature Materials*, **1**, 106.
10. Klimov, V. I., Mikhailovsky, A. A., Xu, S., Malko, A., Hollingsworth, J. A., Leatherdale, C. A., Eisler, H.-J., and Bawendi, M. G. (2000). *Science*, **290**, 314.
11. Klimov, V. I., Ivanov, S. A., Nanda, J., Achermann, M., Bezel, I., McGuire, J. A., and Piryatinski, A. (2008). *Nature*, **447**, 441.
12. Malko, A. V., Mikhailovsky, A. A., Petruska, M. A., Hollingsworth, J. A., Htoon, H., Bawendi, M. G., and Klimov, V. I. (2002). *Appl. Phys. Lett.*, **81**, 1303.
13. Schaller, R. D., Petruska, M. A., and Klimov, V. I. (2003). *J. Phys. Chem. B*, **107**, 13765.
14. Bimberg, D., Grudmann, M., Heinrichsdorff, F., Ledentsov, N. N., Ustinov, V. M., Zhukov, A. E., Kovsh, A. E., Maximov, M. V., Shernyakov, Y. M., Volovik, B. V., Tsatsul'nikov, A. F., Kop'ev, P. S., and Alferov, Zh. I. (2000). *Thin Solid Films*, **367**, 235.
15. Hospodková, A., Křápek, V., Kuldová, K., Humlíček, J., Hulicius E., Oswald, J., Pangrác, J., and Zeman, J. (2007). *Physica E*, **36**, 106.
16. Ledentsov, N. N., Grudmann, M., Heinrichsdorff, F., Bimberg, D., Ustinov, V. M., Zhukov, A. E., Maximov, M. V., Alferov, Zh. I., and Lott, J. A. (2000). *IEEE Selected Topics in Quantum Electron*, **6**, 439.
17. Kirstaedter, N., Schmidt, O. G., Ledentsov, N. N., Bimberg, D., Ustinov, V. M., Egorov, A. Yu., Zhukov, A. E., Maximov, M. V., Kop'ev, P. S., and Alferov, Zh. I. (1996). *Appl. Phys. Lett.*, **69**, 1226.
18. Bimberg, D., Grudmann, M., and Ledentsov, N. N. (1999). *Quantum Dot Heterostructures*. John Wiley, Chichester.
19. Wiersma, D. S. (2008). *Nature Physics*, **4**, 359.
20. Lau, S. P., Yang, H. Y., Yu, S. F., Li, H. D., Tanemura, M., Okita, T., Hatano, H., and Hng, H. H. (2005). *Appl. Phys. Lett.*, **87**, 013104.
21. Chan, Y., Steckel, J. S., Snee, P. T., Caruge, J.-M., Hodgkiss, J. M., Nocera, D. G., and Bawendi, M. G. (2005). *Appl. Phys. Lett.*, **86**, 073102.
22. Puls, J., Mikhailov, G. V., Henneberger, F., Yakovlev, D. R., Waag, A., and Faschinger, W. (2002). *Phys. Rev. Lett.*, **89**, 287402.

Silicon nanophotonics

15

In Section 10.4 we stated that silicon is a material completely unsuitable to become an active laser medium and we also commented on attempts to build empirically a silicon laser, which generally failed.

Ever since the beginning of this century, the effort to fabricate a silicon laser, or at least an efficient silicon-based LED, has intensified. One of the reasons is the so-called interconnection bottleneck threatening integrated silicon circuits. This terms refers to the fact that the ever increasing density of transistors on silicon chips leads to excessive lengths of metal interconnects among them (the overall length of interconnects on a chip exceeds 10 km!). The consequence of such long interconnects can be a slowed-down operation of the integrated circuit and its overheating. Furthermore, one is likely to come across similar problems also in other fields of silicon microelectronics. A possible solution to this problem is to replace at least part of the electrical interconnects with optical ones. In order to realize such a replacement, however, a suitable injection laser, easily integrable onto the silicon chip, is necessary. In an ideal case, the laser or the LED diode would be made of silicon.[1]

A way to sort out this situation which immediately comes to mind is the exploitation of unique properties of silicon nanocrystals, which are, in stark contrast to bulk silicon, very efficient phosphors in the visible spectral range. In this context, one begins to speak of *silicon nanophotonics*. In what follows, we will first focus on the fundamental manifestations of photoluminescence in silicon nanocrystals, then we will discuss optical gain in these nanoparticles and specific properties of active silicon-nanocrystal-doped waveguides and finally we will treat the electroluminescence of an ensemble of silicon nanocrystals.

[1] Alternative approaches are possible, e.g. the incorporation of a III–V semiconductor laser on a silicon wafer (Jones, R., Cohen, O., Paniccia, M., Fang, A. W., and Bowers, J. (2007). *Photonics Spectra*, **41** (January), 54), the generated radiation being distributed by means of a rib Si waveguide surrounded by SiO_2 oxide (so-called silicon-on-oxide technology, SOI).

15.1 Silicon nanocrystals

Efficient room-temperature photoluminescence of silicon nanocrystals[2] was first described by Canham [1] in 1990. Depending on the preparation method, the emission spectrum of these nanocrystals can be situated in nearly any part of the visible region, most frequently, however, red or orange luminescence with the maximum of the band between 650 and 800 nm is reported (Fig. 15.1(a)) [2]. Its decay is relatively slow, in the order of tens to hundreds of microseconds (Fig. 15.1(b)), and it follows a stretched-exponential law

$$I(t) = I_0 \exp\left[-(t/\tau_L)^{\delta}\right]. \qquad (15.1)$$

At the same time, the decay time τ_L decreases with increasing energy of the luminescence photon. Lifetime distribution functions $f(1/\tau_L)$ for wavelengths of 560 and 800 nm are shown in Fig. 15.1(c).

Up to now, the microscopic origin of the luminescence of silicon nanocrystals is not known exactly. It is highly probable, though, that concerted action of several mechanisms comes into play:

- The quantum-confinement effect and the corresponding opening of the bandgap (Section 12.5).
- The small volume of the nanocrystals which, to a large extent, eliminates the non-radiative recombination centres, randomly distributed in the lattice.
- Efficient passivation of the surface of nanocrystals, preventing non-radiative recombination on surface dangling bonds of silicon atoms and, furthermore, constituting possibly a surface-related luminescence channel.
- In the case of very small silicon nanocrystals (1–2 nm) with a strong quantum-confinement effect, a gradual transformation from the indirect bandgap to the direct bandgap as a result of the uncertainty principle $\Delta x \Delta p \geq \hbar/2$ (strong localization in real space $\Delta x \to 0$ leads to delocalization of electrons and holes in reciprocal space $\Delta p \to \infty$, see Fig. 15.2), which implies the occurrence of quasi-direct no-phonon radiative transitions and increases in this way the probability of radiative recombination.

The importance of the individual contributions depends critically on the surface passivation of the nanoparticles. Silicon nanocrystals can be passivated, e.g. by embedding them in a suitable matrix (the most frequent one being SiO_2), but it is also possible to attach chemically various organic species onto the surface of the nanocrystal.

Now we shall mention briefly how the silicon nanocrystals can be fabricated. At present, a large number of methods requiring equipment with various levels of sophistication are available: electrochemical etching of crystalline silicon wafers, pyrolysis of silane SiH_4 combined with chemical post-etching, plasma-enhanced chemical vapour deposition (PECVD) of substoichiometric oxide $SiO_x (x < 2)$ with subsequent annealing at high temperatures $> 1000°C$, electron beam lithography with reactive ion etching and oxidation, implantation of

[2] More precisely speaking, the material in question was so-called porous silicon, which is composed of a large number of interconnected silicon nanocrystals, with sizes ranging between approximately 2 and 6 nm.

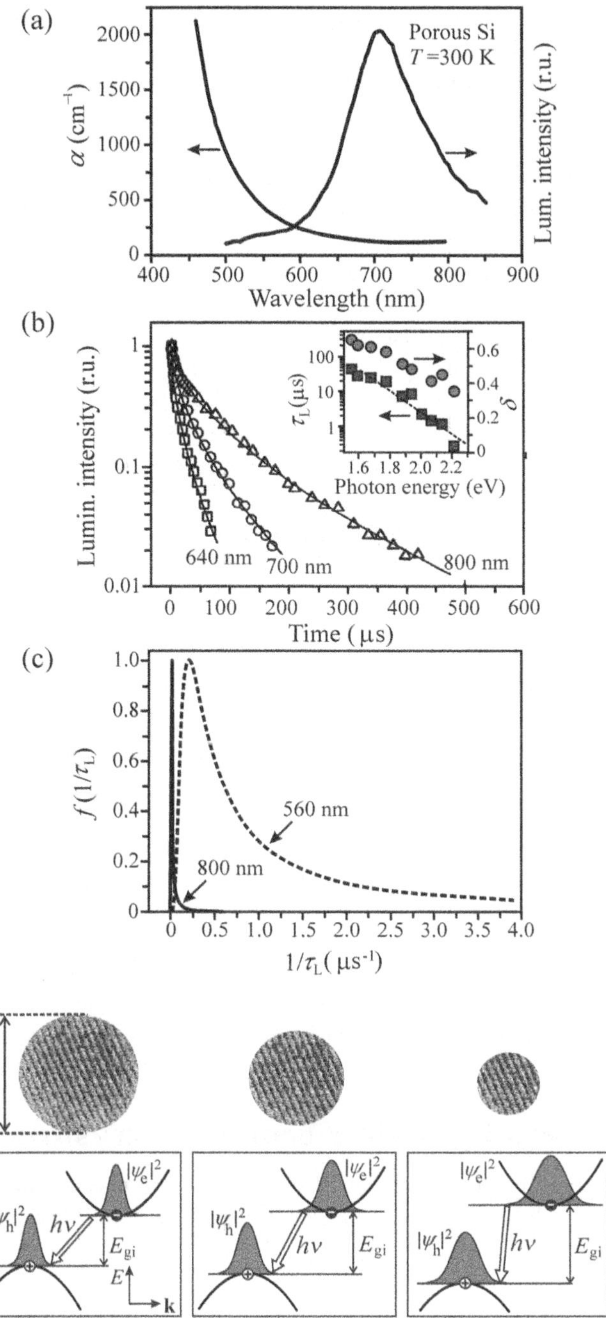

Fig. 15.1
Characteristics of photoluminescence of porous silicon (ensemble of silicon nanocrystals). (a) Absorption and emission spectra; the emission is excited with a 488-nm line of an Ar^+ laser. (b) Decay curves under excitation by picosecond pulses of 532 nm. Solid lines are fits with formula (15.1). (c) Distribution functions of luminescence decay times, determined by Laplace transform of the stretched-exponential function (Section 9.3). Room temperature, after Malý *et al.* [2].

Fig. 15.2
Simplified scheme of a gradual bandgap E_{gi} opening and the transition from indirect to quasi-direct radiative *e–h* recombination with decreasing size of Si nanocrystals. Grey areas represent schematically the probability density of the occurrence of an electron and a hole.

Si^+ ions into the surface region of a quartz slab followed by high-temperature annealing, etc. Two of these methods will be treated in more detail: electrochemical etching (in fact the simplest method) and implantation of Si^+ ions into quartz glass (probably the most demanding one, requiring a big laboratory with sophisticated specialized equipment–large ion implanter).

Electrochemical etching was exploited by Canham in his pioneering work on the surprising luminescence properties of Si nanocrystals [1]. The method consists of using a commercial silicon wafer (with p-type conductivity) as the bottom of a teflon chamber, into which an ethanol-based solution of hydrofluoric acid HF is poured and an inert electrode (usually platinum) is immersed. The silicon wafer is biased to several volts with respect to the inert electrode. Current flowing through the solution then etches narrow pores in the wafer, giving rise to a porous structure. Residual silicon between the pores represents a system of interconnected nanocrystals of size several nanometres. This *porous-silicon* layer is usually several micrometres thick and exhibits intense red/orange photoluminescence under an ultraviolet lamp. Because porous silicon is inherently unstable by itself, the prepared layers are often mechanically removed from the substrate (by means of ultrasonication or pulverization) and the resulting powder containing silicon nanocrystals can be embedded into a solid matrix or form a colloidal dispersion in a liquid environment (water, alcohol, organic solvents).

In the implantation method, the starting material is a polished quartz slab (e.g. Infrasil), exposed to a flux of accelerated Si^+ ions penetrating to a thin subsurface layer, where the ions form an oversaturated solid solution in SiO_2. During the subsequent annealing of the slab to 1100°C, the excess ions precipitate to small nanocrystals of spherical shape, which again exhibit red luminescence under ultraviolet irradiation. The energy of the accelerated ions ranges approximately between 10–600 keV (this parameter drives the penetration depth of ions into the slab) and the overall implant fluences amount to roughly 10^{16}–10^{18} cm^{-2}.[3] Thanks to the discrete energy spectrum of Si^+ ions, the mean value of their penetration depth beneath the SiO_2 surface has only negligible dispersion. Thus, the nanocrystals constitute, just below the surface of the slab, a planar formation with refractive index n higher than that of the surrounding SiO_2 matrix ($n_{Si} \doteq 3.5$, $n_{SiO_2} \doteq 1.45$). This is how the so-called active[4] planar waveguide, exhibiting unique spectral and spatial characteristics of the emitted luminescence, is fabricated. These special luminescence properties will be dealt with in more detail in Section 15.3. The monographs [3–5] offer further reading about specific luminescence properties of Si nanocrystals.

15.2 Optical gain in silicon nanocrystals

As for the possibility of light amplification by stimulated emission, silicon nanocrystals exhibit several features significantly different from those of bulk silicon:

- the high value of the photoluminescence quantum efficiency η_{PL} at room temperature (values in the literature lie in the range $\eta_{PL} = 5$–80% depending on the preparation method and surface passivation);

[3] Interestingly, in order to attain, e.g. a fluence of 6×10^{17} cm^{-2}, the implantation process needs to continue ceaselessly for three (!) days.

[4] 'Active' in this sense means that light is not coupled to the waveguide from an external source but, instead, is generated inside the waveguide.

- bandgap opening with shrinking size of nanocrystals (quantum-confinement effect) from $E_{gi} = 1.1\,\text{eV}$ up to $E_{gi} = 2.5$–$3\,\text{eV}$ (again, the bandgap width strongly depends on surface passivation);
- the corresponding shift of energy of luminescence photons from infrared to the visible spectral region, i.e. shortening of the luminescence wavelength λ, leads to a decrease in the free carrier absorption coefficient, because it follows from (10-21) that $\alpha_{FCA} \sim \lambda^2$;
- gradual transition from indirect to direct recombination with shrinking size of nanocrystals which, most of all, results in an increase of the radiative recombination rate $1/\tau_r$ and, consequently, according to (10–12) also in an increase of the optical gain coefficient $g(\nu)$. In the limit case, this effect can be further enhanced by lifting the conduction band degeneracy (six equivalent minima close to the X point).

It is obvious that all of these steps are heading in the 'right direction', that is towards a higher probability of stimulated emission in silicon nanocrystals. The progressive decrease in α_{FCA} accompanied by the increase in $g(\nu)$ during shrinking of the nanocrystal should lead to a situation when the inequality $g < \alpha_{FCA}$, valid for bulk silicon, inverts. Thus, one should obtain a positive net optical gain $g - \alpha_{FCA} > 0$.

The foregoing discussion, however, has been intentionally simplified: it does not account for the influence of non-radiative Auger recombination. The rate of this recombination will also increase with increasing level of localization of the excited *e–h* pairs in a nanocrystal, i.e. with its shrinking size, as was explained in Subsection 14.3.1. The condition for the onset of stimulated emission is then determined by the critical competition between the rate of Auger recombination $(\tau^A)^{-1}$ and that of stimulated emission (τ_{stim}^{-1}), and can be described by relation (14.5). We interpreted this relation in such a way that it is a transparent matrix of high optical quality with densely packed nanocrystals, accompanied with as strong pumping as possible, that are required in order to achieve positive optical gain. Silicon is no exception to this rule. In addition to these 'standard' requirements, in silicon nanocrystals sizes as small as possible are favourable (because of short-wavelength emission, conditional on increasing weight of the quasi-direct transitions). Moreover, surface passivation of high quality is a must, as it prevents non-radiative surface recombination and consequently boosts the luminescence quantum efficiency, the last condition favouring stimulated emission being a size distribution as narrow as possible.

Detailed theoretical calculations or predictions do not exist at present. The only possible way to assess the occurrence of optical gain in Si nanocrystals thus lies in the preparation of a system of nanoparticles which would satisfy the above criteria as closely as possible, and in subsequent experimental investigation of optical gain. The VSL method, possibly combined with the SES method (Subsection 10.6.1), is the most frequent way to measure such a gain. Figure 15.3 [6] serves as an example of the results of such a measurement. The distinct (positive) deviation of the VSL curves from the integrated SES curves at high excitation fluxes indicates the occurrence of stimulated emission. Results very similar to this one were obtained also in other laboratories (e.g. [7]); it is still not clear, though, whether the experiment is not incorrectly

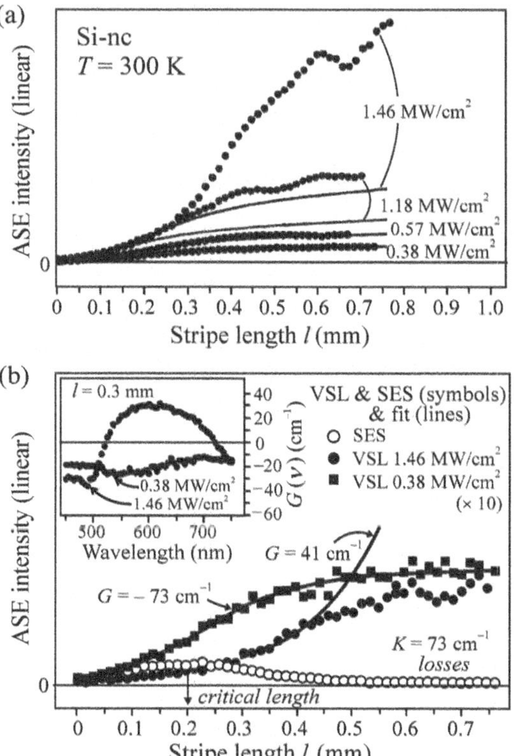

Fig. 15.3
(a) VSL (symbols) and integrated SES (solid lines) curves for silicon nanocrystals embedded with density of $\sim 10^{18}\,\mathrm{cm}^{-3}$ in an SiO_2 matrix. The increasing departure of the VSL points from the integrated SES curves with increasing pump intensity indicates the onset of stimulated emission, i.e. positive net optical gain. (b) Detail from panel (a) for the lowest ($0.38\,\mathrm{MW/cm^2}$) and highest ($1.46\,\mathrm{MW/cm^2}$) pump intensities, demonstrating the 'switching' from negative to positive gain. At the same time, a critical length for minimization of artefacts of the VLS method is defined here. The inset represents the gain spectrum obtained by means of eqn (10.34). Room temperature, excitation with a pulsed XeCl laser (308 nm, $\sim$ 15 ns). After Dohnalová [6].

interpreted; for the time being, no one has succeeded in the fabrication of a real laser, i.e. a resonator with positive feedback containing silicon nanocrystals as an active medium.

15.3 Active planar waveguides made of silicon nanocrystals

A polished quartz slab exposed to Si^+-ion implantation and subsequent annealing contains a layer of densely packed silicon nanocrystals just beneath the surface. An example of the depth in which the nanocrystals are embedded and of the profile of their density (also representing the profile of the refractive index of the layer) is shown in Fig. 15.4. The nanocrystalline layer then constitutes a planar optical waveguide.

When the photoluminescence of Si nanocrystals is excited inside the waveguide, a surprising result appears in some samples: the emission spectra collected from the surface of the sample (directed along the surface normal) significantly differ from those collected from the edge of the sample, see Fig. 15.5(a) [8]. The spectra of the radiation coming out of the polished edge consist of a 'doublet' comprising two relatively very narrow lines. Furthermore, the lines are linearly polarized, perpendicular with respect to each other: the shorter-wavelength line has the electric field vector $\mathscr{E}$

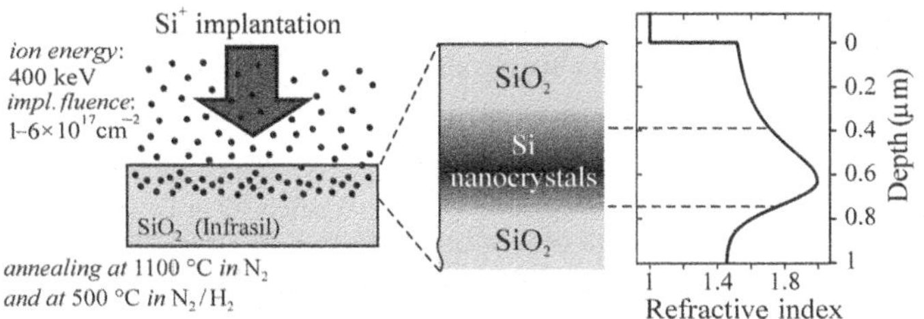

Fig. 15.4
Schematic of implantation of Si^+ ions into a quartz slab (on the left), cross-section of a layer of Si nanocrystals prepared in this way close to the surface (in the middle), and the corresponding profile of refractive index (on the right).

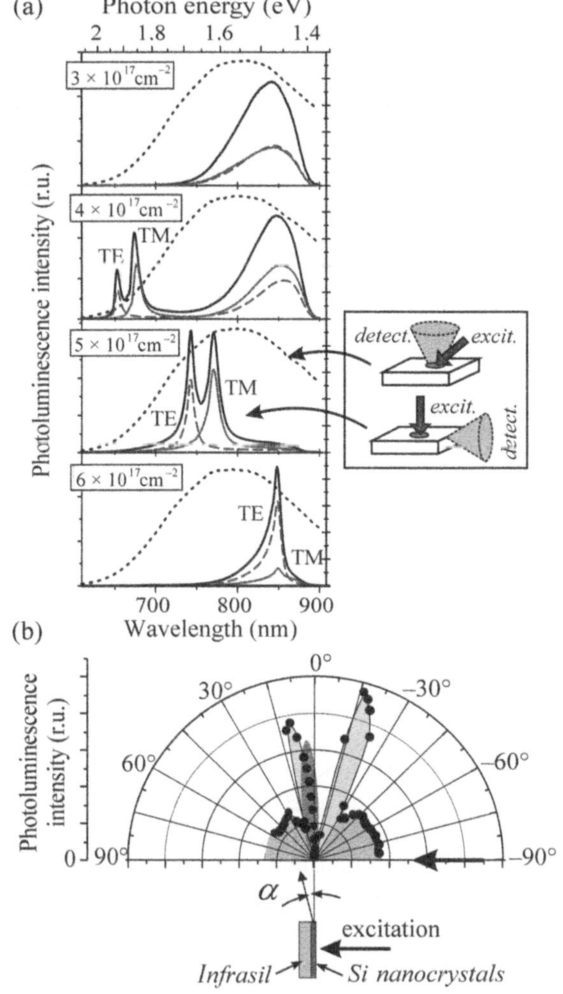

Fig. 15.5
(a) Emission spectra of silicon nanocrystals constituting a planar waveguide. Spectra of four different samples are shown, the voltage accelerating the Si^+ ions during implantation was 400 kV and the corresponding implant fluences are shown by the spectra. (b) Polar diagram of spectrally integrated emission as a function of the angle related to the plane of the sample. A TE/TM doublet is characterized by a sharp lobe at $\alpha \approx 4°$. Room temperature, cw excitation from a HeCd laser, 325 nm. After Pelant *et al.* [8].

parallel to the nanocrystalline layer (TE polarization), whereas $\mathscr{E}$ of the longer-wavelength line lies in the plane perpendicular to the nanocrystalline layer (TM polarization).[5] The doublet as a whole systematically shifts with

[5] The order can be reversed, depending on the particular profile of refractive index across the nanocrystalline layer.

increasing implant fluence towards the longer-wavelength side of the spectrum. Moreover, the doublet appears to be strongly direction-dependent: it can be observed only in a direction close to the plane of the nanocrystalline layer (near $\alpha = 0°$, see Fig. 15.5(b)). On the other hand, luminescence coming out from the surface is not polarized and is characterized by a broad spectrum that is very similar to other systems of silicon nanocrystals (the reader can compare Figs 15.5(a) and 15.1(a)). The position and the shape of the band are practically independent of the implant fluence.

The presence of these spectral features suggests that the luminescence is strongly influenced by waveguiding effects when collected from the edge of the sample.[6] Theoretical simulation together with auxiliary experiments proved this hypothesis. At the same time, however, they showed that these modes—rather than being identified with the standard guided modes of the waveguide—represent so-called *radiative substrate modes*. They are characterized by undergoing a total internal reflection on the upper interface (waveguide/air)—see Fig. 15.6—whereas they impinge on the bottom interfaces at an angle slightly below the critical angle θ_c for total reflection. This means that they are still efficiently reflected back to the waveguide; nevertheless a small part of the radiant power is decoupled from the waveguide and propagates through the substrate, in the direction nearly parallel with the plane of the waveguide. After undergoing a large but finite number of reflections, i.e. after propagating over a certain distance, all the power of the mode 'leaks' to the substrate and in the end it leaves the edge of the sample at a small angle α (Figs 15.6 and 15.5(b)).

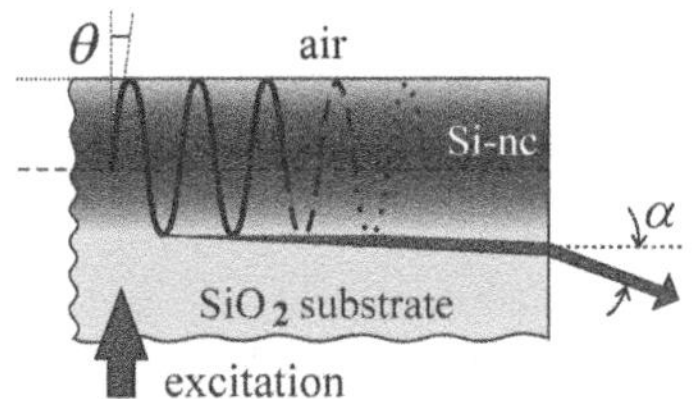

Fig. 15.6
Schematic of radiative substrate modes, manifesting themselves in silicon-nanocrystal-based waveguides.

A glass slab with such an active waveguide thus can, to a certain extent, remind one of a Lummer–Gehrcke plate, used in high-resolution spectroscopy. In this context, it is not surprising that significant narrowing of the emission spectrum of nanocrystals occurs. The splitting into the doublet is then owing to a slightly different phase shift for the TE- and TM-polarized luminescence components, resulting from the total reflection on the upper interface (as stems from the Fresnel equations). It is thus obvious that in order to obtain the resulting effect—the narrow emission doublet—asymmetry of the waveguide is necessary because both the upper and the lower interface come into play, each of them having a different function.

In textbooks on optics, radiative substrate modes are usually considered parasitic and undesirable effects; as a result, they are treated only superficially. Here, the question arises why their manifestation is so prominent just in the case of silicon-nanocrystal-based waveguides. The reason lies in significant losses due to Mie-type scattering of light, occurring in the waveguides under discussion. In this type of scattering, light scatters on inhomogeneities featuring sizes comparable with the order of the luminescence wavelength. (Silicon nanocrystals, much like nanoparticles made of other semiconductors, tend to

[6] It might have occurred to the reader that these characteristics (spectral narrowing, directionality of radiation) are a proof of stimulated emission. Detailed experiments, however, confirmed that stimulated emission is not present. On the contrary, these samples turned out not to be suitable for optical gain, because they have relatively long-wavelength luminescence and considerable losses due to light scattering.

form larger aggregates.) Power transmitted by the 'true' guided modes is thus gradually scattered all around the waveguide and, consequently, lost for further exploitation. The strongly directional substrate radiative modes are eventually the only utilizable part of the output radiation.

This effect definitely has application potential in spite of the losses in the waveguide; after all, they are not extremely high (about 40–50% of the excited radiant power leaves the waveguide as the substrate modes). The effect can be exploited, e.g., for selective tuning of the wavelength in silicon photonics or for sensoric applications.

15.4 Electroluminescence of silicon nanocrystals

So far, our discussion on silicon nanocrystals has covered their photoluminescence. In order to be commercially applicable, however, the final product has to be a silicon-nanocrystal-based light source working on the electroluminescence principle, not under optical pumping. Electroluminescence from a system of silicon nanocrystals might seem at first glance a hardly realizable feat, because luminescent Si nanocrystals are usually embedded into a dielectric matrix, most frequently SiO_2. It will thus not be trivial to ensure the supply of excitation electrical energy into nanocrystals; despite that, various types of silicon-nanocrystal-based electroluminescence structures have already been realized in the laboratory.

The first attempts to fabricate such electroluminescent structures date back to the time shortly after the discovery of porous silicon photoluminescence (1990). They proved that electroluminescence, although weak and unstable in time, could be observed under various conditions, in experiments based either on a simple contact of porous silicon with a metal electrode (Schottky effect) or on a porous p-n junction (see Fig. 11.12(b)). Although these results were very promising, insufficiently defined samples together with the complexity and instability of the porous-silicon structure itself made the interpretation of the results considerably difficult.

A bit later, another type of structure started to be prepared: thin SiO_2 films with embedded nanocrystals for the metal-oxide-semiconductor (MOS) type of electroluminescent structure. In this case, a thin film of SiO_2, grown on a crystalline silicon substrate and containing Si nanocrystals, is prepared and a semitransparent electrode is then deposited on top of the thin film. These structures, which can be easily biased in order to study the emission of light, are much better defined than the 'as grown' porous silicon. Experiments showed that if the oxide layer with nanocrystals is thick (100 nm and more), it is necessary to apply a relatively high voltage (tens to hundreds of volts) to excite electroluminescence. Injection of free carriers into the film then takes place through Fowler–Nordheim tunnelling, charge is transported through the oxide via hot electrons and these generate electron–hole pairs in nanocrystals via impact excitation. Therefore, one deals essentially with high-field electroluminescence. The flow of hot electrons through the oxide then naturally leads to rapid degeneration of the device, as was confirmed by a large number of experiments.

It has thus turned out favourable to realize a system of silicon nanocrystals suitable for the injection type of electroluminescence. Nevertheless, the question arises as to whether it is actually possible to prepare a standard p-n junction in a system of isolated nanometre-sized silicon nanocrystals. Taking into account what was presented in Section 11.3, one comes to the conclusion about the impossibity of the task because a continuous macroscopic slab of material is required in order to create a depletion layer and related diffusion potential. Consequently, another technique to inject electron–hole pairs in a defined way had to be found. With progressive development in the nanotechnology field, the effort focused mostly on structures with a thin (~10 nm) oxide film, making it possible for such injection to be carried out in a defined way. This mechanism has the real prospect of meeting the requirements for a commercial device in the future, which means low power consumption, high modulation frequency and long durability. Now, we present an example of one particular device.

This device, shown in Fig. 15.7 [9], is based on a design very similar to the MOS field-effect transistor (MOSFET). An SiO_2 thin film (only 15 nm thick) thermally grown on a floating gate contains a layer of Si nanocrystals prepared by ion implantation. A semitransparent polycrystalline silicon electrode is deposited at the top. If a positive voltage is applied to the gate, electrons flowing in the conductive channel between the source S and drain D are injected via Fowler–Nordheim tunnelling to the nanocrystals (Fig. 15.7(a)). After an abrupt change of the voltage polarity, holes are injected into the nanocrystals (Fig. 15.7(b)), also through the Fowler–Nordheim tunnelling

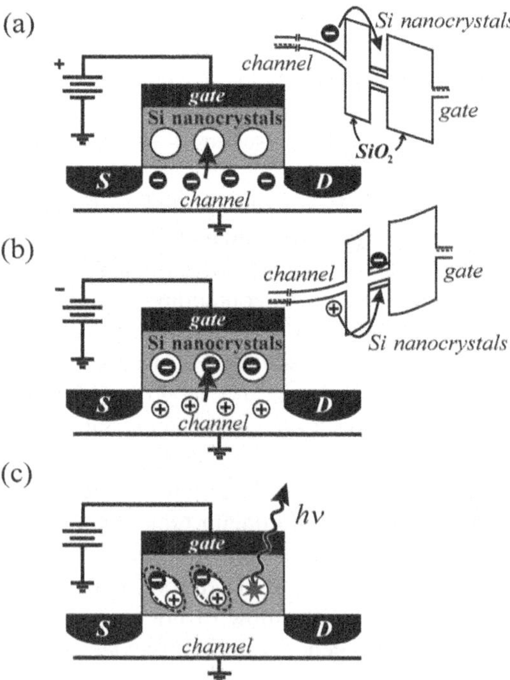

Fig. 15.7
Schematic of the electroluminescence mechanism in a MOS transistor structure with silicon nanocrystals. The injections of (a) electrons and subsequently (b) holes takes place via Fowler–Nordheim tunnelling. (c) The next step involves radiative recombination of *e–h* pairs (excitons) in nanocrystals. After Walters *et al.* [9]. If only a single type of carrier is injected, the device can operate as an optical memory.

mechanism, which is this time enhanced by Coulomb interaction.[7] Finally, the excited nanocrystals, containing one electron–hole pair, emit luminescence photons (Fig. 15.7(c)). In this case, a gradual (sequential) injection of carriers takes place, in contrast to simultaneous bilateral injection in a conventional p–n junction. The authors suggest that this type of luminescence be called a field-effect light-emitting device (FELED).

Figure 15.8 depicts a comparison of photoluminescence and electroluminescence spectra, confirming that the device is based on the radiative recombination taking place in silicon nanocrystals. Temporal traces of the electrical excitation pulse and of the relevant electroluminescence response are shown in Fig. 15.9. Luminescence can be detected only when the polarity of the applied voltage changes, which agrees with the above mentioned mode of sequential injection. A very important point here is that all the operations applied during the fabrication of this device are compatible with silicon CMOS technology. The device exhibits sufficient durability ($> 5 \times 10^9$ cycles); its frequency bandwidth is quite low, however, amounting only to about 30 kHz. Ways to increase it are under investigation (e.g. [10]).

The foregoing discussion has shown that the injection of charge carriers into nanocrystals is by no means a simple task as numerous technical obstacles can be encountered on the way. Surprisingly, it turned out that when these obstacles are overcome, the electrical excitation of silicon nanocrystals can be, in a way, more effective than optical excitation. This can be proved by using the so-called excitation cross-section σ, which expresses the effectiveness of the luminescence excitation in nanocrystals.[8] The cross-section σ satisfies the relation (see Problem 15/2)

$$\frac{1}{\tau_n} = \sigma\phi + \frac{1}{\tau}, \tag{15.2}$$

where τ_n is the time characterizing the transient rise of luminescence after the excitation is switched on, ϕ is the flux of excitation photons (in the case of photoluminescence) or electrons (in the case of electroluminescence) and τ stands for the luminescence decay time. According to (15.2), the luminescence rise time τ_n shortens with increasing excitation intensity ϕ; one can then estimate σ from the slope from an experimentally acquired plot $(\tau_n)^{-1} = f(\phi)$. Figure 15.10 shows the results of relevant electroluminescence experiments

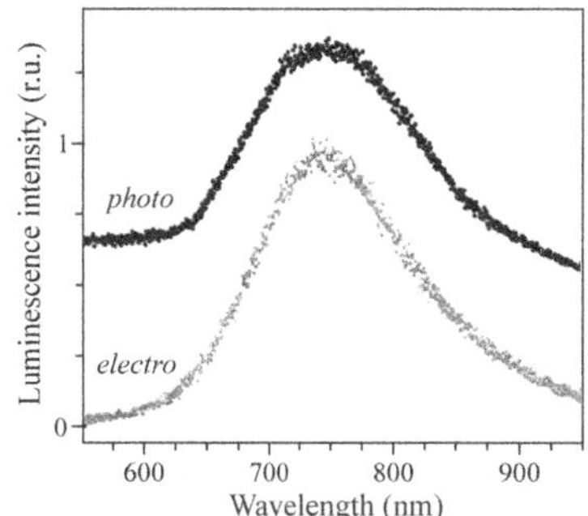

Fig. 15.8
Emission spectra of photoluminescence (upper curve) and electroluminescence of the device shown in Fig. 15.7. Room temperature, after Walters *et al.* [9].

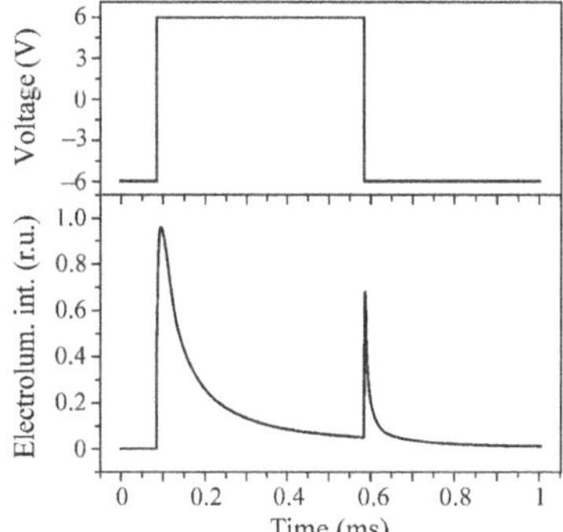

Fig. 15.9
Electroluminescence dynamics of the device from Fig. 15.7 (at the bottom) and the corresponding time development of the gate voltage (at the top). After Walters *et al.* [9].

[7] The carriers tunnel through an oxide layer only ~4 nm thick, which minimizes the undesirable effects of hot-carrier transport. Attractive interaction helps also to remove the so-called Coulomb blockade, which otherwise may complicate the flow of charge carriers through a system of nanocrystals. The Coulomb blockade results from an increase in electrostatic energy of a 'capacitor' represented by the nanocrystal–contact system due to an electron with charge e having tunnelled into the nanocrystal. If C stands for the capacity of the nanocrystal–contact system, then the increase in energy due to nanocrystals charging is $e^2/2C$. Because of the negligible value of $C(\approx 10^{-19}$ F), the increase in energy leads to the formation of a high potential barrier (the order of volts), which makes the charged nanocrystal represent high electric resistance for further carriers.

[8] The excitation cross-section of nanocrystals σ generally differs from their optical absorption cross-section σ_{abs} (Section 13.6) because σ_{abs} can comprise possible transfer of excitation energy to other luminescence centres in the matrix, see Section 15.5. If this is not the case (and if the absorption cross-section is measured by means of the subsequent nanocrystal luminescence with efficiency η), one may consider, in the case of photoluminescence, $\sigma = \eta\sigma_{abs}$.

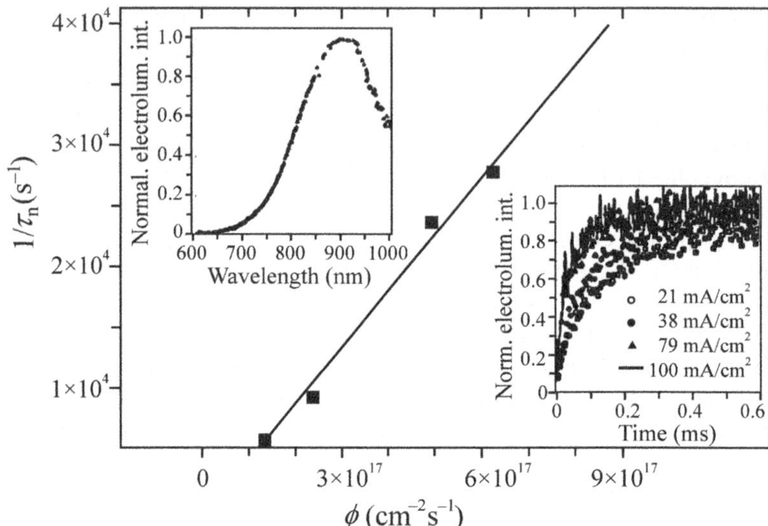

Fig. 15.10
Reciprocal value of the electroluminescence rise time of Si nanocrystals $(\tau_n)^{-1}$ as a function of the excitation electron flux ϕ. The upper inset represents the emission electroluminescence spectrum, and the lower inset then depicts the curves of the onset of electroluminescence in time under different current densities $j = e\phi$. Room temperature, after Irrera *et al.* [11].

performed on samples of MOS structures in which the oxide layer contained Si nanocrystals [11]. The excitation was carried out using rectangular pulses with repetition rate of 55 Hz. The onset time τ_n really shortens with increasing current density, decreasing from the value of 175 μs to 36 μs. The slope of the linear plot in Fig. 15.10 then yields the electroluminescence excitation cross-section of $\sigma_{EL} \approx 4.7 \times 10^{-14}\,\text{cm}^2$. The authors performed an analogous experiment on the same sample under optical pumping (488-nm line from an Ar^+-ion laser) and obtained a value of the photoluminescence excitation cross-section of $\sigma_{PL} \approx 1 \times 10^{-16}\,\text{cm}^2$. Thus, the process of electrical excitation was found out to be by two orders of magnitude more effective when compared to optical excitation, at least for the investigated type of samples

Finally, it is certainly noteworthy that the device in Fig. 15.7 can in principle operate also as a nanocrystalline memory with optical readout. Let us ask what happens if one type of carrier only, e.g. electrons, is injected into the gate. They can be trapped in a potential well (represented by the nanocrystal) for a relatively long time because the barriers of the surrounding oxide are high. This state can be denoted for example as a logical zero; the readout can be carried out by measuring the intensity of photoluminescence excited by an external source. The intensity of photoluminescence in the state when nanocrystals are not charged will then be different; it can serve as the logical one. In this way, logical binary Hi-Lo information on the state of the memory element can be obtained.

15.5 Silicon nanocrystals combined with Er^{3+} ions

Up to now, the luminescence we were discussing in this chapter was spectrally situated in the visible region of the optical spectrum. In silicon photonics, however, wavelengths in the near-infrared region, particularly around 1.3 and

1.5 μm, attract considerable attention. The reason lies in the fact that the best properties—in terms of optical signal dispersion and attenuation—of glass (SiO_2) optical fibres, widely used for the transfer of the signal in telecommunication and cable TV, are situated in this spectral region. In particular the 1.54 μm wavelength falls at the absolute minimum of the glass fibre attenuation and the transmitted optical signal thus achieve maximum range.

Naturally, the idea to exploit this wavelength in the near future also for optical interconnects on the board-to-board or even chip-to-chip level arises. Thus, a unique fusion of optoelectronics and microelectronics could be based on a single material—silicon. The scenario that one could, some time in the future, connect his or her computer through an optical connector to a world-wide optical communication network with very fast access to data and high transmission rate is not totally unrealistic. As the 1.54-μm radiation is not absorbed in silicon, this infrared signal could be guided and distributed inside a computer in the silicon chip itself. Tiny silicon rib waveguides with cross-sections of $\sim 10 \times 10\,\mu m^2$ could be prepared with this objective in mind; such optical components, surrounded by SiO_2 oxide, would then resemble very closely nearly ideal channel waveguides ($n_{Si} > n_{SiO_2}$). All this could be realized in the framework of the SOI approach, which is compatible with the current CMOS technology of Si integrated circuits, being doubtlessly advantageous.[9]

How can one fabricate a silicon-based light source emitting the 1.54-μm wavelength? The answer to this question has, in principle, been well known for quite a long time: dope the crystalline lattice of silicon with erbium Er^{3+} ions. Rare earth ions embedded to almost any matrix maintain nearly unaltered their optical properties thanks to the screening of the optically active partially filled electron shell from the influence of the surrounding matrix (Section 5.6). Thus, the Er^{3+} ions emit their inherent luminescence at the 1.54-μm wavelength, originating in the transition from an excited atomic-like $^4I_{13/2}$ state to the ground $^4I_{15/2}$ state even inside a crystalline silicon matrix. If, in addition, silicon with a 'suitable' electric conductivity is chosen as the matrix, it may be feasible to prepare an electroluminescent p-n junction, in which the excitation energy is transferred from injected electron–hole pairs to the Er^{3+} ions (Fig. 15.11(a)). It thus seems that we have succeeded in finding the 'ultimate' silicon-based light source. Unfortunately, we did not.

Here, one of several problems lies in the fact that the solubility of erbium ions dispersed in a crystalline silicon matrix is low (around 10^{17} cm^{-3}). Above this threshold, the ions start to form clusters, causing quenching of the photoluminescence. Futhermore, backward transfer of energy from the Er^{3+} ions to the host matrix takes place, see Fig. 15.11(a) [12]. Luminescence of the Er^{3+} ions in crystalline silicon is, consequently, too weak to be exploited for a viable light source. In this context, a:Si-H seems to be a better candidate for the matrix; however, other problems arise (degradation of the device).

[9] On the other hand, purely silicon-based photodetectors could not be employed in this case. Most probably, a SiGe-based alloy would be the replacement of choice, possibly posing a problem with lower quality of alloy detectors.

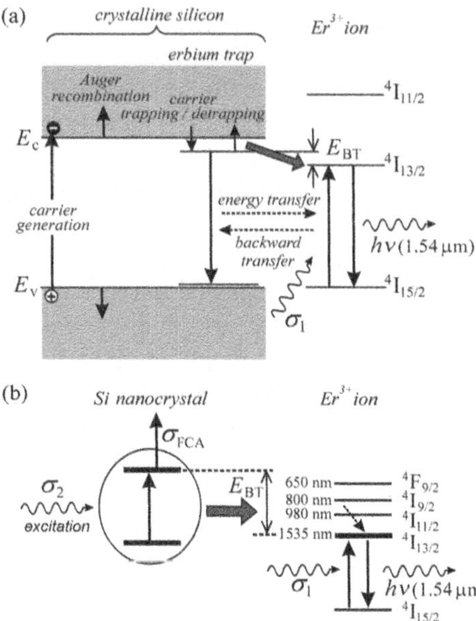

Fig. 15.11
(a) Energy band structure of crystalline silicon with the excitation energy transfer to embedded erbium ions. The quantity E_{BT} represents the activation energy of the backward transfer. After Kenyon [12]. Excitation cross-section σ_1 of direct excitation of the Er^{3+} ion itself is also depicted. (b) Schematic of the process of excitation of the Er^{3+} ions via energy transfer from silicon nanocrystals; both the nanocrystals and erbium ions are embedded in an SiO_2 matrix. $\sigma_2 \approx 10^5\sigma_1$ stands for the overall excitation cross-section of the radiation at 1.54 μm. After Pavesi [13].

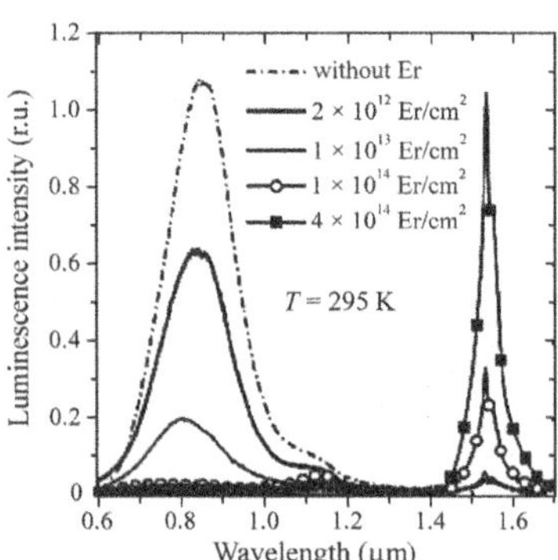

Fig. 15.12
Emission photoluminescence spectra of a 'mixture' of silicon nanocrystals and erbium ions in SiO_2. As a result of the efficient Si nanocrystals → Er^{3+} ions energy transfer, the intensity of the red nanocrystal luminescence (~ 0.8 μm) decreases with increasing concentration of Er^{3+} ions, whereas the intensity of the emission from these ions (~ 1.54 μm) grows. Room temperature, excitation with an Ar^+-laser 488 nm. After Franzò *et al.* [14].

A quartz glass SiO_2 matrix offers better conditions for the Er^{3+} ions: the luminescence spectrum remains unaffected, the concentration of Er^{3+} ions can be considerably higher and the system does not degrade. These advantages have in fact been exploited for a long time in practice, namely in optical fibre amplifiers for optical communications. Such amplifiers are, however, unsuitable to be employed for photonics on the personal-computer level, in principle for two reasons: (a) the Er^{3+} ion luminescence has to be excited with high-power lasers because the excitation cross-section of the Er^{3+} ions is low and (b) such a system cannot be effectively electrically driven.

This is where—finally—silicon nanocrystals come into play: they act as very efficient sensitizers of the erbium-ion luminescence. If SiO_2 oxide is co-doped with both Si nanocrystals and Er^{3+} ions, several noteworthy effects can be observed:

- The nanocrystals are efficient absorbers of both the visible and near ultraviolet excitation radiation and they transfer this accumulated energy to the Er^{3+} ions very efficiently (~70%) and quickly (with a transfer time constant $\tau_{tr} \sim 1\,\mu s$), see Fig. 15.11(b) [13]. In this way, the photoluminescence excitation cross-section of Er^{3+} ions σ_1 can be increased from 10^{-21} cm^2 to ~ 10^{-16} cm^2, by five orders of magnitude!
- Energy back-transfer of the excitation energy from Er^{3+} to the nanocrystals does not occur.
- The luminescence band typical for Si nanocrystals (at 700–800 nm) nearly entirely disappears, see Fig. 15.12 [14].

The high efficiency of the energy transfer (Si nanocrystals)→ (Er^{3+} ions) and negligible probability of the back-transfer stem from the fact that the offset of the energy levels of an Er^{3+} ion with respect to the silicon nanocrystal substantially differs from the case of crystalline silicon. The back-transfer activation energy E_{BT} is much higher in the case of nanocrystals and, as a result, it is much more difficult to thermally activate the back-transfer at room temperature.

Now, an important question arises whether positive optical gain can be achieved in the system (silicon nanocrystals + erbium) at the 1.54-μm wavelength, as this is a prerequisite for the development of a laser in the future. Since this issue is currently hotly debated and still open, we shall confine ourselves to a short comment only. While optical gain was experimentally observed by some authors, others claim the contrary. Naturally, there are loss factors that counteract the signal amplification (as always and everywhere) at 1.54 μm, among the most significant ones belonging 'free' carrier absorption in Si nanocrystals (see the absorption cross-section σ_{FCA} in Fig. 15.11(b)). Perhaps most importantly, an appropriate proportion between the concentration of silicon nanocrystals and the Er^{3+} ions is regarded here as the key problem. If the number of Er^{3+} ions per excited nanocrystal is too high, the extra ions will not be excited and will hamper optical gain via absorption transitions from the ground state $^4I_{15/2} \rightarrow {}^4I_{13/2}$. If, on the other hand, the number of Er^{3+} ions is too low, the resulting weak optical amplification will not overcome the losses due to free carrier absorption in nanocrystals. In any case, only low values of optical gain (several cm^{-1}) have been measured so far. All the effort is aimed at increasing the gain values by optimizing the technology of sample preparation.

15.6 Biological applications of silicon nanocrystals

Nanoparticles applied in the human body and living organisms have promising application prospects in both diagnostics and therapy. It turns out that they can, for example, be loaded with a drug and thus exploited for targeted drug delivery via the bloodstream to a specific part of the body. If these nanoparticles exhibit luminescence, they can be employed as both drug delivery agents and fluorescent markers in order to monitor the drug's propagation through the organism as well as its release in the target. Both nanoparticles of organic molecules and semiconductor nanocrystals have been put forward as potential candidates for this kind of *in vivo* monitoring and imaging. Important criteria these candidates need to meet are as follows:

- Their emission spectrum has to fall into the visible spectral range, it must be easily distinguishable from the autofluorescence of the living tissue ('background') and should propagate through the tissue without substantial attenuation.
- They must not be toxic.
- They have to persist in their active state long enough to fulfil their purpose; afterwards, however, it is advantageous if they decay spontaneously, i.e. undergo biodegradation.

The first candidate for a luminescent drug-delivery agent that comes to mind are obviously nanocrystals of II-VI semiconductors: by choosing a suitable size and surface passivation of the nanoparticles, it is not difficult to achieve a particular emission wavelength, the luminescence intensity being definitely sufficient. Recent studies, however, show that their toxicity is not, due to the presence of heavy metals, negligible. On the other hand, the latest results of experiments carried out with luminescent silicon nanocrystals and mouse fibroblast cell culture (L929) confirmed that Si nanocrystals can enter the cell via phagocytosis while keeping their red photoluminescence and, moreover, that this process is harmless for both the cell and its progeny. Furthermore, recent experiments by Sailor's group in California yielded very promising results on the imaging of a tumour in mice using larger agglomerates of luminescent silicon nanoparticles made of porous silicon [15]. These agglomerates (sized around 150 nm) turned out to be more advantageous for this particular application than smaller agglomerates or even single silicon nanocrystals, because such 'large' particles cannot leave the body too early; the kidneys will remove them only after their partial degradation. Research in this field is very intense, showing rapid progress.

15.7 Problems

15/1. The occurrence of radiative substrate modes, mentioned in Section 15.3, is not limited to silicon-nanocrystal-based waveguides. These modes can be, under certain conditions, a general property of active thin films, both organic and inorganic. Compare Section 15.3 with e.g. the publications [16–18].

15/2. *Determination of the excitation cross-section of silicon nanoparticles.* The excitation cross-section σ (cm^2) is an important parameter describing the efficiency of the luminescence excitation process with light of a particular wavelength or electrons with a given energy. (This is an analogy to the excitation spectrum of a 3D semiconductor.) The cross-section σ can be determined experimentally by means of the temporal shape of luminescence rise $I(t)$ upon switching the excitation, as described by [19]

$$I(t) = I_0 \left\{ 1 - \exp\left[-\left(\sigma\phi + \frac{1}{\tau} \right) t \right] \right\}, \tag{15.3}$$

where ϕ is the flux of exciting photons or electrons ($cm^{-2}s^{-1}$) and τ stands for the luminescence decay time. If the onset rate $1/\tau_n = \sigma\phi + 1/\tau$ is measured for different excitation fluxes ϕ (i.e. excitation intensities), the cross-section σ can be determined as the slope of a linear plot $1/\tau_n = f(\phi)$. In what follows, only the case of photoluminescence will be taken into account for the sake of simplicity.

(a) Show that the formula (15.3) is a solution to

$$\frac{dN^*}{dt} = \sigma\phi(N - N^*) - \frac{N^*}{\tau}, \tag{15.4}$$

where N^* is the concentration of excited nanocrystals and N their total concentration. The intensity of photoluminescence is given by $I(t) = N^*/\tau_r$, where τ_r stands for the radiative lifetime.

(b) Find out the physical meaning of the I_0 constant in (15.3) and its relation to the photoluminescence quantum efficiency $\eta = \tau/\tau_r$.
(c) Is this determination of σ limited to weak or high excitation fluxes, or is it excitation-flux independent?
(d) Formula (15.4) contains, with regard to the experimentally determined character of the photoluminescence decay of silicon nanocrystals, a certain simplification. What is its nature?

15/3. Is the photoluminescence intensity of the optical memory shown in Fig. 15.7 higher for electrically charged or neutral nanocrystals? See also [20].

15/4. An erbium atom has the electron configuration of $[Xe]4f^{12}6s^2$. Its most common ionized state is Er^{3+}, losing both the 6s electrons and one of the 4f electrons. The not fully occupied 4f shell with eleven electrons is shielded from its surroundings by filled 5s and 5p shells, which is why luminescence arising from the radiative recombination of electrons in this shell is virtually independent of the host matrix, the electron–phonon interaction is weak and the emission is composed of atomic-like narrow lines.
(a) Using Hund's rules, derive the Russel–Saunders term ($^4I_{15/2}$) of the Er^{3+} ground state.
(b) Are optical transitions within the f shell allowed or forbidden? What implication stems from this fact for the radiative lifetime τ_r? Is this order of magnitude of τ_r advantageous or disadvantageous when compared to the time constant of energy transfer from silicon nanocrystals to erbium ions, $\tau_{tr} \approx 1\ \mu s$?

References

1. Canham, L. T. (1990). *Appl. Phys. Lett.*, **57**, 1046.
2. Malý, P., Trojánek, F., Kudrna, J., Hospodková, A., Banáš, S., Kohlová, V., Valenta, J., and Pelant, I. (1996). *Phys. Rev. B*, **54**, 7929.
3. Ossicini, S., Pavesi, L., and Priolo, F. (2003). *Light Emitting Silicon for Microphotonics, Springer Tracts in Modern Physics*, Vol. 194. Springer, Berlin.
4. Khriachtchev, L. (ed.) (2009). *Silicon Nanophotonics. Basic Principles, Present Status and Perspectives.* World Scientific/Pan Stanford Publishing, Singapore.
5. Pavesi, L. and Turan, R. (ed.) (2010). *Silicon Nanocrystals. Fundamentals, Synthesis and Applications.* Wiley-VCH, Weinham.
6. Dohnalová, K. (2007). *Study of optical amplification in silicon based nanostructures.* PhD dissertation, Charles University in Prague, Faculty of Mathematics & Physics, Prague, and Université Louis Pasteur, Strasbourg.
7. Ruan, J., Fauchet, P. M., Dal Negro, L., Cazzanelli, M., and Pavesi, L. (2003). *Appl. Phys. Lett.*, **83**, 5479.
8. Pelant, I., Ostatnický, T., Valenta, J., Luterová, K., Skopalová, E., Mates, T., and Elliman, R. G. (2006). *Appl. Phys. B*, **83**, 87.
9. Walters, R. J., Bourianoff, G. I., and Atwater, H. A. (2005). *Nature Materials*, **4**, 143.
10. Carreras, J., Arbiol, J., Garrido, B., Bonafos, C., and Montserrat, J. (2008). *Appl. Phys. Lett.*, **92**, 091103.
11. Irrera, A., Pacifici, D., Miritello, M., Franzò, G., Priolo, F., Iacona, F., Sanfilippo, D., Di Stefano, G., and Fallica, P. G. (2002). *Appl. Phys. Lett.*, **81**, 1866.

12. Kenyon, A. J. (2005). *Semicond. Sci. Technol.*, **20**, R65.
13. Pavesi, L. (2006). *Optical gain in silicon and the quest for a silicon injection laser*. In: *Optical Interconnects* (ed. L. Pavesi and G. Guillot), p 15. *Springer Series in Optical Sciences*, Vol. 119. Springer, Berlin.
14. Franzò, G., Vinciguerra, V., and Priolo, F. (1999). *Appl. Phys. A*, **69**, 3.
15. Park, J.-H., Gu, L., von Maltzahn, G., Ruoslahti, E., Bhatia, S. N., and Sailor, M. J. (2009). *Nature Materials*, **8**, 331.
16. Kiisk, V., Sildos, I., Suisalu, A., and Aarik, J. (2001). *Thin Solid Films*, **400**, 130.
17. Sheridan, A. K., Turnbull, G. A., Safonov, A. N., and Samuel, I. D. W. (2000). *Phys. Rev. B*, **62**, R11 929.
18. Yokoyama, D., Moriwake, M., and Adachi, C. (2008). *J. Appl. Phys.*, **103**, 123104.
19. Priolo, F., Franzò, G., Pacifici, D., Vinciguerra, V., Iacona, F., and Irrera, A. (2001). *J. Appl. Phys.*, **89**, 264.
20. Walters, R. J., Kik, P. G., Casperson, J. D., Atwater, H. A., Lindstedt, R., Giorgi, M., and Bourianoff, G. (2004). *Appl. Phys. Lett.*, **85**, 2622.

Photonic structures

16

Nowadays, we are witnessing the continuous development of novel optical technologies which are step-by-step substituting for electronic communications because the optical signal can be treated much faster and more effectively. In the framework of this strategy, new media, materials and composites capable of tailoring in a desired way the parameters of luminescence radiation are being developed. Throughout this chapter we briefly take notice of several of these new concepts, commonly called 'photonic structures'—photonic crystals, microresonators, microcavities and single photon sources.

16.1 Photonic crystals

Many phenomena in solid-state physics can be explained using *wave* properties of electrons, the attribute connected primarily with optical signals. On the other hand, optics has borrowed the idea of photonic crystals from the solid state: there exist bands of allowed energies for electrons in a solid with a periodic arrangement of atoms; similarly, allowed and forbidden energy states will exist also for photons in a medium with a suitable structural periodicity. This means that only photons with energy ranging in a certain interval will be able to travel through such a 'crystal'; the propagation velocity of photons with other energies will be slowed down or their propagation will be totally suppressed. This idea was, for the first time, expressed in 1987 by Yablonovitch [1] and the corresponding medium is called a *photonic crystal*. It is obvious that if such a medium contains luminescence centres of any kind, their radiative recombination will be influenced by this fact, anyway.

What is the 'suitable periodicity' for that purpose? In analogy with a crystalline metal or semiconductor, where the distance between atomic planes matches the de Broglie wavelength of the electron, one may conclude that the 'lattice constant' or period of arrangement (periodicity of the dielectric constant) in a photonic crystal must be comparable with the wavelength of light. Such photonic crystals then show features similar to electrons in crystals, namely, the occurrence of an energy band structure for photons, including the band of forbidden energies, the concept of Brillouin zones, defect states localized within the forbidden bandgap, and reduced propagation velocity in the vicinity of the band edge.[1]

In nature, a kind of condensed matter with periodic arrangement of its basic building blocks on the scale of hundreds of nanometres or units of a micrometre occurs very rarely. Photonic crystals must thus be prepared in a synthetic way. Contemporary micro- and nanotechnologies offer in this direction a great diversity of possibilities. A variety of structures can be manufactured (see Fig. 16.1): from the simplest 1D photonic crystals (for example alternating dielectric layers with different dielectric constants, the so-called Bragg mirrors or multilayers), via 2D crystals (for instance a grating manufactured from parallel rods or pores) to 3D photonic crystals (e.g. rods arranged in the form of a wooden pile, or synthetic opal—a face-centred cubic or hexagonal lattice with closely packed submicron spheres of SiO_2). From the previous exposition one may deduce that the 'ban' on the propagation of photons in a photonic crystal is not—unlike, e.g., a semiconductor crystal—due to their absorption, but due to reflection and interference of light on lattice planes at certain angles of incidence (causing *opalescence*, the rainbow colour of both natural and artificial opals). One may also tell that the photonic forbidden gap for electromagnetic waves incident on the crystal from outside is due to their total reflection.

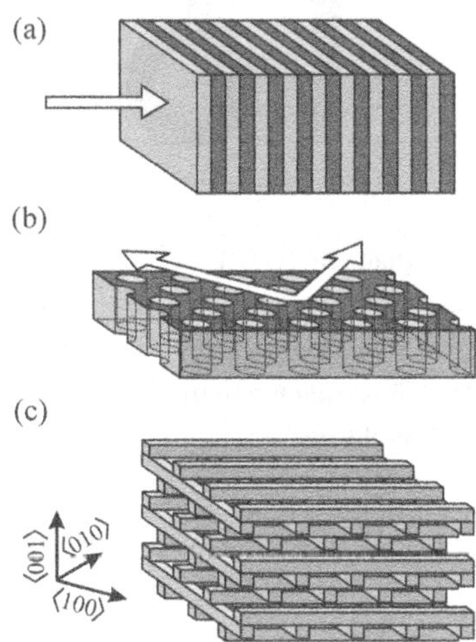

Fig. 16.1
Examples of photonic crystals of different dimension: (a) Bragg mirror (1D), (b) hexagonal lattice of pores—air cylinders in a dielectric matrix (2D) and (c) 'wooden pile' (3D).

Figure 16.2 represents a typical band structure of a 3D photonic crystal [2]. The absence of modes of propagation of the electromagnetic field in the photonic forbidden gap offers interesting application possibilities for optoelectronic components. Besides some straightforward applications like different types of optical filters and polarizers, one may give an example of special optical waveguides, because intentional insertion of a line of point defects into a regular structure of the photonic crystal relaxes the suppression of photon propagation, and light can then propagate along a defined direction. The reader may find a comprehensive treatment on these and many other properties of photonic crystals, including detailed theoretical essentials, in the books [2–4]. In what follows we restrict ourselves to the cases when a photonic crystal shows luminescence or (if not luminescent itself) is impregnated

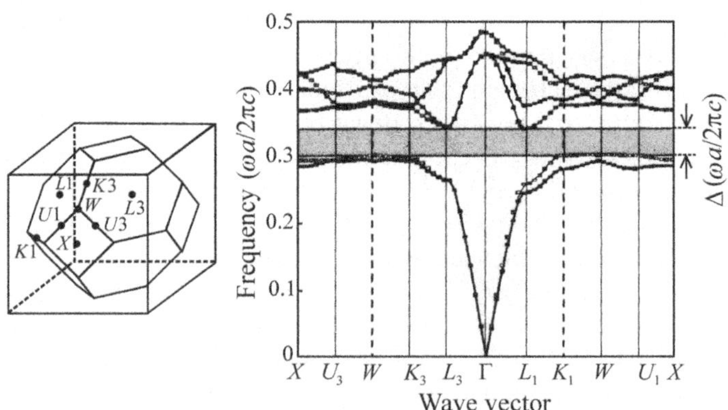

Fig. 16.2
Demonstration of the band structure and Brillouin zone of a 3D photonic crystal, so-called Yablonovite (a special system of holes introduced at different angles in a dielectric substrate). The band of forbidden photon energies (frequencies $\Delta\omega$) of relative width $\Delta\omega/\omega \simeq 0.15$ is marked out. The basic translation period of the lattice is a, light velocity is c, and $(\omega a/2\pi c)$ stands for the normalized frequency. After Lourtioz *et al.* [2].

[1] The existence of bands of allowed and forbidden energies is a general characteristic of any waves propagating in periodic structures, including lattice vibrations—phonons (Chapter 4).

by luminescent dopants—usually by semiconductor nanostructures. Luminescence behaviour is then influenced by the imprinted photonic structure.

16.1.1 Spontaneous emission

Here one deals with the original idea by Yablonovitch [1]: if the energy of the luminescence photon falls within the photonic bandgap, radiative recombination and spontaneous emission theoretically cannot occur at all (inhibited spontaneous emission). Then it would be possible to reduce in this way the threshold pumping power of lasers because spontaneous emission to all modes except the laser mode—where it provides an important triggering signal for laser action—represents in this sense harmful losses. However, the effect of inhibition of spontaneous emission is not so unambiguous in reality, and whether or not it occurs depends on many circumstances.

Primarily, the photonic forbidden gap is of a different 'quality' in different photonic structures. It depends on the particular lattice symmetry and on the magnitude of the reflectivity coefficient on the interface between the two dielectric media. This coefficient is very responsive to the ratio of dielectric constants (refractive indices) and so also is the strength of suppression of photon propagation. At low contrast of the refractive indices the inhibition will be partial only, and then we speak of the so-called *stop-band* instead of the total forbidden gap.[2] This is just the case of synthetic opals manufactured from SiO_2. Spontaneous luminescence therefore is not totally inhibited; however, it is possible to observe a significant decrease in luminescence intensity at wavelengths corresponding to the stop-band. This is demonstrated in Fig. 16.3 for luminescence of silicon nanocrystals incorporated into synthetic opal [5]. The dip in the emission spectrum in panel (c) corresponds perfectly to the increased sample reflectivity measured under illumination with a white source—panel (d). However, to observe the effect, a strict periodic ordering of the SiO_2 spheres is required, indeed, as Figs 16.3 (c) and (d) clearly demonstrate.

An interesting result is the luminescence decay curves displayed in Fig. 16.3(e). One would expect that a decrease in the probability of radiative recombination (in other words, an increase in the radiative lifetime τ_r) will cause a slowing down of the luminescence dynamics, therefore the decay should get prolonged. However, we do remember that the luminescence decay time τ is in a decisive way (except when the luminescence efficiency η is extremely high) driven by the non-radiative transitions. Here, this is the case (η is 'only' several percent) and therefore we cannot expect any marked modifications of the luminescence decay. Panel (e) of Fig. 16.3 is in accord with this fact; nevertheless, one can notice two effects: first, the decay gets moderately faster while after $\sim$ 150 μs it gets prolonged, thus both the fast and slow components seem to be enhanced. The interpretation is based mainly on the partial *increase* in the density of photonic states due to defects in the opal lattice (fast component), and, on the other hand, the slow component reflects

[2] The band structure in Fig. 16.2 with a well-developed forbidden gap was calculated for a high ratio of dielectric constants equal to 13.

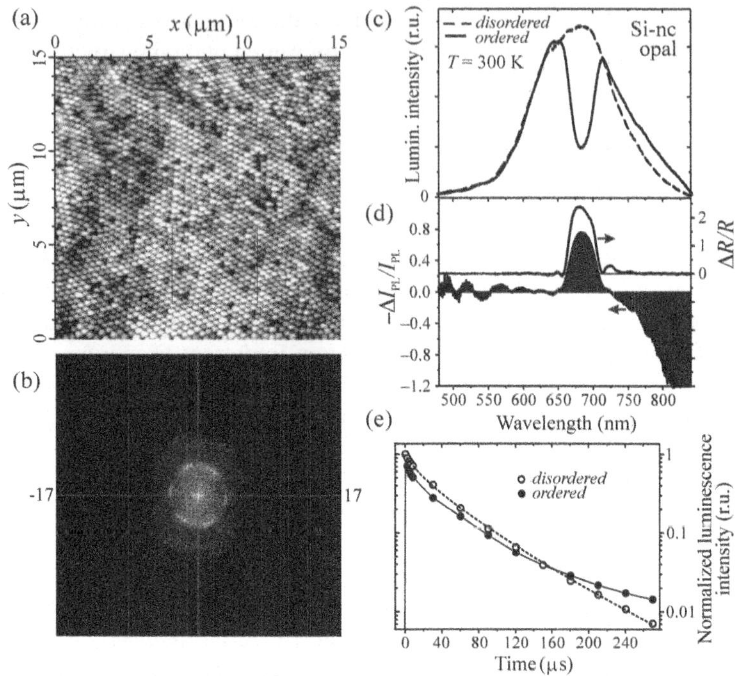

Fig. 16.3
(a) Surface view of a sample of opal ($15 \times 15\ \mu m^2$) in an atomic force microscope (AFM). It is possible to distinguish clearly single spheres of SiO_2. (b) The two-dimensional Fourier transform of the structure displayed in the upper panel shows the opal hexagonal symmetry. (c) Photoluminescence spectra of opal impregnated with Si nanocrystals. The full line is a spectrum taken from a well-ordered area of the sample, and the dashed curve originates in a sector of the sample lacking a periodic arrangement of the spheres. (d) Relative decrease of photoluminescence $\Delta I_{PL}/I_{PL}$ from panel (c) in comparison with the relative change of reflectance $\Delta R/R$ on the same spot of the sample. (e) Photoluminescence decay measured at wavelength of the stop-band: full symbols correspond to the ordered area, open symbols to the disordered part of the sample. Room temperature, after Janda *et al.* [5]. Reproduced with kind permission of IOP Publishing Ltd.

the *lower* density of photon states in the perfectly arranged part of the sample. Spontaneous emission can thus be either inhibited or in principle also amplified (by means of a suitable defect).

The influence of a photonic crystal on the parameters of spontaneous luminescence manifests itself more noticeably if there is a higher constrast of refractive indices, better lattice ordening (fewer defects) and higher luminescence efficiency. All these parameters are met in the next example—a two-dimensional photonic crystal GaInAsP of the same type as shown in Fig. 16.1(b) with an inserted quantum well of 5 nm thickness made of the same compound [6]. The well, acting as a luminescence source, lies in the plane of the crystal slab. The application of state-of-the-art technology of preparation—a combination of epitaxial growth with electron beam lithography and plasma etching—allowed samples of very good quality with regular arrangement of the hole to be fabricated. The constrast in refractive indices with respect to air is 3.27, therefore is significantly higher than that in opals (~ 1.45). Finally, measurements were performed at low temperature (4 K), which implies high luminescence efficiency in this material; therefore the luminescence decay also reflects the radiative lifetime τ_r, or the spontaneous emission rate $R_{sp} = 1/\tau_r$. The whole series of samples was prepared with lattice constant a ranging in the interval 300–500 nm.

The essence of the results is summarized in Fig. 16.4 [6]. Panel (a) shows the emission spectra of 2D GaInAsP photonic crystals with different values of the lattice constant a. Grey shading depicts the photonic bandgap. It is seen that a substantial increase in the emission intensity occurs under spectral

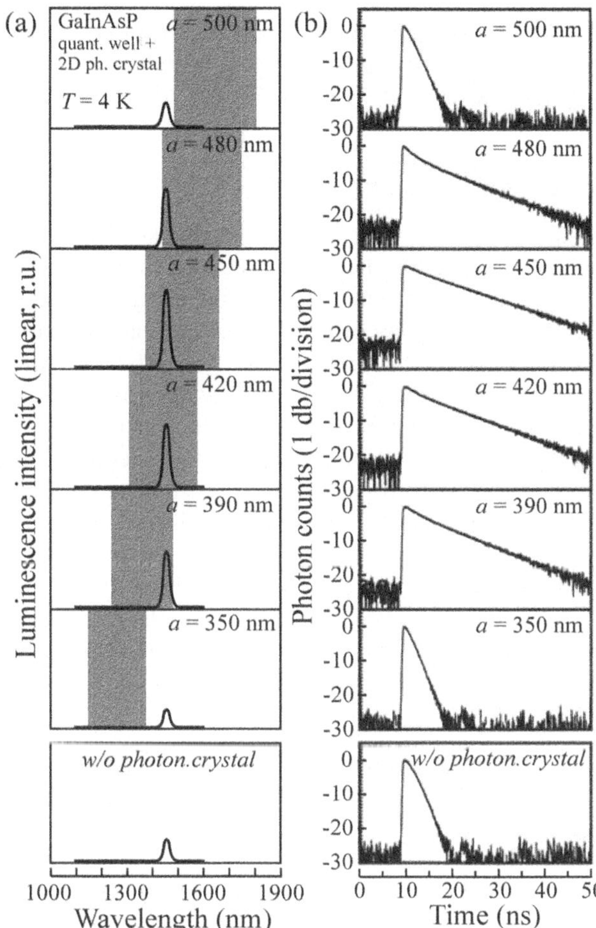

Fig. 16.4
(a) Time integrated emission spectra of GaInAsP quantum wells, incorporated into a series of 2D photonic crystals with different lattice constant a. Photoluminiscence was measured in the normal direction. The shaded area marks the photonic bandgap. (b) Corresponding curves of luminiscence decay. If the spontaneous emission spectrum coincides with the grey area, fivefold prolongation of the decay occurs. Excitation with a pulsed Ti-saphir laser 980 nm, temperature $T = 4$ K. After Fujita *et al.* [6].

coincidence of this band with the luminescence of the built-in quantum well. This is, however, in contradiction with what one might expect; it means a decrease rather than the observed increase; here, it is thus important to specify that the luminescence was measured perpendicular to the sample planes while photonic suppression of the emission works only in the directions of the periodic arrangement of holes, thus within the sample plane. The rate of spontaneous emission in this plane, $R_{\text{sp}}^{\parallel}$, is then strongly reduced and the emitted light power is redistributed: the spontaneous emission rate into the normal direction $R_{\text{sp}}^{\perp}$ becomes enhanced. That is, photons emitted from the quantum well can be effectively coupled out into vertical modes of radiation only. Consistent with this observation is then also a—seemingly paradoxical—total slowing down of the decay, because we may write for the luminescence decay time τ

$$\tau^{-1} = R_{\text{sp}}^{\parallel} + R_{\text{sp}}^{\perp} + \tau_{\text{nr}}^{-1} \tag{16.1}$$

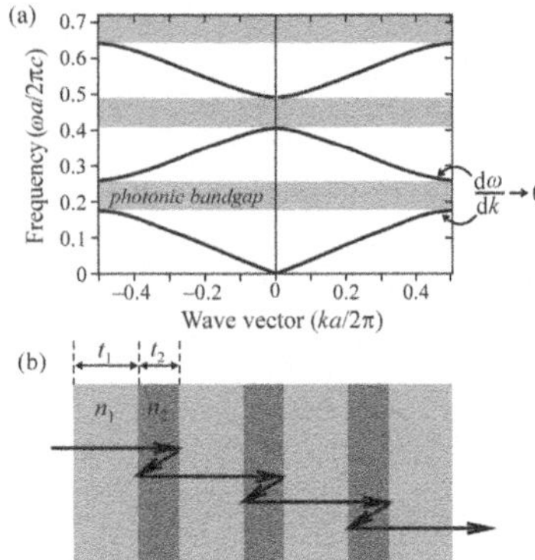

Fig. 16.5
(a) Example of a band scheme of a 1D photonic crystal. (b) One-dimensional photonic crystal, consisting of alternating dielectric layers (with thicknesses t_1 and t_2, and refractive indices n_1 and n_2); one of the layers constitues a medium with optical gain g. The arrows depict schematically prolongation of the optical path of photons due to multiple internal reflections.

and a strong reduction in $R_{sp}^{\parallel}$ leads to a significant decrease of the right-hand side in (16.1), thus to a prolongation of the decay time τ.[3] This is shown convincingly in panel (b) in Fig. 16.4.

16.1.2 Stimulated emission

We have seen that the decreased density of photon states in a photonic crystal inhibits spontaneous emission in certain directions. In the case of stimulated emission, however, the situation may develop markedly differently. Here, conversely, one can take advantage of the decreased photon density of states in the vicinity of the edge of the photonic bandgap, where the group velocity of propagation of the optical signal $v_g = d\omega/dk$ decreases significantly. Theoretical calculations show that the lasing threshold is proportional to v_g^2 [4] and if the amplified luminescence radiation falls spectrally into the region where $d\omega/dk \to 0$ (Fig. 16.5(a)), a significant decrease of this threshold might be expected. In other words, it is possible to enhance in this way the laser action. One can imagine the story also like this: a photon undergoes multiple reflections in the lattice and slowly wriggles its way through the structure, which is for a 1D crystal depicted in Fig. 16.5(b). The effective length of the optical path gets longer and therefore the effective optical amplification in an 'inverted' medium with population inversion increases.[4]

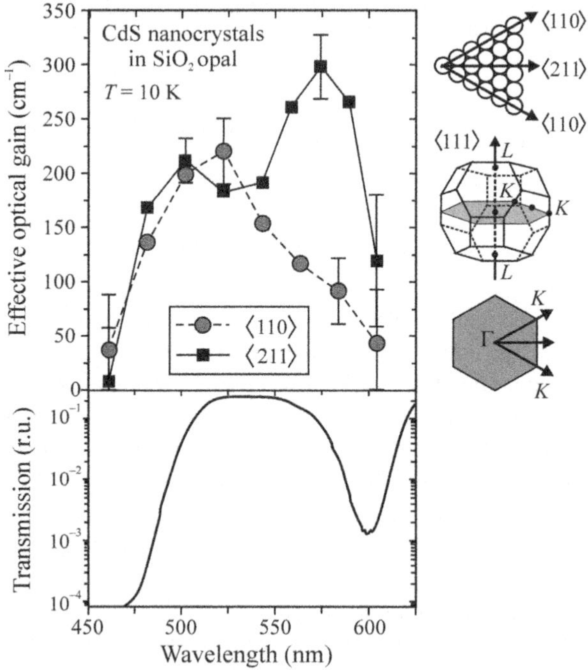

Fig. 16.6
Upper panel: Spectra of optical gain in a photonic opal, impregnated by CdS nanocrystals, as measured in two different lattice directions in the (111) plane. Excitation by nanosecond pulses 308 nm, temperature $T = 10$ K. On the right, the arrangement of the spheres of SiO_2 (cubic face-centred lattice, lattice constant $a = 260$ nm), the corresponding Brillouin zone and its projection into the (111) plane of **k**-space are shown. Lower panel: Spectral dependence of transmittance of the opal/CdS sample in the visible region. The marked minimum in the transmittance centred at ~ 600 nm reflects the photonic stop-band at the L point of the Brillouin zone. After Vlasov *et al.* [7].

[3] An increase in $R_{sp}^{\perp}$ is not sufficient to compensate the decrease in $R_{sp}^{\parallel}$.
[4] A certain analogy with laser distributed feedback may be found here.

Verification of this theoretical idea in an actual 3D photonic crystal—cubic opal SiO_2—is presented in Fig. 16.6, upper panel [7]. Thin films of the opal can be best prepared parallel to the plane of the closest packing of the SiO_2 spheres, thus parallel to the (111) plane. Luminescent nanocrystals of CdS embedded into voids among the SiO_2 spheres were the source of amplified radiation. The amplified spontaneous emission and optical gain in different directions of the (111) plane were measured in these samples by the VSL method (Subsection 10.6.1). It is seen that up to threefold enhancement of the optical gain occurs in the $\langle 211 \rangle$ direction at a wavelength of $\sim 570\,\text{nm}$; this is thus the 'allowed direction' for amplification. The wavelength of 570 nm indeed falls on the edge of the photonic bandgap, as proved in the lower panel of Fig. 16.6.

16.2 Microresonators

Optically active microresonators in the form of a miniature flat disk or ring appear to be other attractive components for optoelectronic applications; besides, they may also find their use elsewhere, e.g. in chemistry and biology. In such devices, light propagates owing to total reflections along the device perimeter and if the condition that there is an integer multiple of wavelength per round is met, i.e.

$$M\frac{\lambda}{n} \doteq 2\pi R_D, \tag{16.2}$$

then stationary modes are established and the device can be used for example as a laser resonator.[5] In eqn (16.2) M stands for a large natural number, λ is the wavelength of light in air, n is the refractive index of the resonator and R_D is its radius, see Fig. 16.7. In analogy with the specific mechanism of propagation of sound in circular corridors and cathedral domes, these modes are commonly called 'whispering gallery modes' (WGM).

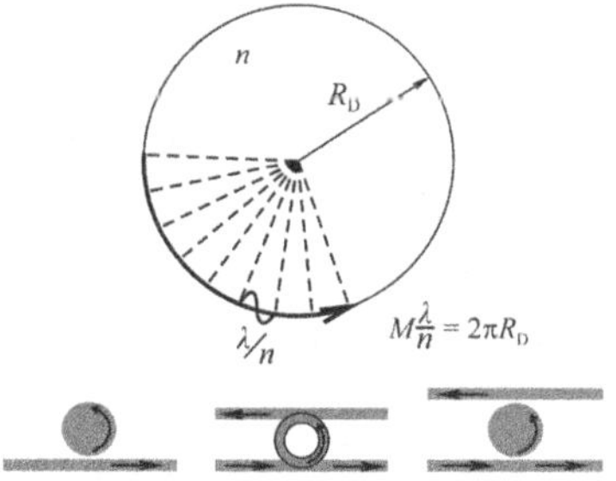

Fig. 16.7
Schematic of a 2D disk microresonator of radius R_D, along the perimeter of which an electromagnetic wave propagates owing to a large number (M) of total internal reflections. Three possibilities of coupling the radiation in or out of the resonator are depicted at the bottom: lateral contact of the microdisk with a waveguide, vertically coupled microdisk, and a microdisk with both input and output waveguides.

The principal possibility of utilization of such resonators in optics has been, of course, known for a long time ago; however, only contemporary advanced microtechnologies have revealed their potential. That is, the WGM can reveal themselves distinctly only when the resonator radius is sufficiently small, of the order of $R_D = 1$–$10\,\mu\text{m}$. Otherwise the spectral separation between modes $\Delta\lambda_{WG} \doteq \lambda^2/2\pi R_D n$ (see Problem 16/2) quickly decreases and the modes merge into one another; in addition, technical difficulties arise connected with manufacturing larger radius resonators of perfectly circular shape and these resonators then exhibit considerable optical losses due to mechanical imperfections.

In order to use the microresonators as miniature photonic sources, a semiconductor quantum well may, for example, be incorporated into a disk resonator (if the resonator material itself is not a good phosphor), or the resonator may be impregnated by luminescent nanocrystals. With respect to small

[5] Relation (16.2) holds true only for flat 2D microresonators with negligible thickness. In the general case the use of Bessel functions is required in order to calculate the modes.

resonator dimensions and low density of modes, such a microlaser can manage with a low pump power, thereby meeting the basic requirements imposed on photonic integrated circuits. Perhaps it is not yet fully understandable to the reader how the optical signal, generated inside the microresonator and circulating along its perimeter, leaves the resonator. This may happen owing to the evanescent wave, penetrating during total reflection into the outside medium. This wave is then coupled to a waveguide being tightly adjacent to the microresonator (Fig. 16.7 bottom).

The disk microresonators can nowadays be manufactured via a combination of photolithography or electron beam lithography with different etching techniques. Microdisks in a 'mushroom' shape are very often manufactured. Microdisk GaN resonators, shown in Figs 16.8 (a) and (b), were prepared by depositing thin circular islands of GaN on a silicon substrate, which was then under-etched, thus creating mushroom formations [8]. Important information is depicted in panel (c) of Fig. 16.8: with increasing intensity of optical pumping, resonant WGMs of low-temperature edge luminescence around 365 nm become strongly enhanced. Moreover, a break in the intensity dependence proves this observation to be the onset of lasing. This result is, among others, important also because it demonstrates the principal feasibility of 'GaN on silicon' as one of the alternatives for future photonic circuits. At the same time, silicon is far from being an ideal substrate for gallium nitride due to the considerable lattice mismatch. This is reflected in the low resonator quality, reaching in this case only a value of $Q \simeq 80$; further technological development is thus necessary for the purpose of practical implementation. Besides, the problem of electric pumping has not yet been solved for microresonators.

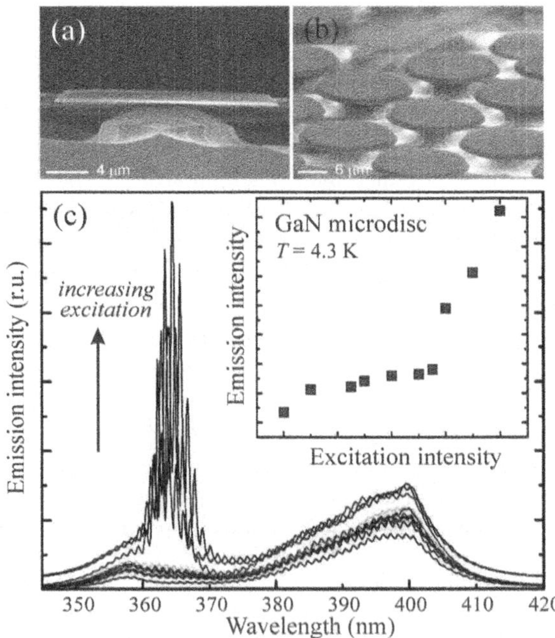

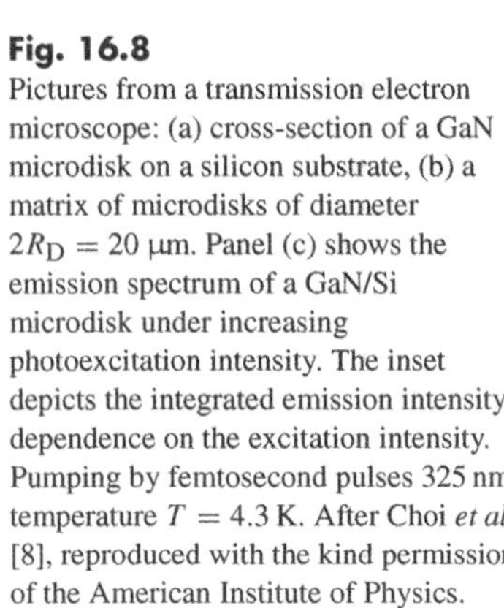

Fig. 16.8
Pictures from a transmission electron microscope: (a) cross-section of a GaN microdisk on a silicon substrate, (b) a matrix of microdisks of diameter $2R_D = 20$ μm. Panel (c) shows the emission spectrum of a GaN/Si microdisk under increasing photoexcitation intensity. The inset depicts the integrated emission intensity dependence on the excitation intensity. Pumping by femtosecond pulses 325 nm, temperature $T = 4.3$ K. After Choi *et al.* [8], reproduced with the kind permission of the American Institute of Physics.

It would be misleading to form the impression that the WGM modes may find application only in stimulated emission. They can be observed experimentally also in spontaneous luminescence spectra and their width, intensity and spectral position respond sensitively to any change in the parameters of the resonator and surroundings. Therefore, also their potential for diagnostic and sensoric applications in physics, chemistry and biology are being studied. Here, the fact that the required amount of the analyte is extremely low is of particular attraction.

16.3 Microcavities

A photonic microcavity may, for example, be a specific defect in a 1D photonic crystal, as depicted in Fig. 16.9: two parts of a chain of alternating semiconductor or dielectric layers of thicknesses L_1 and L_2 (with refractive indices n_1, n_2) are separated by a 'spacer' of thickness L_c with a different refractive index n_c. A source of electromagnetic radiation with wavelength λ_0 is placed within this 'spacer' or microcavity, the thicknesses of the surrounding layers being chosen to fulfil the conditions $L_1 = \lambda_0/4n_1$, $L_2 = \lambda_0/4n_2$. The optical paths in both types of layers are thus equal to one-fourth of the wavelength. An array of such closely packed layers is called a 'distributed Bragg reflector' (DBR).[6] Radiation generated in the microcavity is then returned back (at least partially) on either side by means of these mirrors. That is, a partial reflection occurs at each interface between the layers of the DBR mirrors due to the different values of the refractive index $n_1 \neq n_2$, the wave reflected at the back interface of one layer gaining a path difference $2 \times \lambda_0/4 = \lambda_0/2$ and simultaneously a phase shift π with respect to the wave reflected on the front interface. Reflected waves therefore interfere constructively and if the number of the layers in either DBR mirror is sufficiently large, total reflectivity in the vicinity of wavelength λ_0 approaches unity (corresponding to the photonic stop-band) and a narrow dip appears in the reflectivity at the resonant wavelength λ_0, like in a Fabry–Perot resonator.

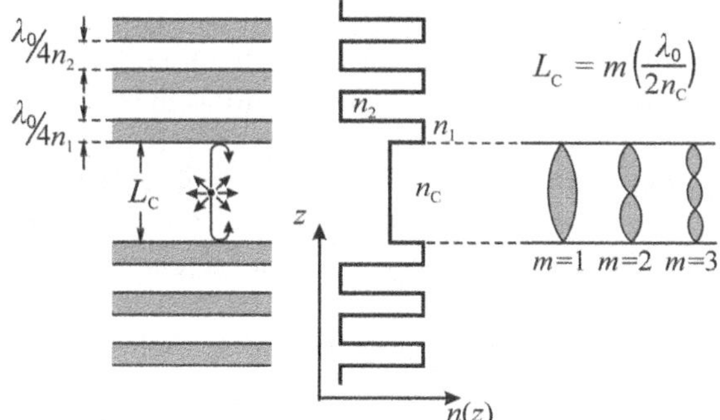

Fig. 16.9
Cross-section of a photonic microcavity of thickness L_c sandwiched between Bragg reflectors DBR. Also refractive index profiles and examples of the electric field amplitude distribution of the light wave are shown.

[6] The term originates from an analogy with the Bragg mechanism of X-ray reflection on crystal atomic planes.

In addition to the specific shape of the mirrors, the microcavity as a linear optical resonator is characterized by the fact that its thickness L_c is order-of-magnitude comparable to the wavelength of the radiation generated inside. Microcavities are designed with the intention of L_c being a (small) integer multiple of a half-wave, and therefore must fulfil the condition

$$L_c = m(\lambda_0/2n_c),\ m = 1,\ 2, 3\dots, \tag{16.3}$$

or, to say it differently, the wavenumber along the microcavity axis reads $k_z^m = m\pi/L_c$. Therefore, a standing wave pattern of resonant modes with a low number of nodes and antinodes (Fig. 16.9) may arise between the DBR reflectors in the microcavity. We speak of a $\lambda/2$-cavity or λ-cavity etc, if we have in mind a microcavity with $L_c = \lambda_0/2n_c$ or $L_c = \lambda_0/n_c$, respectively. The distribution of the electric field amplitude $\mathscr{E}$ is then described by

$$\mathscr{E} \sim \sin(k_z^m z). \tag{16.4}$$

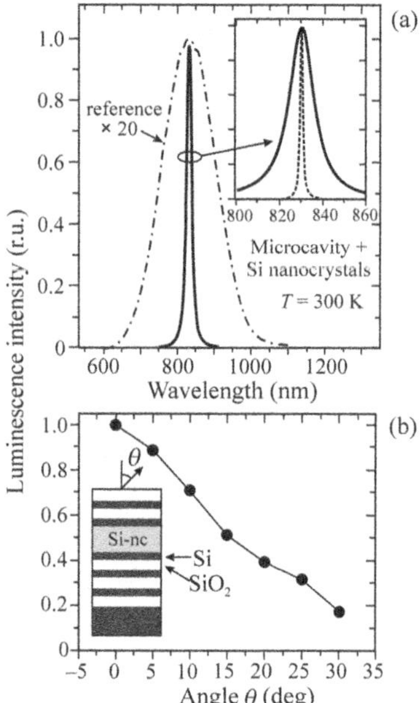

Fig. 16.10
(a) Photoluminescence spectrum of a microcavity embedded with silicon nanocrystals, tuned to a resonant wavelength of $\lambda_0 = 830$ nm. The reference spectrum (multiplied by a factor of 20) originates in the same nanocrystals but placed outside the microcavity. The inset demonstrates, by comparing the dashed and solid lines, how the emission spectrum narrows upon increasing the number of double-layers in DBR. (b) Photoluminescence intensity as a function of the deviation from the microcavity normal. Room temperature, adapted after Iacona *et al.* [9].

A dipole source of radiation placed in the microcavity thus has the possibility of transfering its energy only to a small group of allowed modes in a privileged direction. The directionality of the output radiation is significantly enhanced in this way and, moreover, if the characteristic spontaneous emission spectrum of the emitter is broadened, the microcavity may modify this spectrum into a single narrow quasi-monochromatic line.[7] At the same time, however, it depends critically on how the radiation source (for instance a luminescent nanocrystal) is placed with respect to the nodes and antinodes of the electromagnetic field: in agreement with (16.4), the matrix element or square of its modulus $|M|^2$, characterizing the spontaneous emission rate, will attain its maximum value when the emitter is placed at the antinode, while $M = 0$ is expected for an emitter with z corresponding to a node. In microcavities of good quality (also in microdisks and microresonators of other types) with a high value of Q, small volume V and with luminescent nanocrystals implemented in a well-defined manner, an enhancement of the rate of spontaneous radiative recombination may even occur, namely, owing to the Purcell effect, already mentioned in Section 12.7. This enhancement is characterized by the so-called *Purcell factor* $\mathscr{F} = (3\lambda_0^3/4\pi^2 n^3)Q/V$ [2, 3].

We will illustrate this topic using the example of a relatively simple microcavity containing silicon nanocrystals. The cavity is manufactured by PECVD for a resonant wavelength $\lambda_0 = 832$ nm (Fig. 16.10 [9]). Its DBR mirrors are formed by alternating layers of Si ($n_1 \simeq 3.5$) and SiO_2 ($n_2 \simeq 1.45$); the active layer itself is composed of silicon oxide SiO_2 with incorporated Si nanocrystals, has refractive index $n_c \simeq 1.7$ and thickness $L_c = 241$ nm; it is therefore a half-wave microcavity. Figure 16.10 shows how the emission spectrum of nanocrystals gets narrower with respect to their luminescence outside the microcavity, the quality of the cavity reaching up to $Q = \lambda_0/\Delta\lambda \simeq 550$. The directionality of the output radiation enhances at the same time.

[7] Some of the microcavity modes (16.3), of course, must fall spectrally within the band of spontaneous emission.

Enhanced interaction between electronic excitation and the electromagnetic field in the microcavity may also lead to a significant change of the character of exciton–polariton luminescence. In Section 7.1 we described the luminescence of a free exciton–polariton in direct-bandgap bulk semiconductors as a minuscule doublet in the emission spectrum, composed of the upper polariton branch (UPB) and the lower polariton branch (LPB). This spectral structure is due to accumulation of polaritons in a 'bottleneck' because their thermalization via the emission of low-energy acoustic phonons is slow while the polariton lifetime is short. Therefore, before they can fully relax the polaritons annihilate radiatively resulting in the emission doublet UPB/LPB. After all, the lower polariton dispersion curve that is heavily populated has no local minimum into which the polaritons could 'fall down'.

However, if a light-emitting material (quantum well) is placed in a suitably designed microcavity with strong coupling[8] between excitons and photons, the dispersion relations are modified substantially. A new resonant condition, originating in the reflections on Bragg mirrors leads to the fact that a local minimum, the so-called 'polariton trap', emerges on the lower branch in the vicinity of $k = 0$, at which the polaritons may accumulate. Under the condition of sufficiently high excitation they get there as a result of collisions with other excitons or with free carriers; common thermalization owing to phonon emission is not, like the case of a bulk semiconductor, sufficiently efficient. This is then a specific case of Bose–Einstein condensation. The condensate emits coherent radiation, and because reabsorption does not occur (the trap has a lower energy than the other electronic excitations), it may be easily amplified by stimulated emission. Indeed, laboratory examples of a so-called *polariton laser* with very low pump threshold have already been realized. Investigation of the properties of this type of laser and its possible applications represent nowadays one of the 'hot topics' of photonic research, see the monograph [10].

16.4 Single photon sources

The development of solid-state light sources, emitting pulses containing just one photon (or, in other words, generating single photons on demand), has been initialized rather recently on the basis of the requirements of quantum cryptography. The protocol of secure optical information transfer via quantum measurements on photons is based on the fact that if anybody tries to 'eavesdrop', i.e. to capture photons propagating through the optical fibre, changing at the same time the photon properties, then participants at both ends will become aware of it. To generate single photons, it is hardly possible to use an

[8] The so-called 'strong coupling' of a quantum system with an electromagnetic field in a cavity denotes such a regime in which light can dwell for a relatively long time in a (small volume and very-high-Q) cavity so that a new joint light–matter state develops. In other words, the energy exchange rate between the emitter and the cavity field dominates over the cavity decay rate. This interaction lies beyond the perturbative approach used to describe the 'weak coupling' case where dissipation dominates over the system interaction (Kavokin, A. V., Baumberg, J. J., Malpuech, G., and Laussy, F. P. (2007). *Microcavities*. Oxford University Press, Oxford; Gaponenko, S. V. (2010). *Introduction to Nanophotonics*. Cambridge University Press, Cambridge).

attenuated conventional light source because of photon bunching. This means that in this case the 'light pulse' with very high probability either does not contain any photon or contains two or more photons. The situation seems more promising in the case of an attenuated laser beam (photons of laser radiation do not exhibit bunching), however, losses in the channel of transmission and false (dark) pulses of the detector act adversely in the required regime of low photon fluxes.

A substantial increase in the security of the transmission can be assured only by making use of perfect single photon sources implemented via an isolated quantum system. Semiconductor quantum dots are one of the candidates because they are able—in principle—to met three basic requirements:

(a) The emission wavelength may be tuned—by choosing a suitable material and tailoring the size of the quantum dot—to fit some of the telecommunication windows of optical fibres;
(b) they have high efficiency of emission into a given mode with a defined polarization; and
(c) they can be excited electrically.

Quantum dots made from a wide scale of semiconductor compounds are at present being intensively studied in this respect. It is clear that success will depend critically on the preparation technology of well-defined and sufficiently space-separated quantum dots. In addition, the sample topography together with a suitable experimental arrangement has to guarantee that the luminescence will be recorded from a single dot only. Emission of single photons can be monitored using correlation photon measurements (Fig. 16.11(a)): a beam splitter divides the luminescent radiation into two optical paths in which spectral filtration of luminescence into a narrow band occurs, and then the photons hit the detectors (photomultipliers). Electric signals from the detectors are led to a time $\rightarrow$ voltage amplitude converter, combined with a multichannel analyser. An autocorrelation histogram of the relative delay τ between the detection of photons by photomultipliers 1 and 2 is then constructed using the data from the analyser.

An example of the result of such experiments is shown in Fig. 16.11(b) and (c) [11]. Panel (b) represents the emission spectrum of GaN quantum dots with a labelled monochromatized emission line at 3.49 eV (radiative recombination of an exciton). Panel (c) shows an autocorrelation histogram. The peak at time $\tau = 0$ corresponds to the event when two photons originating from one and the same pulse are detected. Essential in this graph is the fact that for $\tau = 0$ the autocorrelation function has a substantially lower value (0.42) in comparison with the other maxima (1). This means that the probability of simultaneous emission of two photons is 0.42 times lower than in a conventional non-coherent source, thus these particular GaN quantum dots indeed show an increased effect of 'antibunching'. Photon statistics are discussed in more detail in Subsection 17.5.2 and Appendix L.

Nevertheless, the reader probably recognizes that the whole matter still has some weak spots. Primarily, it works so far at low temperatures only, and the antibunching is not absolutely perfect. Next, no photonic structure (incorporation of the dots into a microcavity, etc.), which should potentially

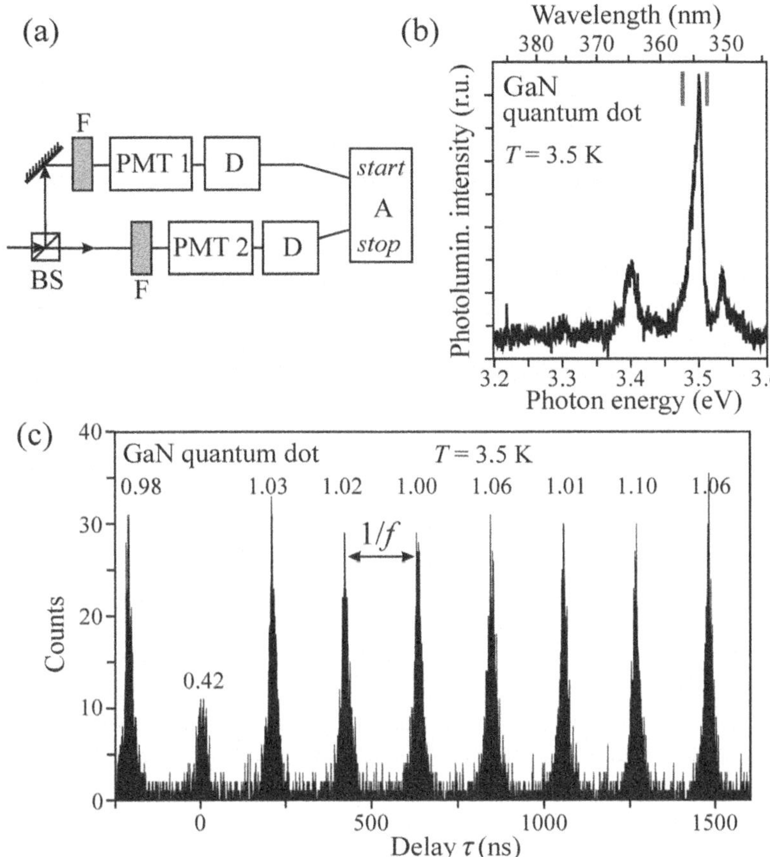

Fig. 16.11
(a) Schematic of the experimental arrangement destined to study single photon emission from GaN quantum dots. BS–beam splitter, F–spectral filter, PMT–photomultiplier, D–discriminator, A–converter time → voltage together with a multichannel analyser (a meter of time interval τ). (b) Photoluminescence spectrum under a pulsed excitation of repetition rate $f = 4.74$ MHz. (c) Autocorrelation histogram of detected pulses. Numbers above the peaks give the normalized peak areas. Excitation wavelength 266 nm, temperature $T = 3.5$ K. After Kako *et al.* [11].

improve the coherence properties of the emitted photons, has actually been used in this example. Technologically this is not a simple task, complicated moreover by problems of electric contacting. We have mentioned the search for single photon sources for two reasons: to draw the reader's attention to the current application pressure in this direction and, secondly, because here we happen to meet the so-called microluminescence technique, used for the study of single quantum dots. The next chapter is fully devoted to this advanced technique of luminescence spectroscopy.

16.5 Problems

16/1. Why, when measuring the optical gain in SiO_2 opal impregnated with CdS nanocrystals (see Fig. 16.6, upper panel), was the $\langle 111 \rangle$ direction omitted, if it is known that theoretical calculations predict the occurrence of a photonic bandgap in the vicinity of the L point of the Brillouin zone? What is the reason for the deep decrease in the sample transmittance at wavelengths below 500 nm (Fig. 16.6, bottom panel)?

16/2. Using eqn (16.2), derive an expression determining the spectral separation of two neighbouring modes of a disk microresonator $\Delta\lambda_{WG} \doteq$

$\lambda^2/2\pi R_D n$. Calculate $\Delta\lambda_{WG}$ for a GaN microdisk of radius $R_D = 10\,\mu m$ and compare the result with the mode structure displayed in Fig. 16.8(c).

16/3. In Sections 10.4 and 15.2 we mentioned the crucial role of free carrier absorption (FCA) on the way towards optical amplification in silicon nanocrystals. It turns out that, in order to quantify FCA (i.e. to determine the relevant absorption cross-section σ_{FCA}) it is possible to use the properties of whispering gallery modes (WGMs) in a microresonator containing Si nanocrystals. The principle of the method consists of experimentally monitoring the width of the WGM modes as a function of the excitation intensity: a higher concentration of the free carriers in the nanocrystals causes higher losses in the resonator, thus decreases its quality or, in other words, leads to mode broadening. Discuss this method after [12].

16/4. What are the principal differences between a microcavity formed by two DBR mirrors and a standard (macroscopic) Fabry–Perot resonator with metallized mirrors? How do these differences manifest themselves in the number of allowed modes? For the cavity depicted in Fig. 16.10(b) is it essential that it is the layers of Si (not SiO_2) that are adjacent on both sides to the active middle layer containing nanocrystals, or it would be possible to reverse the order of the layers of Si and SiO_2?

16/5. Explain why the expression $\xi_m(z) = 2\sin^2(k_z^m z)$ is called the *antinode factor* of the microcavity. Towards what value does the mean value of $\xi_m(z)$ converge in a sufficiently thick microcavity?

References

1. Yablonovitch, E. (1987). *Phys. Rev. Lett.*, **58**, 2059.
2. Lourtioz, J.-M., Benisty, H., Berger, V., Gérard, J.-M., Maystre, D., and Tchelnokov, A. (2008). *Photonic Crystals. Towards Nanoscale Photonic Devices.* Springer, Berlin.
3. Benisty, H., Gérard, J.-M., Houdré, R., Rarity, J., and Weisbuch, C. (1999). *Confined Photon Systems. Fundamentals and Applications.* Lecture Notes in Physics, Vol. 531. Springer, Berlin.
4. Sakoda, K. (2001). *Optical Properties of Photonic Crystals.* Springer, Berlin.
5. Janda, P., Valenta, J., Rehspringer, J.-L., Mafouana, R. R., Linnros, J., and Elliman, R. G. (2007). *J. Phys. D.: Appl. Physics*, **40**, 5847.
6. Fujita, M., Takahashi, S., Tanaka, Y., Asano, T., and Noda, S. (2005). *Science*, **308**, 1296.
7. Vlasov, Yu. A., Luterová, K., Pelant, I., Hönerlage, B., and Astratov, V. N. (1997). *Appl. Phys. Lett.*, **71**, 1616.
8. Choi, H. W., Hui, K. N., Lai, P. T., Chen, P., Zhang, X. H., Tripathy, S., Teng, J. H., and Chua, S. J. (2006). *Appl. Phys. Lett.*, **89**, 211101.
9. Iacona, F., Franzò, G., Moreira, E. C., and Priolo, F. (2001). *J. Appl. Phys.*, **89**, 8354.
10. Kavokin, A. V., Baumberg, J. J., Malpuech, G., and Laussy, F. P. (2007). *Microcavities.* Oxford University Press, Oxford.
11. Kako, S., Santori, C., Hoshino, K., Götzinger, S., Yamamoto, Y., and Arakawa, Y. (2006). *Nature Materials*, **5**, 887.
12. Kekatpure, R. D. and Brongersma, M. L. (2008). *Nano Lett*, **8**, 3787.

Spectroscopy of single semiconductor nanocrystals

17

In Chapter 12, we discussed the basic properties of low-dimensional semiconductors. When measuring the luminescence spectra of such structures in practice, we can apply the very same techniques as for bulk semiconductors (Chapter 2). We might, however, run into a major problem—large *inhomogeneous broadening*, that is, 'smearing' of the spectral lines that can completely obscure any fine structure possibly present in the spectra. The main challenge connected with this technique consists in the preparation method, because it is not possible to produce an ensemble of absolutely identical nano-objects (quantum wells, dots, ...), even using the most advanced techniques (such as molecular beam epitaxy and electron beam lithography). Each object will exhibit individual variations in shape, size, surface states, or possibly chemical composition, defects, and impurities. All these variations then influence the energy levels and consequently also the luminescence properties of the object under investigation. Thus, when measuring an ensemble of such individual nano-objects, what we observe results inevitably from optical properties being averaged over the whole ensemble. Naturally, one has to pose the question: Is it possible to measure the spectra of individual low-dimensional objects and thus to get the hidden information?

High-resolution optical spectroscopy faced a similar problem, in the beginning of the 1970s, at that time connected with measurement of an ensemble of molecules at low temperatures (usually complex organic molecules frozen in an amorphous matrix offering a variety of embedment configuration sites). Challenging spectroscopic techniques had been developed in this field during the 1970s and 1980s, which enabled researchers—to some extent at least—to determine the homogeneous shape of the spectrum within the inhomogeneously broadened bands. The measurement was based on resonant excitation of only a very small sub-ensemble of molecules using the very narrow spectral line of a tuneable continuous laser. Yet, despite their complexity, these techniques of site-selection spectroscopy and spectral hole-burning were only able to provide indirect information and only for some systems (see the monograph [1]). True spectroscopy of individual molecules came out of this breeding ground in 1989 (the pioneering work by Moerner and Kador [2] and Orrit and Bernard [3] of 1989–1990). In its pure form, unfortunately, this technique can only work with a few suitable molecules, at very low temperatures and under excitation with tuneable lasers; however, it yielded beautiful results confirming the quantum chemical description of molecules and the theory of their interaction with light [4]. The focus later shifted towards the observation of individual molecules at higher temperatures, at which the spectra are no longer so interesting; what the experimenters are mainly concerned with is the statistics of the emitted photons [5].

The first announcements of spectroscopic observations of individual semiconductor structures followed shortly after those of individual molecules in 1992 [6]. This opened up an important field in luminescence spectroscopy, established on the techniques of spectroscopy of individual molecules and yielding fundamental knowledge primarily about semiconductor quantum dots—nanocrystals. We can generally include the techniques of spectroscopy of individual nanocrystals in the wider field of optical spectroscopy with high spatial resolution (such as the very widely used technique of micro-Raman spectroscopy). Some authors use the term *local probe techniques* to refer to these methods; see, for example, the lucid article by Gustafsson *et al.* [7]. In this chapter, we shall first outline the fundamental principles of spectroscopy of individual quantum structures, then we shall discuss the experimental approaches currently in use, and finally we shall give illustrative examples of the most interesting results. We shall limit our attention mainly to quantum dots; however, quantum wires will also be mentioned briefly.

17.1 Basic principles

As we know from Chapter 12, low-dimensional semiconductor structures are unique due to their quantum-confinement effect, which sets in roughly for dimensions comparable to several times the Bohr radius of an exciton—say, for sizes below 10 nm. If we observe such nano-objects by means of optical methods, we can take advantage of traditional microscopic techniques, which, however, have their spatial resolution limit due to diffraction of the light wave; i.e. the image of a point-like light source is formed by a system of diffraction rings (the so-called Airy disk). The traditional criterion most frequently used to determine the limiting resolution is the Rayleigh criterion, which assumes two points to be distinguishable if their distance d is equal to or larger than the distance between the centre and the first minimum of the Airy disk (see Fig. 17.1):

$$d \geq 1.22\,\lambda \frac{f}{D} = \frac{1.22\,\lambda}{2\,\mathrm{NA}} = d_{\mathrm{m}}, \qquad (17.1)$$

where λ is the wavelength of the light used for imaging, f/D is the relative opening of the imaging optical system and NA is its numerical aperture. Considering the numerical aperture of the commonly available objectives to be roughly NA $= 0.8$, the limiting resolution amounts to around $d_{\mathrm{m}} \approx (3/4)\,\lambda$, that is, several hundreds of nanometres (Fig. 17.1).

Thus, this is where we meet the fundamental discrepancy, which then forms the basis of our further considerations: *The dimension of the observed*

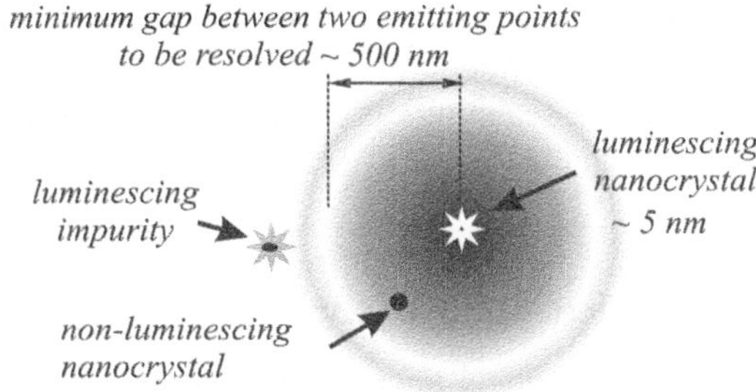

Fig. 17.1
Schematic of the relative dimensions when observing a luminescent nanocrystal through a diffraction-limited optical system. The spectrum of a single nanocrystal can be resolved if there are no other light-emitting objects (impurities, etc.) within the smallest resolvable area, the spectra of which would overlap with the observed object.

object is roughly one hundred times smaller than the limiting resolution of the optical image.

Can we still detect luminescence signal from a single quantum dot?

Yes, we can, much like astrophysicists can and do measure the spectra of stars, objects smaller than the resolving power of their optics. We need to realize that we do not aim at imaging the true size of the object—which is not possible in the optical far field—but what we can do instead is to detect a luminescence signal emitted by an individual quantum dot and distinguish it unambiguously from all other signals.

When measuring luminescence spectra of individual quantum dots the following conditions are to be met:

- The density of the observed objects within the sample needs to be so low that the distance between them is significantly larger than the limiting resolution of the optical image (or that there is at most one active object within the diffraction ring).
- Parasitic light sources (emitting within the wavelength range of the investigated object) inside the diffraction ring must be significantly suppressed so that they do not interfere with the desired signal; primarily, fluorescence from impurities, substrate or matrix and excitation light scattering (Raman scattering) need to be taken care of.
- The spatial resolution and image quality of the optical system must be as high as possible.
- The emitted photons must be collected and detected as efficiently as possible—the solid collection angle of the optical system must be as wide as possible and the quantum efficiency of the detection as high as possible because the measured luminous flux is naturally extremely low.

The first two of the above conditions concern the requirements for the preparation of the sample while the last two concern the imaging and detection system.

17.2 Experimental techniques

Luminescence spectra of individual quantum dots are measured with the use of set-ups consisting of both a micro-imaging system, i.e. a microscope, and a spectroscopic system to detect the very low luminescence signals. If we are to divide the experimental set-ups into groups, the main distinguishing feature is the imaging method: either a wider area of the sample is imaged all at once or the sample is imaged in the scanning, 'point-by-point' mode (by scanning the excitation point or shifting the sample).

17.2.1 Wide-field micro-spectroscopy

The fusion of a traditional optical microscope with an imaging spectrometer and an array detector (a camera), see Fig. 17.2 [8], forms the basis of wide-field micro-spectroscopy. The resolving power of such an apparatus is naturally limited by the light-diffraction ring mentioned above. The imaging spectrometer is usually of a construction similar to the Czerny–Turner monochromator (Section 2.3). However, the first collimating mirror is not spherical but toroidal instead; employment of such a mirror corrects for the aberrations of the spectrometer (astigmatism, in particular), that is, the image created (by means of the microscope) in the entrance slit plane is transformed into a perfect, non-distorted image (usually with 1:1 scaling) at the spectrometer output (there is no exit slit but a multichannel detector, most frequently a CCD camera, see Section 2.2). Of course, in order to obtain a high-quality, wide-angle output image, the input slit needs to be sufficiently wide and the grating has to be adjusted to the zeroth order (direct reflection) or a planar mirror needs to be inserted instead of the grating.

The sample is usually excited by a laser, the beam of which is introduced to the sample either through a gap between the sample and the objective (for transparent samples, it may also pass directly through the sample) or enters through the same objective that collects the emitted light. It is the latter, so-called epifluorescence configuration that makes it indispensable to use several optical filters adequately separating the excitation and emission spectral ranges.

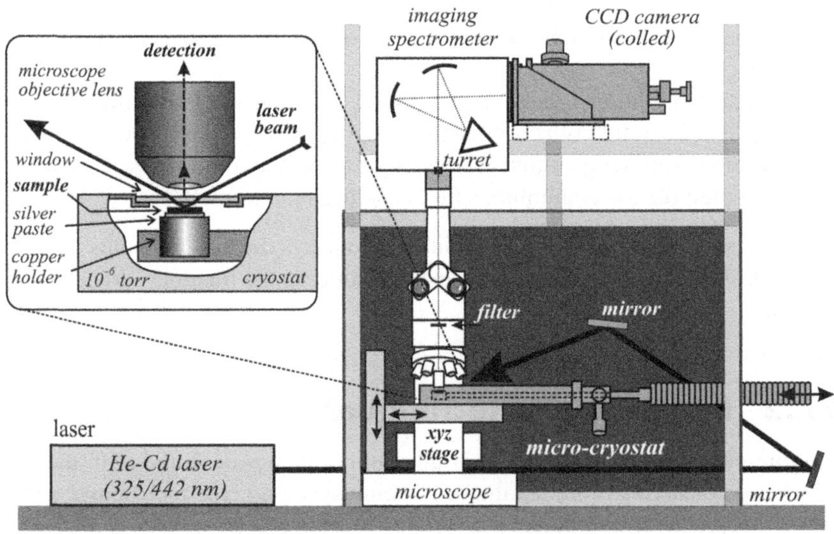

Fig. 17.2
Schematic of an imaging spectroscopic device consisting of a microscope, an imaging spectrometer and a CCD camera. The close-up shows the configuration of the sample and the excitation path in low-temperature measurements with a special cryostat. The excitation laser irradiates the sample at an oblique angle from one side through a gap under the objective (after Valenta and Linnros [8]).

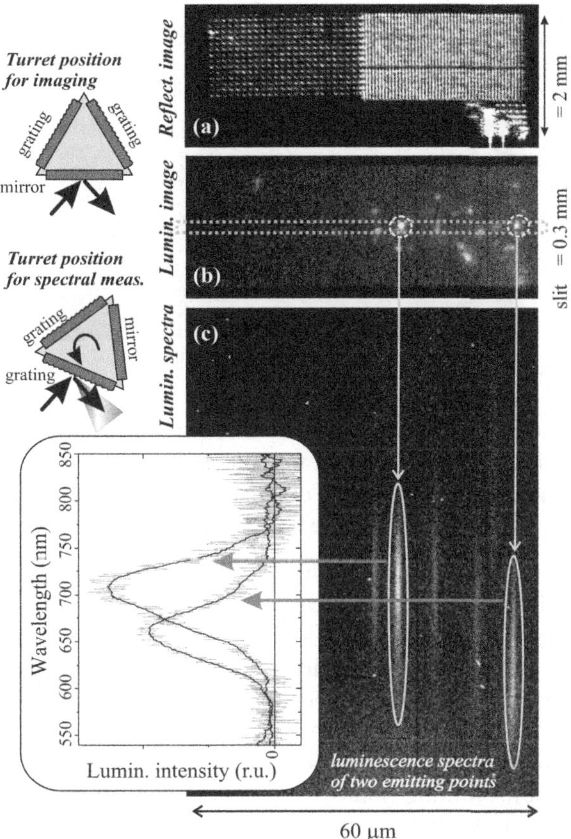

Fig. 17.3
Technique of measuring spectra of individual Si nanocrystals in a lithographic structure. A regular, square matrix with 1 μm (upper left) and 0.5 μm spacing (upper right). (a) The investigated structure is focalized and imaged with the aid of a reflected blue-laser light; the entrance slit is open and a mirror is used in the spectrometer instead of the grating, (b) a photoluminescence image of the same spot on the sample when excited with a UV laser, (c) a spectral image of the luminescent points after applying the entrance slit and inserting (and properly rotating) the diffraction grating in the spectrometer. The spectra of the emitting points can be obtained from the luminescence-intensity line profiles.

The measurement itself then proceeds as follows (Fig. 17.3):

(a) Finding the spot to be studied on the sample—the spectrometer is in the imaging mode, the entrance slit is fully open. To focalize and properly locate the sample, it is appropriate to use reflected or transmitted light (e.g. from a halogen lamp or the excitation laser) for the very first adjustments since the luminescence of the 'diluted' sample itself is very weak. Subsequently, 'promising' objects in the luminescence image are selected and shifted into the centre of the image.
(b) Applying the entrance slit—only light from those objects that are delimited by the slit enters the spectrometer. This step serves not as much to ensure sufficient spectral resolution as towards limiting the number of imaged objects in the direction perpendicular to the slit. We need to keep in mind that in normal microscopic images with a magnification of 100× the diffraction ring is projected to the scale of 0.06 mm (taking into consideration eqn (17.1) for NA = 0.8 and $\lambda = 800\,\text{nm}$). If the entrance slit is closed to, e.g., $\Delta\ell = 0.3\,\text{mm}$, the image of a point source will be several times smaller than the slit width and we shall thus not waste scarce photons.

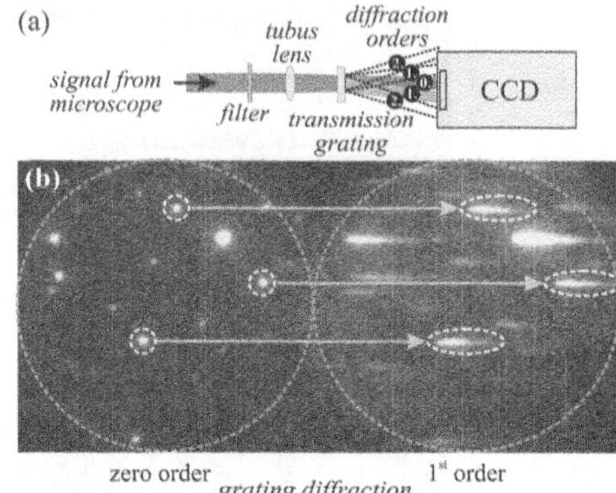

Fig. 17.4
Simultaneous detection of images and emission spectra of individual rhodamine 6G molecules deposited on glass. (a) Sketch of the experiment: the signal from the microscope is imaged (without the entrance slit) directly onto a CCD camera chip. The zeroth and first diffraction orders of the grating are incident within the active detection area. (b) Image of luminescence from a group of molecules (the zeroth diffraction order, left) and their spectral images (the first diffraction order, right). The images and spectra of three molecules are connected by arrows (the reader is encouraged to do the same for the remaining objects). The corresponding wavelengths need to be assigned separately for each object.

(c) Rotating the grating to the dispersion mode and measuring the spectrum—in a direction perpendicular to the slit, a point source will turn into a spectral-trace image and we can obtain the spectrum as an intensity profile of the corresponding rows of the CCD detector.

A noteworthy fact here is that if the image of a point source is smaller than the entrance slit width $\Delta\ell$, the spectral resolution will be determined by the size of the image of the emitting point, multiplied by the reciprocal linear dispersion L^{-1} at the detector (considering $L^{-1} = 4\,\text{nm/mm}$ and the example given above, we achieve a spectral resolution of $\Delta\lambda = 4 \times 0.06 = 0.24\,\text{nm}$, that is, much better than $\Delta\lambda = L^{-1}\Delta\ell = 1.2\,\text{nm}$ for a macroscopic object delimited by the slit). Therefore, theoretically, we could even work without the slit if very few emitting points were present and neither their images nor their spectra overlapped. Nevertheless, in this case we have to correct the wavelength scale—relate it to the zeroth order given by the position of the emitting point's image (Fig. 17.4).[1]

17.2.2 Scanning techniques

Laser confocal microscope

This type of microscope, highly technically advanced, comparatively affordable and quite common nowadays, makes use of excitation by a laser beam focused with a microscope objective into an ideally small diffraction-limited point. The confocal effect consists in the insertion of an iris diaphragm into the imaging plane of the system (a microscope objective and a tubus lens), to allow, for further processing, just the image of such an excited spot that is located both in the relevant object plane and on the optical axis of the system to be transmitted. If the light source is in a different plane or away from the optical axis, its contribution to the detected signal is strongly limited. Scanning is then achieved either by moving the sample fixed on a precise (piezoelectric) x-y-(z) stage or by moving the excitation point using piezoelectrically positioned

[1] One can use a similar configuration to measure changes in the orientation of the dipole moments of molecules when inserting, instead of the grating, a Wollaston polarizing prism, which divides the image into two shifted components with perpendicular polarizations.

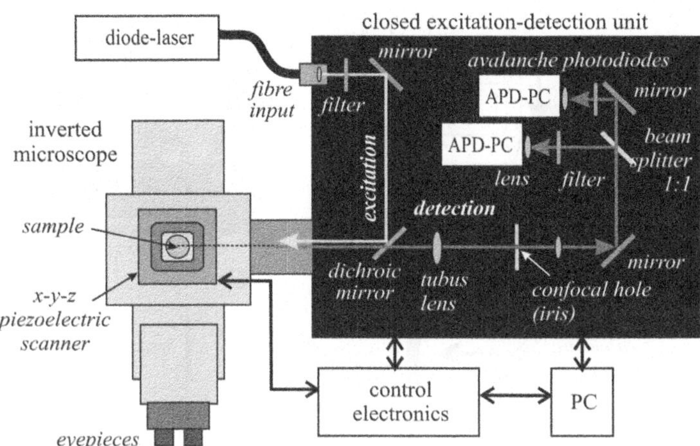

Fig. 17.5
Configuration of a confocal microscope with a scanning table and a detection unit consisting of two detectors—avalanche photodiodes—along with the electronics capable of recording the exact times of detection events at both detectors.

mirrors. The image resolution is limited (in the $\{x, y\}$ plane perpendicular to the optical axis) by the size of the diffraction ring of the excitation laser and is thus somewhat better than for a wide-field microscope where the critical factor is the diffraction pattern of the emitted light (the wavelength of which is longer than that of the excitation light; in the case of two-photon excitation, it is naturally the other way round).

To register luminescence radiation, detectors like those explained in Section 2.2 are used. A point or single-channel detector is sufficient to detect the image in a confocal microscope; usually either an avalanche photodiode (APD) or a photomultiplier (in the photon counting mode) is employed. To obtain a spectrum when using a single-channel detector, a dispersion component—a grating or a prism—has to be inserted in front of the detector. Then, the spectrum can be scanned in the wavelength-by-wavelength mode by rotating the dispersion component, or the entire spectrum can be obtained at once if a multichannel detector (a linear CCD camera or an array of photodiodes) is employed. Using dedicated electronics, it is possible to acquire the statistics of incoming photons or even the correlation function if a pair of APDs is used (see Section 17.5). By applying a high-repetition-rate pulsed laser it is possible to measure the kinetics of luminescence decay using the method of time-correlated photon counting (Subsection 2.9.3). A typical configuration of the confocal microscope enabling detection of photon statistics is shown in Fig. 17.5.

Scanning near-field optical microscope

Up to now, several—mostly rather complicated—methods of somehow circumventing the diffraction limit of optical imaging have been proposed (see, e.g., [9]). In spectroscopy experiments, only those based on near field optical techniques have been used so far. The term near–field refers to the area around a source of electromagnetic radiation closer than the wavelength λ of the emitted or absorbed radiation. Such an optical field can be exploited for imaging via the near-field scanning optical microscopy (NSOM or SNOM) technique [10]. The probe of such a microscope consists of an optical fibre

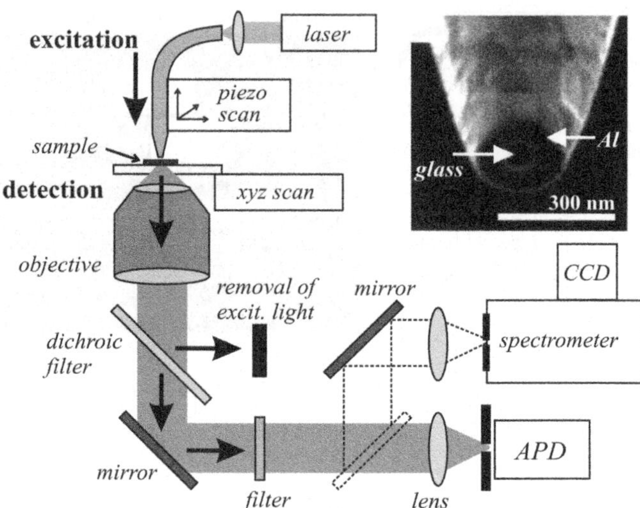

Fig. 17.6
Schematic of an NSOM microscope in the most frequent micro-spectroscopy configuration. The inset image shows an example of an aluminium-coated tip with an opening made using an ion-beam technique. The length of the scale bar is 300 nm. After Novotny and Hecht [10].

tapered to a sharp apex and coated with a metal layer, except at the centre of the apex, where an opening in the metal coating smaller than the wavelength of the light is intentionally left out (with usual sizes between 50 and 100 nm, see Fig. 17.6). If such an aperture is scanned above the sample at a distance smaller than the wavelength of the light, the resolution is approximately determined by the size of the aperture. The best resolution ever demonstrated amounted to $\lambda/40$, but the commonly achieved values range around $\lambda/5$ only. Scanning is realized by means of a piezo-element. The distance of the probe from the sample surface is usually determined by measuring the tunnelling current which is why the fibre is coated by metal. The optical probe can generally be used either in the collecting or exciting mode—that is, it either collects the emitted photons and guides them to the detector, or it transfers the excitation light from the laser to the sample and the emitted radiation is then collected by means of common far-field optics, e.g. by a microscope objective (see Fig. 17.6). The latter mode is used more frequently since it is significantly more sensitive. The higher sensitivity results from the fact that the efficiency of light collection using an aperture smaller than the wavelength is very low (for an aperture sized about $\lambda/10$, the light transmission efficiency is only 10^{-2}–10^{-6} depending on the probe manufacturing method [7]). On the other hand, the collection mode enables one to detect evanescent fields, which do not propagate farther away from the sample. The interaction of the aperture with the near field depends strongly on the distance between the sample and the probe, its shape and surface properties and, therefore, the interpretation of NSOM images may not be a simple task. (The readers will find a detailed discussion of NSOM microscopy in, e.g., the excellent monograph by Novotny and Hecht [10].)

Scanning tunnelling optical microscope

Yet another way to excite emission of light from individual quantum objects is via direct injection of electrons and holes. The tip of a scanning tunnelling microscope (STM) readily springs to mind as a possible source for the local injection of charge carriers. For this purpose, however, the STM system must be modified—supplemented with collection and detection optical systems (which is not an easy task; it is difficult to achieve a collecting angle as high as in optical microscopes due to the limited space around the scanning probe). Most often, a metal tip is applied but a strongly doped semiconductor will also do the trick. The beneficial feature of charge injection from an STM tip is the possibility to tune the energy of the injected electron or hole by adjusting the voltage between the tip and the sample.

The injection mechanisms can be divided into two groups: *inelastic* and *elastic scattering*, that is, whether the tunnelling electron loses part of its energy during the tunnelling process itself or not, respectively [7]. It has been established that in inelastic scattering, particularly in the case of electron tunnelling from the STM tip to a metal, electromagnetic modes of the coupled tip–sample system are created (surface plasmons). Their radiative decay then represents the energy emitted in the form of photons (Fig. 17.7(a)). This process may be denoted also as *inverse photoemission*. In the case of elastic scattering, the carriers are injected without any loss of energy in the tunnelling process, i.e. electronic states of the investigated system (e.g. a semiconductor nanocrystal) are resonantly populated (Fig. 17.7(b)–(d)). If the voltage is set to match the energy difference of a suitable excited state, resonant injection can take place directly into the state from which radiative or non-radiative recombination occurs. If the energy of the injected electron or hole is somewhat larger, the excess energy dissipates via, e.g., phonon emission or Auger recombination. After attaining a certain energy threshold, impact ionization, i.e. the generation of secondary *e–h* pairs, can occur (Fig. 17.7(d)). It should be noted that surface states acting as non-radiative recombination centres may

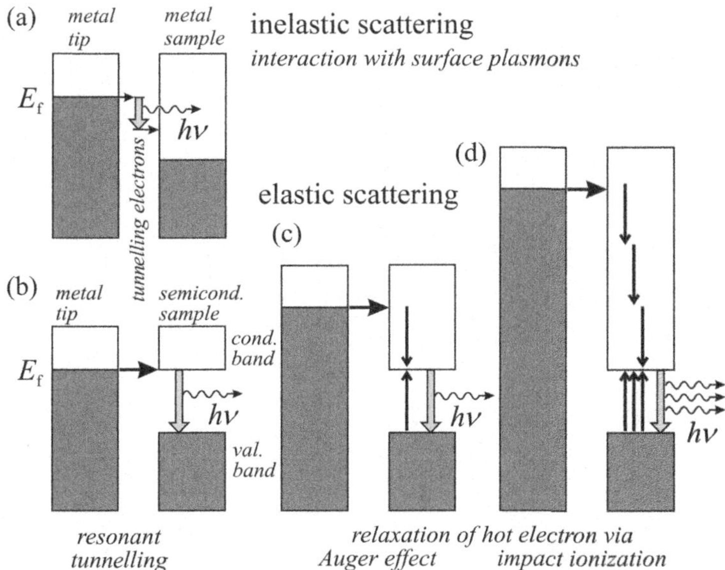

Fig. 17.7
Various processes of tunnelling and radiative recombination between the metal tip of an STM microscope and a metal (a) or semiconductor (b)–(d) sample. Adapted from Gustafsson *et al.* [7].

impede efficient light generation through the STM tip. The influence of these states must be minimized by, e.g., suitable passivation of the surface bonds.

To disperse the STM-tip-induced light emission spectrally, high-luminosity spectrographs must be used, because of the low photon fluxes. For the same reason, only high-sensitive photodetectors like modules with avalanche photodiodes are applied; particularly for recording the spectrally resolved STM-induced emission, CCD multichannel detectors are a must, since they enable one to record the spectrum of an extremely weak luminous flux through extended integration times (Section 2.2).

The STM provides a unique opportunity to characterize the energy levels of individual nanocrystals. Banin *et al.* published an extraordinary example of the experimental determination of the density of states in an individual quantum dot; see Fig. 17.8 [11]. The measurement is based on the fact that the tunnelling current is directly proportional to the local density of states. Under a given preset voltage, tunnelling can only proceed through states with a corresponding or lower energy (including the energy necessary to add a charge to a nanocrystal—the so-called charging energy E_c). As the voltage is increased (when measuring the current–voltage curve), the tunnelling current jumps up every time the next state is reached (see Fig. 17.8(a)). If we take the derivative of the current–voltage curve, we obtain a plot of the density of states (see Fig. 17.8(b)). At the same time, the sample topography revealed by scanning the STM tip can be used to establish the size of the nanocrystal (taking into account the height of the particle with respect to the surface since this value is not much affected by the tip shape—in contrast to the lateral dimensions), suggesting that it is possible to compare the distributions of the density of states for different sizes of nanocrystal (Fig. 17.8(c)).

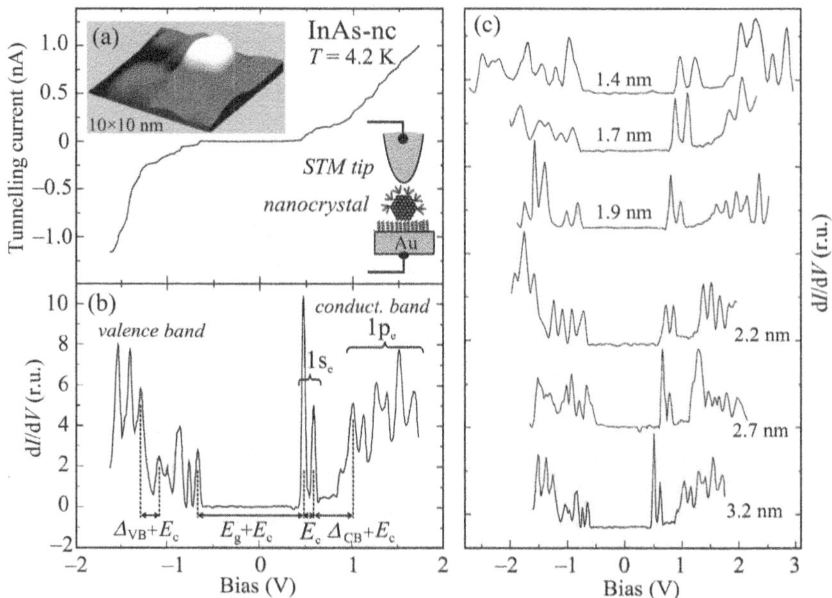

Fig. 17.8
Tunnelling spectroscopy of individual InAs nanocrystals (passivated with organic ligands), deposited on a gold surface: (a) The current–voltage curve and an STM image ($10 \times 10\,nm^2$) of an InAs nanocrystal of radius 3.2 nm. (b) Spectrum of tunnelling conductivity (proportional to the density of states in the nanocrystal) obtained by taking the derivative of the current–voltage curve: Δ_{CB} and Δ_{VB} are energy differences between the two lowest states in the conduction and valence bands, respectively. In the conduction band (unlike a more complicated situation in the valence band), one can clearly resolve lifted degeneracy of the states (a doublet and a hexaplet) caused by the step-by-step occupation of the 1 s and 1 p states following the Pauli exclusion principle. (c) Spectra of tunnelling conductivity for nanocrystals of various dimensions. Adapted from Banin *et al.* [11], reproduced courtesy of Macmillan Publishers Ltd.

17.3 Preparation of samples

Preparation of suitable samples is a key prerequisite for successful spectroscopic measurements of individual nanostructures. In general, two types of procedures are applied; the former uses advanced techniques of electron and ion-beam lithography, whereas the latter is based on producing diluted colloidal dispersions of nanoparticles.

17.3.1 Electron- and ion-beam lithography

Lithography has been the cornerstone of semiconductor technology for as long as several decades. However, optical lithography (albeit in the far UV) does not have sufficient resolution to produce patterns on the tens of nanometres scale and, therefore, the lithographical procedure is carried out by means of an electron or possibly ion beam. Lithography can be applied either directly for the preparation of the nanostructures to be studied or just as an additional tool for some modifications.

A model example of using electron-beam lithography to produce isolated silicon nanocrystals is the technique developed by Linnros *et al.* at the Royal Institute of Technology in Stockholm (Fig. 17.9). The first step consists of etching silicon columns about 100 nm wide and 200 nm high, which can be arranged arbitrarily, e.g. in a square lattice with a period of 1 μm (see Fig. 17.3) so that they can be resolved in an optical microscope. This is followed by a two stage oxidation with the oxide layer being etched away after the first oxidation step. Oxidation parameters must be precisely adjusted so that, ultimately, the initial silicon nanocolumn finally gives rise to an SiO_2 column concealing in its upper part a small remnant of the original silicon material—a nanocrystal. (Here, one takes advantage of the so-called self-limiting oxidation effect, that is, a reduction in the oxidation rate due to the mechanical tension generated in the high-curvature surface area near the top of the column.)

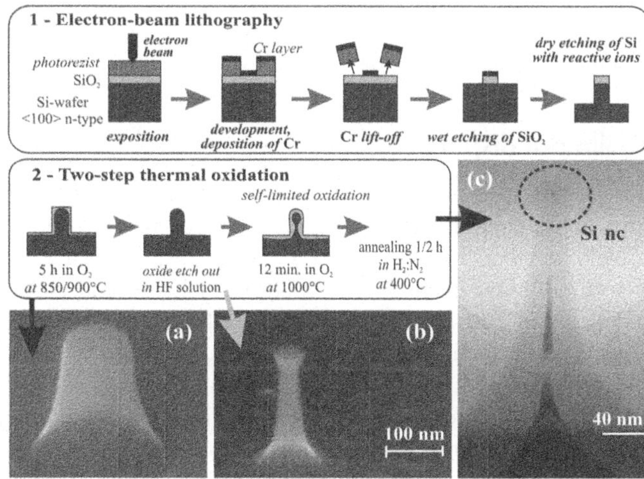

Fig. 17.9
Preparation of isolated silicon nanocrystals via lithography using an electron beam and two-stage thermal oxidation. The bottom images are from SEM (a),(b) and TEM (c) microscopes and show an Si column in various oxidation phases. After Valenta and Linnros [8], reproduced courtesy of Springer Science and Business Media.

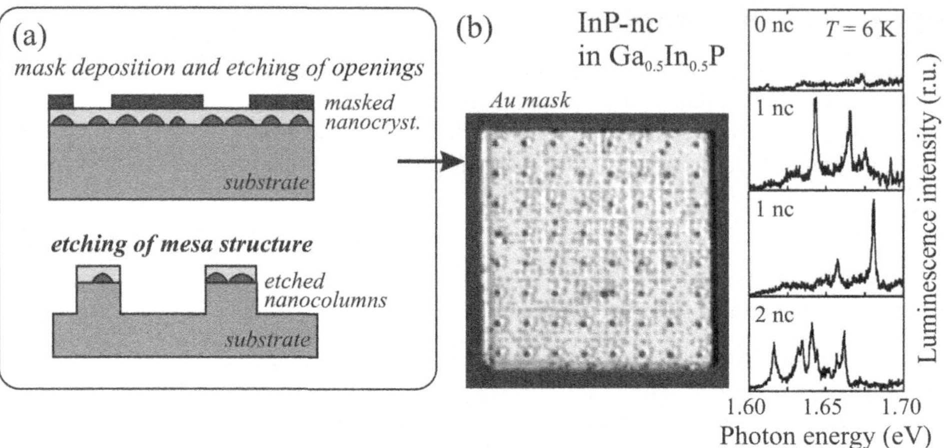

Fig. 17.10
(a) Two possible treatments of a sample with a nanocrystal layer to enable measurement of the spectra of individual nanocrystals. Top—deposition of a cladding (metal) layer into which sufficiently spaced small openings are etched using lithography or a focused ion beam; bottom—etching a mesa structure, i.e. columns on top of which a small section of the original nanocrystal layer is preserved. (b) An example of measuring emission spectra of InP nanocrystals from openings $700 \times 700\,\mathrm{nm}^2$ etched in a deposited gold layer. Right—examples of spectra obtained from four different openings and corresponding to the emission from none, one or two nanocrystals (nc). After Gustafsson *et al.* [7], reproduced courtesy of the American Institute of Physics.

Other preparation techniques produce a layer of nanocrystals arranged irregularly on the substrate, e.g. the self-organized growth of InAs nanocrystals on a GaAs surface (the Stranski–Krastanow method, Subsection 14.3.2). The surface density of these nanocrystals is usually 100–1000 $\mu\mathrm{m}^{-2}$, that is, too high to be resolved optically. Here, electron-beam lithography can be used in two ways: (1) A metal layer can be deposited on the sample and sufficiently spaced small openings can be made in it (Fig. 17.10(a) and (b)) using electron-beam lithography (or directly by a focused ion beam). (2) Almost the entire sample is etched away so that only small, distant columns remain (a so-called mesa structure), similarly to the first phase of preparing Si nanocrystals as shown in Fig. 17.9. Part of the original nanocrystal layer is preserved on top of the columns and, sometimes, it can contain just a single nanocrystal as desired (Fig. 17.10(a)).

17.3.2 Colloidal dispersions

Some of the preparation methods of nanocrystals take advantage of chemical reactions in solution (particularly in the case of type II-VI semiconductors). Or, alternatively, powder-form nanocrystals can be dispersed in a solvent also giving rise to a colloidal dispersion (a mixture of liquid and particles sized from several to about 200 nm, unlike in a true solution, where the mixture is homogeneous even on the molecular level). Then, the preparation of samples for spectroscopy of individual nanocrystals is relatively easy, as it consists just of getting rid of possibly present clusters (e.g. by filtration, sedimentation, and centrifugation) and sufficiently diluting the dispersion. Only a small amount of such dispersion is then deposited on a suitable substrate (quartz glass, silicon wafer, etc.) either simply by dropping a droplet or by deposition techniques such as spin-coating or dip-coating. Naturally, the fundamental requirement here is to keep the solvents, substrate and all laboratory equipment absolutely clean because most problems are caused by impurities luminescent in the same spectral region as the objects under investigation. The most active luminescing objects are usually various organic molecules, whose emission, however, can be quenched, as they are damaged by exposure to ultraviolet irradiation.

17.4 Experimental observation of luminescence from individual nanocrystals

Now, we will present examples of several phenomena observable only if individual nanocrystals are detected (these effects vanish if the measurement is averaged over an ensemble).

17.4.1 Hidden fine structure of luminescence spectra

When compared to an ensemble, luminescence spectra of individual nanocrystals almost always exhibit narrower spectral lines. The spectral line narrowing thus confirms the existence of inhomogeneous broadening due to the variance in properties as well as energy levels of individual nanocrystals in an ensemble. However, at room temperature, when luminescence lines are strongly broadened due to their interaction with phonons, the narrowing effect is rather small (see Fig. 17.11). Only at cryogenic temperatures, does it become possible to observe very narrow lines[2] and quite commonly also fine structure, which gets 'smeared' in the spectrum of a large ensemble. The differences in the phonon structures of the spectra in Fig. 17.11(b) are caused by different excitation methods. In the experiments of single-Si-nanocrystal spectroscopy, UV excitation (325 nm, i.e. 3.81 eV) was applied; whereas the UV excitation results in a direct transition, the resonant excitation of porous silicon requires participation of one phonon. A second phonon can then take part in the emission act—this is why the spectrum involves both one-phonon and two-phonon replicas.

An interesting result on the fine structure in the spectrum of individual $In_{0.6}Ga_{0.4}As$ quantum dots was published by Bayer *et al.* [12], as you can see in Fig. 17.12. These nanocrystals were fabricated by self-organized growth with a density of about 100 μm^{-2} on a GaAs substrate. Subsequently, the sample was etched to give rise to a mesa structure with columns of a diameter of about 100 nm, which were individually excited by a focused laser beam. On average, there was just a single nanocrystal on each column and its spectrum was measured as a function of the excitation intensity. With increasing excitation, several electron–hole pairs were generated per nanocrystal, and in addition to the transitions between the lowest states (s-states), p-state transitions

[2] With low excitation intensities, short detection times, and temperatures below 10 K, extremely narrow spectral lines with FWHM of about 0.1 meV (limited by the resolving power of the spectral system) were observed. This suggests that transitions take place between exactly defined levels, thus justifying the use of the term 'artificial atom'.

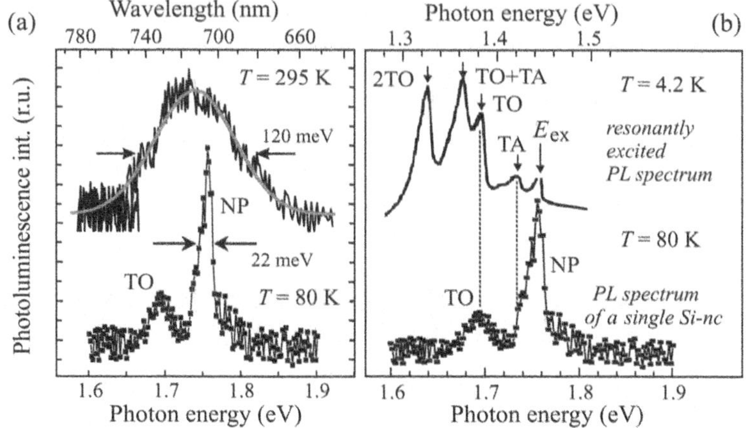

Fig. 17.11
Emission spectra of individual silicon nanocrystals prepared by electron lithography (see Fig. 17.9). (a) Comparison of a rather broad spectrum (FWHM of about 120 meV) at room temperature with a spectrum obtained at a temperature of 80 K where the no-phonon line (NP) and the TO-phonon replica are apparent. (b) Comparison of a low-temperature spectrum of one Si nanocrystal with a resonantly excited spectrum of an ensemble of Si nanocrystals in porous silicon ($T = 4.2$ K). Adapted from Valenta and Linnros [8].

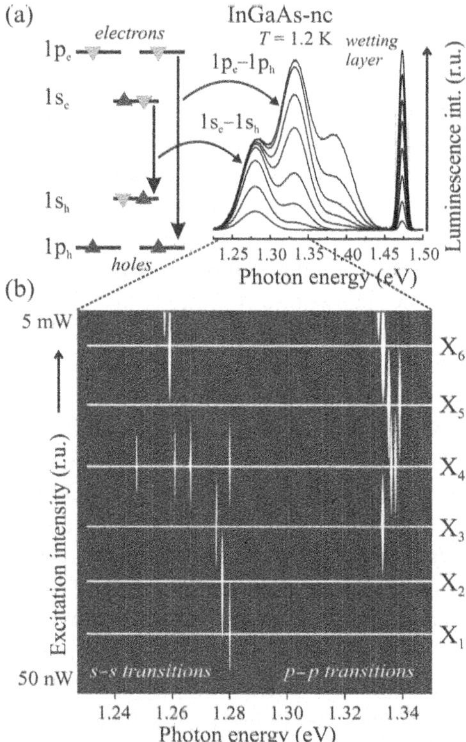

Fig. 17.12
Spectroscopy of individual quantum dots of $In_{0.6}Ga_{0.4}As$ in a mesa structure. (a) The luminescence spectrum of an ensemble of nanocrystals shows only broad bands corresponding to transitions between s- and p-states of electrons and holes (as schematically shown in the left part of the figure). (b) Spectral map representing changes in the luminescence spectrum of one nanocrystal (which can be obtained as a horizontal profile of the graph) as a function of the excitation intensity: the horizontal lines marked X_i show intensities for which the generation of i electron–hole pairs in one nanocrystal is most likely. Very narrow spectral lines enable the observation of spectral fine structure. Adapted from Bayer *et al.* [12], reproduced courtesy of Macmillan Publishers Ltd.

were observed as well. The spectrum obtained from an ensemble contains only broad bands corresponding to the s- and p-transitions (Fig. 17.12(a)), while the spectra of an individual nanocrystal (at a temperature of 1.2 K) reveal interesting fine structure within the bands (Fig. 17.12(b)). The fine structure of the s- and p-bands varies strongly with varying excitation intensity. The symbol X_i ($i = 1, 2, \ldots, 6$) in Fig. 17.12 denotes the number of generated

e–h pairs; similarly to an exciton which was described previously, in terms of the analogy with a hydrogen atom, here, one can look at the multiply excited nanocrystal as at excitonic helium, lithium, beryllium, boron, and carbon. The fine structure results from the Coulomb interaction between the charge carriers in the nanocrystal and from selection rules for the recombination of excitonic complexes (in analogy to the bound multiexciton complexes, see Subsection 7.2.1).

17.4.2 Changes in spectra: jumps, shifts, blinking

The discovery of fine structure in single-nanocrystal spectra, otherwise distorted by inhomogeneous broadening, is basically an expected result of single-nanocrystal spectroscopy. A more surprising phenomenon, however, is the considerable amount of variability in single-nanocrystal spectra with time, which shows up almost always if the measurements are repeated several times (without changing the experimental conditions). Time evolution of single-nanocrystal spectra might bring about

(a) spectral shifts in the band positions—so-called spectral diffusion;
(b) spectral jumps of the band between two or more discrete spectral positions; or
(c) switching on and off of the luminescence band—so-called intermittency or blinking. (Smooth changes in the band intensity usually do not occur.)

These phenomena are so characteristic that they are often accepted as evidence for the observation of a single nanocrystal.

Likely causes of the phenomena listed above are changes in the local electric field around the nanocrystal due to an electric charge trapped at the interface between the nanocrystal and the surrounding matrix or in its vicinity. Moreover, they can also result from a jump between several eigenstates of the nanocrystal, which may have long lifetimes.

As an example of these phenomena, let us mention measurements of single colloidal CdSe nanocrystals performed by Bawendi's group and shown in Figure 17.13 [13]. In this experiment, CdSe nanocrystals (coated/passivated with a ZnS shell) were mixed with PMMA (polymethylmethacrylate) and deposited on a quartz substrate using the spin-coating method. Then, the samples were placed in a cryostat and their spectra were measured by imaging spectroscopy. The luminescence spectrum of a single CdSe nanocrystal consists of a no-phonon line and a one-phonon or even two-phonon replica. If we concentrate our mind on the width of the no-phonon line (Fig. 17.13(a)) we find that it increases both with the spectrum accumulation time and with increasing excitation intensity (Fig. 17.13(b)). This broadening results from spectral diffusion and jumps as can be seen from repeated spectral measurements (Fig. 17.13(c)). The bottom panel in Figure 17.13(c) represents a histogram of the luminescence peak position obtained by processing 150 spectra (with an integration time of 0.1 s). It turns out that two positions of the emission peak are preferred—at 2.051 eV and 2.0535 eV. Obviously, these spectral shifts depend on the total amount of excitation energy supplied during the

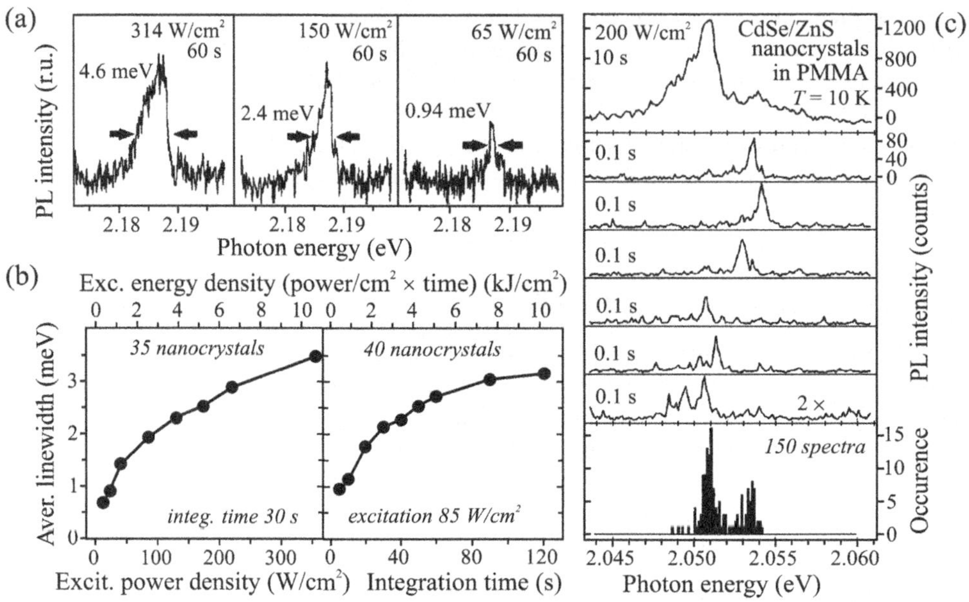

Fig. 17.13
Low-temperature spectral diffusion of individual CdSe/ZnS nanocrystals in PMMA ($T = 10$ K). (a) Photoluminescence spectrum of a 4.5-nm-diameter nanocrystal recorded for a period of 60 s at various excitation intensities. (b) The average width of a no-phonon line as a function of the excitation intensity (left) and of the detection (integration) time (right). (c) Manifestations of spectral diffusion during repeated spectral acquisition of the emission from a single 5.65-nm-diameter nanocrystal—the top spectrum was integrated for 10 s and the six spectra below it were each recorded for a period of 0.1 s. Adapted from Empedocles *et al.* [13].

measurements (the upper axis in Fig. 17.13(b)) as well as on the excitation wavelength (spectral separation from the absorption edge). The interpretation of these phenomena assumes the existence of a local electric field generated by charges trapped in the vicinity of the nanocrystal (the shifts of the nanocrystal's energy levels are then the manifestations of the Stark effect, which will be discussed in the following subsection). The amount of dissipated energy then determines how fast these charges can decay, diffuse, etc.

17.4.3 Stark effect

In order to get a deeper insight into the effects of local electric fields (generated when charges are localized close to the surface of a nanocrystal), it is important to study the influence of an external electric field on the optical properties of nanocrystals. The shift of energy levels due to an external electric field is called the *Stark effect*. Generally, one can observe a Stark shift of energy levels ΔE that depends linearly or quadratically on the electric field intensity F

$$\Delta E = \mu F + \frac{1}{2}\alpha F^2 + \cdots , \tag{17.2}$$

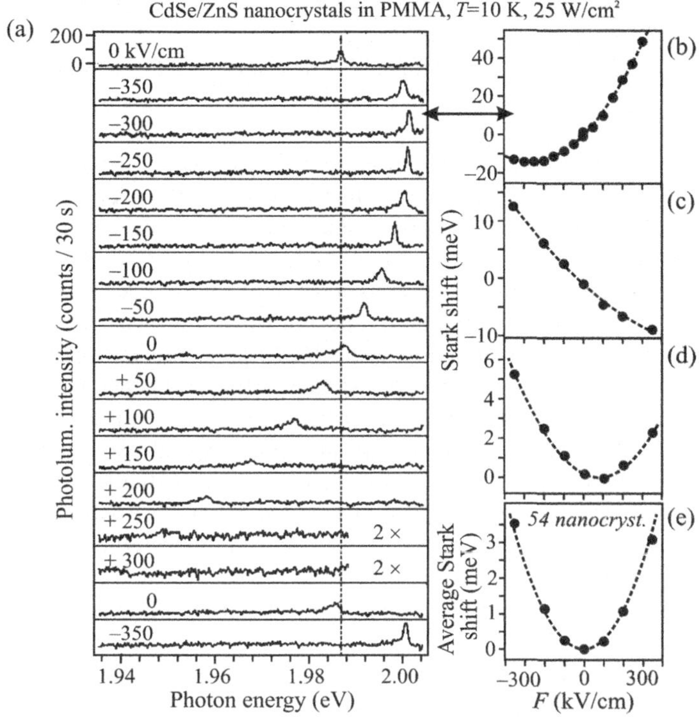

Fig. 17.14
Stark effect observed in the photoluminescence spectrum of a single CdSe nanocrystal (ZnS-coated, the mean diameter in the original ensemble equals 7.5 nm). (a) Repeated measurements of the spectrum of a single nanocrystal with applied external electric field of various magnitudes from –350 to +300 kV/cm. (b) Shift in the spectral position of the peak from panel (a) as a function of the electric field intensity. Panels (c) and (d) then present Stark shifts for two different 5.8-nm-diameter individual nanocrystals. (e) Average Stark shift obtained by processing 54 individual nanocrystals. Cw excitation with an Ar^+-laser, 514 nm, after Empedocles *et al.* [14].

but in some cases it can also contain higher terms. In (17.2), μ and α are the projections of the *dipole moment* of an excited state and of its *polarizability* into the direction of the electric field, respectively.[3]

The linear Stark effect occurs if a dipole is present in the nanocrystal (e.g. when an *e–h* pair is trapped in a surface state of the nanocrystal); the quadratic effect then manifests itself if the electric field polarizes an excited state, that is, an *e–h* pair or an exciton (a delocalized excitonic state in the nanocrystal). What can then be expected apart from the shifts of energy levels is reduced oscillator strength of optical transitions of the (quasi-)free exciton (see Fig. 17.14(a) [14]). The reason behind the reduced strength lies in the Coulomb interaction, as the electron and hole get farther apart and thus the overlap of their wavefunctions decreases. However, due to the spatial localization in a nanocrystal, this effect will be much weaker than in a bulk semiconductor.

As expected, when measuring ensembles of (immobile, fixed) nanocrystals, one observes only the quadratic Stark effect—the possible dipole moments of the excited states in nanocrystals are oriented randomly and, on average, their effects cancel out. On the other hand, measurements performed

[3] To determine the magnitude of the electric field inside a semiconductor nanocrystal in a matrix, we have to take into account the effect of the local field. Therefore, the intensity of the field inside a nanocrystal F_i is not equal to the external field F, but $F_i = 3F/(2 + \varepsilon/\varepsilon_b)$ holds true instead; ε and ε_b are the relative permittivities of the nanocrystal and the matrix (barrier), respectively. In the case of a semiconductor embedded in an insulator (e.g. in glass), the semiconductor permittivity is always higher $\varepsilon > \varepsilon_b$ and the internal field is thus weaker than the external one. See also Subsection 17.4.4.

on single nanocrystals, e.g. CdSe [14], have shown that most nanocrystals exhibit both the linear and the quadratic Stark effects, but, in very different ratios (Fig. 17.14). Experimentally, the Stark effect measurements are performed by depositing a colloidal dispersion on a quartz substrate covered with 5-μm-spaced Ti-Au photolithographically prepared electrodes. Only those nanocrystals that settle down in between the electrodes are then investigated. It is observed from panel (e) that in this case the average shift scales quadratically with the field intensity; the linear term is cancelled out due to averaging.

Comparing spectral shifts measured under application of an external electric field with spontaneous shifts due to spectral diffusion enables us to deduce that spectral diffusion is caused by the local field with an intensity of the order of 10^5 V/cm. This roughly corresponds to the field generated by a single elementary charge localized near the surface of the nanocrystal. Under such a strong field, phenomena which complicate the theoretical description emerge, examples of which are the strong mixing of states near the bandgap edge as well as breaking of the inverse symmetry of the exciton wavefunction. This way one can explain the unexpectedly strong exciton–LO-phonon interaction observed in single CdSe nanocrystals: polarization of excitons in the strong electric field enhances their Fröhlich interaction with LO-phonons in this polar semiconductor.

The entanglement of the phenomena of spectral diffusion and Stark shift is nicely illustrated in Fig. 17.15. Here, luminescence spectra of a single CdSe nanocrystal are successively measured (100 repetitions with a detection time of 30 s per spectrum) while changing the applied fields in a series of values of 0, –350, 0, +350, 0, –350 kV/cm, etc. One can see that the position of the emission peak jumps in a step-like manner at certain moments and, at the same time, the Stark shift magnitude changes. Within the interval between the 26th and 32nd minutes of the measurement, the Stark effect is almost exactly quadratic (both field polarities yield the same peak position), signifying that the projection of the dipole moment of the excited state into the direction of the external electric field is zero. A more detailed analysis shows that the polarizability remains constant during the measurements while the magnitude of the dipole moment of the excited state changes by almost a factor of 50. This leads to the conclusion that both the spectral diffusion and Stark effect are consequences of variations in the induced dipole moment of the excited state, produced by either the local or external electric field.

17.4.4 Luminescence polarization

In addition to the spectral and temporal dependence of the luminescence signal of a single nanocrystal, there is another property which can be characterized, namely, luminescence polarization. It can be determined by inserting a polarization filter (the so-called analyser) into the detection path in front of the detector; we can then rotate the filter and thus acquire polarization-resolved images or spectra. Moreover, we can also rotate the polarization of the

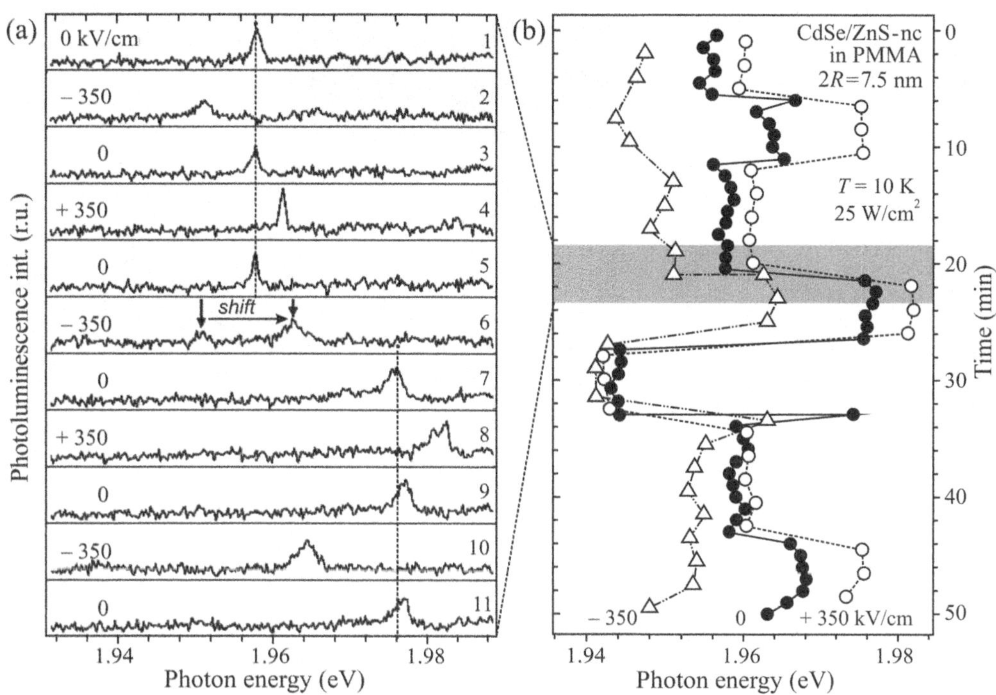

Fig. 17.15
The impact of spectral diffusion on the Stark shift in the photoluminescence spectra of a single CdSe/ZnS nanocrystal. (a) Eleven spectra measured successively (detection time 30 s) under different polarities of the applied external electric field (± 350 kV/cm). A sudden spectral shift occurred in spectrum 6 and it consequently led to a step-like change in Stark shifts. (b) Spectral positions of the emission peak during the entire 50-minute experiment (100 spectra); the grey rectangular area corresponds to the time span of panel (a). Excitation with an Ar^+-laser, 514 nm, after Empedocles *et al.* [14].

excitation light and observe the corresponding changes in luminescence. What needs to be emphasized here is that in all polarization-resolved experiments, the response function of the imaging and spectroscopic system under polarized light needs to be carefully investigated, since the reflectivity of various optical elements (particularly dielectric mirrors and diffraction gratings) can differ significantly for polarization parallel with or perpendicular to the plane of incidence.

Figure 17.16 [15] displays an example of polarization-resolved photoluminescence measurements of individual silicon nanocrystals, as prepared by a lithographic method and introduced in Fig. 17.9. The plot of the luminescence intensity as a function of the analyser orientation greatly varies for individual nanocrystals but it can always be very well fitted using a squared sine function. The degree of polarization p (the shorter term *polarization* is sometimes used [16]),[4] is then determined by

[4] In addition, in fluorescence spectroscopy of organic molecules, another quantity called *polarization anisotropy* is often applied. It is defined as the ratio $a = (I_\parallel - I_\perp)/(I_\parallel + 2I_\perp)$, where $I_\parallel$ and $I_\perp$ denote the intensities measured in the direction parallel with or perpendicular to the orientation of the emitting dipole (or, alternatively, of the polarization of the excitation light, which 'selects' from the ensemble those molecules that have an absorption dipole moment oriented along

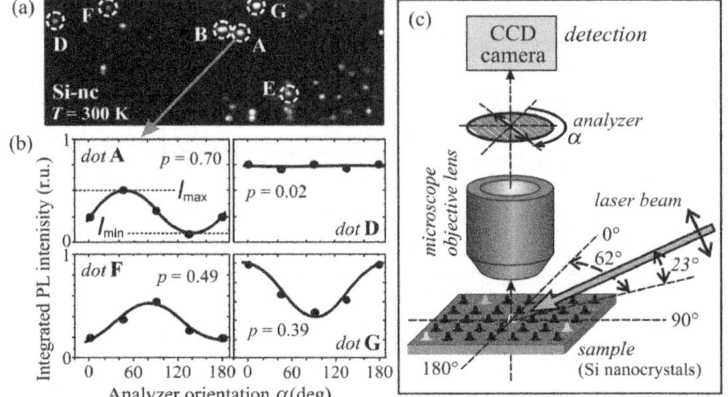

Fig. 17.16
Polarization of photoluminescence from individual silicon nanocrystals. (a) A luminescence image of a group of light-emitting nanocrystals; some of the measured objects are highlighted. (b) Behaviour of the integrated luminescence signal as a function of the orientation of the analyser (inserted inside a microscope, between the objective and the CCD camera); the 0° angle corresponds to the horizontal direction in panel (a). (c) Schematic of the experimental setup indicating also the polarization of the exciting He-Cd laser (325 nm) projected at $\alpha = 62°$. Room temperature, after Valenta *et al.* [15].

$$p = \frac{I_{\max} - I_{\min}}{I_{\max} + I_{\min}}, \tag{17.3}$$

where $I_{\max}$ and $I_{\min}$ are the maximum and minimum intensities obtained by measuring the polarization dependence. The limiting values $p = 1$ and $p = 0$ correspond to a completely polarized and non-polarized signal, respectively.

What is the reason for the partial linear polarization of luminescence of a single nanocrystal? The primary reason usually lies in *breaking of the perfect spherical symmetry of the nanocrystal.* A wide range of preparation methods yield nanocrystals of the shape of a rotational ellipsoid—slightly or markedly cigar- or disk-shaped. The asymmetric shape then leads to an increase in the transition probability along the ellipsoid major axis due to the dielectric confinement effect (see below) or alternatively it can result in a preferred orientation of the transition moment.

The degree of polarization of a single molecule with a given orientation of the transition dipole moment should always be equal to unity. This is because the strength of the optical transition is proportional to the squared product of the transition dipole moment $\boldsymbol{\mu}$ with the polarization vector of the emitted light $\mathbf{P}$, that is $|\boldsymbol{\mu} \cdot \mathbf{P}|^2 \simeq \cos^2\theta \cos^2\phi$, where ϕ is the angle formed by $\boldsymbol{\mu}$ and the sample plane and θ is the angle between the projection of $\boldsymbol{\mu}$ to the sample plane and the polarization vector (oriented in the same direction as the analyser), see Fig. 17.17(a). This implies that the polarization-resolved emission intensity achieves its maximum value $I_{\max} = |\boldsymbol{\mu}|^2 \cos^2\phi$ whenever the direction of $\mathbf{P}$ matches the projection of $\boldsymbol{\mu}$ into the sample plane and, on the other hand, a minimum value $I_{\min} = 0$ for the mutual perpendicular orientations of $\mathbf{P}$ and the projection of $\boldsymbol{\mu}$. Therefore, based on the polarization-resolved measurement, one can only establish the orientation of the projection of the transition dipole moment. Information about the angle formed by the dipole and the sample

the excitation polarization) and correspond to the intensities denoted as $I_{\max}$ and $I_{\min}$ in eqn (17.3). The reason why anisotropy is introduced in fluorescence spectroscopy is that it describes appropriately depolarization of light emitted from an ensemble of molecules excited by polarized light.

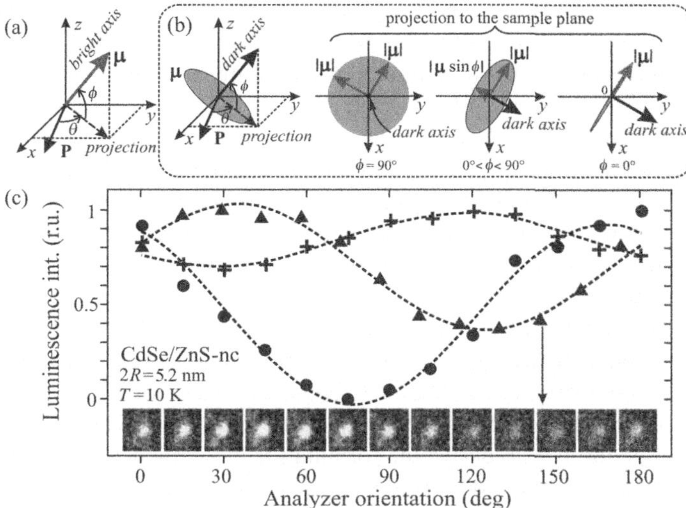

Fig. 17.17
(a) Schematic of the relative orientation of the emission dipole μ and the polarizer **P** in the case of a molecule characterized by a 'radiative' axis (dipole orientation). (b) Diagram of a degenerate transition where the dipole can have an arbitrary orientation within a certain plane and is characterized by the so-called 'dark' axis; three cases with different orientations of the dark axis as projected into the sample plane are displayed. (c) Dependence of the photoluminescence signal of three different CdSe nanocrystals on the polarizer orientation. The bottom row of pictures shows polarization-resolved images of emission from a single nanocrystal (represented by triangles in the plot). After Empedocles *et al.* [17], reproduced courtesy of MacMillan Publishers Ltd.

plane is then hidden in the polarization signal amplitude, the quantitative determination of which is hardly possible. However, an interesting scenario occurs in high-symmetry molecules (e.g. benzene), in which the dipole moment can point in an arbitrary direction within a plane perpendicular to the symmetry axis. This axis can be called the *dark axis* (Fig. 17.17(b)). If we now relate the angles θ and ϕ to the dark axis, the transition strength becomes proportional to $(1 - \cos^2\theta\cos^2\phi)$. The maximum and minimum of the polarization-resolved signal are then equal to $I_{\max} = |\mu|^2$ and $I_{\min} = |\mu|^2 \sin^2\phi$. This means that based on the $I_{\min}/I_{\max}$ ratio, the angle ϕ (more precisely $\pm\,\phi$) can be determined and thus also the 3D spatial orientation of the dark axis. This has been experimentally established by Empedocles *et al.* for CdSe nanocrystals [17]. These nanocrystals have the wurtzite crystal structure and, in addition to that, they are slightly elongated (the ratio of the longer to the shorter axis of the rotational ellipsoid ranges between 1.1 and 1.2). In this case, theory predicts that the emission from the lowest state is spin-forbidden but can occur thanks to the mixing of the lowest state with higher allowed states. Out of three of these states, one has a 'radiative' dipole oriented along the **c** axis of the wurtzite structure and the other two have a degenerate dipole oriented arbitrarily within a plane perpendicular to the **c** axis, resulting in the correspondence between this axis and the dark axis. The experimental results can then be interpreted as a signal coming from the transitions characterized by the dark axis since the degree of polarization of the signal from individual CdSe nanocrystals strongly varies and does not attain unity, see Fig. 17.17(c).

Let us note that the experiments on CdSe nanocrystals clearly indicate that the dipole moment of the excited state, which is involved in the Stark effect (see Fig. 17.14), is not correlated with the polarization of the emission radiation in any way, that is, with the orientation of the emission transition dipole moment. This means that under application of an external electric field of the order of 10^5 V/cm, no changes in the degree and orientation of luminescence polarization are observed [13]. Such a result corresponds to expectations because the dipole moment of the excited state is governed mainly by the envelope functions of the electron and hole wavefunctions while it is the crystal wavefunctions that determine predominantly the dipole moment of the transition, which is, as a result, oriented along one of the crystallographic axes.

However, the interpretation of polarization measurements in Si nanocrystals (Fig. 17.16) cannot be based on the same reasoning as in the case of CdSe nanocrystals since the silicon crystalline structure possesses a high degree of symmetry. Moreover, when using lithographic etching of a crystalline wafer, the crystalline orientation of all nanocrystals in the sample is the same and as such cannot be the source of a difference in the degree of polarization. In silicon nanocrystals, the source of luminescence polarization is the so-called *dielectric confinement effect,* which we already came across in Chapter 13 (a low-dimensional structure characterized by a dielectric constant ε surrounded by a matrix with a dielectric constant $\varepsilon_b < \varepsilon$). In the following, we will give some hints about how to derive the formula for the influence of the dielectric confinement effect on the polarization of (photoluminescence) absorption and emission of silicon nanowires [18].

The basic idea lies in a well-known fact usually taught in the very first courses on electrostatics: at an interface between two dielectric media, the tangential component of the electric field vector is conserved while the ratio of the normal components is governed by the magnitude of the corresponding dielectric constants. Consequently, let us consider a nanowire (cylinder) with a dielectric constant ε surrounded by an external material with a dielectric constant $\varepsilon_b < \varepsilon$. If we put the wire into an external electric field F, the field inside the nanowire will have the component parallel to the wire axis $F_{i\parallel}$ identical to the external field, while the perpendicular component $F_{i\perp}$ will be reduced in the ratio (Problem 17/3)

$$F_{i\perp} = A\,F_{\perp} = \frac{2\varepsilon_b}{(\varepsilon + \varepsilon_b)} F_{\perp}; \tag{17.4}$$

obviously, $A < 1$. Although the validity of relation (17.4) was originally limited to static electric fields, we can assume it will also hold for optical fields if the diameter of the nanowire is very small compared to the wavelength of light (which is obviously satisfied for nanowires of diameter $2r < 100\,\mathrm{nm}$ and the visible spectral region).

The first consequence of the reduction of the perpendicular field component is a reduced absorption probability of light polarized perpendicularly to the axis of the wire, or, in other words, a lower absorption coefficient $\alpha_{\perp}$. Since we can consider the nanowire to be an optically thin material ($\alpha_{\perp} 2r \ll 1$), the photoluminescence intensity scales linearly with $\alpha_{\perp}$ (see Section 2.1). Therefore, the amplitude of the photoluminescence signal when excited by polarized light with $F_{i\parallel}$ will be larger than for the perpendicular orientation of excitation polarization $F_{i\perp}$ and this difference will obviously increase with increasing ratio $(\varepsilon/\varepsilon_b)$.[5]

[5] The ratio of the corresponding luminescence signals satisfies $(I_{\parallel}/I_{\perp}) = 1/A^2$.

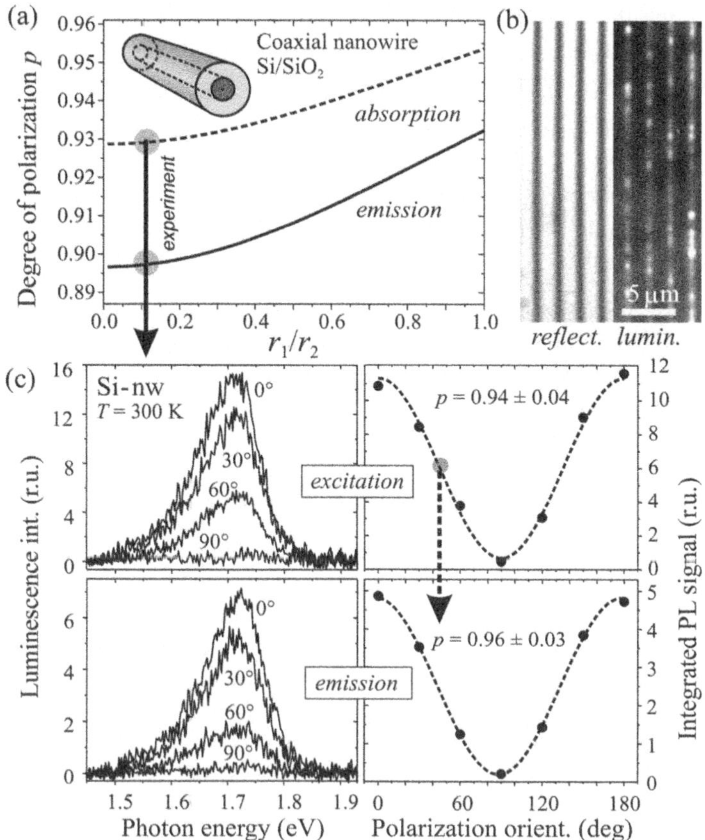

Fig. 17.18
Polarization of light absorption and emission from silicon nanowires passivated with SiO_2. (a) Theoretical dependence of the degree of polarization on the radius of the core and shell, after Ruda and Shik [18]. (b) Reflection and photoluminescence image of several Si nanowires (the spacing of the nanowires is 2 μm). (c) Top row-left: luminescence spectra of an Si nanowire obtained for various polarization orientations of the exciting laser. The angles are relative to the orientation of the wire, zero corresponds to the wire axis. Right: a plot of the integrated intensity as a function of the polarization orientation fitted with the squared cosine function. The bottom panels show spectra and integrated emission intensities for various orientations of the polarization filter (analyser) in the detection path and for the orientation of the excitation polarization fixed at 45°.

The second consequence of the polarization anisotropy discussed above is the high polarization degree of luminescence radiation from nanowires. Ruda and Shik [18] derived an explicit expression for the degree of emission polarization p and extended their calculations to include the case of a coaxial nanosystem composed of an inner ('active') wire with dielectric constant ε and radius r_1, which is encapsulated with a cylindrical shell (passivating layer) of dielectric constant ε_p and radius r_2. The whole system is embedded in a dielectric medium with dielectric constant ε_b. The result of their calculations enables us to compare directly theory with experimental data, as shown in Figure 17.18.

Panel (a) in Fig. 17.18 represents the calculated values of p for both the excitation and emission polarizations in the case of a silicon wire covered with an SiO_2 layer in air ($\varepsilon = 12$, $\varepsilon_p = 2.1$, $\varepsilon_b = 1$), where the ratio of the inner and outer radii r_1/r_2 ranges from 0 to 1 (i.e. the thickness of the shell ranges from infinity to zero). As we can see, the degree of polarization of absorption is larger than the emission polarization for all values of r_1/r_2. The result may be compared with measurements performed on individual lithographically prepared Si nanowires with $r_1/r_2 \simeq 0.1$, photos of which are shown in panel (b). The comparison of theory with experiment is depicted in Fig. 17.18(c). The measured value of the degree of polarization of absorption (excitation) matches the theory very well while emission polarization is slightly higher than the theoretical prediction. The increased polarization degree of emission may have various causes, e.g.

(a) the emission transition moments are anisotropic due to quantum confinement [19] or
(b) a polarization memory effect sets in, i.e. the preferred excitation of dipoles oriented along the nanowire is reflected in the preferred orientation of the emitting dipoles.

In any case, the degree of polarization of luminescence in silicon (as well as other) nanowires is very high and can be significant in applications, e.g. in the detection of polarized light or in achieving stimulated emission (the number of emission modes being reduced).

17.4.5 Luminescence intermittency—blinking

Probably the most surprising and interesting effect related to light emission from individual nanocrystals is so-called *blinking*—random switching between the emitting and dark states under continuous excitation. This *emission intermittency* is observed in virtually all individual emitting quantum objects—molecules, fluorescing proteins, polymer chains, as well as semiconductor nanocrystals and nanowires. The surprising point is that intermittency manifests itself in a very similar way in all these objects, which are indeed very different. The time-scale on which the switching occurs can range over several orders of magnitude, from the shortest measurable intervals (ns) to very long times (minutes) [20].

It is important to realize that intermittency can be of essential importance when considering applications of light-emitting nanocrystals or when interpreting and comparing measurements of individual nanocrystals as well as of nanocrystal ensembles. As an example, let us discuss the effects intermittency has on an image of luminescent Si nanocrystals obtained in a scanning confocal microscope (Fig. 17.19) [21]. The image of a $6 \times 6\,\mu m^2$ area consists of 256 rows scanned successively as the sample was shifted by a piezoelectric scanner with a step of 23 nm. Therefore, the image of one nanocrystal (the size

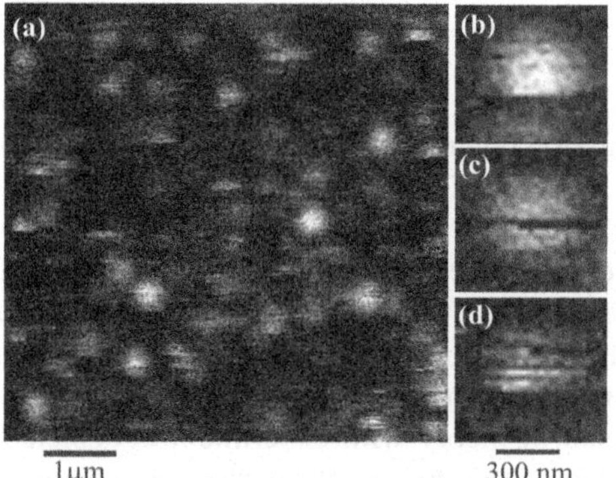

Fig. 17.19
Photoluminescence image of silicon nanocrystals (made from a powder of porous Si and deposited on a quartz glass) obtained by means of a scanning laser confocal microscope (excitation at 514 nm, 40 kW/cm^2). The left panel (a) shows an area of approximately $6 \times 6\,\mu m^2$ imaged successively in 256 rows, each of them with 256 steps. Therefore, one step is 23 nm and the dwell time at one pixel is 10 ms. Panels on the right (b)–(d) are detailed images of three different nanocrystals showing emission intermittency. Room temperature, after Valenta *et al.* [21], with kind permission from Springer Science + Business Media B.V.

of which is determined by the diffraction ring) consists of about 15 rows. If the nanocrystal is blinking, then—in the moment when it is traversed by the excitation beam—it can be either in the emitting state or in the dark state or it can even just be switching over. The image of such a nanocrystal can thus contain some dark rows or even most of the rows can be dark, making the nanocrystal almost unidentifiable (see Fig. 17.19(d)).

Intermittency measurements are performed by simply repeatedly detecting the luminescence signal from an individual object (a nanocrystal) continuously excited with a laser beam. The detection interval should be substantially shorter than the length of the typical intervals between the emission switching over.

It is possible to use a CCD camera as a detector provided it is sufficiently fast; if so, then one can, from a series of luminescence images, extract signals of the individual objects. However, avalanche photodiodes in the photon counting mode are more commonly employed detectors. Now, we will focus on how to process a signal obtained in this way.

Statistical processing of intermittent luminescence from an individual nanocrystal

Let us assume that our detection system (with one or two detectors) records the exact times of the detection events. In order to display the intensity as a function of time, we must divide the duration of the experiment into small time intervals (time bins) and sum up the number of pulses falling into each interval (see Fig. 17.20). If our system shows blinking or jumps between multiple intensity levels, we should recognize this phenomenon in the temporal intensity trace. The optimum choice of the elementary time bin lies in the construction of a histogram of intensity distribution in all elementary time intervals. The best-choice time bin to be used in the processing is then the one whose histogram has the most distinct peaks formed around the preferred emission intensity levels, that is, the background level (OFF emission, dark state) and the level corresponding to the ON emission (bright) state, or, alternatively, other levels between which the system switches (Fig. 17.20(a),(b)). Let us note that if the detection system enables us only to sum up repeatedly the pulses in a given acquisition time window (not to detect the arrival times of individual pulses), the acquisition window should be sufficiently short (at least ten times shorter than the typical time of the ON and OFF states). Later, it is possible to merge the intensities from two or more detection windows in order to improve the intensity histogram, but it is obviously impossible to divide them into shorter segments.

The next step should be directed towards separating exactly the ON and OFF states on the temporal intensity trace. Firstly, we determine the mean values of the individual states and then the boundaries (thresholds) between them. If we quantify the mean values of the ON and OFF states I_{ON} and I_{OFF}, respectively, it is then possible to determine a threshold value I_{th} between them so that its separation from both levels is the same; we can achieve this using the relation [22]

$$\frac{I_{\mathrm{th}} - I_{\mathrm{OFF}}}{\sqrt{I_{\mathrm{OFF}}}} = \frac{I_{\mathrm{ON}} - I_{\mathrm{th}}}{\sqrt{I_{\mathrm{ON}}}}.$$

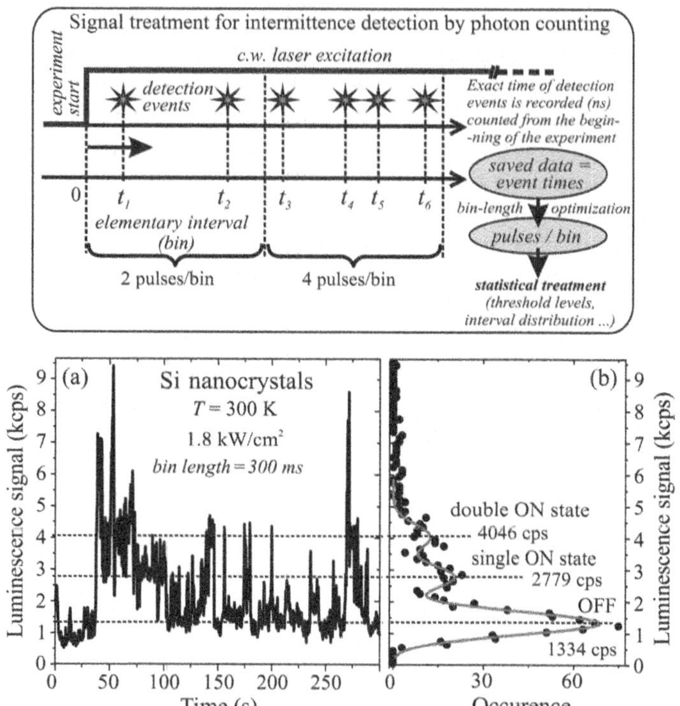

Fig. 17.20
Top: diagram of recording and processing of the individual detection events of luminescence from a single nanocrystal. (a) Photoluminescence intensity for a cluster of Si nanocrystals as a function of time; the curve was obtained by applying an elementary time interval (bin) of 0.3 s on the recorded detection events. The suitability of this choice of time window is apparent from the constructed occurrence histogram of intensities (b) that shows clearly the preferred states of the system and also enables us to determine the boundaries between the states. (The abbreviation kcps denotes kilocounts per second, i.e. one thousand pulses/s.) Excited by a cw 444-nm laser diode at room temperature. Adapted from Valenta and Linnros [8].

The threshold value then enables us to determine whether the system is in the first or in the second state for each elementary interval. Consequently, it becomes possible to determine the system's dwell times in the ON and OFF states, respectively, and, based on these values, produce another histogram, this time a histogram of these dwell times (see Fig. 17.21) [23]. Most often, what is observed is either an exponential distribution of the ON and OFF intervals (corresponding to random switching between the two states—'random telegraph') or a power-law distribution $t^{-\alpha}$, where α usually fluctuates around the value of 1.5. In the case of a cluster of Si nanocrystals (porous silicon grains) in Fig. 17.21 we observe a power-law dependence turning into an exponential law for longer intervals.

The method of intermittency assessment from a sequence of luminescence images is then analogous. We accumulate the signal at the position of the blinking nanocrystal for a pre-selected time period (if the background level changes appreciably over time, e.g. decreases due to bleaching, it has to be evaluated and subtracted) and we treat these values in the same way as the number of pulses obtained by integrating over the elementary intervals—the bins. Figure 17.22 gives an example of such an evaluation of blinking of individual lithographically produced Si nanocrystals from images taken at 15-s intervals. The uppermost panel (Fig. 17.22(a)) shows a temporal luminescence record of one of the nanocrystals, the intermittency statistics of which practically cannot be established because it only emits light rarely and for a short time. Most easily established are the intermittency statistics evaluated in nanocrystals that remain in both the emitting and dark states for approximately the same time

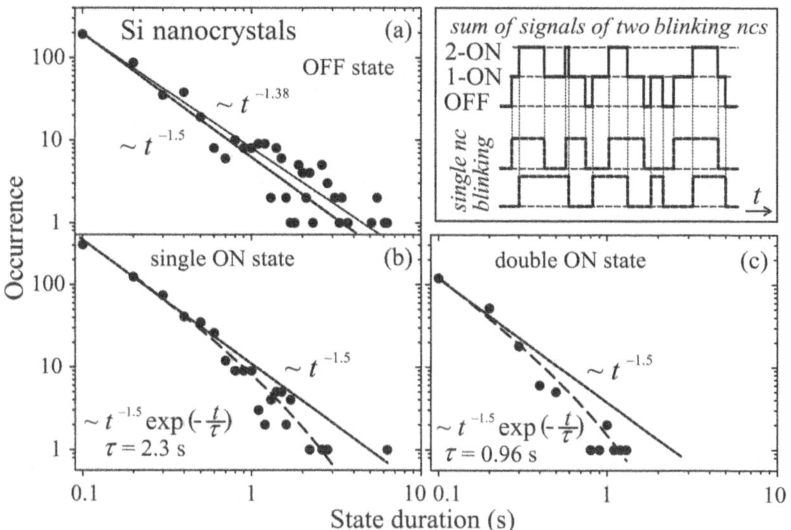

Fig. 17.21
Distributions of dwell time intervals of a Si nanocrystal cluster (from Fig. 17.20) in one of the three states defined by the histogram in Fig. 17.20(b): (a) The OFF state, (b) a single ON state and (c) a double ON state. The top right diagram suggests an interpretation of the double ON state in terms of a superposition of intermittency of two nanocrystals. Such a superposition of blinking of two or more nanocrystals affects (shortens) the statistics of all states. After Valenta *et al.* [23].

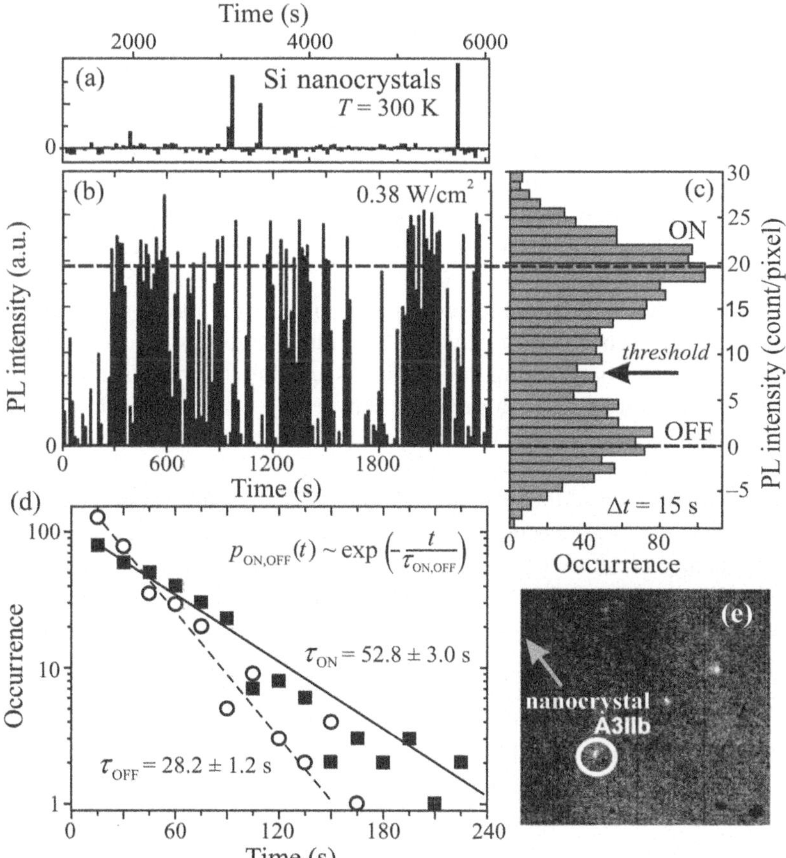

Fig. 17.22
Photoluminescence intermittency of single Si nanocrystals prepared by electron lithography. (a) A nanocrystal emitting light only for short time intervals and only very rarely. The statistics of such intermittency cannot be determined. (b) A nanocrystal staying in both the ON and OFF states for approximately the same time. The background signal was subtracted. (c) Distribution histogram of intensities from panel b. (d) The distribution of the ON and OFF state durations is fitted by an exponential decay with lifetimes $\tau_{ON} = 52.8 \pm 3.0$ s and $\tau_{OFF} = 28.2 \pm 1.2$ s. (e) One of the luminescence images highlighting a nanocrystal, the intermittency statistics of which are shown in panels (b)–(d). Cw excitation with a He-Cd laser (325 nm) at room temperature, after Valenta and Linnros [8].

(Fig. 17.22(b)). For lithographical Si nanocrystals, we get (at room temperature and photoexcitation intensity of $0.38\,\mathrm{W/cm^2}$) an exponential distribution of bright and dark intervals with mean lifetimes of 53 s and 28 s, respectively (Fig. 17.22(d)).

If the intermittency mechanism is to be determined, it is important to analyse its dependence on the excitation intensity and sample temperature. As established empirically, the switching rate into the dark state in most cases substantially increases with increasing excitation while the reverse switching from the dark state into the bright one does not depend on the intensity at all or only weakly. For example, in the Si nanocrystals introduced in Fig. 17.22, the dwell time in the bright state decreases quadratically with increasing excitation intensity while the duration of the dark periods practically does not change [24]. However, the opposite situation was described in the case of individual InP quantum dots covered with a GaInP layer; also these dots exhibit an exponential distribution of the ON and OFF periods [25]. Here, the switching rate (i.e. the inverse of the mean lifetime) into the dark and into the bright state increases in proportion to the second and fourth power of the excitation intensity, respectively (Fig. 17.23(a), (b)). The temperature dependence of intermittency for these InP nanocrystals (Fig. 17.23(c)) shows prolonged OFF intervals with decreasing temperature, which indicates that the process of switching from the dark into the bright state is thermally activated (with an activation energy of around 10 meV). It is difficult to find a simple model of nanocrystal blinking because the intermittency statistics are not necessarily stable in time. For example, in the case of the InP nanocrystals mentioned above, sudden changes in the switching rate can occur (Fig. 17.23(d)), or the intermittency can even stop for some time. These changes may also be accompanied by a change in the intensity of the emitting state (the bottom curve in Fig. 17.23(d)).

As can be deduced from the several examples given above, the manifestations of luminescence intermittency of individual nanocrystals can be diverse. In spite if this diversity, taking into account such cluttered results, it is possible to find some typical—and surprisingly universal—features that do apply not only to nanocrystals but also to nanowires and other nanostructures [20]:

(a) The distribution of the ON and OFF intervals has the form of a power-law function $t^{-\alpha}$, where α usually takes on the values between 1.2 and 2, most frequently close to 1.5. For longer times intervals (seconds and longer), the distribution of the ON intervals sometimes falls off exponentially (see Fig. 17.21(b), (c)). This can be a sign of the presence of a competing mechanism turning off the emission. A purely exponential distribution occurs very rarely. The reason for its appearance in Fig. 17.22 probably lies in insufficient time resolution of the detection channel that was unable to see shorter time intervals, possibly governed by a power-law distribution.

(b) The blinking itself is connected with the very excitation of nanocrystals by light. This is revealed by the ON intervals getting shorter for increasing excitation intensity and the OFF intervals gradually getting longer. Or, to put it differently, the average intensity of emission from a nanocrystal ensemble decreases in time (this is the so-called *bleaching*) due to pho-

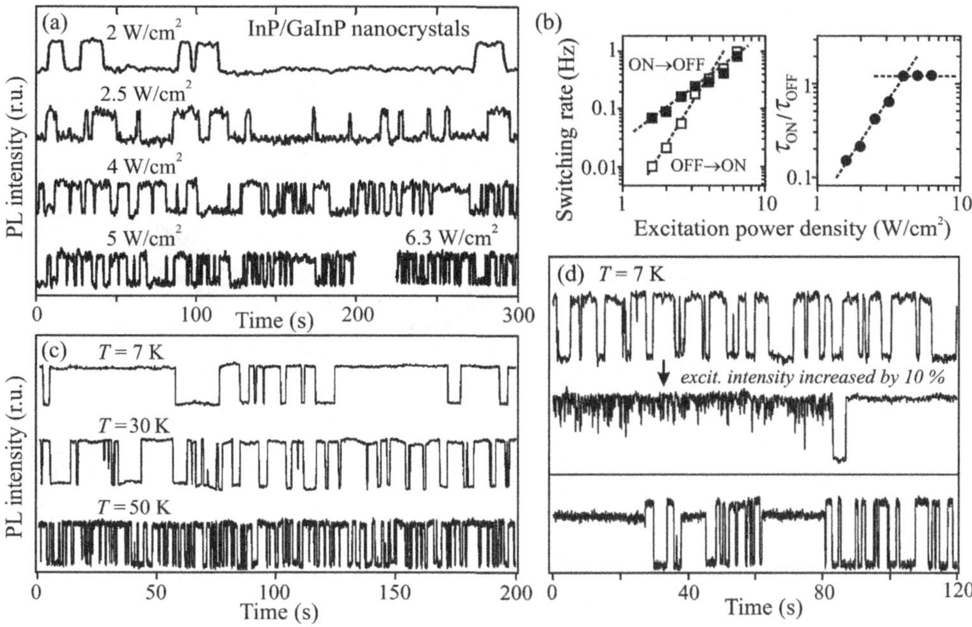

Fig. 17.23
Intermittency of InP quantum dots in GaInP. (a) Variability of the intermittency of a single nanocrystal under varying photoexcitation intensity. (b) Changes in the ON → OFF (black squares) and OFF → ON switching rate with increasing excitation. The right panel displays how the ratio of the average durations of the ON and OFF states varies as a function of the excitation intensity. (c) Changes in the intermittency with temperature. (d) Examples of sudden changes in the switching rate. Excitation with a cw Ar^+-laser (488 nm), $T = 7$ K (if not specified otherwise). After Pistol *et al.* [25].

toexcitation. After turning the excitation off, luminescence slowly returns to its original level.

(c) Intermittency is connected with spectral diffusion (Fig. 17.13). Neuhauser *et al.* showed that large spectral jumps are correlated with emission intensity switching [26]. Therefore, both phenomena may have a common cause in charge redistribution near the nanocrystal surface.

(d) Another quite frequently observed phenomenon is that intermittency does not occur between two exactly defined states but there is rather a distribution of emission intensities and also of lifetimes (both radiative and nonradiative). This means that the quantum efficiency of emission may change in time [27].

(e) Sensitivity to electric fields: Modulation of an external electric field influences the modulation of intermittency, which further confirms the link between the local electric field and nanocrystal intermittency [28].

Efros and Rosen [29] were the first to submit a model explaining nanocrystal intermittency. They saw the cause of the emission being turned off in Auger recombination—the inelastic scattering of light-generated electron–hole pairs. This process enables an electron to gain sufficient energy to become captured in a trap site (acceptor) near the nanocrystal surface. This gives rise to a positively charged nanocrystal, which cannot emit light due to the very efficient Auger quenching of the photogenerated *e–h* pairs. Emission is restored after

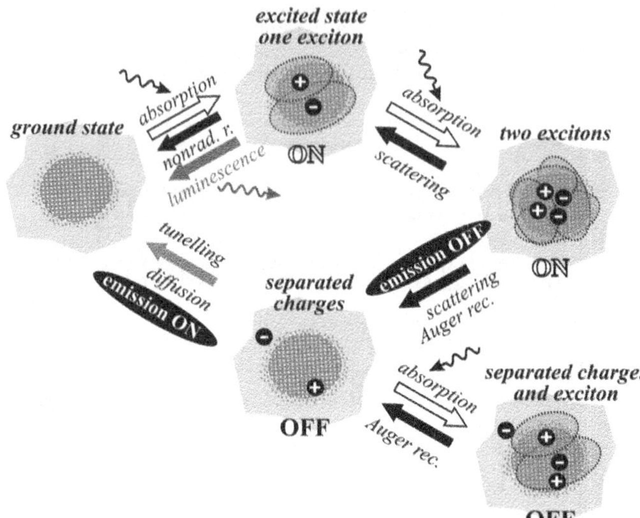

Fig. 17.24
Schematic representation of the mechanism that switches nanocrystal luminescence on and off due to Auger recombination. Adapted from Valenta *et al.* [23].

the free charge is neutralized (see Fig. 17.24) [23]. However, in its simple form, the model only leads to an exponential distribution of the duration of the ON and OFF states. More sophisticated models have thus been proposed subsequently; they all introduce distributions of certain trapping and release parameters of an electron in the trap site. We can list a review of these models according to [20]:

(1) The *multiple trap* model assumes the existence of multiple different trap states near a nanocrystal with a statistical distribution of the trapping and release rates.
(2) The *spectral diffusion* model assumes resonant tunnelling between the nanocrystal and a nearby trap, the energy level of which changes (diffuses) as a result of, e.g., local electric fields, thus modifying the electron trapping and release rates.
(3) The *spatial diffusion* model assumes that an electron ejected from the nanocrystal diffuses all around until it happens to come back. The model anticipates the distribution of the ON and OFF periods to be exactly of the $t^{-3/2}$ shape; one has to consider anomalous diffusion for other values of the exponent.
(4) The *fluctuating barrier* model assumes the depth and width of the potential barrier through which an electron tunnels between the nanocrystal and a trap to be variable. This results in a fluctuating probability of electron penetration through the barrier.
(5) *Fluctuation of non-radiative lifetimes* due to deep surface states also leads to a power-law dependence of the ON and OFF times in the form of $t^{-3/2}$.

Due to the diversity of effects of luminescence intermittency in low-dimensional semiconductors, it is highly unlikely that a single, universal mechanism that would explain all experimental observations could be found. Different optimum descriptions will probably persist for different systems.

Therefore, intermittency of light emission from quantum sources remains a remarkably appealing topic for many theorists as well as experimenters. From the point of view of practical applications of nanocrystals, however, it is important to find a technology that would prevent or highly reduce blinking. The first signs of non-blinking nanocrystals appeared in several works, which pointed to a link between different ways of passivating the nanocrystals and profound modifications of their intermittency. Recently, the first paper on nanocrystals showing constant light emission for up to several hours was published [30]. These were CdZnSe nanocrystals coated by a ZnSe layer, the potential barrier at the interface between the materials being of gradual rather than the commonly used step-like shape.

17.5 Nanocrystals as sources of non-classical photon flux

Photoluminescence of an individual quantum object is triggered by absorption of one photon that brings the system into an excited state and, as it relaxes back to the ground state, there is a certain probability that emission of one photon will also occur (if we neglect nonlinear optical phenomena such as multiple-photon absorption and emission). The time sequence of the emitted photons bears unique information about the internal dynamics of the system. In this section, we shall discuss the techniques of detecting and describing the statistics of the emitted photons and we will show that an individual nanocrystal is the source of a non-classical photon flux. The reader can find a brief summary of the theoretical description of photon statistics in Appendix L.

17.5.1 Measuring photon statistics

Complete information about one luminescence act in an individual quantum object would consist of the absorption time and subsequent emission time of a photon (we disregard now further properties such as the polarization and energy of the photon). It is not possible to establish directly the very instant of photon absorption. We can only restrict the exciting light into short pulses that determine the interval during which absorption could occur, or we can relate the time of detecting a photon to the previous detection event. We can detect the individual emitted photons using photon counters—photomultipliers—or, more frequently, avalanche photodiodes (APD). However, this detection technique has a number of limitations that result in the non-identity between the sequence of the detection events and that of the emitted photons: (a) the efficiency of collecting and detecting the emitted photons is not absolute (even if in APDs the efficiency can attain values as high as 90%), (b) the detector also produces false pulses (the so-called dark counts; for APDs, the typical background level is 20 or more dark counts per second, depending on the quality and price of the detector), and (c) after a particular detection act, the detector becomes 'blind' for a moment (the so-called dead time, several nanoseconds) and, thus, it is not able to detect two photons coming in in rapid succession.

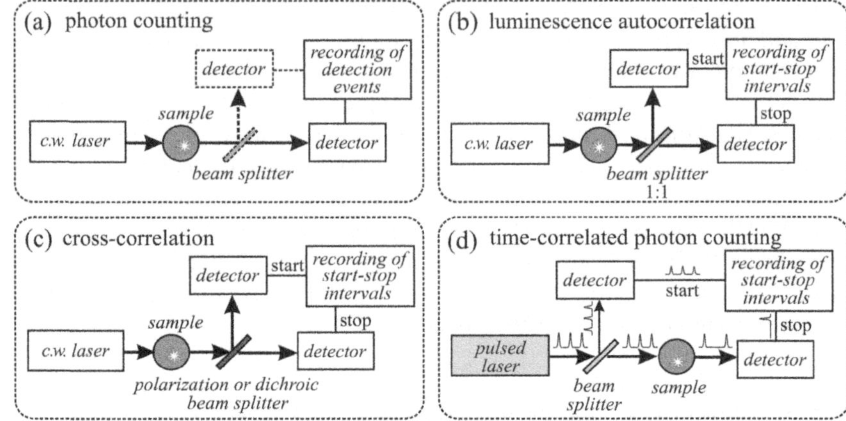

Fig. 17.25
Techniques for measuring photon statistics. (a) Detection of photon arrival times using one or two detectors (to circumvent the effect of the detector dead time). (b) Luminescence autocorrelation measurements using a pair of detectors between which the photon flux is split in the ratio 1:1—the so-called Hanbury Brown–Twiss interferometer. (c) Cross-correlation measurements with the photon flux split by polarization, wavelength, etc. (d) Time-correlated counting of single photons using a pulsed excitation and start–stop electronics. Adapted and updated from Lippitz *et al.* [22].

The basis of an experimental set-up measuring photon statistics of individual quantum objects usually lies in a confocal microscope described in Subsection 17.2.2 (Fig. 17.5), apart from the detection system, whose design is special. Its various versions are analysed in more detail in Fig. 17.25. In the simplest variant, the emitted light is focused on a single detector (APD), see Fig. 17.25(a).[6] Pulses from the detector are processed by dedicated electronics that record the arrival time of the pulse with a high accuracy (< 1 ns). To partially circumvent the problem of the detector dead time, the collected light can be split across to two detectors.

The two-detector configuration is often used to measure the *photon correlation spectrum* or *luminescence autocorrelation* (we correlate one photon flux split into two parts), see Fig. 17.25(b). Here, one detector provides the start signal and the other the stop signal. Special electronics measure the duration of the start–stop intervals, and the frequency of various interval occurrences can be evaluated.[7] We can split the detected light between the detectors in the ratio 1:1, but there are also other ways to split the detected signal: by polarization (polarization beam splitter) or by their wavelength (dichroic filter); then, the so-called cross-correlation is measured (see Fig. 17.25(c)), which is used, for example, for monitoring energy transfer in molecular systems. Naturally, we can calculate the correlation function subsequently from the recorded detection times of the individual photons on both detectors; however, the reverse case is not possible. The photon arrival times contain the most detailed experimentally available information on the distribution of the emitted photons.

Measurements of the start–stop intervals can also be employed in connection with pulsed excitation. In such measurements, the start pulse comes from a reference detector monitoring the excitation pulse whereas the stop pulse

[6] The confocal iris diaphragm can sometimes be omitted since the small detection area of the APD detector itself does its job.

[7] The first to perform and describe the measurements of intensity correlation using an interferometer (Fig. 9.25(b)) and a pair of photomultipliers were Hanbury Brown and Twiss (Hanbury Brown, R., and Twiss, R. Q. (1956). *Nature*, **177,** 27). They tested the principle of the experiment with a classical light source (a lamp) and then they applied it to astrophysics. This work laid down the cornerstone of the field of measurements of the coherence and statistics of light.

is derived from the detector recording the emission. Thus, the electronics record the statistical distribution of the delay of emitted photons after the absorption event. We have already described this technique of time-correlated single-photon counting (TCSPC) in Subsection 2.9.3 in connection with the measurements of the kinetics of luminescence decay. Here, the corresponding experimental setup is presented in Figure 17.25(d). If one wants to apply TCSPC to detect individual quantum objects then what should be taken into account is the fact that one excitation pulse generates at most one detected photon (in practise, less than one-tenth of the excitation pulses results in a photon detection event). If the accumulation of sufficient statistics is not to take too long, it is necessary to use pulsed laser excitation with a high repetition rate. If, however, the lifetime of the excited state is long (of the order of tens of μs and longer) or the quantum efficiency of the photon emission and detection is very low, the TCSPC technique basically cannot be applied at all.

17.5.2 Experimental manifestation of non-classical light emitted by a single nanocrystal

In a correlation experiment (Fig. 17.25(b)), we measure the *distribution of the waiting times* between two detection events $C(\tau)$, or alternatively the *conditional probability density* $G(t+\tau, t)$ of detecting a photon at time $(t+\tau)$ assuming that a photon was detected at time t. This, in fact, is the non-normalized alternative to the *intensity correlation function*, $g^{(2)}(\tau)$. In the classical description of light, the correlation function $g^{(2)}(\tau)$ must comply with several conditions: It is a non-increasing function of τ, at zero lag time $\tau = 0$ it has to be larger than or equal to unity, $g^{(2)}(0) \geq 1$, and $g^{(2)}(\tau) \geq 0$ holds true for all τ. In contrast, the quantum description imposes a single condition, namely, $g^{(2)}(\tau) \geq 0$ for all τ (see Appendix L).

Correlation measurements of the luminescence signal from individual semiconductor nanocrystals clearly show that these objects behave similarly to other types of quantum emitters (individual molecules, defect centres, etc.) and are the source of a *non-classical* photon flux. An example is shown in Fig. 17.26 which displays a correlation function obtained by detecting the emission from individual quantum dots of InAs produced by self-organized growth [31]. One can see that at short delays, $\tau \rightarrow 0$, the correlation function $g^{(2)}(\tau)$ drops below unity (even considerably below 0.5), which is in contradiction with the classical description of light. The cause of this phenomenon can be easily understood: After absorbing one photon, the excited nanocrystal has to return back to its ground state, whether radiatively or non-radiatively, after a certain average lifetime. Therefore, if the nanocrystal has emitted a photon, it goes back into its ground state and it needs some time before a new excitation (through photon absorption) and radiative recombination can occur. Two photons thus cannot be emitted in rapid succession or, in other words, the probability of radiating a photon at time τ if a photon has been emitted at time 0 drops to zero for $\tau \rightarrow 0$. This effect is called *antibunching* and we came across it already in Fig. 16.11.

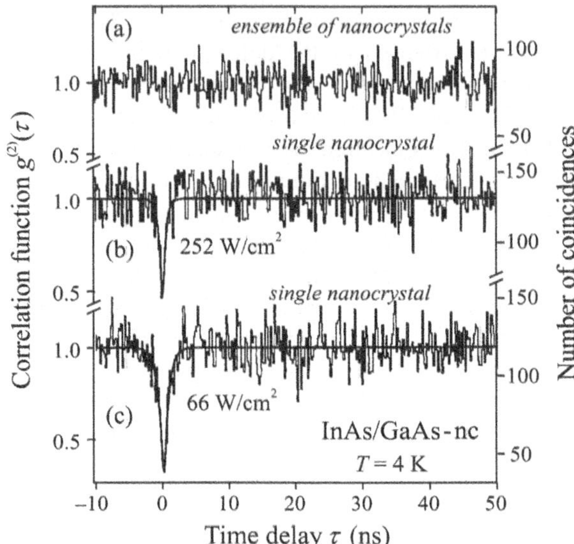

Fig. 17.26
Luminescence autocorrelation function of light emission from InAs quantum dots (at the wavelength corresponding to the radiative recombination of one exciton). (a) Signal when detecting a larger number of nanocrystals, (b),(c) signal from a single nanocrystal under excitation intensity of 252 and 66 W/cm^2, respectively. The noisy curve represents the measured number of coincidences for the given delay; the smooth line is a fit by a model correlation function described by eqn (17.5). Excited with a cw laser, $T = 4$ K. Adapted from Michler *et al.* [31].

However, the experiment represented by Fig. 17.26 shows that $g^{(2)}(\tau)$ does not attain a zero value for $\tau = 0$. This is due to the imperfections of the experiment—detector noise and emission background (scattered light, impurity fluorescence, etc.). Nevertheless, the considerable decrease in the correlation function $g^{(2)}(\tau)$, which drops below unity around $\tau = 0$, provides sufficient evidence of antibunching and it is widely accepted as the most straightforward way proving that the detected signal really comes from a single nanocrystal.

In Fig. 17.26(b),(c) we notice that under stronger excitation the variations in the correlation function around $\tau = 0$ get faster. The form of the correlation function can be fitted by

$$g^{(2)}(\tau) = 1 - a \exp(-\tau/t_d), \tag{17.5}$$

where the coefficient a expresses the effect of the detection background and $t_d = 1/(\Gamma + W_p)$ is the antibunching time constant, in which Γ and W_p represent the spontaneous emission rate and the pumping rate into the excited state (from which the radiative recombination occurs), respectively. The decrease in the antibunching time constant with increasing pump intensity is due to an increase in the pumping rate W_p that depends on the absorption cross-section of the nanocrystal and on the excitation photon flux. The spontaneous emission rate should not depend on the excitation intensity at low excitation levels. Measuring t_d for several excitation intensities enables one to determine a very interesting parameter—the absorption cross-section of one particular nanocrystal (see, e.g., [32]). To evaluate the time constant t_d reliably, one has to take into account the temporal resolution of the experiment (0.42 ns in the above case of InAs nanocrystals) and, if necessary, to deconvolute the measured curve.

Correlation and cross-correlation measurements at the wavelengths of different spectral lines can significantly contribute to the interpretation of the lines

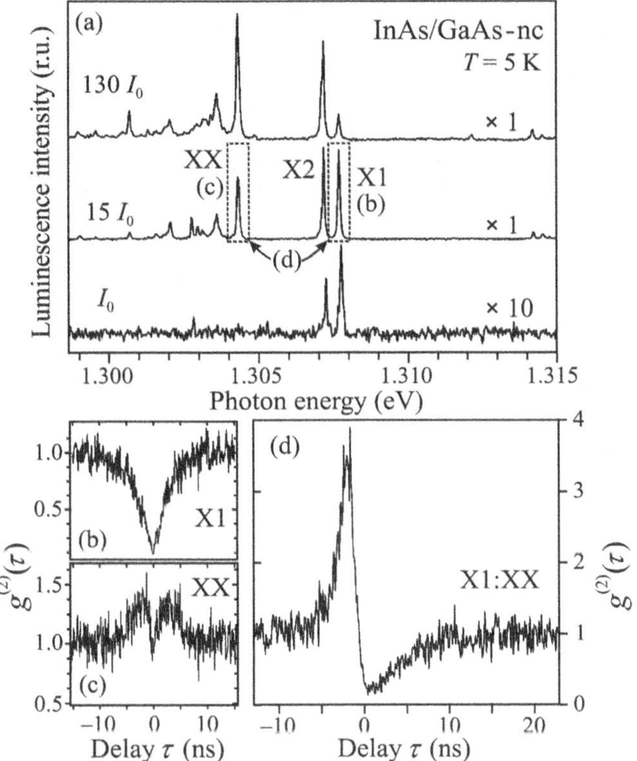

Fig. 17.27
Spectrally resolved measurements of luminescence correlation and cross-correlation of an individual InAs nanocrystal. (a) Three luminescence spectra obtained at various excitation levels (multiples of I_0). The bands X1, X2, and XX are due to the radiative recombination of one exciton, trion (charged exciton) and biexciton, respectively. The dotted rectangles denote parts of the spectrum used in the correlation measurements. (b) Luminescence autocorrelation of the X1 exciton line, (c) luminescence autocorrelation of the XX biexciton line, (d) cross-correlation of the X1 and XX lines. Excited with a cw laser, $T = 5$ K. Adapted from Kiraz *et al.* [33].

and the determination of some material parameters, such as lifetimes of certain states, absorption cross-sections, etc. Let us quote one more example of an experiment performed on individual InAs quantum dots [33]. Figure 17.27(a) displays a luminescence spectrum of a single InAs nanocrystal at various excitation intensities. The dominant spectral lines are denoted X1, X2, and XX and they correspond to the radiative recombination of a single exciton, a charged exciton (trion) and a biexciton, respectively. Using a monochromator or narrow band-pass filters (FWHM of 0.5 or 1 nm), one can select a narrow part of the spectrum and perform correlation measurements using a Hanbury Brown–Twiss interferometer. Panels (b) and (c) show luminescence autocorrelation measured on the excitonic (X1) and biexcitonic (XX) lines. While the exciton correlation clearly exhibits antibunching similarly to Fig. 17.26 as expected, the biexcitonic correlation is an interesting combination of bunching and antibunching. Bunching manifests itself as a wider band around $\tau = 0$ and it occurs here because, after the biexcitonic emission act (at time $\tau = 0$), an exciton remains inside a nanocrystal which increases the probability of generation of another biexciton and subsequent emission of another photon. The fall off of the bunching band really corresponds to the exciton lifetime (3.6 ns). The time constant of the antibunching notch then corresponds to the biexciton lifetime (2.6 ns). Cross-correlation, shown in Fig. 17.27(d), was then

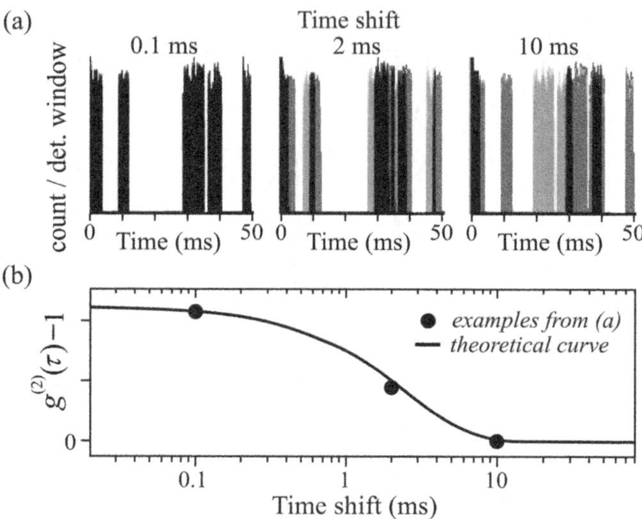

Fig. 17.28
Connection between intermittency and photon bunching: (a) Autocorrelation of the intermittency signal for delay times of 0.1, 2, and 10 ms. The dark and light grey colours denote the original and shifted signals; the black area then means an overlap of both correlated signals. (b) Luminescence autocorrelation of the intermittency signal from panel (a). After Lippitz *et al.* [22].

measured by sending the signal of the X1 line to the start detector (an avalanche photodiode) and the signal of the XX line to the stop detector. The curve is of a markedly asymmetric form—antibunching dominates the positive-delay side because after the excitonic emission act the nanocrystal goes back to the ground state, while significant photon bunching observed for negative delays results from biexcitonic emission, which leaves an exciton behind; as the remaining exciton can rapidly emit a photon, the biexciton and exciton photons are very likely to come shortly after one another. The correlation experiment here clearly confirms the interpretation of the luminescence lines and, above all, it unambiguously proves they come from a single object—the nanocrystal.

Finally, let us note that, in addition to antibunching, the luminescence correlation function of the signal from one nanocrystal often shows *bunching* as well (which, as is known, exists for classical light sources too, and it thus does not represent a manifestation of a non-classical photon flux); in most cases, bunching occurs for much longer delay times than antibunching. Bunching results from luminescence intensity fluctuations over time, which can have various reasons, most frequently intermittency (discussed in Subsection 17.4.5), but also spectral diffusion, bleaching, etc. [22]. Therefore, luminescence autocorrelation measurements can also be used to study these relatively slow processes.

An illustrative explanation of the connection between intermittency and bunching is given in Fig. 17.28, showing the correlation of a model intermittency signal for three different time shifts and yielding the form of the second-order intensity correlation function.

17.6 Problems

17/1: Estimate the level of the detected luminescence signal from one nanocrystal if you know the radiative lifetime, photoexcitation power density and detection efficiency.

17/2: What theoretical limiting resolving power R could be achieved in luminescence spectroscopy of a single nanocrystal when using a double grating monochromator with reciprocal linear dispersion $L^{-1} = 0.5\,\text{nm/mm}$? Assume emission wavelength of 800 nm and a size of the nanocrystal image (diameter of its diffraction ring) of 60 μm. What prevents us from realizing this idea in practice? Do we need such resolving power in reality?

17/3: Derive relation (17.4) for the perpendicular component of the electric field of a light wave in a nanowire. Follow, for example, a procedure analogous to calculating the Lorentz internal field in Dekker's textbook [34], Section 6.6-5.5.

17/4: *Electroluminescence of individual nanocrystals.* Throughout this chapter, we have been discussing photoluminescence as a sole experimental means of spectroscopy of individual nanocrystals. However, researchers have also succeeded in studying the luminescence of individual Si nanocrystals driven electrically. Discuss the relevant experimental realization of the method according to [35] and compare the emission electroluminescence spectra obtained in this way with photoluminescence spectra.

References

1. Moerner, W. E. (ed.) (1988). *Persistent Spectral Hole-Burning: Science and Applications*. Topics in Current Physics, Vol. 44. Springer, Berlin.
2. Moerner, W. E. and Kador, L. (1989). *Phys. Rev. Lett.*, **62**, 2535.
3. Orrit, M. and Bernard, J. (1990). *Phys. Rev. Lett.*, **65**, 2716.
4. Basché, T., Moerner, W. E., Orrit, M., and Wild, U. P. (eds) (1997). *Single Molecule Optical Detection, Imaging, and Spectroscopy*. Verlag-Chemie, Weinheim.
5. Gell, C., Brockwell, D., and Smith, A. (2006). *Handbook of Single Molecule Fluorescence Spectroscopy*. Oxford University Press, Oxford.
6. Brunner, K., Bockelmann, U., Abstreiter, G., Walther, M., Böhm, G., Tränkle, G., and Weimann, G. (1992). *Phys. Rev. Lett.*, **69**, 3216.
7. Gustafsson, A., Pistol, M.-E., Montelius, L., and Samuelson, L. (1998). *J. Appl. Phys.*, **84**, 1715.
8. Valenta, J. and Linnros, J. (2009). *Optical spectroscopy of individual silicon nanocrystals*. In *Silicon Nanophotonics: Basic Principles, Present Status and Perspectives* (ed. L. Khriachtchev), p. 179. World Scientific/Pan Stanford Publishing, Singapore.
9. Zhang, X. and Liu, Z. (2008). *Nature Materials*, **7**, 435.
10. Novotny, L. and Hecht, B. (2006). *Principles of Nano-Optics*, p. 121. Cambridge University Press, Cambridge.
11. Banin, U., Cao, Y. W., Katz, D., and Millo, O. (1999). *Nature*, **400**, 542.
12. Bayer, M., Stern, O., Hawrylak, P., Fafard, S., and Forchel, A. (2000). *Nature*, **405**, 923.
13. Empedocles, S. A., Neuhauser, R., Shimizu, K., and Bawendi, M. G. (1999). *Adv. Mater.*, **11**, 1243.
14. Empedocles, S. A. and Bawendi, M. G. (1997). *Science*, **278**, 2114.
15. Valenta, J., Juhasz, R., and Linnros, J. (2002). *J. Luminescence*, **98**, 15.
16. Lakowicz, J. R. (1983). *Principles of Fluorescence Spectroscopy*. Plenum Press, New York.
17. Empedocles, S. A., Neuhauser, R., and Bawendi, M. G. (1999). *Nature*, **399**, 126.

18. Ruda, H. E. and Shik, A. (2005). *Phys. Rev. B*, **72**, 115308.
19. Allan, G., Delerue, C., and Niquet, Y. M. (2001). *Phys. Rev. B*, **63**, 205301.
20. Frantsuzov, P., Kuno, M., Janko, B., and Marcus, R. A. (2008). *Nature Physics*, **4**, 519.
21. Valenta, J., Linnros, J., Juhasz, R., Cichos, F., and Martin, J. (2003). *Optical spectroscopy of single silicon quantum dots*. In *Towards the First Silicon Laser* (eds. L. Pavesi, S. Gaponenko, and L. Dal Negro). NATO Science Series, Vol. 93, p. 89. Kluwer Academic, Dordrecht.
22. Lippitz, M., Kulzer, F., and Orrit, M. (2005). *Chem. Phys. Chem.*, **6**, 770.
23. Valenta, J., Fucikova, A., Vácha, F., Adamec, F., Humpolíčková, J., Hof, M., Pelant, I., Kůsová, K., Dohnalová, K., and Linnros, J. (2008). *Adv. Funct. Mater.*, **18**, 2666.
24. Sychugov, I., Juhasz, R., Linnros, J., and Valenta, J. (2005). *Phys. Rev. B*, **71**, 115331.
25. Pistol, M.-E., Castrillo, P., Hessman, D., Prieto, J. A., and Samuelson, L. (1999). *Phys. Rev. B*, **59**, 10725.
26. Neuhauser, R. G., Shimizu, K. T., Woo, W. K., Empedocles, S. A., and Bawendi, M. G. (2000). *Phys. Rev. Lett.*, **85**, 3301.
27. Schlegel, G., Bohnenberger, J., Potapova, I., and Mews, A. (2002). *Phys. Rev. Lett.*, **88**, 137401.
28. Park, S. J., Link, S., Miller, W. L., Gesquiere, A., and Barbara, P. F. (2007). *Chem. Phys.*, **341**, 169.
29. Efros, A. L. and Rosen, M. (1997). *Phys. Rev. Lett.*, **78**, 1110.
30. Wang, X., Ren, X., Kahen, K., Hahn, M. A., Rajeswaran, M., Maccagnano-Zacher, S., Silcox, J., Cragg, G. E., Efros, A. L., and Krauss, T. D. (2009). *Nature*, **459**, 686.
31. Michler, P., Imamoglu, A., Kiraz, A., Becher, C., Mason, M. D., Carson, P. J., Strouse, G. F., Buratto, S. K., Schoenfeld, W. V., and Petroff, P. M. (2002). *phys. stat. sol. (b)*, **229**, 399.
32. Lounis, B., Bechtel, H. A., Gerion, D., Alivisatos, P., and Moerner, W. E. (2000). *Chem. Phys. Lett.*, **329**, 399.
33. Kiraz, A., Fälth, S., Becher, C., Gayral, B., Schoenfeld, W. V., Petroff, P. M., Zhang, L., Hu, E., and Imamoglu, A. (2002). *Phys. Rev. B*, **65**, 161303.
34. Dekker, A. J. (1963). *Solid State Physics*. Prentice-Hall, Englewood Cliffs, N.J.
35. Valenta, J., Lalic, N., and Linnros, J. (2004). *Appl. Phys. Lett.*, **84**, 1459.

Appendices

A Convolution

For the purpose of this book we define the convolution of two real functions f_1 and f_2 of a real variable a by means of the integral (t is a real number)

$$K(t) = \int_{-\infty}^{+\infty} f_1(a) f_2(t-a) \mathrm{d}a. \tag{A.1}$$

Let us consider only functions f_1, f_2 that are 'reasonable' from the physical point of view, i.e. continuous, finite and complying with the conditions $\lim_{a \to \pm\infty} f_1(a) = 0$, $\lim_{a \to \pm\infty} f_2(a) = 0$. Let us discuss immediately two simple properties of convolution (A.1). Firstly, if $f_2(t-a) = \delta(t-a)$ (Dirac's delta function), we have

$$K(t) = \int_{-\infty}^{+\infty} f_1(a)\delta(t-a) \mathrm{d}a = f_1(t). \tag{A.2}$$

Secondly, by substituting $t - a = b$ the definition (A.1) can be written as

$$K(t) = \int_{-\infty}^{+\infty} f_1(a) f_2(t-a) \mathrm{d}a = \int_{-\infty}^{+\infty} f_1(t-b) f_2(b) \mathrm{d}b. \tag{A.3}$$

The operation of convolution is thus commutative.

Now we are interested in whether one can give a visual demonstration of the convolution $K(t)$.

Let us consider a particular case: two Gaussian curves of the same shape $f_1(a) = \exp(-a^2/\sigma^2)$, $f_2(t-a) = \exp[-(t-a)^2/\sigma^2]$; 2σ means the width of the curves at the 1/e value. Figure A.1 displays the dependences of these functions on the variable a for several discrete values of the parameter t. It is observed that the variation in t implies a shift of $f_2(t-a)$ along the axis of the integration variable together with a gradual overlap of f_1 and f_2. Because $K(t)$ has a non-zero value only if the product $f_1 f_2$, is non-zero, obviously $K(t) \approx 0$ whenever the Gaussian functions are sufficiently separated from each other.

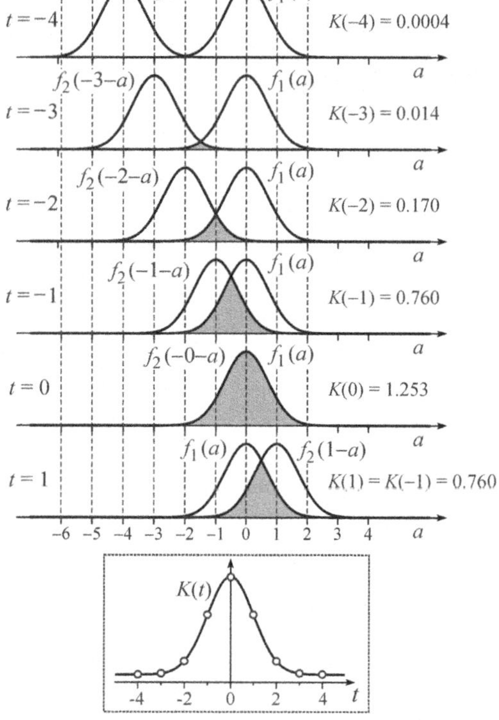

Fig. A.1
Visual demonstration of the convolution $K(t)$ of two Gaussian curves $f_1(a) = \exp(-a^2/\sigma^2)$, $f_2(t-a) = \exp(-(t-a)^2/\sigma^2)$. For the sake of simplicity $\sigma = 1$ has been chosen. The resulting convolution shape is shown in the bottom frame.

Only in a relatively narrow range in the neighbourhood of the origin $a = 0$ does the product $f_1 f_2$—and thus also $K(t)$—acquire sufficiently large non-zero values, as shown schematically by the shadowed areas in Fig. A.1. The resulting shape of $K(t)$ is at the foot of the same figure. It follows from this discussion that convolution can be realized as the mutual 'crossing' of both functions with simultaneous integration of their product. Moreover, we can see that in numerically evaluating the integral (A.1) it is possible to replace its infinite limits by convenient finite numbers; in our specific case it should be sufficient to integrate, e.g., over the interval $\langle -6\sigma, +6\sigma \rangle$.

Another example of convolution is shown in Fig. A.2, which shows the convolution of a bell-shaped function $f_1(a)$ having non-zero values for $a \in \langle 0, +\infty)$ only (and whose analytical expression is not of immediate importance), with the function

$$\begin{aligned} f_2(t-a) &= \mathrm{e}^{-(t-a)/\tau} && \text{for } a \le t, \\ f_2(t-a) &= 0 && \text{for } a > t. \end{aligned} \tag{A.4}$$

Again a similar shifting of the function f_2 is observed, the resulting convolution shape being indicated at the bottom of Fig. A.2. In this case it is interesting to note that in order to evaluate $K(t)$ numerically, it is sufficient to perform the process of integration from 0 to t, therefore (A.1) can be written as

$$K(t) = \int_0^t f_1(a) f_2(t-a) \mathrm{d}a. \tag{A.5}$$

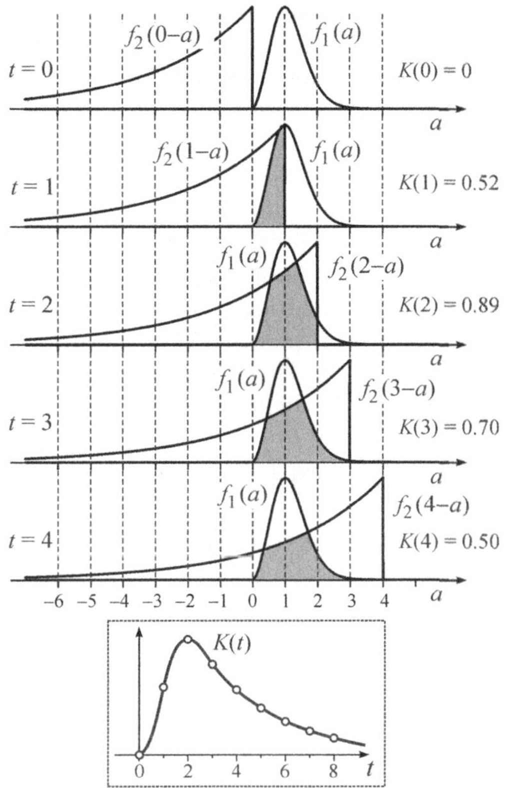

Fig. A.2
Visual demonstration of the convolution of two functions f_1, f_2 having at $t = 0$ a common boundary point ($a = 0$) of otherwise disjunctive domains of non-zero values.

Hence the last expression holds if the functions f_1 and f_2 are non-zero only on the following domains:

$$
\begin{array}{ll}
f_1(a) & \text{on the domain } \langle 0, +\infty), \\
f_2(a) = \exp(-a/\tau) & \text{on the domain } \langle 0, +\infty), \\
(\text{or } f_2(-a) = \exp(a/\tau) & \text{on the domain } (-\infty, 0\rangle).
\end{array}
$$

Outside these domains f_2 is zero by definition and f_1 is not defined.

B Emission spectrum of free excitons including phonon broadening

The free exciton theoretical lineshape is driven by the Maxwell–Boltzmann distribution (7.16) or (7.13)

$$I_{\mathrm{FE}}(h\nu) \approx \sqrt{(h\nu - E_{\mathrm{P}})}\exp(-(h\nu - E_{\mathrm{P}})/k_{\mathrm{B}}T) = \sqrt{(h\nu)'}\exp(-(h\nu)'/k_{\mathrm{B}}T), \tag{B.1}$$

where E_{P} stands for the low-energy threshold of the emission spectrum, i.e. $E_{\mathrm{P}} = E_{\mathrm{gi}} - E_{\mathrm{X}} - \hbar\omega$ or $E_{\mathrm{P}} = E_{\mathrm{g}} - E_{\mathrm{X}} - 2\hbar\omega$. The line is narrow (FWHM $\sim 1 - 10\,\mathrm{meV}$) and usually of low intensity, such that in experiments one has to adjust wider monochromator slits. Line broadening then inevitably takes place, as described by the apparatus function (2.58).

It appears, however, that in some cases not even the application of (2.58) is able to reproduce satisfactorily the experimentally acquired lineshape. Then one introduces an additional phenomenological broadening factor—due to the final lifetime of participating phonons. This is also referred to as broadening by phonon collisions and is expressed by the Gaussian function

$$F(E' - (h\nu)', \sigma) = \frac{1}{\sqrt{2\pi}\sigma} \exp\left(-\frac{(E' - (h\nu)')^2}{2\sigma^2}\right), \tag{B.2}$$

where 2σ (a parameter independent of temperature) denotes the width of the Gaussian curve at the points of inflexion. This parameter is connected with the FWHM of the Gaussian curve σ_F through relationship $\sigma_F = 2\sigma\sqrt{2\ln 2}$.

Taking into consideration that the slit broadening is also described via a Gaussian function (2.58), which will be written now as[1]

$$S(E - E') \cong \exp\left(-\frac{4\ln 2(E - E')^2}{\sigma_S^2}\right), \tag{B.3}$$

we arrive through a combination of (B.1)–(B.3) at the following double convolution, required for fitting the experimentally observed emission spectrum of a free exciton:

$$I_{FE}^{\xi}(E) \cong \int \left[\int \sqrt{(h\nu)'} \exp(-(h\nu)'/k_B T) \exp\left(-\frac{(E' - (h\nu)')^2}{2\sigma^2}\right) d(h\nu)'\right] \times \exp\left(-\frac{4\ln 2(E - E')^2}{\sigma_S^2}\right) dE'. \tag{B.4}$$

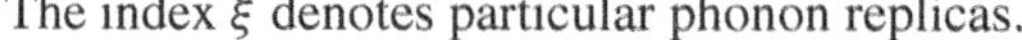

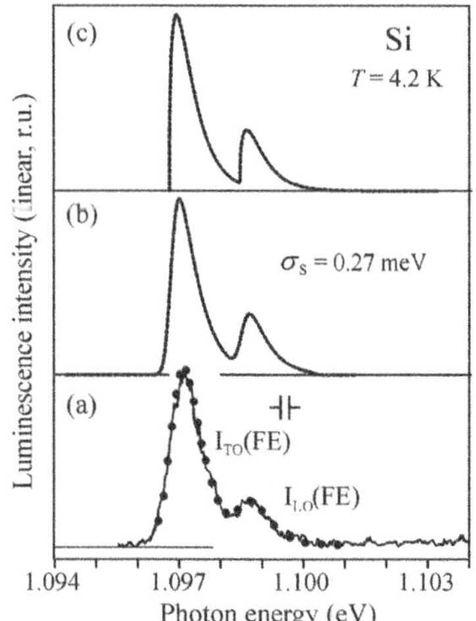

Fig. B.1
(a) Emission spectrum of a free exciton in silicon (LO- a TO-replicas) at $T = 4.2$ K. The full line stands for experiment, the symbols represent a theoretical fit according to (B.4): $\sigma = 0.22$ meV, the slit width of the monochromator was 0.25 meV, corresponding to $\sigma_S \approx 0.27$ meV. After Pelant *et al.* [1]. (b) Theoretical spectral shape when taking into account the slit broadening (B.3) only. (c) Theoretical shape of the Maxwell–Boltzmann distribution (B.1) without any broadening.

The index ξ denotes particular phonon replicas.

We show in Fig. B.1 how the various corrections fit the observed free exciton emission spectrum in crystalline silicon. Panel (a) represents the experimental observation (full line) together with symbols resulting from the double convolution (B.4). Agreement between the numeration and the experiment is perfect [1]. There are three fit parameters: σ ($\approx$ 0.2–0.3 meV, sample dependent), the amplitude ratio $I_{LO}(FE)/I_{TO}(FE)$ and the effective temperature T of the exciton gas (here we may, provided the excitation is not too strong, identify the effective and lattice temperatures, owing to the longevity of free excitons (>1 μs) entailing their full thermalization). Panel (b) exhibits the calculated lineshape that takes into account finite slit widths (B.3) only; potentially worse agreement with experiment is evident, especially on the low-energy wing. Panel (c) depicts the simple Maxwell–Boltzmann shape (B.1) without any corrections.

Nevertheless, how things stand has to be considered from case to case. In another indirect-bandgap semiconductor, AgBr, the parameter σ is smaller ($\sigma \approx 0, 1$ meV) [2] and, moreover, the free exciton lifetime is much shorter

[1] σ_S means the FWHM of the curve. We use this parameter here in contrast to (B.2) because the shape (B.3) is being derived from measurement of a narrow spectral line (Section 2.8), and determination of the full width at half maximum of the experimentally acquired line is easier than determination of its σ.

(10–100 ns) owing to efficient non-radiative processes. The effective temperature is then always higher than the lattice temperature, thus the emission line is predominantly temperature broadened (B.1), possibly also slit broadened (B.3). Phonon broadening then need not to be taken into account mostly.

References

1. Pelant, I., Dian, J., Matoušková, J., Valenta, J., Hála, J., Ambrož, M., Vácha, M., Kohlová, V., Vojtěchovský, K., and Kašlík, K. (1993). *J. Appl. Phys.*, **73**, 3477.
2. Kawate, E. and Masumi, T. (1988). *J. Phys. Soc. Japan*, **57**, 1826.

C Luminescence of an excitonic molecule

We shall derive, by making use of the quantum mechanics formalism, expressions for the emission lineshape pertinent to the radiative decay of an excitonic molecule; we shall treat separately the cases of direct and indirect bandgaps.

Direct bandgap

A general formula covering the emission lineshape relevant to the radiative decay of an excitonic molecule (into a photon and a recoil exciton) can be written down as

$$I_{\mathrm{sp}}^{\mathrm{M}}(h\nu) \sim \int \mathrm{d}^3\mathbf{K}_\mathrm{i} \int \mathrm{d}^3\mathbf{K}_\mathrm{f} f(\mathbf{K}_\mathrm{i})|M_{\mathrm{if}}(\mathbf{K}_\mathrm{i},\mathbf{K}_\mathrm{f})|^2 \delta(E_\mathrm{i}-E_\mathrm{f}-h\nu) \times \delta(\mathbf{K}_\mathrm{i}-\mathbf{K}_\mathrm{f}-\mathbf{K}_\mathrm{p}). \quad \text{(C.1)}$$

In this formula the index i relates to the initial crystal state (excitonic molecule or biexciton) and the index f to the final state (exciton). $E_\mathrm{i}(E_\mathrm{f})$ is therefore the biexciton (exciton) energy, $\mathbf{K}_\mathrm{i}(\mathbf{K}_\mathrm{f})$ denotes the biexciton (exciton) wavevector and $\mathbf{K}_\mathrm{p}$ the wavevector of the emitted photon. Two delta functions ensure the conservation of energy and wavevector, M_{if} is the transition matrix element and $f(\mathbf{K}_\mathrm{i})$ stands for a distribution function of biexcitons with wavevector $\mathbf{K}_\mathrm{i}$. The double integration over $\mathbf{K}$-space indicates involvement of contributions from all initial and final states meeting the condition $\mathbf{K}_\mathrm{i} = \mathbf{K}_\mathrm{f} + \mathbf{K}_\mathrm{p}$.

The biexciton quasi-momentum is carried away by both the created photon $\mathbf{K}_\mathrm{p}$ and the recoiled exciton $\mathbf{K}_\mathrm{f}$. In the radiative recombination under discussion then, all of the biexcitons can participate, regardless of their $\mathbf{K}_\mathrm{i}$. To a good approximation we can therefore assume $|\mathbf{K}_\mathrm{i}|$, $|\mathbf{K}_\mathrm{f}| \gg |\mathbf{K}_\mathrm{p}|$ and thus neglect polariton effects. (A discussion can be found in [1].)

In the classical limit, which describes very well most of the current experimental results, the biexciton energy distribution is given by the Boltzmann function

$$f(\mathbf{K}_\mathrm{i}) \equiv f_\mathrm{M} \approx \exp(-E_\mathrm{i}(\mathbf{K}_\mathrm{i})/k_\mathrm{B}T_\mathrm{M}), \quad \text{(C.2)}$$

where T_M is the effective temperature of the biexciton gas. Under the above assumptions we can re-write (C.1) as

$$I_{sp}^{M}(h\nu) \sim \int_{-\infty}^{+\infty} d^3\mathbf{K}_i \int_{-\infty}^{+\infty} d^3\mathbf{K}_f \exp(-E_i/k_B T_M)\delta(E_i - E_f - h\nu)\delta(\mathbf{K}_i - \mathbf{K}_f), \tag{C.3}$$

where we have adopted, for the direct allowed transitions, $|M_{if}|^2 = \text{const}$ and put it outside the integral, in full analogy with radiative recombination in a direct bandgap under weak excitation. The wavevector conservation law further enables us to perform the integration in (C.3) over $\mathbf{K} = \mathbf{K}_i = \mathbf{K}_f$ only:

$$I_{sp}^{M}(h\nu) \sim \int_{-\infty}^{+\infty} d^3\mathbf{K} \exp(-E_i/k_B T_M)\delta(E_i - E_f - h\nu). \tag{C.4}$$

Now we have to consider the particular shapes of the biexciton and exciton dispersion curves. According to Fig. 8.3(a)

$$E_i = 2(E_g - E_X) - E_B + \frac{\hbar^2 K^2}{4m_{exc}} = 2(E_g - E_X) - E_B + \epsilon_M,$$

$$E_f = (E_g - E_X) + \frac{\hbar^2 K^2}{2m_{exc}} = (E_g - E_X) + \epsilon_X, \tag{C.5}$$

where ϵ_X and ϵ_M are the kinetic energies of the exciton or biexciton, respectively. Then (C.4) transforms to

$$I_{sp}^{M}(h\nu) \sim \int_{-\infty}^{+\infty} K^2 dK \exp\{-[2(E_g - E_X) - E_B + \epsilon_M]/k_B T_M\}$$

$$\times\delta\{E_g - E_X - E_B + (\epsilon_M - \epsilon_X) - h\nu\}.$$

In addition, we have converted the integration into polar coordinates (K, ϑ, ϕ) in $\mathbf{K}$-space and, on condition that the distribution of $\mathbf{K}$-vectors is fully isotropic, maintained the operation of integration with respect to a single variable $K = |\mathbf{K}|$ only.

Now only the formal treatment follows. Because either $(E_g - E_X)$ or E_B do not depend on K, because $\epsilon_M - \epsilon_X = \hbar^2 K^2/4m_{exc}$ and because $\delta(cx) = \delta(x)/c$, we obtain

$$I_{sp}^{M}(h\nu) \sim \int_{0}^{\infty} K^2 dK \exp(-\hbar^2 K^2/4m_{exc} k_B T_M)$$

$$\times\delta\{K^2 - (E_g - E_X - E_B - h\nu)4m_{exc}/\hbar^2\}$$

and if we make use of $\int f(t)\delta(t^2 - c^2)dt = f(c)/|c|$ (which is valid for an even function $f(t)$), we have

$$I_{\rm sp}^{\rm M}(h\nu) \sim \frac{(E_{\rm g} - E_{\rm X} - E_{\rm B} - h\nu)4m_{\rm exc}/\hbar^2}{\sqrt{(E_{\rm g} - E_{\rm X} - E_{\rm B} - h\nu)4m_{\rm exc}/\hbar^2}}$$
$$\exp\{-[(E_{\rm g} - E_{\rm X} - E_{\rm B}) - h\nu]/k_{\rm B}T_{\rm M}\}$$
$$\approx \sqrt{(E_{\rm g} - E_{\rm X} - E_{\rm B}) - h\nu}$$
$$\exp\{-[(E_{\rm g} - E_{\rm X} - E_{\rm B}) - h\nu]/k_{\rm B}T_{\rm M}\}. \qquad \text{(C.6)}$$

Therefore, we indeed get the inverse Maxwell–Boltzmann distribution (8.6). An upper bound of this distribution, i.e. the maximal energy of emitted photons, is equal to $h\nu_{\rm max} = (E_{\rm g} - E_{\rm X} - E_{\rm B})$. Let us note that the joint density factor $\sqrt{(E_{\rm g} - E_{\rm X} - E_{\rm B}) - h\nu}$ appears in (C.6) 'automatically', as a consequence of the parabolic dependence of $\epsilon_{\rm X}$ and $\epsilon_{\rm M}$ on K in (C.5).

Indirect bandgap

In the indirect bandgap the formula for the spectral line of biexciton luminescence (C.1) reads

$$I_{\rm in}^{\rm M}(h\nu) \sim \int {\rm d}^3\mathbf{K}_{\rm i} \int {\rm d}^3\mathbf{K}_{\rm f} f(\mathbf{K}_{\rm i})|M_{\rm in}^{\rm M}(\mathbf{K}_{\rm i}, \mathbf{K}_{\rm f})|^2\delta(E_{\rm i} - E_{\rm f} - h\nu - \hbar\omega), \qquad \text{(C.7)}$$

where, this time, of course, $\mathbf{K}_{\rm i} \neq \mathbf{K}_{\rm f}$. In conformity with Fig. 8.3(b) let us denote $\mathbf{K}_{\rm i} = \mathbf{K}_{\rm M}$, $\mathbf{K}_{\rm f} = \mathbf{K}_{\rm X}$; we see also that $E_{\rm i} - E_{\rm f} = E_0 + \epsilon_{\rm M} - \epsilon_{\rm X}$. At the same time $\hbar\omega$ is the energy of a phonon participating in the process. Relation (C.7) holds for low temperatures and thus takes into consideration phonon emission only. In comparison with (C.1) the delta function of $\mathbf{K}$ is missing because here wavevector conservation is met automatically at any difference of $(\mathbf{K}_{\rm i} - \mathbf{K}_{\rm f})$ owing to the phonon participation.

As we have explained in detail in Subsection 8.2.1, the matrix element $M_{\rm in}^{\rm M}(\mathbf{K}_{\rm i}, \mathbf{K}_{\rm f})$ is no longer independent of the integration variables. According to the theoretical model [2] it reads

$$|M_{\rm in}^{\rm M}|^2 \cong \left| \frac{1}{[(\mathbf{K}_{\rm M}/2 - \mathbf{K}_{\rm X})^2 + a_{\rm M}^{-2}]^2} \right|^2. \qquad \text{(C.8)}$$

In order to convert (C.7) into a form enabling numerical evaluation, we shall introduce, referring to Fig. 8.3(b), new wavevectors $\mathbf{Q}$, $\mathbf{q}$ defined with regard to the abscissa pertaining to the minimum of the dispersion curve of biexciton $\mathbf{Q}$ and exciton $\mathbf{q}$, respectively:

$$\mathbf{K}_{\rm M} - 2\mathbf{K}_0 = \mathbf{Q} \Rightarrow {\rm d}^3\mathbf{K}_{\rm i} \equiv {\rm d}^3\mathbf{K}_{\rm M} = {\rm d}^3\mathbf{Q},$$
$$\mathbf{K}_{\rm X} - \mathbf{K}_0 = \mathbf{q} \Rightarrow {\rm d}^3\mathbf{K}_{\rm f} \equiv {\rm d}^3\mathbf{K}_{\rm X} = {\rm d}^3\mathbf{q} \qquad \text{(C.9)}$$

or

$$(\mathbf{K}_{\rm M}/2 - \mathbf{K}_{\rm X}) = (\mathbf{Q}/2 - \mathbf{q}).$$

In eqn (C.7) we now replace the integrations with respect to $\mathbf{K}_i(\equiv \mathbf{K}_M)$ and $\mathbf{K}_f(\equiv \mathbf{K}_X)$ by those with respect to $\mathbf{Q}$ and $\mathbf{q}$, introducing in each valley spherical coordinates:

$$\begin{aligned} d^3\mathbf{K}_M &\equiv d^3\mathbf{Q} \to Q^2 dQ \sin\vartheta_Q d\vartheta_Q d\phi_Q, \\ d^3\mathbf{K}_X &\equiv d^3\mathbf{q} \to q^2 dq \sin\vartheta_q d\vartheta_q d\phi_q. \end{aligned} \tag{C.10}$$

The kinetic energies of the participating quasi-particles are then

$$\begin{aligned} \epsilon_M &= \hbar^2 Q^2/4m_{exc} \Rightarrow Q^2 = 4m_{exc}\epsilon_M/\hbar^2,\ Q^2 dQ \approx \sqrt{\epsilon_M}\, d\epsilon_M, \\ \epsilon_X &= \hbar^2 q^2/2m_{exc} \Rightarrow q^2 = 2m_{exc}\epsilon_X/\hbar^2,\ q^2 dq \approx \sqrt{\epsilon_X}\, d\epsilon_X. \end{aligned} \tag{C.11}$$

Taking into account (C.8)–(C.11), we pass in part from integration in **K**-space to integration in energy space and the lineshape (C.7) then reads

$$\begin{aligned} I_{in}^{M}(h\nu) \sim & \int_{\epsilon'_M}^{\infty}\int_{0}^{\infty} \sqrt{\epsilon_M}\sqrt{\epsilon_X}\exp(-\epsilon_M/k_B T_M) \\ & \times \left\{ \int_0^{2\pi}\int_0^{2\pi}\int_0^{\pi}\int_0^{\pi} \left| \frac{1}{[(\mathbf{Q}/2-\mathbf{q})^2 + a_M^{-2}]^2} \right|^2 \sin\vartheta_Q \sin\vartheta_q d\vartheta_Q d\vartheta_q d\phi_Q d\phi_q \right\} \\ & \times \delta(E_0 + \epsilon_M - \epsilon_X - h\nu - \hbar\omega) d\epsilon_M\, d\epsilon_X. \end{aligned} \tag{C.12}$$

The meaning of the lower bound ϵ'_M is discussed in Subsection 8.2.1. The expression to be integrated, which arises from the matrix element, we arrange easily using (C.11):

$$\left| \frac{1}{[(\mathbf{Q}/2-\mathbf{q})^2 + a_M^{-2}]^2} \right|^2 = \left| \frac{1}{(a - Qq\cos\alpha)^2} \right|^2, \tag{C.13}$$

where $a = (m_{exc}\epsilon_M/\hbar^2 + 2m_{exc}\epsilon_X/\hbar^2 + a_M^{-2})$ and α is the angle contained by $\mathbf{Q}$ and $\mathbf{q}$.

The formulas (C.12) and (C.13) tells us that we integrate over all possible mutual orientations of $\mathbf{Q}$ and $\mathbf{q}$ in both valleys. From the physical point of view the result will, of course, be the same even when shifting one of the vectors in such a way that $\mathbf{Q}$ and $\mathbf{q}$ have a common origin; the angle α remains unchanged. Moreover, we can fix one of the vectors, say $\mathbf{Q}$, to be parallel with the z axis, $\mathbf{Q} = (0, 0, Q)$, and let the remaining vector $\mathbf{q} = (q_x, q_y, q_z)$ take arbitrary positions—see Fig. C.1. In this way all possible mutual orientations are still covered. Then, of course, $\alpha \equiv \vartheta_q$ and

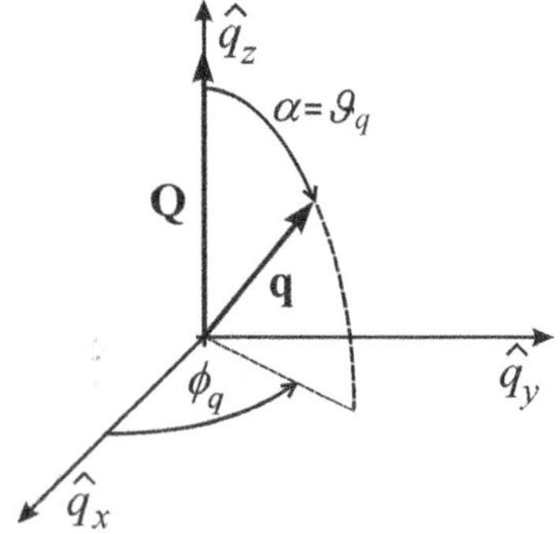

Fig. C.1
Calculation of $I_{in}^{M}(h\nu)$.

$$\mathbf{Q}\cdot\mathbf{q} = Qq_z = Qq\cos\vartheta_q,$$

which considerably simplifies the calculation, since in the inner integral (C.12) only one couple of integration variables $\vartheta_q \equiv \vartheta$ and $\phi_q \equiv \phi$ is left.

The integral (C.12) then, when applying (C.13), reads

$$I_{\text{in}}^{\text{M}}(h\nu) \sim \int_{\epsilon'_{\text{M}}}^{\infty} \sqrt{\epsilon_{\text{M}}}\sqrt{E_0 + \epsilon_{\text{M}} - h\nu - \hbar\omega}\, \exp(-\epsilon_{\text{M}}/k_{\text{B}}T_{\text{M}})$$

$$\times \left\{ \int_0^{2\pi} \int_0^{\pi} \frac{\sin\vartheta \,\mathrm{d}\vartheta}{(a - Qq\cos\vartheta)^4} \mathrm{d}\phi \right\} \mathrm{d}\epsilon_{\text{M}},$$

which, by putting $\overline{h\nu} = h\nu + \hbar\omega - E_0$ and using (C.11), gives

$$I_{\text{in}}^{\text{M}}(h\nu) \sim \int_{\overline{h\nu}}^{\infty} \sqrt{\epsilon_{\text{M}}}\sqrt{\epsilon_{\text{M}} - \overline{h\nu}}\, \exp(-\epsilon_{\text{M}}/k_{\text{B}}T_{\text{M}})$$

$$\times \left\{ \int_0^{\pi} \frac{\sin\vartheta \,\mathrm{d}\vartheta}{(a - m_{\text{exc}}2\sqrt{2}\sqrt{\epsilon_{\text{M}}}\sqrt{\epsilon_{\text{M}} - \overline{h\nu}}\cos\vartheta/\hbar^2)^4} \right\} \mathrm{d}\epsilon_{\text{M}}.$$

The dimensionless substitutions $z = \epsilon_{\text{M}}/k_{\text{B}}T_{\text{M}}$ and $y = \overline{h\nu}/k_{\text{B}}T_{\text{M}}$ transform the last expression to

$$I_{\text{in}}^{\text{M}}(h\nu) \sim \int_y^{\infty} \sqrt{z}\sqrt{z - y}\, \exp(-z) \left\{ \int_0^{\pi} \frac{\sin\vartheta \,\mathrm{d}\vartheta}{(A - B\cos\vartheta)^4} \right\} \mathrm{d}z, \qquad \text{(C.14)}$$

where

$A = a(\hbar^2/2m_{\text{exc}}k_{\text{B}}T_{\text{M}}) = \Gamma + 3z/2 - y$, $\Gamma = \hbar^2/2a_{\text{M}}^2 m_{\text{exc}}k_{\text{B}}T_{\text{M}}$ and

$B = \sqrt{2z(z - y)} \geq 0$.

It appears that of basic importance now is the integral in curly brackets, which is certainly analytically integrable:

$$\int_0^{\pi} \frac{\sin\vartheta \,\mathrm{d}\vartheta}{(A - B\cos\vartheta)^4} = \frac{1}{3B}\left[\frac{1}{(A - B)^3} - \frac{1}{(A + B)^3}\right] = \frac{2}{3}\frac{(3A^2 + B^2)}{(A^2 - B^2)^3}.$$

We thus finally get from (C.14)

$$I_{\text{in}}^{\text{M}}(h\nu) \sim \int_y^{\infty} \frac{Be^{-z}(3A^2 + B^2)\mathrm{d}z}{(A^2 - B^2)^3}. \qquad \text{(C.15)}$$

This expression was applied (with $\Gamma = 4$) for calculating the theoretical fit of the M line in Fig. 8.4(b).

References

1. Peyghambarian, N., Chase, L. L., and Mysyrowicz, A. (1983). *Phys. Rev. B*, **27**, 2325.
2. Cho, K. (1973). *Optics Commun.*, **8**, 412.

D Kinetic model of exciton condensation

An illustrative model of exciton condensation into electron–hole drops was introduced by Pokrovskii and Svistunova [1]. Let us consider a steady-state regime with a photoexcitation rate G (cm^{-3} s^{-1}). Let us assume that dynamic equilibrium takes place between the drops, containing the condensed phase or electron–hole liquid (EHL), and the surrounding gas of free excitons with concentration n_X. Let each drop of volume $V = (4/3)\pi r^3$ contains an EHL with an electron–hole pair density n_0. If τ stands for the free exciton lifetime and τ_r^{EHL} denotes the radiative lifetime of the EHL inside the drops, then the basic kinetic equation for exciton generation, recombination and condensation reads

$$G = \frac{n_X}{\tau} + I^{EHL}, \tag{D.1}$$

where $I^{EHL} = (n_0/\tau_r^{EHL})\, N\, V$ is the rate of radiative recombination in the condensed phase.[2] N denotes the density of electron–hole drops and, consequently, $NV \leq 1$ represents the filling factor that describes the degree of occupation of unit volume by the liquid.

Under steady-state conditions the total flow of quasi-particles from the gas phase to the droplets $n_X v N \pi r^2$ must be equal to the sum of the radiative recombination rate I^{EHL} inside the drops and the reversed flow of carriers evaporating from the drops, $4\pi r^2 N A T^2 \exp(-\varphi/k_B T)$:

$$n_X v N \pi r^2 = \frac{n_0}{\tau_r^{EHL}} N \frac{4}{3}\pi r^3 + 4\pi r^2 N\, A T^2 \mathrm{e}^{-\varphi/k_B T}. \tag{D.2}$$

Here v stands for the mean thermal exciton velocity, φ is the EHL binding energy and A (cm^{-2} s^{-1} K^{-2}) is the Richardson constant familiar from the thermionic emission of electron from solids. Eliminating n_X from eqns (D.1) and (D.2) yields

$$G - \frac{4AT^2 \exp(-\varphi/k_B T)}{v\tau} = \frac{4}{3}\frac{n_0}{v\tau_r^{EHL}}\left(\frac{1}{\tau} + N v \pi r^2\right) r. \tag{D.3}$$

Hence to create drops with radius $r \geq 0$ it is necessary to satisfy the condition $G - (4AT^2/v\tau)\exp(-\varphi/k_B T) \geq 0$. In other words, the condensed phase will appear only when a threshold photogeneration rate G_{th} connected with a threshold temperature T_{th} is achieved:

$$G_{th} = \frac{4A(T_{th})^2 \exp(-\varphi/k_B T)}{v\tau}. \tag{D.4}$$

The lower the temperature, the lower is the required threshold photogeneration rate or the threshold excitation intensity. (Expression (D.4) makes sense for $T_{th} <$ the critical temperature T_c, though.)

If G is sufficiently small, we can assume that $N v \pi r^2 \ll 1/\tau$ holds, or that the probability of condensation is negligible with regard to the exciton decay

[2] For the sake of simplicity, the authors assume 100% luminescence efficiency in the condensed phase which, however, does not limit the general validity of the discussion.

rate in the gas phase. In this case the solution of (D.3) reads $r \sim (G - G_{\text{th}})$. Since the intensity of the EHL luminescence is proportional to the total volume of all drops, $I^{\text{EHL}} \approx r^3 N$, then the intensity dependence of this luminescence takes form

$$I^{\text{EHL}} \sim (G - G_{\text{th}})^3. \tag{D.5}$$

Shortly after attaining the threshold conditions the EHL luminescence intensity is thus proportional to the cube of the excitation intensity. However, at sufficiently high excitation $G \gg G_{\text{th}}$ (and sufficiently low temperatures), the majority of the non-equilibrium carriers is concentrated in EHL drops, so that we can assume $N v \pi r^2 \gg 1/\tau$. Then (D.3) implies

$$I^{\text{EHL}} \sim G, \tag{D.6}$$

thus the EHL luminescence intensity grows linearly with excitation intensity.

Equations (D.5) and (D.6) fit very well the experimental observations, e.g. in very pure germanium, indeed. However, if additional channels of radiative recombination (impurities, excitonic molecules) are present in the material, the simple starting equation (D.1) is no longer justified and experimental results may differ substantially from (D.5) or (D.6).

Let us note that from (D.2) also the temperature dependence of the EHL luminescence can be derived, namely [2]

$$I^{\text{EHL}}(T) \sim \left\{1 - \exp\left[-\frac{\varphi}{k_{\text{B}}}\left(\frac{1}{T} - \frac{1}{T_{\text{th}}}\right)\right]\right\}^3.$$

References

1. Pokrovskii, J. E. and Svistunova, K. I. (1970). *Fizika i technika poluprovodnikov* (*Physics and Techniques of Semiconductors*), **4**, 491.
2. Kaminskii, A. S., Pokrovskii, J. E., and Alkeev, N. V. (1970). *Zhurnal exp. teor. fiz.* (*Journal of Exper. and Theor. Physics*), **59**, 1937.

E Bose–Einstein condensation

The maximal number of bosons in a given ensemble can, based on (8.34), be expressed as

$$n_{\text{CB}} = \int_0^\infty \rho(E) f_{\text{BE}}(E) \mathrm{d}E = \frac{(2m)^{3/2}}{4\pi^2\hbar^3} \int_0^\infty \frac{E^{1/2}\mathrm{d}E}{\exp\left(\frac{E}{k_{\text{B}}T}\right) - 1} \tag{E.1}$$

and the task of finding this number can therefore be reduced to evaluating an integral of the type

$$I(x) = \int_0^\infty \frac{x^{1/2}\mathrm{d}x}{\mathrm{e}^x - 1}. \tag{E.2}$$

The decision of whether or not Bose–Einstein condensation may occur is thus intimately related to the very existence (as a finite number) of this integral. A closer look at $I(x)$ reveals that the existence of this integral is not obvious at all, though. The integrand $f(x) = x^{1/2}/(e^x - 1)$ in (E.2) has, among others, a singularity at zero, $\lim_{x\to 0} f(x) = +\infty$; this may be inferred for instance from Fig. 8.21(a) for the case $\alpha = -\mu/k_B T \to 0$ or checked by an easy calculation. To investigate the existence of $I(x)$ we can proceed, e.g., by writing

$$I(x) = \int_0^\infty \frac{x^{1/2}\mathrm{d}x}{e^x - 1} = \int_0^1 \frac{x^{1/2}\mathrm{d}x}{e^x - 1} + \int_1^\infty \frac{x^{1/2}\mathrm{d}x}{e^x - 1} = I_1(x) + I_2(x)$$

and investigating separately the existence of I_1 and I_2.

(1) The integral $I_1(x)$. For $x \in (0, \infty)$ evidently $e^x - 1 > x$ or $(e^x - 1)^{-1} < x^{-1}$ holds. From this it follows that

$$\frac{x^{1/2}}{e^x - 1} < \frac{x^{1/2}}{x} = x^{-1/2}. \tag{E.3}$$

Evaluating $\int_0^1 x^{-1/2}\mathrm{d}x$ yields

$$\int_0^1 \frac{\mathrm{d}x}{x^{1/2}} = \lim_{\delta\to 0} \int_\delta^1 \frac{\mathrm{d}x}{x^{1/2}} = \lim_{\delta\to 0} [2x^{1/2}]_\delta^1 = 2, \tag{E.4}$$

and consequently this integral exists and according to (E.3) also the integral $I_1(x) = \int_0^1 x^{1/2}(e^x - 1)^{-1}\mathrm{d}x$ must exist and, moreover, $I_1(x) < 2$ will hold.

(2) The integral $I_2(x)$. A criterion analogous (E.4) cannot be applied here since $\int_1^\infty x^{-1/2}\mathrm{d}x$ diverges. Now let $g(x) = x/(e^x - 1)$. Then the function $f(x)$ can be expressed for $x \in \langle 1, \infty)$ as

$$f(x) = \frac{x^{1/2}}{e^x - 1} \le \frac{x}{e^x - 1} = g(x). \tag{E.5}$$

It is known that $\int_0^\infty g(x)\,\mathrm{d}x = \pi^2/6$. The definite integral $\int_1^\infty g(x)\,\mathrm{d}x (< \pi^2/6)$, all the more so, has to exist and (E.5) implies that also $I_2(x) = \int_1^\infty f(x)\mathrm{d}x$ does exist, at the same time satisfying the inequality $I_2(x) < \pi^2/6 \approx 1.645$.

We have thus found not only that $I(x)$, as a sum of two definite integrals $I_1(x)$ and $I_2(x)$, does exist, but we also obtained the estimate $I(x) < 2 + 1.645 = 3.645$.

The final evaluation of $I(x)$ cannot be done analytically but with the aid of special functions [1]

$$\int_0^\infty \frac{x^{p-1}\mathrm{d}x}{e^{ax} - 1} = \frac{\Gamma(p)}{a^p}\left[1 + \frac{1}{2^p} + \frac{1}{3^p} + \ldots\right] = \frac{\Gamma(p)}{a^p}\zeta(p), \tag{E.6}$$

where $a, p > 0$, p need not to be an integer. $\Gamma(p)$ stands for the gamma function whose values are tabulated, $\zeta(p)$ is the so-called Riemann zeta function, tabulated as well, e.g. [2]. The integral $I(x)$ represents the particular case with $p = 3/2$, $a = 1$; therefore we get from (E.6) for $\Gamma(3/2) = \sqrt{\pi}/2$ and $\zeta(3/2) \approx 2.612$ [2] the result

$$I(x) = \Gamma(3/2)\zeta(3/2) \approx 2.31$$

in line with the above estimate.

Finally to obtain n_{CB}, the use of (E.1) with $a = 1/k_B T$ gives

$$n_{CB} = \frac{(2m)^{3/2}}{4\pi^2\hbar^3} 2.31(k_B T)^{3/2},$$

which is (8.35).

References

1. Dwight, H. B. (1961). *Tables of Integrals*. The Macmillan Company, New York.
2. Janke-Emde-Lösch (1961). *Tafeln Höherer Funktionen*. B. G. Teubner, Stuttgart.

F Emission band due to strong electron–phonon interaction

In Chapter 4 we introduced the concepts of the configurational coordinate and of the Franck–Condon principle and presented a qualitative description of the emission or absorption spectrum of a localized optical centre. We found that in the case of strong electron–phonon coupling the spectra acquire the form of a broad Gaussian band (phonon wing), completely lacking the no-phonon line and exhibiting a Debye–Waller factor equal to zero: $u_{DW} = 0$. Now we derive an explicit expression to describe the relevant band shape [1].

In Fig. F.1 we reproduce, with a minor modification, the previous Fig. 4.8(b). Referring to this figure, we introduce a simplifying notation: $q = Q - Q_{g0}$, $Q_{e0} = Q_{g0} + b$, $Q_{g0} = 0$. The energy of the electronic excited state (4.12) then becomes

$$E_e(q) = E^* + Aq^2 - Bq,$$

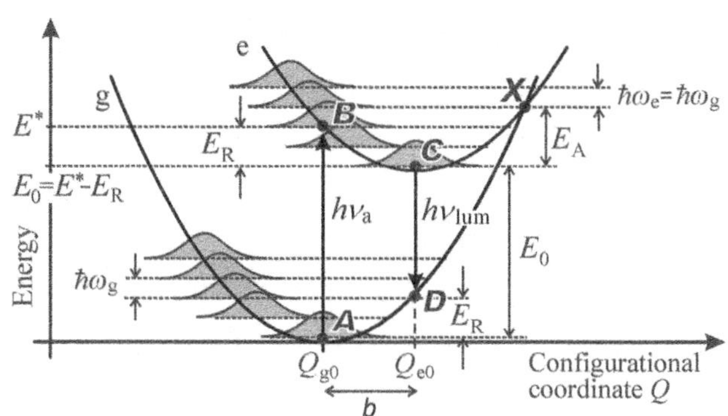

Fig. F.1
Franck–Condon principle of optical absorption ($h\nu_a$) and emission ($h\nu_{lum}$) within the configurational coordinate model in the case of strong electron–phonon interaction.

where $E^* = E_{e0} + (f_e b^2)/2$, $A = f_e/2$ and $B = f_e b$; B is proportional to the strength of the electron–phonon interaction. The relaxation energy is given as $E_R = S\hbar\omega = (f_e b^2)/2$ and, in addition, $A/B^2 = (4E_R)^{-1}$ holds.

Since the energy of the ground electronic state is $E_g(q) = Aq^2$, the emitted photon energy as a result of the radiative downward transitions is

$$h\nu = E_e(q) - E_g(q) = E^* - Bq. \tag{F.1}$$

The excited state minimum is characterized by $q = b = B/2A$ and we have $h\nu_{lum} = h\nu(q = b) = E^* - 2E_R = E_{e0} + (f_e b^2)/2 - 2E_R = E_{e0} - E_R \approx E_0 - E_R$.

Let us consider low temperatures ($k_B T < \hbar\omega_e = \hbar\omega_g = \hbar\omega$), when the wavefunction of the system in an electronically excited state but in a ground vibration state, being the wavefunction of a linear harmonic oscillator with frequency ω, has the form [2]

$$\psi(q) = \text{const} \exp(-Aq^2/\hbar\omega). \tag{F.2}$$

If B is large enough, this function controls the emission band shape $I(h\nu)$. In this case, the emission rate $P(h\nu)$ of the photon (F.1) is proportional to the vibration wavefunction $\psi(q' = q - b)$ and upon substituting from (F.1) into (F.2) we obtain

$$\begin{aligned} I(h\nu) \approx P(h\nu) &= \text{const exp}\left\{-\frac{A(E^* - h\nu - Bb)^2}{B^2\hbar\omega}\right\} \\ &= \text{const exp}\left\{-\frac{[h\nu - (E_{e0} - E_R)]^2}{4E_R\hbar\omega}\right\} \\ &\approx \text{const exp}\left\{-\frac{[h\nu - (E_0 - E_R)]^2}{2\sigma^2}\right\}, \end{aligned} \tag{F.3}$$

if we consider that $q = q' + b$. The emission band is therefore Gaussian with FWHM equal to

$$\sigma_F = 2\sqrt{2\ln 2}\sigma = 4\sqrt{\ln 2 E_R \hbar\omega}. \tag{F.4}$$

Increasing strength of the electron–phonon interaction entails an increase in $E_R \sim B^2$, thus the FWHM (F.4) also grows and, simultaneously, the band undergoes a red-shift, thereby increasing the Stokes shift $2E_R$. Everything has been summarized in Fig. F.2.

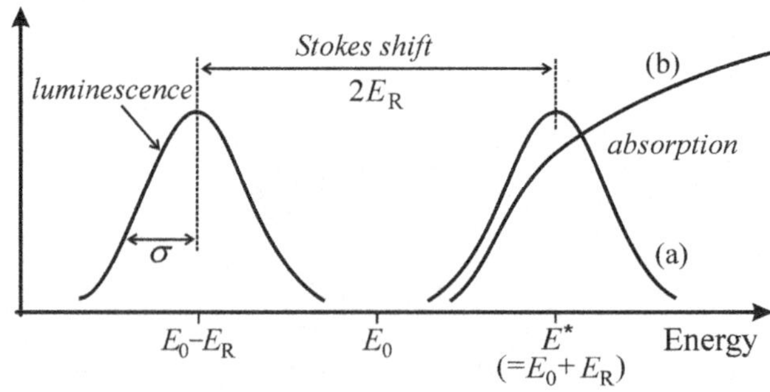

Fig. F.2
Spectral location and shape of the absorption and emission bands corresponding to the case depicted in Fig. F.1. Curve (a) stands for an optical centre with its own absorption band in a transparent matrix, while (b) describes the optical excitation of the centre via interband absorption in the host. After Street [1].

References

1. Street, R. A. (1991). *Hydrogenated Amorphous Silicon*. Cambridge University Press, Cambridge.
2. Curie, D. (1960). *Luminescence Cristalline*. Dunod, Paris.

G Fitting the optical gain spectral shape in the model of k-relaxation

Although the **k**-relaxation model is not theoretically well-founded, it may be of convenience to fit the spectral shape of the optical gain due to radiative recombination in an electron–hole plasma (EHP). This shape is described by eqn (10.18) which reads

$$g(\nu) \approx \int_0^{\overline{h\nu}=h\nu-E_g'} \sqrt{E}\sqrt{\overline{h\nu}-E}[f_c(E) - f_v(\overline{h\nu}-E)]\mathrm{d}E, \qquad \text{(G.1)}$$

where E_g' stands for the reduced bandgap and $f_c(E) = \{[\exp(E - F_e)/k_B T_{ef}] + 1\}^{-1}$ and $f_v(\overline{h\nu} - E) = \{[\exp(\overline{h\nu} - E - F_h)k_B T_{ef}] + 1\}^{-1}$ denote the Fermi–Dirac occupation functions for electrons in the conduction band and holes in the valence band, respectively. The fitting procedure is quite simple and yields reliable values of the important plasma parameters: the density N_0 of electron–hole pairs and the effective temperature T_{ef}. We recall the relevant scenario [1].

Because of the overall electro-neutrality of the EHP, the concentration of electron–hole pairs N_0 equals the concentrations of both electrons N_e and holes N_h. The electron and hole Fermi energies F_e, and F_h, respectively, and the effective temperature of the EHP then control the concentration N_0 by means of a sum over all occupied states in the bands

$$N_0 = N_{e(h)} = \frac{(2m_{e(h)})^{3/2}}{2\pi^2\hbar^3}\int_0^\infty \sqrt{E} f_{c(v)}(E)\mathrm{d}E$$

$$= \frac{(2m_{e(h)}k_B T_{ef})^{3/2}}{2\pi^2\hbar^3}\int_0^\infty \frac{\sqrt{x}\mathrm{d}x}{\exp(x - F_{e(h)}/k_B T_{ef}) + 1}, \qquad \text{(G.2)}$$

where $m_{e(h)}$ stands for the electron (e) and hole (h) effective mass, respectively.

To fit an experimental gain spectrum with the theoretical curve (G.1), N_0 and T_{ef} appear as adjustable parameters. To begin with, we do the first estimates of T_{ef} (always higher than the bath temperature; one way to asses T_{ef} quite reliably will be given towards the end of this appendix) and F_e, whereupon we calculate N_e from (G.2). Then we perform an analogous computation using the same T_{ef} and an estimated value of F_h, thereby obtaining N_h (knowledge of the carrier effective masses is required, of course).[3] If we are lucky, we get $N_e = N_h (= N_0)$ and, substituting this value (through F_e and F_h) together with the estimated T_{ef} into (G.1), we obtain the $g(\nu)$ spectral run in relative units. Upon normalizing $g(\nu)$ with respect to the experimental curve maximum, comparison of experiment with theory becomes feasible.

However, more likely we obtain $N_e \neq N_h$ and, in this case, iterative enumerations of (G.2), based on successive variations of F_e and F_h, are in order, until we get equal densities $N_e = N_h$. Afterwards we proceed as mentioned above. If the calculated curve $g(\nu)$ comes out apparently broader than the experimental curve, in the next step we have to reduce F_e and F_h while keeping their proportion; if the calculated curve exhibits a correct low-energy shape but its high-energy wing is broadened, then probably the chemical potential μ' is too large and thus T_{ef} needs to be reduced, etc. In this way we reach, sooner or later, an optimum fit and acquire the values of N_0 and T_{ef}. Then, through settling the theoretical curve on the photon energy axis, we can determine E'_g and $\mu' = F_e + F_h + E'_g$. Nevertheless, the values of E'_g and μ' might not be quite reliable owing to **k**-relaxation.

The preceding discussion should not be taken literally, though. It describes the principle of the procedure only. In fact the fit procedure may be more straightforward, which is connected—paradoxically to a certain extent—with the fact that the integral in (G.2), sometimes called the Fermi or Fermi–Dirac integral,[4] cannot be solved analytically using elementary functions. Either tabulated values [2] or analytical approximations have to be applied. For instance Zarrabi and Alfano [3] put forward the series

$$\frac{F_{e(h)}}{k_B T_{ef}} \cong \ln\left(\frac{N_{e(h)}}{N^C_{e(h)}}\right) + 0.353\left(\frac{N_{e(h)}}{N^C_{e(h)}}\right) - 4.95 \times 10^{-3}\left(\frac{N_{e(h)}}{N^C_{e(h)}}\right) + \dots, \tag{G.3}$$

where

$$N^C_{e(h)} = 2\left(\frac{2\pi m_{e(h)} k_B T_{ef}}{h^2}\right)^{3/2}.$$

From this it follows that for the determined temperature T_{ef} and a selected density $N_0 = N_e = N_h$ we can make use of (G.3) to calculate the relevant Fermi energies F_e, F_h and then, with the aid of (G.1), to obtain $g(\nu)$. The overall fit routine thereby simplifies.

When studying time-resolved emission spectra (for instance Fig. 10.3), it is of course possible to apply the above procedure for various delays τ of the detection window behind the excitation pulse. This enables us to acquire interesting information on the temporal development of the density $N_{e(h)}(\tau)$ and the temperature $T_{eh}(\tau)$, therefore for instance gaining an insight into the

[3] To generate starting values of F_e, F_h at the estimated temperature T_{ef}, a simple proportion $F_e/F_h = m_h/m_e$, valid for $T = 0\,K$ and F_e, $F_h > 0$ may be used, or density-of-states effective masses m_{de}, m_{dh} can be introduced if needed; see eqn (8.27).

[4] More specifically, the Fermi–Dirac integral is usually considered to be of the form $\int_0^\infty x^{1/2}[1 + \exp(x - \mu)]^{-1}dx$. Some authors denote the expression $(2/\sqrt{\pi})\int_0^\infty x^{1/2}[1 + \exp(x - \mu)]^{-1}dx$ as the Fermi–Dirac integral.

efficiency of carrier cooling in the bands (rate of energy loss) due to phonon emission.

Finally we are going to describe the commonly used method to estimate $T_{\rm ef}$, namely, by means of the high-energy tail of the EHP spontaneous emission spectrum. That is, the spontaneous emission—unlike the optical gain spectrum, see Fig. 10.9(a)—also includes photons coming from recombination of the most energetic electron–hole pairs. The relevant electrons and holes always find themselves in the Boltzmann tails of the respective Fermi–Dirac distributions, and the description of the high-energy wing of the EHP spontaneous emission line ($h\nu > \mu' = E'_{\rm g} + F_{\rm e} + F_{\rm h}$) can be obtained if we split the convolution integral (8.25), describing the spontaneous lineshape $I_{\rm sp}^{\rm EHP}$, into two parts (A) and (B):

$$I_{\rm sp}^{\rm EHP}(h\nu) \approx \int\limits_0^{\overline{h\nu}=h\nu-E'_{\rm g}} \frac{\sqrt{E}\sqrt{\overline{h\nu}-E}{\rm d}E}{\left[\exp\left(\frac{E-F_{\rm e}}{k_{\rm B}T_{\rm ef}}\right)+1\right]\left[\exp\left(\frac{\overline{h\nu}-E-F_{\rm h}}{k_{\rm B}T_{\rm ef}}\right)+1\right]}$$
$$\approx \int\limits_0^{(F_{\rm e}+F_{\rm h})} ({\rm dtto}){\rm d}E + \int\limits_{(F_{\rm e}+F_{\rm h})}^{\overline{h\nu}} \frac{\sqrt{E}\sqrt{\overline{h\nu}-E}{\rm d}E}{\exp\left(\frac{E-F_{\rm e}}{k_{\rm B}T_{\rm ef}}\right)\exp\left(\frac{\overline{h\nu}-E-F_{\rm h}}{k_{\rm B}T_{\rm ef}}\right)}$$
$$= (A) + (B). \tag{G.4}$$

The first integral (A) defines the value of the emission spectrum at $\overline{h\nu} = F_{\rm e} + F_{\rm h}$, alias for $h\nu = E'_{\rm g} + F_{\rm e} + F_{\rm h}$. Those electrons and holes, providing photons $h\nu > E'_{\rm g} + F_{\rm e} + F_{\rm h}$, are supposed to be already driven by the classical Boltzmann statistics, which is expressed through the integral (B). Evaluating this integral yields

$$(B) = I_{\rm sp}^{\rm EHP}(h\nu \geq E'_{\rm g} + F_{\rm e} + F_{\rm h}) \approx {\rm e}^{-\overline{h\nu}/k_{\rm B}T_{\rm ef}} \int\limits_{F_{\rm e}+F_{\rm h}}^{\overline{h\nu}} \sqrt{E}\sqrt{\overline{h\nu}-E}{\rm d}E$$
$$\approx {\rm e}^{-h\nu/k_{\rm B}T_{\rm ef}}(h\nu - E'_{\rm g})^2 r(h\nu); \tag{G.5}$$

the function $r(h\nu)$ can be shown to be limited in the interval $h\nu \in \langle E'_{\rm g} + F_{\rm e} + F_{\rm h}, \infty)$, and $r(h\nu) = 0$ holds at $h\nu = E'_{\rm g} + F_{\rm e} + F_{\rm h}$. Therefore, at sufficiently large $h\nu$, the exponential drop $\exp(-h\nu/k_{\rm B}T_{\rm ef})$ predominates and the approximation (G.5) starts to reflect the emission lineshape very well. The overall scene is sketched in Fig. G.1. By plotting $\ln I_{\rm sp}^{\rm EHP}(h\nu \gg E'_{\rm g} + F_{\rm e} + F_{\rm h})$ against $h\nu$ we can then extract from the slope of the graph a reliable estimate of $T_{\rm ef}$.

Let us note that we can asses in a similar way the effective temperature of the exciton gas, namely, from the high-energy wing of free exciton luminescence under weak excitation. An example of the determination of $T_{\rm ef}$ with the aid of this technique is shown in Fig. G.2 [4].

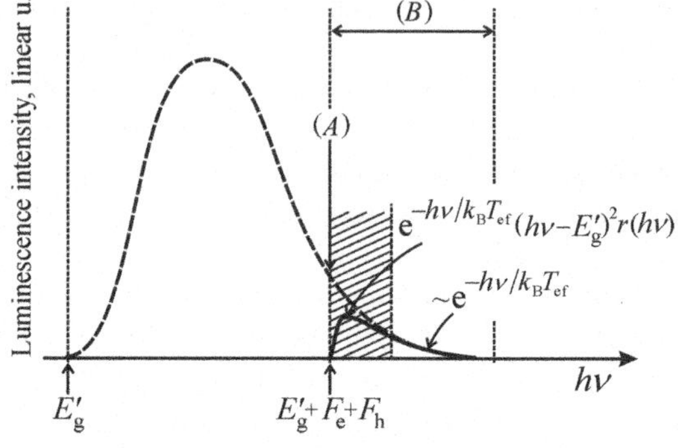

Fig. G.1
Outline of the EHP spontaneous spectrum. The labels (A) and (B) correspond to the integrals (A) and (B) in eqn (G.4). The dashed area indicates a region where the approximation (G.5) does not properly fit the actual spectral shape.

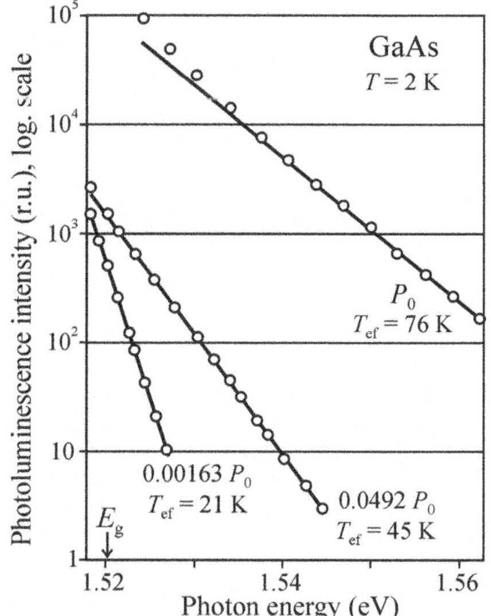

Fig. G.2
High-energy tail of the spontaneous emission spectrum of GaAs, at a bath temperature of 2 K and under Ar^+-laser excitation (514 nm, 2.41 eV). The lines determine the photocarrier effective temperature T_{ef} that increases with increasing excitation intensity. P_0 means the maximum absorbed power. After Shah [4].

References

1. Tanaka, S., Kuwata, T., Hokimoto, T., Kobayashi, H., and Saito, S. (1983). *J. Phys. Soc. Japan*, **52**, 677.
2. Blakemore, J. S. (1962). *Semiconductor Statistics.* Pergamon Press, Oxford, Appendix II; Fistul, V. I. (1967). *Highly Doped Semiconductors*, Appendix 3. (In Russian: *Sil'no legirovannyje poluprovodniky.*) Nauka, Moskva.
3. Zarrabi, H. J. and Alfano, R. R. (1985). *Phys. Rev. B*, **32**, 3947.
4. Shah, J. (1978). *Solid-State Electron.*, **21**, 43.

H Reabsorption of luminescence in semiconductors

Semiconductor luminescence is usually situated in the spectral region of material transparency, where the photon energy is considerably less than the width of the forbidden band. Therefore, the effects of reabsorption of the luminescence radiation on the emission spectrum shape can be neglected in most cases. The edge emission, or luminescence lines due to radiative recombination of free or bound excitons, may represent exclusion, though. Here we show how spectral distortion caused by such reabsorption can be corrected [1].

Cadmium telluride CdTe is a direct-bandgap semiconductor. In Subsection 7.1.2 we discussed thoroughly, in the framework of the exciton–polariton concept, the luminescence behaviour of free excitons in semiconductors of this type. We have shown that the low-temperature emission line X is very often—especially in II-VI and III-V semiconductors—underdeveloped, almost invisible, and composed of two parts, the upper polariton branch (UPB) and lower polariton branch (LPB). In what follows we shall see that such a lineshape can be, all the same, mimicked even in much simpler way: through resonant luminescence reabsorption.

Figure H.1 displays absorption and emission spectra of CdTe in the exciton region at $T = 80\,\mathrm{K}$ [1]. The experimentally acquired emission spectrum $i(h\nu)$ indeed contains two components that could be identified with UPB and LPB. The absorption line, pertinent to the creation of free excitons with principal quantum number $n = 1$, is located at $h\nu = 1.5855\,\mathrm{eV}$ and, what is remarkable, the minimum observed in the emission spectrum corresponds exactly to the absorption curve maximum. This clearly indicates the possible involvement of reabsorption.

In order to asses the influence of reabsorption in a quantitative way, we have to consider the depth x which the luminescence arises from. This means determining the concentration profile $n(x)$ of photocreated excitons beneath the sample surface. By multiplying this profile by the depth-dependent absorption term $\exp(-\alpha(h\nu)x)$, summing over x and considering also a reparation to the sample interface reflectivity $R(h\nu)$, we create a correction factor

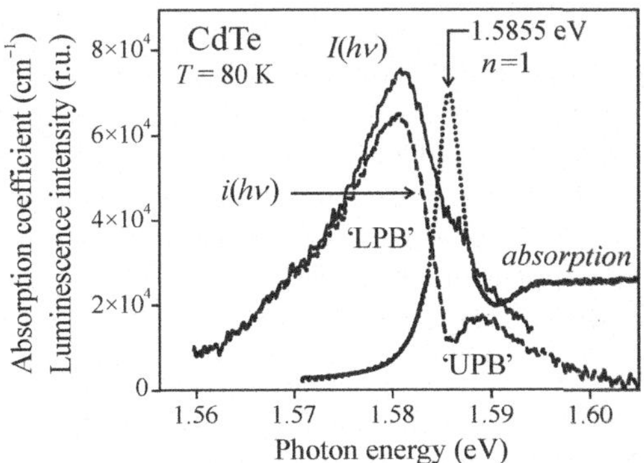

Fig. H.1
CdTe optical spectra around the free exciton energy at $T = 80\,\mathrm{K}$: absorption line to $n = 1$ free exciton state, measured emission spectrum $i(h\nu)$ and the same spectrum corrected for reabsorption $I(h\nu)$. After Horodyský [1].

$$K(h\nu) = (1 - R(h\nu)) \int_0^\infty n(x) \exp(-\alpha(h\nu)x) \, dx, \tag{H.1}$$

where $\alpha(h\nu)$ stands for the absorption coefficient as a function of the luminescence photon energy $h\nu$. This correction factor acts upon the true emission spectral shape $I(h\nu)$, transforming it into the experimentally observed spectrum

$$i(h\nu) = K(h\nu)I(h\nu). \tag{H.2}$$

Alternatively, a reciprocal recipe is applied to correct for reabsorption and to obtain the true spectrum $I(h\nu) = i(h\nu)/K(h\nu)$.

To determine $n(x)$, we can start with the diffusion–recombination equation (6.1)

$$D\frac{d^2n(x)}{dx^2} - \frac{n(x)}{\tau} = -G = -I_{ex}\alpha(h\nu_{ex}) \exp(-\alpha(h\nu_{ex})x), \tag{H.3}$$

where I_{ex} is the excitation photon ($h\nu_{ex}$) flux density impinging on the sample from an external source (laser), $\alpha(h\nu_{ex})$ stands for the corresponding absorption coefficient, D denotes the exciton diffusion coefficient and τ the exciton lifetime. Solving eqn (H.3) requires application of some boundary conditions. One of them is the surface exciton concentration $n(x = 0)$. To keep the problem simple, we suppose the rate of surface recombination infinitely large; in this case the first boundary condition reads $n(x = 0) = 0$. The second condition requires $n(x \to \infty) = 0$ because the sample thickness was much larger than the exciton diffusion length $l_{exc} = \sqrt{D\tau}$.

Under these conditions eqn (H.3) can be solved analytically. Substituting the resulting $n(x)$ into (H.1) then provides the correction factor

$$K(h\nu) = \frac{\alpha(h\nu_{ex})}{1 - D\tau\alpha^2(h\nu_{ex})} \left(\frac{1}{\alpha(h\nu_{ex}) + \alpha(h\nu)} - \frac{1}{\alpha(h\nu) + (D\tau)^{-1/2}} \right).$$

Finally, if according to (H.2) we divide the experimentally acquired emission spectrum $i(h\nu)$ by this $K(h\nu)$, enumerated for characteristic values of the CdTe parameters,[5] we obtain the corrected spectrum $I(h\nu)$ as shown in Fig. H.1. This spectrum indeed highlights a smooth curve, while the observed minimum in $i(h\nu)$ has originated owing to the strong narrow reabsorption event.

It is worth noting, however, that the peak of the corrected emission curve is red-shifted with respect to the absorption line. Now, the question arises as to how to interpret the microscopic origin of the luminescence line $I(h\nu)$. Is this really a manifestation of *free* exciton luminescence? Does the simple $I(h\nu)$ shape mean that the sophisticated model of polariton luminescence is no longer valid? Is the dip in the exciton luminescence due to the elastic scattering of exciton–polaritons from residual neutral donors in insufficiently pure (unintentionally doped) samples (Subsection 7.1.2)? The answer is not

[5] $\alpha(h\nu_{ex}) = 3.5 \times 10^4 \text{cm}^{-1}$ (excitation with a He-Ne laser $h\nu_{ex} \approx 1.96\,\text{eV}$), $D = 0.1\,\text{cm}^2/\text{s}$, $\tau = 0.5$ ns and the $\alpha(h\nu)$ spectral shape is defined by the absorption curve in Fig. H.1.

easy; in the authors' opinion, this problem has not been unambiguously solved yet. All of the three mechanisms (i.e. polariton effects, elastic scattering by residual impurities and reabsorption) may happen to contribute, depending upon the semiconductor itself, particular the sample purity and temperature of measurement [2]. The goal of this appendix has been just to demonstrate, through a suitable example, how to realize a possible correction for reabsorption in luminescence experiments.

References

1. Horodyský, P. (2006). *Optical properties of* $Cd_{1-x}Zn_xTe$. PhD Thesis, Charles University in Prague, Faculty of Mathematics and Physics, Prague.
2. Sermage, B. and Voos, M. (1977). *Phys. Rev. B*, **15,** 3935.

I Oscillator strength

If there are several resonant frequencies ω_{0j} in an optical material, we can write its complex relative permittivity in the familiar form (Lorentz model) [1]

$$\varepsilon(\omega) = 1 + \frac{Ne^2}{\varepsilon_0 m_0} \sum_j \frac{1}{(\omega_{0j}^2 - \omega^2 - \mathrm{i}\gamma_j \omega)}, \tag{I.1}$$

where N denotes the number of atoms per unit volume, m_0 is the electron mass and γ_j stands for a damping factor. Based on (I.1), every oscillator (electron) in the system would contribute to the optical absorption or emission with equal strength, while, e.g., atomic absorption lines, as is well known, may be of various intensity. This is reflected in classical physics by prescribing to any resonance a dimensionless number f_j, the so-called oscillator strength. Equation (I.1) then becomes

$$\varepsilon(\omega) = 1 + \frac{Ne^2}{\varepsilon_0 m_0} \sum_j \frac{f_j}{(\omega_{0j}^2 - \omega^2 - \mathrm{i}\gamma_j \omega)}.$$

Importantly, the sum rule $\sum_j f_j = 1$ holds for any electron. We can imagine that a particular electron takes part in several transition events in parallel, and the 'absorption strength' or absorption (emission) intensity is distributed over all these transitions.

Quantum mechanics comes with a logical explanation: different transitions have different values of the relevant transition matrix element. The oscillator strength is then defined as a dimensionless quantity that is proportional to the squared modulus of the matrix element

$$f_{ij} = \frac{2m_0\omega_{ij}}{3\hbar} \left|\langle j|\, \mathbf{x}\, |i\rangle\right|^2. \tag{I.2}$$

Equation (I.2) has been written for dipole-allowed transitions ($e\mathbf{x}$) between non-degenerate levels i and j.

A similar routine can be applied also in the case of Wannier excitons, namely, by starting with eqn (I.2), and completing it with a factor $|\phi(r = |\mathbf{r}_e - \mathbf{r}_h| = 0)|^2$, which defines the probability of finding an electron and a hole at the same lattice point. Taking into account the analogy between a hydrogen atom and an exciton, we come to the conclusion that this probability should be proportional to the squared modulus of the wavefunction of the hydrogen atom. This means that with increasing principal quantum number n (which indicates excited states of the exciton) the value of $|\phi(r = 0)|^2$ will diminish because for increasing n the Wannier exciton radius increases and, consequently, $|\phi(r = 0)|^2$ must decrease. The general shape of the hydrogen wavefunction $\Psi(r, \theta, \phi)$ is rather complicated and is expressed with the aid of Laguerre and Legendre polynomials; however, it is not difficult to verify that if $r = 0$ and, for instance, $n = 2$, $l = m = 0$ (2s state) we get $|\Psi|^2 = 1/8\pi a_B^3$ [2], where a_B is the Bohr radius of the hydrogen atom. In the more general case of $r = 0$ and $n \neq 0$, $l = m = 0$ we obtain $|\Psi|^2 = 1/n^3\pi a_B^3$. This means that for a dipole-allowed s-exciton the oscillator strength, normalized to unit cell volume Ω, reads

$$f_X(n) = \frac{2m_0\omega}{3\hbar\pi a_X^3} |\langle n, l = 0| x |0\rangle|^2 \frac{\Omega}{n^3}.$$

Absorption lines related to transitions to higher exciton states $n = 2, 3, 4, \ldots$ thus rapidly fade away as can be clearly seen, e.g., in Fig. 7.2. If the intensity of one of these lines becomes enhanced for any reason, it follows from the sum rule that the intensities of the remaining lines must be reduced. This occurs for instance in quantum wells.

References

1. Fox, M. (2001). *Optical Properties of Solids.* Oxford University Press, Oxford.
2. Beiser, A. (1969). *Perspectives of Modern Physics*, Table. 9.1. Mc-Graw Hill, New York.

J Fitting with a double exponential (Kočka's summation)

The dynamics of luminescence is often characterized by a decay curve being a mixture of two exponential components with amplitudes a_1, a_2 and decay times τ_1, τ_2, see (3.8),

$$i(t) = a_1 e^{-t/\tau_1} + a_2 e^{-t/\tau_2}. \tag{J.1}$$

The first summand in this equation usually represents a fast component ($\tau_1 < \tau_2$), which fades away rapidly with increasing t. In a coordinate system comprising a logarithmic intensity scale and a linear time scale, both summands in (J.1) are represented by straight lines as depicted in Fig. J.1. If $a_1 \gg a_2$, which is a frequent case, the overall curve will tend towards a linear shape for

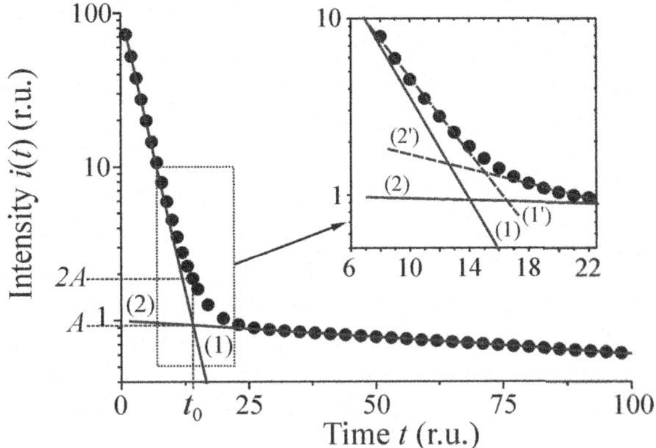

Fig. J.1
Model curve of luminescence decay $i(t)$ (symbols) formed by two components $i(t) = 100\exp(-t/3) + \exp(-t/200)$, which are depicted by straight lines (1) and (2), respectively. The inset demonstrates an example of an incorrect fit via straight lines (1′) and (2′).

both $t \to 0$ and $t \to \infty$; this is shown in Fig. J.1 for $a_1 = 100$, $a_2 = 1$, $\tau_1 = 3$, $\tau_2 = 200$.

Let both lines cross one another at time t_0. Obviously, the relation

$$a_1 e^{-t_0/\tau_1} = a_2 e^{-t_0/\tau_2} = A$$

must hold, or according to (J.1),

$$i(t_0) = 2A. \tag{J.2}$$

In other words, the resulting overall curve at $t = t_0$ is situated above the point of intersection of the straight lines and the ordinate of the resultant curve is a double of that pertinent to the point of intersection (i.e. is equal to $2A$).

Everything seems self-evident and trivial; nevertheless, in the literature one can encounter the following incorrect technique when evaluating experimental results. Provided the decay curve has been measured over a insufficiently wide time interval (see the inset in Fig. J.1), authors tend to fit the experimental points with two straight lines so as to minimize the deviation of their point of intersection from the experimental data, see lines (1′) and (2′). Then the condition (J.2) is not fulfilled and the decay times τ_1, τ_2 extracted from the slopes of the lines do not correspond to reality, of course.

A similar problem occurs—maybe even more often—when one fits double- or multi-exponential dependences of other types, as for instance those resulting from the investigation of the temperature behaviour of luminescence or semiconductor electric conductivity, aimed at determining the relevant activation energies.

K Absolute quantum yield of luminescent materials

The absolute quantum yield (efficiency) of luminescence η is a very important quantity, especially for evaluating application prospects of a particular phosphor. The only way to asses η is by experiment and this experiment is, as already stressed several times, difficult and delicate (like all absolute measurements of similar type). Specialized laboratories are mostly dedicated to this purpose. The intensity of luminescence in current experimental basic research is measured in relative units only. Nevertheless, even then occasionally there is a need to evaluate η absolutely. It might thus be of importance to outline the principles of the measurement of luminescence efficiency.

There are basically three experimental techniques for evaluating η for *photo*-luminescence: (1) absolute optical; (2) relative optical; and (3) photocalorimetric one. The first method basically evaluates the exact amount of both the absorbed light energy and the energy emitted by the sample. The second one, relative optical, can be realized more easily, because it relies upon a comparison of the measured sample with a luminescence standard. It may, however, be less accurate because of the uncertainty related to η of the applied standard. The photocalorimetric or photoacoustic method consists in the transformation of the absorbed excitation energy into heat and takes into account the processes of non-radiative recombination. Its implementation is difficult owing to the necessity to ensure good thermal isolation of the sample from its surroundings and to detect very small temperature variations (mK).

Also measurements of *electro*luminescence efficiency can be classified as absolute or relative, the latter relying on a known standard. In principle, the assessment of η in the case of electroluminescence is easier than that for photoluminescence because the supplied excitation energy can be evaluated quite smoothly through measurement of the excitation electric current.

We shall outline three particular experimental approaches: an absolute photoluminescence method, a relative photoluminescence method and an absolute electroluminescence method.

Absolute determination of photoluminescence η [1]

A schematic of one possible experimental set-up is shown in Fig. K.1. The central component here happens to be an integration sphere, which is a hollow spherical solid with inner surface covered by a non-luminescent, strong diffusion reflecting film (MgO). The sphere has four ports, located in the equatorial plane: an excitation port for input of the excitation beam, a port destined to detect the excitation radiation (photodetector A), a port to detect the emitted luminescence radiation (photodetector B) and the sample port. A sample holder (quartz tube) can contain both a liquid sample and a solid one (powder). The excitation radiation is scattered by a diffuser upon entering the sphere. Suitable glass filters are located in front of the photodetectors in order to let through only the excitation (photodetector A) or photoluminescence (photodetector B)

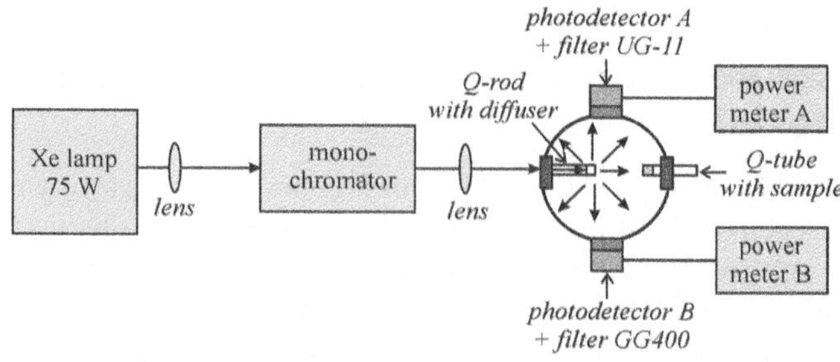

Fig. K.1
Experimental set-up for absolute measurement of photoluminescence quantum efficiency of solid and liquid samples. After Rohwer and Martin [1].

radiation, respectively. A simple spectrometer constitutes an independent part of the set-up to measure the spectral composition of the photoluminescence in relative units.

The absorbed optical power P_{abs} is evaluated by comparing two sets of data from the photodetector A, recorded with and without the sample (a reflecting blind flange is fixed in the relevant port instead of the sample) inside the sphere.

The emitted radiation power P_{em} is determined by making use of similar data from the photodetector B (measurements with and without the sample). Knowledge of the spectral sensitivity of both detectors as well as of the spectral sensitivity of the sphere is indispensable. The photoluminescence power efficiency $\eta_P = P_{em}/P_{abs}$ can then be assessed; hence, the quantum efficiency η will subsequently be determined with the aid of the relation $\eta = (\overline{\lambda_{em}}/\lambda_{ex})\eta_P$, where $\overline{\lambda_{em}}$ is the wavelength of an average energy luminescence photon. As for the details, the reader is referred to the original work [1]. The accuracy of fixing η in this manner may be relatively high (error $< 1\%$).

Relative determination of photoluminescence η

This method is based on comparison of the emission intensity of the material under study (in relative units) with an analogous measurement performed on a luminescence standard whose photoluminescence quantum efficiency is known. An identical excitation wavelength λ_{ex} is applied in both cases, and similarities in the spectral position and width of both emission spectra appear advantageous for eliminating additional corrections. Of basic importance is then to ensure that the same amount of excitation energy is absorbed in both cases. This is feasible to arrange in *liquid samples* (solutions) of phosphors only, when the absorbed optical power can be written as

$$P_{abs}(h\nu_{ex}) = P_0(h\nu_{ex})(1-R)\left[1-\exp\left(-\varepsilon(h\nu_{ex})cl\right)\right], \qquad \text{(K.1)}$$

where R stands for the reflectivity of the cell filled with the investigated solution, l is the cell length, $\varepsilon(h\nu_{ex})$ is usually called the molar absorption or extinction coefficient and c denotes the substance concentration (number of moles per unit volume). By varying the concentration c we can thus easily tune the same value of P_{abs} in both cases. Furthermore, there is no need to know the values of the molar coefficients ε, but we are able to achieve the desired result in a purely empirical way—through a suitable choice of both concentrations so that the excitation beam has the same intensity upon passing through the cells in both cases. It may appear that for semiconductors as solids the method is useless; nevertheless, it can be nicely applied to colloidal dispersions of semiconductor nanocrystals, as will be mentioned below.

The experimental arrangement of both the excitation and the detection paths may then be rather arbitrary, provided that in the course of measurement all experimental parameters remain fixed, apart from exchanging the cell with the measured solution for that containing the luminescence standard. An identical geometrical localization in the cell of the luminescent spot as well as an identical spatial radiation pattern can be reasonably supposed in both cases.[6]

A solution of quinine sulphate dihydrate in 1N sulphuric acid can serve as an example of a widely used international luminescence standard. It is a long-term stable, non-oxidizing liquid standard with emission maximum at $\lambda_{em} = 450$–460 nm and photoluminescence quantum yield $\eta = 52\%$ (under excitation wavelength λ_{ex} ranging from 224 to 390 nm and at a concentration of $c = 10^{-2}$ M) [1]. However, for practical applications various organic dyes as luminescence standards appear to be more easily applied; they are available in powder form from several manufacturers (e.g. Exciton, Inc.) and are soluble in common solvents like ethanol and methanol. By way of example Rhodamine 6G in ethanol ($\eta \approx 94\%$, $\lambda_{em} = 560$–580 nm, concentration $c = 10^{-7} - 10^{-2}$ M, room temperature) can be quoted. For the details the reader is referred to the literature [1–3]. It should be pointed out that the values of η for organic dyes, cited in the literature, fluctuate somewhat [4, 5] (η depends on purity of the dye, the solutions are not fully photostable in the long term, measurement temperature varies, etc.), which makes this method less reliable by comparison with the foregoing case.

What may constitute a difficulty with this method is residual light scattering in the sample. For instance, in a colloidal dispersion of luminescent Si nanocrystals (of size of a few nanometres) gathering of the nanocrystals into large agglomerates of about 100 nm in diameter occurs, which leads to Mie scattering of the excitation radiation. Given these circumstances, even if we prepare the standard and the colloidal solutions so that the transmitted excitation intensity coincides in both cases, eqn (K.1) does not straightforwardly give the absorbed power, because the exciting beam is attenuated in the colloid—in addition to useful absorption—also by light scattering. The problem can be solved by introducing artificial scattering into the standard solution, for instance by means of polystyrene beads of diameter similar to that of the nanocrystalline aggregates [6]. Figure K.2(a) presents an emission spectrum of a colloidal dispersion of Si nanocrystals (black curve) that comprises also a narrow line at 480 nm owing to elastic scattering of the excitation beam. The figure also contains an emission spectrum of a luminescence standard solution (Rhodamine 6G in ethanol, R6G grey curve), to which polystyrene beads with a diameter of 68 nm were added; a pure R6G 'genuine' solution does not exhibit any light scattering, see the grey shading. The number of beads was chosen so that the scattering line at 480 nm exhibited an amplitude equal to that observed in the Si colloidal dispersion; this compensated for the effect of light scattering.

By comparing the areas below the curves it become obvious that the quantum yield of the Si-nanocrystal dispersion is smaller in comparison with η of R6G. Quantitative evaluation of the areas at varying excitation intensities then

[6] It might not be like this if the concentrations of the solutions differ substantially. Complications arising due to possible reabsorption of luminescence radiation may be avoided by choosing as low solution concentrations as possible (but still compatible with an acceptable value of the signal-to-noise ratio).

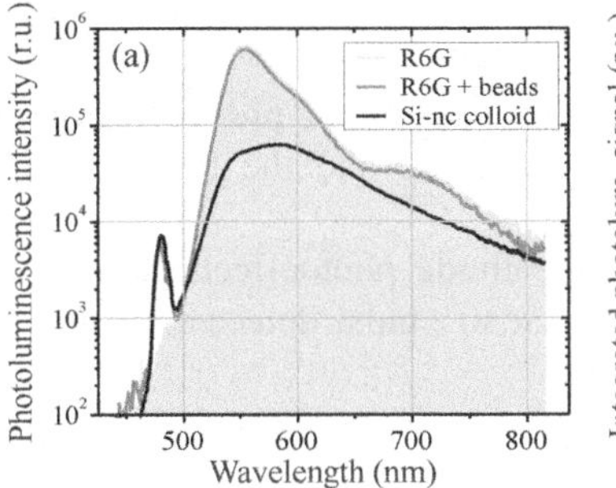

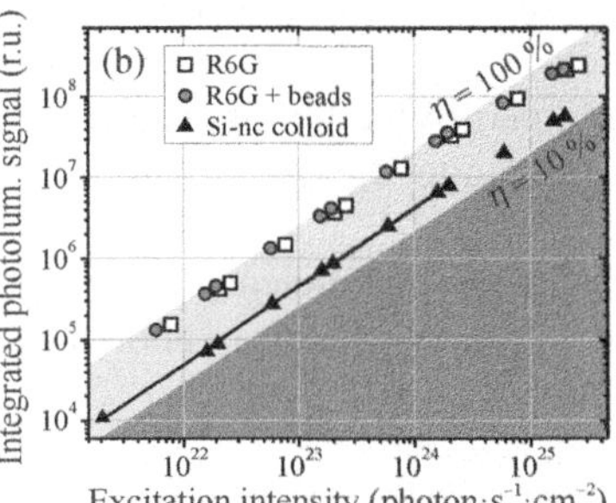

Fig. K.2
(a) Comparison of photoluminescence emission spectra of Rhodamine 6G (grey shading), Rhodamine 6G with polystyrene beads (grey curve) and a colloidal dispersion of Si nanocrystals (black curve) at a cw photoexcitation with $\lambda_{ex} = 480$ nm. The amount of excitation energy is the same in all three cases. (b) Spectrally integrated luminescence signal of the same samples as a function of excitation intensity. Room temperature, after Kůsová *et al.* [6].

leads to the curves shown in Fig. K.2(b): over four orders of excitation intensity $\eta(\text{Si-nc}) \approx 0.22\eta(\text{R6G}) \approx 20\%$ holds.

Absolute determination of electroluminescence η

We shall deal with the total (external) efficiency of injection electroluminescence η_{tot}, which was defined in Subsection 11.2.1 as η_{tot} = (number of photons emitted from sample surface/number of injected electron–holes pairs). The number of injected *e–h* pairs can be established from the excitation current density $j = I/S$, where I is the electric current flowing through the p-n junction, therefore a quantity easily measurable, and S denotes the junction cross-section. This can be commonly evaluated on the basis of the sample geometry; for instance in a sample in the form of a thin film deposited on a non-transparent conductive substrate, S means the area of a semitransparent electrode from which the electroluminescence radiation emerges. The number of injected *e–h* pairs per second is then

$$n_{e-h} = j/e = I/Se, \tag{K.2}$$

where e is the electron charge.

We continue to suppose we have a planar sample with vertical radiation through the semitransparent electrode. Let us suppose further that our light detection system consists of a photomultiplier tube in connection with a photon counter, which provides data on the number of detected photons per second, n_p. We are then faced with a task defined basically by the experimental geometry: how to deduce the total number of photons emitted from the sample surface, n_{p0}, knowing the number of detected photons, n_p.

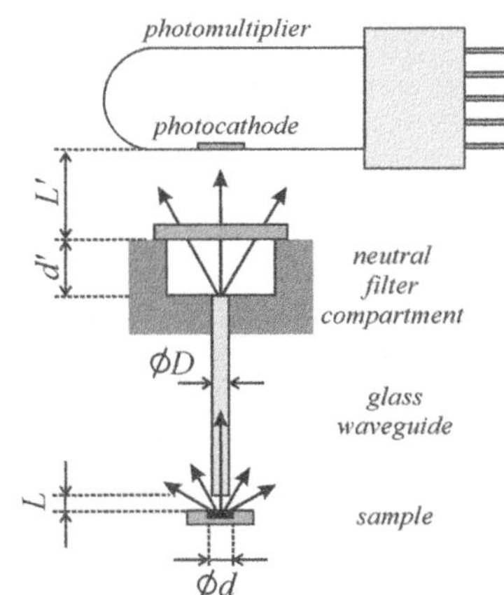

Fig. K.3
Measurement of electroluminescence efficiency [7].

An example of a particular experimental set-up is shown in Fig. K.3 [7]. The sample is fixed in a light-proof chamber and excited with current pulses from a pulsed generator. The luminescence radiation is collected closely above the sample by a 'light pipe' (a glass rod with polished ends). The rod faces a compartment of neutral glass filters (aimed at avoiding saturation of the photodetector) on which a photomultiplier is mounted. Obviously, n_p and n_{p0} are linked through the proportion $n_p = \xi n_{p0}$, where the coefficient $\xi < 1$ comprises the effect of five factors:

1. Only part of the emitted radiation is coupled to the glass rod (factor c_1).
2. The rod itself causes a certain attenuation (factor c_2).

3. Only part of the radiation leaving the rod enters the filter compartment (factor c_3).
4. Only part of the luminescence beam impinges on the photodetector, owing to the finite dimensions of the photocathode and the beam divergence (factor c_4).
5. The quantum efficiency of the photocathode photoeffect η_p is less than unity, because not every photon gives rise to a pulse detected by the counter (factor η_p).

We can thus write

$$\xi = \prod_{i=1}^{4} c_i \, \eta_p.$$

Taking into account eqn (K.2), the electroluminescence efficiency can then be transformed into a form suitable for numerical evaluation:

$$\eta_{tot} = \frac{n_{p0}}{n_{e-h}} = \frac{n_p}{\xi n_{e-h}} = \frac{n_p Se}{\prod_{i=1}^{4} c_i \eta_p I}. \tag{K.3}$$

The photocathode quantum efficiency η_p along with the photocathode spectral sensitivity are listed in manufactures' data sheets. The coefficients $c_i \leq 1$ can be found by combining effortless experiments with easy computations. Thus, the part c_1 from the total luminous flux emitted from a circular electric contact of diameter d and coupled to a light guide of diameter D at a height L above the sample (Fig. K.3) can be calculated by anticipating the cosine character of the radiation; the result reads [7]

$$c_1 = \frac{1}{2d^2}\Big(d^2 + 4L^2 + D^2 - \sqrt{16L^4 + d^4 + D^4 + 8L^2d^2 + 8L^2D^2 - 2d^2D^2}\Big).$$

The transmission coefficient of the light guide (c_2) can be found experimentally if a flat circular light source is installed instead of the sample. The circular source (e.g. a flat LED diode) should have its emission spectrum and spatial radiation pattern similar to those of the investigated sample. A photodiode fixed closely behind the opposite light guide end then measures the amount of transmitted light. The c_3 coefficient can be determined again from experiment, namely, by measurement of the angular distribution of light—emitted by the above mentioned LED—at height d' above the light guide exit. This intensity distribution is subsequently compared with the limiting dimensions of the neutral filter compartment. Finally, the c_4 coefficient is to be calculated starting from the measured spatial radiation distribution at height d' (which can be approximated by a Gaussian curve with dispersion $w^2/2$), if we take into consideration the lateral dimensions of the PMT photocathode, its location at height L' and the divergence of the luminescence beam. The result for a rectangular photocathode $2a \times 2b$ reads

$$c_4 = \mathrm{erf}\,(a'/w)\mathrm{erf}\,(b'/w),$$

where $a' = a\,(d'/(d' + L'))$, $b' = b\,(d'/(d' + L'))$ and $\mathrm{erf}(x) = (2/\sqrt{\pi}) \int_0^x \exp(-t^2)\mathrm{d}t$ is the so-called error function. For the details, the reader is referred to [7].

In view of the approximations involved and of possible experimental errors the accuracy of this method cannot be overvalued; the relative error makes up to $\sim 30\%$. However, even this value is helpful when assessing novel phosphors. Of course, if a reference standard light source (e.g. a LED diode calibrated in absolute units by the manufacturer) is available, the precision of the method may be increased considerably.

References

1. Rohwer, L. S. and Martin, J. E. (2005). *J. Luminescence*, **115**, 77.
2. Kubin, R. F. and Fletcher, A. N. (1982). *J. Luminescence*, **27**, 455.
3. Fischer, M. and Georges, J. (1996). *Chem. Phys. Lett.*, **260**, 115.
4. Abrams, B. L. and Wilcoxon, J. P. (2009). *J. Luminescence*, **129**, 329.
5. Rohwer, L. S. and Martin, J. E. (2009). *J. Luminescence*, **129**, 331.
6. Kůsová, K., Cibulka, O., Dohnalová, K., Pelant, I., Matějka, P., Žídek, K., Valenta, J., and Trojánek, F. (2009). *Mater. Res. Soc. Symp. Proc.*, **1145**, 1145-MM04-13.
7. Luterová, K. (1996). *Transport and photoelectric properties of luminescent forms of silicon* (in Czech). Diploma thesis, Charles University in Prague, Faculty of Mathematics and Physics, Prague.

L Basic description of statistics of light from classical and non-classical sources

In this appendix we shall indicate some approaches used to describe the statistics of light. These fundamental concepts are needed for understanding the non-classical photon statistics of individual luminescent semiconductor nanocrystals (Section 17.5). For in-depth reading, the reader is referred to Loudon's classic textbook [1].

A *classical light source*, as for instance a light bulb or any other incandescent lamp, is composed of a huge assembly of emitting species, radiating without any phase correlation. The resulting emission from such a source is characterized by a chaotic temporal development of the electric field intensity $\mathscr{E}(t)$ and therefore also of the light intensity (Fig. L.1(a)) [2].

We can define the normalized *first-order correlation function*

$$g^{(1)}(\tau) = \frac{\langle \mathscr{E}(t+\tau)\, \mathscr{E}^*(t) \rangle}{\langle \mathscr{E}(t)\, \mathscr{E}^*(t) \rangle}, \tag{L.1}$$

which, in principle, describes the interference pattern that we obtain in an interferometer—i.e. by splitting the signal into two parts and subsequently by reuniting them after introducing a relative variable temporal shift τ. The angle brackets in (L.1) denote an ensemble average and the asterisk means the complex conjugate. At $\tau = 0$ obviously $|g^{(1)}(0)| = 1$ holds, and $g^{(1)}(\tau)$ approaches zero for $\tau \to \infty$, as shown schematically in Fig. L.1(b). The absolute value of $g^{(1)}$ thus ranges between zero and unity and represents

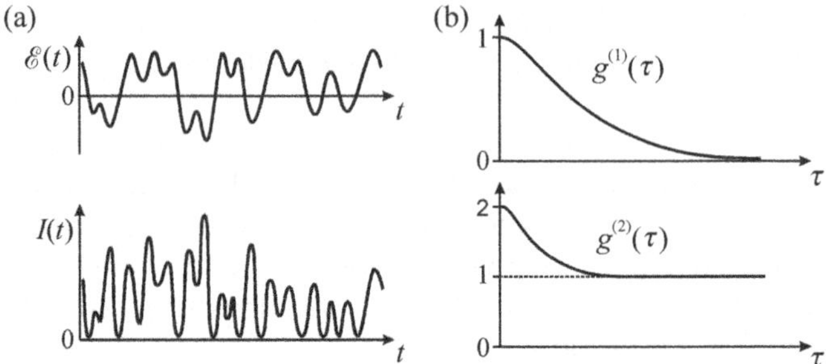

Fig. L.1
Classical chaotic (thermal) light: (a) temporal course of electric field and light intensity, (b) corresponding first-order and second-order correlation functions. The Fourier transform of $g^{(1)}(\tau)$ can be used to yield the intensity spectrum of the source. Adapted from Orrit [2].

the degree of *coherence of light*—it determines how two light waves shifted in space or time (spatial or temporal coherence) will be able to interfere provided they are linked together. Perfectly coherent light $|g^{(1)}| = 1$ would be represented by a harmonic wave of a single frequency, which would create an interference pattern with maximum contrast. Of course, light with a finite spectral width will always be incoherent for sufficiently long times $\tau \to \infty$, therefore $\lim_{\tau\to\infty} |g^{(1)}(\tau)| \to 0$.

The normalized *second-order correlation function*

$$g^{(2)}(\tau) = \frac{\langle \mathscr{E}^*(t+\tau)\,\mathscr{E}(t+\tau)\,\mathscr{E}^*(t)\,\mathscr{E}(t)\rangle}{\langle \mathscr{E}(t)\,\mathscr{E}^*(t)\rangle\,\langle \mathscr{E}(t)\,\mathscr{E}^*(t)\rangle} = \frac{\langle I(t+\tau)I(t)\rangle}{\langle I(t)\rangle^2} \tag{L.2}$$

describes the correlation of intensities, experimentally attainable via intensity correlation experiments in the Hanbury Brown–Twiss interferometer (see Fig. 17.25). Calculation of $g^{(2)}(\tau)$ can be visualized as the computation of the overlap integral of the temporal intensity profile with the same profile but shifted in time by τ (see Fig. 17.28), the result being normalized to the square of the average intensity.

It can be shown that in the case of *classical*, chaotic light the intensity correlation function has the following properties [1]:

$$\begin{aligned} &1.\ g^{(2)}(0) \geq 1,\\ &2.\ g^{(2)}(\tau) \leq g^{(2)}(0),\\ &3.\ g^{(2)}(\tau) \geq 0. \end{aligned} \tag{L.3}$$

This means that $g^{(2)}(\tau)$ is a non-increasing function of τ and simultaneously at $\tau = 0$ it cannot be less than unity. For chaotic light it can be shown that [2]

$$g^{(2)}(\tau) = 1 + |g^{(1)}(\tau)|^2,$$

see Fig. L.1(b). A characteristic feature of classical light sources is that the intensity correlation function achieves its highest value close to $\tau = 0$. This signifies that two photons have higher probability to be detected (or emitted) together (or one shortly after another) than with a long delay in between. This property is called *photon bunching*.

The quantum description of light (see for instance [1]) imposes only one limitation on the normalized second-order correlation function, namely,

$$g^{(2)}(\tau) \geq 0,$$

which corresponds to the third property valid in the classical description of light (L.3). If there are light sources not complying with the classical conditions 1 or 2 in (L.3), or even with both conditions simultaneously, they can be called *non-classical light sources*. It turns out that such sources violating the classical conditions—in particular at short times τ—are single quantum emitters such as molecules, nanocrystals, etc. The reason for this behaviour is that after emitting a photon, the emitter rests in the ground state and a subsequent photon can be emitted after a certain time period only; during this period the emitter becomes re-excited via absorbing an excitation photon and relaxes afterwards to a light-emitting level with a finite mean lifetime. In other words, the probability density of detecting two consecutive photons separated by a short time delay τ is low and decreases with decreasing delay. This effect is called *photon antibunching*. Experimental observation of antibunching in the correlation signal is taken as the main evidence that a single nanocrystal (molecule) is the origin of the detected photon stream (see Sections 16.4 and 17.5).

Now, let us discuss how the intensity correlation function can be established experimentally [3]. We shall consider a simple experimental set-up employing cw photoexcitation and a single ideal detector (i.e. a detector possessing 100% detection efficiency and no noise). A basic observable, enabling us to characterize the photon stream, is the distribution of delay times between the arrivals of two consecutive photons (this means a distribution acquired in a start–stop experiment). Let us call this function the *delay distribution* $C(\tau)$. The probability of detecting the next photon arrival with a delay occurring in the time interval $(\tau, \tau + \mathrm{d}\tau)$ is equal to $C(\tau)\,\mathrm{d}\tau$. The function $C(\tau)$ thus represents in fact a probability density that should obey the normalization condition

$$\int_0^\infty C(\tau)\mathrm{d}\tau = 1.$$

In a similar way we can introduce a *conditional probability density* $G(t+\tau, t)$, related to the probability of detecting a photon at time $t+\tau$, provided one previous photon was recorded at time t (detection of other photons between t and $t+\tau$ is not excluded). The function $G(t+\tau, t)$ is in fact the *second-order non-normalized intensity correlation function pertinent to* $g^{(2)}(\tau)$, thus

$$G(t+\tau, t) = \langle I(t+\tau)I(t)\rangle = g^{(2)}(\tau)\,\langle I(t)\rangle^2 .$$

In case we detect single photons rather than the instantaneous intensity, we can rewrite eqn (L.2) in the form

$$g^{(2)}(\tau) = \lim_{T\to\infty} \frac{T}{N^2} \sum_{i>j} \delta\left[\tau - (t_i - t_j)\right],$$

where N means the total number of photons recorded within a time interval of duration T and all possible photon pairs, delayed by τ, have been taken into account.

Let us now investigate whether there is a relation between $G(t+\tau, t)$ and $C(\tau)$, and if so, what this relation looks like. The total conditional probability density $G(t+\tau, t)$ is obtained by summing the probabilities for a photon, detected at time $t+\tau$, to be the first, the second, the third one, etc. arriving after the photon recorded at time t. Therefore,

$$G(t+\tau, \tau) = C(\tau) + \int_0^\tau C(\tau')\, C(\tau-\tau')\, \mathrm{d}\tau' +$$

$$\int_0^\tau \int_0^{\tau-\tau'} C(\tau')\, C(\tau-\tau')\, C(\tau-\tau'-\tau'')\, \mathrm{d}\tau'' \mathrm{d}\tau' + \cdots$$

This is a complicated relation and can be solved with the aid of Laplace transforms, because convolutions are replaced by simple products of Laplace transforms of the relevant functions. The above equation then converts into a geometrical series whose summation, performed by Reynaud [4], yields

$$g(s) = \frac{c(s)}{1-c(s)}, \tag{L.4}$$

where $g(s)$ and $c(s)$ stand for the Laplace transforms of $G(t+\tau, t)$ and $C(\tau)$, respectively.

A fundamental example of photon statistics is a coherent light source generating a photon stream that follows a Poissonian distribution

$$p(m) = \frac{\langle n \rangle^m}{m!} e^{-\langle n \rangle},$$

where $p(m)$ is the probability of recording m photons in a given time interval and $\langle n \rangle$ stands for the mean number of detected photons per interval. This might be, for instance, a perfectly stabilized cw laser. In this case the delay distribution $C(\tau)$ appears to be an exponential function of the delay: $C(\tau) = a \exp(-a\tau)$. Its Laplace transform reads $c(s) = a/(s+a)$, therefore by making use of eqn (L.4) we get $g(s) = a/s$. This signifies that the intensity correlation function G is a constant independent of τ (see Fig. L.2). In other words, a particular detection event tells us nothing about the probability of the next photon arrival–the coherent light is neither bunched nor antibunched and a Poissonian source therefore produces an absolutely random stream of photons.

Real (chaotic) sources always exhibit larger mean quadratic deviation (variance) of the number of detected photons in comparison with a perfect Poissonian source. The deviation of real photon statistics from the ideal Poissonian case can be characterized by the *Mandel parameter* $Q(T)$

$$Q(T) = \frac{\langle n^2 \rangle_T - \langle n \rangle_T^2}{\langle n \rangle_T} - 1,$$

which is based on a typical property of the Poissonian distribution, namely, that the variance in the number of photons $\langle n^2 \rangle_T - \langle n \rangle_T^2$, detected in the interval T, is equal to the mean photon number $\langle n \rangle_T$ detected in the same

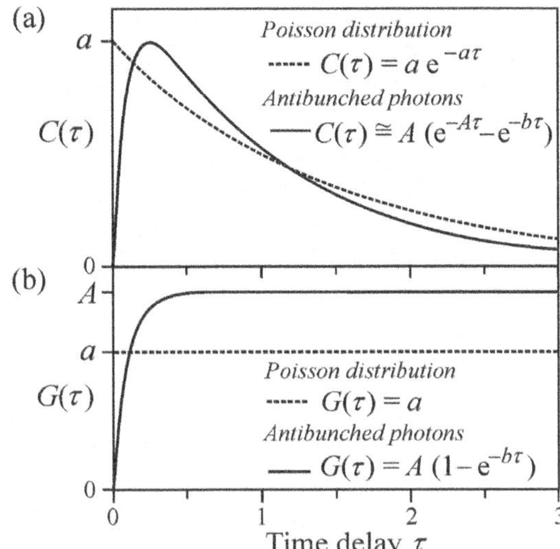

Fig. L.2
Comparison of (a) the delay distribution function $C(\tau)$ and (b) the non-normalized intensity correlation function $G(\tau)$ for two simple examples: a Poissonian source emitting an absolutely random photon stream (dashed curves) and an antibunched photon stream originating in a non-classical light source (full curves). After Verberk and Orrit [3].

interval. A Poissonian source is thus described by $Q(T) = 0$. A negative value $Q(T) < 0$ is characteristic of a so-called *sub-Poissonian* photon distribution, while the photon distribution with a positive value $Q(T) > 0$ is said to be *super-Poissonian*.[7]

In order to interpret real measurements of photon statistics, it is important to know how these measurements can be affected by delayed absorption, reduced quantum detection efficiency (less than 100%), and by the occurrence of noise. Like in the previous discussion, we shall follow the approach of Verberk and Orrit, and for further discussion we refer the reader to their paper [3].

Between a photon emission and the subsequent excitation event due to another photon absorption, a random delay may appear, which manifests itself especially when the absorption probability is low, no matter whether this is due to low excitation intensity or to a small transition absorption cross-section. Let us suppose there is no correlation between the waiting times of the absorption and of the emission, the two random processes being independent. Let $D(\tau)$ be the distribution of these additional absorption waiting times and $d(s)$ its Laplace transform. The overall distribution of waiting times between consecutive photons will be given as a convolution of the absorption-related and emission-related distributions $D(\tau)$ and $C(\tau)$, respectively; in Laplace space this convolution turns into the product $d(s)c(s)$. The Reynaud relationship (L.4) thus transforms into

$$g(s) = \frac{d(s)c(s)}{1 - d(s)c(s)}.$$

Let us now denote through η the efficiency (probability) of photon detection, i.e. the ratio of the number of detected photons to the total number of emitted

[7] Photon antibunching and the sub-Poissonian photon distribution are sometimes confused or misused. It is true that both cases are manifestations of non-classical photon statistics and often both effects indeed coincide. However, it does not always have to be like this, and the exact definitions of both effects do matter.

ones. This photon detection efficiency includes the efficiency of the emitted photon collection, losses in the optical system and the photodetector detection yield (η is usually less than 20%). The distribution of waiting times between the detected photons is obtained by adding up the individual probabilities of detecting two of these photons, provided we consider a supplementary condition, namely, that $0, 1, 2, \ldots$ etc. photons arrived in between without being detected. Verberk and Orrit arrive at the following equation for the Laplace transform $\gamma(s)$ of the non-normalized correlation function of the detected photons $\Gamma(\tau)$:

$$\gamma(s) = \frac{\eta\, c(s)}{1 - c(s)} = \eta\, g(s).$$

A very important message for the interpretation of the experimentally acquired photon statistics therefore is as follows: If the efficiency of photon detection η less than 100%, only the contrast of the (non-normalized) correlation function is modified, not its overall shape. The normalized correlation function remains unmodified, independently of η. This is, of course, good news for the experiment; nevertheless, with low photon detection efficiency the experimental determination of the correlation function will be very difficult.

Finally we briefly assess the impact of the noise background on the detected photon statistics. If we take a constant Poissonian background into account with a mean value B, then only the contrast of the normalized correlation function $g^{(2)}(\tau)$ becomes modified (multiplication with a constant term). That is to say, applying the ergodicity of the field $\langle I(t)\rangle = \langle I(t+\tau)\rangle = I$, the new correlation function that considers the background becomes

$$g_B^{(2)}(\tau) = \frac{\langle (B + I(t))\,(B + I(t+\tau))\rangle}{\langle B + I(t)\rangle^2} = 1 + \frac{1}{(1 + B/I)^2}\left(g^{(2)}(\tau) - 1\right).$$

Unlike the correlation function, however, the distribution of delays will be affected by the noise background in a rather complicated way [3].

References

1. Loudon, R. (1983). *The Quantum Theory of Light*. Oxford University Press, Oxford.
2. Oritt, M. (2002). *Single Molecules*, **3**, 255.
3. Verberk, R. and Orrit, M. (2003). *J. Chem. Phys.*, **119**, 2214.
4. Reynaud, S. (1983). *Ann. Phys.* (Paris), **8**, 315.

M Behaviour of multi-component spectral mixtures: the isostilbic point

The reader may have noticed throughout the book (see for instance Figs 2.28, 7.19 or 8.1) that luminescence spectra can be composed of several emission bands which may be either well separated, or partly overlapping, or sometimes even badly resolvable. Such composed spectra can be regarded as *spectral mixtures* which have certain interesting properties that are worth mentioning here.

In order to get an insight into the problem, let us consider an example. In Fig. M.1 we show time-resolved luminescence spectra of AgCl crystalline foils (polycrystals rolled into a thin sheet) doped with Cd^{2+} ions (500 ppm) [1]. The spectra were accumulated at various temporal detection windows after the end of the excitation pulse. Time resolution was obtained using a pair of synchronized mechanical choppers; this is the so-called phosphoroscope dedicated to measuring slowly decaying emission signals. Two spectral components are readily recognized. A faster component is centred in the blue-green region around 510 nm and a slower orange component around 590 nm. The short-wavelength band is due to recombination of self-trapped excitons (STE) in the AgCl crystal; the thermally activated energy transfer from STE sites to Cd^{2+} substitution centres (producing long-wavelength emission) takes place on the scale of tens of ms [1]. The presented spectra are normalized to have the same area. The most striking feature here is that all the spectra intersect in one point around 555 nm. This point is called the *isostilbic point*.[8]

The reason for the existence of the isostilbic point is purely mathematical. Suppose we have n different emission centres, each of them contributing with a single emission band $g_i(x)(i = 1, 2, \ldots, n)$ to the overall emission spectrum of the material. Here x denotes any spectral variable, e.g. wavelength, photon energy, wavenumber, etc. Let each of the spectral components be normalized to unity at the interval of measurement $<x_L, x_H>$ (that should cover the whole extension of the spectral bands), i.e.

$$\int_{x_L}^{x_H} g_i(x)\mathrm{d}x = 1 \qquad i = 1, \ldots, n. \tag{M.1}$$

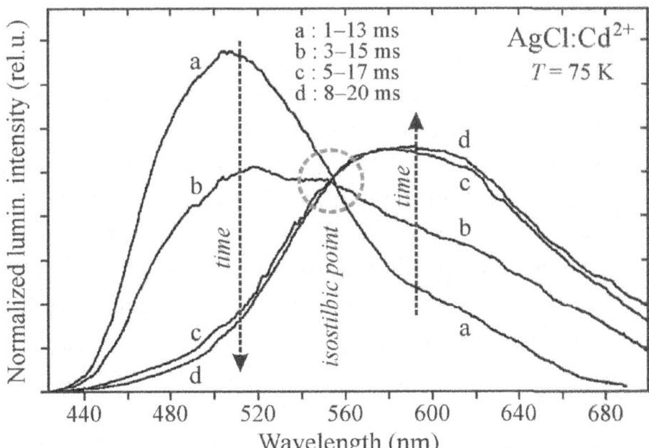

Fig. M.1
Time-resolved photoluminescence spectra of AgCl : Cd^{2+} (500 ppm) crystalline foils at $T = 75$ K (excited by a cw Hg lamp and detected by a PMT with a lock-in amplifier). The detection time windows are sequentially shifted to longer delays going from spectrum (a) to spectrum (d). The corresponding delay values are indicated in the upper part of the figure. Each spectrum is normalized to equal integral intensity and the isostilbic point is indicated (dashed circle). Adapted from Valenta *et. al* [1].

[8] According to IUPAC terminology (McNaught, A. D. and Wilkinson, A. (1997). *IUPAC Compendium of Chemical Terminology.* Blackwell Science), the isostilbic point denotes the wavelength at which the intensity of emission of the sample under consideration does not change during a chemical reaction or physical change. The term derives from the Greek words meaning 'the same brightness'. Sometimes alternative terms like 'isoemissive' or 'isolampsic' are used. The equivalent point occurring in absorption spectra is called the *isosbestic point*.

Suppose we performed a series of measurements (e.g. time-resolved luminescence at various delays after excitation), obtained several spectra $h_j(x)$ and normalized them to unity at $<x_L, x_H>$. Each of these spectra is a spectral mixture of the pure components, or

$$h_j(x) = \alpha_1 g_1(x) + \alpha_2 g_2(x) + \ldots + \alpha_n g_n(x) = \sum_{i=1}^{n} \alpha_i g_i(x), \qquad \text{(M.2)}$$

where α_i plays the role of the 'concentration' of the $g_i(x)$ component in the spectral mixture. Then the set of these concentrations (α_1, α_2, ..., α_n) can be regarded as a concentration vector $\boldsymbol{\alpha}$. As all $g_i(x)$ and $h_j(x)$ spectra are normalized to unity, integration of eqn (M.2) by taking into account eqn (M.1) immediately yields

$$\sum_{i=1}^{n} \alpha_i = 1.$$

This means that all concentration values α_i range between 0 and 1 and their sum is equal to unity.

Let us suppose that all the normalized experimental spectra $h_j(x)$ intersects at one (isostilbic) point, i.e. they share the same value $h_j(x_C)$ at spectral coordinate x_C. This condition is equivalent to the invariance of the function $h_j(x)$ with respect to the vector $\boldsymbol{\alpha}$ at the point x_C. Considering the special values of $\boldsymbol{\alpha} = (1, 0, \ldots, 0)$, $\boldsymbol{\alpha} = (0, 1, \ldots, 0)$, ..., $\boldsymbol{\alpha} = (0, 0, \ldots, 1)$—which correspond to the pure spectral components—then the existence of an intersection point is equivalent to the condition

$$h_j\,(x_C) = \sum_{i=1}^{n} \alpha_i g_i\,(x_C) = g_k(x_C), \quad k = 1, \ldots, n. \qquad \text{(M.3)}$$

In other words, condition (M.3) means that once all the experimentally acquired 'mixed' spectra $h_j(x)$ cross each other at one point, then also all *pure* (normalized) spectral components must intersect at one point whose abscissa is x_C. Besides the trivial situation associated with a single spectral component ($n = 1$), this condition may be generally satisfied in a two-component mixture ($n = 2$) only. Any two overlapping bands must have at least one (possibly two) point of intersection. For spectral mixtures with more than two components ($n > 2$) it is extremely unlikely that there will be any common crossing point.

In Fig. M.2 we illustrate the existence of the isostilbic point by calculating spectral mixtures composed of two Gaussian bands for three different values of a relative band shift. Note the existence of *two* isostilbic points for two close bands (Fig. M.2(a)) and, on the other hand, an almost disappearing isostilbic point for bands separated by more than the sum of their FWHM (Fig. M.2(c)).

In conclusion, whenever an experiment shows a set of (normalized) spectra sharing one (or more) isostilbic point, this observation practically proves that: (i) one is dealing with a spectral mixture of just two spectral components, (ii) the spectral shape of both (pure) spectral components does not change during

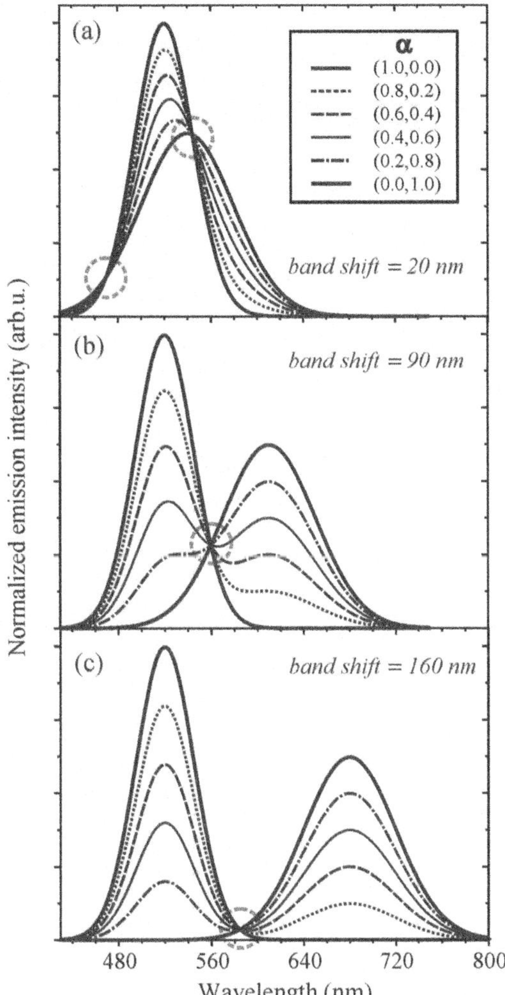

Fig. M.2
Numerical simulations illustrating the presence of isostilbic points in spectral mixtures of two Gaussian bands. All spectra are normalized to a unified area. The two pure Gaussian components (bold full lines) have FWHM of about 60 and 95 nm. While the position of the left bands is fixed at 520 nm, the right bands are peaked at 540, 610, and 680 nm at panels (a), (b), and (c), respectively. The spectral mixtures described in the inset of panel (a) have the value of concentration vector $\boldsymbol{\alpha}$ = (0.8, 0.2), (0.6, 0.4), (0.4, 0.6), and (0.2, 0.8), where the first and second components relate to the left and right bands, respectively. Isostilbic points are indicated by grey dashed circles.

the measurement and (iii) there is a mechanism transforming the emission from one to the other centre (e.g. an energy transfer in the case of time-resolved measurements or some kind of photochemical change). A note may be in order: obviously, the above discussion does not apply to a single luminescence centre that manifests itself via several emission bands (e.g. phonon replicas); however, such a case is not difficult to recognize on the basis of experimental behaviour.

References

1. Valenta, J., Pelant, I., Kohlová, V., Bradnová, V., Trchová, M., Klimovič, J., and Hála, J. (1995). *J. Phys.: Condens. Matter*, **7**, 433.

Subject index

Material index

Conversion of spectroscopic units

The most frequently used units:
Wavelength: nm
Photon energy: eV
Wavenumber: cm^{-1}
Frequency: THz

The basic relations for conversion of units:
$\lambda = hc/E = 1/\nu^* = c/\nu$
$E = hc/\lambda = hc\nu^* = h\nu$
$\nu^* = 1/\lambda = E/hc = \nu/c$
$\nu = c/\lambda = E/h = c\nu^*$

Indispensable constants:

Planck constant $h = 6.626176 \times 10^{-34}$ Js
Speed of light in vacuum $c = 2.99792458 \times 10^8$ m/s
Speed of light in air $c = 2.99709199 \times 10^8$ m/s,
where the relative refractive index $n = 1.0002778$ was considered—this value is valid for dry air under standard conditions (15 °C; 101.325 kPa) for a wavelength of 550 nm
Elementary charge $e = 1.602189 \times 10^{-19}$ C

The following ***conversion table*** is obtained by inserting constants into the above basic relations (again, the speed of light in dry air under standard conditions is assumed)

Quantity	*Label* (unit)	*Conversion relations*			
Wavelength	**λ (nm)**	1	1239.51 / E	$1 \times 10^7 / \nu^*$	299709 / ν
Photon energy	**E (eV)**	1239.51 / λ	1	0.000123951 ν^*	0.0041357 ν
Wavenumber	**ν^* (cm^{-1})**	$1 \times 10^7 / \lambda$	E / 0.000123951	1	ν / 0.0299709
Frequency	**ν (THz)**	299709 / λ	E / 0.0041357	0.0299709 ν^*	1
		λ (nm)	**E (eV)**	**ν^* (cm^{-1})**	**ν (THz)**

Note: When transforming the spectral axis from, for example, wavelengths to photon energies, one has to keep in mind that a spectral device is measuring the spectral density of radiation. Therefore, the perfectly correct transformation of spectra should include a correction of the spectral shape, see Section 2.7.

Semiconductor materials

The commonly fabricated and investigated semiconductor crystals (as well as several less common semiconductor-like materials mentioned in this book) are plotted in coordinates of lattice constant and bandgap energy (both related to room temperature). The lines connect semiconductors that are used to prepare ternary or quaternary alloys. The grey rectangles show approximate groups of materials that can be combined in heterostructures as they have similar lattice constants; see Chapter 12 for more details.

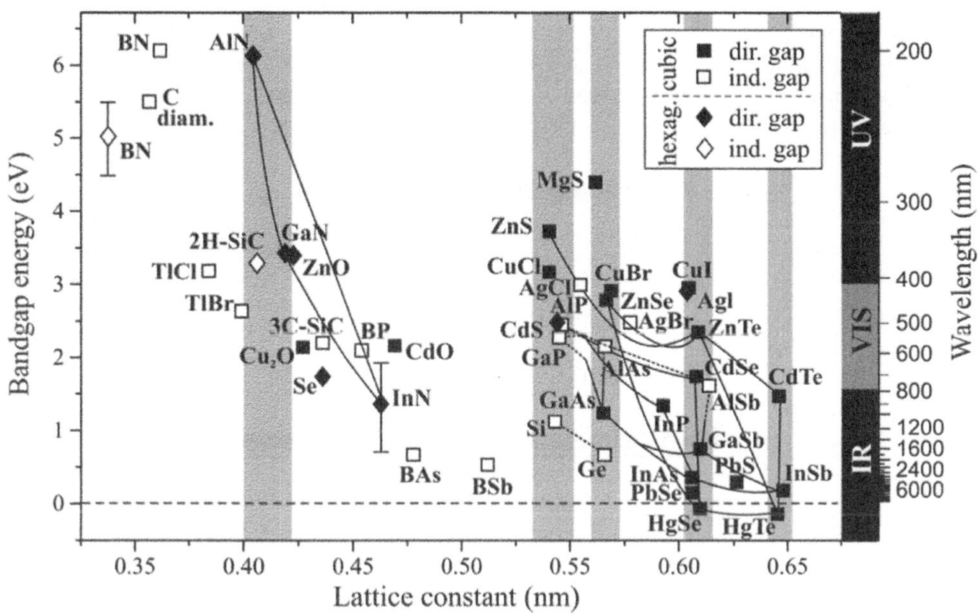

Black (■) and white (□) symbols indicate materials with direct and indirect bandgaps, respectively.

□ cubic crystals

◇ hexagonal (wurtzite) crystals (here the average value of *a* and the *c*-axis lattice constant is used for the plot)

Primary data are taken from: Martienssen, W. and Warlimont, H. (ed.) (2005). *Springer Handbook of Condensed Matter and Materials Data*. Springer, Berlin.

Landolt–Börnstein: Numerical Data and Functional Relationships in Science and Technology–New Series. (1996). Springer, Berlin.

Overview of the most common excitation sources for luminescence spectroscopy

Nowadays, laser diodes and diode-pumped solid-state lasers (DPSS, mainly Nd:YAG) are becoming the most important excitation sources for spectroscopy. They have a number of advantages: excellent stability, low energy consumption, and user-friendly control. Chemical excimer lasers and dye lasers are becoming obsolete due to the instability of their emission and potential health risks.

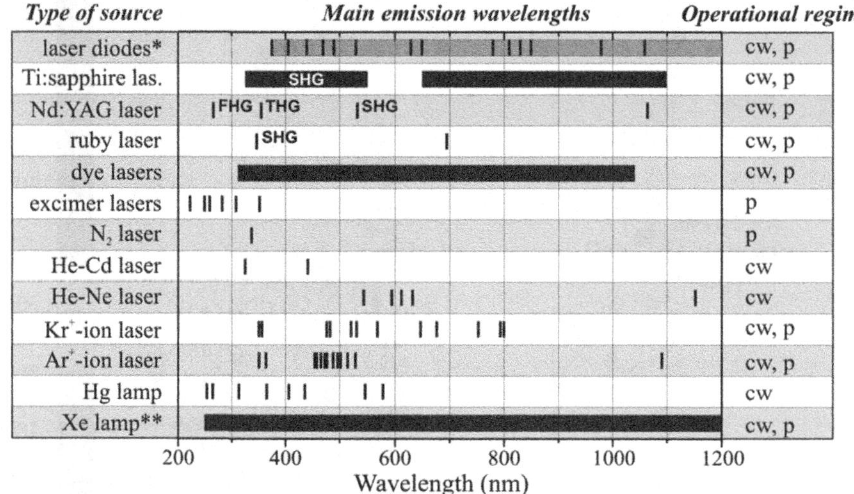

SHG, THG, FHG—generation of the second, third, and fourth harmonic frequency, respectively, from the fundamental laser wavelength using a nonlinear optical crystal

Operational regime: cw—continuous wave, p—pulsed regime

* Laser diodes are currently the subject of rapid innovations and the range of available wavelengths and powers is steadily increasing. Therefore the presented data should be considered as temporary

** Powerful Xe lamps are used mainly in connection with monochromators to measure luminescence excitation spectra

Detectors for optical spectroscopy

The table gives an overview of the most common detectors used in luminescence and other optical spectroscopy techniques. These detectors can be divided into four main groups depending on the physical principles used to convert light into an electrical signal: CCD, PMT, photovoltaic and photoconductive detectors. The indicated detection ranges should be taken as illustrative because the exact range depends on many parameters of the specific device, like the operational temperature, the material of the detector window, the antireflection coating, alloy composition, etc.; see Chapter 2 for more details. Data are taken from materials provided by producers.

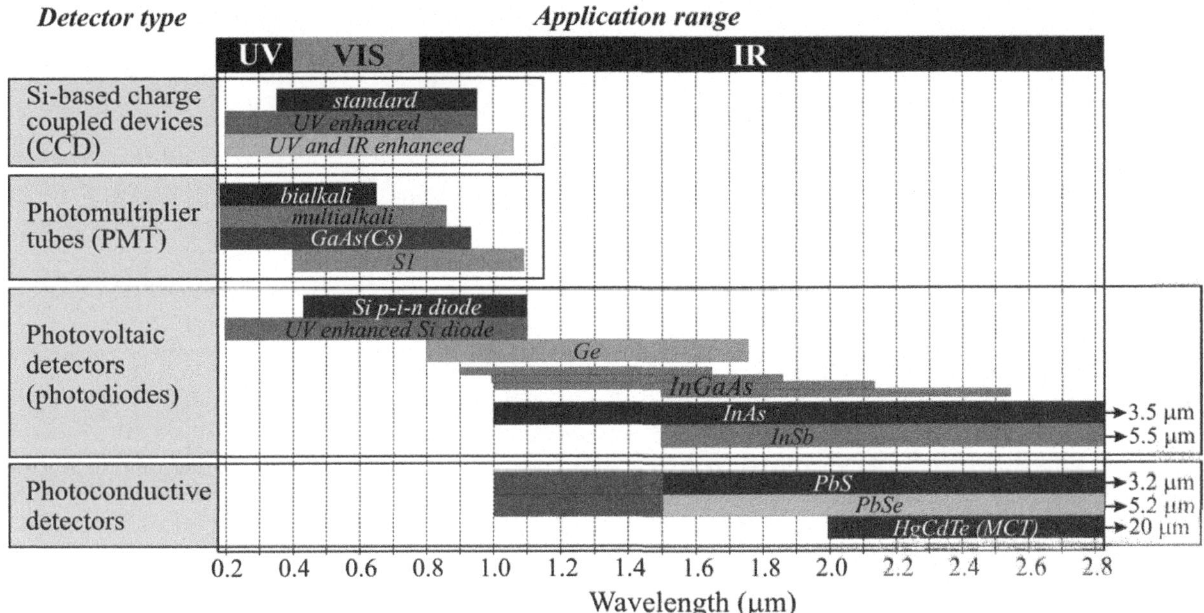